Mud, mud, glorious mud,
Nothing quite like it for cooling the blood.
So follow me, follow, down to the hollow
And there let us wallow in glorious mud.

Flanders and Swann, 'The Hippopotamus Song'

Published by Echo Books.

Echo Books is an imprint of Superscript Publishing Pty Ltd, ABN 76 644 812 395
Registered Office: Registered Office: PO Box 669, Woodend, Victoria, 3442.
www.echobooks.com.au

Creator: Brunskill, Gregg, Author
Title: Glorious Mud: A Scientist in Search of Adventure
ISBN:978-1-922603-22-7 (paper back)

NATIONAL LIBRARY OF AUSTRALIA

A catalogue record for this book is available from the National Library of Australia

Book layout and design by Peter Gamble, Canberra
Set in Garamond Premier Pro Display, 12/17, and MinervaModern .

echo BOOKS

Glorious Mud

A Scientist in search of Adventure

Gregg J. Brunskill

Dedication

This book is dedicated to my strong and good wife Sheila Margaret Brunskill, who helped me keep sane during most of my career. And yes, also my good parents Gregg and Virginia Brunskill, who brought me up right.

Contents

Introduction

There is more than one way to set fire to a haystack.

In spring 1958 several of my high school cronies (who shall remain nameless for good reasons) built a rocket out of some old rusty pipe and plumbing connections, filled with some magnesium filings and sugar, a lot of sulphur, and a fuse made out of waxed butcher paper. The one mete long tube was propped up on the water hand pump behind our ranch house, which sat on a concrete slab over a cistern that held our household water. I lit the fuse and ran for cover, and a large explosive roar filled the hot dry air of western South Dakota. We ran back to watch the rocket rise high in the sky, crossing the nearby Interstate Highway 90 into a field of my father's neatly stacked prairie grass hay from last year. The rocket vanished for a while. Then we noticed that the rocket had made a large black coating on the concrete slab, and the hand pump casing had been tipped over. While we fixed the hand pump and scrubbed the black powder off the concrete, someone shouted that the haystack was on fire across the highway. Being crack rocket scientists, we all knew that we had done this, and that a prairie fire was a bad thing. Within ten minutes the fire siren in town blew and the fire truck was soon beside the haystack, hosing it down. Some time after supper that night I heard my father talking on the phone with the town sheriff, wondering how that haystack caught fire. They discussed how some glass might have caught the sun's rays, or maybe someone tossed a cigarette too close to the haystack.

This juvenile episode was really the fault (gift) of our science teacher, Mr. Charles Staudenbauer, who had started his lecture in sophomore chemistry class with some photographs of the Russian rocket that carried the first-ever satellite into space. *Sputnik*, an 84 kg beach ball sized satellite, launched 4 Oct 1957 from the Tyuratam Rand in Kazakhstan, into earth orbit at 600 km, orbiting the earth every 96 minutes. He described the physics and chemistry of rocket lift-off, and the proportions of fuel needed for the mass of rocket cargo.

Later in our chemistry lab, we combined some common lab ingredients to make small explosions, with great sulphurous smoke rising to the ceiling. The windows were wide open, fortunately.

I'm writing this book to show young people that science is not boring and incomprehensible, and that white lab coats are not required. In fact, in my life it has been a series of wild outdoor adventures, some of them life-threatening and dangerous, and some of them hilarious.

Serendipity led me into a life of adventure in Bermuda, New York City, bush camps in NW Ontario, the big lakes of northern Canada, along the Mackenzie River and Yukon to the Arctic Ocean, visits to the high Arctic islands near the North Pole, the tropical coast of northern Australia, New Guinea, Indonesia, and Borneo.

Along this path, I have attended conferences all over North America, Europe and Asia, where I met other curious folks who were trying to discover how the world works, where the best restaurants were, where the good music and dancing was happening. This life of discovery is open to all who wish to wander down this path, as long as you are willing to open your eyes, work hard toward this goal, and continue to have a stubborn curiosity. I was fortunate to have good family support and school education, but the main thing is to be able to focus and the desire for adventure, in both a mental and geographic way.

If video games are your escape and adventure from the boring real world, please consider how the computer and game software was imagined, constructed, tested, manufactured, advertised, and presented to the initially sceptical public. Too many young people are either ignorant or frightened of the vast number of books and journals, facts, equations, language contained in print and digital libraries, implying that everything is already known and explained and there isn't much left to be discovered.

> 'All the good music has already been written by people with wigs and stuff.' Frank Zappa said this.

We are stumbling around in a dark room, trying to find a candle that might not be there at all. It seems that 95% of the mass of the universe has not yet been found. It is important to have some appreciation of the extent of our ignorance on this planet and how we should recognise, take advantage, and seek to reduce our ignorance in important aspects of our lives. Some universities are now giving classes about the useful recognition of ignorance, scientific mistakes and serendipitous adventures that led to useful discoveries. It is hard to teach serendipity and innovation and creativity, but it is very easy to discourage these abilities in our family and educational systems. As computers and robots take over many of our routine jobs, young people need to find interesting and rewarding careers far in advance of the binary and quantum codes in computers.

As we stumble through life's choices to find our careers, do not underestimate your abilities. Most people are poor judges of their own abilities and capacity for brilliance and innovation.

Science is a method of seeking a rational way to understand and explain how the world works, without reference to supernatural powers, miracles, myths, and legends. It is not so much all the books and journals, facts and equations, but rather it is the creative process of trying to find an explanation for how one part of how the world works. Science should not be left to technocrats and jargon laden science journal papers. Most scientific methods and results (like music and the other arts) can be understood by everyone, with a little effort and imagination. In fact, science requires a great deal of creative imagination, in many ways similar to the artistic world. Indeed, in scientific research, we often create/imagine mental scenarios or models of how some natural process might work and then devise methods to test (or falsify) the model. And so scientific research is much more than logic, reason, and machines that go 'ping'. Science also includes a kind of devotion, a sense of wonder and awe, a naive desire to create order out of apparent chaos, an awareness and lack of fear of the unknown, and some vague appreciation of potentially dangerous uncertainty. This understanding is very important for citizens who wish to take advantage of new scientific findings and vote intelligently on governmental actions on local, national, and global problems. Scientific methods are good for discovering and testing useful new information and ideas, which sometimes leads to improvements in social, cultural, and technological lifestyles. There exist many forms of anti-intellectualism, denial, hoaxes, and cult beliefs that mislead some of us, and indeed it is sometimes difficult to determine the better source of information. There probably has always been a 'knowledge war'. Most of the urgent threats to our relatively happy quality of life in this century require knowledge, wisdom, and action from citizens to ensure a sustainable use of the world's limited resources for us and our descendants. Left to their own pleasures and profits, the businessmen and politicians will certainly not do this for us.

Some of these threats are with us now. Cattlemen in Argentina, Africa, Australia and North America need to know that runoff from seasonal rain is now changing, and the growth of grass and grain might not be similar to that of the past 10,000 years. Retail shops and the construction industry should prepare their inventories to satisfy demand for low energy consumption products and living spaces. The manufacturing sector needs to seek low energy ways to provide essential goods and services, and to recycle materials used in repairs and new products. They could even manufacture stuff that was designed for easy recycling! The transport industry will have to find ways to move people and freight without use of coal and petroleum products, leading to a wondrous step into new technology and industry. The insurance and finance industries must find ways to encourage mitigation and adaptation to a hotter planet with rising sea levels and greatly reduced natural plant and animal diversity. The excessive wealth of the multinational business corporations must be invested in the alleviation of poverty, disease, inequity and planetary pollution. Medical, social, and political organizations need to address ways to limit population growth, to ensure adequate resources for a happy self-sufficient life for our grandchildren. All of this needs to be done in concert with continued exploration of the arts, music, dance, history and literature, where human nature can be enriched and enhanced in our small damp corner of the cosmos.

Mom's apple pie, you say? The next generation can worry about this, because it is too expensive now and we are saving to buy a new Hummer? The previous paragraph is not wishful future thinking, but a plain statement of what is happening today. These issues are linked to survival of human communities, agriculture, business, and the avoidance of war and pestilence. Some of us are comfortable now, but we see the stress coming (and so did our grandfathers). Over two-thirds of the world's human population lives in poverty, hunger, and disease, and they are definitely not comfortable. It is appropriate, ironic and cruel justice that the most complex social and economic systems of the developed world are probably the most vulnerable to natural and technological disasters, whereas the few remaining pockets of so-called 'primitive societies' would likely survive very well without electricity, banks, computers, oil and coal.

Anyway, here is my story of a mad curious scientist wandering the world in search of adventure, pleasure, knowledge, and perhaps seeking wisdom without knowing where to look.

I didn't win a Nobel Prize, I made no earth-shaking discoveries, just a lucky prairie boy puzzling about how the world works. I was lucky to stumble into science during a time when there was an international environmental support movement happening. I was lucky to be associated with some good academic and research teams involved in important work. I was lucky to have a stable, richly rewarding family life and I was lucky I didn't get killed during some of these adventures. I thought I was carrying a banner for logic and reason, wanting to do good with the tools of science, but I was pushed and shoved by political pressures, newspaper headlines, irrational public concerns, and the powerful forces of economics and megacorporation wealth.

It is no secret that a lot of scientific research is fun. Can you imagine getting paid to map a high Arctic lake or ride a research vessel up a tropical river?

I also give you some of the startling world events that happened along the way, examples of the powerful forces of good and evil that push us in directions we might not have considered under normal circumstances.

Well, hell, during these times, there were no 'normal circumstances'.

Chapter 1. What Happened When I Was a Kid

'You can take the boy out of the country, but you can't take the country out of the boy.'
– Unknown

I was born two months before the bombing of Pearl Harbour on 7 December 1941. The Hitler Nazi army laid siege to Leningrad during the month that I was born. In the year of my birth, Nobel Prize winner Glenn Seaborg synthesized plutonium by neutron bombardment of uranium. Out in the Black Hills, Gutzon and Lincoln Borglum ran out of money in October 1941, so the carving of four presidents on the hard Harney Peak granite (some 1.6 billion years old) stopped before the designed sculpture could be completed, soon to be called Mount Rushmore. James Augustine Aloysius Joyce died in 1941, one of the most enigmatic and lyrical poets and novelists of this century. Enrico Fermi and his colleagues built the first nuclear reactor (CP-1) underneath the abandoned west stands of the Stagg Field Stadium in Chicago in December 1942. The Manhattan Project began, one year after I was born. In the following three years, Robert Oppenheimer and many young physicists/chemists built mathematical models and then real test bombs at Los Alamos, New Mexico. At 5:30 a.m. on 16 July 1945, the world's first man-made nuclear explosion turned four hectares of sand into glass at the Trinity Site in the Jornada del Muerto Desert of southern New Mexico. Scientist and leader of men, Oppenheimer watched the nuclear bomb tests and sadly commented on the role of physics and mathematics in this war 'Now I am become Death, the destroyer of worlds.' Halfway around the world, the Japanese Divine Wind religious pilots bombed Darwin and Townsville, northern Australia in 1942.

Thomas Watson, Chairman of IBM, said in 1943 'I think there is a world market for maybe five computers.' In December of 1943, the UK turned on the Colossus, the world's first digital programmable electronic computer, which operated from long paper tapes of instructions, to help decode Nazi messages during the war. In July 1945 a US B-25 Mitchell bomber aircraft lost its way in foggy weather and crashed into the Empire State Building in New York City. Colonel Paul Warfield Tibbets helped end World War II by dropping 'The Little Boy' five ton enriched uranium bomb from the B-29 *Enola Gay* on Hiroshima on 6 August 1945. 73,884 Japanese citizens died quickly, and 74,909 others were injured from a plutonium bomb dropped on Nagasaki at 11:02 on 9 August 1945 from another B-29 named *Bockscar*. All in all, over sixty million people died in the cold mud of Europe and the Asian Pacific theatre of World War II.

On Christmas day in 1948, Thomas Jukes of Lederle Laboratories in Pearl River, New York, discovered that feeding the antibiotic aureomycin to chickens caused them to double in weight gain, thus beginning the massive and dangerous use of antibiotics in agriculture. In October 1949, Mao Zedong and the Chinese Communist Party came to power, pushing the Nationalist KMT Chinese and Shiang Kai-shek onto the small island of Formosa (Taiwan). The USSR tested their first atomic bomb in August 1949 in the deserts of Kazakhstan. During these times, there were no commercial jet airliners for public travel. The human population of earth was just over two billion.

It is hard to imagine my father in a white shirt with a necktie, but I guess he must have trundled off to work with a lunch bucket at a Philips 66 oil refinery office in Kansas City, Kansas during the war years. He was a 1939 mechanical engineering graduate from the State University of Iowa, and this job was a war-critical industry. He took his favourite horse with him, and I have some photographs of him and Blaze in a small pasture near the city. I was born there 10 September 1941. I suspect my father got sick of working in the city, and he moved back to the White River family ranch in western South Dakota when I was a baby.

Two of my uncles came home shortly thereafter from the war with heavy trunks filled with mysterious objects of great interest to me. My mother's brother Raymie was in the Navy in the South Pacific, dropping depth charges on Japanese submarines from a cheap plywood PT boat, and he also played lead trumpet in the US Navy Concert Touring Band. His trunk was full of cowrie shell bracelets, grass skirts, coconut shell brassieres, carvings from the South Pacific islands, and heavy cardboard music stand folders full of jazz and classical printed music for lead trumpet. It seems very likely that he took his rest and recreation time in a little Australian airport town called Townsville, Queensland. My father's brother Gerald was at the European Front, an Infantry Sergeant leading his men across the cold and muddy human wreckage of Belgium, France and Germany. I can remember my uncle Gerald walking into the ranch house kitchen in uniform, with his army duffel bag and trunk, and my grandmother making the coffee with tears in her eyes. His trunk contained a German Luger pistol, some sporty calendars in French, and a long Viennese smoking pipe with a decorated ceramic bowl. A few years later, my mother gave me a library copy of Anne Frank's diary and firmly told me to read it. I could not understand why the German people would kill over a million Jewish and Gypsy citizens, and it haunted me for many of my young years.

In 1944, before World War II was over, the Allied Nations met in Bretton Woods, New Hampshire, at the Mount Washington Hotel. They proceeded to make plans for rebuilding the world economy and created the Marshall Plan, which was wildly successful in Germany and Japan over the next two decades. In 1944, a young plant pathologist named Norman E. Borlaug accepted a position in Mexico to improve strains of wheat; twenty-six years later he got the Nobel Prize for his 'Green Revolution' in food production. In 1945, a British Royal Air Force radar technician by the name of Arthur C. Clarke published a short article on 'Extraterrestrial relays' in the *Wireless World* magazine. His ideas led to the development of satellite communications several decades later, as well as a lot of mind-bending science fiction. The Soviet Army won the Battle of Berlin in 1945, Adolf Hitler and his mistress Eva Braun committed suicide, and their bodies were burned two days later. In January 1945, Russian troops liberated Auschwitz, which had been abandoned by Nazi troops. In 1945, in the wreckage of bombed-out Tokyo, Masaru Ibuka and Akio Morita launched their radio repair shop, under the brand name Sony. Also in 1945, the brilliant polymath Hungarian-American John von Neuman scribbled down some notes and equations on a train between Philadelphia and Princeton, New Jersey, providing the basic principles for computer programming of the ENIAC.

Also in 1945 President Franklin Delano Roosevelt commissioned Vannevar Bush to make a report on the usefulness of science in post-war USA, and a new chapter in science and technology began. Winston Churchill gave a speech in Fulton, Missouri in March 1946, indicating that '...an iron curtain has descended across Europe', and the Cold War began. It wasn't until 1946 that penicillin became generally available to the public. On July 8, 1947, witnesses claimed that an unidentified flying object with five aliens aboard crashed onto a sheep and cattle ranch just outside Roswell, New Mexico. While I was in grade school, Willard Libby discovered (1947), and invented a way to measure, carbon-14, which allowed estimation of the age of organic materials. On 16 December 1947, John Bardeen, William Shockley, and Walter Brattain of Bell Laboratories in New Jersey built the first transistor, and the modern electronic age began. Also during 1947, Thor Heyerdahl took a bullrush and bamboo raft from Peru to French Polynesia in the Pacific Ocean. George Gamow, Ralph Alpher, and Robert Herman published in 1948 their theory of the creation of our universe, the big bang, and nucleosynthesis of the lighter elements. Gamow was supposed to have said 'Give me hydrogen, and you can have the periodic chart'. Arthur Miller's drama *The Death of a Salesman* opened on Broadway in February 1949, a tragedy of the common man.

Mao Zedong's 1949 Cultural Revolution in China began. Biochemist Melvin Calvin at the University of California in Berkeley published papers describing the mechanism of photosynthesis, perhaps the most valuable chemical reaction on earth, for the production of oxygen and plant biomass, for food for everybody. The Korean War was happening in 1950-53, killing perhaps three million people, ending in a ceasefire, not a treaty, and nothing was resolved. In 1951, Gerard Kuiper created the idea of the existence of lots of icy comets and small planets outside of the orbit of Neptune, based mostly upon theory. In the next decades, over 1500 such objects were found, and this astronomical zone became called the Kuiper Belt. Glenn Seaborg was awarded the Nobel Prize in 1951 for synthesising plutonium and nine other elements beyond uranium (americium, curium, berkelium, californium, einsteinium, fermium, mendelevium, nobelium and element 106, which, while he was still living was named seaborgium in his honour), extending the periodic table. The University of Chicago gave him the original 2.77 microgram plutonium sample to keep, encased in a transparent plastic box. Edmund Hillary and Tenzing Norgay climbed to the top of Mount Everest in May 1953, the highest point on earth. James Watson and Francis Crick (with important help from Rosalind Franklin) made an imaginative model of desoxyribonucleic acid (DNA), and a completely new direction in biology began in 1954.

Increased stress on DNA was provided by the Castle Bravo dry fuel hydrogen-lithium fusion bomb (15 megatons TNT) on the Marshall Islands (Namu Island, in the Bikini Atolls) in the Pacific Ocean on 1 March 1954. The explosion fireball was 7 km high in the atmosphere, and the fallout contaminated 7,000 square miles of the surface ocean, including many small oceanic islands. It was supposed to be a secret test, so the native islanders were not evacuated until 3 days after the explosion, and many suffered radiation burns and sickness. At that time, there was little or no knowledge of the effects of nuclear radiation on biological organisms, including humans. Albert Einstein died in April 1955 in Princeton, New Jersey. His brain was preserved in formaldehyde in a big pickle jar and parts of it were sliced for histological sections, seeking unique signals for his brilliant mind. In 1956, IBM got the first hard disk drive to work, and it was the size of a cow. In 1953-59 a young graduate student named Stanley Miller and veteran physics professor Harold Urey published science journal papers on the laboratory synthesis of amino acids under primitive (pre-biological) earth (& other planetary) conditions, starting the dimension of space biology. He irradiated a sterile glass vessel with UV light, heat and electrical sparks, which caused the ammonia, methane, hydrogen and mineral water solution to form all sorts of organic compounds, some of them being components of living organisms.

Despite this warlike, explosive and innovative environment, I had a peaceful childhood, growing up on a cattle ranch on the floodplains of the White River in western South Dakota. I knew little or nothing of the tumultuous events listed above, but this was the kettle of ferment into which I was born. We didn't have electricity, and the wood stoves in the kitchen and living room burned cottonwood and coal in the winter time. I can remember in a childhood reverie the sight of my mother putting little clamping candlesticks on the juniper/cedar Christmas tree, and then magically lighting the candles for a brief period only. My father would make a special trip to town to get fresh oysters, and my mother would make oyster soup for New Year's Eve supper. Our cattle would make icy hard bottomed walking paths in the deep snow of the river bluffs and timber, and these hard-packed pathways would be the last to melt in the spring. My father would bring in new-born March-April calves to warm up behind the kitchen stove, when there was still hard snow on the ground outdoors. In late winter/early spring, the river ice would begin to move, and ice jams would cause localised over-bank flow onto the floodplain. Sometimes these floods covered our river bottom alfalfa fields, and even came into the house a few times. River silt was deposited inside the central pedestal of our big kitchen table on more than one occasion. When spring came, the upland prairie briefly turned green, and usually by early summer the grass turned brown. During spring runoff, high flow in the river dug deep holes beside the cut banks, where the big catfish would remain during low water of the rest of the year. Later in the hot summer, we drifted a fish trap (tarred netting over a series of hoops) baited with rotten mouldy cheese over these deep holes, to get hundreds of pounds of tasty channel and white catfish. Mother would roll these catfish in peameal or cornmeal and fry them in lard. The ranch house kitchen always smelled of cinnamon and nutmeg, and there was no shortage of oatmeal cookies. We had a big black and white house cat called Tinkerbell and dozens of barn cats that got fed fresh milk from a tame Hereford cow. Our two border collies, Jiggs and Greasy, would sometimes move cattle in directions not desired by my father. My mother read all of the Laura Ingalls Wilder *Little House on the Prairie* books to us, and the Will James books about cowboys.

In early summer, my parents planted two gardens: mother's garden by the house for berries, salad greens, onions, beans and such, and my father's larger garden up beside the river bank, which was for watermelon, muskmelon, sweet corn, black squash and potatoes. The gardens were irrigated from windmill pumps and from the White River. Wheelbarrow loads of fresh-picked sweet corn were shucked, and mother cut off the soft kernels with a sharp knife, briefly blanched them, and they were preserved for the winter by canning in soldered tins. After a long hot summer putting up hay for the cattle, it was a special pleasure to pick chokecherries from wild shrubby trees down by the river breaks, with two five gallon buckets tied across the saddle horn of a gentle horse. My father had a special thing about chokecherries, and he would happily spend days picking buckets full of them. He always said that there might not be any next year. My mother would briefly complain that we already had lots of chokecherry juice from previous years, but she would get out the giant cast aluminium pot to cook them down and squeeze out the juice from the seeds by twisting them inside a cheesecloth sack. And indeed, there were rows and rows of quart glass jars of chokecherry juice in the basement, some of them labelled a decade ago.

After spring runoff, on hot days, the White River would flow milky white, dimpled by the wind, like a river of pearls. Shitepokes (sandhill cranes) would walk carefully through the shallow water over the sandbars and cottonwood fluff would slide down the river in the current. On holidays and family gatherings, the old hand-cranked White Mountain ice cream freezer bucket would get soaked for two days in the corral water tank, and then we would make the best ice cream in the world, with salted ice in the wooden bucket. It was almost too much to bear when grandmother would put crushed garden strawberries on top of the freshly made ice cream.

The Brunskill brothers on the hill above the White River ranch, 1950s

My father hired John Paul High Pine to help with the haying. He was a tall proud Oglala Sioux Indian who was a careful man on a tractor and could do magic with horses. He would help me saddle Old Sugar, so I could ride with him to check the fences. Another hired hand was Donavan Wildman, who got his ear nearly torn off trying to saddle a bad biting horse in the barn. On the other side of the White River from the ranch house where I grew up was Sioux Indian reservation land. Grandpa said it was Two Kettle land. Just as our region was settled by Norwegians, Swedes, Danes, Poles, Czechs, Dutch, Scots and English, we were really using the land of the Arikira and Sioux Nations. We knew some of the Oglala, Hunkpapa and Lakota people, and I learned that they didn't call themselves Sioux (that name came from French fur trappers), but another more difficult name that meant 'the only true people'. There were two large reservations nearby, the Rosebud, and Pine Ridge, where Catholic schools tried to convert the proud plains Indians to white man's culture. I got to know Howard, Minerva, and Willard Bordeaux in my school, all good Indian students, and all went on to good careers.

My father was determined that all of his sons would get swimming lessons, so he hauled all of us to Pierre for lessons in the big swimming pool. After the chilly lessons we would have a picnic lunch on the green grass of the nearby park, and sometimes a croquet game would be played. He would also haul us to the local rodeo in White River or Fort Pierre, where the world saddle bronc riding champion Casey Tibbs would perform. Tibbs would drive into the rodeo grounds in his lavender Lincoln, with his spurs hanging on the rear-view mirror. We visited the Falcon Café in Pierre, where Casey had all his prize saddles, photos and belt buckles on display.

On one of our family expeditions to Pierre we hiked to the top of the Pierre Shale bluffs above Fort Pierre and the Missouri River, to see the lead plaque buried there by the sons of Pierre Gaultier de Varennes, Sieur de La Vérendrye, in 1742. I think we were told they were wandering French trappers who got lost in the Rocky Mountains hunting for the Pacific Ocean. I think it may have been a copy of the original plaque. Several decades later I would find that La Vérendrye's sons had come from up north of the Canadian city of Winnipeg, their Fort Maurepas, and that they had travelled all the way to the Little Bighorn Mountains in Wyoming. The French connection is not widely appreciated in South Dakota, and our pronunciation of the name of our capital city causes French speaking people to burst into laughter.

When the first frosts came, my father and his sons would walk through the melon patch, eating the hearts out of the big frost-touched watermelons. We would haul wagon loads of potatoes, squash, and beets from the river garden to a root cellar beside the ranch house, which would keep vegetables unfrozen for most of the winter. Big bull snakes were often living in the earth and hay roof of this root cellar, fat from eating mice. Our water came from a sandpoint well, which was especially sweet and good year around.

I grew up towards the end of the transition between the horse and the tractor, which was a bit delayed in west river South Dakota, compared to the rest of USA. When my father was in high school during the dirty thirties dust bowl and depression, the White River dried up and the family drove the cattle and horses down to Chadron, Nebraska, where there was water and feed. My father went to school there for a few years and then helped drive the cattle back to the White River ranch. Grandfather Brunskill was loudly belligerent about his dislike of lawyers and the banks, especially the banks. Apparently he had borrowed some money from the bank in Murdo and couldn't pay it back promptly. So he cut out the equivalent value in beef cattle, drove them to town (some twenty miles of rough country), down Main Street in front of the bank, and marched into the bank to demand a receipt for his payment.

My grandfather Brunskill had several teams of big black Percherons which were used for some years of my early childhood, mainly in places where they didn't want to use the tractors. In the winter time, when snowdrifts made tractor use difficult, four-horse teams would pull large hayracks on skis to where the cattle were wintered down in the timber beside the river. I can remember sitting on the cold metal seat of the hayrack front while my father used a pitchfork to toss down the hay for the cattle, as the horses slowly moved the hayrack down the fence line. I vaguely remember a horse (and later tractor)-powered belt-drive thrasher for wheat/oats/barley, and two men would gather the straw into hayracks to haul to the barn loft (used for bedding). The grain spout went into a green wooden wagon, which was emptied by human scoop shovel energy into a bin in the barn.

When I was very young, living at the ranch on the White River, my parents would pack us kids all up with lunches, pillows and blankets, and we went to the Saturday Night Dance at the Westover Community Hall, a one-store/post office rural community beside the White River bridge. They would have a local dance band (fiddle, guitar, C melody saxophone, piano, drums) and a caller would lead the square or round dance. Kids would play along the side and eventually go to sleep under the tables. They had 'box cake events' where by draw or quiz or some sort of game you could win a box, usually with cake or candy or toys inside. At the back of the hall was the food table, with big platters of hot and cold food, drinks, desserts and a giant coffee urn. They would sprinkle cornmeal on the dance floor to make it more slippery. In the hot summer the men would go outdoors to smoke cigars and pipes. In the winter they had a big pot-belly stove but the heat from the dancers was enough to keep people sweating. My memory probably tricks me but there must have been 30-40 people dancing, whooping and hollering. There was no electricity, so there was no amplification of the band and lights were from kerosene lanterns hanging from the rafters.

That White River bridge got washed out during an ice jam flood in the early 1950s, Westover general store finally collapsed when the old trapper Tom Wheeler died, and you can still see the foundations for the community hall. On the south side of that White River bridge and up on Prosperity Flats was the ranch of my Aunt Margie and uncle Bernard Strait and Bernard's father Walt Strait. When I was a young kid Old Walt powered his house and barn with windmill electricity and you had to turn off everything else in the house if you wanted to use the toaster. He was well liked by the old Sioux people along the river as he could speak a little of their language, and he would give them a steer to eat and medical help when times were tough. He had several old rusty milk pails full of arrow and spearheads down by his windmill water tank. Walt Strait took me deer hunting with him one chilly fall day in his old beat-up pickup truck. I was sitting between Walt and his son Bernard, with the gear shift between my legs. Walt took this old pickup into places he should not have, steep gullies and foothill shale slopes, and he had his rifle out his window, ready to shoot. He drove right over the top of big yucca cactus and badger mounds and small cedar trees. He didn't want the big bucks but he took two young does on that trip, gutted them where they fell with a big Green River knife, and hauled them back to his barn to hang on a singletree for a few days. We would have heart and liver and kidney suppers for the next few days.

Every now and then my Grandpa Brunskill would pick out a nice colt from the upland horse pasture, rope and lead him down to the river bottom barn. Within weeks, the colt would be begging for treats, wearing a halter and a heavy saddle blanket around the corral. Grandpa Brunskill's method for breaking a colt was to make it easy to do the right thing, and hard to do the wrong thing. He fed them treats if they did the right thing, sometimes a pinch of Copenhagen snuff tobacco. Grandpa himself was rarely without some snuff under his lower lip, or a chaw of plug tobacco in his cheek, or a half chewed cigar in his mouth. There must have been at least a dozen barn cats, which got fed cow's milk and kitchen scraps and all the mice and rats they could catch. I can still smell the barn, the bitter disinfectant, the sweet smell of horse manure, old harness shining with neatsfoot oil, the smell of saddle blankets saturated with horse sweat, saddle soap in tins and the cloying smell of old hay.

Grandpa B also had a special horse-drawn winter sled for hauling ice from the river to the icehouse, where 50 cm thick slabs of ice were stacked in sawdust. Cut blocks of this stored ice would last till late summer and trimmed blocks were used in a big icebox cooler in the house. Some of the old leather horse harness gear hung in the barn loft long after the tractors took their work, but they have vanished now. I think one of my brothers saved some of the buckles and hanes but the leather had dried up and cracked beyond repair. My grandfather and father and uncle Gerald had a terrible soft spot for horses, and they kept a herd of 20-50 horses on the upland pasture for many years. Over time many of them were sold off to rodeo stockmen.

After uncle Gerald died, my brother Obie took several truck loads to his ranch, where a few of the colts proved useful when broke to saddle. My father and uncle Gerald converted some of the horse-drawn equipment (harrows, disks, ploughs, cultivators for gardens, sickle mowers, dump rakes) to a tractor hitch by the time I was in high school. Grandpa Brunskill and my father had a fetish about keeping tools sharp and the wooden handles oiled. I clearly remember them both sitting on stumps in the barnyard with a file to sharpen hoes, shovels, mower sickles, cultivator blades and disks. Ploughshares were kept shiny and smooth by applying axle grease after use.

Grandpa B (and later my father) had domestic pigs gone wild down in the timber by the river. When he wanted some pork he'd go down on horseback with a rifle, shoot two or three and drag them back to the house with a rope to the saddle horn. He'd use a wagon single-tree on a block and tackle to hang the pigs for butchering. This was done very quickly in the chilly fall season. Same for beef, deer and antelope. After the spring floods Grandpa B would supervise the setting of the big black tarred hoop fish trap in the deep holes beside the cut bank of the river. Grandpa would drive a truck to Pierre to get a barrel of rotten cheese and a shovel full of this would go into a cheesecloth sack, tied inside the upstream end of the fish trap. The trap was held in place by a long steel post hammered into the hard sand upstream from the hole, and the trap was drifted over the deep hole. We would get several catches of 50-100 fish, maybe 2-10 lb channel, mud, and white catfish. These were hauled in tubs to the barnyard windmill and stock tank, where they were nailed to a plank for skinning and gutting. Some were salted and preserved in brine in soldered tin cans but we also ate fresh catfish (rolled in cornmeal and fried in butter) for several weeks. Our Norwegian neighbour told us we should plant our sweet corn only during the full moon but castrate our bull calves only during the Saracen moon. My father snorted at this and said it was nonsense. The land on the other side of the White River was Two Kettle land, on the Rosebud Indian Reservation. My father hired John Paul High Pine to help with the haying season and he was very good at soothing rough horses. He taught me a few Sioux words, showed me a few places on river meanders where arrow heads could be found, and he said these were places where big powwows happened in the olden days.

That Norwegian neighbour's ranch was just down the river from us and Anfin Haugland was a little hard to understand when he got excited. He hated prairie dogs, badgers, and coyotes, because of the damage they did to his hayfields and cattle. He had a little Cessna airplane and like several ranchers in this region used his airplane to hunt coyotes and deer. Apparently Anfin could fly his airplane with one hand and his feet and shoot a rifle at a coyote with the other hand. He died in an airplane crash while doing this. Just up the river was Don Hight's ranch, where he decided to raise buffalo and sell the meat to high-end restaurants on the east and west coast. He also had a herd of Texas longhorn cattle which looked right at home in the river bluffs. In fact, there were several big herds of buffalo running on the prairie between the White River and up towards Cheyenne-Eagle Butte.

Grandmother Edna May Gregg Brunskill was a champion basketball player when she was in high school in Haywarden, Iowa. She could put her hands flat on the floor without bending her legs, even when she was old. She would put a pan of soapy water beside the kerosene lamp on the kitchen table to collect the miller moths during the hot summer days. She kept a straw tick mattress on the porch, for guest sleeping in the summer. She was Scots-Presbyterian and claimed her ancestors were run out of Scotland by English lords and sheepherders. Her maiden name was Gregg, which would probably originally have been MacGreggor, a banned clan in Scotland during the clearances. I still remember her walking from the kitchen down to the corral beside the windmill, with her apron full of potato peelings, saying 'waste not, want not' as she fed them to the big work horses, Clydesdales and Percherons.

Her nephew John Bailey Gregg, a medical doctor from Sioux Falls, would often fly his Apache airplane out to our ranch to hunt pheasant and grouse. When I was young, he loaded his passenger seat with cardboard boxes of grouse, our muskmelons and two big watermelons, and took off to go home. He didn't gain altitude fast enough and ran into one of our alfalfa haystacks. My father brought our Chevy grain truck up beside the stack and we placed a ladder from the truck box to the airplane wing. I held the ladder while my father crawled across to the airplane, and pulled him out of the cockpit. He was unconscious and had part of the aircraft instrument panel embedded in his forehead. The cockpit was a stinking mess of grouse feathers and guts and shattered melons. My father drove him to the hospital in Pierre and he was not badly injured. Dr. John Bailey Gregg was a graduate of Sioux Falls High School and the University of Iowa School of Medicine. He practiced medicine in Sioux Falls for twenty-five years. He also regularly conducted medical clinics on the Pine Ridge and Rosebud reservations and conducted medical clinics in Southern Mexico. Dr. Gregg derived great satisfaction from teaching and taught medicine at the University of Iowa, University of South Dakota and University of Nebraska. His avocation was physical anthropology. He enjoyed studying Plains anthropology and native and early American medical remedies. He also taught medical anthropology, including diseases of the early native American population, at the University of Tennessee. He was the author of numerous publications relating to medicine and physical anthropology.

Dr. Gregg was an avid hunter and a pilot. For a number of years he conducted examinations and gave certifications for pilots in Sioux Falls. He spent many hours studying family genealogy and his ancestral homeland of Scotland.

My father taught me how to operate the Farmhand hay stacker on a Model A John Deere tractor. I was proud to operate the Farm Hand, lifting buckloads of alfalfa hay up to my father on top of the stack with his pitchfork, smoothing the stack into an egg shape that would shed rain and snow for several years. My father and grandfather smoked cigars and pipes but they couldn't smoke while haying, due to the fire danger. So my father would chew plug tobacco or put snus under his bottom lip while he was on the haystack. More than once I watched him spit tobacco juice at the head of a rattlesnake while up in the haystack, then used his three-tined pitchfork to toss the snake away from the stack.

I ran a little Ford tractor mounted post hole auger to dig the holes for the telephone and electricity pole lines that came down to the river ranches in the early 1950s. It was a rural shared telephone line with each home having a different number of short and long rings. I helped my father pull a cross-ways calf out of a cow after sticking my arm inside the cow to straighten the calf around. My father patiently taught me how to plough a straight furrow and I learned how to use one of the John Deere wheel brakes to make square corners for our 9 foot number 9 John Deere sickle mower in the alfalfa fields. I experienced the relentless boredom of driving a tractor day after day in 115° F summer heat and the wonderful satisfaction of seeing a job well-done at the end of a month's work. But also there were equipment breakdowns and I got to watch my father and uncle Gerald fire up the forge and an old Lincoln arc welder to make new parts or weld a failed piece of metal.

Sometimes my father would have to go to town to get new tractor parts and mother would come along to get some groceries. We all looked forward to these rare trips because we would always stop at the Dairy Queen soft ice cream shop for double chocolate double malt shakes on the way home. I learned how to handle horses and cattle, milk a cow, take care of chickens and hunt mule deer in the White River bluffs. These blue shale bluffs contained abundant Pierre Shale fossils, ammonites and various species of *Baculites*, and lots of rattlesnakes. I probably did not fully appreciate the wonderful sense of freedom and peacefulness that came while walking amongst the river bank cottonwoods and willows and riding up into the foothills on Old Sugar, the only safe horse for kids.

One chilly spring day, a car full of university 'eggheads' came to the ranch house door, asking permission to look for rattlesnakes in the river bluffs. My father told me to take Old Sugar and ride to the bluffs to show them the location of the rattlesnake dens in shale washout caves along the river. I watched them untangle the cold snakes from their hibernation balls and milk them into rubber diaphragm topped vials. They explained that they would use the venom in their research to find new medical applications of the hundreds of compounds in the poisonous liquid.

My mother and grandmother England were piano and organ teachers with heavy allegiance to Bach, Beethoven, Brahms, Mendelssohn, Liszt, Chopin and the hymnals of the Methodist Church. I learned the major, minor and modal scales more by ear osmosis than practice on the piano. Grandmother Gertie England had piano students from a large area of western South Dakota and she would drive through blizzards to attend to her students in nearby towns. There was a special room in her house for the piano and her music collection and the piano and bookshelves were covered with little white busts of the major composers. There was a large collection of 78 rpm records in Grandma England's house, including the major operas and most of the symphonies of Beethoven, Brahms, Mendelssohn, choral music from the Mormon Tabernacle Choir and Fred Waring. There were also several old books that gave the story plots and photographs of singers and stage sets for operatic productions in Vienna, London, Moscow and La Scala in Italy. Even out in the dusty prairies of western South Dakota, 23 year old classical pianist Van Cliburn (from Texas!) was a cultural hero for winning the 1958 Moscow International Tchaikovsky Competition. His real name was Harvey Lavan Cliburn. In my mother's photo album there is a photo of me pretending to be Van Cliburn on the piano during one of our school drama plays, all dressed up with a bow tie.

But grandmother England's piano room also had a closet full of old 78 rpm records, some of them not your typical classical music. Here were post-world war II recordings of Benny Goodman, Glenn Miller, Cab Calloway, Les Brown, Duke Ellington, the big swing bands. I eventually got some 45 rpm records of Louis Armstrong and Ella Fitzgerald and all this jazz completely expanded my concept of music, even if my grandmother did not approve of it. It was about then that it occurred to me music was a lot of fun! Later on, I saw the 1954 Hollywood film about Glenn Miller and his US Army Touring Band, with Jimmy Stewart along with some very good jazz musicians of the time (Louis Armstrong, Ben Pollack, Gene Krupa, Lionel Hampton).

My mother had several large Renaissance and Classical art books in the house but we had prints of Harvey Dunn's prairie paintings on our walls. In addition to the Sears and Roebuck Catalogue, the pinnacle of shopping was at the J. C. Penny Store in Pierre, where they had pneumatic money transfer gadgets flying on wires across the rooms. In spite of all this culture the men in the family set the breakfast time radio station to WNAX, Yankton, So. Dak., where the cattle prices and weather reports were interspersed between cowboy songs by George B. German, Jimmie Rodgers, Bob Nolan and The Sons of the Pioneers, Bob Wills and his Texas Playboys, Tex Ritter and Gene Autry. More than once I was found guilty of having a wild imagination: one of my grade school teachers apparently reported me to my mother, saying I had fabricated a disturbing story about a herd of giant man-killing buffalo up in the White River bluffs. I don't remember this at all but I do remember my father and uncle Gerald telling completely ridiculous fantastic and scary stories to entertain the kids down beside the stock tank and windmill, rummy dummies they called us, while doing rope tricks. I do remember enhancing the Paul Bunyon and his blue ox Babe story to fit the South Dakota badlands and several versions of 'Pecos Bill' stories that deviated considerably from the original.

My mother's brother Raymie was a very good US Navy Concert Band trumpet player, and I learned some simple tunes on his old silver Conn horn. After a while, my grandparents bought me a new Olds Raphael Mendez trumpet, which was easier to play. I played in the school band and even won some high marks in local music competitions. My grandmother England hauled me away to summer music camps in Gunnison, Colorado and the Peace Gardens in North Dakota, where I discovered that a lot of young people were serious about music and I found myself in symphonic concert bands and *a cappella* choirs. The mind of a South Dakota prairie boy must be very plastic, as I was singing the Latin mass '*Kyrie eleison, Christe eleison, Hodie Christus...*'. I was truly transformed, elevated into another sphere of musical communication and I was often deeply moved such that I couldn't speak for the swelling in my chest and throat. The music thrilled me but I had no interest whatsoever in the church teachings. Sometimes music is presented as something serious that you have to work hard and practice for most of your life but here were kids like me who simply enjoyed the music, had fun and took performance seriously!

Back home, I helped to form a school jazz band (The Hungry Seven) so we could play for local dances in our regional small towns, much to my grandmother's distress. I combed my hair into a ducktail, joined the Jazztone Record Society, subscribed to the *Jazztone Magazine* and learned how to jitterbug to Bill Haley and the Comets and Elvis Presley's 'Hound Dog', 'Don't Be Cruel' and 'Blue Suede Shoes'. I had 45 rpm records of Les Paul and Mary Ford, the Everly Brothers, the Four Freshmen, and the Platters. My jazz idols were Chet Baker and Gerry Mulligan of the West Coast 'cool jazz' style, Louis Armstrong, and the big swing band of Duke Ellington and his screech trumpet player Cootie Williams, who my parents hastened to explain were all drunks and drug addicts and 'no good at all'.

I also played Taps with the local Veterans of Foreign Wars Riflemen and Color Guard at dozens of small town lonely cemeteries on Memorial Day. It was a sad and odd feeling to send the bright sound of the trumpet across these desolate gravestones and the empty rolling prairie, not knowing how these young prairie men had died in war on foreign lands. These living survivor riflemen made sure they had a bottle of whiskey and a big sack of cheese sandwiches in the back seat of our car.

1959

In the 1920-30s, my grandfather Ray England and his cousin Jimmy ran a road building company powered by mules and horses and a few kerosene tractors. During this time, there were several England and England dirt moving companies, all of them cousins, uncles, or relatives of my grandfather. They built dirt roads all over the Dakotas, Nebraska, Montana and Wyoming.

My mother and uncle Raymie grew up with the summer road gang, while my grandmother and her sister Hazel did all the cooking and washing for thirty men. My mother described helping with the wash for these men, shopping for groceries for thirty men, making bread and pies and cakes for them and running the first aid kit to injured workers. My grandmother England grew up near Minneapolis, was a well educated piano teacher and she held a very dim view of those dirty workmen who played banjos and fiddles and sang rough songs about life in the west

When I was very young all that was left of this great adventure was dozens of work horse harnesses hanging in the barn, dump wagons, water tank wagons, mobile tool sheds and bunk cabins for the road crew. These were parked behind the England barn and I liked to explore them. They were eventually given to museums. During the years of my childhood some of the old time mule skinners would come to Christmas dinners at my grandfather's house and they would tell stories and smoke cigars way past midnight.

Grandpa Ray England had been all over the Midwest building roads and he was a very gentle, patient man in most situations. He did get a bit riled when 'city slickers' commented on the barren emptiness of West River South Dakota. One of his favourite comments was:

> 'When city people visit the country they come home and wash their hands to get rid of the dirt from out there in the country. When country people visit the city they come back home and wash their hands to get rid of the filthy city dirt.'

Grandpa England experienced a varied career in his life. He left high school in Illinois to work on the railroad gangs as a teenager. After his career building roads in the Midwest was torpedoed by the Great Depression he ran a dairy for a while, operated a Texaco fuel station in town and helped his piano teacher wife Gertrude run a bed and breakfast operation for tourists in the summer time.

The road camp kitchen days also provided my grandmother with gigantic pots and pans, pitchers and big spoons, these were much later used to make cottage cheese and butter when Grandpa England had a dairy for some years. Butter was made in a large churn with hand-cranked wooden paddles and the cold iced buttermilk left over was a special treat for Grandpa on hot days. Grandmother also raised two hundred or so chickens in the summer, some for sale and some she canned for the winter days. During the war years the big England house was used as an overflow home for the local wounded soldiers. I suspect some of my respect for water came from my grandmother England, as I had to help her carry many buckets of wash water to trees and gardens in her yard. Grandmother England also kept bees, centrifuged the honey combs in the basement, packed a lot in fancy jars for sale and made sewing beeswax lumps and candles from the wax. She also made her own soap in large washtubs and I helped her cut it into small bricks. She would walk down to the stock dam with a bamboo pole, string and big fish hook, with a bucket of worms and return with a stringer of bullheads and perch. She also put bricks in the flush toilet tank to reduce water consumption, this being a water limited country. One of my jobs was to haul water from the nearby town water tower to the England home cistern, using a little Ford tractor and an old road camp water wagon. There was a flat-on-the ground double door into the coal chute in the England house where the coal truck would dump a ton of anthracite into the coal room, beside the big iron furnace in the basement.

When my Grandpa and Grandma England retired, they packed up a horse trailer with some of their goods and moved to a trailer house park in Rapid City in the Black Hills. When they left the Murdo house, the last words I heard my Grandpa England say were 'This country sure looks good to me.' He died shortly thereafter in 1961, probably bored to death in the city, this man who had built most of the roads in the upper Midwest, sitting in a plastic chair twiddling his thumbs. Grandma England was a tough lady so she dragged her trailer back to the home ranch and kept on giving piano lessons until she had a stroke in August 1981.

I grew up with three younger brothers and lots of interaction with aunts and uncles, cousins, two sets of grandparents and family friends. The extended family would gather at the White River ranch for the 4th of July celebrations when roman candles were launched in roof eavestroughs propped against the barn and we would have fried catfish, roast pheasant, grouse and venison and home-made ice cream. My Aunt Marjory helped the town Book and Thimble Club write a book on the pioneer family history of everybody in the region (*Proving Up*). The local Masons, Lions Club, Veterans of Foreign Wars and Ladies Aid were socially active. Everybody knew each other in town and the local Town Sheriff was a friend of our father, who gave us driving lessons and the test to get our driving license. The Sheriff would ask us to drive to the local coffee shop where he would buy a cup of coffee and put it on the dashboard of the car or pick-up. If you spilled any coffee, you failed the test.

The High School band played concerts on the porch of the Jones Country courthouse in the summertime, there was a fall town fair where 4-H heifers, chokecherry jelly and home-made butter were judged and the homecoming parade and football game was a big event. When the Murdo High School band marched down Main Street with baton twirlers, snappy uniforms, the brass blowing out John Philip Sousa's 'Star and Stripes Forever' with the flutes and piccolo descant, it gave shivers down your spine. We listened to The Lone Ranger, Hopalong Cassidy, Roy Rogers, Dragnet and Gene Autry on the radio. We were all enthralled by John Wayne cowboy films and Jane Russell's cleavage. Town people later had TV sets and we were lucky to be invited to watch the Ed Sullivan Show, Steve Allen and Simon and Schuster's comedy show. If you played badly at last night's basketball game, or failed a math test, the town barber and the garage mechanic might tell you how to do better next time. Kids played marbles in the dirt and your marbles were kept in a little leather bag attached to your belt. Harold Lathrop's barber shop on Main Street was a tiny wooden shack with faded framed prints on the walls, paintings of cowboys in action by Frederick Remington and Charlie Russell. Old timers sat on crude wooden benches in the barber shop waiting their turn, talking about the weather, the price of cattle, good horses and how crooked the politicians were.

Family friends came to supper with us, played cards and sang popular tunes around the piano. The Poet Lariat of South Dakota, Badger Clark, came to our town school in a black cape, high lace boots and a high-crowned hat to recite his prairie/cowboy poems. Traveling black gospel quartet singers gave concerts at the Methodist Church and the Assembly of God bible thumpers had tent revivals, complete with speaking in tongues and fainting spells during testimony. Here was 'Dem Bones, Dem Bones' and 'Rock-a-My Soul in the Bosom of

Abraham' sung with joy and hand clapping. I remember sitting on a cold, hard church bench for Sunday School and being told about Satan and Lucifer, Sodom and Gomorrah, the end of times Apocalypse, Armageddon, the Whores of Babylon, the sulfurous pits of Hell and eating the flesh and blood of Jesus.

To a young mind there was a dangerous confusion between military, patriotic and religious fervour which I didn't recognise until many decades later. There was a hymn 'The Caissons go Rolling Along', and the Battle Hymn of the Republic 'Mine eyes have seen the glory of the coming of the Lord, He is trampling out the vintage where the grapes of wrath are stored, He has loosed the fateful lightning of His terrible swift sword. His truth is marching on.' While my Grandpa Brunskill was sitting on an upside down five gallon pail down by the windmill, sharpening a garden hoe, I asked him about Satan and Lucifer and the Apocalypse/End of Times. He looked a bit startled at this question from a small boy but then he spat a long string of chewing tobacco juice and said 'That's a goddamned whole lot of hooey. Don't worry about it.' And that settled it for me, right there and then. Grandpa Brunskill did not like preachers, lawyers, or bankers, no matter how much they smiled.

I can still remember the 'clank clank clank' of the metal chains rattling on the big steel A-frame of the flagpole and the playground swings at the grade school. At recess, we would play 'Pom Pom Pull Away' and sometimes the big kids would run right over the little kids. During my father and mother's time at Murdo High School, basketball was the main sport and the gymnasium was located in the high school basement. There was little room for spectators and the local town Fathers decided to build a big auditorium. I suspect this was done with local labour and goods and services donated by town businesses. My father brought his John Deere tractor with the Farmhand hay fork to help lift packets of asbestos shingles up on the huge curved roof. And of course the various ladies groups would organize lunch for the working men, served in the hot sun on the gravel pad beside the new auditorium grounds, served on 4 X 8 sheets of plywood supported by alfalfa hay bales. These ladies groups were pretty damned fierce once they figured out a project. They helped put on dramatic plays and musical productions in the schools, expanded and stocked the high school library and raised money for uniforms for the high school band. It was the sort of town where our first grade teacher could stand at the exit of our high school graduation, greet us by name and give us a pencil kit present in a zippered plastic bag.

In school I discovered that I liked to learn new stuff and I was fortunate to have good teachers. In grade school I rebelled at the arbitrary nature of the rules of arithmetic and I wondered if it was possible to create many completely different kinds of number systems, in the same way that there are many different languages. Years later, after university courses in calculus and science philosophy, I found that mathematicians had done exactly that. My father tried to convince me that zero was a useful number, being the average of -1 and +1, but to his credit he implied that it was OK to question what everybody else accepted as true. I startled myself (& my classmates) by correcting our high school chemistry teacher about the arrangement of double bonds in benzene on the blackboard. I had good help from my father (a mechanical engineering graduate) for physics and mathematics and he would slide rule the answer before I could do the calculations with pencil and paper. My mother was knowledgeable about literature, history, art and music, and had a good library on these subjects. I enjoyed Mrs. Evelyn Kuhrt's world history and some of the classics of English literature. Mrs. Kuhrt had a conniption in front of the principal's office because she wanted boys to take typing instruction (normally only girls took typing). I benefited from this conniption and decades later I was grateful to be able to touch-type on computer keyboards (which were not even invented then). The typing classroom had dozens of ancient typewriters that went 'ding' at the carriage return and on the classroom walls were formal photographs of Murdo High School graduation classes back to the 1920s, including my father and mother, Aunts and Uncles. Whenever I had a term paper to write, my mother would drive to the biggest nearby library in Pierre and return with a trunk full of books on the subject; for example, C. W. Ceram's book on archaeology, which I read more than once.

One of the traditions in the Murdo High School is that the juniors would decorate the school gymnasium for the Senior Prom dance in the spring. My junior class decided to decorate the gym in a jungle theme. So I borrowed my father's cattle truck and two of my classmates came with me down to the floodplains of the White River, where we cut down two dozen 10-12 m high cotton wood trees, maybe 20-30 cm in base diameter. We were hauling them to town but we were intercepted by the town sheriff Ben Arndt in his sheriff car with sirens and flashing lights on top. Apparently we didn't have the correct red reflectors or flags on the trees sticking out of the back of the truck box. We told Ben why we were doing this, he nodded and grinned and got a big red flag out of his car trunk, tied it on the longest tree and then led us back to the school gym with his sheriff car flashing lights. These trees were wired to the wall of the gym, leaning over the dance floor, and draped with coloured paper chains, tinsel, and blinking Xmas tree lights to give the appearance of forest vines hanging down above the heads of the dancers. Some time later my father quietly gave me a little pamphlet giving the rules for truckers hauling over-sized cargo on the highway. I can just imagine the conversation at the local coffee shop amongst the old timers, my father, and the sheriff.

I was on the high school football and basketball teams but at 98 pounds I was not a heavy hitter. After basketball practice the team would gather at Jack Mowell's Drug Store for chocolate or lemon cokes from the soda fountain and sit at white ceramic round tables in the back, listening to Jack's wild stories of cowboy life in his youth, reading comic books and telling jokes. My parents allowed me to clear out the basement in our home so that we could host rock and roll dance parties after basketball games in town. Bill Haley and the Comets, Elvis, and Chet Atkins 45 rpm records were stacked up on my mother's old record player, turned up to full volume. In the summer time I helped put up hay with my father and brothers, did some chores and played with the town baseball team at first base.

All of this was wonderfully peaceful but I remember sitting in my upstairs room, looking out the window at the prairie hills, wondering what was going to happen to me. I wanted to cross the Pacific Ocean on a raft like Thor Heyerdahl on Kon-Tiki, or hunt seals with Vilhjalmur Stefansson and the Eskimoes. I read a National Geographic article about Frenchman Auguste Piccard and his bathyscaphe, which could safely carry a man to 4000 m depth in the ocean. I poured myself into these books, lived these wild experiences, in the same way that I would swagger after watching a John Wayne cowboy film at our local theatre. I read J. D. Salinger's *Catcher in the Rye* and I could not identify with the bitter, city-bred, angry Holden Caulfield at all, I thought it was a terrible book. But during these times a new generation of rebellious youth were gathering their privileged, coddled and well-fed spirits to change the world. And so my brothers and I learned from our parents and grandparents and school teachers, especially when they were not teaching, and especially when we didn't know we were learning. We learned by watching and helping with the work.

After working for Ken Foster, a neighbouring rancher, putting up hay and windrowing oats all summer, I saved enough money to help pay for my first year at Augustana College, about three hundred miles east, a world away from the dusty prairie grass of home. My Grandmothers and mother tried to whip me into shape for university education. Grandma England used to say that 'Goodness gracious land sakes alive, you won't amount to a hill of beans unless you get your long hair cut'. I was a part-time jazzman playing trumpet with dance bands in small town auditoriums on the weekends and I combed my hair in a duck-tail style, remotely related to Elvis and Chet Baker. When I went to Harold Lathrop's barbershop on Main Street, Murdo, and asked for a crew cut, Harold said '...and Jesus wept!' Harold was one of my father's classmates in Murdo High School of 1935. He kept a long leather strop attached to the barber's chair so he could sharpen his shaving razor (whop whop whop). He had a large ceramic bowl for the shaving cream on his counter along with all the liniments, perfumed lotions, hair lotions and rubbing alcohol squirt bottles. He also had a semi-outdoor shower at the back of the little shop. Some old time cowboys would come in for their yearly shower, haircut and shave and they would go to the barber's chair with their big hat onto their long greasy hair. Harold would say 'Sweet Jesus, I can't cut hair with your hat on!'

My mother bought me a used school Remington typewriter and a dictionary. In the fall of 1959 my father helped me load my books, one cardboard suitcase of clothes, bedding, LP record player, my jazz albums and an aluminum laundry shipping box full of banana bread and tea bags into his grain truck and hauled me to Sioux Falls, where the next chapter of my life began. I must have looked a little uncertain standing at the doorway to Solberg Hall on the Augustana College campus; as my father left, he simply said 'Pay attention, be a gentleman, and you'll be fine.' My father went on to Sioux City (Iowa) to get a truck load of corn so he could grind it with his own oats for his overwintering calves. I figured I would take an academic major to support an easy career in music or maybe nuclear physics, but I really wanted to be an astronaut.

Off I went, with Will James's *Lone Cowboy*, the archaic exuberance of Shakespeare, the world-weariness of T. S. Eliot's 'J. Alfred Prufrock', the King James version of the Bible and the wild rhythmic midwestern poetry and songs of Carl Sandburg and Badger Clark in my head. No, maybe not an astronaut, but perhaps a fossil hunter in exotic lands, like Roy Chapman Andrews in the Gobi Desert discovering dinosaur eggs. But fundamentally important was the dictionary from my mother, and my father 'loaned' me his slide rule.

For unexplained reasons a signature sound amidst my upbringing was the almost human call of the big crows/ravens down by the White River bluffs. Whenever I visited home over the next several decades that was one sound I wanted to hear.

Chapter 2. Off to College in Cowboy Boots

A little learning is a dangerous thing;
Drink deep, or taste not the Pierian spring:
There shallow draughts intoxicate the brain,
And drinking largely sobers us again.

....

But more advanced, behold with strange surprise
New distant scenes of endless science rise!

Alexander Pope, Essay on Criticism (1711)

Ah, yes, but then 'If a little learning is dangerous, where is a person who has so much as to be out of danger?'

Thomas Henry

My career in music was thwarted by New Orleans jazz. Well, I did begin as a music major at Augustana College, taking classes in first year music theory, composition, choir, brass ensemble and symphonic orchestra, philosophy, history and literature. I must have been a distinctive sight in my West River high-heeled cowboy boots, Levis and pearl snap button shirt and leather vest. But early in my second year I met a pretty girl who was a teaching assistant in the biology department. She told me that the teaching assistants in biology were travelling in a bus to New Orleans over Easter break. So I joined the biology and chemistry departments quickly, helping to set up labs in invertebrate zoology for Professor Will Rosine.

We really did travel to New Orleans in a bus, having a ball on the way down the Mississippi River, eating pecans collected from the road ditches. We did some botanical collections and bird watching in the swamps around Lake Pontchartrain and toured the biology and chemistry departments at Tulane University. We discovered the French Quarter, voodoo parlours, hookers and pan-handlers, Preservation Hall Dixieland jazz, Pete Fountain's emporium, K-Paul's Louisiana Kitchen, blackened catfish, gumbo and tiny shrimp, poor boy sandwiches, hot pralines from street vendor hotplates, strong chicory coffee and beignets at the wonderful Café du Monde down by Jack's Brewery on the levied banks of the Mississippi River delta channel. This was mind candy for me and everything Mark Twain said about Tom Sawyer and Life on the Mississippi became real to this dry dusty prairie boy. The girl who snatched me away from a career in music was Elaine Pearson, a pre-medical student whose laugh you could hear a half mile away. I took her to the Court of Two Sisters for supper and we jitterbugged to live Dixieland jazz.

In my waspish upbringing in the high prairies I had heard people talk about zombies and voodoo in a jokey sort of way. I was startled to find that in New Orleans Voodoo was one of the African/Creole folk-religious ways of life and that zombies were probably real people drugged with low doses of brain poisons. I went into the voodoo shops thinking to buy some souvenirs but I found these shops were serious (and a little scary) sources of religious paraphernalia, dried roots, symbolised wooden canes and some books. I bought a book about Marie Laveau who was one of the most powerful Voodoo Queens in the 1830s and who seemed to combine very interesting pagan African/Creole beliefs with European Catholicism, Islam, nature worship, jazz (especially drums) and good food. She worked to help the hungry, poor and sick in her neighbourhood and she also cast spells, handled snakes, gave potions for love sickness and helped slaves escape bad masters.

During the hot summer months I returned home to help my father with haying. I took a job with Ken Foster, a nearby rancher, who needed a hand who knew how to run a hay stacker. He had an antique yellow Minneapolis Moline tractor with a Farm Hand stacker mounted on it and my job was to make hay stacks all alone. On our home ranch I always had my father in the stack with a pitch fork to round and smooth the tops but here I had to do it with the hydraulic push-off buckteeth. I also helped windrow oats and move the winter cattle barn manure out onto the fields with a manure spreader. The food there was very good, as Mill Foster was an excellent cook and Ken enjoyed long talks on the porch in the evening with melons, strawberries and hot tea. I can't remember how we got onto the subject of horse feet but I made the statement that they were similar structure to all other mammal feet, according to Darwinian evolution. This caused some upraised eyebrows and Mill pursued me on this topic for several weeks, as she was sure Darwin was the Devil.

In the next years I learned genetics and anatomy from Professor Sven Froiland, cell physiology from Professor Will Rosine, botany and ecology from Professor Dilwyn Rogers and chemistry from K. K. Kintner. During my second year summer holidays I made a collection of invertebrate animals from my father's ranch and tried to identify them, pinned in neat rows in plastic foam boxes. The ecology class also had a field trip to the western part of South Dakota, my home ground, and they stayed overnight camped in the cottonwood and willow timber beside my White River. I could proudly lead them into the foothills to peer at rattlesnake dens, locations where honey bees were inside dead branches of cottonwood trees and where to find certain dung beetles underneath the abundant mature cowpies. Seminar classes included some difficult books such as Ernst Mayr's *Animal Species and Evolution*, Pierre Teilhard de Chardin's *Phenomenon of Man* (of which the Catholic Church held a very dim view) and C. P. Snow's *Two Cultures*. These books gave us the possibility of going to New Guinea to study birds of paradise or going to Chinese deserts to discover dinosaur nests full of fossilised eggs. Teilhard's several books proposed a teleological evolution to a planetary and solar system convergence (the Noosphere and the Omega Point) which is vaguely similar to Ray Kurzweil's logarithmic and computational/cultural evolution towards a 'singularity' which will happen about 2025 or thereabouts. In 1962-66, Claude Levi-Strauss published *The Savage Mind*, a cultural anthropology paradigm change, which suggested that most primitive cultures of ancient and present times were not savage but highly civilised. We were also thrust into the atomic bomb test question, the cold war and the significance of on-going bomb fallout from tests in Nevada and the Pacific Islands. President Dwight Eisenhower warned us all in 1961 about the acquisition of unwarranted power by the military-industrial complex. We read several social and political papers on the 'military-industrial complex' and some of the early experiments on the biological effects of bomb radiation, which scared us witless as we were doing our *Drosophila* (fruit fly) genetics experiments. We didn't know it then but we were in the midst of the maximum fallout of bomb radionuclides (1962-64) all around the world.

This being a liberal arts college, I was also able to enjoy several courses in the history of different religions, art and philosophy and seminar classes that required reading in evolution, religion, anthropology and the history of science. I discovered that I was embarrassed I couldn't pronounce 'avant-garde', *weltanschaung* and 'angst' with a casual foreign accent. A wonderful Norwegian lady taught us the German language and I was forced to take courses in public speaking (thankfully now). I took an advanced course in cultural anthropology and we read scholarly accounts of life in the Trobriand Islands in the South Pacific and Arnhem Land in north Australia. I took a paying job as a laboratory assistant in biology and chemistry, washed dishes in the campus cafeteria and got paid for being an usher in the Sioux Falls concert series.

I had fun playing in brass ensembles, the local 'Town and Gown' symphony orchestra, playing the high trumpet in Bach cantatas (*Magnificat, Ein feste Burg*) and singing in a Renaissance motet group. Leo Kucinsky, conductor of the Augustana Town and Gown Symphony Orchestra, often asked his trumpet section to sound like 'sweet angels in butter', or sometimes 'angry lightning', or sometimes 'dignified announcement of the Queen'. One of the guest conductors for the orchestra was a young Daniel Barenboim who led us through a complex Middle Eastern influenced suite where trumpet and oboe were in a recursive theme statement. My friends in the audience commented that I just about fell off my chair, twisting and turning and counting the irregular and complex Middle Eastern rhythms. I was a musician's union member and got paid well for playing lead trumpet in an off-campus local big swing band trying to play Stan Kenton and Duke Ellington charts. These times were near the end of the acoustic dance band era and I was lucky to listen and watch the big jazz bands of Artie Shaw, Cab Calloway, George Shearing and Count Basie play in large Sioux Falls ballrooms where hundreds of people were dancing and enjoying the music.

I also went on a Christmas break midwest tour with Arnold Running's very good Augustana Choir, playing the high trumpet parts to J. S. Bach's cantata *Magnificat*. This particular piece of music made my skin shiver and my throat get tight. When everything went well on some nights, when everybody did their parts well, it felt so good that I was playing my part with tears rolling down my cheeks. I think you get this feeling only rarely, and only if you are in the middle of the music. For me, this music was about as good as it gets. One of the organ teachers asked me to play some of the high trumpet solos with organ accompaniment and we performed two recitals of Spanish, German, and Italian Renaissance music. We also did Bach's *Ein feste Burg* and I almost blew my gaskets playing the high trumpet parts on a little piccolino trumpet. Sometimes it felt like the top of my head was going to explode, not just because of the trumpet air pressure but also because all the different lines of music weaving together made my blood boil. I felt like this was one of peaks in my life and if I did nothing else in my life, these performances made it all worthwhile. The cohesive ensemble elation and warm camaraderie after a good performance were better than drugs or booze.

T. S. Eliot said it exactly in his *The Dry Salvages*:

> Music heard so deeply
> that it is not heard at all
> but you are the music
> while the music lasts.

We made a pile of money playing Medieval and Baroque tower brass music for the local churches on religious special days and playing the long herald trumpets for official events, under the guidance of our trumpet/brass teacher Jim Ode. One of our specialties were the 24 Contrapunctus from J. S. Bach's *The Art of the Fugue*. I was powerfully affected by local composer Fred Teuber, who wrote several pieces for our brass and percussion ensemble for special events. Teuber wrote a brass, percussion and pipe organ ensemble piece for a local Lutheran church that was very dissonant, loud and gloriously sonorous, and we got asked to not perform this piece but we all enjoyed it very much. I think Teuber took pride in being kicked out of the church. Prairie writer Herb Krause taught a wonderful course about Shakespeare and a history of lesser-known prairie writers (most of which were already in my mother's library at home).

All this went by as mostly fun but there were other forms of education happening here. I used to take naps in the attic of the lab preparations room and would observe certain people sneaking in to drain off a cup of lab ethanol for non-scientific uses. My genetics professor began crying during his lecture about anomalies in human development and we later found out that his second child had been stillborn the previous day. He also took us to a nearby institution for deformed and deranged people, where we observed microcephalic and hydrocephalic children, mentally disturbed people and people badly deformed from accidents and war. My jazz band experiences included sax players stoned on marijuana, LSD and booze. One of my jazz band leaders got in a fight on the sidewalk at three am in downtown Sioux Falls and he got his arm and jaw broken (which ended his musical career). I watched Gerry Mulligan and Chet Baker try to play jazz at a local dance hall and they were so looped/stoned that they couldn't keep the beat steady for the dancers. Swiss chemist Albert Hofmann was seeking medicinal uses of extracts of a wheat fungus in 1943 when he was exposed to small amounts of lysergic acid diethylamide (LSD) and his brain went on a trip of fantastic pictures with garish colours. He did experiments on himself, hoping that this substance would be useful in research on mental disorders. He was dismayed by the manufacture and use of LSD for recreational purposes during the hippie years of the 1960s. All of this broadened considerably my understanding of the human condition.

My senior thesis work in organic chemistry was to study the terpenoid compounds in pine pitch and to think how this might relate to the genesis of amber. I think I started this research because I saw several arm/ankle bracelets of smoky amber that I found in my uncle Raymie's US Navy trunk in the attic at home. I had read about fossil insects and bird feathers preserved in amber in Willey Ley's book. These lumps of amber probably came from the South Pacific islands, where my uncle served during World War II. I wrote a letter asking for help from a professor at Cornell University who had some experience with the biochemistry of fossil organic material. This was Jack Vallentyne, a kind person who sent me reprints and long letters of advice, accompanied by the forms for application to graduate school at Cornell. I had only the vaguest idea of what happens in graduate school, I knew nothing about Cornell University, but it sounded interesting. Serendipity was happening.

Chapter 3. Piled Higher and Deeper: East Coast Ivy League University Days

'The thing the sixties did was to show us the possibilities and the responsibility that we all had. It wasn't the answer. It just gave us a glimpse of the possibility.'

John Lennon

After another summer of stacking White River floodplain alfalfa hay with my father's John Deere tractor and Farmhand stacker I could tell that this next chapter in my life would be a more difficult experience. I suspected that large east coast universities would expect me to have a lot of brains, and I might not be up to the challenge. In the fall of 1963 my parents hauled me down to North Platte, Nebraska where I boarded the train for Chicago and the east coast. As the train pulled out, both of my parents had tears in their eyes and I was more than a little scared.

The train rolled through grimy industrial Chicago, Cleveland and Toledo, to Rochester, New York, where I had to find a cheap hotel for the night. I didn't like these big dirty noisy cities, the traffic and the busy loud people, and I worried that I might have made a mistake coming east. I took the bus to Ithaca, down past Cayuga Lake, dotted with apple orchards and small farms. This was much more to my liking and I began to relax a bit. I took a taxi to Cornell University, finally dumping my two heavy trunks and my aluminium laundry box on the steps of Stimson Hall.

I wandered the halls of this old creaky building for a while, finding the Zoology Department Head in his office, but the doorway was blocked by a very large, drooling sleeping black Newfoundland dog. A very amused Lamont Cole gave me directions to the office of Professor John Reuben Way Vallentyne, where my life in scientific research began.

His office was partly occupied by two tall metal cabinets which held the latest automated analytical machines to measure amino acids. After brief introductions and curious examination of my South Dakota boots, he began to explain how the machines worked and showed me the latest results. Only later did he explain that the samples being analysed were groundup meteorites, and he was trying to find biochemical evidence for life from outer space. He was also doing experiments with amino acid spiked quartz in ovens, to determine their thermal stability, which could be extrapolated to determine how long these components of proteins could be expected to survive in ancient terrestrial (and other planetary) rocks.

After a while I came to understand that Professor Vallentyne had been a graduate student at Yale under George Evelyn Hutchinson, an English polymath scholar of the old school. Hutchinson was one of these almost mythical characters who could speak dozens of languages and had worked in Tibet, South Africa, Pacific Islands, Greenland and Siberia. His literary essays were full of references to medieval saints and Renaissance students of natural history and he wrote a large treatise on the Biogeochemistry of Vertebrate Excretion, amongst other strange subjects. I could now understand why Professor Vallentyne entitled some of his journal papers 'Love and Pyrite Spherules', or 'The Romance of Mud'.

I was introduced to the local graduate students, most of them being east coast people who were greatly interested in a South Dakotan boy in Levis and a leather vest. They helped me find a share house to rent, and showed me the university bookstore and coffee houses. Roger Green demanded that I observe his methods for making roadkill venison stew. Stu Hurlburt took me into the forest looking for salamanders, and Andy Sheldon took me to Oswego Creek to sample stream invertebrates. According to Andy I kept saying 'It sure is green!' I was startled to hear about Tom Pliske's thesis work on the hair pencil penis of *Danaeus gillepus*, a butterfly that lived in the swamps of Florida. Tom worked in the Everglades, where Cuban gunrunners in speedboats and float airplanes were a hazard to research. Roy Horst worked with Bill Wimsatt on diseases of bats in Mexico and one of his helpers died from a fer de lance bite in a bat dung cave. Stuart Ludlam was trying to construct a several millennia history of Cayuga Lake, based upon his long Kjullenberg sediment cores.

These strange people studying very strange things were my friends and helpers during my time at Cornell University.

I was roped into being a teaching assistant in the department and worked for Bill Leonard in the endocrinology and histology labs. For this I got $3400 for the academic season, which was enough to live well by my standards (I ate a lot of beef hearts and potatoes). I managed to pass my examination in the German language on the first attempt, but the French language was my downfall. Many times I climbed the dingy stairs to a cubbyhole apartment where my

French tutor lived, and I didn't pass my French test until my fifth attempt. Calculus, physical chemistry, statistics, mathematical ecology, soil microbiology and biochemistry courses were difficult but interesting, but I really enjoyed invertebrate palaeontology and geology with John Wells.

Wells was a small, darkly tanned man with wrinkly eyes, who had studied the coral reef ecology of the thermonuclear bomb sites (Bikini Atoll, Eniwetok) in the tropical Pacific, before and after the bomb tests in the late 1950s and early 1960s. He was interested to hear that I'd collected *Baculites* sp. and other ammonites from the shale foothills of the White River in western South Dakota. His lecture slides about coral reef research in SE Asia and the Great Barrier Reef of Australia contained a lot more than photographs of marine organisms and fossils, and I fantasised about seeing the tropical seas.

This kind of graduate school learning was very different from my high school and undergraduate studies, as most of the Cornell University lecturers assumed you had read and understood the assigned textbooks. Their lectures were mostly about their own new research results, some of which conflicted with textbook information, indicating that you shouldn't believe everything you read in science books. Some of the lecturers expected class discussion, and arguments on alternate points of view sometimes extended hours after the class was over.

Professor Vallentyne's graduate seminar class regularly met in the back of a beer hall in downtown Ithaca, where we debated aspects of cosmic and planetary evolution, pheromone ecology, biodiversity and genetics over pitchers of beer and hot sausages. Some of the best lecturers would bait us with obviously false information or logical traps, and some brave students would have to stand up to challenge the professor in midstream. Sometimes we failed to recognise these false points and we would be chewed out properly in the next lecture for being sleepy or dumb.

Guest lecturers at the Cornell Colloquium Series included some of the stars in science and philosophy (Carl Sagan, Hans Bethe, Thomas Gold, H. T. and Eugene Odum, Ernst Mayr, Jerry Brown). Sagan was then calculating the greenhouse warming of Mars and some of the moons of Saturn and Jupiter and wondering if there were oceans of methane and water on these planets. Bethe was still involved in the cold war debate about the deadly logic of nuclear war and intercontinental missiles. The Odum brothers were making ecological history by showing how plant and animal communities could be represented by electronic diagrams of resistors, capacitors, batteries and energy flow. Mayr was trying to figure out how new species of plants and animals came to be, and why some were now extinct. Jerry Brown gave a series of lectures on South American ant colony evolution, with all of the complexity of their division of labour, community work habits and reproduction ethics.

In spite of total immersion in science I didn't lose track of the humanities, and we had opportunities to attend rehearsals of the great operas, some symphonic and chamber orchestras and concerts of Renaissance and Baroque period music.

One of my favourite poets, Robert Frost, died in 1963 and I treasured my mother's copy of one of his books of poems, *A Boy's Will.*

In this rarefied atmosphere, I was becoming aware of major breakthroughs in big science.

In Edinburgh, Peter Higgs published a series of papers beginning in 1964 which predicted the existence of an energetic field pervading the universe that somehow endows mass to energetic waves during the big bang. This prediction became known as the 'Higgs boson', which was ridiculed initially and finally accepted into the options for a standard theory of physics. In 1964 Rachel Carson died of cancer, having ignited the environmental movement. Arno Penzias and Robert Wilson of Bell Laboratories in New Jersey built and calibrated a horn antenna to measure radio waves and accidently discovered cosmic microwave background radiation, the heat energy of the big bang expansion, supporting several hypotheses from Albert Einstein's theories. In 1965 Gordon Moore was working in the semiconductor industry, and he wrote an article that claimed the number of components of integrated electronic circuits (transistors, resistors, diodes, capacitors) would double every two years into the future. More simply stated, this means that computing power would double every two years. This became 'Moore's Law' and was true for several decades into the future.

A lot of this was going on around me and I was completely unaware of the revolutions happening in the physical sciences until years later.

I shared a rental house in Ithaca's college town with Peter Sobel, who was in graduate school in aeronautical engineering. He was the son of a wealthy building contractor in New York City. Pete drove an XKE Jaguar and had a little Cessna airplane so he could fly back to NYC whenever he wanted. Pete knew personally the best restaurant cooks in town and he introduced me to good Italian food. Pete really liked my mother's homemade peanut brittle and almond toffee and he phoned her to ask for a special shipment, with a special note inside indicating that it was to be shared.

The Ithaca region had lots of apple orchards and in the fall there were lots of roadside stands selling cider. If you walked behind the vendor stand there were usually several gallon jugs of cider than had 'gone off', and they would give these away if you asked for them. Knowing a little about fermentive chemistry, I devised an anaerobic bubble release system to allow complete fermentation to about 14% alcohol. This stuff was really good, so Pete and I gathered about a hundred gallons of 'bad cider', poured it into a wooden barrel in our backyard in late fall, and let it freeze in early winter. At the core of the frozen barrel was a small volume of liquid applejack, sort of like an apple brandy. Goodness gracious, this was good stuff.

I was his best man at his wedding and that is when I discovered he was from a Jewish family. This was something I knew nothing about. So I stepped on the light bulb at the New York City wedding, learned about blintzes, bagels and lox and yarmulkes. Pete took me to NYC several times to visit the Cloisters, smoky jazz joints, the Museum of Modern Art, Mama Leonie's Panzerotti Palace, the top of the Empire State Building and Greenwich Village, full of hippies and musicians and beggars. I was reduced to a wide-eyed shivering pulp by an early performance of *The Fantasticks* at the Sullivan Street Playhouse, a cheap off-Broadway production. Pete's parents took us to a posh theatre where we watched a high-powered and menacing *West Side Story* with a very hot big jazz band in the pit. We also watched a Greenwich Village production of Samuel Beckett's *Waiting for Godot*, which left us puzzled. Being introduced to New York City by a cynical native was wonderful, and I would have been frightened to wander the city alone.

I had to study hard to pass most of my classes, and I was supposed to be selecting a Ph.D. thesis problem as well. I ended up with a minor in Geology at Cornell University, primarily due to Professor John W. Wells, who taught me invertebrate palaeontology and showed a lot of field research photographs of his team working on the Great Barrier Reef of Australia. One of his graduate students, Ken Reed, an Australian, insisted that I also learn some taxonomy of the *foraminifera*. Ken also loved to recite Australian bush ballads, including the *Man from Snowy River*.

One of my geology class field trips included a visit to nearby salt mines where I collected some large perfect crystals of halite with liquid inclusions inside them. These salt deposits were from Silurian times and I decided to try to extract the liquid and measure the amino acids inside these salt crystals. This might provide evidence that there were living bacteria inside these Silurian salt crystals, in cyst form, or after having decayed millions of years ago.

This was of interest at the time, as other laboratories were trying to find evidence of living organisms in moon dust, extraterrestrial meteorites and ancient rocks. In 1953 Stanley Miller and Harold Urey (University of Chicago) published experiments showing that with appropriate kinds of energy, inorganic compounds (water, methane, ammonia, hydrogen) would form organic compounds similar to that found in living organisms. Based upon earlier speculations from A. I. Oparin (USSR), Medewar, Harold Urey, and Carl Sagan, thus began the field of experimental exobiology, which included simulations of how life might have begun in space, the early days of the formation of the earth, and looking for signs of life in extraterrestrial rocks and soils from space. Part of my education from Dr. Vallentyne was to analyse the amino acid composition of one single fingerprint on the inside of a super clean glass vial. This was to serve notice that tiny contamination could ruin my results.

After disinfecting the exterior of selected perfect large salt (NaCl) crystals with alcohol, flames, and vacuum I extracted the liquid and was able to find all of the common amino acids (components of living protein), plus a few rare acids commonly found in microbes and fungi. The dense salty brine inside the salt crystals probably pickled the organic molecules and slowed their rate of natural decomposition. Now I was expected to write up these results for a science journal.

Chapter 4. My First Encounter with Warm Seawater

My thesis professor was dumbfounded that I'd never seen the ocean. So he sent me off to a three month course of training in marine science at the Bermuda Biological Station during the summer of 1964. Up to this time I had very little experience with personal navigation in cities and airports, and my progress from the tiny rural airport near Ithaca to New York City was accompanied by some trepidation. By some miracle I found the BOAC terminal at La Guardia Airport and found my seat on this big airplane. As would become usual, I went to sleep shortly after takeoff but was awakened by a pretty stewardess who handed me a menu. I was hungry and thirsty but I didn't have very much money, so I decided I couldn't afford this fancy meal and would wait until I got to the Biological Station where food was provided.

Nobody told me that the airplane food and drink was included in the price of the ticket.

Somebody picked me up at the Bermuda Military Airport in an old truck that smelt of fish and tar. Bermuda is a British colony, full of the exaggerated accent of colonial nations, and I was startled to see people walking around in shorts and cars driving on the wrong side of the road. I arrived in the late afternoon and the first thing I saw at the pastel pink Biological Station was Ivy the cook hanging conch shells on the clothes line. She happily explained that these conchs had been brought in by scuba divers a few hours ago. She was hanging them by the opercula (the closing lid) and the shells would fall off when the conch muscles relaxed. These fresh de-shelled conchs then went into a stew pot with a lot of strange greens and half a bottle of black Jamaican rum.

I was near starving to death and it smelled and tasted very good to me, along with homemade bread and canned butter and chicory coffee. I was introduced to my classmates, who happened to include Stephen Jay Gould from Harvard, Maryann Meehan from Columbia University, Gary Greiner from Case Western University and Don Gordon from Dalhousie

University. I was assigned a motorised bicycle and a bunk bed and a lot of reading material for my scuba lessons coming the next day.

After some preliminary scuba training in a shallow lagoon they took us out past the reefs for seawater experiences. I cannot describe how I felt when I first saw the sea stretching away to the horizon, me being a prairie boy who used to put bricks in the toilet water reservoirs to conserve water. I'd never been swimming in warm water before and it was a wonderful shock to discover how easy it was to flipper through the clear water along the outer reef face.

I think they sensed that I knew nothing about boats and safety at sea, so the Station Manager's daughter Robin took me out in her sailboat and talked about storm waves, tides, how to hold a course on a compass at sea and how past students had gotten into serious trouble with boats and scuba diving operations. That night we all rode our motorbikes to a nearby hotel for beer and conversations till late.

Then began three months of lectures by professors I'd never heard about but much later discovered that they were amongst the top marine chemists and biologists. Bob Garrels, Harold Helgeson, Keith Chave and Fred Mackenzie taught us all about how plants and animals produced most of the reef and lagoon sediment and about the cycles of the elements dissolved in seawater. Phil Abelson was interested in the very early geological history of the earth but his main love was getting us all out to the volleyball court. Conrad Newman was busy building a mechanically driven tidal box for experiments. Visiting lecturers included Gordon Riley from Halifax in Canada (who spent a lot of time bubbling air through filtered seawater to generate particles of organic matter), and D. W. Sutcliffe, who enjoyed dumping bushel baskets of yellow pollen out of airplanes onto atmospheric and seawater windrow convection cells offshore from the Bermuda reefs.

Routine lab classes demanded understanding and precision in measurement of oxygen, alkalinity titrations, pH and salinity of seawater. We were all expected to have our own research projects and I decided to explore the faint bands of algal/fungal growth inside carbonate beach rock and cliffs of coralline rocks above seawater. I shaved thin layers of rock into tiny vials to determine the pigments and other biological compounds that would tell me about their growth rates.

Steve Gould worked on terrestrial, marine and fossil snail speciation and the mathematical (allometric) codes by which their shell was formed and coiled. He worshipped the order and symmetry in biology but was brought to tears when someone showed him a chunk of *Vermicularia spirata* reef, composed of dead Oculina corals covered with an apparently chaotic uncoiling snail. Several years later he sent me a reprint of his paper about how he discovered that the 'uncoiling' was actually all right handed and orderly, depending upon their growth substrate.

Margaret Meehan (from the Columbia School of Dentistry) worked on the phosphatic minerals in some mollusc radula (grinding teeth). Don Gordon worked on the chemistry of dissolved organic matter in seawater. We also team collected and dissected many sea cucumbers (*Holothuria*) to get enough of their anal sphincter rings, to determine their composition and mineralogy. Sea cucumbers are fat bread loaf-sized globs of gristle and jelly that lie on the reef lagoon seafloor, chewing up the sediment to digest the microorganisms and excreting the waste in characteristic stringy faeces. I knew I was never going to be able to explain this particular project to my father or my brothers back in South Dakota.

And here was serendipity pushing me along in interesting useful directions, myself not completely understanding the usefulness of this kind of education. Bob Garrels smoked his pipe while lecturing on carbonate equilibria and the state of carbonate saturation with respect to the carbonate minerals formed by snails, corals, starfish, fish and human bones. He and Harold Helgeson drew 'stability diagrams' on the blackboard, showing what minerals were stable under variable conditions of oxygen, alkalinity, pH, salinity in Precambrian, Cretaceous and present day seas. Phil Abelson told of his experience with uranium fluoride in the Oak Ridge (Tennessee) centrifuge building during World War II and some of his post-war work on defining the reaction chemistry of radionuclides produced by Oak Ridge nuclear reactors. This brief education in radiochemistry gave me a leg-up when I tried to build an alpha and gamma spectrometer laboratory to measure natural ^{210}Pb and nuclear bomb fallout ^{137}Cs for sediment core chronology work in the Canadian Arctic.

At that time I didn't know that my Ph.D. thesis would lean heavily on Garrels and Christ's book about carbonate equilibria and I would use Helgeson stability diagrams to sort out the different forms of arsenic in mine-polluted lakes in Yellowknife, NWT. The Woods Hole research vessel *Chain* anchored offshore briefly and we were all given a tour of the ship and its scientific equipment. This somehow spoke to my sense of adventure and I enjoyed imagining a life at sea out of sight of the coast. Many years later I did get that chance.

I eventually bought a pair of Bermuda shorts and hoped to hell that my brothers in South Dakota would never find out about this. We rode our bikes around the island, exploring the mangrove coasts, the sea caves and cliffs, hotel pubs and candy stores. We ate frog legs, lobster and abalone. We discovered a small boat load of sharks being butchered on the docks. The dorsal fin muscles were being cut into little round cylinders with a sharp 'cookie cutter', to be sold as scallops. Bermuda cedars were almost gone when I was there and I bought some hand-carved decorative wall hanging cedar sailboats for my parents. I had my Fort Pierre Stetson hat with me and I've kept a Bermudian one pound note in the hat band ever since this time. I became aware that I was strangely at peace on the *Panulirus* (the small Biological Station ship), that I didn't get seasick and I was good at handling scientific equipment on a working ship. I think there was a connection between the endless South Dakota prairie grass waves in the wind and looking out to sea over the wind waves.

GJB, Robin Sutcliffe and Stephen Jay Gould. Bermuda Biological Station, 1964

It was sad to leave all these good friends at the end of the summer. As each person left for the airport we siphoned the fuel remaining out of their motorbikes for our own use. Gary Greiner died of cancer about a year later. Maryann Margaret Meehan went on to use tetracycline to track the growth of teeth in a variety of mammals. Don Gordon wrote many papers on the chemistry of dissolved organic matter in the seas of the world. Steve Gould wrote several papers on the Bermuda land snails (living and fossil), many pleasurable polymathic essays and books on natural history, evolution, and the history of science as well as being a member of the Bach Choir in Boston. He survived for two decades after being diagnosed and treated for peritoneal mesothelioma, from inhaling too much asbestos in the old Zoological Museum back rooms of Harvard University. During his chemotherapy he smoked marijuana to alleviate pain and nausea and he testified in court cases about the medical benefit of pot. He died in 2002 from a metastatic adenocarcinoma of the lung, completely unrelated to his dose of museum asbestos.

Chapter 5. Across the Atlantic to Europe: An Innocent Abroad, Cold Toast and Warm Beer

These were the days when scientific research was relatively well funded. Somehow Professor Vallentyne found enough money to send me to a big science conference in Warsaw, Poland in the fall of 1965. I was to report on my research into the amino acids found inside the Silurian salt crystals. I gratefully got help from my fellow graduate students on how to make 35 mm slides of the data and carefully tailor the duration of my accompanying speech. In the meantime, my mentor and thesis advisor Professor Vallentyne organised my itinerary for this trip, including visits to several European scientific institutions along the way, to inspect and order scientific equipment and to gather advice about my thesis work.

It is hard to imagine a more naive and unprepared person being thrust into the chaos of international travel, the crush of humanity in a hurry, in taxis and airports, the babble of many languages over garbled booming loudspeakers, the brief instructions (in garbled accents) on how to find the right boarding gate at the boarding time. Somehow I stumbled upon the right path and developed travel habits that insulated me from becoming frightened or lost. I enjoyed people watching, noting the exotic logos for foreign airlines, surveying the useless but attractive items for sale in airports and desperately trying to find good coffee and pastries. I also discovered a great talent for sleeping in hard chairs waiting for my next flight.

In an accelerating scale of madness I progressed from peaceful Ithaca to Binghamton in a small plane, from there to JFK airport in a slightly larger plane, then struggled to find the TWA terminal through concrete tunnels and anxious non-English speaking people with piles of luggage. New Yorkers might as well have been speaking in a foreign language and it took me several runs to understand them. Gulls circled the terminal in the dirty sour air and little boys traced airplane take-offs with their fingers on the glass of the floor to ceiling windows in the TWA lounge. This time I knew enough to take the airplane food from the stewardess and enjoyed a small steak and several beers before falling asleep over the Atlantic Ocean.

I woke up over the Orly Airport on the outskirts of Paris the next morning and I was happy to see fields of grain and small farms with red tiled house and barn roofs. Of course I couldn't find a breakfast of bacon and eggs and was forced to eat cold croissants with some very good grape jam. I took the train to Geneva, where I wanted to visit several research institutions on Lac Leman. The Chief Scientist at the Thonon-le-Bains research station fed us a lunch of deep-fried tiny perch with crusty bread and wine, while we sat on the veranda overlooking Lac Leman.

Later in the day I took a paddle wheel ferry to Lausanne, where Auguste and Jacques Piccard's *Mesoscaphe* was offering a submarine tour of the lake bottom. This was magic to actually see the surface of the lake mud at 100 metres water depth and watch fish digging in the soft mud. For some reason this affected me deeply and I began to be interested in the 'romance of mud', or the millennial history that could be gained from studying the changes in mud chemistry and microfossils in long cores of the lake mud. On the ferry returning to Geneva I met a group of Canadian girls who helped me eat garlic butter steak and enticed me to visit a Geneva night club called Gay Ba-Ta-Clan, where a jazz quintet was playing. Over the next few days I visited museums, art galleries, attended an evening of ballet, an opera and a concert of symphonic music. My journal indicates that all of this cost less than \$30/day, with the fine hotel room being only 25 francs/day. Throughout this adventure I tried very hard to send appropriate postcards to my parents and brothers back in South Dakota. I'm sure they thought that I was crazy, and would get lost.

I took a train in the rain from Geneva to Zurich to catch a Polish airline flight to Warsaw the next day. The plane was full of international scientists going to this conference in Warsaw and it was wonderful to meet some of the giants in the field that I had read about in journals. The Poles served a large lunch on the airplane, supplemented with generous servings of vodka and beer (*piwo*), and it was like a large scientific party up in the air. Absolute chaos happened upon arrival in Warsaw, with lost baggage and buggered hotel reservations being common to our party. I joined Dr. Vallentyne and a group of his friends in a hotel supper and found out that they had spent the last week in gambling casinos in London. We ate lots of cabbage and pork and watched old people dancing to a C melody sax, accordion and out-of-tune violin.

And here was the crushing sadness of history. I walked the quiet, dark sad streets of Warsaw, where war had clearly left its mark. Sprays of bullet holes could be seen on stone buildings and some sections of the city had not been rebuilt. Piles of housing rubble blocked some of the streets. There were few cars, mainly diesel trucks carrying water and vegetables to the markets and hotels. The city was grey and dark brown and I was well aware that we were behind the Iron Curtain. The conference organisers put us on a bus the next morning and we toured down to Krakow, an even older capital city. Krakow smells old, it smells of stale beer, wood smoke, disinfectant, urine, bakery goods, sweat, horse manure, fresh-cut grass, musty cathedrals and pigeon droppings. It was also a cultural city, with Arthur Miller's *After the Fall* drama being performed, the movie theatres featuring Paul Newman's *Hud* and the musical *Oklahoma* on billboards, in addition to Polish theatre and entertainments. The theme for the 'Pink Panther' was on the radio in my hotel room.

While sitting in the Krakow old town square drinking hot tea, Polish friends tried to explain the story of the tower of St. Mary's Basilica, where a 13th century trumpeter tried to warn the city of Mongol invaders. The trumpeter sounded the warning signal (*heynal*) but was struck in the throat by an arrow in the midst of the signal, and ever after the warning trumpet signal stopped at the same place. We heard a long lecture at the Jaegiellonian University (established 1364!) and had a banquet at the medieval Wawel Castle in the midst of walls covered with Flemish tapestries, Italian art and French medieval armour. European people sitting beside me at the banquet table could speak easily in four or six different languages and I found that my university German and French language skills were nearly useless. The oppressive weight of history here began to crush my optimistic and naive American ego, and this was probably a good thing.

We toured on down to Zakopane, on the edge of the mountains near the Czech border. Agricultural activities here were very basic, with horse-drawn wagons hauling coal, bricks, timber and hay. Many of the field workers were bright head-clothed women, hand-scythe cutting, raking and stacking wet hay, or knitting while watching herds of goats and cattle. Many of the barns and animal shelters were decorated with floral wreaths and crucifixes. We walked hiking trails up to Morskie Oko and other small lakes in the Tatra Mountains, and it was good to get some clean country air in my lungs.

That night our group had a good meal of *golabki* (stuffed cabbage leaves) and beer in a cabaret, where a troupe of gypsy dancers were performing. My Italian friend Guiliano (the Tiger of Pallanza), Sir Charles Goldman (the Caliphe of California), Sikko Parma (a mad Dutchman) and I managed to snag dances with some of the gypsy troupe and they came to our table to eat up our leftovers and ask for more booze. One of these painted ladies in black stockings attached herself to me and claimed she had read Kafka, Hemingway and Sartre while she was a drama student in Prague. She was also familiar with Duke Ellington, Louis Armstrong and Miles Davis and wanted to know if I had any J. F. Kennedy half dollars to spare. She hoped to work in Paris with a jazz dance group.

Back in Krakow the severe problem of breakfast was explored with a Swedish lady scientist, who was given a plate with some strange grey lumps. She wondered aloud...'It is some type of egg. I wonder if it should be an omelette?' It turned out to be an excellent scrambled egg with mushrooms, but we couldn't find any coffee or toast. We visited a large salt mine in Wieliezka with a Polish geologist, who explained that some salt miners lived underground for parts of their lives, carving homes into the salt. Polish people also believe that breathing the salty air was good for their health. We also visited Bialowieza Puszcza, a forest reserve little disturbed since medieval times, and it was hard to imagine that most of Europe was covered with massive oak forest before the advent of agriculture. A few of Przewalski's horses (*Equus przewalskii*) were said to be living in this remnant forest. Charles Goldman, Bob Wetzel and I visited Auschwitz, to see dark rooms full of eye glasses, shoes and false teeth. We were very quiet for the rest of the day.

Most of the week (back in Warsaw) was taken up with scientific sessions, where hundreds of short papers were presented on all topics related to freshwater, lakes, streams, reservoirs, swamps and groundwater. Heated discussion sometimes followed, in a welter of languages that I couldn't follow. I soon developed a team of European companions who would translate softly in my ear, chortling about ignorant Americans. I expected Professor Vallentyne to give me some encouraging advice on presenting my first paper at an international science meeting: he simply said 'Check that your zipper is up...' with a big grin on his face.

I boldly stepped up to the podium (shaking) to present my paper, speaking slowly and loudly as instructed by my mentors. I was wearing my high school graduation suit and a necktie and felt very uncomfortable. It all had to do with how long we should expect organic molecules of dead organisms to persist in the geological record of rocks and whether this would tell us something about how earthly and other planets' microbes evolved over long periods of time. The amino acid proportions I had found in the New York Silurian salt crystal liquid inclusions were slightly different than modern microbial amino acid proportions. I was happily surprised to be swamped with questions and discussion, as many eastern European scientists were doing similar research, including samples from the Wieliezka salt mines nearby. I was invited to apply for post-doctoral positions in Sweden and Russia.

These meetings were held in a large Palace of Culture and Science, with samovars of hot tea and platters of little cakes scattered amongst the labyrinthine hallways. Mid-day lunches were lavish displays of hot sausages and cold salty meats, pickled mushrooms, stuffed carp, duck eggs in gravy, pressed beef tongue, headcheese, dark breads, cheeses, beer, vodka, wine, separate tables of fruits, desserts and the inevitable large brass samovars of steaming tea. Toilets were controlled by little ladies in headscarves, sitting behind a table just inside the door, expecting a tip for keeping the toilets clean. And there were many long evening feasts in restaurants, where the science discussions continued with passion, vodka and beer, cigars and pipe smoke. My friend Giuliano, the Tiger of Pallanza, would plug his ears so he wouldn't have to tip the musicians. A gigantic banquet was held on the final evening of the sessions, where representatives of all the counties would stand and drink a vodka toast to Poland and the organizers of the meetings. One of the favourite vodkas had a stem of buffalo grass in the bottle, this being one of the locations of the now-extinct European bison (wisent, *Bison banasis*). I had to fake many of the toasts as I couldn't handle so much vodka, me being a puritan prairie boy a long way from home.

And here was also a shiver of recognition, a glimpse of an international community of scientists and friends that somehow jumped above the level of language and race, culture, nationality, past wars and the Iron Curtain. This warm circle of congeniality also included art, music, history, philosophy, sport, practical jokes and a surprising temporary negation of national patriotism. This had a heavy impact on my understanding of how the world works.

But my prairie family pursued me, even here behind the Iron Curtain. A distant cousin of my mother was a diplomatic officer in the American Embassy in Warsaw and she invited me to supper. We were accompanied by several of her multilingual Polish friends, who tried to educate me on the history of this part of the world. This included small glasses of *slivovitza*, an overproof plum brandy much loved by the Poles. This supper also included many stories about Polish princes who had French wives and Russian mistresses in the 16th and 17th centuries, which explained why there were so many French paintings and sculptures in Polish castles. These polite and wickedly fun ladies took great pleasure in extracting stories about my life in South Dakota, hooting at the idea of hunting deer on horseback and jitterbugging to Elvis Presley records. I had to take off my cowboy boots so they could examine them closely. Second cousin Cherry Avery took me to see a musical drama (*The Three Musketeers*) and a visit to Chopin's home, where I spent my last *zlotys* on a pile of early recordings of Chopin nocturnes for my grandmother the piano teacher, who had a white ceramic bust of Chopin on her piano.

One of these well educated Polish ladies asked me if I'd ever heard of *The Good Soldier Schweik*. She explained that this was a book written after World War I by a Bohemian from Prague. Schweik was a dumb country boy patriot who went to war with great enthusiasm but managed to sabotage every action he touched. The book satirically presents the worst of military behaviour, illustrating the inevitability and futility of war and the trivial events that start wars. The book was banned by the Nazis during World War II. The Polish lady said that this national character perfectly accounts for Polish reluctance to honour military power and the many conquerors that have ruled their weary land. And likewise, most Poles have little respect for the divine right of kings and queens. This evening supper conversation was worth more than a full university course in European history.

I was strangely sad to leave cloudy dark Warsaw. My hotel tea ladies knew me by first name and would present my preferred pastries and a cup of tea as I walked past their table. The hotel piano man played Polish folk songs and New York Broadway tunes and the radio broadcasts in English were berating President Lyndon Johnson for his actions in Viet Nam and the Dominican Republic. I wanted to go to Uppsala in Sweden to visit a famous lake research lab and the Linnaeus museum. So I flew across the Baltic Sea to Malmo and Copenhagen and whammo, I was in a different, loud, colourful world of neon lights, automobile traffic and bright clothing in the warm September evening.

I found a cheap student hostel with open bunk beds and a large cheerful dining hall with Miles Davis records being played. After some beer and good sausages I was dragged along by a group of pretty girls to the Student's Club, where lectures were in progress (in Danish) on the philosophy of Soren Kierkegaard, followed abruptly by a folk band featuring Danish versions of Joan Baez songs, a Norwegian key fiddle (*nyckelharpa*) and hurdy-gurdy band, a Greenlander group in mukluks and walrus skin trousers that sang hunting songs, a sled dog whip demonstration (five beer bottles broken in five attempts) and finally the National Anthem at 3 am, after which we ate pickled herrings in sour cream and drank more beer. The next day I walked the Hans Christian Anderson Boulevard of Kobenhavn, visiting Tivoli, a classical music festival in the park, an agricultural fair and had some wonderful pastries with good strong coffee. I boarded the noon train for Stockholm and was surprised to see the train drive onto a ferry boat shortly after Helsingor (of Shakespearean note). Marijuana and other mind-expanding drugs were being sold and consumed in an open manner. About half the passengers got seasick and the rest of us ate tax-free cold meats, cheese, hard bread and drank beer all the way to Sweden (which wasn't very far away).

On the electric train from Malmo to Stockholm I could see ploughed fields, barns with roof-tops covered with soil and grass, church spires and fishing boats pulled up on land. The train station was clean and efficient but there were bums begging and drunks fighting on the street toward my hotel. I was familiar with the hippie movement in USA but the Swedes seem to have a great abundance of long-haired dirty tie-dyed young people on the streets, with apparent abundance of booze and drugs to support their freedoms. The next morning I took a bus further north to Uppsala, where Professor Rohde welcomed me to his Limnological Institute. One of my future options was a post-doctoral fellow position in his lab and we were well suited to each other, liking classical music, jazz, pipe tobacco and good food. He took me on a tour of lakes and rivers of this region, inspected some runes, Dag Hammarskjold's grave and we ate more pickled herring in sour cream from small town street vendors. We visited Carl Linnaeus' summer home, where the system of scientific names was invented (in the 1730s) and applied to all known plants, animals, dragons, mythical beasts, daemons and trolls, which must abound in Sweden. His sensual description and naming of the reproductive organs of plants is still in use today. Again the weight of history was on my shoulders, oppressive sometimes, and it made me feel naive about history.

My next task was to visit a research lab in northern England. So I boarded a train in Stockholm with a sack of cheese and hard sausage and little tins of juice. This was because I was running out of money and train food was expensive. I was reading Ian Fleming's James Bond paperback books along the way. I met a Swedish oil field engineer who was going down to London, where he would fly back to Saudi Arabia where he worked. He had come back to Sweden to find a bride but had failed, he was miserable and wanted to borrow some of my pipe tobacco. After several switches to ferries, trains, and airplanes, I found myself in Prestwick, Scotland, and rode a bus down to the Ferry House on Lake Windermere. Here were Ayrshire cattle and sheep grazing on sparkling wet grass encircled by winding stone fences up and down the hills. This was the home ground of the romantic lake poets William Wordsworth and Samuel Taylor Coleridge of the late 1700s, whose works I had read in high school back in South Dakota. Now I could imagine the opium-dazed Coleridge trying to capture his exotic dreams of Kublai Khan and the 'Rime of the Ancient Mariner'.

The exposed slate-grey rocky hill tops were just like I imagined from reading Emily Bronte's *Wuthering Heights* and I expected the dark skinned Heathcliff and some gloomy dogs to be found walking across the morose scree in the cold rain. Small villages of brick and stone houses with smoke coming out of multiple chimneys appeared as if in children's story books about medieval England. And old and exhausted it looked to my North American prairie boy eyes. This country looked like it had been chewed down by centuries of too many sheep, the hills were bare of trees and all the tiny rock fenced pastures looked like every blade of grass was managed by the farmer's hands. Hundreds of generations of Picts, Celts, Saxons, Danes, Romans, Normans and now modern tractors had left very little natural landscape. It was easy to see why most European and English people consider North America to be a wild and savage place even today.

And I found myself in one of these old stone buildings shortly, an Ambleside bed and breakfast inn with a crotchety old Scotch doorman, who I asked for information on the schedule of the ferry that crossed Lake Windermere. In a language that I could barely understand, he said 'Fellow what runs the ferry don't go by schedule. He's either on the other side, or on this side, or in the middle, or in his pub. He don't have high blood pressure.'

The kind old lady that ran the accommodations took great pains to explain how to flush the toilet with a chain hanging from the ceiling, that soap cost extra and there were extra heavy wool blankets and four pillows in the room closets. She acted like I was a wild savage from the other side of the Atlantic Ocean. The next morning was a wonderful experience as there on the sideboard were steaming hot platters of bacon, fried eggs, sausages, rolls and bowls of butter. I had trouble with the European 'continental breakfast' over the past weeks and this was just what I needed. The only trouble was the toast. They had these diabolical little metal trays that held the toast vertical, apparently to cool the toast so that it wouldn't melt the butter. Another trouble later on was the experience of drinking warm beer and ales. I couldn't figure out why such a chilly damp country would keep their beer warm.

I took the cable ferry across Lake Windermere to the Ferry House. For a freshwater scientist this is equivalent to approaching the Vatican. The Ferry House was established as a centre for aquatic research in 1931 and many very important research papers came from this wonderful old building. I finally found John Mackereth in the basement workshop, threading steel pipe to fit his continually upgraded mud coring device. We had lunch on the grass near the shore of Lake Windermere, watching the sailboats sliding by with the gentle wind. Mackereth was a kindred spirit, as he liked to make new gadgets to measure what happens in lakes. He had also cored the deep mud of lakes of this region to mud depths of 8-10 meters, to discover the history of the lakes back into the glacial times. He showed me some of the oldest sections of one sediment core that was annually laminated, indicating that it might be possible to determine ecological history on a yearly basis, for mud that was 15,000 to 20,000 years old. I asked him to give evidence for the annual nature of the laminations and he smiled warily, saying 'That would be a good Ph.D. thesis for you.'

He was also a kindred spirit because he would slip the conversation into medieval history of this region, stories about the 'Witch of Westmorland', folk tales about the Lake District, how the 'wallers' developed the skill of building stone fences and homes without mortar and how the high tarns and peat bogs were being destroyed by farmers. He also told me to look up my name in the local telephone book, as there were Brunskill families in this region. I phoned one of them but they were not interested in distant relatives from USA. I later found that my Brunskill ancestors probably originated a few miles east of Lake Windermere in the upper reaches of the Swale River (Swaledale), around Grinton and Muker.

Part of my reason for visiting this famous lab was to order Mackereth's newly developed oxygen electrode on seventy meters of cable. Of course he insisted that we wait six months for the new version of this electrode to be shipped to Dr. Vallentyne at Cornell University. In the afternoon some of the graduate students and staff took me to the Far Sawrey pub, where they proceeded to get me drunk on frothy hard cider. I thought cider was a weak drink and consumed several litres of the stuff. I could hardly get up from my pub bench to walk to the ferry and my bed and breakfast room. I went to sleep dizzy and forgot to eat supper. Near Far Sawrey is where Beatrix Potter lived and at my mother's request, I visited her home, which was a Nature Conservancy and tourist site. The whole place looked like the illustrations in the Peter Rabbit and Jemima Puddleduck books.

While in my bed and breakfast room in Ambleside I got two letters from my family in South Dakota. My youngest brother John told me that the local rodeo was a success and he'd got bunged up in the first high school football practice for the year. My mother's letter told about one of the local bank managers leaving his family and running off with a younger lady. But my father had got the hay up, the stock dams were full, and the cattle were looking good.

I took the train to go north to Edinburgh, through Carlisle country sparsely dotted with sheep that had numbers painted on their backs. I was going to the Edinburgh International and Fringe Festival to meet some of my musical friends from the USA. Upon arrival I found that the city was booked solid, and no accommodations were to be had at the hostels or YMCA. My personal good fairy must have been near, as I made acquaintance with a group of fellows in the street who invited me to stay with them. It turned out this was a musical comedy troupe called 'The Pedestrians', London's Hopping Madmen, although most of them were from Cambridge. Their leader, John Reagan, had rented a small cottage on the outskirts of the city. He gave me the living room couch to sleep on (beside a coal-fed fireplace), but little sleep was obtained as everybody stayed up till the wee hours playing music and telling jokes, smoking pot and drinking whiskey. I met the group Pentangle at this house and they were never far from their instruments, rehearsing till the wee hours. Also in this house were several fiddlers and ballad singers, the conductor of an Oxford opera company, some stage sound technicians, a theatre drama producer, several children and a bearded fellow who claimed to be a street poet. This was a wonderful escape from the scientific culture that had claimed my soul over the last few weeks.

Edinburgh was sparkling wet, happy, usually raining, and the city was crawling with musicians and artists of every kind. Venues for new dramatic productions, string quartets, lectures on Scots fiddle music history, folk music, medieval pageant plays, Renaissance motet concerts, sessions of poetry and spoken ballads, experimental (& erotic) films, cathedral organ and choir recitals, electronic music, symphony orchestras, drum and bagpipe corps, country and western bands (complete with oil rough-neck Texas drawls), modern dance and of course the Blackwatch Pipe and Drum ceremony on the Esplanade at Edinburgh Castle. The bagpipes of the Edinburgh Tattoo gave me the shivers, tears were running down my cheeks and I wondered if there was some distant genetic Scots blood in my veins. Maybe so, as my father's mother was a MacGregor. Tickets were cheap and I listened to Sir John Barbirolli conduct the New Philharmonia Orchestra playing several new compositions by Benjamin Britten (the violin soloist was from Warsaw, interestingly). The Swansea University Players presented Dylan Thomas' play *Under Milk Wood* in a lyrical language that was closer to singing than speech. I discovered some incredible traditional folk music being played by Pentangle and I spent some of my last money on two of their LP records. I could look down from Princess Street to see the Sir Frances Scott Memorial while I was eating a bowl of meaty oxtail soup. There were bagpipers and fiddlers busking on the streets and poets lustfully reciting on street soapboxes. Late at night the jazz joints below street level were busy, with comedians and storytellers in the intermissions. It is hard to describe how all of this creative expression felt to a South Dakota prairie boy. I was ready to give up a career in science if someone would teach me how to play the fiddle.

Completely tuckered out from lack of sleep and too much culture, I rode the bus to Glasgow and the Renfrew Airport outside of town. I only woke up to eat on the airplanes flying to London and New York City and I probably spent my 25th birthday asleep over the North Atlantic Ocean. Groggily I found a taxi to get to La Guardia Airport, through slums and dirty industrial parks, a smaller plane to Binghamton and an even smaller plane to Ithaca. Now I had to get back to my lab and get to work on my research thesis.

Chapter 6. The Romance of Mud

Bill Harmon was a one-eyed retired Navy Seal demolition diver and he was my new roommate in our rented house in College Town, up on the hill above Cayuga's waters. He was after a Ph.D. in biology, studying the taxonomy and ecology of clams and snails in the Finger Lakes region. He could identify the species of big river clams when you threw them up in the air. I learned to identify most of the aquatic snails in the area from him. He had grown up in this region, had a wife and a home in a small town north of Ithaca and several nice wooden boats. He was determined to teach me how to waterski, something that had never occurred to me to be interesting or useful. I'm sure it was a pitiful sight to watch this skinny South Dakota cowboy, wobbling and trembling on the skis, after several days of trial and error attempts. He also played the five-string banjo and knew some old time ballads. He got paid for collecting road kill deer which he delivered to the parasitologists over on the Cornell University Agricultural Campus, but we often kept the loin and haunch for us to eat if the kill was fresh. In the spring we would go smelt fishing with dip nets on Cayuga Lake and bring home laundry baskets full of small fish to freeze. They were very good deep-fried whole or pickled in vinegar, salt and green chilli sauce. On special days Bill and I would walk down the street to Johnny's Big Red Bar and Grill to feast on manicotti, the cheapest dish on the menu.

I had accomplished most of my academic lecture courses (with the exception of that cursed French language) and it was time to get serious about my Ph.D. thesis work. Stuart Ludlam and I visited Fayetteville Green Lake, up near Syracuse, to collect water samples for Ed Deevey at Yale who was studying the isotopic fractionation of sulphur by anaerobic and photosynthetic bacteria. The bottom half of this lake was devoid of oxygen, stinking of hydrogen sulphide, and these exotic bacteria were thick (and pink/purple in colour) at the interface between the oxygenated surface water and the deep sulphurous water. We spent a day bringing this pink water to the surface and carefully trying to suck it into nitrogen filled glass vials without allowing any oxygen into the vial. It is a wonder we didn't get sick from inhaling hydrogen sulphide from the pipettes we were using. Stuart was interested in lake sediments so at the end of the day we quickly took a dredge sample of this stinking deep soft mud, put it in a bucket in the trunk of the car and drove back to Ithaca.

We had forgotten about the dredge sample of deep lake mud and it lived in its bucket in the corner of the lab for several months. Stuart finally opened the Ekman dredge in a fume hood and the dried block of sediment plopped out. To our amazement, the sediment was neatly layered into white and dark horizontal layers. Stuart had observed glacial varves (annual couplets of silt and clay) in long sediment cores from Cayuga Lake, but these were from ten to fifteen thousand years ago during times of pulsed melting of the North American glaciers. Within a few days I had my hypothesis for my Ph.D., that these light and dark layers were currently forming and my task was to identify the mechanism for their formation.

Professor Vallentyne gave me grant money to buy an old Chevy station wagon for $200, as I needed to haul equipment back and forth between Cornell University and Green Lake. This sturdy car was affectionately called 'The Limnorocket'. I was given Stuart Ludlam's pontoon coring boat (photo), as he had finished his thesis work. This was a flat working platform boat with a large square hole in the middle, over which a strong wooden plank A-frame and hand winch was perched. We called it the *African Queen*. This allowed us to lower sampling gadgets and measuring tools into the water from the centre of the boat. The Green Lake State Park officials kindly allowed me to keep some equipment in their tool shed and I even slept overnight there many times. Every few weeks I would sit on my pontoon boat in the deepest part of the lake making measurements of the vertical profiles of temperature, conductivity (salt content), oxygen (using John Mackereth's new probe from Ferry House in England) and dissolved ions and filtering the water to obtain the suspended particles. I found that late spring and early summer surface warm water was filled with tiny crystals of $CaCO_3$ but these crystals were not found in surface or deep water in fall or winter. We used snowshoes to walk on the thin ice, dragging a large sled behind us, to obtain water samples under the ice. Several near disasters occurred when I fell through the ice in early spring and I was wet to the neck in cold water. Fortunately the Park Office had a hot shower and I learned to wear a scuba wetsuit and carry a change of clothes on these sampling trips. We often worked past dark on the lake and would stagger into Sergio's Restaurant in Canastota for a plate of garlic-rich Italian food at midnight. We would then pick up a box of fresh hot donuts and refill our thermos with hot coffee for the drive back to Ithaca, with the old Chevy station wagon loaded to the roof.

The African Queen *research pontoon boat on Fayetteville Green Lake, N. Y., in 1966.*

With the help of other graduate students and the engineering shops I invented an underwater filtering device that could be used by a diver to obtain undisturbed large samples of these magic crystals. We also built some 'sediment traps' (funnel topped tubes supported on a frame, which was deployed into the lake on an anchored wire) to catch falling particles over a year. This work showed that the tiny glittering white $CaCO_3$ crystals were formed as the water warmed in spring and early summer and these crystals settled to the bottom to form a neat white layer in the bottom mud. As the deep water was lacking in oxygen, no bottom animals lived there to stir and eat the organic matter in the mud. This was the mechanism for the formation of annual white layers in the mud, with the fall and winter seasons being represented by a dark silt and clay layer. My job was to measure how all this happens and to prove the annual nature of the white and dark layers in the lake mud. If this were true, the lake mud then becomes a history book for the region, preserving annual signals of past changes in land use, runoff, climate and vegetation.

I gave several lectures on this research at local conferences and attracted the attention of some important scientists who influenced my career in science. Bill (W. H.) Diment, from the US Geological Survey in Reston, Virginia, was hunting for locations to record geothermal heat flow and he showed up at Green Lake with a semi-trailer of complex gadgets for measuring deep earth heat flux into the isolated bottom water of Green Lake. He found year-round heat convection cells in the deep water, evidence that the heat from the earth was moving into the deep lake water and into the atmosphere. Two German microbiologists, Hans Trupper and

Holger Jannasch at Woods Hole, wanted to see the pink layer of water in this lake and came with special containers to sample these pink and purple sulphur bacteria.

Drs. Taro Takahashi and Wally Broecker (from Lamont-Doherty Geological Observatory, Columbia University) appeared with another truck full of gear to measure the partial pressure of carbon dioxide and radioactive carbon-14 in the deep water. From the absence of thermonuclear bomb ^{14}C in the deep water they could claim that most of the lake water came from deep spring inflow, not surface runoff. They also took a long core of the deep lake sediment and decided that the white-dark layers in the mud could not be annual. This bothered me a lot but did not deter me from continuing to seek evidence for my hypothesis. Wally measured the outflow of the lake by throwing an orange in the stream, timing the rate of movement and measuring the cross section of the stream. We sat on the jetty eating corned beef sandwiches and beer and talking about how to figure out how the world works. Wally and Taro talked about the world ocean as if it was just another lake in front of us. Wally was only ten years older than me and had come from a conservative religious college near Chicago but he had certainly broken the bonds with the Midwest. He was one of a few geochemists that realised the thermonuclear bomb tests in the Pacific had labelled most of the surface of the earth with radioactive tracers, so that we could figure out how fast the ocean circulated and where the organic matter in soil and the ocean was going. He was running the radiocarbon laboratory at Lamont, dating samples from all over the world and trying to figure out the changes in the global water cycle over the past several glacial cold periods and warm interglacial periods. His style of thinking and calculating ballpark data in the field stuck with me later in life. I flew into New York several times to consult with him and he taught me how to keep several hypotheses open in your mind, while gathering evidence to show that some of them were false.

Large reef-like growths on the margins of my study lake were called bioherms and no one knew how or why they formed. I tried to culture the blue-green algae that coated these reefs and went to Frances Drouet at the Philadelphia Academy of Science for help on identifying and culturing these algae. This expert on these algae bluntly told me that putting names on these crazy algae (now classified as cyanobacteria) was a waste of time, as they changed shape and physiology depending upon where they happened to live. But he did show me how to use a special microscope to see the mineral coating on the outside of the cells. It seemed likely that the algae formed $CaCO_3$ crystals in their slimy cell walls and this gradually formed the large reefs around my Green Lake. But I gave up this line of investigation, as algal taxonomy and physiology of this group seemed to be hopeless. While I was in Philadelphia a folk festival was happening and I wandered past many stages of fiddle and banjo players, Appalachian ballad singers and many clones of Bob Dylan and Joan Baez.

During all this focused research I also managed to escape into the world of experimental films, music, drama and literature. My journals are full of programs of recitals, concerts and festival programs, which I kept to show my mother and grandmother at home. Duke Ellington, Larry Coryell and Gary Burton played jazz at Cornell and I met and shook hands with the Duke in College Town the next day. Cornell had a museum of ancient musical instruments and weekly student recitals of medieval and Renaissance music. Dr. Jim Ode, my brass ensemble leader from Augustana College, was now teaching at Ithaca College and I got to have Thanksgiving supper with him. Professor Vallentyne's wife Anne was a cellist in the local symphony orchestra and string ensembles and she invited me to play the trumpet part in Alan Hovhaness' *Prayer of St. Gregory*, which allowed me to escape science for several weeks of intense practice. I was in correspondence with my Augustana classmates, some of whom were in medical school, other graduate schools and a few in Vietnam. Several new friends I had met in Europe visited my lab, telling stories of good times at the Ba-Ta-Clan nightclub in Geneva, to my embarrassment. I joined an intramural basketball team and played through the winter. I remember being disappointed by the *Mariner 4* fly-by of Mars, taking photos of that arid, crater-pocked planet, lacking the expected canals and seas.

I managed to go home for Christmas this year. I wore my Fort Pierre Stetson hat and polished my cowboy boots. Upon arrival at home I was immediately taken to the barn to help my father sew up the prolapsed uterus of an old cow with a new calf, using crimping pliers and short pieces of baling wire. My brother Obie was breaking a colt. We dowsed for water along the river gravel beds and put a sand point in for winter water for the cattle. We got a little juniper tree for the house Christmas tree and opened presents as a family, feeling very good to be home. Obie got a deer up in the foothills, a young doe, very tender meat. And then a round of visits and feasts with all the grandparents, aunts and uncles and family friends. The feast tables were loaded with venison, home-grown beef, wild turkey, prairie chicken, candied sweet potatoes, cranberries, hot buns and home-made ice cream with crushed strawberries. My grandfather Brunskill asked me about my studies and announced that he didn't think much of the emerging space industry, or east coast academic eggheads. Brother Obie's high school basketball team was very good and they proceeded into the State Tournament this year. Obie and I got two deer with our father's 300 Savage rifle and carried them on horseback to the barn for butchering.

My father took me into town after supper for coffee at the local highway gas station/ coffee shop, where I was grilled by his friends:

'I hear you've been over behind the Iron Curtain?' says Ken Foster.

'Yup. I was in Poland for a big science meeting.' Says I, proudly.

'What's it like over there, did you have any trouble?

'Nah, it was all very civilised and our hosts took good care of us. But you sure could tell they haven't yet fully recovered from the war. Some parts of Warsaw had piles of building rubble from the war and there were still bullet marks on some buildings.'

'Did they feed you well?' asked Helge Anderson.

'They eat a lot of sausage and cured meat, pickled eggs and boiled milk.'

'I'll be damned. And why were you there?'

'Well, I gave a talk about my research on organic material inside salt crystals.' Says I.

There was a long silence, stirring of coffee, nodding of heads.

'And why on earth would you study shit in salt crystals?' They asked, greatly puzzled at this queer business.

'So we could figure out how long the organic material would last. These salt crystals were many millions of years old.'

And then there was an even longer silence and we all ordered some pie.

'But didn't the Communists spy on you over there in Poland and watch where you were going?'

'Sure, they probably did. We had 'Intourist guides' that went with us all the time, helped us with translations in shops and kept us from going into certain parts of the city.'

'What sort of crops do they grow?'

'We drove through the country where they had milk cows, they were putting up hay in little doodles and there was some corn fields, grain in shocks. Little old ladies were collecting mushrooms in the forests. Lots of cabbage in their yard gardens. They like to wrap cabbage leaves around meat and cheese and potatoes.'

'Well, I'll be damned. They can't be all bad if they got cows.' And they nodded their heads approvingly.

As I packed up to return to the East Coast my mother and Grandmothers packed my aluminium laundry box full of Christmas fruitcake, banana bread, Christmas almond toffee and a hand-knitted wool sweater from my grandmother England. My father drove me down to Sioux Falls, 350 miles on icy roads in his old white Chevy pickup truck. We had lunch at the Stockyards Café and an awkward goodbye at the airport. My father was not comfortable in a big city but he was curious about me riding in a big airplane. And then it was Chicago, Detroit, Buffalo (Stan Getz playing 'Girl from Ipanema' and Ramsey Lewis playing 'Movin' Easy' in my headphones) and back to Ithaca, a different planet it seemed.

It took some effort to regain my enthusiasm for my research work, as Ithaca was dull grey, damp and cold in January. I had a wonderful basement lab in Stimson Hall, with steam heat pipes running over the ceiling. Even deeper in the sub-basement of this building were dirt-floored corridors and rooms used for storage. I found a collection of dated and named human heads preserved in formalin in giant pickle jars that had been used by some Cornell scientist in a mid-1800s study of the relationship between head shape and criminality, and some skeletons of middle American peccaries. I had a fume hood in my lab, which I mainly used for a hot plate cooker of omelettes for late night suppers. My small fridge did have a few carefully labelled calibration solutions in Erlenmeyer flasks but most of the fridge space was for hard cider and beer. I could get WWVA Wheeling, West Virginia on my little lab radio, so I could listen to bluegrass music and late night jazz. I had a complete LP set of the Beethoven symphonies and Bach's *The Art of Fugue*, which I could stack onto my little portable record player for late night work. I had inherited an old lean-back chair from another departing graduate student and I often took long naps in this chair with the door locked and the phone unplugged. My task was to measure the tiny amounts of $CaCO_3$ crystals on the filters that I had obtained on the lake over two years of sampling. I used an old gas chromatograph (abandoned a decade ago by another graduate student) to measure the CO_2 evolved when I acidified these filters in a nitrogen atmosphere. There were many hundreds of these filters and I could do about ten analyses per day, so it took over a year to complete this exacting and boring work. Fellow graduate students and new Ph.D. Stuart Ludlam taught me how to operate the new Olivetti Programma 101, the latest development in small computers, so we could do detailed calculations about the solubility of $CaCO_3$ in water of seasonally changing temperature.

While all this work was grinding on Professor Vallentyne left Cornell on a Guggenheim Scholar year's leave in Pallanza, Italy, with the Tonolli family at Lago Maggiore. I was allowed to use Dr. Vallentyne's office and I enjoyed the creaking wooden floor, his shelves full of curious objects and books in French, Russian, Italian and German languages. Things sort of fell apart for a while. There were pot and LSD parties in College Town and the flower children of the new rock and roll generation were finding their voices. The April 1964 issue of *Scientific American* was all about hallucinogenic drugs, especially marijuana, LSD and psilocybin, none of which were illegal at this time. Logical arguments were made that drug-induced experiences were beneficial and inspirational. I was led astray several times but I found pot and booze simply put me to sleep, with no special pleasures as promised. Because I was hanging out with a lot of chemical/ecological graduate students I was aware that several labs were measuring the potency of the different grades of marijuana. They found that delta-9-tetrahydrocannabinol concentrations (1-5%) in the weed varied greatly and these results were openly posted on bulletin boards in the biochemistry labs.

Students protesting the Vietnam War blockaded the Cornell Student Union, the police were called in and there was considerable unrest on the campus. Wannabe revolutionary students walked around campus with copies of Chairman Mao's 'Little Red Book', Karl Marx's *Das Kapital* and Che Guevara's *Guerrilla Warfare*. Yellow-robed Hari Krishnas walked around handing out free *Bhagavad Gita*s and Indian gurus peddled peace, sex and personal wealth. I was too busy in the lab and taking courses to pay much attention to all this. Many graduate students couldn't take the academic pressure and went home to find other careers. A friend of mine from India who was in astrophysics with Professor Thomas Gold, Chilukuri Krishna, committed suicide by jumping off a cliff above one of the gorges near town. He was not on drugs at the time, according to the autopsy. I and another graduate student arranged to have his body shipped back to India, using money granted by the Student Union.

I found it hard to keep my focus and I would escape to downtown Ithaca for films and concerts. I was drafted for military service and I took the mental and physical exams in Syracuse. I was declared fit for service. Walking naked through those cold, dark, dirty corridors with a sheaf of papers in my hand, a plastic numbered bracelet on my wrist, I wondered if I was entering a hellish new life. And here was the military command 'Lean over and spread your cheeks', just like Yossarian said. I wondered what they expected to find down there? These people handled us like slabs of beef in a meat processing plant. I had been injected and inspected all over my skinny body and they were especially interested to know that I was a deer hunter in my earlier days. The cold lonely night ride back to Ithaca in a dirty bus allowed my anxiety to increase and for once I couldn't sleep while travelling. The Zoology head of department (Lamont C. Cole) wrote an apparently good enough letter to my draft board asking for deferment, explaining that my research was in the national interest. For the life of me I couldn't figure out why my research was more important than military service, but I didn't question this twist of fate.

Meanwhile, back home in South Dakota, my younger brothers John and Gerald were riding bulls and bareback broncs in the local rodeos and Gerald bunged up his knee. This kept him out of the Army, so he could study metallurgy at the South Dakota School of Mines in Rapid City. My brother Obie was graduated from South Dakota State College with a degree in Animal Husbandry but he also spent too much time riding saddle broncs in rodeos. He broke his tailbone finally, which slowed down his rodeo career somewhat. This didn't keep him out of the Army, as he was drafted shortly after graduation. He did boot camp in Fort Huachuca (Arizona), where he showed that he was an excellent marksman with a rifle.

His buddies explained what happened to sharp-shooters (they ended up as jungle snipers in Vietnam) and suddenly his marksmanship scores declined. He phoned me from Fort Dix in New Jersey as they were shipping him off to Germany to serve as a supply clerk in Berlin for two years. This was the second generation of my family in Germany in a military uniform, my uncle Gerald having marched across the ravaged land of Belgium and Germany in 1945-46 and my uncle Raymie having served in the Navy in the Western Pacific. Several of my Augustana classmates wrote letters from Vietnam, in harm's way mentally and physically. One of them was a conscientious objector who ended up in the military medical corps, hauling dead bodies out of the jungle. He survived Vietnam but his brain was warped afterwards and he died of a rare kidney infection shortly after returning to South Dakota. I always wondered if his kidney infection was the result of being close to Agent Orange sprayed on the jungle.

My grandmother Brunskill died in March 1967 at 72 years of age and I was not able to return to South Dakota for the funeral. She was a small thin athletic lady who could put her hands flat on the floor without bending her knees. She was a point guard on her high school basketball team. She was always telling me to stand up straight because good posture makes the brain work better. She loved gardening and always had long rows of flowers interspersed between her beans, peas and sweet corn. I always enjoyed staying with her when I was young because she would make little log cabins out of fingers of cinnamon toast for me, when the men were having their coffee break.

With Professor Vallentyne away in Italy I got scientific help from Wally Broecker at Lamont and Bob Berner at Yale, good people to have on your side. I was left to finish up my field measurements and lab work and help with lab preparations for several other courses at the University. An older graduate student of Vallentyne's, Stuart Ludlam, who had his new Ph.D. was hired to teach Vallentyne's courses. He helped me a lot during this difficult time. We got the loan of a Mackereth pneumatic corer from a colleague at the University of Waterloo in Canada and tried to raise a long core history of my study lake. We were successful in driving the corer five metres into the deep sediment but we couldn't get it to come up. We cranked the winch to nearly sink the boat, swam to shore and left it overnight. The next day we cranked the winch further and late that afternoon the large corer device rose to the surface connected to a large tree. Apparently a tree had fallen into the lake and slid into the deep bottom mud. The core tube was bent, so we had to find money to hire a workshop fellow to install a new steel core barrel. This nearly emptied the funding remaining in my research grant. Unbeknownst to me, out in California in 1965 Gordon Moore, a young physicist from Silicon Valley, announced that computer chips would diminish in size and increase in speed in an exponential manner, with a doubling speed time of less than two years. We were at the beginning of the computer age.

Dave Culver and Jim Russell helped me try to measure the pigments and growth rates of the exotic photosynthetic green and purple sulphur bacteria at the interface between the oxygenic and anaerobic water in Green Lake. We used radiocarbon spikes in light and dark bottles, the normal method at the time. This didn't work very well because there was enormous and rapid uptake and release of the ^{14}C in the dark and in the light. This was because there were a large variety of microbial organisms living at this interface and they were able to use oxygen, sulphur and iron in their metabolism. Some of these microbes did not need light to grow, which confounded our measurement methods. Dave was also trying to sample the depth distribution of zooplankton in the lake and he found that nearly all of the crustacean biomass was feeding on the sulphur bacteria at 18-20 m water depth. These two fellows were good friends, Dave having worked in radiochemistry at the Oak Ridge National Laboratory, and Jim was an unreconstructed hippie from Buffalo who lived on Planter's Peanuts and marijuana oatmeal cookies. During these times we developed several fetishes and occult rituals while working on the lake. One was to cast a small pinch of tobacco on the water before difficult research operations, to satisfy the glowering kraken that lived in the sulphurous deep water.

Immediately upon Professor Vallentyne's return to Cornell in fall of 1966 we drove my old Chevy station wagon from Ithaca to a Gordon Conference in New Hampshire, where Jack met with a group of experimental aquatic scientists who had some important planning for a new project that I knew nothing about. This science meeting was a wild affair, with much time spent in the local bar. Here I met Bob Harriss, who later came to work with me on Green Lake, Philip Abelson (an old friend of Jack Vallentyne), Werner Stum from Switzerland, Fred Mackenzie, who I had met in Bermuda, Bob Garrels, Enrico Bonatti and Cesare Emiliani, who later did some of the best work on foraminiferan oxygen isotopic tracers of climate change in long ocean sediment cores. All the way home, driving in near blizzard conditions, Vallentyne talked about this new project, which was to use remote lakes as experimental test tubes, adding chemical and exotic species of fish and plankton to test hypotheses about lake evolution. I must admit that I thought this was a 'pie in the sky' dream.

In June 1967 Stuart Ludlam and I drove the old '53 Chevy station wagon to Madison, Wisconsin for a major symposium on eutrophication. This big word means 'the feeding of nutrients to a lake' and this subject was to consume a large chunk of my life in the next decades. We heard magnificent presentations from G. E. Hutchinson, David Frey, Richard Vollenweider, Fred Lee, Charles Goldman and Wilhelm Rohde, all major stars in the lake research community. Vollenweider was a Swiss scientist who developed a mathematical model of the relationships between lake nutrient (nitrogen, phosphorus) supply rates, lake area and average depth and water flow-through time. This was a major breakthrough that caused a lot of us to think bigger than usual. A topic of discussion around the beer and supper table was the need for direct experimental evidence for the mechanism of lake enrichment in nutrients and the unpleasant consequences that were becoming more obvious in lakes of Europe and North America. As usual, Hutchinson told elegant and tragic stories about how this was not a new problem and that sediment core histories from Italian lakes showed the fouling of lake water supplies in Roman times, causing disease epidemics, due to agricultural and city wastes.

I met Hutchinson twice at science meetings when I was a humble grad student. Vallentyne introduced him to me at a post-conference supper table. He was certainly aloof, arrogant, lofty, sure of himself, eloquent and scarily intelligent. He would mumble scatological curse words about somebody, clear his throat, loudly castigate in Latin and then translate it for us nitwits. He would then quote some 19th century German philosopher to clinch the argument. I was introduced to him as someone from South Dakota, as if that were the dominant part of me. He immediately asked if I knew about the vetch *Astragalus* sp. that accumulated selenium and how it gave western prairie cattle the blind staggers. Actually, I did know about this species and he grinned ever so slightly. This was my contact with one of the great gods of my science world. We also were treated to the views of Ernst Mayr, Howard and Eugene Odum and Carl Sagan, who were invited to our graduate student seminar discussion group.

It was the custom at the time to have a 'comprehensive exam' for Ph.D. students, to determine if the student was broadly educated enough to begin writing the Ph.D. thesis. My exam was hastily called together in January 1967, consisting of an algal taxonomist (John Kingsbury), a geochemist (Bob Berner from Yale), a coral reef palaeontologist (John Wells) and Jack Vallentyne. I was asked to explain (in biochemical reactions) how a steer would decompose on the South Dakota prairie and how this would happen differently in the Arctic or tropic regions. There were mathematical puzzles, questions about astrobiology and cosmology, deep sea biology, fossil stratigraphy, speciation in Galapagos finches, short questions in German and French. Most of this was fun, even if I didn't know the direct answer, I could identify how I would find information about the issue. The exam committee was impressed enough to sign the form and take me to the faculty lounge for whiskey sours and a full supper. In the next week I finally passed my French exam on the fifth attempt.

During these Cornell years I had learned how to assimilate information and reject less promising directions of study and research. I had stumbled through several alternate lines of investigation, learned methods that did not work well and wasted an enormous amount of effort in unproductive reading and lab work. Here I was, with over two years of detailed physical and chemical data on one small lake in upstate New York, and I was expected to make a Ph.D. thesis out of it and publish the stuff in peer-reviewed science journals. I had sketched a rough outline of the thesis but was anxious about the quality of this information and why anybody would want to read about it. I was also a hopeless romantic with a relentless interest in poetry, drama, music, literature and history; these interests often dragged me away into another world when I should have been focusing on science. I certainly enjoyed the intellectual challenge of graduate school, research and interpretation of information about how the world works but I had experienced no intellectual firestorm of inspiration or revelation of the secrets of the universe. More than once I was ready to give it all up and return to South Dakota to be a cowboy. Some combination of determined persistence and luck allowed me to slowly put small chunks of information together, about how a small part of the natural world operates according to the seasons, years, centuries and millennia.

Well, the Gordon Conference idea caught intellectual fire and within a few months Jack Vallentyne quit Cornell University, moved to Winnipeg, Manitoba, Canada, to help start a new institution to begin his pet project with the Canadian federal government behind it. He offered me a stipend in Canada to help with this new project and I was assured I could write my Ph.D. thesis there. I had other options for post-doctoral research at several US and European universities but these positions were all in big cities. The lure of an adventurous life in the wilds of Canada appealed to me. It took me about a year to finish all my measurements, pack up my lab and notebooks, LP record player, my Remington typewriter and my old aluminium laundry box full of cheese, sourdough bread and hard sausage. Cornell University sold me the Chevy station wagon for $1 and in October of 1967 I started driving north to Canada. I had about $100 in my wallet, a new USA passport and a sleeping bag in the car. I knew almost nothing about Canada, but I must have had a great faith in serendipity.

The fall tree colour was spectacular on the highway beside Cayuga Lake and the Cornell-Colgate football game was on the radio. The corn crops were being harvested and deer and bird hunters in orange vests with dogs could be seen at the edges of the patches of forests. This rural upstate New York country was very peaceful and some sadness and nostalgia hit me hard, as I had four good years in this part of the world. I had friends in Syracuse and Rochester where I could overnight. The next morning my colleague Stuart Ludlam phoned to let me know he had wrecked his car somewhere east of Syracuse, which had rolled into a farmer's barn manure pond, with all his precious notebooks and Ekman dredges full of beautifully laminated Green Lake mud. I picked him up at a nearby coffee shop, desolate and discouraged, helped him rescue his possessions and he sold the car for $50 to the farmer. We got a shower at a motel, a big supper and then I drove him to Amherst, Massachusetts that night. After another good feast, some whiskey and the Hummel trumpet concerto on the stereo, I returned to the highway to try to find Canada.

Chapter 7. My First Job in Science... Canadian Style

Canadian immigration officials in Niagara Falls were sufficiently impressed with my letter of employment with the Fisheries Research Board, a crown corporation of the Federal Government. They gave me landed immigrant status sometime in late October 1967 and off I went into the wilds of Canada, looking for the RCMP, Sergeant Preston and his Huskies and the Eskimos. What I found was Quebec French radio stations, donut shops, shopping malls and four lane highways across southern Ontario. The further northwest I drove, the better it got. I was happy to discover the little town of White River, Ontario, where a large sign declared that it was the birthplace of Winnie the Pooh.

Here began snow and ice covered roads, with lots of moose hunters in the road ditches along the spruce forest north shore of Lake Superior. I had a little magnetic based smoking pipe holder on the dashboard, a box of cigars in the glovebox, the aluminum laundry box full of food and four gallon jugs of hard cider in the passenger footwell. I became expert in one-handed roll-yer-own cigarettes at 70 mph. I slept in my car somewhere between Nipigon and Port Arthur when I couldn't see the road anymore, due to blowing snow. I was feeling the freedom of the road, having read Jack Kerouac and Allan Ginsberg, and this was certainly new country to explore. Somewhere along this boreal forest road the radio presented a folk song by Wade Hemsworth, the tune called 'Little Blackflies'. It stuck in my head and I was to become much more familiar with Canadian black flies in the future. I also sang every cowboy song I ever knew to keep awake on this long road (all from WNAX Yankton, South Dakota, Livestock Market Report at 0600 - 'The Strawberry Roan', 'Windy Bill', 'Work's all Done This Fall' etc.) With great relief I rolled into Kenora on a near-empty gas tank, had a wonderful lunch with cream cheese pie and strong coffee and looked at the map to discover the location of Winnipeg further west. Even greater relief was felt as I drove west from Kenora, as the Canadian prairie opened up before my tired eyes. Cattle, haystacks, grain elevators and lots of open space and clear skies indicated I was back in the Midwest again.

I rolled into the big city of Winnipeg at dark on 10 October 1967 and spent some frustrating hours trying to find Professor Vallentyne's house along the banks of the Assiniboine River. I found his sons at home and they gave me some supper and a bed. I slept till noon the next day. After sorting out Canadian social security I found a small apartment above a garage down near the University. I had a one-burner hot plate, a little kerosene space heater and a nice big bed. Later that winter it got very cold but I had my mother's home-made quilt and some Hudson Bay blankets to keep me warm at night. I still had some of my grandmother's banana bread, which would freeze when I was at work, but I could warm it up in a buttered frypan on the stove.

I was immediately thrust into the exciting world of the formation of a new science institution, with round-table meetings with visiting scientists from Germany, England, Japan, Italy, Austria, Switzerland, Poland and the USA. I remember sitting at a good supper table with Domenico Povoledo (Italy), Holger Jannasch (Germany) and Kazimerz Patalas (Poland), who discussed the post-war scientific progress of Europe from very different perspectives. The big task of all these visitors was how to locate, select and operate an experimental lake area so that we could treat whole lakes like test tubes in the lab, to manipulate lakes to discover how they respond to increased nutrient supply, contaminants, changes in fishing pressure and water flow-through rates. Being the youngest scientist in the group, I didn't realize what else was happening at these meetings. The director of the new Fisheries Research Board (FRB) institute, Wally Johnson, and Jack Vallentyne were hiring staff from around the world. It couldn't have been an easy job, as Winnipeg was not exactly near the centre of the cultural or scientific world, and it felt like one of the coldest places on earth in winter. Nevertheless, they were very successful and scientists, technical staff, librarians and workshop metal artists were arriving with their families almost every week. Plans were being drafted for a new building for the institute but in the meantime we were all huddled in cramped offices and labs of the University of Manitoba campus. Part of the reason that the field work got accomplished was that quite a few of us were young, unattached and from upbringing in the country. We were not afraid of roughing it in the bush and we enjoyed the hard physical work that most PhDs would not consider.

In November we were flying into lake districts on the east shore of Lake Winnipeg and around the Lake of the Woods in Ontario, using a helicopter on floats to measure the depth, take water and sediment samples and a net tow for plants and animals in the water. My new partners in this survey work were John Flannagan and Willy Burton (Scotland), Ole Saether (Norway), and Andy Hamilton (a Canadian). We spend parts of every day playing cards in cold hotel rooms in small towns, waiting for the weather to clear. During the winter these data were graphed, analysed and evaluated against logistic and scientific criteria. Amongst other exciting biological discoveries Ole Saether found a new species of a midge larvae and named it after me, as I had filled out the collection data sheet. This was *Chaoborus brunskilli*.

I managed to escape all this scientific excitement to go to my South Dakota home for Christmas, driving the old LimnoRocket through blizzards and icy roads southward, to where Stetson hats and high-heeled boots were common. I found my father asleep in front of the TV, with a big bowl of home-made Xmas candy beside him. I managed to get a deer from the White River bluffs and let the cut and wrapped venison freeze on the back porch, so I could take it back to Winnipeg with me. Grandpa Brunskill had changed a lot, as Grandma B. had died and he didn't like batching. I managed to play some trumpet tunes with my grandmother England on the piano but we were both out of practice. Brother Obie was still in the Army in Germany, Brother Gerald was learning mining engineering at South Dakota School of Mines in Rapid City and youngest brother John was just out of high school, and wanted to be a wealthy rancher with lots of horses and cattle. Local gossip was that I had gone to Canada to escape the Vietnam war draft. While this was not true, my mother said it would probably be safer for me to live in Canada as she didn't approve of this war. I sadly left my family, with my Chevy station wagon loaded up with venison, some buffalo meat, my aluminum laundry box full of banana and cranberry bread, almond toffee and an assortment of kitchen equipment that my mother knew I needed but would never buy for myself. My father gave me an old Army parka and insulated trousers in case I got stuck on the roads and had to overnight in the ditch.

During this winter I was also struggling to focus on my trunk full of notebooks and data sheets from my Ph.D. thesis work. This being times before computer graphics, I used pencil and graph paper to plot out the seasonal changes in Fayetteville Green Lake water chemistry and gradually supported the hypotheses in my head. And here was another glimmer of the excitement, pleasure and reward resulting from the hard work of collecting scientific data. Out of this puzzling mess of field notebook records, notes on other scientific research in this area, graphs and tables emerged a credible story of how this little lake in upstate New York managed to produce annual laminations in the mud. I was lucky to have help from the FRB graphics department and a wonderful secretary (Joyce Verbeke) who typed my thesis from my smudged and edited Remington typewriter paper. After several drafts were read and criticised by Professor Vallentyne and my Cornell University thesis committee I sent the Ph.D. thesis back to Cornell in the spring of 1968, where it was eventually accepted. A few weeks after this I was offered a proper research scientist position with the Fisheries Research Board, at the amazing salary of $12,000/yr. I had never had this much money in my life, so I found a nicer (warmer) apartment (sublet from a jazzy airline stewardess, who left her black and white TV for me to use). I also bought a big reel-to-reel tape recorder and some Wharfedale speakers so I could have good music sound at home, both for the J. S. Bach *Magnificat* and the cowboy songs of Ramblin' Jack Elliott. One of the startling unanticipated consequences of becoming a Fisheries Research Board employee was that they asked me to sign a form giving my allegiance to the Queen of England, this being a Commonwealth country. Being an obstreperous Yankee I refused, much to the consternation of the administrative officers. I tried to explain that a few of my ancestors fought a war to get out from underneath the thumb of divine rulers. About a week later they found another form that allowed me to swear that I would not sabotage the government of Canada, which I happily signed.

During the winter of 1967/68 we had decided on the location of our lake research area and we hauled several ATCO trailers through Kenora into some lumber camp roads SE of Kenora, near Hillock Lake. We began setting up a working field camp as the ice went off the lakes, with a crude kitchen, lab and sleeping bunks. We were using small noisy petrol generators to provide electrical power to scientific equipment and our chemistry lab (being created by F. A. J. Armstrong and Mike Stainton). I spent a lot of time trying to adapt and calibrate a little Fischer-Hamilton gas chromatograph to measure the extremely low levels of alkalinity and CO_2 in this very dilute lake water. Mutsuru Sakamoto was trying to measure radiocarbon plankton productivity but his most appreciated skill was his use of a wok to cook the evening meal for our crew. We used Pete Peterson's Kenora-based helicopter to drop light canoes or boats into a group of nearby lakes. I hired Jim Russell, a Ketchican Alaskan fisherman I had met at Cornell, to help me run the depth sounder so we could make bathymetric maps of these lakes. We would copter into a lake, cargo hook-drop a canoe or small boat into the lake, set up two tents and use the boat to make transects across the lake with a big Raytheon depth sounder burning a trace on the chart paper. We nailed aluminum pie plates to trees on prominent shore features (that could be identified from air photos) so we could try to cross the lake on straight lines. It took us several days to do one lake and we had some problems with bears raiding our groceries and ripping up our tents. So I began to learn how to operate in the Canadian bush, hanging the food bags up on tree branches, thirty metres away from the tents. Jim Russell was very good at keeping track of the sounder traces and he used air photos to trace the shoreline of the lake and make a good map of the depth zones of dozens of lakes.

Again I have to emphasise what a strange experience this lake exploration was for a

South Dakota prairie boy, whose home range had only muddy stock dams. We saw moose, bear, beavers and large patches of blueberries. In winter we could see the slides of otters down the snowbanks beside the open flowing water of the lake outlets. The land was all rock, gravel and sand and I could imagine my father saying that it wouldn't grow any potatoes, cattle would have a hard time finding grass to eat, and all that stony gravel would cause foot problems for horses.

One of the new scientists arrived at our field camp with his wife, two kids, two Gordon Setter dogs and supplies for camping. This was David Schindler, who became the research leader of this experimental lakes project. He built a cabin on an island near our field camp for his family and dogs and immediately took charge of the scientific program. He was from a North Dakota/Minnesota family of farmers, was a North Dakota State University football and wrestling competitor and had a D.Phil. from Oxford University in England. Under his leadership we moved the main field camp further into the bush, with a lot of preliminary building, plumbing and a better shed for a new big Caterpillar diesel generator.

The winter of 1968/69 was spent fooling around on Bombardier skidoos, trying to map the catchment areas of the many lakes and making trails to lakes that we wanted to sample on a regular basis. I was guilty of damaging several skidoos by hitting boulders or stumps under the snow and I was chewed out for reckless behaviour by the Winnipeg administrative pinstripe suits more than once. Hydrologists Bob Newbury and John Cherry were advising us on how to construct water budgets for these lakes. David Schindler and I hauled several tonnes of sodium nitrate and phosphate fertiliser on big sleds into one small lake that was to be our first experimental treatment. This is easier said than done. We got stuck in lake blow-hole wet snow several times and had to remove some trees to make hard-packed snow trails up some of the hills. We also had great winter feasts back in the camp kitchen trailer in the evening and the inevitable pranks began. We baked a wonderful black lake mud pie for the birthday of Staffan Holmgren, a Swedish expert on algal taxonomy.

During February 1969 I built a large piston corer and ice tripod to handle it. I was inexperienced with the little electric winch we bolted to the tripod, as I'd never used one before. We lowered the half tonne corer and it buried itself in the sediments of Lake 239. The little winch could not lift the corer up. So we disconnected the battery from the winch and went back to camp kitchen for coffee to talk it over. When we came back the winch had free-spooled all the cable out and it was only a friction tie at the end, so all the cable went down the hole. We got Bob Newbury to fix the position of the ice hole with his surveying equipment and left the scene of the crime.

The next summer we returned with diving gear and Noel Burns (from CCIW, with a big UW flashlight) and I went down to 32 m to tie a bigger steel cable to the top of the corer. We had two ELA technicians in the boat to handle the cable. As you might guess, it was cold and dark and we did fit a big snap hook onto the corer and were ready to go up. One of the old cable ends hit my facemask and mouthpiece, dislodging my mouthpiece, and I couldn›t find it with my cold hands. So I free ascented all the way up, blowing air out to the surface. I coughed blood for a few days. But we secured the cable to a buoy and next winter we did retrieve the corer with a winch powered by a gas engine, operated by Willy Burton. But my Woods Hole core catcher failed and all the soft mud dribbled out the end of the 4 m core tube.

In the spring of 1969, as soon as the ice was frazzled we began monitoring temperature, chlorophyll, nutrients on a small lake, because we wanted to use this lake for a big experiment on the physics, chemistry and biology of excessive nutrient addition (eutrophication). Later in June we began dosing this lake with fertiliser nitrogen and phosphorus, at supply rates predicted to cause algal blooms by Richard Vollenweider's OECD model. The hydrologists began a program of drilling through the thin glacial gravel/sand till around our lakes, to install wells (piezometers) to measure groundwater flow and to build water flow monitoring weirs on the inlets and outflows of several lakes near the main camp. Regular sampling of fish, plankton and bottom animals began. Brian Graham and I began a survey of vegetation, rock and soil types in this region and sampled all the lakes for mud for chemical analyses. Most of these early adventures at the experimental lake project were published in the *Journal of the Fisheries Research Board.*

Back in Winnipeg the administrative brains of our operation were in high gear, as the Canadian government decided to build another freshwater research facility to study the St Lawrence Great Lakes in Burlington, Ontario. Our Winnipeg office promptly provided notice to the Ottawa bureaucrats that there were a few other 'great lakes' in Canada such as Lake Winnipeg, Lake Athabasca, Great Slave Lake, Great Bear Lake, all of which were of similar area and volume to the St Lawrence Great Lakes. Without appreciating the immensity of the task, we proposed to survey the chemical and biological status of these northern great lakes. One of the reasons for this public and governmental interest was the appearance of bubbly foam on Ontario rivers flowing to the Great Lakes and the appearance of smelly masses of rotting algae on the lake shores near cottages and boating facilities. Most of the foamy material proved to be caused by soapy detergent and sewage, which was rich in phosphorus. A bitter debate erupted between the freshwater scientists and the soap and detergent industry about the need to remove phosphorus from detergents. The soap and detergent industry (Procter and Gamble, Lever Brothers, powerful opponents) claimed that phosphorus in detergent was harmless and that the supply of carbon dioxide controlled the growth of algal blooms. The scientific community were sure this was false, that phosphorus and nitrogen were the prime agents that feed algal blooms and the supply of carbon dioxide was in easy supply by diffusion from the atmosphere. All of a sudden the fertilizer additions to the experiment lake in our bush camp became very important and we needed to further design experiments to test the combined and separate effects of carbon dioxide, nitrogen and phosphorus supply to lakes.

Chapter 8. Great Slave Lake Adventures

When I review my mud-stained journal pages for this part of my life, it seems that too much was happening too fast. During our exploration and establishment of the experimental lakes project in NW Ontario in 1968, our headquarters bosses were also pushing for exploration of the northern great lakes of Canada. These remote lakes had previously been surveyed by aquatic biologists and fisheries scientists to determine their suitability for commercial fisheries, but there was little information on the nutrient supply rates, chemistry and thermal structure of these big lakes. The FRB had a field station at Hay River, Northwest Territory, on the SW shore of Great Slave Lake.

Upon our arrival at the Hay River Fisheries Office in early June 1968 the Chief Fisheries Officer did not believe I was a scientist, as I was too young and dressed for the bush. I had to get him to phone our Research Director in Winnipeg to verify that I was a *bona fide* agent of the FRB. After shaking his weary head at this green southern visitor, the Chief introduced me to Con Haight, a large jovial ship captain of the FRB research vessel *Stenodus*. He took us to visit the ship and there it was, a steel-hulled 22 m grey and white rusty beauty, pulled up on top of several logs on the beach. Beyond the beach towards the lake were piles of ice cakes as far as the eye could see, the ice being pushed along the south shore by the chilly NW wind. Con assured us that we were about a month too early for safe travel on the lake but we took the old *Stenodus* out for a run in the ice-free parts of this big lake.

During this time, while we were waiting for the ice to clear, we explored the sheds behind the Fisheries Office. One of these sheds contained hand-sawn thick plank shelves to the ceiling, with neat rows of labelled sample bottles full of preservatives, containing plankton, bottom animals, fish and fish stomach contents, birds and eggs and a fetal moose. These were the samples from D. S. Rawson, a pioneer aquatic biologist from the University of Saskatchewan, who had worked on Great Slave Lake in the 1940s. He must have been a tough man as most of his work was done from a small fishing boat with a hand winch, camping on islands across this large lake. These sheds were due to be destroyed and the scientific samples had no home, so we took a few of the sample bottles. I kept the fetal moose in my office for the next twenty-three years and he would mysteriously rise and fall in the preservative, apparently depending upon the changes in weather. This was another personal lesson about science: I gained a lot of respect for earlier scientists who had struggled with lack of money, ships, equipment and logistic support. Despite all their troubles they published their work in the science journals, so that we could benefit and advance to the next level of information about how the world works.

Later in July Con arranged for Dave Schindler and me to accompany the Canadian Department of Transport buoy tender *Eckaloo* when they put out the navigation buoys on the lake over the next few weeks. This was a much larger freighter ship that could withstand icy conditions with its reinforced hull. We had our crates of science equipment hauled to the *Eckaloo*, made a primitive lab in one of the bunkers below deck and installed some makeshift davits and blocks to run 600 meters of wire for our reversing thermometers and water sampling bottles, plankton nets and bottom sampling gear.

After a week's dithering about the weather, the Captain finally ordered the hawser to be cast off from the Hay River dock and we spent the next two weeks steaming around the lake on the *Eckaloo*, avoiding thick ice patches, setting the navigation and warning buoys for the lake fishermen and freight ships. With the powerful winch on this ship, we could quickly sample water and plankton all the way down to 600 m depth in Christie Bay. Our lab was not so nice and David grumbled about spending too many long evenings filtering water, preserving biological samples and measuring chemical properties of the surface and deep water. We were in a rusty cold wet hold with one light bulb and a low ceiling but we had lots of rusty bolts on the floor to tie down our bench and equipment for the rough weather patches. Our bunks and the galley food were great but this ship was not designed for science.

Now can you imagine a South Dakota prairie boy out on a big ship on a huge lake with rolling waves? I had to learn the language: pulleys were blocks, a rope was a line, a cable was a wire, the kitchen was a galley, the toilet was a head, and then of course stern, aft, forward, bow, port and starboard, wheelhouse, hold, hatch, hawser, knots and nautical miles. But I knew how to tie a bowline and a clove hitch quickly, which impressed the deck hands. There was no shower but we could heat a large pot of water on the galley stove once a week to wash ourselves, whether we needed it or not. We had primitive radar and could fix positions from points of land, but no GPS. Slippery decks, gumboots, wool socks, wet weather gear...but cowboy leather gloves were still very useful.

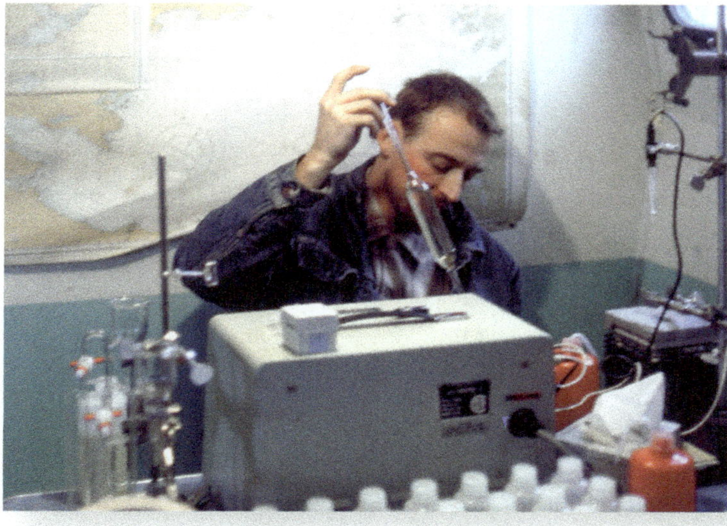

GJB in the lab on the Eckaloo, Great Bear Lake, 1968

Zooplankton biologist Kaz Patalas, Dave Schindler and I returned to Great Slave Lake twice that summer and enjoyed the better services of Captain Con Haight and the good FRB ship *Stenodus*. *Stenodus* is the genus of the inconnu (*Stenodus leucichthys*), a salmonid fish which occurs in cold water regions of northern Canada, Alaska and the USSR. The weather was very rough in August and we had to seek shelter in the Slave River Delta for nearly a week. As we were running out of groceries, we shot a woodland caribou to stock the meat larder. We ate huge meals, probably because it was cold. There were woodland buffalo in this area too but we didn't see any of these. On the way back out of the delta, in thirty knot wind, trying to reach the sheltered east arm of the lake, a rubber fitting caught fire in the aft hold and sour black smoke was rising from the hatch cover. Con shut the engine down, discharged the hold fire extinguishers, keeping the ship's bow into the waves with a bow drogue. Since I was the only deck hand that was not seasick, I went into the smoky hold to remove the burnt fitting and replaced it with a metal pipe. I was dizzy and sick for two days afterward from inhaling the fumes in the hold.

But we did make it into the narrow wind-protected bays of the East Arm of the lake and all was better, the sun was shining and the air felt much warmer. Great Slave Lake lies on the boundary between the northern boreal forest and the taiga, the land of little sticks. We were close to the sub-Arctic tree line to the north of us, where only low shrub-like willow grows on the rocky land. This part of the lake featured some spectacular scenery, with high cliffs looming over crystal clear cold deep water. It was here in Christie Bay that D. S. Rawson's soundings showed over 600 m water depths and we wanted to see if the chemical properties of the deep water were suitable for bottom animal life, with adequate oxygen concentrations. Our winch wire did reach 600 m but it took two or three hours to get water and mud samples up from that depth. There was lots of oxygen in the deep water. Water and plankton samples were taken for analysis back in Winnipeg labs but we did alkalinity, pH, filtrations and nutrient measurements on the galley table on the ship, usually late at night after supper. We were also taking blood and tissue samples of fish for fish biologists back in the Winnipeg labs, and we set nets for whitefish and lake trout. We caught lots of fish and after sampling, we took the cheeks and tongues for our supper. We had saved a carboy of cold 600 m water for mixing with whiskey for our after supper drinks, and this made us feel a little like scientific pioneers playing around with a nearly pristine ecological environment.

Historical interjection: British explorer Captain George Back, camped at the eastern extremity of McLeod Bay, Great Slave Lake, 25 December 1833, from his journal:

> Christmas day was the appointed time for opening a soldered tin case, the gift of a lady at New York; but our companion Mr. McLeod being absent, we thought it fair to postpone the gratification of our curiosity till he could participate in it; and Mr King and I made a cheerful dinner of pemmican. Happiness on such occasions depends entirely on the mood and temper of the individuals; and we cheated ourselves of our friends at home, as if we had partaken of the roast beef and plum pudding which doubtless 'smoked upon the board' on that glorious day of prescriptive feasting.

George Back's party nearly starved to death that winter, growing weak from eating nothing but boiled whitefish.

We took turns cooking meals in the galley and on one of my turns I tried to make bakin' powder biscuits. For some reason these turned out rock hard and even soaked in coffee they didn't taste good. So we pitched the whole batch out the galley porthole. These biscuits floated around the ship for some time, even the gulls couldn't crack them, and Captain Con Haight promptly named this little bay 'Biscuit Bay'. Con was a trickster and he collected some dried moose turds (piles of little round black droppings), coated them with olive oil and put them in a pretty dish on the supper table as Spanish olives. A few of us bit into them. He also had his own recipe for home-made pickled/fermented fish, which was bloody awful. Most of the time Con was at the skipper's wheel but he had to give up the wheel to his mate Hugh Trudeau when he visited the head (the ship's tiny toilet, crammed in the bow rope and chain bin). On one particularly rough day the *Stenodus* hit a large wave, Con was bucked off the toilet seat and came somersaulting down the walkway past the bunks and galley, bare bum in the air, shouting 'quarter on, quarter on, not head on to the waves!' He was instructing the steersman Hugh Trudeau to put the bow of the ship 45° to the biggest waves, not directly into the wave.

FRB Research Vessel Stenodus, *Great Slave Lake*

Our work confirmed previous surveys which suggested that these large cold deep sub-Arctic lakes had lots of old (40-60 years) fish and that high fishing pressures could not be sustained. We hypothesised that nutrient supply rates were low, due to the short open water season in this sub-Arctic region. The photosynthetic algal productivity of the lake was relatively low, especially in the east arms of the lake, and the food chain could support only limited numbers of fish. The big fish survived by eating most of the young fish. Years later my colleague Mike Healey would test this hypothesis by experimentally netting all the big carnivorous (Lake Trout, Pike) fish out of some small lakes north of Yellowknife, to see if herbivorous whitefish productivity would increase by removing the large carnivorous Lake Trout. And yes, the whitefish productivity did indeed increase when their predators were eliminated.

Chapter 9. Hookah Pipes, the Jordan River and Amsterdam Hookers

Steven Weinberg's Dictum: 'With or without religion, good people behave well and bad people can do evil; but for good people to do evil...that requires religion.'

Within a week of returning from Great Slave Lake I was on an airplane travelling to Israel for another big aquatic science conference. It was a tour of airports from Winnipeg to Montreal, London, Rome, Athens, with miserable airplane/airport food and no good coffee. I guess I was dreaming of those wonderful fresh caribou loin chops on the *Stenodus* on Great Slave Lake. We arrived in Tel Aviv, on the shore of the turquoise Mediterranean Sea, and were hauled across the barren rocky land to Jerusalem in a convoy of military vehicles to guard our dark tinted window/air-conditioned buses. They had the 'Six-Day War' here in June '67, with Israel fighting Egypt, Syria and Jordan. The military was still nervous and we were rarely allowed to move around without guards or Israeli escorts. Apparently during the previous week several small bombs had gone off inside garbage cans outside movie theatres. There were armed Israeli soldiers at the hotel doorway and we had to show our passports to them. Somebody forgot to tell me that this science meeting was being held in a war zone.

Once inside the halls of the scientific meeting however, it was as if the tension outdoors vanished and we were amongst international friends gathered from many countries of the world. The official commencement of the plenary sessions was in Hebrew, Arabic, several European and Slavic languages and finally in English. At this time the president of the scientific society (SIL) that sponsored the meeting was Dr. Wilhelm Rohde, who I had met in Uppsala, Sweden. His opening speech mentioned the 'brain drain to Winnipeg', the establishment of the Experimental Lakes Area in Canada and a new era of whole lake ecosystem experimental science. This made all of us busting proud to be associated with this project and it also indicated the international networking activities of Jack Vallentyne, who was head-hunting staff for this new research operation.

After these grand opening ceremonies there was a week of fifteen minute presentations from all of us, from eight in the morning to six in the evening, spread over twelve different rooms with 35 mm slide projectors and overhead projectors blazing away concurrently. Here we learned more about lakes and rivers of eastern Europe and the Middle East, lots of very salty hot water lakes, most of them having been stressed from city and industrial waste pollution from thousands of years of human habitation. I was interested to hear accounts of the ecology of the wetlands of the Tigris-Euphrates River valley in Iran/Iraq, the Fertile Crescent (which Mrs. Kuhrt taught us about in Western Civilization History back at Murdo High School), water budgets for the Jordan River, the Sea of Galilee and the Dead Sea. Some of these aquatic ecosystems had been used for irrigation and domestic water supply for six thousand years, and it sure looked like it. The Jordan River was mostly siphoned off to grow oranges and the remaining trickle was polluted with agricultural and domestic rubbish. There were 'No Swimming' signs along the Jordan River, near the location where Jesus was supposed to have been baptised. You could eat tiny perch from the Sea of Galilee, cooked whole in hot oil. The Dead Sea was evaporating away, declining in water level steadily over the past decades. Freshwater was precious in this region and the engineers were building structures to harvest atmospheric condensation water, water that had evaporated from the Mediterranean Sea. But we also heard presentations about salt lakes in Turkey, Lake Baikal in Siberia, brine salt harvesting in India and volcanic, acidic caldera lakes in Indonesia. With a little more confidence this time I gave my paper on upstate New York lakes that were permanently stratified, with dense salty deep water and lighter freshwater on top. I was still wearing my high school graduation suit and a necktie borrowed from my father's collection (which probably came from his University of Iowa days). Usually I would get overloaded with information every day and escape to the balcony to get some fresh air and talk with the soldiers on guard there. They offered me Egyptian and Turkish cigarettes and unfortunately I got hooked on nicotine from that time on, standing on the balcony overlooking the Muslim minaret towers and golden onion domes of the Orthodox churches.

My room-mate was a proper British gentleman scientist, J. D. H. Strickland, co-author of one of the chemical methods books we all used. Even in the heat he was dressed to the hilt, vest and top coat and tie, golden watch chain across his belly, polished shoes. I was fortunate to go to breakfast with him, as he loudly brow-beat the Israeli waiters to give us something equivalent to bacon and eggs and strong coffee. They tried to feed us sour cream and cucumbers at first but he beat them down, giving us some sort of ground chicken patties with hard-boiled eggs. He would time the settling of the coffee grounds in his Turkish coffee cup with his vest pocket watch and he would complain about the coffee if there wasn't a thick coating of coffee grounds all over the bottom of the cup. We made a strange group, proper British chemist Strickland, the microbiologist Holger Jannasch from Germany and a few Dutch friends and I would walk to the Arab Quarter of Jerusalem to find some good meat to eat. We would sit on pillows around a low table, eating with our fingers and taking whatever the waiter brought to us (as we couldn't speak their language). We ate a lot of camel and goat meat. Holger told stories about his grandfather being a Moravian missionary in Newfoundland and the Belcher Islands of southern Hudson Bay in the mid 1800s. Apparently the missionary society shipped dozens of trombones into the Belcher Island communities so the converted Inuit could play hymns in church. It boggled my mind to think of choirs of Inuit trombonists playing the doxology on an icy island in Hudson Bay! Against the advice of Strickland, Holger and I smoked a hookah (tobacco, I think) pipe after one of these sinful Arab meals and we paid for it dearly all night.

Our hosts bussed some of us to visit Kinneret and the Dead Sea (more salty than seawater), where we could float easily in the very dense soapy foul-smelling water. We visited Herod's fortress mountain Masada, Beersheba and down through the Negev to a kibbutzim dedicated to growing vegetables in the desert, using brackish well water. We went on south to Ein Gedi, several date palm wadis with camels wandering around, and Eilat on the Gulf of Akabah. We could see the hills of Sinai, Jordan and Saudi Arabia from the salty shore of this desolate gulf. I had never seen true desert dunes and the heat was severe. I began to wonder why so many wars were fought over this desolate, windy desert.

I remember expecting Biblical history to prove itself here. Abraham of the Jews led his Israelites here, the Christians believe Jesus walked, talked, and was crucified here and the Prophet Muhammad was transported by a flying horse to Jerusalem, where he ascended to heaven. But the land looked exhausted and ravaged, plagued by the same consumerism and greed of western nations. Extreme wealth existed next to extreme poverty in Jerusalem. Most of the pre- and post-biblical six thousand year history of this region was mainly a series of vicious battles between various sects of Arab Muslims, Jebusites, Mamelukes, Ottoman Turks, Jordanians, Jews, Persians, Greeks, Romans, Egyptians, the European Christian Crusades and then World War II. Even in this day each of these religious groups dehumanised their enemies, considering them demons. Jerusalem was like a Jewish military garrison where even the theatre ushers were armed, certainly not a city of peace and light. Why would people fight over this rocky desert for millennia? Muslims, Jews, orthodox and western Christians and the pagan Bedouins consider this ground holy, but I couldn't see the cedars of Lebanon anywhere. A group of us visited Bethlehem to find the village pathways lined by souvenir and fast food shops and plastic Jesus-in-the-manger figurines made in China, some with glue on the pedestal for sticking on the dashboard of your car. My childhood Methodist Sunday School pictures were not the same as what I saw here and the military guards never put down their rifles.

Exhausted from too much ancient history and modern science, we sleepily said goodbye to our colleagues and boarded our airplanes to return to the western world. On the way home I visited the Geobotanische Institut (ETH) in Zurich, Switzerland, where one of my scientific idols had done excellent research on the laminated mud of Zurichsee, the big lake beside the city of Zurich. I had cited his papers in my Ph.D. thesis and admired his succinct German writing style. I asked the receptionist if anyone here knew Fritz Nipkov, who had studied sediment cores of Zurichsee in the 1920s. She promptly phoned an old fellow who had long been retired but worked with Nipkov before he died. Hans Siegl rode up to the Institut on his bicycle and led me up a ladder to the attic of this large building. And there, on the bare timber ceiling joist, were Nipkov's original sediment cores in metal half tubes, with labels in old-style handwritten 'Switzerdeutsch', carefully documenting where samples had been taken from this annually laminated mud. I took flash photos of these old cores, and complimented Herr Siegl on keeping these valuable records of Zurich history. Nipkov was a Zurich Apotheke (the Swiss word for a pharmacist) who developed a hobby of looking at the microscopic hard parts of algae (diatoms). He realized that changes in these microscopic plants would happen as the lake water chemistry changed and that he could determine millennia of lake history from looking at these diatom remains in cores of the lake mud. Herr Siegl invited me to his home for supper, which turned out to be a large artist's loft full of paintings by himself and his wife. His English was poor so I tried to use my college German. He immediately wanted to know why an American would speak German with a Norwegian accent. So I had to explain that my Augustana College German teacher Ruth Stenseth was a Norwegian lady and that my French was much worse. After a good supper of onions, dumplings and pork, followed by dark Swiss chocolate coated pastries and espresso, he showed some of his (and his wife's) recent paintings. I bought one that attracted me, a modernistic Picasso-like distorted image of an old church organ, with gleaming white keyboard like teeth. He invited me to take a bed for the night, as he could tell I was a wandering student without too much money because I had haggled a bit about the price of the painting.

For reasons I can't remember I had a two day layover in Amsterdam, where several of my new scientific friends lived. Pieter Schrovers and Sikko Parma were also young and free, and they provided a tour of the city that was hard to forget. After a quick tour of the science institutions and the ocean-going research ships we went directly to the Red Light District. And here were all the ladies of the night, sitting provocatively in their boudoir windows, garishly illuminated by red and blue lights, playing cards and smoking cigarettes and beckoning to the street walkers. Street vendors were selling hot sausages and condoms. There were marijuana cafes, tattoo parlours, all very sinful to my naïve prairie boy eyes. My scientific colleagues revealed to me that prostitution was a lucrative professional business for both females and males of the species. Once again my puritanical upbringing was assaulted by the realisation that most Dutch families were good hardworking honest people who tolerated this biological service profession, rather than driving it underground with the criminal elements. But this seedy district of Amsterdam was neat and clean, with flower gardens, canals and a giant hospital and police station, ready to handle any emergency. My hosts did take me to a few art galleries and a smoky jazz joint loft where some very good Dutch musicians were playing, and the Dutch beer was very good. We must have spent over four hours at a good Dutch/Indonesian restaurant eating about thirty courses of a Riz Tafel.

While in Europe I visited many ornamental gardens, parks and forest preserves. These were beautifully manicured, immaculately ordered, designed and groomed and attended by professional caretakers with a life-long attachment to each pocket of protected serenity. Many of them were centuries old, having been founded by some king or queen. My mother would have loved them. But I found that my wild prairie genetic code was rebelling against all this divine order, where nature had been suppressed to the will of human desires. In some North American savage way I recognised that this rebellion was also against the kings, queens and gods of European culture. I suspect my father would have been uncomfortable in these neat gardens, as I longed for the open prairie or the ragged forests of northern Canada.

Again having had little sleep, I boarded my airplane and slept my way back to Montreal and Winnipeg, following the sun, back into a very different Canadian winter life. During most of this past year's travel I was on government money and my bank account back home was rising. I'd never had so much money before, so I went to Hudson's Bay and bought a sofa for my tiny apartment and a big reel-to-reel tape recorder. I also traded in my old 1953 Chevy station wagon for a used 1965 Cougar, a sporty car with lots of zip on the road. With my old Remington typewriter I edited down my Ph.D. thesis to more concise and separate pieces suitable for the science journals, and sent them off to the journal editors. We also spent more time at the Experimental Lakes Area camp, skidoo hauling fertiliser through a series of frozen lakes and the spruce bush into our test lake.

One of the FRB biologists, an expert on larval (aquatic) midge flies, Andrew Hamilton, invited Dave Schindler and me to come on his family moose hunt on Torch River in northern Saskatchewan. I purchased a Browning 30.06 semi-automatic with a good 4-8X scope and two five-shell clips for $300. I still had my Boy Scout compass and signalling mirror. I rubbed Arctic Dubbin on my boots and prepared a little canvas day pack to hold my Aladdin thermos bottle and lunch and some survival gear. Andrew's parents, brothers, uncles, cousins and old school friends arrived at the little Saskatchewan farm house with trucks and pick-ups, all primed for a meat harvest. This was not sport hunting but was a serious business of providing meat for their families. We all slept on the living room floor in sleeping bags, took turns doing dishes and house chores and made our own lunches. Andrew's father Ellis (an expert old-time trapper and hunter, now occupied raising leaf-cutter bees) showed us air photos of the hunting area across the river, with some key land features to use if we got lost. This was scrubby snow-covered spruce and pine forest with no roads but we could always get back to the river using our compass bearings. The style of hunting was to find recent tracks in the snow and fresh moose droppings, and follow them quietly and quickly. We walked all day, from dawn to dusk, tracking moose and trying to select young cows or yearling bulls. Most of us were successful, the moose was gutted where it fell, the ribs propped wide open with a stick and left to freeze under a pile of spruce boughs, the outside perimeter of which we peed on to ward off wolves and wolverines. After a week of this, on Sunday (no hunting) Andrew's father Ellis brought a small John Deere caterpillar across the river ice and pulled the moose carcasses down to a ramp where trucks could be loaded. Some of these northern Saskatchewan farmers went home with two or three moose, some deer, elk and sometimes a woodland caribou, tagged with licences that included their wives' and brothers' names. They would hang the frozen moose in their barns, turn on the oil heaters to partially thaw the hide, skin them and then take the meat off the bone for home use. David and I brought home two moose, which we skinned and butchered in his heated garage. I sure enjoyed eating this meat for the rest of the year...some of the steaks would cover an entire plate. We had to put some pig fat into the ground meat for sausage, as it was very lean.

I proudly drove my Cougar to South Dakota for Christmas and my brothers were very impressed with this sporty car. All the grandparents, aunts and uncles and other relatives gathered for feasts and it was good to be home. My father usually invited me to come along to visit his friends at the local Interstate Highway fuel station and restaurant, a hang-out for the 'good old boys'.

The conversation went something like this:

'You been overseas again?' says Ken Foster.

'Yup, I was in Israel for a science meeting', says I.

'So what's it like over there?', Ken stirring his coffee.

'Pretty dry and hot, they irrigate grapes, orange trees and garden vegetables in little patches in the parched desert. Didn't see any cattle or horses. A lot of dry sand and rocks, no grass. Well, they do have camels and goats.'

'So what do they eat then?'

'Who, the goats? They mostly nibble on piles of rubbish.' say I.

'No, the people in that country.'

'Lots of cottage cheese and cucumbers. And goats too. You can't get bacon and eggs for breakfast over there.' Says I.

There was a long silence, shaking of heads, sipping coffee, maybe a bite of sour cream raisin pie.

'They still fighting those Arabs over there?' asks Helge Anderson.

'Well, there were guards with machine guns at the doorway to our hotel. Some bombs went off in garbage cans outside movie theatres. But we did go to the Arab Quarter of Jerusalem to find some meat to eat, and we smoked a hookah pipe.'

'Well, I'll be damned. I hope the Christians win the fight', says Helge.

'No, it's not the Christians doing the fighting, it is the Jewish Israelis. Israel is a Jewish nation plunkered down on top of Arab Palestine in the aftermath of World War II. I don't think anybody will win, as they've been fighting since long before Jesus.'

There was another long silence, stirring of coffee.

'What about the Holy Land? Is there a special part that is the Holy Land?' asked Wayne Sanderson.

'Well, they got parts of Jerusalem that are holy for the Jews, parts that are holy for the Muslims, parts that are holy for the Greek and Russian Orthodox Christians and parts that busloads of tourist USA Christians go to, like Bethlehem. A lot of these people are still mad about the crusades.' Says I.

'Bethlehem? You mean it is still there?' says Wayne.

'Yup, you can buy a plastic Jesus (made in China) for the dash board of your car, on the street leading to the manger.' Says I.

'So what was wrong with the crusades?' asks Ken Foster.

'Well, the English and European Christians slaughtered a whole lot of Jews and Arabs about a thousand years ago, anybody that was an enemy of the Roman Pope. They talk about it as if it had happened only a few years ago and they are still mad about it, arguing about who won the battles.'

'Well, I'll be damned. I thought the Crusades fellows were knights who rode around England on big Lipizzaner horses with armour and heavy saddles.' Says Wayne.

It was very good sour cream raisin pie, almost as good as my mother's, but the coffee was weak, like dishwater.

Chapter 10. Lake Winnipeg Adventures and the CGS *Bradbury*

I questioned the wisdom of our Fisheries Research Board administrative leaders when they asked me to lead a study on Lake Winnipeg during the winter of 1968-69. On the other hand, none of the staff at our institute had experience working on big ships on big lakes... except our marine chemist F. A. J. Armstrong, who was too smart to get involved in this risky and unpleasant field work. This ambitious impulse was due in part to the rising enthusiasm of the environmental movement of the times, and the liberal Trudeau Canadian government was pushing for more action in this direction. This was the legacy of Rachael Carson, Harrison Brown, Paul and Anne Erhlich, Meadows and Meadows, Club of Rome, all enthusiastic environmentalists. During the 1960-1980s there was an increase in non-military science funding in Canada and the USA. I was lucky to enter the profession at this time, as 1980-2008 funding for scientific research declined or was flat over the following decades.

Now, Lake Winnipeg is one of a series of large lakes in an arc from the St Lawrence Great Lakes (discharging water to the Atlantic Ocean), west and northward to Lake of the Woods and Lake Winnipeg (discharging water to Hudson Bay), Great Slave and Great Bear lakes in northern Canada discharging water to the Arctic Ocean. Lake Winnipeg is over 400 km long, with a surface area of 23,750 km^2, but it is relatively shallow (mean depth = 12 m) compared to other large lakes. This means that the wind can blow the water up and down the lake in a vicious wave climate and push several metres of water into the wetlands and river deltas around the lake. This lake was an important 'inland sea' for early explorers because the tributary rivers feeding the lake allowed canoe travel from Montreal and the St Lawrence Great Lakes westward to the Pacific and northward to the Arctic Ocean. The Frenchman Sieur de la Vérendrye was probably the first to build a fur trading post on the southern river delta region (Fort Rouge, on the Red River) of this lake in 1738. (This rang a bell in my mind, as la Vérendrye had also ventured southward and planted a leaden plaque on a hill near the present site of Fort Pierre, South Dakota. This is drummed into the heads of South Dakota school boys and girls). British explorer John Franklin and a party of Metis men had paddled large Hudson Bay freight canoes from Montreal, through the Great Lakes, Lake of the Woods and into Lake Winnipeg in the early-1800s, en route to the Mackenzie River and the Arctic Ocean. At their camp on the northernmost shore of Lake Winnipeg they got an explanation as to why this big lake was so muddy.

According to the local Cree Indian people of that region, the magic trickster Wisakedjak (sometimes a bird, sometimes a shaman) had tipped over cooking kettles and ripped open food bags during one long winter. The people caught the naughty trickster magpie and put him in a small willow cage behind their tent camp. The cage became filled with faeces and in the spring the cage was muddy and foul, and so was Wisakedjak, and he begged to be let out of the cage. As the summer heat melted the ground, one of the willow branches of the cage came out of the ground and Wisakedjak escaped. He flew to Lake Winnipeg and washed himself for several days, and the water has been muddy ever since then. The name 'Winnipeg' means 'muddy water' in the Swampy Cree language.

Around 1875 Icelandic people settled along the western shore of the southern part of the lake, and they have influenced many of the names and uses of the lake up into the present time. Increasing catches of sauger, whitefish and walleye reached the markets of Chicago and New York City in the 1900s. Some of the glory days of the Icelandic fishermen are captured in the songs of Sol Sigurdson, and many old photographs of the fishing camps and ships are on display at the Icelandic Museum in Gimli, Manitoba. The first good chemical and biological research on Lake Winnipeg was done by Alexander Dimitri Bajkov, who earlier in life had graduated from the University of Moscow, served in the Russian Artillery and survived capture by escaping a Bolshevik prison camp in World War I. He migrated to Canada in 1926 and by 1931 he had his Ph.D. from the University of Manitoba. He established his field camp in Gimli in 1929 and used the schooner *Breeze* to sample the biological communities of this large lake over 1930-32. After he finished his work on the lake and had moved on to the next project, the *Breeze* and two crew members were lost in a late fall storm in 1933. This lake can come up rough in a short time and has claimed the lives of many fishermen.

Being naive and ambitious, we hired an Icelandic fisherman to take us out to the centre of the south basin ice of Lake Winnipeg in March 1969. The 318 Chrysler engine bombardier had a sled-trailer with most of our sampling equipment in insulated and padded boxes and a thirty inch Sigurdsson ice augur (with a 3 m shaft) attached to the rear chassis. We imagined sliding along blissfully over the snow-covered ice but this was not to be. The surface was extremely rough, with jagged 1-2 m floes of ice sticking up in long rows and the blowing snow obscured any landmarks and the shore. We navigated by compass and dead reckoning based on travel times. It was only about -25°C outside but the inside of the bombardier was over heated and full of exhaust fumes. We were also to learn that the ice was 1-1.5 m thick and it took about thirty minutes to take our water chemistry and biological samples, during which times our horse-hide and wool liner mittens would get wet and freeze to any metal object. Our plastic water bottles would crack in the cold and the plankton nets would freeze and break when hauled out of the ice hole. This was so much fun, we made two more of these bombardier trips in April, when the weather was just a bit kinder to us. My scientific colleagues were impressed that I could go to sleep during the long, very rough bombardier ride back to our shore camp.

Coring on Lake Winnipeg Ice

We used a helicopter to sample the under-ice properties of the north basin of this large lake and we promptly realised we were completely out of our comfort zone. After a few augur holes in the centre of this northern basin the weather turned vicious and we were in a white-out. We were about 35 km from any shore, the wind-blown snow obscured the horizon, and we couldn't tell up from down. The helicopter pilot asked that one of us walk on a compass bearing to Grand Rapids, at the mouth of the Saskatchewan River. The helicopter pilot could then cautiously fly behind the walking person, about 20 m above the ice. John Flannagan, a benthic biologist, and I took turns walking into the teeth of this strong wind. Just before dark, the pilot could see the lights of Grand Rapids bridge and we made a beeline for the hotel restaurant, where many good stories were told over beer and full bellies. These winter expeditions on the lake told us that the water under the thick ice was well oxygenated, that the plants and animals in the water and on the bottom were in good shape despite huge river supply rates of nutrients and rich organic prairie soil during the previous spring/summer season.

We found that one of the government ships on the lake was not busy, so we arranged to have the Coast Guard Ship *Bradbury* available for monthly cruises on the lake during the open water season of 1969. I visited the ship on dry dock in late winter, just to get a feel for the situation. It was a wonderful old ship, 46 m long, 8 m beam and draft of 2 m, with a cruising range of 1200 miles. She had oak and maple furniture, polished wooden stair railings, an elegant wheelhouse with large oak map tables, gleaming spic and span engine room and the cabin furnishings reminded me of Victorian England, with a porcelain water pitcher and washing basin fitted securely into the wooden wardrobe. This ship had been prefabricated in Sorel, Quebec in 1914, shipped in pieces by rail to Selkirk, Manitoba, and welded together on the banks of the Red River in 1915. They steam-fitted oak planks, which were bolted to the wrought-iron hull, to provide more protection as they wanted to use the ship as an icebreaker. During the flu epidemic of 1917 the ship broke 30-50 cm of ice to reach remote northern settlements, to supply doctors and medicines around the lake. She served as a fisheries patrol vessel, a lighthouse and navigation buoy tender, harbour dredge tender and a lake freighter, burning coal as a steam ship up to 1935. In 1952 the ship was refitted with Marconi Radar, Raytheon sonar, gyro compass navigation gear and two twenty tonne Kelenberg diesel engines.

We got advice from the St Lawrence Great Lake scientists in Burlington, Ontario, who supplied some scientific and ship equipment that we didn't have. The FRB metal workshop fellows, especially Willy Burton, built us some swinging davits that could be bolted onto the gunnels of the ship, with hand-operated rope winches for shallow water work. FRB Institute carpenters created a ship's laboratory underneath the forward deck, with fridges, freezers, lab benches bolted to the wooden floor and wooden bench rails and eyebolts to lash down lab equipment in rough weather. Most electronic ship lab equipment was fitted into heavy-duty padded transport boxes that could be partially opened for use on the ship and quickly snapped back together if rough weather happened. The walls of the lab were covered with navigation charts of different parts of the lake and a few Playboy fold-out beauties. We also built a deck incubation tank frame on the upper deck, where we could get full sunshine for estimating algal growth rates in the water. These tanks also had lids with fluorescent light banks inside so we could use artificial light to do these measurements at night. We had barely finished these preparations in early June when the ice cleared in the south basin of the lake. I went around the institute offices, promising luxury cruise facilities, fun and games, good food, begging for a dozen competent people to do the biological, chemical and physical work on the ship for our first cruise.

CGS Bradbury *at Grand Rapids MB, 1969*

After loading two full trucks of equipment on the ship we departed the Selkirk jetty at dawn in early June, when the Red River was still high from spring runoff. It was difficult to see the deep channels as the flood water covered the adjacent deltaic swamp land and the foggy mist obscured the few navigation signal towers. Captain Kris Thorsteinson was calm, having three decades of experience, sipped his coffee and steered the ship slowly through the delta into the big lake and the land shore vanished. There was deep anxiety in my belly, as I didn't know how our scientific crew would perform all this sampling and measurements in this short nine day cruise. But the sun came out, warm on our skin, even though the lake water was cold, and the wind was mild. We managed to obtain our samples in the southern part of the very muddy lake in the first two days and proceeded northward at ten mph through the Narrows to the much larger North Basin. We off-loaded supplies to the community of Berens River one rainy night, a surprisingly large shipment of beer, tobacco, Spam, corned beef and tinned peaches. We encountered large ice floes just north of Berens River but the ship could find open water channels to reach our desired locations for water and biological samples. It was cold here, the deckies were wearing winter parkas, but we had to use our bare hands to operate the meters, water sampling bottles and bottom dredge samplers. We kept a hand towel on our belts to dry our cold hands before putting on our leather covered wool mittens. After three days of exploring this large basin we had to skedaddle southward, as the weather report predicted strong winds from the NW.

With this initial experience under our belts and safely back in the FRB workshops, we improved our shipboard equipment in anticipation of ice-free water in July. Sure enough, the sun was hot and the work was fast and pleasant on the little south basin of the lake. When we reached the north basin the ice was gone and the water was flat calm for a few days. The water was deep pale green from diatom blooms and we established about twenty new sampling stations. But then the NW wind came up, the ship began to pitch and roll, the ship's galley fridges and cutlery bins fell down and our laboratory tables ripped loose from the oak floor screws and were sliding across the lab room. Our big upper deck tanks for incubation of algal samples contained about 1 m^3 of water each (= 1 tonne) and as we were draining them they ripped themselves apart. All precious science lab equipment was sent to bed with each person on our crew and it was a miserable night for all. Benthic biologist Phil Chang and my sidekick Brian Graham reported that the rivets were popping out of the walls of their bunk cabins (and ricocheting across the room) during the worst of the storm. The next morning the wind was even stronger but the ship had found anchor in a little bay of George Island. We tried to rebuild our lab and some damaged deck gear but we didn't have the proper tools for this work. Mike Stainton repaired the light bulb fitting on our little spectrophotometer (for nutrient measurements) with a sardine can lid and a soldering gun. So we steamed across rough water into the little town of Grand Rapids, at the mouth of the Saskatchewan River. There was a new bridge across the river, the First Mate Merv Sinclair had too much engine power on the screws while going under the bridge, and the top mast light was knocked off by the bridge struts. Captain Kris was calm and explained that all these disasters was the result of recently taking the rusty sheet metal and rotted oak planking off of the hull, which lightened the ship and made it ride higher in the water, and less stable. It was with great relief when we stepped onto the Grand Rapids dock but we found the land was bobbing around underneath our feet, as our ear bone sense of balance had adjusted to the motion of the ship. We hired a pick-up truck to obtain some hardware, tools and a load of 2 x 6' planks to repair our deck and lab gear.

While all this exciting rough weather was happening, several of our science crew turned pale green and vanished to their bunks. The ship cook was a sadist who persisted in cooking sauerkraut and pork hocks during the worst of the weather and the strong smell pervaded the entire ship. Our expert Swedish algal taxonomist, Staffan Holmgren, was found two days later in the ship crew bathtub, moaning miserably and cursing my name. We had to physically carry him to the showers and then put him to bed. John Flannagan could only eat bread soaked in milk for several days and he claimed he was cured of ever working on big lakes again. We certainly had damaged the morale of our scientific crew and I had to conceal the interesting fact that I didn't get seasick, and I had happily slept through the worst of the storm. I had a big reel-to-reel tape recorder roped firmly underneath my bunk so I could listen to Bach violin partitas, Bob Dylan and the Lake Winnipeg ballads of Sol Sigurdson through headphones, warm and dry in my bunk.

After a busy day (tied up at the dock) rebuilding our lab and deck equipment we all trooped to the local hotel bar in Grand Rapids, where we found some delicious food that was not sauerkraut and wieners, drank too much beer and got involved in a shuffleboard contest with some local truckers. The leader of our team was our First Mate Merv Sinclair, a Métis, who was a very good player. There was some dispute over one game, a fight broke out, the furniture was broken and the truckers attacked Norm. So of course the ship crew went to his aid. During the melee our science crew was on the margins, trying to keep new arrivals from joining the fight. One persistent fellow pushed too hard to join the fight, so my colleague David Schindler (a North Dakota State University wrestling team champion) put him in a headlock. We later found that this fellow was an out-of-uniform Royal Canadian Mounted Police officer who was trying to break up the fight. When things quieted down we had one more beer and walked in a very tight group back to the dock and our ship bunks.

JRW Vallentyne and DWS Schindler at the algal growth measurement tanks on CGS Bradbury, *1969*

After this difficult and bruising research cruise experience, we again improved our equipment and shipboard activities to allow safer operating conditions during rough weather and instructed our good Captain Kris to head for shelter when the wind came up. We ran research cruises in late July, August, September and watched the muddy brown lake water warm up and become pea soup green with massive blooms of bluegreen algae (now called bacteria, since they have no nucleus). Some massive rafts of decomposing algae looked like islands from the ship's wheelhouse, with birds perched on them, and the ship slowed down before proceeding, uncertain that they might contain logs or the remains of winter ice fishermen shacks. John Flannagan, Ole Saether and Andy Hamilton adapted the ship's lanterns to collect mayflies and other lake emerging insects, and we could have collected barrel-full quantities of them. By October the frazzle ice was beginning to form, and we retrieved the navigation buoys for the end of the shipping season.

After the first ice formed, wind would blow the ice into ridges and valleys where snow would accumulate and the lake surface was very rough. We used a bombardier to get to some of our summertime sampling stations and an ice augur to open a hole to the water underneath. We were trying to determine if the ice cut off the supply of air and oxygen to the water, because deoxygenation would be deleterious for fisheries and other biota. We hired Icelandic fishermen and their bombardier with ski-mounted fuel and cabin trailer to go out on the south basin ice and the two or three hour ride was very jarring. Of course I would sleep through all of this rough ride and my scientific colleagues would be impressed that I could wake up just before we arrived at our sampling station. The north basin of Lake Winnipeg was too big for quick bombardier trips so once again we hired a helicopter to reach a few of the central sampling stations. We continued to take water samples (for chemical analyses) from the inflowing rivers up to 1974, to help average the variation in annual inflow of nutrients and some trace elements to the lake. For the next decade my biological colleagues at the Freshwater Institute were publishing papers on various aquatic insects, crustaceans, clams and algae from this lake.

While all of these wild adventures on ships and helicopters were happening, my colleagues Jack Uthe and F. A. John Armstrong discovered that the fish from Lake Winnipeg contained too much mercury to allow sale on the market. This closed the commercial fishery for several years and much urgent research was done to locate the source of mercury inflow and what could be done to reduce the mercury content of fish. It was discovered that most of the contaminant Hg was coming from pulp and paper plant cathodes upstream on both the Winnipeg and Saskatchewan Rivers. The medical profession then discovered that the incidence of mercury poisoning of fish-eating Indian village people in the vicinity of Grassy Narrows (part of the

Winnipeg River) was serious and the pulp and paper plant was temporarily shut down until they could find a different technology to make paper. Some of our lake mud cores taken from the CGS *Bradbury* were used to determine the history of background and contaminant mercury accumulation in the lake mud. This indicated that Hg was considerably enhanced in the top twenty cm of lake mud, representing perhaps the previous twenty-five years of the pulp and paper industry. From our estimates of the mud accumulation rates we considered that it would take a decade to bury this contaminated mud, out of reach of resuspension by storms. However, Uthe and Armstrong searched far northern lakes, remote of any industry, to find that some big old fish had high concentrations of mercury in undisturbed lakes. Some of this mercury may have come from Flin Flon and Sudbury refineries and coal combustion and atmospheric transport from eastern North America, but some of this bioconcentration was probably a natural process.

Our sediment cores also revealed an episode of history that was to become more important in the coming years. About 1-1.5 m below the lake bed mud surface in the south basin of the lake we discovered an unreasonably dry, conchoidally fractured clay horizon that appeared to be bound together with root hairs. Upon consulting with University of Manitoba geologists Jim Teller and Bill Last, we learned that this dry clay probably represented a time in the past when the climate was more arid and this lake basin was dry and covered with prairie grass. Many years after our work on the lake geological seismic records showed several of these crumbly dry horizons in the deeper sediment, probably representing series of millennial episodes of dryer climate with less runoff.

About this same time, proposals for massive dams on the Churchill and Nelson Rivers were made, with the intent to divert water from the northerly Churchill River into the Nelson River to generate more hydroelectric power for the expanding needs of southern Canada and the USA. The proposals called for water flow control structures on the outflow of Lake Winnipeg, to moderate the annual variation in seasonal water flow to the turbines downstream. This regulation of outflow of Lake Winnipeg would lead to reduced flushing of the lake during spring runoff and more regulated lake level which would disadvantage the deltaic and swampy wetlands around the lake. This led to a decade of research before and after construction of the Churchill River dam, led by my colleague Bob Hecky and several other scientists. The damming of the Churchill River flooded large areas of muskeg and jack pine forest and greatly reduced the ability of the local Indian villages to harvest fish for domestic consumption and for the commercial market. To add to the problem, fish from the new dead tree-studded reservoir was found to have high Hg concentrations. It was later found that the Hg came from natural sources and was being converted to an organic mercury form that was more readily taken up by fish-food organisms and fish.

Over the next few years I wrote series of reports based upon the 1969 CGS *Bradbury* cruises on Lake Winnipeg and our measurements of the chemistry of the inflowing and outflowing rivers. And again out of this tangled mass of data tables, graphs and lake maps of physical, chemical and biological data came a relatively simple story of how the lake does its work over the seasons of thick ice cover, spring melt, the warming summer, the wind- stirred fall and the retreat to still water under two metres of rough ice and snow. Colleagues Paul Campbell, Susan Kasian, Brian Graham and Mike Stainton created order in the data files, graphs and tables that allowed the seasonal picture to appear. There was intense pleasure and excitement around the emerging picture, which could be captured in simple diagrams and photographs for presentation at scientific meetings. The main message was that the lake was well supplied with nutrients under pre-European times and that modern agriculture, city sewage and a few industries were adding even more nutrients and trace elements. The lake was shallow, so it warmed and cooled quickly, and it acted as a very well-designed sewage digester for a large agricultural watershed. There was much concern about pollution from sewage from the City of Winnipeg (about half a million people then) but I had to remind the policy wonks that 95% of the river-borne nutrients came from upstream of the city, from erosion of the rich prairie soil, hog and chicken barns, crop fertiliser and other agricultural wastes. The local newspapers shouted about the 'toxic algal blooms' rotting on the shores of popular beaches and resort towns but I had to remind them that Captain Huyshe of the British Army Rifle Brigade had canoed from the Winnipeg River to the Red River, through mats of floating algal blooms in 1871. His troops were on the way to Ft. Rouge to put down Louis Riel and the Métis Republic of Manitoba.

All this was teaching me that everything is connected to everything else. And while I was frantically busy trying to keep the ship operations running smoothly, the world around me was becoming increasingly bizarre.

As soon as there was good early winter tracking snow in northern Saskatchewan we took two weeks off work in early November for a successful moose hunt. I managed to get my Ph.D. thesis carved into five separate science journal articles and I awaited fame and fortune, proud to see my name in print. I was supposed to be writing up the results of the Lake Winnipeg cruises but I kept getting interrupted by new international developments in science. The International Biological Program (IBP) had begun in 1964, to study large-scale aspects of biological productivity and human welfare around the world and new research camps were proposed for the Canadian and Alaskan Arctic regions. Professor Frank Rigler at the University of Toronto and Jaap Kalff at McGill University invited some adventure- seeking scientists to a planning meeting in Montreal that winter. We were to evaluate and plan research at Char Lake on Cornwallis Island in the Canadian High Arctic. After this intense three day meeting we had a banquet at a good restaurant, where a skinny little fellow scurried around the banquet table after we had finished, gathering up the baked potato skins, steak gristle and fat and uneaten desserts. After gathering up a large amount of leftover food he sat down and ate the lot. This was Buster Welch, the Chief Scientist for the IBP Char Lake research effort. He was to become a good friend and scientific colleague for many years. On the way back to Winnipeg I stopped to visit Buster and his family at their home on Toronto Island.

The other big project brewing had to do with the ARCO and Exxon discovery of a large oil field in the vicinity of Prudhoe Bay in Alaska in 1968. IBP had established a research camp at Prudhoe Bay, helped along by the oil industry. There were proposals to deliver the crude oil to the USA via a heated pipeline down the valley of the Mackenzie River, to the existing oil pipeline network around Edmonton, Alberta, about 3800 km. The Canadian government wanted to scope some of the social, cultural, environmental, fishery and engineering problems that would accompany a hot oil pipeline crossing permafrost soils, rivers, Inuit and Dene communities and I attended a blizzard of planning meetings for this effort. In the spring of 1970 I travelled to Alaska to visit with the oil industry and ecological scientists in Fairbanks, and on to Dead Horse and Prudhoe Bay to see the drilling rigs and the ecological research effort on the cold desolate coast of the Alaskan Beaufort Sea. Inside the camp kitchen the food was good and I found Augustana colleague Larry Tiezsen there, working with gas chromatographs trying to estimate productivity of the Arctic grasslands and permafrost melt ponds. Other scientists were already worrying about the effect of elevated or buried hot oil pipelines on migrating caribou and other wildlife. I also found puffins walking along the tinkling frazzle ice of the coastline, piles of gigantic whale bones and frozen lakes full of asphalt-like oil. I flew over to the Mackenzie Delta, the little towns of Aklavik and Inuvik and down the huge Mackenzie River valley to Edmonton. This eye-aching vast ecologically diverse area of the northern Rocky Mountains, extensive river floodplains and boggy frozen swamps was almost impossible to grasp in my mind. I did write a report on this trip, which became the basis for a large new Mackenzie Valley pipeline project for my institute for the next decade. I began to collect and read everything I could find on the history, geography, hydrology, culture and economics of this western Arctic river, including a first edition of the journals of Alexander Mackenzie.

Chapter 11. Romance and Rhabdomancy

Sitting in my little office on the fifth floor of our new federal government research building on the southern fringe of Winnipeg, my head was near to exploding in the spring of 1970. There was just too much going on all at once and as would become common during the rest of my career, new projects were beginning before I wrote up the results of the last piece of field work. I should have been writing papers on Great Slave Lake and Lake Winnipeg surveys. We were also in the initial stages of planning a large field research program in the Mackenzie River valley, stimulated by the Prudhoe Bay hot oil pipeline proposal, with four field camps (accommodations, kitchens, labs, warehouses) in Yellowknife, Fort Simpson, Inuvik and Old Crow. Logistic meetings were held in Alaska, British Columbia, Alberta and of course Ottawa. The Canadian IBP research was steaming ahead, demanding ideas and participation in the Char Lake Project on Cornwallis Island in the high Arctic. The media were running hot with stories about pollution of the St Lawrence Great Lakes and a big debate was emerging about the importance of phosphate (an essential algal growth nutrient) in detergents as being an important cause of scummy algal growths in lakes. The soap and detergent industry denied this and Dr. Jack Vallentyne of our institute went to battle with them. This also meant that our remote field station experiments, with additions of nitrogen and phosphorus to whole lakes, was much more important to the Canadian government and the International (USA-Canada) Joint Commission, which was trying to referee this argument. One of the proposed replacements for phosphate builders in detergents was nitrilo-triacetic acid (NTA) and a newly hired microbiologist, Dr. Robert Hamilton, wanted to test this chemical compound in ELA lakes to determine its toxicity and nutrient capacity.

All this caused a blizzard of urgent meetings, some new research initiatives and much excitement for our new young research group. The leader of the Experimental Lakes Area

(ELA), Dr. David Schindler, went to Europe in the spring of 1970 for some science meetings, leaving me with the opportunity to escape my city office to be a temporary chief scientist at ELA in the remote western Ontario bush country. Aha, says I, this would be a good opportunity to peacefully write up science journal papers on my work on Great Slave Lake and Lake Winnipeg. Southern Manitoba is in the flat rich soil of the prairies, where busy farmers and livestock operations were everywhere. ELA was five hours drive east into the Precambrian Shield rocky jack pine country, where the air smelled of pine and spruce, there was no traffic and no distractions. So before the ice was out I arrived at the ELA camp with boxes full of field data, my father's slide rule, my high school Remington typewriter and boxes full of graph paper. Nobody warned me what was going to happen this summer.

The first glimmer of trouble was the news that the bosses back in the Winnipeg office had hired some university women to work as 'summer students' this year. The camp had been all male for the past years and this caused a flurry of construction and sanitation of the camp to make it suitable for the ladies. A separate ladies accommodation ATCO trailer was installed, the kitchen dining area was expanded and a whole new sewer system was laid down in the frozen gravelly sand. I had to go back to Winnipeg for some meetings and there I met some of these women as they were raiding my NWT camper truck toolbox. Two male and four female summer students arrived at ELA camp in early June, when it was still snowing. One of these ladies put a snowball down my neck. This was Sheila Dennett and I knew she was going to be trouble. Sheila was assigned to work with Dr. R. D. Hamilton on the new NTA project. Hedy Kling was an algal taxonomist in training with Staffan Holmgren, Norma Kritsch was working with Mike Capel in the field chemistry lab and Doreen Middleton was an assistant to John Flannagan. Liz Frost worked with Russ Schmidt on algal productivity.

For some unknown reason the ELA camp became aware that spring of the opportunity to have fun in the bush after the day's work had been done. We built a volleyball court and established one of the ATCO trailers for recreational pursuits, such as table tennis and card games. Guitars and folk music suddenly appeared after supper. That feisty Sheila Dennett, who tried to put a snowball down my neck, turned out to be a good singer and she sang a haunting version of the medieval ballad 'Barbara Allen'. Sheila also had a tame grackle at this time which would follow her everywhere, even into my radon lab

The multinational scope of our field staff was displayed by the national flags of Scotland, Wales, Germany, USA, Sweden, Britain and Norway outside the labs of the respective occupants. Some senior administrator must have noticed this and ordered that a very tall flagpole was to be installed, with the Canadian flag above all the others. One of the legendary highlights of this year was the 'Gunfight at the OV Corral' during one holiday weekend.

OV was a popular Canadian beer, OV being for 'Old Vienna', and the guns were water pistols, sometimes loaded with dilute ammonia or blue dye. Other events were the orange ravishing contest, races with 30 kg sacks of fertiliser on your shoulder, wine and cheese parties, table tennis, darts, the ceremonial presentation of the blue balls award and of course a volleyball tournament. Honest John Shearer was the referee and MC for most of these events. Fast Eddie the cook got very upset when some of us raided the kitchen late at night to make grilled cheese sandwiches. Padlocks were finally installed on the kitchen door. Fast Eddie had a snazzy yellow Torino car and he tried to get some of the girls to come with him to have drinks in Kenora.

Gunfight at the OV Corral, ELA, 1970
Left to right are GJB the gun slinger, Chemist Mike Capel, and zooplankton counter Richard Barnes.

And I have to tell you about Jim Russell. He magically appeared as a technical helper during my Ph.D. thesis work at Cornell University. When I needed help in water depth mapping of the ELA lakes I knew he would be useful, as he had helped make our maps of Fayetteville Green Lake in New York. I wrote to him in Buffalo, New York and within a month he appeared in Winnipeg, ready to go to work. He was brilliant at keeping the Raytheon sonar depth sounder working in difficult field conditions (rain, snow, heat, part failures) because of his experience on fishing ships in the Gulf of Alaska. He calibrated the sonar signal using a very large Secchi Disk on calm days, allowing accuracy to about 10 cm in water depth. He kept a miniature electronics toolbox during our work, keeping the stylus clean so it would burn the paper to indicate depth. Now, despite Jim's expertise and careful map drawing skills, he was not a tidy individual. He had long unkempt hair and rarely took a bath, usually was dressed in black clothing from the Salvation Army and he almost always had a toothpick in his mouth. During one of his ELA camp visits the Chief Financial Officer from FRB in Winnipeg, Mr. Stan Schick, presented himself at the supper table after some dirty field work. Stan was a good man, keeper of the finances, but he was also fastidious, dressed in pin striped suits with a tie and polished shoes. Upon returning to the head office in Winnipeg, Stan declared that there should be a dress and cleanliness code at our field camp, in accordance with some government policy. Jim Russell took offense at this memo and promptly shaved his hair off, sitting on a chair in the middle of the camp accommodation trailer circle. He then rubbed hair remover over his head and polished his shining bare head with a towel. During rare visits to Winnipeg, Jim would wander innocently into the administration wing (mahogany row) of the Freshwater Institute, where Stan's office was situated, and would stand there admiring the modern art on the wall in a full black suit, white shirt and tie, with his polished bald head. The secretaries all knew what he was doing and they would loudly applaud, much to the embarrassment and amusement of Stan Schick.

During working hours we were busy with regular measurements and sampling of dozens of selected experimental lakes, during the dangerous period of melting ice. Bob Hamilton's crew were building 10-20 m long tubes, about 2 m in diameter, made from PVC plastic sheets. These tubes were hung from jack pine trees during construction and were later deployed into the lake after ice-out, to isolate a column of water for experimental treatment with NTA and other chemical stressors. Hydrologists Bob Newbury and John Cherry from the University of Manitoba were drilling to bedrock and inserting plastic pipes to monitor the flow of groundwater through the porous sandy gravel into the lakes. They also built inflow and outflow weirs to measure the flow of water through the lakes and with the help of their supertechnician Ken Beatty, built a Class A weather station on a hill above the camp.

During the previous winter we had hauled hundreds of kilograms of sodium nitrate and small drums of phosphoric acid into a small lake about an hour away from camp, using skidoos and large sleds. We were waiting for ice-out to apply the fertiliser and making careful measurements of the carbon and nutrients in the lake water held over during the winter dormant period. Sampling during ice-out is an adventure, as the lakes usually had a narrow 1-4 m moat of liquid water around the margins, with variable thicknesses of ice over the deeper part of the lake. We would push a small canoe into the moat and gingerly step from the canoe onto the centre ice, dragging the canoe (with our measurement and sampling equipment inside) with us in case we fell through the weak ice. Summer student Jim McComiskey and I were the lucky first winners of the 'blue balls award' that spring, as we went through the ice several times.

When David Schindler returned from Europe and the lake water had warmed a bit, we began the fertiliser experiment that year by mixing the sodium nitrate fertiliser and phosphoric acid with lake water in a boat, then driving the boat rapidly around the lake with the aft drainage bung open. Within weeks the lake turned pea soup green, in contrast to other natural lakes nearby that were clear blue or tea coloured (from forest soil organic matter). With abundant supplies of nitrogen and phosphorus the algae grew so fast that they depleted the supply of CO_2 in the water by mid-day and then slowed their growth rate until the next morning. We reasoned that the morning burst of algal growth was supplied with CO_2 from plant respiration, decomposition and the invasion of atmospheric CO_2 into the lake water. This was uncomfortable news because the soap and detergent industry had postulated that CO_2 would limit algal growth, not N and P. By mid-summer we had clearly shown that N and P additions could cause mats of smelly algal blooms in the lake, even with severe afternoon draw-down of the naturally supplied CO_2.

We needed to find a way to measure the transfer of atmospheric CO_2 into the lake water, so I wrote a letter to Wally Broecker at the Lamont-Doherty Geological Observatory (LDGO) of Columbia University in New York, describing this problem. Wally had been developing methods to measure gas exchange between the atmosphere and oceans, using radon, the biologically and chemically inert radioactive daughter of radium. Radon gas is naturally produced from radium in rocks, soils and water and diffuses rapidly into the atmosphere, where it decays rapidly to radioactive lead and polonium. Wally had measured the radium and radon content of the surface ocean water and knowing the deficiency of radon (relative to its parent radium) that had escaped to the atmosphere, he could estimate most of the parameters needed for the exchange rate for most other gases, such as oxygen and CO_2. It must have been a boring summer at LDGO because within a month I picked Wally up at the Kenora airport and we had a week to review the options for the measurements of gas exchange on our severely nutrient stressed lake.

We finally decided to spike the lake with lots of radium, to make it easier to measure radium and its atmosphere-seeking daughter radon. This was facilitated through the nearby Pinawa branch of the Atomic Energy Commission of Canada by a Mr. Cecil Plunket, who arrived at camp with a very heavy lead casket of high activity radium and a full plastic suit for the transfer work. He used long steel tongs to break the small glass ampoule of high activity radium into a barrel on a raft in the lake. We then used an outboard motor boat to pull the raft around the lake and the spike radium was added to the lake surface water via a small spigot hose. Once diluted in the lake, the radium was no longer dangerous and the water could have been drinkable, except for the smelly green algal bloom. This experimental treatment with a radioactive tracer probably could not be done easily and quickly today, with all the health and safety regulations and the need for many permits and questionnaires. But it helped that ELA was on Crown land and was far away from any towns or settlements.

Of course, Wally now had to teach us how to sample and measure radium and radon in the lake water and LDGO Ph.D. graduate student Steve Emerson spent ttime setting up the alpha scintillation counting gear and teaching me how to do the measurements. And thus began a fruitful and productive collaboration between Wally's LDGO graduate students and ELA staff for the next fifteen years. These students taught us the value of using radioactive tracers in ecological field research, as the feedback was extremely rapid. From our direct measurements with alpha and gamma spectrometers in this remote field camp we could have good data to examine one day after the sampling period and decide what to do next. Our spike radium isotope (^{226}Ra) has a half-life of 1600 years but it decays to one of the several isotopes of the inert gas radon (^{222}Rn) which has a half-life of only 3.8 days, which decays by emission of alpha particles to isotopes of polonium, bismuth and lead. We were mainly interested in measuring the degassing of radon from the lake into the atmosphere as a tracer of how fast carbon dioxide might be adsorbed from the atmosphere or released from the lake to the atmosphere. We found that radon was exhaled from the lake rapidly, that algal uptake of carbon dioxide was rapid (depleting the lake water of CO_2 every morning in two hours) and exchange with the atmosphere was rapid, equilibrating in seconds to minutes.

Sampling lake water from surface to deep for the noble gas radon and its parent radium required special equipment. Since radon is almost insoluble in water, it would degas quickly when exposed to the atmosphere. Using Wally Broecker's methods, we would create a vacuum in twenty litre flint glass bottles, firmly wedged into a wooden crate. These large evacuated glass bottles were dangerously fragile to transport in boats and canoes and backpacks. After gingerly carrying these bottles to the lake, we used the vacuum inside the bottles to suck water from varying depths, using long plastic tubing wound around a metered rope. After filling five to eight of these bottles at the lake and carrying them back to the lab (over several other lakes and portages) the samples had to be processed and analysed immediately. The radon gas was collected by bubbling the glass bottle with argon, and the gas stream was cleaned and frozen in a copper U-tube immersed in liquid nitrogen. After a lengthy bubbling time the frozen U-tube was disconnected from the line and hooked up to a separate line that led to a magic black box (a Lucas cell photo-scintillation detector) that measured the timed rate of radioactive alpha particle disintegrations of radon that came from the twenty litre glass bottle. After being trained by Steve Emerson, who shortly left to return to LDGO, I spent half the summer doing these measurements, with just a few interruptions.

Just a few interruptions, yah, sure, you bet. I was invited to the ELA Benthos Lab for a wine and cheese party by Sheila Dennett and even though I don't like wine, I happily attended to listen to her folk songs and guitar accompaniment. I was doomed, entangled, inveigled, hornswoggled and attracted to her. One thing led to another and we began a torrid romance that completely negated any after-hours effort on my Lake Winnipeg writing assignments. On one of my trips back to Winnipeg she asked if she could ride with me in my flashy Cougar, as she wanted to buy a new guitar with her summer student pay check. She had planned to stay at the Winnipeg YWCA but it was closed, so she came to my apartment for the weekend where I fed her moose steaks and fried potatoes and we listened to Ramblin' Jack Elliott on the tape recorder. She bought a 00018 Martin steel string guitar for $300 and had no money left for a case, so I bought one for her. To celebrate we went to a fancy restaurant, smoked cigars beside the Parliamentary grounds fountain and visited a soft ice cream parlour. I was seriously hooked, besotted, hornswoggled and discombobulated too. Back at ELA at work, she decorated my raincoat with Tolkien runes and proceeded to label me as 'The Wizard'.

While Wally Broecker was at the ELA camp with us the camp sewage system got plugged somehow and the Camp Manager Hermann Schneider got out his dowsing rods to determine which buried sewage pipes were running and which were plugged and empty. The logical scientist Broecker loudly stated that this was nonsense and that dowsing rods could not work. Hermann and his assistant Phil Michaelis (nick-named 'Hammer') asserted that ignorance resided in some scientists and the war began. Wally proposed that we perform an experiment to test the ability of dowsing rods to detect water. Inside one of the ATCO office trailers we filled a washtub with water, so we could drag the tub into any room. Hammer and Hermann were on the roof of the trailer with their bent welding dowsing rods, walking back and forth waiting for the rods to cross. As I recall, the dowsing rod people declared that there wasn't enough water for the rod to detect and the experiment was abandoned with no clear winner. The scientists ended up in my little office drinking single malt whiskey, during which time I played my reel-to-reel tape recorder version of Arlo Guthrie's 'Alice's Restaurant' story to them. I nearly died laughing at the expression on Broecker and Schindler's faces, these people not being associated with the folk music scene of this time.

Another interruption was a trip to visit the Char Lake Project on Cornwallis Island where Buster Welch was leading a multidisciplinary group of scientists that were trying to disentangle the carbon and nutrient cycles and the food chain of a simple Arctic lake. The base camp was in the settlement of Resolute, where Inuit families, Canadian and US military personnel and the Canadian government Polar Continental Shelf Project were sited. Buster and his good wife Cathy lived in a house constructed out of lumber, insulation and furniture collected largely from the dump and abandoned housing projects. The project scientists were using newly designed double layer wetsuits to collect algae and animals living on the underside of the ice. Having my scuba diving certificate, I volunteered to help with this work. This involves getting into the base camp shower under very hot water and filling the space between the two layered wetsuits with hot water. We then quickly walked outside into the wind and blowing snow to get into a heated Bombardier snow machine, which carried us along a rough road up the hill to Char Lake. There was a trap door at the back of the Bombardier so that the driver could position the Bombardier right over the 1 m diameter ice hole, to allow the divers to swim downward into the black water. We had head lights and a flashlight attached to our wrists and a safety line was down the hole. Nevertheless, it was scary and I wasn't of much use to the scientific team, as I got numb in about ten minutes trying to help scraping the green scum off the bottom of the ice. My real role here was to help the group figure out the importance of the accumulation rate of mud (including the burial rate of carbon, nutrients, and microfossil history) at the bottom of the lake, and I collected a few short sediment cores for this purpose.

Upon returning via Montreal and Winnipeg to ELA after the Char Lake trip, I drove my Cougar up to my field camp office trailer and was immediately assaulted by this young lady Sheila, barefoot, dressed in an old pair of my Levis and a South Dakota cowboy shirt tied off above her belly button. Her passionate greeting reminded me that there was more to life than science. We spent a lot of time picking wild blueberries in the bush, avoiding the feasting black bears, and Sheila made some good blueberry pies. I made zero progress on my plans for peaceful data analyses and writing of the Lake Winnipeg manuscripts. As the fall season came around I was still measuring radon and radium in the spiked lake, finding that both isotopes were disappearing faster than anticipated. Also with the fall season, the summer students began to vanish, to return to their university courses. One chilly late night in the radon lab Sheila was baking a gingerbread cake in the lab oven, to have with Labrador tea, and the tame grackle was sitting on top of the radon counting 'magic box'. Sheila was fussing about when to leave camp to return to her studies at the University of Waterloo. I turned around from setting the clock on the radon counting board and told her that I'd like to marry her, so she wouldn't have to leave. After the explosion of emotion and Sheila's teary agreement with this proposal, I duly recorded this incident in my radon lab notebook and we had a nice bitter cup of Labrador tea and fresh baked gingerbread. This spontaneous proposal surprised me greatly, as I had assumed I was going to be a lifelong bacheloR and I had never had any vision of married life. As would become more common, this serendipitous action was the best decision of my life.

GJB adjusting radon board at ELA with a grackle on his back, 1970

Shortly thereafter we phoned our parents about this development. Sheila's mother was not surprised, as she had already guessed that the situation was serious from Sheila's letters to home. My mother was surprised, and she seriously instructed Sheila to think twice about this deal. My mother told Sheila that her Brunskill boys were reckless, ornery and not easily tamed. But...but... all of my three brothers were already married. We went to Winnipeg to buy wedding rings that would fit over our swollen ELA volleyball knuckles and drove to Sheila's family home in Ontario (with the pet grackle in the back seat in a cardboard box) to surprise the Dennett family. In short order we arranged for the Salvation Army Chaplain to marry us, despite Dennett family friends trying to tell us to slow down. Sheila's father (president of the Canadian branch of Decca Radar) went down to Toronto to buy a silver tea set for us and the ceremony was done on 8 September 1970. I was introduced to the Dennett family chaos, where a large python in a glass cage was being fed raw chicken legs on the kitchen table, a kinkajou lived in the laundry room, an iguana lived under the living room sofa, a tame skunk lived in the electronics lab and antique rifles were stacked in several corners of the house. Sheila's mother tried to feed me kippers in milk for breakfast. I figured that this must be some sort of test of my suitability for marriage to Sheila, so I ate the damned stuff with great difficulty. The pet grackle was given over to Sheila's mother who understood how to talk to birds, giving them cigarette smoke baths and feeding them chewed food from her mouth.

Sheila and Gregg on their Wedding Day, 1970

We returned to Winnipeg, where Sheila arranged to finish her biology degree at the University of Manitoba and was whisked off to a university field station at Delta Marsh, on the southern shore of Lake Manitoba. I returned to the ELA field camp to continue the radon/radium measurements. Some honeymoon. My radium measurements indicated that we were losing radium from the lake somehow and I thought my analytical measurements must be wrong. A few years later, further research showed that much of the radium had been taken up by the algae along the shoreline. At Christmas we drove to my parents' home in South Dakota, through the blustery bleak winter snow-covered prairies. I suspect my parents and brothers were very curious about this soft-voiced English-accented lady that I had married so suddenly. All the aunts and uncles and grandparents came to see her (...'a prize catch', said Grandma England, admiringly), and we feasted on roast prairie chicken, venison, pheasant, candied yams, hot buns from the oven and home-grown beef.

Since Sheila was involved with a mammalogy course with Dr. Pruitt at the University of Manitoba, she took this South Dakota visit as an opportunity to collect some unusual skulls for her collection. She managed to obtain and clean skulls from some coyotes, prairie dogs, gophers and a buffalo skull from Don Hight's back yard. We also found a dead frozen road-kill skunk, which she kept frozen for dissection and cleaning later. At this time I was driving a Mercury

Cougar, a little sporty car that my brothers much admired. On our way back to Canada the smell of the dead skunk got too strong from the trunk, so we dropped it in a garbage container on the main street of Pierre. As usual, we arrived at the US-Canada customs and immigration gates very late, which was fortunate, as we were carrying a lot of South Dakota boned and wrapped venison and pheasants, as well as Sheila's skull collection and an uncured salted buffalo hide. The tired inspector took one look at our luggage and waved us into Canada.

There were giants among us during these times and of course we didn't recognise this until several decades later. We had Jack Vallentyne and Wally Johnson at our back in the mahogany row administrative offices. Cliff Jones, Willy Burton and staff of the institute workshop set up our field lab and accommodation ATCO trailers, made new scientific sampling instruments and fixed stuff that broke. David Schindler was a good leader and pushed us all to think science options around the ELA kitchen table. John Cherry, Bob Newbury and Ken Beatty saved our bacon many times over by providing the essential hydrological and atmospheric measurements. Wally Broecker and a dozen of his geochemical graduate students (Ray Hesslein, Tom Torgersen, John Crusius, Steve Emerson and Paul Quay, among others) enriched our ELA camp, way beyond the boundaries of normal limnological methods. Domenico Povoledo, Richard Vollenweider, Staffan Holmgren and John Rudd were also involved. F. A. J. Armstrong and half a dozen of his technical support staff (Mike Capel, Mike Stainton, Norma Kritsch, Sheila Newsam and Greg Morden) kept our chemical methods and results on the leading edge. The ELA camp could run itself for months due to supertechnicians Honest John Shearer, Russ Schmidt, Phil 'Hammer' Michaelis, Jim Russell, Brian Graham, Hedy Kling, Alex Salki, Paul Campbell and others. Steady inputs from biologists Andy Hamilton, Ole Saether, John Flannagan and Kaz Patalas kept a good perspective on what to do next. It was a rare and very lucky conglomeration of people that mostly got along well together and accomplished many difficult tasks that enriched the whole group. Many of these young graduate students and scientists were to become well-known in research institutions and universities around the world in the following decades.

Chapter 12. Seeking Yellowknife and the Mackenzie River

The winter of 1970/71 went by in a blur of meetings all over western Canada and the Northwest Territories. Sheila was busy at the university, taking graduate courses in biology. She was on a late winter field camp study session, where cross-country skis were used to check the live-trap lines and make observations of small mammal habitat under the snow cover. She developed a distaste for tea on this expedition, which was a bit strange. We were living in my tiny Wellington Crescent apartment on the Assiniboine River bank in Winnipeg. I was doing a lot of office work that I was not trained for, such as hiring new scientific and technical staff, ordering ATCO trailers and fittings for research camps for the new Mackenzie Valley Pipeline Study, assembling scientific equipment and bargaining for land, sewer and water supplies to our planned lab and accommodation headquarters in Yellowknife and purchase order arrangements for groceries and hardware. We were also planning smaller field camps in Fort Simpson, Inuvik, and downstream from Old Crow in the Yukon. All of this was being coordinated with various fisheries agencies, forestry and water survey departments, the NWT government, the wildlife service and geological survey and the Indian Affairs Office. The Canadian government was trying to give us a budget for this large operation, with the money oscillating between a quarter to three quarters of a million dollars and thirty to forty person-years of new positions.

By early May we had traded my bachelor snazzy Cougar for a big Dodge Econoline van, in anticipation of commuting between Winnipeg and Yellowknife. We packed our household goods in storage and loaded the essentials in our van, which technician Jim McComiskey would drive to Yellowknife. Before we left Winnipeg for the NWT Bob Hamilton, the scientist that had initially hired Sheila for ELA work last year, invited us to his home for supper. While sitting on their couch being gracefully social, he looked at Sheila and told her she was pregnant. This surprised both of us greatly but it was later confirmed. Sheila and I took the train from Winnipeg to Jasper, Alberta, where yet another planning meeting went on for several days. We got to hike some trails near Jasper, smell the high mountain terpenoid piney air and watch the elk. We flew to Yellowknife in early May, where there was cold wind and snow on the ground and we were met by Hugh Trudeau, one of the Fisheries deck hands on the good FRB ship *Stenodus*. Yellowknife is the capital of the Northwest Territories, perched on a northern bay of Great Slave Lake. This small town was plagued with infestations of transient government employees, gold miners, some fishermen and trappers of the Dogrib Indian nation. The town had the normal complement of a Chinese restaurant, two hotels, hardware and grocery store, fuel depot, liquor store, hospital, airport, mechanical repair shops and a diesel electricity generating station. The Polar Bear Lounge in Yellowknife had posh fittings, where town drunks, mining executives and fishermen coexisted easily. During the winter, supplies came from the south over an ice road across the lake. During the open water season, a ferry carried vehicles across the western outlet of Great Slave Lake, the beginning of the Mackenzie River.

Mike Healey (and his offsider Bill Woodall) was in charge of a Fisheries Research Board experimental fishing study on some lakes north of Yellowknife. He and I were the scientific leaders at our new field HQ and within a month we had four accommodation ATCO trailers and a lab trailer on site and hooked up to water, sewer and electricity (with the help of Cliff Jones of FRB Workshop and some local tradesmen). These ATCO trailers were very well furnished and perched well above the ground to avoid melting the permafrost soil underneath. Outboard motors, tin boats, Hudson Bay freight canoes and lab equipment arrived by semi-trailer trucks almost every week. Sheila was cooking for everybody at $1/person-day and I'm sure she was happy to see Mike's wife Judy arrive in late May.

Feeling a little guilty, I left my new wife waving at the Yellowknife jetty and stepped into a fully loaded de Havilland Twin Otter on floats, for our first sampling trip on the Mackenzie River at the end of May. The lake ice was still firm but parts of the river were open. We were flying with experienced staff from the Canadian Water Survey Agency, the main field man being Herb Wood, who taught me how to live comfortably in a cold airplane. Woody's job was to retrieve water flow records and service flow stations on the major tributaries and the Mackenzie River. This allowed my group to obtain water samples (for chemical analyses) at locations where water flow data was available for several decades. We flew into Fort Liard in the Nahanni Mountains, a snow-covered and gorgeous country. But before we could land, the left propeller engine stopped, causing some concern in the cockpit. It was restarted, coughing and sputtering, and the plane limped back to Fort Simpson, mostly on one engine. Controlling the aircraft on floats in a strong river current with only one engine was difficult but we finally made it to the Ft. Simpson dock safely.

It was going to take a day to get an aircraft mechanic to us and at least a day to repair the faulty fuel line pump, so we used boats to visit tributaries of the southern Mackenzie River. Most of the Water Survey stations had been damaged by ice blocks being pushed onto the shore line and all of these required repairs and calibration. Johnny's Hotel and Restaurant got a roaring business from us. The 'Special Menu' was written with felt-tip pen on a brown paper bag on the wall, and the menu didn't change during the time we were there. After supper we walked the river bank to watch the ice move downstream and there was a bagpiper in a kilt playing Celtic warpipe tunes to the river and the ghost of Alexander Mackenzie himself. We got attacked by Indians, too. They were six to ten years old, whooping and giggling, throwing mudballs at us from behind a pile of driftwood.

Three days later we flew back into the Nahanni Mountains to service the flow station and obtain water samples. The Fort Liard Forestry Service and RCMP allowed us to use their home-made long riverboats to obtain samples of this river of mud. We flew further into the mountains to sample tributaries to the Liard River, stopping for lunch at the Nahanni Hot Springs, where Gus and Mary Kraus had their cabin. This place smells of hydrogen sulphide from the hot springs, some of which are just behind Gus's cabin, at 37°C, bubbling sulphide and methane, with a white scum on the surface of clear bluegreen water. Pink, green and red sulphur bacterial coatings on the muddy margins of the steaming ponds appeared in paisley and oriental patterns. The hot springs, plus the southern exposure to the spring sun, made this little pocket in the mountains a very warm microclimate and the vegetation was already green, despite there being some ice still on the river and snow on the nearby mountain slopes. Mary had her tiny garden planted already, with some potatoes already coming up. The river flowed through a 1200 m high canyon just downstream from Gus's cabin, the walls being sheer white limestone. Gus and Mary made us bacon and eggs with fried bannock, tinned butter and thawed fermented high bush cranberry jam. The coffee was good but there was Nahanni River silt at the bottom of the cup. The cabin was chinked with sphagnum, with shelves to the ceiling full of little bottles of curious plants and animals preserved in ethanol, unusual mineral specimens, bird eggs and feathers and lots of bird books. The yard around the cabin featured two gardens, several secure food caches up in the trees, three small diesel generators in a sound-insulated shed, two Hudson Bay freight canoes upside down on high pole racks, tarp-protected split wood piles, fish and pelt drying racks and sheds for his trapping and hunting gear. Gus sure liked to talk and Woody encouraged him to tell many adventurous tales of his early days. His wife Mary told the story about how she shot a young Gus with a .22 rifle because she thought he was fooling around with another woman.

The bullet went through his cheek and almost took his ear off and it took about a week to get him to the Yellowknife hospital, where he spent three months recovering. Mary stayed beside his hospital bed the whole time, because '...he shouldn't sleep alone.' Gus and Mary were packing up their possessions as the National Park Service had bought them out. They were moving to Little Doctor Lake on the eastern flank of the Nahanni Mountains, where this seventy-two year old trapper was going to build another cabin and set up a tourist/fishing resort.

At the Water Survey cabin on the South Nahanni River just below Virginia Falls we stashed two barrels of turbofuel for future twin otter emergencies because we expected to visit this region over the next few years. Some of the interior mountain rivers were said to flow partly open all winter, as some were spring fed with warm groundwater. It was strangely warm here, black flies and mosquitoes were out in force and the river bank trees were budding green already. This cabin was designed to resist bears and wolverines and was stocked with emergency supplies and dried/tinned food in case of aircraft or weather troubles. We made it back to Fort Simpson before dark, the menu at Johnny's Restaurant had not changed and the bagpiper was again playing tunes to the river at sunset. We helped refuel the Twin Otter, 500 lbs of turbofuel at $25/45 gallon barrel, 50 cents/gallon, costing us $2.50/hr to ride along, flying at 110-120 knots, maximum cargo plus passenger weight being about 2000 lbs.

The Fort Simpson Indian people gillnet suckers, burbot and pike from the muddy Liard and Mackenzie to feed their dogs. Good eating grayling, inconnu and whitefish could be found only in the clearer water of the eastern tributaries of the Mackenzie and upstream Mackenzie clear water (the outflow of Great Slave Lake). The trappers took beaver ($11), marten ($35), and lynx ($60) pelts and some made $5000/yr. Moose were abundant further up the Liard River and black bears were a real pest at any bush camp. Later on we found that a roasted hindquarter of smoked bear was as good as a hog ham.

Over the next two weeks we flew northward along the Mackenzie River, servicing the Water Survey stations on the large roaring muddy western mountain tributary rivers and the smaller clear water eastern tributaries. Where settlements had accommodations we enjoyed the luxury of soft beds. In some locations we camped in trapper's cabins near our sampling sites, ate a lot of bacon and bannock and made strong coffee for our Stanley thermos bottles for the card games at night. The Twin Otter maintained radio contact with Yellowknife and Fort Smith and we often visited RCMP stations along the river for local news. Most of the settlements along the river were residual Hudson Bay posts for gathering furs over the past century, but Norman Wells settlement was the location of some sweet crude oil wells. The little muddy town was thick with geologists, oil riggers, black flies and mosquitoes, a tough town, but we found an old ATCO trailer for our bunks for the night. Forestry Service friend Steve

Zoltai (who had helped me with ELA soil descriptions) and Owen Hughes of the Geological Survey of Canada were at the fancy Mackenzie Mountain Lodge, which was out of beer, but they had a lot of rum on the shelf. They told stories of finding sink hole lakes and oil seeps in their explorations to map soils and glacial deposits of this region. There was no beer because the ice road had melted and the river barges were not yet running. There was a rope running from a shoreline tower out across the river ice to another tower, where the rope was attached to a bell. The big town game was to bet on the day and time that the bell would ring as the ice moved downstream. Woody was a clever fellow and he knew how to make good friends along the river. We emptied our Twin Otter of all science gear and flew to Inuvik on the further north Mackenzie River Delta, where they had lots of beer and where the Mackenzie River water meets the Beaufort Sea and Arctic Ocean. The aircraft returned loaded to the ceiling with beer, and we were applauded loudly as heroes when we entered the Norman Wells Hotel that night. We would benefit from this act of kindness for the next few years.

Further north, we visited the Arctic Red and Peel Rivers, above the tree line now, the western mountains there being covered with a soft green taiga vegetation. The eastern flank of the mountains became muskeg swamps and wetlands with few trees. Even though the ice was just melting, the rivers and lakes were full of plant and animal life. The western mountain rivers were a muddy rust-red colour with islands of foamy debris and whole trees moving at several metres/second, making our water sampling trips in an inflatable rubber boat a bit dangerous. We tried to visit Great Bear Lake but the ice had been pushed westward against the outlet, where Woody's flow station was perched, and we could not land on floats. We did land on a narrow patch of open water near the Great Bear Lake Settlement and I got a local person to take me to the outlet for my water samples on a skidoo across the rotting lake ice.

Woody had a terrible habit of wanting to depart at five a.m. so we could arrive at our first sampling location at first light. In all these travels, after returning to our temporary overnight base in the late evening I would get out my water filtering equipment to process the samples, recover the muddy suspended silt on filter paper and measure a few parameters in the filtered water. My field book indicates that I often finished this work between midnight and two a.m., so I got in the habit of sleeping during the daytime longer aircraft flights. These processed samples were labelled and packed in sturdy transport boxes for shipment back to the Yellowknife lab and/or Winnipeg, where clever chemists Greg Morden, F. A. J. Armstrong, Mike Stainton, Rudy Wagemann and Paul Campbell did further chemical analyses and tabulated the data.

These Arctic spring and early summer aircraft sampling trips were conducted in the 'land of the midnight sun'. This meant that we could work twenty hour days, until we collapsed.

I suspect my later ability to sleep amidst loud noise, diesel fumes, rough travel, smelly cargo and confusing schedules was learned here in the cabins of various Twin Otters, turbo-Beavers, helicopters, bombardiers, canoes and airport lounges. Some of my colleagues were very impressed that I could sleep through violently rough air travel. Neurophysiologists years later discovered 'sleep spindles' that are able to shut down parts of the brain, to allow sleep to proceed undisturbed.

The question that our research team was pursuing had to do with the ecological effects of construction and operation of a hot oil pipeline, to run down the Mackenzie Valley somewhere. The surface ground here melted and froze seasonally, but below the ~half metre melt layer was permafrost, ground that had been frozen for many millennia. We worried that construction activities would open the ground to summer warming, increased erosion would occur with the mud flowing to the rivers and this would deleteriously affect the river plants, small riverbed animals and fish. In addition, a hot oil pipeline, either above ground or insulated and buried below ground, would surely change the heat budget of the soil and if the ground melted, the pipeline foundation would be endangered and broken pipelines might occur, with oil spills on the land and at river crossings. From this initial field sampling trip I could already see that the greatest danger of increased erosion and mud supply would be to the clear or tea-coloured water of the eastern tributaries. The mountainous western tributaries were already carrying enormous loads of mud and sand, with large erosive energy to carve new channels. This would cause engineering problems for pipeline river crossings. This was the beginning of my understanding of the problem and the beginning of my appreciation of the logistic problems of working along this mighty river.

At the same time other environmental and social research was beginning, as the money flowed in from Ottawa and the oil pipeline design and geographic plans became available. Wildlife ecologists were mapping the movements of the caribou herds, studies on the fur-bearing animals were begun, several fisheries study groups were formed, soil and forestry experts were mapping the region, studies on the commercial and subsistence economy of this vast region began and surveys of the social impact on aboriginal communities were started. The bureaucrats in Ottawa created a task force to manage this effort, with the usual gigantic tangle of procedures for task proposals, evaluation, funding and workshops to be held in the small communities along the river valley.

We returned to our base in Yellowknife in early June, where the north arm of Great Slave Lake was melting rapidly. We were tired, bruised and our hands were cut and sore from loading and unloading heavy gear from the aircraft with wet hands. Sheila gathered me up, fed me and put me to bed, where I slept for about a day and a half.

Sheila had been staking out a garden near our ATCO trailer, scraping off the organic topsoil to let the permafrost melt and dewater, then covering it up again with rich soil. Her garden grew huge radishes, onions and oversized zucchini in the short summer period, when there was sunlight from four a.m. to midnight. She was also running a small mammal live trap line as part of her Master's Degree project, catching mostly voles and mouse-sized creatures. It was wonderful to have home-cooked meals, tea-time breaks with hot donuts and to sleep late without the howl of the Twin Otter engines in my ear. Sheila collects pets like a magnet attracts iron filings. At this time, she had a Brazilian Matamata side-necked turtle in a wading pool in our living room, a small red-neck turtle in an aquarium and I had brought her an unusual frog from the Johnnie Hoe River, which liked to eat mosquitoes fortunately. Sheila and I took a small Chestnut canoe up Yellowknife Bay to portage into Prosperous Lake, where we camped for the weekend on a high shoreline ledge near a large eagle nest in a tall tree.

During this down time our summer students were using an old Raytheon sounder to produce bathymetric maps of six small lakes near Yellowknife, some of them polluted with city sewage and mine tailings waste water. Brian Graham and Greg Morden got the Yellowknife chemistry lab running smoothly and we invented a better method for measuring arsenic in lake water and plankton. We were happy to receive the attention of our CEO Wally Johnson, who visited several times to see how we were getting along. This kind of executive support was to become rare in the coming years and he was demanding quick results to show to Ottawa, as we were consuming large amounts of money and man-power.

I flew with Woody and my sampling crew all along the Mackenzie River in July and August of 1971 in our specially outfitted Twin Otter. I gained huge respect for the de Havilland Otters and Turbo Beaver aircraft on floats, wheels and skis and the pilots that flew them through the mountains and boggy floodplains. We worked out of Ft. Simpson, Norman Wells and Inuvik/Tuktoyaktuk, and I slowly learned all the radio call codes for the remote stations along the river. There were forest fires, flood-damaged sampling stations, aircraft mechanical problems, fuel supply delays and we would breathe mosquitoes and blackflies while sampling and filtering our water samples. We would blow our noses into facial tissues and find a black insect sludge on it. We ate some interesting meals at various field camps: curried moose tongue, bear liver with caribou kidney fat, smoked and candied whitefish, fried beaver tail, ptarmigan, muskrat soup with lots of tea and pilot biscuits. As the high western mountain snowfields melted, the rivers flooded in sequence going northward. The outlet of Great Slave Lake is the beginning of the Mackenzie River and it is clear blue lake water. As the Mackenzie flows northward, it gains input from the brown or rust red muddy mountain rivers from the west. And it gains clear tea-coloured water from the muskeggy swampy eastern rivers. These different coloured water streams would persist downstream in the big Mackenzie River for hundreds of miles, with muddy opaque water on the western shore and clear or tea-coloured water on the eastern shore of the river. We repeatedly visited Keith Kepke's garden up in the headwaters of the Liard River, where he had luxuriant cucumbers, tomatoes, rhubarb, peas, beans, sweet corn, squash, onions and potatoes growing well during the very short but long-lighted summer. Gus Krause's abandoned cabin on the Nahanni River was invaded by Swiss tourists and mineral prospectors. We gained skills in avoiding fights and hookers in the Inuvik pub (affectionately known as 'The Zoo').

Benthic biologist David Rosenberg began research on river bed biota in the Fort Simpson region, Norm Snow was hired for Mackenzie Delta and estuary studies out of the Inuvik research laboratory, Gian Vascotto was hired to scope the stream and river biology of the Old Crow region, Porcupine River, tributary to the Yukon River. Good friend and stream biologist John Flannagan also toured the Mackenzie River with us. This was an insane time of rapid expansion and I was supposed to be on top of it all, including four widely dispersed research camps and thirty person-years of field and lab research personnel. We really didn't have enough money to support all of this activity but Ottawa kept telling us to go ahead and do the work, the money would be found.

I had received news that a scientific society had granted me $800 toward travel costs to attend a big science meeting in Leningrad in the fall. Sheila had some of the Russian language and her M. Sc. thesis advisor Professor Pruitt had contacts with Russian small mammal ecologists, so Sheila was anxious to visit the Russians as part of her thesis program. Sheila and I packed up our possessions, her Martin guitar and turtles and we returned to Winnipeg for the next adventure.

Chapter 13. Babushkas, Hot Crotch, and the Wizlet

Upon returning to civilisation in Winnipeg, we had a week to find a new apartment that would be suitable for our new baby and to prepare for our trip to Russia. This included a briefing from two RCMP officers in full uniform and I was startled to find they had a complete file on my history, with fingerprints and letters of recommendation. I also gave some seminars, slide shows and pep talks to my institute bosses, confirming that we had done some useful work during the Arctic summer. And as was to become normal in the coming years, the institute administration was going through a reorganisation and we lost the scientific leadership of Jack Vallentyne, who became one of the chief scientists at a bigger institution in Ontario and began his work with the International Joint Commission on the Great Lakes.

By mid August 1971 we were flying east via Toronto, New York, Amsterdam, Stockholm, finally landing in Leningrad in a dark gloomy empty airport. After several hours of hassles with multiple languages, passports and gaining some roubles, our party of US, Canadian and European scientists were hauled to a special hotel for westerners in an Intourist bus. And then we had another series of hassles with passports, getting a room key sometime after midnight, with Sheila sleeping wearily on a bench in the hotel lobby. Of course we slept in the next morning and got on an afternoon bus for a tour of the town (Pushkin Theatre, Zoo, several museums, Lenin statues). We then we waited two hours to be served supper (sour malt beer, chicken in sour cream, tomatoes in sour cream).

I attended the science meeting sessions all day while Sheila toured the Zoological Museum, with letters of introduction from Professor Pruitt at the University of Manitoba in Winnipeg. I was happy to meet Professor O. A. Alekin, whose books on Russian Arctic River hydrology and chemistry I had read. All of my old European friends were here and we walked downtown to find a bookshop, tobacco shop and lunch (meat and sour cream blinis and tea).

Most of the presented science papers were in the Slavic languages that day so I left early to join Sheila for supper and we attended a wonderful concert at the ornate Pushkin Theatre of Stravinsky's *Petrushka* and *The Rite of Spring*.

Being pregnant, Sheila was desperate for whole fresh milk and eggs, so we tried to find some in the drab grey Nevski Prospekt shops. The best we could find was boiled milk and hard-boiled eggs. We also purchased some children's books in Russian, as Sheila had some of the Russian language. We found a balalaika, some Russian books on small mammal ecology and a wonderful Russian tea cosy. At our hotel supper table we met a Spanish lady named Doloros Planas, who had her young son and daughter with her at this meeting. She recognised Sheila's pregnancy and immediately began harassing the waiter for special fresh food for her with some success, because she could speak Russian. They were trying to feed me hard-boiled eggs and cold bread for breakfast but I finally got them to provide fried eggs and ham by the middle of the week. Our good Dutch friends Sikko Parma and Jaap Kaalf found a street vendor who would provide a fresh bottle of milk for Sheila on a daily basis at the hotel door, but only if we paid in US dollars. Also at the hotel door were ragamuffin children trying to sell fake ancient Russian church icons, black market roubles for US cash and hand-carved nested babushka dolls.

The science papers being presented were from all over the world, including problems with fisheries in African lakes, saline water in the high mountains of Chile, Scandinavian acid lakes, Indonesian sinkhole lakes, the Mississippi River and Lake Tahoe. I tried to use my halting German language sometimes, to little effect. My French language was worse, bringing laughter from the French speakers. The science congress organisers provided session translators and our Intourist guides could also speak a dozen languages. The 35 mm slide projectors broke down regularly, the schedule was half a day late and some sessions continued on past nine p.m. My paper on Lake Winnipeg went well, with some penetrating questions from Charles Goldman on the futility of trying to control nutrient supply to a lake that was naturally well-fed with nutrients.

One evening Livia Tonolli and Han Golterman gathered Sheila and me for a special expedition to a good fish restaurant. This expedition led us down into the Leningrad Metro, a subway train deep beneath the city, with neat and clean shops and rest stops and lavishly decorated marble arches. Livia appointed herself to be Sheila's grandmother and gave her little scolding lectures about taking it easy in her pregnancy, avoiding long walks and crowds and eating more cheese, all this in heavily accented north Italian language. She obviously didn't trust me to take care of her. And here again was the international family of science, crossing language and cultural barriers easily.

The usual science conference dinner was held in an old palace, with giant samovars of hot tea, vast quantities of wine and vodka, cold fatty sausages, iced buttermilk, pickled fish, bushels of whole tomatoes in sour cream, watermelon soup, fruit and giant loaves of hard dark Russian bread. They played 'Hey Jude' and 'Hello Dolly' over the speakers, with some people dancing. Later on a live band appeared with accordions, balalaikas, zithers and an electric guitar, blasting out Russian folk songs across the hall. All the countries attending had special representatives who toasted the organisers with large glasses of vodka, pepper grains sprinkled into the glass, each toast being translated into five languages. Sheila had been visiting science institutions in the city all day and her feet were tired. So she took her shoes off and we sat on the steps of a winding staircase above the merry crowd. One of the security men came up to us and in a sparklingly British accent asked us kindly to not steal the cutlery.

I was privileged to talk with some good Russians who had excellent English and a long grasp of history. They were puzzled that the USA and USSR were enemies in this cold war, with thousands of nuclear warhead missiles pointing at each other.

'Why does USA hate USSR? We were allies during World War II, stopping the Nazi war machine. Why do we even have a cold war?'

The 'cold war' was a strange logic developed by the military-industrial complex of a group of countries, but mainly the USA and USSR. The point was to avoid nuclear war and to make 'conventional war' a game to be played to allow the balance of power to adjust. Most of the people of the USA and USSR are not stupid or wicked but somehow they insanely paid their governments for thousands of nuclear warheads and missiles, in addition to larger and larger land, air and sea military forces. Such discussion went on into the cool night air. The Russians had many sly ironic jokes, for which I had no come-back:

'During Stalin's time the Russian economy stood at the precipice of disaster. Since then, we have made great strides forward.'

But we were all saddened to hear that a faulty valve in the Soviet space capsule *Soyuz 11* caused the cabin to depressurise at an altitude of 168 m above the earth, killing three cosmonauts.

During our last day off from the science meetings Sheila was met by her interpreter Tanya and we ventured through the Metro to the Arctic and Antarctic Museum to meet several Arctic small mammal experts who were friends of Professor Pruitt. They gave Sheila some decorated wooden spoons and several maps of Arctic white fox distribution in the Russian Arctic. Leningrad (also called St. Petersburg) is quite far north, in the taiga zone, so there were several research enclosures for scientific studies, containing taiga rats, ground squirrels, unhappy foxes and box cages of small vole/mouse-like animals. We walked across the Neva River to board a fast hydrofoil boat to visit Peterhof, the castle/estate for the Russian royalty house of the Czarist Peter the Great, up on the Gulf of Finland. This estate was built in the early 1700s and consisted of numerous gardens, fountains, ponds with useless bridges, ornate gold-plated palaces, chapels, onion-domed towers and sheds full of gold-plated horse carriages and buggies. Most of this had been damaged by the Germans during World War II and was still being renovated and repaired when we were there. Despite all the gold-plated buildings what was missing was a toilet. We were finally directed to a small wooden building behind some hedges, which held an earthen pit, much to the disgust of the European ladies in our group. After a complicated series of kopek- fuelled buses, boats, trains and Metro rides, we made it back to our hotel for a gigantic pot roast in the dining room, surrounded by all of our scientific friends, saying farewell until the next meeting.

After frantic packing the next morning we took a taxi to the airport, where we had to give up our remaining kopeks and roubles and were subjected to careful luggage inspections by the Russians. Upon walking to the Finnish airplane we were met by a sparklingly colourful attractive smiling stewardess with a basket of perfectly crisp apples. The Finns certainly did know how to distinguish themselves from the Russians. Flying into Helsinki in the evening, we were almost blinded by the riots of neon-coloured lights, in contrast to the dour darkness of Leningrad streets. Our hotel provided fresh green salads and fruit, fresh milk, excellent cold beer and grilled fish and rare reindeer meat, which was eagerly devoured by us. We also were introduced to Lakka liqueur, made from sub-Arctic cloudberries, which we liked very much. We spent two days exploring Helsinki in the rain. I bought a winter beaver-skin hat and Sheila bought a wonderfully soft wool blanket for the trip home. We visited several parks, monuments to Sibelius, attended some concerts and loafed. We flew from Helsinki to London, New York City, and Toronto, where Sheila's father met us at the airport and we were whisked away from the big city to the comfortable arms of the family at their country home north of Toronto.

Sheila stayed with her family while I returned to Winnipeg for urgent meetings about the reorganisation of the science groups at our institute. Paul Campbell and I moved our household stuff from storage to a new apartment in south Winnipeg, a cold empty place without my Sheila. I unpacked just enough for me to cook, eat and sleep, and left the rest in cardboard boxes.

All sorts of urgent reports had to be written and I had Pipeline Project Task Force meetings in Edmonton and Vancouver and one more fall aircraft survey trip from Yellowknife to Inuvik on the Mackenzie River valley in mid-September. Keith Kepke at Fort Liard in the Nahanni Mountains was digging his garden potatoes and getting ready for the long winter.

There was fresh moose meat at the Fort Simpson restaurant. Some Swiss spelunkers had been living in Gus Krause's cabin for most of the summer and they reported on several new caves they had discovered. The riverboats were pulled up above the ice damage level on shore at Norman Wells. Most of the tributary rivers were at low discharge and the rapids at San Sault on the Mackenzie were very bad. Jim McComiskey and I swamped a zodiac boat in these rapids because we misjudged the two metre standing waves over the big boulders. We managed to capture the upside-down rubber inflated boat below the rapids, right it, start the outboard and collect all of our gear with no loss, thanks to our water-tight gear boxes. After taking our water samples we returned to shore to light a big fire to dry our clothing, while the Twin Otter returned to Norman Wells for fuel. We had double caribou steak for supper in Inuvik that night, after a long hot shower. After returning to our HQ in Yellowknife we started preparing our accommodations and lab facilities for the winter. This meant we spent some time crawling underneath the ATCO trailers to drain the water pipes and putting anti-freeze in the toilet systems. Some equipment was packed for shipment back to Winnipeg, for calibrations and repairs, and the canoes and machinery were stashed in a rented warehouse. It was getting cold and windy and I was grateful that Sheila had left some tubs of frozen rich moose meat chilli and home baked bread in our cabin.

By late September I was back in my tiny cubicle office in Winnipeg, trying to clear my desk of mountains of urgent mail and demands for more data reports and cost estimates for the Mackenzie River project. Most of these paper demands were from desk-bound administrators and task force managers in Ottawa who wore suits and neckties every day, worked nine to five, whose main job was to keep track of the money. They really didn't care about anybody getting dunked in the cold Mackenzie River from an overturned zodiac boat. They couldn't tell a Beaver aircraft from a Twin Otter in the daylight and they had little concept of the eyeball-aching vastness of the Mackenzie Valley. Nor could they understand the mind-jarring transitions between rubbing elbows with great scientists in Leningrad castles and then immediately being a field man in the Mackenzie River valley.

Our new institute in Winnipeg was still very small and our administrative support officers, Section Leaders and Director were personally involved in evaluating and pushing the scientific effort. Our Director Wally Johnson took interest in our project and personally rammed some of our cost estimates through the bureaucratic tangles in Ottawa. I enjoyed doing the rough field work and extracting the information from the piles of data sheets from all the chemical and biological analyses. But I had little patience for the administrative aspects of project management and more than once I was guilty of spending too much money to complete some scientific task.

I was rescued from this office insanity by the arrival of my dear young wife Sheila, who had spent some time with her family in Ontario, making baby blankets, deer hide moccasins and elderberry jelly. After a visit to the doctor to inspect the baby in the belly she was immediately involved in coursework for a Master's degree in small mammal ecology with Professor Pruitt. We bashfully went shopping for nappies, a crib, baby clothing and some furniture for our new apartment. Disgusted with the fancy stuff in the stores, we found everything we needed in a dusty second-hand shop, including an old pedal-bellows pump organ (from a high-class Winnipeg whorehouse that had sadly gone out of business).

My grandfather George Brunskill died on 29 September '71, snoozing in an easy chair with his pipe. We decided to attend the funeral, flying through lightning storms via Minneapolis to Pierre, where my father and two of my brothers were waiting for us. There was the usual big extended family dinner at the big house at home, only this time with sad eyes and quiet talk. Relatives from Haywarden, Iowa joined us at the funeral in the Methodist Church, where a Minister who didn't know him spoke about life on the prairies. This religious talk did not fit my grandfather, who had little respect for the church. I remember him talking about preachers as being people who make a living from other people's fear and guilt. A good selection of White River ranchers carried his coffin to the graveyard. Then we all returned home for more coffee, date-nut bread, donuts, pies and cakes and home-made ice cream with strawberry puree. Grandfather would have been happy to know that the talk then turned to cattle prices, the good rain, the stock dams being full and a few outrageous stories about old George's most famous exploits. And of course, he was the great-grandfather to our unborn child. In my journals I recorded a long account of his influence on me, including his love of gardening, horse training, windmill repairs, chewing tobacco and John Deere tractors. He had cleared the river floodplain land with a team of horses and a lug-wheeled McCormick-Deering kerosene tractor in the early days and he regarded the White River as an adversary that had to be fought every time it flooded. I remembered him sitting in the barn, rubbing linseed oil into bridles and saddles, sharpening garden hoes with a file and trimming horses' hooves with large nippers and a frog knife. He didn't finish grade school and was in the cattle business at age fourteen. Even though he didn't think much of advanced education he sent generous cheques to me through my college and graduate student years. One of his famous quotes was 'There can never be too much bacon!'

The big England house west of town was where my mother and uncle Raymie grew up. Now Grandpa and Grandma England decided to move to Rapid City in a trailer park, leaving the ranch, barns and house to my father and mother. I suppose this was a good family decision but it didn't seem right to me. Both of them were prairie people, had come through the 1930 Depression building roads and dams, and bought that big house to raise their family. I remember that quiet gentle man Grandpa England looking out across the barn and cattle pasture, saying 'This country sure looks good to me.' And then they drove off to Rapid City for retirement and a life of boredom in the city.

Back in Winnipeg in mid-October I was trying to insulate our Dodge van, to make it warm for travelling during the - 30°C winter. Sheila called me in to time her labour pains and it was our first child ready to come out. We went to the hospital and Sheila spent two painful days in labour before the doctors decided to do a caesarean delivery. At that time I wasn't allowed to be with Sheila, so I spent most of this time very uncomfortably on the metal benches in the dreary noisy hospital hallways. Both of us were a long way from our family network of support during this stressful time. A network of good friends fed me, mainly Paul and Margaret Campbell, John Flannagan, Roger and Mary Green, David and Kathe Schindler. Hand-knitted baby blankets, rocking cribs and baby toys showed up at our apartment. A healthy baby boy was born early morning 19 October and the kind clucking nurses whisked him away, leaving Sheila exhausted and drugged. Like most bewildered fathers I just waited for her to wake up, worried and equally exhausted, sitting in a metal chair watching her breathe. But all was well eventually and we came back to our apartment in a few days, where I enjoyed watching our little boy nurse on his good mother's breast. I wanted to name him Alexander Mackenzie, after the explorer of the Mackenzie River, but Sheila's mother cautioned me not to go outside my Scots ancestral clan. So he was called Ian MacGreggor and his baby nickname became 'Wizlet'. The next few weeks I spent cooking a lot of moose steaks and washing nappies, going to the shop for more nappies and taking care of my tired Sheila, as the first winter snow began to accumulate.

Within a month Sheila hired Margaret Campbell as a babysitter, so that she could return to her Master's Degree program at the University of Manitoba. My brother Gerald came to visit in early November, which made it possible for me to return (with Andrew Hamilton and David Schindler) to the northern Saskatchewan Torch River country to get a moose for our freezer. I was involved in many planning meetings across the country to get ready for our next year's research activities in the Mackenzie River valley, and we were hiring some permanent staff for our research stations in Yellowknife, Fort Simpson, Inuvik and Old Crow.

With the help of my brother Gerald we had insulated our van, installed a large sleeping platform behind the rear seat, a baby bottle warmer plugged into the cigarette lighter, installed battery, engine block and oil pan heaters. A heavy wooden box was floor-bolted and strapped to the rear seat, for the comfort of the baby Wizlet. This was a komatiq box, a specially designed box for fitting neatly on an Arctic sled pulled by a skidoo. Sheila had found some mushroom stickers which were applied to the outside of our van and it became known as 'The Mushroom Van' for the next decade. We drove down to South Dakota to be with my family for Christmas and to show off our new son. All the aunts and uncles, nieces and nephews and family friends clucked over the new baby. Two of my brothers were married and had small kids, so the house was full of small hooligans (rummy-dummies, as my father called them) tearing around the house. The family feast table was groaning under a giant turkey and whole ham, baked potatoes, candied yams, corn and the inevitable horrid green Jello salad with olives and grated carrots suspended and quivering. Sheila had brought one of her family Christmas cakes (a mature aged dark cake full of brandied citron and cherries), which my grandmother England liked so much she tried to steal it. My mother figured I was incapable of taking proper care of a wife so she had made a dress for Sheila and a large thick wool quilt for our bed. Tears came to my eyes when my mother gave a large box of my own ragged children's story books to our Wizlet. My parents also gave us a White Mountain ice cream maker, a wooden bucket with a gallon metal can inside, hooked up to a crank for triple-action mixing during freezing of the milk, sugar and crushed vanilla beans. As in the case of this Christmas, one of my family traditions for important feasts was to make home-made ice cream, and Sheila and I used this one for several decades before it rusted out. My brothers were all hunters and we must have seen over a hundred deer along the White River, of which we got several white-tail and mule deer and one antelope. This was all butchered, wrapped and labelled and divided amongst the brothers and hung in the barn rafters in gunny sacks to freeze in the -25°C weather. Sheila had collected the antelope skull, an antique braided leather bridle, several shed buffalo horns and one entire hindquarter of the biggest mule deer for her family gifts and New Year's feast in Loretto, Ontario.

And yes, there was the usual grilling from the local South Dakota ranchers at the local coffee shop, with my father quietly smiling at the edge of the group. Remember now, this was in the midst of the cold war and the Iron Curtain was a scary reality.

'I hear you been over there with the Rooskis?' says Wayne, suspiciously, squinting at me.

'Yup, we were in Leningrad for a week.' says I.

'Why on earth would you want to go over there with those communists anyway?', asks Wayne.

'Well, a US science society gave me travel money to go to a big international scientific meeting to give a presentation on my research in northern Canada. A lot of the country in the north of Canada is very much like that in USSR.' says I.

'I don't know, but if someone gave me travel money I'd go somewhere else. Did you get into any trouble over there?' asks Ken.

'No, no real trouble that I knew about. They had interpreters and guides that kept us in safe places. They didn't want us taking too many pictures of things.'

'What kind of horses and livestock to they have over there?' asks Helge.

'We were mostly in the city, but we did see big work horse-drawn wagons sometimes along the country roads, usually hauling hay. And they had a lot of milk cows. And there were some old tractors and farming equipment too.' says I.

'Did you have to go through the Iron Curtain?' asks Wayne.

'No, we flew from Amsterdam into Leningrad and they had the normal customs inspection at the airport.'

'I'll bet you ate a lot of that beet soup...what do the Russkis call it?' asked Ken.

'That would be borscht, and yes, I ate some of that but I didn't like it much. We ate a lot of boiled meat and potatoes and tea, and they expected us to drink lots of vodka before and after the meals.'

'Did you drink lots of vodka?' asked Ken, worried about my personal habits now that I was so far away from the safety of home and the First Methodist Church.

'No, I don't like the stuff. But I had to drink a little shot glass of it when they made toasts and speeches. Sometimes I sneakily dumped the shot glass in the potted plants, if I could get away with it.' says I, aware that they took a dim view of strong liquor.

'Well, did you like it over there?' asks Helge.

'It was ok. I met some interesting people from all over the world and I made a few Russian friends. The Russian people were very warm, gracious and polite and I understand a little bit of their history better now. But I'd rather live here.' Says I.

'You'd rather live here? Well then, what are you doing up there in Canada with the Eskimos? When are you going to come home?' says Ken.

'Well, I'm a little bit busy having fun exploring northern Canada, where there is even more wild country than around here. It is sort of like being Lewis and Clark, exploring the prairies, except that we get to use airplanes and helicopters.' Says I.

'Why is Canada paying you to fool around up there? It is too cold to grow crops or cattle, isn't it?' says Wayne.

'Well, the Canadian government is worried about the consequences of putting a hot oil pipeline across the frozen ground and rivers up there. They are worried about what happens if the ground melts, the pipeline breaks and oil and mud flows into the rivers and lakes.'

'Well, so what? There is nobody up there.' Says Wayne.

'Oh yes, there are Indian people living there and some of them live on fish, moose, caribou and have fur trap lines. And there are a few white pioneers trying to make a living from the fisheries, growing potatoes, cattle and small town businesses.'

'That must be a hard way to make a living.' says Ken.

There was quiet stirring of the coffee as they mulled this over, and hopefully they decided that I wasn't a Russian spy.

We drove back to Winnipeg in a blinding blizzard but our big van was warm and had good traction on the packed snow road. Two days later we packed up and flew down to spend New Year's Eve with Sheila's parents in a small town north of Toronto. And all of the above paragraphs happened all over again. Sheila had six siblings and the family had lots of friends dropping in for the celebrations and feasts. Sheila's sister Mary was in Ghana, Africa, and she sent her a knitted shawl. Sheila's mother presented us with a heavy cast-iron cornbread pan, as she knew I liked cornbread. Sheila's father had made a small sleigh for the Wizlet, with brass runners and mushroom symbols on the sides. He and Sheila's brother Ian also made a devilish red box with knobs, switches, dials, flashing lights, sirens and whistles. This buzz box lasted for many years. The Wizlet was happy to sleep in an open dresser drawer at night. The New Year's Eve tradition in this family was to assemble all the guns in the house and shoot them off as the New Year rolls in. They had a lot of guns in that house, many of them antique muzzle loaders, flintlocks, revolvers and World War I rifles. The New Year's Day feast was focused on the haunch of South Dakota venison, with lots more of the family Christmas cakes and endless cups of tea. It must have been here that I discovered Rusty Nails (50:50 mixture of whiskey and Drambuie) and I also began writing down the two family genealogy, going back at least eight generations. Our baby son got to visit his great-grandmother Dennett (87 years old) during this time. Sheila stayed with her family while I went to Ottawa for a week-long series of Mackenzie Valley Project meetings and presentations at the University of Ottawa, the National Research Council and various government offices, all bilingual. Flying back west, I met Sheila and the Wizlet at the Toronto airport and we returned to our apartment in Winnipeg, where my brother Gerald was holding the fort. We were a bit puzzled at his sartorial elegance until he told us he had a job selling neckties at a fancy shop downtown. He made quite a sight in a suit and tie in his tiny Triumph TR6 convertible. He made us promise not to tell our family back in South Dakota.

Chapter 14. Life in the Northwest Territories

The next 4 years (1971-74) went by in a blur of family events and science activities and my journals record an almost impossibly busy time. During the winter we lived in Winnipeg and I was consumed by too many trips to Ottawa, Calgary, Denver, Fairbanks, Washington and Vancouver for Task Force planning meetings, meetings with the oil pipeline companies, where I was trying to gain the funding necessary to conquer the Mackenzie Valley Project. I also helped found a Geolimnology course at the University of Manitoba and gave lectures along with Professors Jim Teller and Bill Last.

Sheila decided to give up her Master's Thesis research because our Wizlet started calling Margaret Campbell the baby-sitter 'Mama.' We were able to attend many of the cultural events in town, the Royal Winnipeg Ballet, symphony orchestras, folk and jazz concerts. Sheila was becoming well-known amongst the folk music community in town, with her good Celtic singing voice and guitar. Sheila and I also became experts at making John Bull home-made beer, both the dark and light varieties.

During the short summer open water season we lived in ATCO trailers in Yellowknife, with the rest of my field crew. We have photographs of our young son Ian MacGreggor trying to turn the handle of our ice cream maker under the midnight sun in Yellowknife. Our son also got used to riding in a canoe and he loved to put his hand on the outboard throttle and feel the vibrations.

My Mackenzie Valley Project team was greatly enriched when we hired Clive Nicol (Nic) and Roscoe Dalke, who took charge of most of the problems with our field camp facilities, scientific equipment purchases, staff survival and field training and costs and budgeting. Greg Morden and Sheila Michaelis were running the Yellowknife chemistry lab, Nic, Jim McComiskey, Phil Michaelis and Dennis Bottomley were tough and reliable field technicians, Brian Graham, Paul Campbell, Susan Kasian and Rudy Wagemann kept the Winnipeg labs and data analyses running smoothly. It was all I could do to keep on top of the data reports coming in weekly and trying to make sense out of the big picture. The big picture here was to determine the natural rate of physical and chemical weathering of the river catchment soils and the resulting particle and dissolved elements that were being carried by the rivers to the Beaufort Sea and Arctic Ocean. We thought that road construction and operation of a heated oil pipeline in these permafrost soils would cause greatly increased erosion and would alter the chemistry and food chain of the rivers and wetlands of this region. We were trying to quantify the natural erosion rates (before the pipeline was installed) so that we could state the tonnes km^2 yr^{-1} of dirt, organic matter and dissolved elements that were being carried off the land into rivers and the sea. This required that we have samples of water and suspended matter for the low flow periods in fall and under ice in winter, as well as the short flood season in summer.

Nic had longer experience in the Arctic than I did and he was an expert in survival techniques in cold environments. He trained us in many aspects of safety and survival skills in the front yard of our Yellowknife encampment. We took our 30.06 rifles and over and under .410 shot/.22 guns apart, cleaned them, using graphite lubricant, as cold weather and seawater salts made hydrocarbon lubricants sticky. We practiced some emergency first aid procedures, built some spruce sapling lean-to shelters, quinsies and snow-shelters, checked our double-walled tents and sleeping bags and became familiar with the contents of our emergency food boxes. He even gave us some lessons in defensive fighting tricks, which came in handy for several barroom brawls along the river. Nic was also an expert in several Asian styles of fighting techniques and he worried that he would kill any drunk who attacked him. He had a brass fighting trident that he practiced with late at night in the yard, grunting and thrusting with terrifying swiftness.

On the first day of spring in March 1972 we began our winter survey, again as a partner with Woody and the Water Survey crew in a Twin Otter on skis. We carried a small skidoo and sled, ice augur with three one metre extensions, a heavy needle bar for chipping ice, our scientific and survival gear, compact Black's sleeping bags, a small emergency box of food and first aid supplies, several bottles of overproof Old Sam Navy Rum and a small Herman Nelson propane heater and fan to warm the aircraft engine in the mornings. We sampled all the river stations from Fort Simpson in the south to Inuvik in the northern river delta. Much of the time was spent waiting for decent flying weather in chilly hotel rooms in damp sleeping bags or waiting for aircraft fuel and/or repair parts. I gained the admiration of my colleagues by being able to sleep soundly in the metal/canvas Twin Otter seat, with diesel fumes, noise, cold drafts and rough weather. Upon arrival at a suitable landing site for the Twin Otter on skis we would push the skidoo and sled out the cargo door and jump into the soft snow beside it, sometimes up to our armpits. Where the snow was soft we had to use snowshoes to pack a trail to get the skidoo up on top of the snow and also for the aircraft to take off. Some of the landing sites were wind-packed snow and jagged chunks of ice, making the landing and skidoo travel very rough. After checking the Water Survey station we would try to find a deep channel in the river, clear away the snow cover and augur 1-2 m through the ice to obtain water samples. The weight of the snow and ice above the river water was large and sometimes the water would gush up the hole when the augur broke through, causing us to be wet to the knees for much of the time. The greatest danger was small patches of open water near rapids, where ice would not form despite very cold weather. Some of these open water patches were covered by thick snow cover and we had to learn how to spot the slightly blue-grey colored wet snow that signalled these locations. There were a few hot and saline spring-fed lakes and rivers that remained open all winter. For these sites we used an Alouette II helicopter, which could land near the open water. The Alouette pilots were all Vietnam veterans who enjoyed following moose trails and watching wolf packs approach moose calves. The Nahanni Hot Springs near Gus Krause's cabin would allow comfortable swimming when the air temperature was -30°C. Downstream, the Nahanni River flows over the spectacular Virginia Falls even during winter and the steam rises over the falls. On the Great Bear River we had to scramble over 3-4 m groaning and creaking ice blocks to find a low spot to augur our sampling hole, and even there the ice was over two metres thick.

After a week of this work we arrived at the northern end of our study area, in the Mackenzie Delta and the settlement of Inuvik. This was the biggest town on the river and we enjoyed hot showers, good food and contact with other scientists at the Inuvik Research Centre. They had a small lab where I could use their equipment to measure oxygen, suspended sediment and pH in my preserved water samples. There were wildlife scientists watching the caribou herds migrate from the Yukon coastal plain into the Mackenzie Delta and bird scientists trying to figure out what they ate during the long dark winter. Permafrost soil scientists were looking at curious 'hot spots' of geothermal energy flow that kept some patches of ground warm underneath the snow. The oceanographers were tracking the flow of winter water in the delta channels, out into the Beaufort Sea. At this time the oil industry was building artificial islands offshore from the delta, upon which they hoped to build drilling rigs. One bulldozer operator was eaten by a polar bear that had climbed over the back of the Cat while the operator was eating his lunch. His partners found the Cat still running, with only his mittens left in the cab.

Geomorphologists were trying to map the distribution of pingoes on the floodplain and delta, these being conical elevations in otherwise flat areas, caused by ice formation underground. We also spent some time in 'The Zoo', a wild bar in town, where tied off condoms half full of warm beer were twirled over your head and thrown into the crowd. We would sit and listen to pilots talking about their bad adventures and narrow escapes from disasters. There were far more skidoos and tracked-ski bombardiers parked on the main street than wheeled vehicles. On our return to our headquarters in Yellowknife we got lost in a white-out and flew on compass bearings and dead reckoning for much of the way. This frightened my field staff greatly but I was fast asleep until we landed.

After fixing the frozen water pipes in the YK chemistry lab we processed all the new samples in two days. Nic and our field crew stayed in YK to ready our camp for the summer activities and I flew back to Winnipeg to move out of our apartment one more time and prepare to bring my family back to the NWT for the open water season.

In uncomfortable snowy early May in Winnipeg we packed our Mushroom Van with complete camping gear, supplies for baby Ian, several pet turtles, two chinchillas, a pair of Surinam toads and Max, the Brazilian matamata side-neck turtle. With my brother Gerald helping along the way we drove through Saskatchewan and Alberta (stopping in Edmonton to buy an autoharp for Sheila) and then north on gravel roads to Hay River, NWT, on the south bank of the outlet of Great Slave Lake. The ice road crossing of the lake was closed due to water flowing over the ice, so we left the van in Hay River and flew to Yellowknife, with all the turtles and chinchillas snugly packed in insulated komatiq boxes.

Sheila soon had our trailer house warm and cosy and full of the smell of roast caribou and home-made bread. Max had his own corner wading pool and a good supply of minnows to eat. The Surinam Toad eggs (laid on the jelly-like back of the male) had hatched during our travel and we had several dozen baby toads zipping around in a small aquarium. Gerald had several job offers within a week and helped my crew get the plumbing and electrical systems working for our encampment. Sometimes a cold wind would come off the ice of Great Slave Lake but when the wind was low the spring sun warmed us all. Our little garden was turned over and planted again, on top of the hard permafrost deep soil. With over twenty hours of spring sunlight the radishes grew to the size of softballs and potatoes, zucchini and rhubarb grew very large very fast.

The small lakes around the town of Yellowknife warmed up quickly and we began a regular series of measurements of temperature, salt content, plant and animal surveys and measurements of the common nutrients. A few of the lakes were polluted with town sewage and arsenic from the gold mine tailings ponds. I got into a little trouble with the city fathers after a newspaper reporter interviewed me and I commented on the dangers of arsenic pollution and accumulation of trace metals in fish. However, within a year the polluted lakes had big signs indicating that it was not safe for fishing and boating.

By early May 1972 the Liard and southern Mackenzie Rivers were discharging their load of ice, gouging the river banks, docks and unprotected fishing shacks. Shipments of equipment and supplies were airlifted to us from Winnipeg and new staff were arriving every week. Our grand research plan was coming together now. Before everybody departed for the field camps further north we held a big party on Mother's Day. This meant we cooked several large roasts of moose and Gerald cranked three gallons of home-made over-ripe strawberry ice cream. We had sixty litre drums of home-made beer and lots of greasy frying pan bannock. There were guitars and harmonicas pulled out and we discovered that Nic could sing and tell wild stories of his times in Ethiopia. There was a volleyball tournament and horseshoe throwing contests until 0130 in the morning. We put 4 x 8' sheets of plywood on sawhorses to make tables for about thirty people and feasted under the midnight sun. To add to the festivities there was a military air show happening on the same day and the roar of fighter jets and Hercules overhead added to the mad confusion. A similar spring party tradition was held here for the next few field seasons.

Gian Vascotto and helpers arrived to launch his research camp on the Porcupine River in the Yukon. He used natural landslides as experiments to determine the short-term effect of massive sediment addition to the ecological community of the streams and rivers. When I visited his camp later that year there were abundant blueberries and we ate huge amounts of blueberry pancakes. Gian was careful to explain his research to the small Indian settlement of Old Crow and he hired some of the young people to teach them the navigations skills needed for freight canoes on the river. The Old Crow/Porcupine River Indian people (The Native Brotherhood) were fiercely protective of their land and claimed to be independent from the Canadian government. They understood and approved of Gian's work on natural landslide mud in streams but were puzzled that we wanted to dump oil into their small stream. Gian's work indicated that excess fine sediment addition to clear streams reduced the bottom animal abundances that were important for fish food but these organisms recovered quickly when the landslide sediment was washed away by continued stream flow. A small barrel of crude oil was slowly leaked onto one small stream and Gian's crew monitored the response of the stream bottom animals using drift nets. Other team members measured stream flow rate and took water and sediment samples for chemical measurements of the fractions of the oil in the water. We tried to collect the experimental oil addition below the test area of the stream, using floating oil traps, but logs across the stream worked just as well. It was very hard to burn the collected oil, even with additions of gasoline. The immediate response of the streambed animals was to leave the spill area for several weeks. After a month the stream flow had cleared most of the oil and the fauna returned to near normal, compared to a nearby control stream. There was some anti-government research feeling in the Old Crow community and they blamed us for the late arrival of the salmon that year. One of Gian's helpers was Norma Kritsch, who had worked at ELA during the fateful summer of 1970 when I met Sheila, and Norma married Gian in the fall of '72.

David Rosenberg and helpers arrived to begin his work on benthic biology from a research camp at Fort Simpson, on the southern part of the Mackenzie River. He quickly began experimental work on the effects of sediment additions to naturally clear rivers on the east side of the Mackenzie. They assembled natural stream bed stones in barbeque basket cylinders, staked them to the stream bed and dumped fine mud into the upstream section of the experimental area. They also dunked these BBQ baskets into Norman Wells crude oil, staked them into the stream bed and followed the colonisation of the oil-coated stones, compared to clean baskets of stones upstream. He found that muddy water caused some groups of bottom animals to leave the stream bed and migrate to clear water sections of the stream. Oiled stones were less healthy habitats for most clean water animals but the stones also grew bright green from plant growth after several months and eventually this oil-stimulated plant growth was grazed by stream bed animals and fish.

Norman Snow and his team worked out of the Inuvik Research Lab on the Mackenzie Delta. The delta is large, a confusing tangle of twisting and anastomosing channels, oxbow lakes, perched lakes, pingoes, ice-pushed ridges of dirt and piles of tree trunks from upstream. During the summer floods the brown muddy Mackenzie water pushed the seawater north into the Beaufort Sea, well past Tuktoyaktuk and the old whaling camps along the north coast. Norm's crew tested the effect of oil spills on the delta lakes by carefully adding several barrels of Norman Wells sweet crude oil onto two small lakes that were adjacent to two control lakes of similar area and depth. It was a real mess to do this work, as there had to be two different sets of sampling equipment and clothing, as everything was covered with oil from the test spill site and we had to keep the control lake free from hydrocarbon contamination. I remember particularly the difficulty of rolling my cigarettes with my oily hands and my face covered with DEET insect repellent. I almost got used to the peculiar sour taste of these cigarettes after a week of this work.

Snow found an initial mortality of most surface water organisms from the low molecular weight components of the oil. After several weeks the volatile fraction had evaporated into the atmosphere and a tar-like residue remained on the shore and the shallow lake bottom. After a month, the oiled lakes turned green, as if the oil acted like a carbon-rich fertiliser for the plants in the water. Most of this plant material, and the animals that grazed upon it, were different from the flora and fauna of the unoiled control lakes. Several years later, however, you could still find oily tarry residues beneath the fresh mud along the shoreline of this experimental delta lake. Interestingly, Snow's research group found that the greatest toxicity came from one of the detergent/emulsifiers that was to be used to help clean up real oil spills. One of Norm Snow's technicians was Doreen Middleton, who was one of Sheila's good friends from the summer of 1970 at ELA.

Since we were rapidly becoming Arctic experts in the ecological consequences of oil spills, Norm Snow and I were flown to Resolute on Cornwallis Island in late July '72, where one of the large diesel and aviation fuel tank farms had sprung a leak. We were supposed to tour the site and help write a report on the environmental consequences of the fuel spill. During hours of meetings with the oil companies, Polar Continental Shelf Project staff and the military managers, we were told of a long history (1950s to present) of fuel leaks into Resolute Bay. The airport here was a major staging area for government and military aircraft operations and the oil, avgas and diesel fuel tank farm was supposed to be bermed and impervious to fuel leakage. The recent larger leak had been discovered by University of Toronto and IBP scientist Buster Welch (who later became a good family friend). After being briefed by those responsible for the spill, Buster took us on a walking tour of the coast line, where seals, cod, walrus and whales were taken by the local Inuit. We watched an Inuit lady skinning a seal with her ulu at the shoreline, within ten metres of a pool of diesel fuel on the water. We took several bottom samples, which revealed abundant and diverse marine fauna (starfish, brittle stars, long-necked clams, crustaceans and polychaete worms). Buster told stories about barges full of fuel being crushed by ice, having sunk in this region.

Buster and his wife Cathy were living in Eskimo Village and they fed me some wonderful baked Arctic Char and seal meat. We also walked by the foul-smelling Meretta Lake, where all the sewage from the camp was dumped via pipeline. During our visit there was light fuel, oil mousse and tarry residues on the shore, ice blocks and pools of fuel floating on the open leads of seawater. The Canadian Navy had flown a container full of various sorbents and skimmers to catch the oily fuel on the water. Some of these worked fairly well but they disposed of the collected oil emulsion in the local rubbish dump, which was only 4-6 m above sea level. Despite the foul smell and bad appearance, the local ecological system had apparently tolerated the fuel spills over the past decades. With Buster and Norm Snow's help I wrote a report on my observations for Navy Commander Greenham, before flying back to Yellowknife in bad weather, arriving just as Sheila was taking a batch of cinnamon buns out of the oven.

Phil Michaelis and GJB on Mackenzie River, NWT, 1973

As previously described, my role in this adventure was to lead the project, keep the field stations supplied with money, equipment and technicians, and to keep a steady stream of project reports in front of my bosses in Winnipeg and the Ottawa bureaucrats. Roscoe Dalke and Paul Campbell tried to keep me out of trouble with the budget and manpower regulations. In addition, with the help of Clive Nicol, Jim McComiskey, Dennis Bottomley, Phil Michaelis and Con Haight, we rode Twin Otters, helicopters, freight canoes and the venerable FRB good ship *Stenodus* down and up the Mackenzie at least monthly during the open water season, collecting samples of tributary river water and sediment. We were 'running the ranges', large shore towers in pairs, with diamond shaped black and white markers, which you lined up in order to follow the thalweg of the river channel. The river sandbars would change during big floods, some of the big ship barges would get stuck and they would have to wait for tugs to come along and pull them back into the channels. The big twenty-four foot Hudson Bay freight canoes had transoms reinforced with triple sheets of ¾" marine plywood, so they could support two twenty HP long shaft outboard motors. Phil Michaelis built clever wood and canvas spray covers for the forward part of these canoes so we could make tea and soup while we were underway.

We were trying to measure the rate of chemical and physical weathering of the river watersheds and the rate of delivery of this dissolved and particle load to the Beaufort Sea. We could stand beside these raging rivers and watch the powerful forces of nature carve the land into new shapes and move organic matter and nutrients to the Arctic Ocean. Many of the western Rocky Mountain tributaries to the Mackenzie are large powerful rivers. We painted some 30-50 cm diameter riverbed boulders barn red and lined them up across one smaller river in the low flow fall season. After the next spring flood we couldn't find any of these painted boulders for miles downstream. We also participated in some engineering experiments, where pipeline engineers buried a 1 m diameter steel pipe, heated it with electricity, watched the permafrost melt and the pipe floated away on liquid mud. This was not reassuring at all.

We discovered some strange people along the river, including a potato farmer, somebody trying to raise Hereford beef cattle in the spruce forests and little crude camps of hippies trying to live off the land. My crew became well known at the small settlement hotels on the river and we would often get grudging special treatment for meals and lunches when we arrived late at night, wet and cold. Trying to manage all of this work was an insane job, where equipment broke down in the middle of nowhere, aircraft schedules were compromised by bad weather, bad fuel or mechanical failures. A few of the summer student help were not suited for hard work in the far north and had to be replaced. Although we were struggling to do our jobs, the southern media thought this was good TV material and we had to entertain CBC and BBC film crews several times during this project. It was difficult to keep a straight face when we enacted simple river sampling procedures in front of the cameras, making scientific sounds now and then, under the direction of a film crew with slippery city shoes and inappropriate clothing. They also filmed some of the oil spill experiments done by David Rosenberg and Norman Snow.

Gerald had been helping with all of this and he had worked at Mike Healey's Chitty Lake camp up north of Yellowknife. They were building log cabins and tent camps for their summer experimental fishing research. Gerald helped Bill Woodall cut, peel, notch and pin the logs. Sheila and I flew a Beaver to Chitty Lake for a weekend of grilled fish and rice. My brother Gerald had applied to the Giant Yellowknife Mining Company for a job and they realised he was over-qualified as they hired him. Gerald had a degree in mining engineering from the South Dakota School of Mines so he wrote home to have his textbooks sent to him. He was a good shot with the .22 and he brought us many ptarmigan to cook. Scientific friends from all over began to arrive to see our operation. My colleague David Schindler from ELA came up for a week and we took him in our Chestnut canoe to our Eagle Rock camping site for the weekend. ELA aquatic plant specialist Hedy Kling visited all of our Mackenzie River camp locations, returned to Yellowknife exhausted and slept for two days to recover.

Sheila and I and Wizlet would escape for weekends, canoeing or driving out of town to find some peaceful place to explore. We had a South Dakota tanned buffalo skin for a camping blanket and Trapper Nelson packsacks for our short canoe portages. We could catch fish for supper, made bannock and tea and Sheila usually had a secret stash of home-made shortbreads or fruitcakes for dessert. Little Ian MacGreggor explored the bush confidently and would eat anything he could put in his mouth. Gerald and I, Sheila and the Wizlet went on a moose hunting trip on the Rabbitskin River in the fall and found an old trappers cabin with a wood stove and bunk beds for our comfort. Gerald and I walked upstream to follow moose tracks and droppings for several days and failed to get close enough for a shot. Meanwhile, Ian MacGreggor was eating rose hips and high bush cranberries as fast as he could gather them back at the cabin and his droppings turned pink. Sheila calmly watched a cow and calf moose walk by the cabin, nibbling on the low spruce branches. On one of these trips a year later we drove the Mushroom Van from Yellowknife to Fort Simpson, crossing the Liard River on the ferry. We were greatly surprised to find that the ferry pilot was my brother Gerald. He invited us to the wheel house where he had a bottle of beer in his hand, a rusty 30.30 rifle in the corner and a whole bear liver in the fridge, which was his food for the next few days.

SMB washing nappies at trapper's cabin, Rabbitskin River, NWT, 1972

On our canoe return to cross the river to Fort Simpson I was nabbed by the RCMP, who needed someone to help with an urgent case. (It doesn't take long to become an 'old hand' in the NWT!). A trapper had reported that a hippie family had camped near Gus Krause's cabin up in the Nahanni Mountains all winter and now there was a crazy man shooting a rifle at anything that moved. I guided them to Gus's cabin and we flew around trying to find their camp site. We finally found the fellow, gaunt and yellow-skinned, unconscious, surrounded by piles of rotting fish and empty corned beef tins, his rifle empty beside him. We gave him some water and he said that his wife had died giving birth and the wolves had taken his baby son. The RCMP collected the remaining scattered bones of his family in prospector's canvas sacks and we hauled the whole sorry lot back to Fort Simpson.

In other years I hunted with Con Haight, a big brash good-hearted veteran of living in the north, and captain of the FRB vessel *Stenodus*. One of his hunting camp methods was to find a large pile of ice- or flood-pushed trees and brush near a river and set the whole pile alight. This would generate vast amounts of heat while we put up our tents and in an hour we could use the red-hot coals at the edge of the fire to make bannock and cook some fresh grouse breasts and moose ribs. The fire would still be burning in the morning for our breakfast cooking pleasure. David Schindler, Phil Michalis, Con and I also hunted from Hudson Bay twenty-four foot freight canoes in the swampy outlet region of Great Slave Lake, where we found an old trapper's cabin. We took an enormous number of ducks and geese and two moose on that trip, but most of the ducks were muddy flavoured. We also set gill nets for whitefish, which were very good eating. We built a spit out of scrap metal so we could cook fresh moose ribs over the fire and we had moose brains mixed in with our scrambled eggs for breakfast. David's specialty was moose steak stew and I was famous for my skillet cornbread. Phil made some very good baking powder biscuits and Con made bannock (full of currants and chopped dried apricots) for us to chew on during the canoe hunting. We went a bit berserk trying out moose calls, using moose skeleton scapulae to knock on tree branches, a large coffee can with a wetted string pulled through it and even blowing into disassembled shotgun barrels to make mournful honking trumpet sounds. Some of these ideas came from Albert Bonnetrouge's Metis family, living in a cabin just down the river from us.

Con, Phil Michaelis (The Hammer) and I went with some Dogrib Indian hunters to find the caribou herds north of Yellowknife one other January. We used a Bombardier tracked vehicle with a big trailer to carry supplies and fuel for a week, whereas the Indian hunters used snowshoes and dog sleds. After several days of sitting on high hills with field glasses we finally spotted the herds moving north. It was -25 to -50°F and it was necessary to keep the guns outside so they wouldn't collect moisture and freeze up. We got a lot of caribou and most of them were government airplane shipped back to the Dogrib settlement freezers for Detah community food. The Indian people were very respectful toward the caribou before butchering them; they gave them a drink of mouth-melted snow and spoke to them with reverence. We ate a lot of caribou liver and tongue in camp, dunking our hot bannock in the meat juices. The Dogrib people like the caribou head, which had some choice bits of tender meat. I took two for our own use, carefully boning and labelling the meat, and froze it in komatiq boxes.

I took the boxes of frozen meat to the airport on the morning of our departure to return to Winnipeg but the airplane was delayed. These boxes sat in the holding area of the over-heated airport for six hours and by the time they were to be loaded onto the plane there was blood dripping from the boxes. Never mind, they froze again in the cargo hold of the plane travelling south. Airport security in these days was lax in the north but several times I was slowed down when the Edmonton or Winnipeg airport inspection revealed several clips of 30.06 shells and a skinning knife in my big government briefcase.

We usually drove our Mushroom Van back to Winnipeg as the ice was forming on the Mackenzie River in late October/early November. We often stopped at Ellis Hamilton's farm home in northern Saskatchewan to join in the extended family moose hunt. Ellis had kindly saved cleaned skulls of mink, otter, beaver, lynx and weasels for Sheila's collection. In Winnipeg we stayed with friends until we could find a new apartment and get our household goods out of storage. We were in culture shock for a few weeks, seeing the glaring colour green, colourful flower beds, people dressed in suits and ties and sidewalks that were not covered in ice. It is a good thing we were young and flexible. We also moved out of our temporary offices at the University of Manitoba and moved into our own new five story Freshwater Institute. Brian Graham, Paul Campbell and I now had offices and two labs on the fifth floor. Sadly, our first Director Wally Johnson, a big supporter of our Mackenzie Valley work, left the institute in December 1972 and our institute began a slow spiral into a management-led research institution.

My journals record too many trips to big cities in Canada and USA to present slide shows of our work on the Mackenzie Valley rivers. One of the more amusing trips was funded by the US Army Corps of Engineers, the Cold Regions Research and Engineering Laboratories in New Hampshire. They had been involved in Alaskan oil pipeline studies while we were working in the Yukon and NWT. The meeting was in Santa Barbara, California, in a hotel by a Pacific Ocean beach. What a place to discuss Arctic environmental research! Upon our arrival in the Santa Barbara airport a group of orange-robed Hare Krishnas approached our group, proceeded to put necklaces of flowers around our necks and gave us copies of the *Bhagavad Gita*. Throughout our three day sessions we wore this necklace of fading flowers, earnestly presented our findings and made predictions about the effects of hot oil pipelines in permafrost terrain and river crossings. Just before we were to depart to return to our respective home institutions we gathered up some strings of kelp, jellyfish tentacles and oily strands of fragments of fish nets and wove them into necklaces. Upon arrival at the airport there were the Hare Krishnas, bowing politely and chanting, as we put these stinking seaweed necklaces around their necks. They didn't pause a bit in their chanting as we walked away to the restrooms to wash our hands.

Over the next few years our team wrote many reports for the Mackenzie Valley Pipeline Project, Fisheries Research Board technical reports, science journal papers and book chapters about our work. Not all of this was peaceful collaboration. There were strong disagreements amongst the project scientists on the emphasis, interpretations, uncertainty, extrapolations and long-term reliability of our predictions of the effects of a heated oil pipeline. The administrative leadership of our home institution was changing and some of our research support began to dwindle. There were criticisms of our rough and ready field experiments and some in-house institutional envy of the large amounts of money and man-power we had used in this vast geographic research area.

The oil spill experiments done by Rosenberg, Snow and Vascotto's teams showed that most ecological damage was done by the short-lived volatile low molecular weight fraction of the crude oil, most of which evaporated to the atmosphere within several months. Surface water skimming organisms (both aquatic and terrestrial) were immediately and deleteriously affected by the oil film but most plants and animals in the water column and bottom sediments of the Mackenzie Delta lakes were little affected. Months to years after the experimental oil spills, residual tarry oil remained in the near shore sediments but the lakes and streambed stones responded to enrichment with the oily organic matter by growing different species of green plants. Some of the toxic fraction of the heavy tarry oil residues (polycyclic aromatic hydrocarbons, PAHs) were found in biological tissues and this probably stimulated several physiological mechanisms to excrete or detoxify these substances. Stream bed biota left the immediate site of experimental oil spills but this biological evacuation was short lived and the site was recolonised rapidly (within years) after cessation of the oil input. We didn't do any winter under ice experiments and this might have led to more severe impacts, as the volatile oil fraction would be contained in the water longer by the surface ice cover.

Vascotto and Rosenberg found that experimental additions of fine mud to clear streams caused the streambed animals to leave the site temporarily but these sites recovered quickly after the next year's flood cleared the streambed stones of fine mud. In general, clear water streams and rivers had higher abundance and diversity of plants, bottom invertebrates and fishes, compared to the bigger muddy rivers coming off the northern Rocky Mountains. Snow's work in the Mackenzie Delta lakes also showed that higher elevation clear water lakes were more productive and biodiverse than the muddy lakes near the main channels of the delta. In general, we felt that construction activities and a hot oil pipeline would cause local severe erosion and release of dissolved and particulate material from the permafrost soils over a very small fraction of the big river catchment areas. For the big tributary rivers draining the western cordillera of the Mackenzie River Valley, this would likely have small and very local impacts on river biota. I got in trouble by exaggerating this point at an Ottawa workshop (attended by the press), saying that ten bulldozers pushing frozen mud into the Liard River would have little ecological effect because many hundreds of tonnes of natural erosion sediment was flowing past every day. However, for the clear water tributary rivers of the eastern lowland sub-Arctic forest, these kinds of human disturbances to the ecosystem would have longer lasting effects further downstream. Most of this biological research is summarised by David Rosenberg in his chapter of the book in the Further Reading section.

My own studies revealed that the annual physical and chemical weathering of catchment soils and rocks of the Mackenzie River tributaries was well within the variation observed in much better known Temperate Zone rivers. Since these sub-Arctic and Arctic rivers were largely frozen for six to eight months of the year (and underlain by permafrost), this means that summer monthly water runoff and erosion rates were very rapid. Indeed, some of the western mountainous tributaries in summer flood were rivers of mud, trees and even boulders bouncing along the river bed. In contrast, the gentle muskeg terrain and hard-rock substrate of the eastern tributary rivers resulted in lower runoff of clear or tea-coloured water and low concentrations of dissolved elements. We were directly measuring how many tonnes of calcium, magnesium, potassium, sodium, sulphate, chloride, bicarbonate, copper, lead, zinc, organic matter and other trace elements were being removed from the land and carried to the Beaufort Sea and Arctic Ocean. Exotic exceptions to these generalities were the small saline hot springs that occurred throughout the region which steamed into the winter air, supporting unusual flora and fauna year-round. Our measurements indicated that the Mackenzie River delivered about 333 km^3 of water and 118 million tonnes of suspended sediment to the Delta and Beaufort Sea in an average year.

To gain perspective on these huge numbers I had to review the data for most of the large rivers in the world. In doing this I became addicted to seeking an understanding of the grand cycle of the tectonic uplift of the mountains, the relentless continental erosion by wind and rivers, sediment and salt delivery to the ocean and the growth and decay of organic matter by plants and animals. I learned that just measurements of chemical concentrations and abundances of plants and animals was not as useful as the annual rate of movement of the chemical elements from the land to the sea, or the annual rate of growth and decay of the plants and animals. I could simplify this grand cycle to a simple inorganic chemistry laboratory acid and base titration, something we had done in high school. I had expected that these sub-Arctic and Arctic ecosystems would be fragile and slow growing, but I found that these regions had robust and productive ecosystems that did not appear to be limited by nutrient supply rate or severe climate. This speculation has been confirmed by subsequent research (see the paper by Squires et al., 2009).

There was a glowing internal pleasure in our brains, seeing that we could gain some understanding of how this wild, large, sometimes violent and sometimes deep frozen part of nature could operate so regularly and smoothly.

One side project was the local Yellowknife lakes, which had a curious combination of municipal sewage and gold mine tailings inputs. The local 'greenies' in town got upset about the 'poisoned fish' in Kam Lake and the possibility of poison arsenical gas coming from the tailings ponds and nearby lakes. This was all contrary to the observations from our colleagues Mike Healey and Bill Woodall that showed that the fish of these polluted lakes were fat and happy. Our chemical studies showed that algal biomass was high and arsenic was all in the inorganic arsenate form. Radon techniques were used to sample deepwater gases, which were air shipped to a specialist lab in Toronto for methylarsenic gas measurements (none were detected). I designed an experiment to test the hypothesis that high loadings of nutrients, organic matter and arsenic would cause anaerobic deep water and conditions for the formation of highly toxic methyl-arsenic gases.

Winter sampling, Mackenzie River

We did these experiments at ELA during the winter/summer of '72-'73, with the help of John Rudd, Bob Flett, Phil and Sheila Michaelis and ELA staff. We used one metre diameter steel culverts for tubes in a shallow lake. Rudd and I determined that microbial oxidation of organic matter was inhibited by high concentrations of As, compared to the control in winter, but there was no effect at all during summer. This applied to algal biomass, algal species composition and several other colligative parameters. Doloros Planas from Spain helped do some of the algal work. Doloros had arrived from Spain unannounced with two children and little money, so she lived with Sheila and me for several months. Rudd and I eventually published this arsenic work, indicating that it was a grossly overrated poison, at least with reference to bacteria, algae and fish.

While all this intense field activity was happening the Canadian government appointed British Columbia Justice Thomas R. Berger to review the environmental, social and economic aspects of a Prudhoe Bay heated oil pipeline (and the associated roads, airports, settlements) through Canadian land and water. He was a very good man who created a series of inquiries, hearings and workshops along the river and in southern Canada, to hear the people speak. We scientists were frequently commanded to appear, often on short notice, for hearings in Arctic Red River, Norman Wells, Great Bear Lake Settlement, Fort Simpson, Edmonton and Ottawa. We were not invited to speak about our work but rather to listen to the discussion and when called upon, to answer questions from the public. Our reports were made available to the northern settlement people at these hearings and small truckloads of these reports appeared for distribution.

This consultation went on for five years and the debate between the energy/oil industry consultants, fishing and trapping associations, government scientists, environmentalist NGOs, economists and social workers was in high gear. After all this consultation and deliberation, Berger's final report recommended that the Canadian government delay or reject the oil/gas pipeline development in the Yukon and Mackenzie River valley, much to the surprise of the pro-business bureaucrats in Ottawa and the oil industry. He found that this development would not greatly benefit the native people of the north and in any case, the Canadian government had not settled the land rights issue with the Indian (Dene) and Inuit. He called for a Canadian Native Charter of Rights and the entire Department of Indian and Northern Affairs went berserk for several years.

The oil industry calmly responded by proposing the Trans Alaska Pipeline, which was then built over the next decade, carrying hot oil from Prudhoe Bay across the mountains and tundra muskegs to Port Valdez on the Gulf of Alaska. We were treated to many photographs of Alaskan caribou crossing underneath elevated sections of the pipeline over the next years. Sadly, the liberal government of Pierre Elliot Trudeau abandoned the cause of aboriginal rights in the Mackenzie Valley in 1977. Justice Thomas Berger resigned from the bench in protest of this lack of action and disregard for aboriginal rights.

A 2002 update on the environmental and social situation for the Beaufort Sea region is given in the Ayles and Snow paper, which indicates that the oil and gas industry is still trying to promote oil and gas pipelines in this region.

All this life in the NWT bubble happened while the rest of the world staggered on in search of civilization.

Wally Broecker and colleagues began a series of papers in 1975 in the science journals, worrying about the effect of greenhouse gas warming of the planet on ocean circulation. We were living in a world of cheap fuel, abundant cheap food, exponential growth of technology and apparently endless resources to support this good life. The 'limits to growth' and the 'population bomb' didn't seem to apply to us, up north in the vast tundra and taiga of the Western NWT.

Ah yes, but there was plenty of drama in the NWT as well. A Canadian bush pilot, Martin Hartwell, who worked out of Yellowknife, tried to fly a pregnant Inuk woman, a fourteen year old boy (David Pisurayak Kootook, suffering from appendicitis), and a government nurse (Judy Hill) from Cambridge Bay to the hospital in Yellowknife in November 1972. In bad weather, his aircraft hit a hill near Hottah Lake and the nurse and the Inuk lady died. The pilot had broken both of his ankles so the boy kept the fire going at – 37°C and made a tent and a stone windbreak. The boy died on the 20th day after the crash and the pilot survived by eating the leg of the nurse. He was rescued on the 31st day after the crash. Balladeer Stompin' Tom Connors wrote a ballad about this, tactfully not mentioning Judy Hill's leg. Twenty-six years later, the Canadian Northern Transportation Company named one of their ships after David Pisurayak Kootook.

Chapter 15. A Place of Our Own

'...the oldest task in human history: to live on a piece of land without spoiling it.'

Aldo Leopold.

Over these years of the Mackenzie Pipeline Project, Sheila and I wanted to buy a house and live permanently in Yellowknife. We found several nice houses perched on the shores of Yellowknife Bay but we could not get the Canadian government to commit to a long-term field station in the north. We were getting tired of moving from our winter apartment in Winnipeg to our summer trailer house in Yellowknife. Both Sheila and I had come from snug warm families in a secure home in the country. If we were going to raise our own family, it wouldn't be good for our children to be changing home base every six months. After several years of campaigning for a long-term institutional field station in Yellowknife, it became frustratingly obvious that there was no support for this in Ottawa, at least partly because of the high cost of doing business in the north. We continued our Mackenzie Valley research up to 1975, with smaller and smaller funding every year.

In cold windy March 1973 we bought a small farm house and yard in the country south of Winnipeg. The larger surrounding farm fields were purchased by a local Mennonite farmer who didn't want the old buildings, so we got the house, barn, and four sheds on thirteen acres for $14,000, which was about equal to my annual salary at that time.

It was sad to see old Pete and Agnes Friesen leave their old family home and prepare to live in a big city apartment. Agnes said to Sheila upon leaving the house '...fill it with the happy laughter of kids.' They had a yard sale in May, where Pete sold his combine, tractors and implements to his neighbours for very low prices. Sheila bought some canning jars, a few chairs and an old butter mould. Nobody wanted to buy Pete's little 1952 8N Ford tractor because it wouldn't start, so I got it for $50. Pete and Agnes had their first holiday in Hawaii and Agnes later said that Pete didn't enjoy it much, as he kept worrying about the crops in southern Manitoba.

This small hobby farm was near a one building store/gas station/post office called Tourond, which revealed the early history of French settlers in the area. But most of the neighbours were Mennonites and the nearby towns to the north and east were strong Mennonite communities that originated from a dozen Ukrainian immigrants to this region in the 1850s. To the south of us was the Roman Catholic French-speaking village of St. Pierre, a much more sinful and happy bunch that tolerated dancing, a liquor store and a movie theatre.

Our scary leap into home and land ownership had an electrifying effect on our respective families. Sheila's older brother Ian (an executive in a computer software firm) flew out from Ottawa to help us move from our Winnipeg apartment to our empty farm house. The farm house looked so big when it was empty but it filled up quickly when we unloaded our books, kitchen table and whorehouse pump organ.

Within a month my mother and father drove an old Chevy pickup into our yard, loaded with a disassembled Number 9 John Deere sickle mower adapted to fit a Ford tractor and some old farm tools that we needed. My father built a porch and steps for the house, fixed an abandoned antique lawn mower and mowed the grass, trimmed the trees and helped me fix some shingles on the roof. We got the Ford tractor tuned up, rebuilt an old horse wagon and used an old horse cultivator to turn over a large garden plot, where we planted sweet corn, strawberries, raspberries, currants, asparagus, parsnips, peppers, peas and beans, onions and leeks, pumpkin and squash, potatoes and melons. We made nest boxes for one room in the barn and got some laying hens. We had new bathroom facilities installed upstairs. My mother helped clean the upstairs rooms, which had not been used for at least one generation of Friesens, and made bedrooms for our family, including her own home-made heavy wool-bat quilts for the cold winter.

We moved the old outhouse to the edge of the garden and my mother painted it barn red, to serve as the 'garden office'. An old bathtub was sunk in the outhouse hole, to be used for a summer home for Sheila's collection of turtles. Max the matamata took up residence in the downstairs bathtub for the next eighteen years. Sheila started baking bread in our new kitchen and my mother made a bunch of rhubarb pies. There was nothing like the smell of baking bread to make a house feel like a home! It was rapidly becoming our home.

Sheila's younger brother Alistair arrived with a friend in a 1971 Porsche sports car. He had an alto saxophone with him and not much else for luggage. He said he was escaping a clutching girl friend and he had to go north where she wouldn't find him. During feasts around our kitchen table, one of the South Dakota pearls of wisdom was: I mentioned that during castration and branding of calves, the testicles are cooked on the branding iron fire and are considered a delicacy by the working cowboys. These fatty tender morsels are called 'Prairie Oysters'. A few years later Alistair used this name for his rhythm and blues band in Toronto.

I gave him Con Haight's phone number in Yellowknife and he hitchhiked there in about a week. We next heard from him as he was painting the FRB ship *Stenodus*, working for Con and getting the ship ready for Mackenzie River sampling trips with my crew. He was also playing drums with Inuvialuit musician Louie Goose's band at the Yellowknife Inn. After one trip down the Mackenzie River on the *Stenodus* with us he took a job in Inuvik on the *Pilot II*. On the *Stenodus* Alistair did much of the cooking and he was very good with roast chickens and mince meat pie (in which Con would put half a cup of Old Sam overproof rum).

Sheila's father arrived next, having been at some business meeting in Winnipeg in a full suit and tie. Without changing clothes, he clambered up a rickety ladder in his slippery city shoes to fix the TV antenna on our new house roof. He greatly admired our 'summer kitchen' shed, which contained a rusty old wood stove and a giant iron cauldron that must have been used to render lard in the old days. He also took our earthen-floored basement oil furnace apart, which hadn't been serviced for a long time. I couldn't understand how he knew how to do all this, and he explained that he had grown up in Winnipegosis to the northwest of us, as a Manitoba farm boy. Sheila's father was called 'Hada' by the family and he helped Sheila buy a St. Bernard puppy, which she named Sophie. My scientific and moose hunting colleague David Schindler apparently had too many dogs, because he gave us a female Gordon Setter. Old Jenny gave us several litters of purebred puppies, two of which lived with us for over a decade, Nancy Whiskey and Banjo.

Most of my Mackenzie River science crew would show up at our door when they passed through Winnipeg and Sheila would provide feasts for all. Scientists and technicians from Yellowknife, Fort Simpson, Inuvik and Old Crow became regular guests, and our kitchen table usually had all the extensions fitted for ten to fifteen people. Our garden was amazingly productive of fresh vegetables and strawberries and our home-made ice cream bucket was kept wet with salty water. Gian Vascotto and Norma Kritsch got married and now had a little girl, which they had to show us proudly. My brother Gerald reappeared, after having worked in Florida on a roofing crew for several months. He proceeded to build us a large bookcase and stereo cabinet, full wall-to-ceiling, in our cluttered study room.

In the fall I dug up five large gunny sacks of potatoes and slung them into the basement for winter storage. We had purchased two used large chest freezers to keep our summer garden vegetables, venison, moose, chickens and ducks. Sheila blanched wheelbarrow loads of sweet corn, peas, beans and we had large quantities of strawberry, raspberry and currant jams and jellies. Squash, carrots, onions and pumpkin were in our basement for winter storage. Our little copse of trees revealed several dense patches of chokecherries and I enjoyed both the picking and eating of this essential prairie jelly. We were warned by neighbours to put flax bales over our septic tank, to keep it warm over the winter. Storm windows were installed and our fuel oil tank was filled. As winter approached I was surprised how rapidly Sheila and I adapted to the tasks of country living, with skills that we didn't know we had, probably absorbed unconsciously from our parents and grandparents.

Sheila was now hugely pregnant and she was taking seriously supplemental food to support the new baby. Sheila's older sister Mary (having returned from Ghana) came to help with the birthing and little Jesse George was born on 27 November 1973. Well, he wasn't very little, he was a large chubby baby, so the extra milk must have done its job. We had purchased an old fashioned rocking chair so that Sheila could breastfeed Jesse and rock him to sleep with a full belly. Our little house on the Manitoba prairie was feeling more and more like a home.

We soon found out that our farm house let the cold wind come inside during the worst blizzards. It was probably built in the early 1900s, with an uninsulated double wall construction. The house water came from a well in the barn. The fuel oil furnace required electricity to operate. Our first few winters in this house taught us the value of warm clothing and thick wool blankets. Snow would accumulate on the inside of the window sills and some of the chokecherry jelly froze in the kitchen cupboards. One Christmas Eve the upstairs bathroom water pipes froze and burst and the water dripped through the floor onto our Christmas tree. It was hard to get a plumber to fix the problem during the holidays.

During the winter of January 1975 the power went off during a big blizzard, so we bundled up little Ian and Jesse (and Max the matamata) and pushed/pulled them on a Hada sleigh (covered with a South Dakota buffalo hide) over to our neighbours, who had a wood and oil-burning stove. During each blizzard our yard buildings attracted the blowing snow and we had one to three metre deep snowdrifts barring the driveway. During the next few years we bought a good wood stove for the living room and began replacing the old loose single pane windows with triple-pane windows. We insulated the house with foam urethane, based upon government advice and rebates. This also meant that we spent most fall weekends cutting and hauling standing dead timber to the yard and stacking four to six cords of firewood in the barn. We installed electrical outlets in a shed so we could power our block, oil pan and battery heater in the Mushroom Van in shelter from the wind. But the strongest blizzards would fill the shed with hard packed snow. Both Sheila and I bought cross-country skis, which were handy after the big blizzards because we could ski to the little Tourond Store for supplies.

Tourond Blizzard

During these years Sheila and I became serious followers of the 'back to the earth' philosophy of self-sustainability and we had copies of the *Whole Earth Catalogue*, the *Mother Earth* Magazines and lots of books on how to survive in the country without going to town very often. Sheila recovered many of her mother's recipes for old-fashioned slow food, using our fresh frozen garden produce and home-grown chickens, venison and moose meat in the freezer. I bought a chimney sweep tool to keep the wood and coal soot level down on our old brick chimney. While the snowdrifts were still in the yard, we ordered garden seeds and by late April or May we had young tomato and melon plants in a cold frame attached to a shed. As soon as the spring warmth returned I would fork the frozen chicken and duck manure out of their shed and spread it on the garden. We used the old Ford tractor to turn over the garden soil and a borrowed roto-tiller to pulverise the big clods. When it was ready for planting the soil was so soft you would sink up to your ankles just walking over it. This glacial Lake Agassiz prairie soil is very rich and deep. Sheila and her sister Mary would also plant rows of flowers amongst the vegetables. In addition to the normal vegetables and fruits we also grew comfrey, Swiss chard, oyster plants, gourds, chilli peppers, green and yellow peppers, salsify and manglewurzel. We had a scientific visitor from Switzerland, who had never heard of Swiss chard. The manglewurzel was for the pigs but they didn't like it at all, which was odd as it was a giant sweet sugary white beet/turnip like root.

We rapidly expanded our meat production and at full swing we had about fifty Cornish Giant meat chickens, a dozen laying hens, a dozen Emden geese, twenty Pekin ducks, a dozen turkeys and four weaner pigs by early June. Neighbouring farmers usually cleaned out their wheat, oats, barley and corn bins in the spring and we could have the sweepings free if we hauled it ourselves. Excess garden produce went to the poultry and pigs and it was quite a sight to watch four pigs devour wheelbarrow loads of fall frost- touched tomatoes and melons. Sheila would always name the pigs and sometimes they were named Porkofiev, Hamadeus, Ribsky Korsakov and Fred. Our sons grew up helping with the chores, watering and feeding the livestock, gathering eggs, picking strawberries and raspberries, digging potatoes, shucking sweet corn and the inevitable fall butchering activities.

By the mid-1970s Sheila's sister Mary, sister Heather and family and brother Alistair would participate in our chilly weekend chicken/duck/goose plucking parties, after which hot buttered rum would be served with fresh baked bannock. Later in colder weather we butchered our pigs, letting them hang in the barn for a few days and boning the half-frozen carcasses on the house kitchen table. The meaty bones were stashed in a barn fridge, for treats for the dogs in midwinter. We borrowed a large meat grinder to make our own sausage from the tougher cuts of deer and moose, combined with fatty pork trimmings and garden spices.

This home-grown domestic bliss surprised me greatly, as I thought I was going to be a bachelor and I hadn't imagined being a husband, father and weekend farmer. More than once I was moved to tears when reading Laura Ingalls Wilder books to my sons on our sofa with a South Dakota buffalo robe thrown over it, before bedtime, much the same as I grew up. Some of these *Little House on the Prairie* books were from my mother, dated when I was five years old. And here was my own copy of Will James' *Lone Cowboy*, so I could teach my sons what happens to cattlemen on the prairies. Often times Jesse's big Newfoundland dog Tifty was asleep at our feet, near the wood stove. More than once during this lifestyle transition I was emotionally overwhelmed by all this home place creation. For over a decade I had been a bachelor, living in apartments in cities, travelling the world, spending more time in the lab, offices and airports than at home. I never expected to marry and have children during this time and it was not part of my life plan. But here, suddenly, I had a family, a home and a patch of land that needed care and attention. This was like coming full circle somehow and I felt a special kinship with my father and mother, who were trying to remind me that I should know how to do all this farm and family work.

I was away on science business too much and Sheila often had to fight the blizzards to get our sons to the school bus and keep them amused during the long hot summer days. She also took over the bank account, paid all the bills, got the groceries, took the boys to their dental and medical appointments, music lessons, sports events, made all the school lunches and cooked for major feasts and music parties. Sheila was a magnet for wild animals and over these years we had a pet fox in the house, two pet horned owls, a Canadian goose, cockatiels, a few snakes and many turtles. The fox would sleep with our sons, underneath the covers. In May '78 Sheila rescued two baby Great Horned Owls that had fallen out of their nest in our little woods, into a cold puddle of melt water. They grew up in our living room, usually perched on the back of our sofa. They ate lots of chopped kidney and liver and grew attached to Sheila, following her around the yard, waddling like penguins. One of them would sit on my shoulder during supper. They often fell asleep on the stairs. When they got big they lived in a granary shed and ate frozen chicks (chicksicles). When little they were called 'The Hoolets' but later they were called 'Piffers'.

Jesse and Ian on stairs with owls, Tourond

Sheila bred our purebred Gordon setter, got a set of puppies and entertained the idea of establishing a kennel for show and hunting dogs. After a few contacts with the snobbish show dog people she gave up this idea. Sheila and her sister Mary planted many rows of gladiolus flowers in our garden and little Jesse became the Zinnia farmer. Gladys the Canada goose would follow Sheila down the garden row and pull up whatever she planted, so she had to be confined during garden planting time.

Sheila was a good singer of Celtic folk ballads and we attended many gatherings of musicians in Winnipeg. These music parties were patterned after the Gaelic ceilidh, where the musicians sit in a rough circle in a room and everybody takes their turn singing a ballad, playing a fiddle tune or telling a story. There were penny whistles, harmonicas, banjo, fiddle, guitars and mandolins, Appalachian and hammered dulcimers, both Scots highland and Irish uilleann pipers.

I had been trained in classical music by my mother and grandmother and I didn't understand how all these people could play instruments and sing complex harmonies without any music on paper. On a trip to Steinbach to buy water dispensers for the laying hens, Sheila pushed me into a music shop and encouraged me to buy an El Degas banjo and a Pete Seeger instruction book on how to play it. She also gave me a 1973 Christmas present of banjo lessons from a fellow named Mitch Podolak in Winnipeg. Mitch's idea of giving lessons was to play a tune on the banjo and then hand it to me to see if I could copy his style. This was all done in the ramshackle rear of a crowded little music store, where Mitch was also planning a folk festival, eating his lunch and talking on the phone. Mitch was a wild-eyed Trotsky-Lenin disciple and a very good organiser of folk music events. We attend his first festival in Winnipeg in the early summer of 1974 and we were hooked on this international 'people's music' event.

Over the next fifteen years Sheila and I were on the festival staff as stage crew and the Board of Directors of what became the Winnipeg Folk Festival. This festival grew almost too fast, with later events attracting over 30,000 happy folkies to hear performers from Canada, USA, UK, Australia, the Balkans and France. Here were Celtic fiddlers, Zydeco bands from Louisiana, North Carolina clog dancers, Scots highland pipers, blues and protest singers, Maritime Canadian balladeers, Balkan and Moroccan ensembles, Tex-Mex bands, Welsh and French harpers, Australian story-tellers and some local folks too. Amongst the wonderful pleasure of hearing old and new music at the festival, I also learned a lot about finance, troubles with security and crowd health and the enormous power of a crowd fuelled with musical ecstasy. Careful scheduling of the musical events, as done by Mitch, was partly designed to defuse high emotions as well as invigorate the audience. As the event got bigger and bigger, the finances and business plan got more complex, and I opted out in 1979.

During one of our visits to our South Dakota family in February 1974 my grandmother Gertrude England gave Ian MacGreggor a toy airplane and gave me her family fiddle. This fiddle was probably made by her great-grandfather Nathaniel Butterfield, who was a Vermont cabinet maker. He may have made the fiddle in the early 1800s and it had a repair mark inside it, dated 1874 in Minneapolis. Good musical friend and instrument maker John Weier repaired this old fiddle, to find it had a warm sweet tone. Within a few years I was able to accompany Sheila on some of her ballads on the fiddle and/or banjo and I learned how to play the penny whistle and harmonica. John Weier eventually built a viola for me, which had a dark, rich, deep tone.

With our Danish philosopher friend Hans Hansen we formed the folk band 'Meadowlark'. As time went by, this band sometimes included Sheila's sister Mary (Celtic harp, autoharp) and brother Alistair (bodhran and jaw harp, and his wife Susan on fiddle), as well as good friends David Ediger and Walt Lysack. We played at local festivals, fairs, weddings and concerts in southern Manitoba and Ontario for over a decade and had a lot of fun. We were also called the 'Soft Touch Band', as we often didn't charge for our services if it was a worthy cause, and we depended upon having a lot of good food before any music began.

I also got involved in the Canadian Traditional Music Society, which tried to sponsor the performance and study of folk music from all of the cultural groups. I was on the Board of Directors, briefly President, and wrote several articles for their magazine. One of these articles was about Papaluk, an Inuit lady that I met in our research camp north of Chesterfield Inlet, on the NW coast of Hudson Bay. She was a large jolly lady who played old-time dance tunes on the button accordion, with a unique rhythm and style. This folk music society had its annual meetings in cities from the west to east coast and I was fortunate to hear old-time French/Celtic fiddle tunes for dances in Quebec City, Ukrainian mandolin groups in Winnipeg, Russian (Doukobor) folk songs in Saskatchewan and cowboy songs in Alberta.

Papaluk playing her button accordion at Saq, with Buster and Cathy Welch

Our sons caught some of this musical fire, mostly by cultural diffusion. We would turn them loose on the festival grounds with 'This is a Brunskill Boy: If Found, Please Return to Stage 3' written on the back of their T-shirts. They became groupies of their favourite performers and were often found backstage eating with them. Years later they would often correct the words to ballads that we had forgotten. We had an old upright piano and the whorehouse pump organ in our house and Ian quickly learned keyboard skills, even though he really liked Iron Maiden and Metallica much more than Bach and Beethoven. He was lucky to have a very good band director in high school, Fraser Linklater, and he enjoyed playing percussion. He went to the Universities of Calgary and Toronto to study percussion. Jesse played my old Raphael Mendez trumpet in band, took violin lessons for a few years and learned to play 'Home on the

Range' on the harmonica.

Our boys were probably odd fellows in a largely conservative Mennonite school. They came home once with a survey questionnaire which, amongst other things, asked for our religion. I told them that I was a druid, their mother was a wildlife biologist, but I didn't know what religion they might have. I suspect this Laodicean attitude did not go down well with the school administration. Jesse became famous for being able to put a protractor and hand calculator in his mouth. He and his buddies once taped the science room door shut so the teacher could not come in the room. Jesse was mainly interested in mechanical repairs to his motorbike and skidoos. Ian was in a rock and roll band and they all wore black leather jackets winter and summer. They rehearsed in one of our sheds at our home and the neighbours could always tell when Sheila and I were not home, because the PA volume went to the max and neighbours two sections west could make out the words to the songs.

Sheila was a fierce leader in the local parent-teacher organisation, fought for better playground equipment and supported 4-H and the band boosters group in the local school. She and I became Boy Scout leaders and we led many canoe trips into the local rivers and lakes and taught winter survival skills to these farm boys. We had traded two dressed pigs for a home-made cedar strip/fibreglass canoe, made by our friend Buster Welch. We were proud to watch these sullen, cool rock and roll kids get interested in canoe races, portage races, fire-making, handling sharp knives and axes, camp cooking of fresh-caught fish and mastering the bowline and clove hitch. We helped our sons in science fair projects, including demonstrations of rat nutrition, effects of acid rain on fish, anti-freeze in wasp galls and the oscilloscope harmonics of brass and violin instruments. Both boys were delighted when they got a Commodore 64 computer for Christmas and the inevitable computer games.

During a 1975 Thanksgiving visit to South Dakota my parents gave us the old round kitchen table from the White River ranch house where I grew up. This heavy wooden pedestal table had been hauled by rail from my grandmother Edna Brunskill's home in Haywarden, Iowa in the early 1900s, to the rail head in Chamberlain. She drove a truck with the household possessions from Chamberlain to the site of their ranch, while my grandfather George Brunskill rode the rails with his horses and cattle, farming machinery and tools. My father, aunt and uncle did their school homework on that table and many haying crews ate massive meals around this table. I can remember my grandmother rolling out cinnamon buns with a kerosene lantern in the middle of this table. During several White River floods water came into the house and deposited silt in the empty compartment of the pedestal. When we got this table back to our Manitoba prairie home the river silt was still there. We gathered material for a time capsule to put inside the pedestal as we were assembling the table in our own kitchen.

This table had a powerful effect on me, as it implied some responsibility for me to carry on the family tradition.

We had many house parties where the South Dakota family kitchen table had all its extension leaves in and fifteen to twenty people sat around the table eating home-grown smoked hams and pork chops, honey-garlic moose ribs, haunches of bear, venison, antelope, roast ducks, geese, chickens and giant turkeys. Sheila, her sisters and friends would have 'dessert wars' and the parade of gut-busting cheesecakes, rhubarb-cream and blueberry pies, Sachertortes, giant butter tarts, garden strawberries/raspberries on home-made ice cream, yellow and red watermelon and muskmelon halves filled with whipped cream laced with Drambuie. It makes me hungry just to write this.

Later in the evening these guests would pull out their guitars, banjo, mandolin and fiddles and tunes and ballads would float through the air into the wee hours of the morning.

We entertained forty visiting scientists from an international (SIL) conference in Winnipeg in August 1974, coming to our home for roasted caribou haunch and ribs, home-brewed beer in 120 litre barrels, Sheila's famous bannock and almost endless supplies of fresh-picked sweet corn and melons. Mike Healey and I took two dozen of these mainly European conference scientists on a tour of the Mackenzie River valley and they got to see Mackenzie Delta pingoes, bear and moose, Great Slave Lake fishing boats and many dirty hotel rooms above noisy bars. Years later, scientists from Europe and Asia would be telling stories about our farmhouse party and Arctic field trip, the taste of Old Sam Overproof Rum, the atrocious weather, most of them roughly true.

Throughout all of this happy rural family farm life I was also travelling between Winnipeg and Yellowknife regularly, to maintain our Mackenzie River geochemical sampling program and to assist the scientific staff at our research camps in Fort Simpson, Inuvik and Caribou Bar Creek in the Yukon. This meant many nights in noisy hotel rooms and tents in damp sleeping bags and long days in Twin Otters on floats, flying into all the Water Survey discharge stations, as previously described. Sheila and the boys stayed home, so my field journals mention many meals at the Yellowknife Gold Range Café, where the cook Hung Dung made very large full meal egg rolls (similar to the Italian panzerotto). But kind Sheila did supply me with two weeks supply of oatmeal cookies to have with tea from my Stanley thermos bottle in the cold aircraft.

I was also involved in many NWT Water Board environmental and health meetings in Yellowknife, all about the mine tailing arsenic and sewage pollution of the local lakes. The City of Yellowknife wanted to use Kam Lake as a sewage lagoon, whereas we were trying to explain that the lake wasn't big enough to metabolise and dilute the sewage nutrients and mine waste arsenic. With the help of Brian Graham, John Rudd and Doloros Planas (a Spanish post-doctoral student), we had performed an experiment at ELA to simulate the addition of sewage and arsenic to a lake and showed that arsenic slowed the microbial degradation of organic matter in winter. At an April 1975 science meeting in Toronto, Wally Broecker gave a lecture on the threat of increasing greenhouses gases in the atmosphere and the observed warming of the world oceans. As usual, he was way ahead of most of us on this problem.

Airport Security staff in Edmonton and Winnipeg began to recognise me and routinely asked me to take my hunting knives, rifle shells, steel cables and harmonicas out of my briefcase and put them in my duffel bag. We were also hunting every fall, sometimes north of Yellowknife, sometimes in northern Saskatchewan, and sometimes to the north of Lake Winnipeg. We met some Bloodvein Indian hunters in the Narrows of Lake Winnipeg who showed us how to make tin can moose callers. This was just a big tin can, open at one end, and a bootlace was threaded through a small hole in the other end. When the bootlace was wetted, gripped hard by the fingers and pulled rapidly, it emits a dreadful croak-squawk that is apparently attractive to moose during the rut. Often times in the winter after two weeks in NWT, I would fly into Winnipeg, drive out to our long farm driveway and find my cross-country skis and poles stuck in the snowdrift-blocked road. This meant Sheila had skied to the end of the driveway and left my skis for me to get home.

Both Sheila and I were getting an education in how to handle large groups of people, expeditionary operations and to coordinate social forces to obtain our objectives. We had planned and executed Boy Scout Camporees, helped with a large folk festival, organised family and scientific expeditions to remote locations, research ship cruises, workshops and helped with school music, recreation and educational activities. Sheila wrote a handy manual on this, with an exceptional list of alternate options for when Plan A went wrong or the weather was bad. A simple list of our goals or purposes, the gear required, logistics, food and supplies, cost estimates and contingency plans became supper time conversation. As the years went by I began to appreciate these early lessons, as we had to plan and execute many expeditions that required the use of Plan B, C and D.

Chapter 16. Winding Down and Starting Up New

By 1975 my Mackenzie River research group was disintegrating from lack of support from the head office and Ottawa. We hadn't really finished our oil spill and sediment addition experiments, which would have required another few years of monitoring, but we were running out of money. During the peak of Mackenzie Valley research we had thirty staff working in Winnipeg, Yellowknife, Fort Simpson, Inuvik and Caribou Bar Creek, so it was an expensive operation. We were pushed into planning an Eastern Arctic Pipeline Study, but this didn't work out as the cost was also enormous. Norm Snow and Gian Vascotto drifted away to other jobs and our technical staff dismantled the labs and accommodations in Yellowknife, Fort Simpson, Inuvik and Caribou Bar Creek. Morale amongst institute scientists had declined after our semi-independent Crown Corporation Fisheries Research Board status was changed to full bore civil service government employees and the departure of our best scientific leaders Jack Vallentyne and Wally Johnson. Thus began a dreadful trend of increasing management and dwindling interest in the advancement of science.

I had hauled a pickup load of good prairie soil out to ELA so David Schindler could grow tomatoes on a small patch of the poor gravel soil of ELA for his family. David and I got pissed off at institute administration for some reason so we hauled a pickup load of ELA beach sand into Winnipeg, carried it into the front lobby of our new office building, poured it into the fancy display map case and put sandbox toys with the Director and Chief Financial Officer's name on the bucket and shovel. During this time the administrative wing (mahogany row) of the institute had some modern paintings hanging on the hallways. Andy Hamilton and I got Willy Burton to weld scrap metal into the shape of a fish skeleton in the workshop and stick it on a slab of plywood painted black. This was hung in place of one of the paintings and we hid the original one in Andy Hamilton's Section Leader office. No one noticed for several months and then we all got a stern letter demanding the return of the rented art piece. Willy Burton of the metalworking shop was also making wonderful replica pistols during this time, Colt Peacemakers, cowboy pistols, and he would strut into our offices to show them off, causing some concern amongst the secretaries and senior administrators.

After the Mackenzie Valley program collapsed in 1976 I returned to ELA again, grateful for a stable, focused and well-oiled research program led by David Schindler. By this time ELA was thickly populated with Wally Broecker's Columbia University graduate students. We had two computer supported gamma spectrometers counting whole lake spike radionuclides at camp and the level of science was many orders of magnitude higher than during my early years in '68-'70. I helped Paul Quay and Wally Broecker design 'Ragapos', a system for distributing radium and tritium in the thermocline of Lake 227. It was truly exciting to see the transformation of ELA from a difficult bush camp struggling to do elementary limnology to an advanced, high-tech superscience facility. It was possible to imagine an experiment over beer after supper, construct the necessary equipment, perform the experiment the next day, monitoring the sample results over a week and graph the data for discussion, all at camp. Computer power and gamma spectrometry allowed much of this to happen so fast. It was extremely important to have Broecker graduate students at ELA, as they brought new methods and ideas that never would have occurred to limnologists. They were trained in mass balance logic, physics and chemistry, and used box models to obtain first approximation explanations for processes that were obscure. It was also ironic that the thirty plus year outcome of some of the experiments indicated the most sensitive changes resulting from experimental manipulation (nutrients, metals, acid, organic contaminants) were plant and animal species changes and food chain linkage gaps. Many of our high powered physical and chemical measurements (easier to measure, easier to show changes) could not detect conclusive biological effects. And a bear tore off the camp kitchen door to get into the stove grease trap and ate a lot of bread.

The ELA program was a roaring success and we managed to convince our bosses that we should have a similar research facility in the Arctic. Many northern mining, oil, road and pipeline developments were demanding attention; the conflict between Inuit habits of fish and wildlife harvesting and the rules set by southern white bureaucrats was festering away. In the summer of 1976 Buster Welch, David Schindler, Bob Newbury and I began looking for a northern ELA site. We were looking for a site that had many small lakes (suitable for whole lake experimental tests) and an estuary that was suitable for study with small boats. We toured the west coast of Hudson Bay and northward to Cornwallis Island and Char Lake region. We listened to the village elders at Eskimo Point as they described their kin family territory for hunting and fishing over many past generations. We used a helicopter to explore from Rankin Inlet northward and northward of Chesterfield Inlet, where we camped in the midst of old stone rings near Saqvaqjuac, which was eventually selected as our field camp HQ. We were windstorm bound there for a few days so we shot a caribou, caught grayling and lake trout, ate a lot of porridge, macaroni and tea. Our temporary camp site was an old Inuit fishing camp, where snow knife handles, stone tools, fish spears, fish weir traps, copper wire, komatiq bone runners and Inukshuks were abundant on most islands. We built a stone wall for wind protection, piled large boulders on top of the tent pegs and doubled the tie-down ropes on the helicopter.

Buster Welch, GJB and David Schindler at the outlet of Saqvaqjuac River

We finally made it back to the settlement of Chesterfield Inlet where we were given accommodation in the old Catholic school and dormitory, which had been empty for several years but still kept clean and the smell of wax and cleaning solvents was strong. Under the questioning eyes of the French Catholic administrators we unrolled our damp sleeping bags on wrestling mats in the gym. This had been an elementary school for Inuit children from all over the Keewatin for the past thirty plus years and most of the English-speaking settlement Inuit graduated from this institution, the smart ones shipped to Churchill for high school education. This institution (and the settlement of Chesterfield Inlet) was a post-World War II product of the Church and Canadian government efforts to bring the Inuit off their nomadic ways on the land and sea ice, into organised coastal settlements, where health and education services could be provided. To the older generation of Inuit this was a mixed blessing and many parents did not like the strict orderly nuns and the loss of traditional legends and hunter knowledge. Young Inuit kids grew up using telephones, watching TV, eating Hudson Bay store food and wearing Chinese manufactured jackets and shoes. But on the other hand, all of the Keewatin leaders (lawyers, business men and women) who established Nunavut and ran their own government were educated at Chesterfield Inlet and Churchill.

Over the next decade my colleague Buster Welch (and his family) and technical staff built laboratory and staff accommodations on the protected internal estuary at Saqvaqjuac and began experiments on nutrient additions to several small lakes. Buster hired some of the Chesterfield Inlet Inuit as research assistants and they did very well. Since Buster and Cathy Welch had considerable knowledge of Inuit culture and language, they got along well with the local people who came to Saq for hunting and fishing.

I helped Paul Campbell get a M.Sc. degree from his work on iron cycles in a meromictic (permanently stratified) lake at ELA (whimsically entitled 'The Ferrous Cycle'), and Bill Buhay did a good Ph.D. thesis on carbon and oxygen isotopic signals in tree ring cellulose at ELA. Unfortunately he cut down a favourite meditation pine tree, beloved by the ELA cook, and this caused a large uproar at camp. I was briefly banned from ELA for being involved in this deforestation.

In August 1976 I was asked to interview for a leadership job at a much bigger St Lawrence Great Lakes institute in Burlington, Ontario. I had visited this institute several times and I knew that working there would require living in a big city, but the work would have been exciting on the big lakes. I decided against this upward career move while I was peacefully picking watermelons in our Tourond hobby farm garden.

I was a speaker and session convenor at a big science meeting in Amsterdam in September of that year. All my good-crazy Dutch friends made sure I visited the best Indonesian and Turkish restaurants and made me evaluate their Dutch beers against Canadian beer. We went to several folk music bars, where I was surprised to hear songs written by Canadian Ian Tyson being sung and Michael Smith sang a very sad song about an old retired Dutchman sitting beside the canal with his wife. There were Turkish belly dancers with music in 7/8 time signature at the traditional Dutch beer gardens. A few of us took the bus to Utrecht to visit other friends and the countryside was full of Holstein milk cows, vegetable gardens, church spires, fruit tree orchards, fields of tulips, greenhouses, old stone barns and modern concrete apartment buildings, medieval towers and geese and ducks on the canals. There were bermed polder fields and houses below the water level of the nearby canals carrying big freight ships and barges. Our hosts took us out to the delta works, where new protective dykes were being built five metres above sea level.

Scientists from Plon gave a paper about the difficult problem of freshwater supply for Germany and Holland and the looming problem of industrial pollution and acid rain. Most of the Dutch and German scientists looked to the USA and Canada as being very fortunate to have lots of room and clean freshwater. At the airport, upon departure, there were Dutch ladies in headscarves selling tulip bulbs but my Dutch friends told me that the bulbs really came from Israel. Despite the good science conference and meeting good European friends again I was happy to arrive at the Winnipeg airport, where Sheila and our sons were there to greet me. I had purchased some wooden Dutch shoes for each of them but they were not very comfortable. Also some Delft egg cups, which we used for many years.

I was hired to be a reviewer for the environmental research programs at the Oak Ridge National Laboratories near Knoxville, Tennessee in September. I was on a panel with Cliff Mortimer, Phil Abelson and three oceanographers from Woods Hole. This was a very large federal lab with many layers of military-style security and they had to open my banjo case at the airport to inspect it for dangerous weapons. I was cautiously interviewed by the local security staff, as my passport indicated I had been behind the Iron Curtain several times and I was not a resident of the USA. No cameras or radios were allowed on the ORNL grounds and there were armed guards at the entrance to most of the laboratory buildings that I visited. Some of these old labs had been involved in the plutonium production program for the Manhattan Project during World War II. This large complex of industrial, military and scientific laboratories was in an evolving struggle to adapt to peaceful uses of nuclear technology. Ecological research was a small branch of the ORNL complex but it had a powerful effect on terrestrial and freshwater ecology, due to their mind-boggling analytical tools and lots of money. But there was always tension and conflict between the military-industrial groups and the ecologists and environmental chemists. Several lakes in the area were loaded with twenty-five years of lab activity waste radioactive elements, allowing some spectacular research opportunities in exotic radionuclide pathways through the natural food chains.

The food was great and we ate enormous amounts of catfish, fatty pork and baked beans and cornbread. I tried to order a local beer with my meal and the waitress said 'No suh, honey, this here is moonshine country and we don't mess around with beer.' She brought me a Heineken with a look of distain. They brought a complimentary bottle of Jack Daniels sour mash bourbon whiskey to my hotel room every night. This strangely contrasted with the Bible Belt conservative attitude in most of rural Tennessee.

Several of the ORNL ecological scientists were also bluegrass musicians and we had some good sessions on the back balcony of one of the visiting scientist accommodation quarters. This was at least partly fuelled by several bottles of the complementary whiskey from my room, as I couldn't drink much of their sour mash bourbon. These working scientists had little respect for the military security on the grounds, and they actually stole a security pickup truck to haul me out to one of their experimental field stations at Walker Branch Watershed. They had built an eighty foot high tower above the sweet gums, oaks and hickory tree canopy, to catch rainfall as it moved downward through the forest. The stream draining this little forested watershed was highly gauged so that they could measure the chemistry of surface runoff, particles and groundwater, as well as the annual cycles of plants and animals in the soil and stream bed. With the enormous analytical capability of ORNL, they could measure nearly all of the elements in the periodic chart. One of their projects was about the natural and pollution load of atmospheric mercury from nearby coal-fired electricity generation plants. Most of the active research scientists did not know any of the higher managers at ORNL, who were remote and powerful people. This divide between the bosses and science workers was my main criticism of the place, and it was an omen of the future for many scientific laboratories around the world.

On the way home from this ORNL review work I visited Roger Green in New York and we rescued my dear wife Sheila at La Guardia airport. She brought her Martin guitar, which had been damaged, and we drove to the Martin Guitar Company factory at Nazareth, Pennsylvania, to put a new neck on her guitar. We visited the Lamont-Doherty Geological Observatory (part of Columbia University) on the west bank of the Hudson River, where we made arrangements for my year long sabbatical there, so I could learn some radiochemistry.

As usual, the outside world was more and more confusing and exciting.

Chapter 17. Radioactive Adventures in New York

Radioactivity makes people nervous. The general public will happily inhale automobile exhaust, drink alcohol and smoke cigarettes, eat meat and vegetables loaded with artificial hormones and pesticides, tolerate coal dust and oil spills, have X-rays in their hospitals, consume dangerous drugs, but they don't like the idea of even tiny amounts of silent radioactive stuff in their environment. This attitude probably comes from the fear of the nuclear bombs that ended World War II and the nuclear power station accidents at Chernobyl and Three Mile Island. The global load of radioactive elements from these power reactor sources was trivial, compared to the megatonne military nuclear bomb tests that were detonated between 1950 and the mid-1970s. The USA blew up many of the Marshall Islands in the north Pacific; the UK chose to blow up their dirty nuclear bombs in Maralinga and the Monte Bello Islands of Australia. The French used their colonial islands in the South Pacific for their bomb tests; China was testing nuclear bombs at Lop Nur from 1964 to 1996. Most of these locations of bomb tests are still too hot for human habitation. All of this was a terrible waste of national resources and scientific expertise during the cold war, being acts of political power seeking and world military dominance. The immediate land and sea locations of these bomb tests were made dangerous for human occupation and local plants and animals were subject to increased mutation rates. Sheep grazing near bomb test sites in Nevada and Australia were found to have very high levels of ^{90}Sr in their bones and milk and radioiodine in their thyroid glands.

People are often surprised to learn that all of the natural uranium series elements can be found in the air, water, soil, food, and the tissues of plants and animals, before nuclear bomb testing, uranium mining and nuclear power stations. Some of most serious radioactive exposure threats come from natural radon gas in domestic home basements and groundwater. There have been natural fission reactors millions of years ago, when sufficient concentrations of uranium were brought together by natural processes. These natural underground reactors produced all of the isotopes that the bomb tests produced, but they were apparently retained close to the underground site of the heat/energy production. Nearly all of the energy on the surface of the earth comes from the radiation and heat of nuclear fission power of the sun.

From the technology of military nuclear bomb building came the controlled and safe domestic production of radionuclides for use in biological research and medical procedures and the construction of nuclear power plants that provided relatively CO_2 free electrical energy. Many other less controlled nuclear explosion uses were considered. Project Plowshare began in the USA in 1957, was funded to 1977, and proposed to use peaceful nuclear explosions to create harbours in northern Alaska, to excavate rail and roadways across the Bristol Mountains in the Mojave Desert, to excavate a new canal from the Gulf of Mexico to the Pacific Ocean across Nicaragua, to excavate and heat buried oil shales and natural gas deposits in Colorado and to liberate water and steam from underground water reservoirs in New Mexico and Arizona. Just to prove that they could move a lot of dirt by these methods, they detonated a 104 kiloton bomb on the Nevada Yucca Flats in July 1962. This explosion displaced twelve million tons of dirt 3.7 km high into the atmosphere and this radioactive cloud of dirt was detected east of the Mississippi River within a week. Over 1960-73 there were 839 underground nuclear test explosions (mostly in Nevada, New Mexico and Colorado) and the radionuclides resulting from these bomb tests remain buried at these test sites. None of these large-scale dirt moving projects resulted in useful commercial products. Similar research on peaceful uses of nuclear explosions was done in the USSR and China over these same years.

These nuclear bomb tests sprinkled low-level radioactive fallout of fission nuclides all over the world, from polar to tropical zones of the earth. While this was an act of global vandalism, it also provided time clock marker labels on all of the ecosystems of the earth. Biologists, chemists and physical scientists have since used these bomb tracer nuclides to measure the speed of atmospheric and oceanic circulation, plant growth and many other ecological and geochemical parameters of earth science. These short and long-lived isotopes could be used as ecological tracers, time clocks of geochemical and biological processes. Particle reactive nuclides attach themselves to organic matter and clay and could be used to date the accumulating thickness of lake and ocean mud. Soluble isotopes could be used to track water circulation and the speed of uptake by organisms. By the early 1970s instruments for the measurements of low-level radioactive nuclides (gamma ray and alpha particle detectors) were commercially available and married to computer monitoring systems, making it easier for ecologists and chemists to use radiochemical methods in natural science studies.

The above essay on the nuclear age might seem out of place in a book about the fun and adventure of science. But many times I have had to explain all of this to non-scientific friends and research colleagues who are terrified of radioactive isotopes of any kind, and they assume that radioactivity is evil and deadly.

No one at our institute had much knowledge of radiochemical methods, other than low dose radiocarbon, radiophosphorus and radioiron as measured by liquid scintillation. We had observed the efficient and rapid gain of data and knowledge from Wally Broecker and students at ELA, who used radiochemical tracers in whole lake experiments to determine the exchange rate of CO_2 between the atmosphere and lake water and the accumulation rate of mud and organic matter at the bottom of the lake. I applied for a sabbatical year of leave in the late winter of 1976, to learn some radiochemical methods in Wally Broecker's Geochemistry Laboratory, Lamont-Doherty Earth Observatory, part of Columbia University in New York City, on the west bank of the Hudson River. I was successful in this application and we planned to depart for the east coast of the USA in the fall.

This meant that our usual home farm garden and meat production system had to be shut down for the summer. In the late fall of 1976 we began packing our household possessions that we didn't need in New York and carted it to a storage facility in Winnipeg. Both Sheila and I were saddened by the disaggregation of our familiar household mathoms, books, record collection, kitchen and musical instruments. In a few short busy years we had assembled the clutter of family treasures in our little prairie home, and when the house was stripped of these items it was no longer home. I took Sheila and our two sons to the airport in early December so they could fly to Toronto, to be with Sheila's family for a month, before we went to New York. I stayed behind at home to finish three science journal papers and cleaned up the house. Our two Gordon Setter dogs and several house cats were upset about all this, as their loving family had vanished. Con (the *Stenodus* ship captain) and Jane Haight and their sons Cliff and Gordon arrived in mid-December to live in our house for the winter and John and Pat Flannagan would run our home and garden in the next summer.

I slept in my office for the next week, furiously writing up past research while Max the matamata side-necked turtle lived in my large lab sink, where he had more live fish than he could eat. After finishing my writing work just before Christmas, I packed Max in a Styrofoam box, took a taxi through a blizzard to the airport and convinced the airline hostess that I was a Toronto Zoo employee transporting a rare South American turtle and that Max should be allowed an empty seat. I did have some excess baggage, as I had my banjo, fiddle, a duffel bag and a big komatiq box full of Christmas presents. Sheila's father rescued me in the Toronto airport and whisked me up north to their rural home. There my sons squealed at my arrival and

Sheila anxiously took Max away to coo at him and gently put him into a big warm aquarium. And then the normal madhouse family Christmas and New Year's feasts proceeded, giant roasts of beef, large white and black fruitcakes, and the firing of antique rifles at the beginning of the year 1977.

A friend drove our Mushroom van from Winnipeg to Toronto, loaded with our family gear for our stay in New York. After tearful goodbyes from Sheila's family we drove through the guts of Toronto to the Fort Erie Customs House, where Sheila was declared 'a visitor for purposes of pleasure'. Our little sons were declared triplicate citizens of Canada, the USA and the UK, Sheila having been born in the UK. We got lost in Buffalo in a heavy snow storm but managed to reach the Palisades and the Lamont-Doherty Geological Observatory (LDGO) campus at dark the next day, early January 1977.

This wonderful small wooded campus of Columbia University was once the 1920s weekend residence of a wealthy financier/banker Thomas W. Lamont, and his widow donated the land and buildings (Torrey Cliffs) to Columbia University. Our house was the stable master's residence. The first thing our two sons demanded from our packed household goods was their grandmother Brunskill's homemade quilts for their beds. I built a kennel for our Gordon Setter dogs, Jenny and Banjo, which arrived at J. F. Kennedy Airport within a week. They ran away during our first outing to gather firewood and it took three days to find them, running along the Hudson River below the Palisades cliffs. Like me, they were not accustomed to confinement. I rapidly became addicted to that unique New York City pastrami, pita breads stuffed with crab meat and the spicy apple strudels made from local apples. Our minder and fixer in this transition was Ellen Cox, who was the Geochemistry Secretary and also a good friend.

We found a school for first grader Ian MacGreggor at the Tappan Zee Elementary, and his teacher (Mrs. Izzo) turned out to have a husband who played the hammered dulcimer. This ancient instrument intrigued Sheila, so Mr. Izzo found an old antique dulcimer in NYC for her and I spent a weekend learning how to put new piano strings on it. Several of the Lamont graduate students were also musicians, so we were happy to have some music parties. Old friends from my graduate school days kept appearing for feasts, including Roger Green and family and Stuart Ludlam and family. Many of Wally Broecker's graduate students had worked at ELA, so they knew us as being hippie flower children of the Winnipeg Folk Festival. Wally would visit us often, wanting to hear stories about the Mackenzie River work.

Stuart Ludlam was now an associate Professor at the University of Massachusetts in Amherst. He had worked with me on the annual layers of lake sediment when I was in graduate school at Cornell. We considered the option of trying to determine the bomb fallout radionuclides in these annual layers of mud, to further prove their annual nature. Taro Takahashi was a carbon dioxide geochemist at LDGO and he had modelled the carbon chemistry of Fayetteville Green Lake (where my Cornell Ph.D. thesis work was done). He claimed that the layers in the lake mud could not be annual, so we had a conflict which could be settled by experiments and measurements of time markers in these layers. Stuart agreed to carefully scrape the millimetre thin black and white layers of mud cleanly into vials, so I could try to measure bomb fallout ^{137}Cs, some of the plutonium isotopes, and the natural uranium series radioisotope ^{210}Pb in each mud layer. If these man-made radionuclides in the layered mud matched the known fallout history from bomb tests, then we could claim that we had won the argument. Easier said than done. It took me eight months to determine these nuclides in the tiny samples and my data did agree with the known bomb fallout history (none in the mud layers before 1950, highest in the mud deposited during 1963-64 and declining rapidly by the 1970s). During this time, I was able to take my family back to Cornell University and Fayetteville Green Lake and we took some large water samples to determine these radionuclides in the surface and deep lake water. Tsung-Hung Peng helped me make a mathematical model of the annual cycle of fallout deposition and incorporation of these nuclides into the annual lake sediment layers and the model worked well qualitatively, but not quantitatively.

GJB, Fayetteville Green Lake sampling, 1977

Lamont lab technicians, mainly Bob Trier, taught me how to do the chemical extraction procedures to isolate plutonium and evaporate the small bubble of solvent on a small steel disk for measurements by alpha spectroscopy. This procedure involved a lot of nasty strong hot acids and some foul organic solvents and I was guilty of setting a few fires in the fume hoods. Several of the Lamont graduate students (mainly Curtis Olsen) taught me how to run the computers that operated the gamma spectrometers, so I could measure radioactive caesium in my mud and water sample extracts. Guy Matthieu taught me more details of radon and helium isotope sampling and measurements. It was wonderful and I felt like I was an undergraduate student again. I took Columbia University courses in radiochemistry, physical chemistry, chemical oceanography and geochronology and listened to most of the Lamont seminars by visiting scientists, staff and students. These seminars were not for the faint of heart, as they typically erupted into heated arguments and direct challenges about methods and alternate interpretations, sometimes for several hours after the lecture was over. But then the combatants retired to the LDGO director's house for beer and nibbles, and everybody was friends again. I was pushed to my limit of knowledge by Wally Broecker's request that I serve on several of his Ph.D. thesis committees, and I became familiar with the work of Tom Torgersen, Paul Quay, George Kipphut and Curtis Olsen. These young men went on to become good scientists.

Many of the staff, technicians and students were just returning from oceanographic cruises on most of the oceans of the world, so I got to hear the stories of seagoing boredom and bad weather, gear break-downs in the Indian Ocean, exotic diseases, ship repairs in Chile or some small port in Africa and exciting new discoveries of how the ocean worked. Mathoms collected by the oceanographers migrated between labs and some were displayed in a glass cabinet at the entrance to the Geochemistry Lab. One such curious artefact was a mummified forearm in an evacuated glass cylinder, labelled as being from a South American Saint Matilda. Years later an honourable oceanographer told me he had collected it from a fresh bulldozer road cut on the outskirts of Lima, Peru. I have concealed his name, as he is still afraid of Peruvian justice. He figured it was from a woman, as the hand held two balls of spun wool, so he named her Matilda. Many of the outbuildings around the Geochemistry Lab contained leftover materials from past research projects. I blundered into one shed that contained thousands of carefully labelled whole and ground bone samples, which I later was told were human bones. These bones came from Columbia Medical School cadavers and they were used to estimate the burden of ^{90}Sr and ^{239}Pu in humans during the worst of the bomb fallout in the late 1950s and 1960s. The sediment core lab was an enormous warehouse full of steel pipes full of sediment from all around the world ocean. These sediment cores contained millions of years of history of earth processes, as measured by down-core changes in minerals, organic matter, microfossils and stable and radioactive isotopes. I got to know Gordon Jacoby and Ed Cook in the tree ring lab (dendrochronology), where they were trying to decipher many centuries of climate change information, droughts and floods, from tree ring widths and the isotopes contained in the cellulose.

Sheila and I took kids and our mushroom van into New York City for concerts, plays, museums and art gallery trips. To a New Yorker it would have been amusing to see us in white-knuckle driving terror through these busy skyscraper canyon streets. We were fortunate to hear the music of the New Lost City Ramblers, Doc and Merle Watson, Tom Paxton, Pete Seeger, John Prine, Steve Goodman, Gordon Bok (from Camden, Maine) and Jean Redpath (from Scotland), which we already knew about, and many other new musical groups.

At a Tommy Makem concert at Hunter College, Makem announced that the first row of seats was of course reserved for the clergy, and that the IRA fundraising table at the entrance was not approved by him. During one long, complex, sung and spoken ballad ('Peter Kagan and the Wind', written by Gordon Bok), performed by Tommy Makem and Liam Clancy, our six year old son Ian MacGreggor was writhing in his chair as if he was in pain. We took him into the lobby to ask what the matter was and he explained that the song was too sad. We hadn't realised that he actually understood the words to the song. This was one of many wake-up calls in the parenthood learning curve for us. We had lots of music parties at our house and in the hot muggy summer Sheila would make sixty litre deldrums of ginger beer, which was very popular with the LDGO graduate students. Our sons would often quiz us, after these parties, for the meaning of some of the words in bad blues songs or traditional British ballads.

Wally wanted me to visit the University of Rhode Island for a conference and to inspect the new MERL experimental tanks, where we met the Swiss biogeochemist Peter Santschi and Taiwanese geochemist Yuan-Hui Li. We ate fresh scallops and quahog chowder with biscuits and gravy and gawked at the big sailing ships in the harbour. Sheila and the boys came along on this trip and we also visited with friends at Cape Cod, Woods Hole (John Rudd, Holger Jannasch, John Hobbie), the Universities of Massachusetts and Connecticut. Our young sons got to see the ocean for the first time, gleefully shouting at the crashing waves and picking sea shells from the beach sand. Sheila found a forty-five inch Clement Quebec-built jack loom that she had read about, so we bought it at a shop in Rhode Island. Within a month she was making attractive wall hangings, harmonica sashes, warm winter shawls and heavy wool blankets for our boys.

We were in Montreal in early July for the wedding of Sheila's older brother. We left our two sons with their Auntie Mary in Ottawa and proceeded west to Winnipeg to attend to institute business as well as the Winnipeg Folk Festival. By now we were senior stage managers at this festival, and many of the performers recognised us. Sheila performed some of her Celtic ballads on the traditional song stage. I especially enjoyed the old time banjo workshops and the endless fiddle tune medleys at the hotel parties after the evening concert. We didn't get much sleep at all and returned to Lamont with groggy headaches.

Later that year, in October we travelled to Ontario for another of Sheila's sister's wedding. I spent much of this time practicing the fiddle and cutting firewood for the fireplace. On the way back to LDGO we passed a garage sale in New Jersey, where I found a fiddle made in 1842 in a home-made wooden case, which I bought for $20. It had to be fixed by luthier John Weier, but it had a loud raucous tone.

I was still doggedly trying to apply my newly learned radiochemical methods to the dozens of large water sample extracts and hundreds of tiny sediment samples from the annual layers of mud. I was running out of time, as we were scheduled to leave LDGO in December. By November I had enough data to give joint seminars at LDGO and Woods Hole with Stuart Ludlam on this subject. During this time we also let our sons shout at the Atlantic Ocean again and had good feasts with friends at Woods Hole (Tom and Jennifer Torgersen, Roger Green and family, John Rudd, Michael Bacon). By the time there was ice on the dog water bucket we could buy good hard cider at the roadside stands. After several farewell feasts and music parties with all of our new friends, we packed up our household goods (including taking the big loom apart) in mid December and shipped them back to Winnipeg. We shipped our sedated Gordon Setter dogs in big cages to Pierre, South Dakota. All musical instruments and precious possessions were carefully packed in komatiq boxes in our van, which was driven to South Dakota by a hire driver service. Sheila and I and our two small sons boarded a jet at La Guardia in the city rain and diesel smog. Our sons took full advantage of the stewardesses, who responded with toys and presents and a tour of the cockpit. For some reason our sons developed an interest in the tiny toilets on the airplane, and wanted a photograph of the blue flush water. My father and mother met us in Pierre with a truck and a car, to haul our family, luggage and dogs to the Murdo family home just before Christmas.

Now some people would be affected by culture shock, with an abrupt transition between the big city bustling sky-scraper canyons of New York City and the silent wide open blustery winter prairie of West River South Dakota. Our sons, aged four and six, were up early to ride in the Bronco with their grandfather down to the dam to chip open a line of water holes in the ice, so the cattle could have a few hours to drink for the day. Sheila and I gratefully slipped into the family preparations for Christmas. Brother Gerald and I went to town to find presents for Sheila (goose down vest, Scotch cap). We also bought a bottle of apricot brandy for Grandma England, who always claims she uses it in cooking the desserts. In the few days before Xmas the kitchen became very busy, making steamed (plum) pudding, thawing the turkey and pheasants, making brandied yams, several cakes and rhubarb pies and peanut brittle candy. A little juniper tree from the White River foothills was decorated with precious old ornaments and Sheila added some of her hand-made *ojos de dios* and some New York sweetgum tree seed balls.

And then we had a Christmas feast that couldn't be beat and the children of the extended family tore the present wrapping paper with glee. All of my three brothers were married and had kids now and there were twelve kids and fifteen adults around the big dining room table. A lot of time was spent in the cold windy yard cranking the ice cream bucket. Ice from the cattle watering holes was collected in gunny sacks and bashed with a sledge hammer to make crushed ice for the ice cream bucket. Great-grandmother England sat watching the kids play in the big house, sipping apricot brandy, with a soft shawl (made on the loom by Sheila) on her lap. Later I played the fiddle with Grandma on piano, with Sheila singing some of the Laura Ingalls Wilder songbook tunes. Grandma would tell stories about her grandfather, who played fiddle at socials and dances around Long Lake, Minnesota, and she could even remember the names of the dance tunes (Money Musk, Irish Washerwoman, Turkey in the Straw, Redwing, Arkansas Traveller, Red River Valley, Buffalo Gals). She also told stories about the fiddle, banjo and accordion music played during her 'road gang' years, and some of this she did not approve of, as there were bawdy songs. My mother chimed in with stories of the 'box social' country dances down on the White River most Saturday nights. Seeing that Sheila and I were interested in old time dance music, they sent us up to visit the local jeweller (Ellis Beckwith), who had eight fiddles at home, several old guitars, a Martin tiple, mandolins, a cello and some antique four string banjos.

Gerald and cousin Chuck were running a pig operation down on the White River at this time. Their present to us was a 200 lb hog, which was quickly butchered, wrapped and frozen. Brother Obie got an antelope for us up north of Philip and this was added to the frozen meat pile. I guess they knew we were returning to our Canadian farm, where the freezers would be empty. We also got three whitetail deer and some pheasants from the White River foothills, so we were rich in meat coming home.

New Year's Eve meant oyster soup for my father and myself, and some of the family that would eat those slimy things. This was a family tradition that I never questioned, until later when I found out how hard it must be to get fresh oysters so far from the sea. This was followed by another giant roast beef supper, more sour cream raisin pies and large trays of cinnamon buns. New Year's Day we all went skiing and sledding on the local stock dam grade. Most of the small children could not be identified by name, as they were bundled in winter coats with hoods, covered with snow and shrieking with glee. The kids also discovered a gunny sack in their Grandpa's Bronco, with a wheel of frozen longhorn cheese and some very hard salami, which became insanely delicious between romps in the deep snow. Later in the afternoon we had tea and steamed pudding while watching football games on the TV. My father would sit in his big chair with bowls of Xmas liquorice and peanut brittle beside him, often drifting away into slumber.

Another conversation at Skelly's coffee shop in Murdo:

'What on earth did you have to go back to the east coast for? I'd a thought you'd a had your belly full a that place when you were in school there.' Says Ken Foster.

'I was trying to learn how to measure the radioactive stuff that came from the atomic bomb tests.' Says I.

'What's that good for? We don't have any of that stuff around here, do we?' says Helge.

'Yah, you do have these test bomb radionuclides in the dirt, dam water, in the air, in your bones and cattle bones. They are everywhere, in the Arctic and in the tropics.' Says I.

'Are they gone to hurt us? Damn, maybe that's why Elmer Gilliland had a few deformed calves last year?' asks Wayne Sanderson.

'Probably won't hurt you, because the concentrations of bomb fallout here are much lower than natural radionuclides in the dirt. But the bomb test fallout did bother sheep that were grazing close to the bomb test sites out in Nevada.' Says I.

'Well, that's ok, sheep are no damn good anyway.' Says Helge.

And then the discussion went back to the price of last year's baled alfalfa, the latest stockyard sales prices, and whether or not the river would flood in the spring. The coffee and pie wasn't as good as at home.

I was expected to be back at work in early January, so we packed up our mushroom van for the trip back home to Canada. We had to conceal large boxes of wild meat and another tanned buffalo hide. Mother gave us several large jars of chokecherry and plum jelly. My father oiled and polished little Ian and Jesse's Christmas cowboy boots and told them not to scuff the toes. We loaded our Gordon Setters Banjo and Jenny last and with tearful goodbyes and swollen throats we drove away into another normal Dakota blizzard. The van had an elevated sleeping platform behind the rear seats so that kids and alternate drivers could sleep *en route*. Sheila always worried about customs going home, with our frozen hog, venison and antelope buried in the back of the van. We hit the US/Canadian Customs and Immigration station at 1:30 a.m. and the sleepy agent looked at our packed gear and two dogs, and waved us on to the road going north. We arrived at our Tourond home in the blowing snow and thankfully our long driveway had been recently opened by a neighbour with a very large John Deere tractor snowblower. Our house was chilly and empty and bare. The dogs very happy to be home and immediately checked out the barn and their old kennels, closely followed by our two sleepy sons. And thus began another wild chapter in our lives.

Meanwhile, the earth was spinning around the sun as usual, and the Milky Way paid attention to its gravitational business.

Chapter 18. Back in Winnipeg, *Cosmos 954* Nuclear Satellites and Hudson Bay Adventures

Between blizzards we managed to put our household back together, get our sons in school and re-establish our connections with the musicians in Winnipeg.

Upon return to my office at the institute in mid January 1978, I found it full of giant weather balloons, a welcoming home gesture by my colleagues Brian Graham, Paul Campbell and Susan Kasian. Brian was leaving for a new job in Ottawa and I was looking to hire a radiochemical technician to run a new radiochemical lab. I was furiously ordering the latest models of alpha and gamma spectrometers, fresh from my new knowledge from working at LDGO in New York. We also had to obtain the proper licence to handle radioactive sources in our lab, which meant applications and meetings with Atomic Energy of Canada in Pinawa and Chalk River. One of the recent Ph.D.s from Wally Broecker's lab was hired to continue the radiochemistry at ELA and Ray Hesslein became a valuable partner in setting up this new lab. As with most new ventures I was spending a lot of taxpayers' money to build this lab, and I was concerned to show some useful results as soon as possible.

Andy Hamilton, Ray Hesslein and I were invited to a meeting in Ottawa on 22 January 1978, to discuss Canadian nuclear waste disposal options. Our contribution was to show how natural, bomb fallout and spike radioactive nuclides move through the food chains and eventually get buried in lake mud. This was routine science. But two days later, while I was at home snowbound during another blizzard, the Russian thermonuclear-powered satellite *Cosmos 954* scattered debris from the eastern basins of Great Slave Lake to Baker Lake and western Hudson Bay in the NWT. I got a phone call from my bosses, asking me to get to my office even if I had to use my skis. The search for hot debris was done by Canadian and US nuclear experts, using helicopters based in Yellowknife and Chesterfield Inlet. Since I was familiar with this remote region, I was brought in as a consultant to help plan the logistics of the search. My institute bosses were also concerned about the probability of radioactive contamination of the winter fishery on Great Slave and Baker Lakes, even though they were covered with two metres of ice. All of a sudden radiochemistry was high priority at our institute and we ordered more gamma spectrometers before the excitement wore off.

Over the next months many Indian and Northern Affairs meetings were held in Ottawa and Yellowknife to bring information together to reassure the public and the northern fishermen that we knew what was happening, when we really did not know much at all. We had calculated that the satellite reactor would have contained about 7,000 curies of fission products that it would have been thermally hot from re-entry, and would have melted a hole in the ice. Ray Hesslein and I had half a ton of fresh caught Christie Bay and Baker Lake fish shipped to our Winnipeg laboratories, where we could use our new detectors to determine if any of the exotic reactor nuclides were present in the fish. We could not find any of the reactor nuclides in fish flesh, but we did detect most of the natural radionuclides and a suite of trace elements in fish from this region and control areas. AECL scientists used balloon supported atmospheric gas and particle samplers to show that most of the radioactivity was in small hot uranium and fission product particles in the upper (10-30 km) atmosphere.

I joined nuclear chemistry experts from the Canadian Atomic Energy Ltd. in a series of lectures at small town halls from Great Slave Lake to western Hudson Bay, trying to explain the purpose of the Russian communications satellite *Cosmos 954* and why it may have crashed on their lakes and trapping and hunting land. The TV and print media were at some of these meetings and we were coached to be very careful of what we said. We had to invent some new language to express the concepts of radioactive decay and the possible toxicology of radioactive particles and large metal fragments on the ground. I had my fiddle with me on these long northern lecture tours and it was interesting how many old fishermen and trappers would know the names of the fiddle tunes, would dance, and a few could play the fiddle. After a few fiddle tunes in the hotel lobby they would shyly begin to ask questions about the consequences of radioactive fallout on their land and water. In this quiet atmosphere, with a few beers, more information could be delivered safely than in the bright lights of the formal hearings.

As is irritatingly common in government institutions, reorganizations happened frequently, hopefully for the purpose of terminating old research programs and beginning new directions. The Experimental Lakes Area (ELA) program was now ten years old and it was being questioned about how much longer we wanted to play with nutrient additions to lakes to discover the mechanisms of eutrophication. New proposals for whole lake experimental studies on acid rain were successful in 1975, with the added dimensions of using radioactive tracers for many of the trace elements that would be liberated with acidification from industrial and mine waste acids. As a consequence of this new direction, a group of us attended a NATO scientific session in Toronto to decide how to attack the acid rain problem, with many good papers from our European colleagues who had suffered industrial acidification and metal pollution of the atmosphere, land and water for a century (in some places several millennia, such as in Spain and Italy). ELA was at the front line of this discussion, as we had already spiked lakes with acids, radiochemical tracers of metals and radiocarbon. Under the leadership of David Schindler and Wally Broecker, the ELA camp was swarming with LDGO graduate students with radiochemical and analytical abilities not commonly found in freshwater field camps. This was science in high gear for the next decade.

Another new direction came from a spin-off of our Mackenzie River valley oil pipeline study. New discoveries of natural gas in the central and eastern Arctic led to proposals for gas pipelines along the western margin of Hudson Bay. No one knew what the consequences of a buried gas pipeline break in the muskeg and lake-strewn regions of this huge area would be. Buster Welch proposed to do whole lake experiments on a few small lakes at Saqvaqjuac mainly to see if the addition of methane would cause the bacteria in the cold lake water under the ice to rapidly degrade this new organic matter, causing the lake to become deficient in oxygen and the fish might die. Buster already had several years of background information on Saqvaqjuac lake chemistry and biology and his crew had fertilised a few lakes. They had also done work on removal of the big predatory lake trout, to see if removal of this predation would stimulate growth of the Arctic char. Within a month of funding this new study began during the long winter ice cover season. Big tanks of methane gas were sledded out to the experimental lake and rigged to deliver methane gas to diffuser hoses on the lake bottom, so that the methane would dissolve into the water. Early observations suggested that the methane eating bacteria needed more nutrients than were present in this cold clear lake water. So it was necessary to study the nitrogen and phosphorus cycle to understand the fate of the added methane.

In July 1978 Buster called for help to study the regeneration of nutrients and methane from the lake mud. As Ray Hesslein and I were the local experts in coring lake sediments and extracting the water from the mud for chemical analyses, we packed our equipment in the standard red komatiq boxes and prepared for another expedition. The little experimental lakes were ice-free by the second week in July, with maximum runoff in late June and maximum warmth in August, the surface water now being 8-12° C. Having spent five years of my life in the Mackenzie River valley work I was now about to learn about the west coast of Hudson Bay. This was a rocky and barren taiga country, full of snow and ice for eight to ten months of the year. But up close it was beautiful, eerie and quiet, and you could see for many miles, as there were no trees. Interspersed between the rocks and gravel soils were little patches of moss, cotton grass and tiny willow trees.

Travel in the Arctic is a crap shoot and there is not much use for an itinerary. Leaving Winnipeg in August, where green grass and trees were plentiful, we were stranded in Churchill for several days due to ice fog drifting in from Hudson Bay. The mosquitoes were terrible and the hotel room was over-heated. We were using a mosquito repellent called Repex, about 90% DEET, which would dissolve plastic pens, sunglasses and steering wheels. Rolling my cigarettes with DEET-soaked hands resulted in an unusual tasting smoke. Churchill was full of Inuit kids and young men in flight jackets, motorcycle helmets on skidoos and four wheelers, mangy barking dogs, dirty polar bears at the dump, half- butchered seals on the beach, government trucks with nervous white southerners driving slowly and fishing boats pulled up on the rocky beach. We learned that the beluga whales at the Mystic (Connecticut) and Coney Island (New York) Aquaria come from the Churchill region, trapped under permit from the Canadian Fisheries and Marine Service. We also heard about a bad man who recently took tourists out in freight canoes to shoot belugas with high powered rifles, but he got arrested as this was illegal. The belugas are friendly, will surface beside canoes, and they like their backs to be scratched with a paddle.

We finally escaped Churchill in a Twin Otter airplane on floats, with stops in the Hudson Bay communities of Rankin Inlet and Chesterfield Inlet, where local freight, mail and southern groceries (soft drinks, potato chips, cheese balls, candy) were distributed. We flew into Sacvaqjuac with the Twin Otter heavily loaded with diesel fuel drums, 300 lb methane tanks and our equipment stacked to the ceiling. I held my fiddle in my lap as we bounced along the estuary to find Buster's camp docking platform. Buster's crew were there with wheelbarrows and trolleys to quickly move the gear into the camp and to reload the airplane with outgoing freight destined for our Winnipeg labs. After the Twin Otter left I opened one special box of treats for the camp, including large piles of fresh vegetables from our Tourond garden, comic books and newspapers, fresh fruit, candy bars and several dozen home-grown eggs from our barn.

Unloading plane at Sacvaqjuac, 1978

Ray and Mike Capel and I set up our equipment in the lab and proceeded to take sediment cores from several of the small experimental lakes. The core tube was light plastic, threaded into a heavy steel driving weight and tripping mechanism. This apparatus was lowered over the side of the boat and allowed to gently penetrate into the soft lake mud. Once retrieved and brought back to the lab, the 50-150 cm column of mud was pushed upward into a plastic ring, which allowed 1 or 2 cm slices of the vertical column of mud to be captured. These core slices were squeezed to express the water and treated to preserve the concentrations of methane and nutrients for later analysis. We could also measure the pH and salinity of the water at the lab, and the squeezed mud was put in labelled sacks for radiochemical analysis, to determine the rate of accumulation and age of the mud slices. From this information we could estimate the rate that the mud released nutrients into the deep water of the lake. And this would allow Buster to estimate the capacity of the microbes to digest methane and consume oxygen from the lake water. Most of the bottom mud and water column animals in these lakes were accustomed to oxygen-rich water and any diminution of oxygen would probably cause problems with lake productivity.

This went on for several weeks, often working late into the night. We greatly appreciated that Buster's inventive son Scott built a solar water heater for the shower at camp. We often returned from coring covered with mud and mosquito repellent, and a warm shower was heaven. But some nights were left open for pleasure, and the Saqvaqjuac Symphony was created. After caribou steak suppers, Mike Capel brought out his dobro, Ray had his clarinet and harmonicas and I had my fiddle. Several of Buster's crew could play guitar and all could sing, so there were many choruses of 'You Are My Sunshine' and the 'Tennessee Waltz'. John Le Gault and Art Dalton played some mean blues on guitars, with Ray doing very well on Howling Wolf-style harmonica. Liz, the Inuit cook, demanded to tape our music session and years later this cassette tape was still being copied up and down the coast of Hudson Bay. Liz cooked good bread, a wonderful caribou soup and some of the weirdest pizzas...tinned smoked oysters and sardines, pickled mushrooms and tiny shrimp, bits of caribou and char and sprigs of sorrel from behind the camp. Atomic Cherries were sometimes served with dessert puddings but I didn't realize the full significance of these until later. Buster showed me that one wall of the sleeping room was built from maraschino cherries in bottles in cardboard boxes. These cherries were dumped into a ten litre Erlenmeyer flask, filled with lab ethanol, diluted with the cherry juice and some water. The cherries absorb most of the alcohol and they taste wonderful on top of chocolate pudding.

Many of the scientific samples from the camp biologists were also good to eat. We had clams, geese and ducks, caribou and lots of Greenland cod and Arctic char. Buster's wife Cathy was studying the parasites of fish and birds, and it was a bit disturbing to watch her take small tissue samples from the fish we were going to eat that evening. August is the warmest month of the year and there were flowers and birds galore, seals playing in the estuary, sik sik (big fat ground squirrels), ptarmigan and sandhill cranes wandering around the bare stony camp grounds. Buster and Cathy created a warm family feeling for the whole camp and everybody helped wash the dishes, sweep the floors, fuel the stoves and haul out the toilet throne bucket. We almost felt guilty to leave the camp when we packed up our gear to go back to Winnipeg.

Of course, going home was not simple. We were stranded in foggy Rankin Inlet for a day and stranded for two days in Eskimo Point (Arviat), where we stayed at the Padlei Coop. This was a madhouse, with a large screen TV showing the Arctic Games, US sitcoms and US pro basketball games, all watched intently by young and old. The Arctic Games included a few sports that I had never seen, such as trying to kick a bone hanging from a string about 1.5 m high. Also featured were string figure speed games and judging of seal fat basted bannock and muktuk tasting. I wondered how they could understand the big city jokes on the US TV programs *Laugh In* and *As it Happens*. We visited the Inuit Cultural Institute, where they had taped and translated the old people talking about their lifestyle, hunting grounds, winter/summer migrations and food consumption (and lack of it). All night long there were kids roaring around on ATVs, playing a motorised form of soccer, shouting when anyone scored. Dog teams were chained to boulders around the town, the winter sleds (komatiq) were upside down on top of boulders, covered with caribou skins, and the rocky beach was littered with fuel drums and fishing boats overturned for repairs. Also stranded here was an old fellow from Sydney River, Nova Scotia, who could speak the old Gaelic, sang Celtic ballads, and he demanded fiddle tunes from me. He was a carpenter, going home to his family after a house building job in town. The fog cleared and we finally got a flight south to Churchill and Winnipeg. My seat mate was a Baptist missionary, violently anti-Catholic, and he had been called from on high to convert the Inuit pagans. He was a thoroughly repulsive person who had almost zero understanding or appreciation of Inuit culture and lifestyles.

I was grateful to be in the arms of my bright and shining Sheila at the airport and I was dumbstruck by the colour green outdoors, my eyes having been accustomed to the rocky barrens of Saqvaqjuac. After delivering our half ton of freight to the institute, we went home to the green grass, the garden rich in good food, two small boys running like crazy animals loose on the earth, dogs, Gladys the goose, two owls and four pigs in their barn pen. Now was the time of sweet corn harvest by the wheelbarrow load, picking chokecherries, digging potatoes and putting the onions and squash away for the winter. Not all was bliss however, as a neighbour's dog killed half of our meat chicken flock, so I shot the dog and returned the collar to the owner. On my birthday we drove down to wetland swamps south of Roseau, to release the piffers (tame horned owls). They were transported in dog cages and they hesitated with us before flying away. Sheila was in tears, we had a picnic there and we left some chicksicles hanging on the trees for the first night on their own.

As the ice began to form on the ELA lakes many of the US and European scientists came out of the bush to Winnipeg to go home. But most of them stopped by our house for a feast and some music. Most of the LDGO graduate students knew they could get a good feed and overnight bed at our house. Joerg Bloesch (Swiss) visited and we had him helping split firewood. Joerg told us about his father trying to re-establish storks in Zurich, where he kept ten of them in pens with clipped wings for breeding.

In September I was invited to give a lecture at my undergraduate Augustana College in Sioux Falls, South Dakota, where I met again my old biology and chemistry professors. They had built a new science centre, with a huge amphitheatre for guest lecturers. I didn't feel like a hotshot scientist in front of all those students but they were obviously impressed and jealous that I got to run around in northern Canada in Twin Otters, having fun doing science. It was also fun to participate in a panel discussion with staff and students about how to find new directions for teaching and research in science. One of my recommendations was to keep the liberal arts curriculum in place, so that science students had to take courses in the arts, literature, music and history, and that the humanities students had to have a taste of science and mathematics.

In later years I appreciated more and more having had this kind of liberal arts education before I dived into the specialisation of graduate school thesis work at a much bigger university. Most of my graduate school and institute colleagues in science had little interest in or appreciation of T. S. Eliot, J. S. Bach, the Greek tragedies, the ancient history of the Middle East Fertile Crescent and China, Leonardo da Vinci's artistic concepts in science, Renaissance brass tower music or comparative history and psychology of religions. Revisiting my undergraduate college was an important event in my life, as I could see better where I had come from and where I might be going in the future.

In October I gave a paper on the chemical budgets of the elements for ELA and Saq lakes and streams at a Geological Society of America meeting in Toronto. Lots of my old friends from around the world attended. Norm Guinaso (from Texas) tried to teach me how to interpret the effects of bottom animal mixing of the mud on radioactive lead profiles in sediment cores. Most of this was done on hotel restaurant napkins and the backs of menus, which I kept in my journal. Guinaso was actually more interested in my experience at raising pigs for good pork, as he was trying to live a sustainable life on a hobby farm. John Diebold from LDGO appeared in my hotel room with his guitar and proceeded to play some of his own blues tunes with me on harmonica. Spanish algal scientist Doloros Planas needed a shoulder to cry on, as her husband and two children had returned to Spain without her, but she had a good job at a Quebec university.

I was off to Montreal next, to give a lecture at McGill University and discuss Arctic lake research plans with colleagues Frank Rigler, Jaap Kalff and Rob Peters. Rigler was thin and pale, and I later heard he was dying of cancer. Alain Penn trapped me into attending a meeting with the Grand Council of the Cree and to assist him in predicting the effects of big dams for hydroelectric power production on the Quebec rivers flowing into James and Hudson Bay. This study continued over the next two decades, with the ecologists and hydrologists trying to protect the lifestyles of the Hudson Bay Inuit that live near these rivers and the marine coast. The Quebec generated electricity was mainly for sale to NE USA, not for local people.

One the way home I visited Sheila's family home, up north of Toronto. Sheila's mother was busy in the upstairs bathroom brewing vats of ginger wine, rice wine, cider, elderberry wine and dark stout. I was able to play some trumpet/flugelhorn duets with Sheila's youngest brother Bill, who was becoming a very good musician. Sheila's mother ceremoniously gave me her English riding saddle, thinking that her daughter might use it out there in the Wild West, living with a mad South Dakota cowboy. She finally indicated that I wasn't such a bad sort of person after all, even though I was an American.

After many versions of job descriptions and salary negotiations with balky management nabobs I finally managed to hire Paul Wilkinson in March 1979, to run the Winnipeg radiochemical lab. Both he and Ray knew more about the electronics of measurements of radiation than I did, and our lab quickly became more competent and versatile. Paul and Ray collected radiotracer reference material for uranium, thorium, radium, polonium, lead, plutonium and caesium, so that we could calibrate our new detectors. Most of these reference isotopes were gifts from friends around the world who produced or used these nuclides, an example of the scientific community internal help ethic. That summer, Paul and I cleaned up ELA radwaste storage and took the waste to Pinawa AECL. Paul arrived with an old Gibson guitar, so he was immediately incorporated into the Saqvaqjuac Symphony Orchestra.

There was a quick trip back to New York in April so that Stuart Ludlam (at University of Massachusetts) and I could recover some sampling gear we had intentionally left suspended in Fayetteville Green Lake. Stuart took me on a tour of Amherst and nearby Stockbridge, where Officer Obie was still working with the Police Department, and Alice's Restaurant was now in the train station. [You have to listen to Arlo Guthrie's 'Alice's Restaurant' to understand the deep significance of this previous sentence.] Stuart took me to Cape Cod and Woods Hole, giving lectures to this ignorant prairie boy about the historical importance of this part of the world. I irritated him by commenting that the farms were awfully small and they couldn't support many cattle. We were at Woods Hole for an international meeting about the ecological troubles caused by oil spills and I told the story of Norman Snow's experimental oil spills in the Mackenzie River Delta lakes. Many of my colleagues from the Prudhoe Bay (Alaska) research lab were there and I helped write a summary of our recommendations. We also ate a lot of spicy crab cakes, lobster and cod cheeks.

In late April '79 the Red River flooded most of the flat agricultural land south of Winnipeg. Good friends Paul Campbell and David Schindler had their homes in the flooded area near the river and we spend many days sandbagging around their homes. For several days, Sheila and Susan Kasian took over the cooking responsibilities at Paul Campbell's house as there were seventy people working on his dykes, most of them colleagues from our institute. Our two sons scampered along the tops of the dykes, holding rods for the surveyors to mark the height. Most of the people and livestock near the river were evacuated. Our home was not affected, other than the normal snowmelt water flowing into our dirt basement as usual.

Throughout the rest of the year I was flummoxed and too busy with multiple bundles of messy projects that defied simple solutions. A series of workshops were held in Calgary (the oil capital of Canada) about how to avoid ecological damage from commercial developments in the Mackenzie River valley and the Beaufort Sea, especially the tar sands, oil spills, road construction, melting of permafrost and the resulting erosion of the landscape. Some of these workshops were held at Fairmont Hot Springs, up in the Rocky Mountains west of Calgary, where we could see mountain goats, travertine hot springs, and of course we could soak in the hot springs after our meetings, with the cold clear mountain air above us. (I brought back some water samples from these hot springs, to determine the radioactive components of the high dissolved salt load. We found that radium and radon were very high and off scale for our low-level calibrated instrumentation.) I was involved in a series of meetings in Ottawa and Chalk River AECL to organise the writing of a National Research Council document about natural and anthropogenic levels of radioactivity in Canada, which apparently had never been surveyed before. This involved visiting uranium mining sites, the Canadian nuclear power reactors and their waste holding pools, research and medical treatment radioisotope production facilities and the history of thermonuclear bomb fallout in the remote northern part of Canada. The chairman of this group publication, Ivan Ophel, died shortly after this was published and we never knew he was fighting cancer.

We also were involved in the early stages of International Joint Commission consideration of the effects of intrabasin water transfers. The closest example of this was the intent of North Dakota irrigation engineers to dump salty irrigation water from the Missouri River catchment into tributaries of the Red River, flowing northward into Lake Winnipeg and Hudson Bay. In addition to the salt and nutrient inputs, this action would also allow the transfer of some undesirable aquatic organisms (fish, invertebrates, microbes) that were not known to occur in Canadian waters. The reverse case was also considered, as the US Bureau of Indian Affairs, Nevada and California wanted to import large volumes of cold clear British Columbia water into their water stressed states.

I was also helping to teach a University of Manitoba course in geolimnology with Jim Teller and Bill Last, and I was on thesis committees for several Ph.D. students. I commandeered an empty lab room to hold all the file cabinets, maps and document folders that came from these activities, but I was drowning in a sea of multitasking civil service bureaucracy.

Meanwhile, the real world was proceeding in bizarre and chaotic ways, history and science gone mad. All of these confusing incidents were swirling around us as we were trying to focus on science and our families.

In a much more sane aspect of the world, the first versions of Douglas Adams' *The Hitchhikers Guide to the Galaxy* were published in 1979. Adams wrote:

> 'I'm a scientist and I know what constitutes proof. But the reason I call myself by my childhood name is to remind myself that a scientist must also be absolutely like a child. If he sees a thing, he must say that he sees it, whether it was what he thought he was going to see or not. See first, think later, then test. But always see first. Otherwise you will only see what you were expecting. Most scientists forget that.'
>
> Wonko the Sane on the beach in California, p. 734

Adams also wrote about Marvin the robot, with a brain the size of a planet, and all they let him do is open and close doors. When I read this it sounded like they were talking about most research scientists, who know a lot about how the world works, but are generally ignored by the public and politicians. Adams' only entry in *The Hitchhikers Guide to the Galaxy* for the planet Earth was 'Mostly harmless.' His writing gave some curious (and devious) comfort to a rational person trying to understand why the world behaves as described in the long paragraph above.

But Wonko the Sane hit me directly between the eyes. He was writing about how we indoctrinate our children and university students into seeing and understanding things in the currently approved way and not necessarily the way they really exist. I've always suspected that each of us sees colour, texture, shadow and light differently, and that we all hear different components of language, jazz, symphonic music, thunder and waterfalls. All of these sensations are sorted and embellished in our own culturally warped brain, which has been preconditioned to accept some views and not other new ones. When a scientist analyses his or her data, makes graphs and tables, we sort the information in such a way to falsify or support a hypothesis or to illustrate how we think this part of the world works. Often, a large part of the scientific observations are not used in the journal article or book chapters because they are not useful to enrich the world scene that we want to illustrate. With Wonko's help, I could see better the different views of my wife and sons and my parents, all of which paid attention to different aspects of life than me. This affected my manner of giving lectures, as my preferred interpretations of science data began to include options for outlandish and non-traditional stories of how things might work. Some of these were entertaining and easy to falsify, but some ridiculous ideas persisted and could not be shown to be false. I began to appreciate the alternate views of autistic savants, brain damaged people and those of us who live on the margins of sanity, being able to see the world in a different way.

And the siege of computers began about this time. A 286 computer box and monitor screen appeared on my office desk and a pile of those big floppy disks for storage of files, as we didn't have much memory space. I stumbled into Wordstar and several elementary spreadsheets and Susan Kasian provided computer skills in graphics, data crunching and map work for our publications. Within a year there were internal Canadian government communication channels opened and all sorts of international communication networks were established, such as Omnet, Bitnet and ScienceNet. Bob Heinmiller and his wife Susan Kubany (operating their 64 kb memory servers from the third floor of their Boston home) organised scientific bulletin boards on Omnet for oceanographers and other aquatic scientists. I didn't realise the significance of the act of putting my old high school Remington typewriter underneath a lab table. Now I could type directly to computer documents, add data and text quickly, edit them easily and produce a near-perfect document to submit to a science journal. It took a while for the institute administration to figure out that secretaries were no longer needed to type manuscripts. I was now even more thankful to my high school teacher Mrs. Evelyn Kuhrt, who insisted that boys should take typing as well as the girls.

While all of this science was ballooning into unmanageable tangles, our home life was also happily becoming more complicated. Our two sons were now in school and I was reading J. R. R. Tolkien's *The Hobbit* for their bedtime story. Our Gordon Setter dogs would walk them down our long driveway to catch the school bus. The dogs looked wistfully at the driveway all day, waiting for them to come home. We were taking care of poultry and pigs, a large garden and becoming more involved in the folk music circles in Winnipeg. I was secretary (& briefly President) for the Board of Directors of the Winnipeg Folk Festival and we hosted many interesting year-round fund-raising concerts by touring bards Utah

Philips, Michael Cooney, Tom Paxton, David Essig, Archie Fisher, Gordon Bok, Robin and Linda Williams with Peter Ostrouschko and Stan and Garnet Rogers. For the July '79 Winnipeg Folk Festival Sheila and I butchered six 80 lb pigs and these were spit-roasted backstage for the musical performers and stage crew. This horrified the vegetarian greenies but the carnivores enjoyed it very much. By this time Sheila was performing on the ballad workshop stages, sometimes with her sister Mary. More important were weekly ceilidh house parties with our musical friends in town. Sheila was rapidly learning the fiddle tunes on her hammered dulcimer and my ability on the Butterfield fiddle was progressing slowly. During one January blizzard I spent the weekend making dulcimer hammers out of old wooden venetian blinds. Sheila had her Quebec loom in full production during these cold winters and warm wool shawls, blankets and vests were made. Sheila was also involved in teaching 4-H classes on cooking and how to handle home-grown meat and fresh garden produce. She knew these skills because every fall we put four hogs, thirty to forty chickens and a dozen ducks and/or geese into our freezers. Our friend Hedy Kling brought us a fox kit in May 1979, rescued from a burning shed at their farm. The fox rapidly became tame, grew up with our sons, slept in their beds and ate cereal with them at breakfast time. Sheila's father visited during this time, and it was touching to watch this big man in a business suit sit down on the kitchen floor to cuddle and croon at the little fox. The fox liked to sit on the sofa beside me when I was struggling to play fiddle tunes and he would whine along, as if he were singing to the fiddle tunes. We bought a seventy year old Mason and Reich (Toronto) piano because Ian MacGreggor decided he wanted piano lessons. We were dumbfounded to watch him quickly go through the elementary lesson books and we concluded that this boy might be a musician. And of course every fall there was a moose hunt in northern Manitoba, usually with some success.

In August '79 we drove down to South Dakota for my parent's 40th wedding anniversary and my mother's birthday party. All four Brunskill sons were there to toast our parents for a job well done. Somehow they had kept us relatively healthy, got us through four years of university and kept us in photograph albums of our younger years. There were two rodeos happening this weekend, with town parades, a circus with seven elephants, street dances, baseball games and a real prairie small town celebration. We had fresh White River apples, sweet corn and melons from my cousin Chuck. My brother Gerald and his new wife were about to leave for Arizona, Gerald having a new job as a metallurgical engineer in a big copper mine. Brothers Obie and John were still interested in cattle and horses and the South Dakota prairie grass. And of course my father, quietly and proudly, took me down to Skelly's roadhouse coffee shop, where all of his buddies were stirring their coffee and eating pie...

'So when are you coming home, Gregg John?' Says Ken Foster, who wondered if I could still operate a Farmhand haystacker, now that I was supposedly an egghead scientist.

'I dunno, it is more fun wandering around Hudson Bay than stacking prairie hay.' Says I, grinning.

Many grunts and guffaws at this, and 'But you can't raise cattle up there where they don't even have grass! What do those Eskimos eat, anyways?' says Helge.

'They eat a lot of fish, seals, geese, ducks, caribou and sometimes whale meat. Not a lot of vegetables up there. The kids living in the little villages eat potato chips and candy bars and Coke, just like around here.' Says I.

'Well, that sounds like good grub, and you don't really need vegetables anyway...'cept maybe for sweet corn and melons. Did you eat any whale meat?' Says Wayne, his eyebrows significantly elevated.

'No, I didn't taste whale meat, but I ate some seal, fish and ducks up there. The seal meat tastes like very oily fishy pork. Caribou is good, like deer meat only darker and richer. The eider ducks were big and fat, delicious.' Says I.

'But why is the government sending you up there to Hudson Bay? It must cost a lot of money to go up there?' asks Ellis Beckwith.

'There are proposals to build giant natural gas pipelines from further north in the Arctic, down to the Canadian/USA border. We were supposed to find out if these gas pipelines would damage the lakes and rivers along the construction pathway. The Inuit people that live there are concerned that this industry might affect their harvest of wildlife, the fish and seals.' Says I.

'But what do they need so much gas for, up there in Hudson Bay?' asks Ken Foster.

'They don't need very much in the little villages along Hudson Bay. Most of the gas will be sold to the USA market.'

'Ah, well, that's ok then.' Says Ken, quietly stirring his coffee and asking for another piece of apple pie with whipped cream (NOT DREAM WHIP!) on top, and a chunk of Longhorn cheese.

Then they went back to telling stories about when they were young, and all the trouble they narrowly avoided. There was the price of cattle and grain to discuss, and the usual argument about whether John Deere tractors were easier to fix than International tractors. Other old wind-burnt ranchers would join the circle to get the local gossip and tell of the success of their kids who went off to university and had jobs in the big city. Somebody got hurt trying to shoe a horse and one old fellow claimed he lost a calf to coyotes.

The waitress took a long look at me, and said: 'You are Gregg's oldest boy, aren't you! Where did you get those whiskers? You'd better get your hair cut!', and she calmly walked back into the noisy kitchen. My father grinned and paid the bill.

Chapter 19. Hudson Bay and Radioactive Walrus, *Ayurngnarngmat*

Thank heavens for Kiggavik! The bloody senior management at our institute seemed to take great joy in reorganising the research teams, much to the distress of the scientists who do the work and write the science reports. They were confusing the role of business management with scientific leadership. Ottawa bureaucrats and economic consultants were invited to our meetings, to help discover the economic value of the ecological resources we were studying. And of course they discovered that the economic value of fish and wildlife was much, much less than the money we were spending in our research. David Schindler was so disgusted with the situation that he went out to hustle external support (from AECL, US EPA, EPRI) for his acid rain research at ELA.

The CEO and his henchmen were busy reassigning staff and terminating some projects in early 1980 when the newspapers were suddenly filled with articles about Urangesellschaft's (a German-French uranium mining consortium) plans to build a CDN $200 million dollar mine and yellowcake plant on a patch of land WNW of Baker Lake in the Canadian Northwest Territories. They had found a rich uranium deposit (called Kiggavik, estimated 300 million lbs of U_2O_3), and they proposed to dig up the ore and make yellowcake on site, to fly it out to Yellowknife for truck delivery to the southern market. There were no roads or services in this region, which means they would have to ship many tonnes of equipment, vehicles, materials, fuel and personnel into this remote area by air or winter cat trains. The mining company's environmental impact study suggested that the negative effects on the regional ecological systems and Inuit hunting and fishing would be zero. The NWT government and the National Department of Indian Affairs were supportive of this new industry, but the local Inuit were not so sure that they wanted their fish to glow in the dark. Misinformation flowed both ways between the pro-development businessmen and clean environmental greenies, but the political savvy of the Inuit was underestimated. They knew that mining development in their region would mean some jobs for them and perhaps a better school, hospital and improved communications for their communities. They also knew that most mining camps left behind polluted soil, lakes and streams that were no longer useful for food production. Their local village governments and Nunavut lawyers called upon the Canadian government for information on the natural levels of the radioactive elements in their local environment before they would give approval for any uranium mine developments in their hunting and fishing grounds.

This proposed uranium mining site was about three hundred miles west of Saqvacjuac, Buster's field research camp north of Chesterfield Inlet where he was busy trying to complete his methane gas and nutrient addition experiments in small lakes. Ray Hesslein and I had already proposed a research project on the radiochemistry of Saq plants and animals, water and mud, partly as a result of our efforts in writing our survey of radioactivity in the Canadian environment. But now here was a more compelling reason to do this work and the money flowed in from Ottawa. This saved Buster's field camp from termination and Hesslein, Wilkinson and I scrambled to assemble a realistic field sampling and laboratory analytical plan to determine the geochemical cycles and biological accumulation of the uranium series nuclides (uranium, thorium, radium, the radon gases, polonium, lead) in the terrestrial and freshwater ecosystem and the estuarine environment of Saqvaqjuac Inlet.

We went to ELA to practice our sampling methods in the early spring of 1980, when the lakes were just beginning to melt the winter ice cover. There were moats of soft ice and frazzle around the edges of the ice pan, so we had to push a small boat (loaded with our sampling gear) from the shore, across the frazzle ice, onto the thicker ice in the centre of the lake. Paul Wilkinson and I took 200 litre water samples from the fast flowing snowmelt inflowing streams, the outlet and the surface and deep water. This allowed us to calculate a uranium series element budget for the lake, to know the fraction that flows through the lake to the outlet and the fraction that is trapped in the lake and in the bottom mud. While doing this we created a new sport which we called 'ice dancing', which was mainly a manner of gently and quickly walking to avoid falling through the soft ice.

So here we go again, the roulette game of shipping science equipment and supplies to a remote Arctic field camp on the northwest coast of Hudson Bay. It took us two months to get all the gear from our HQ in Winnipeg into Saqvaqjuac, by air and by SeaLift barge. And on 20 July 1981, Mike Capel, Ray Hesslein and I went through the travel portals of Churchill, Eskimo Point, Rankin Inlet and Chesterfield Inlet (with many attendant delays and adventures), to be met by Buster's son Scott and Guy Amorak. They had a big Lake Winnipeg yawl (with two 20 hp Merc outboard engines) to carry groceries and us visiting scientists to their luxury holiday lodge three hours further north. The cold salty sea spray in our faces actually felt good compared to the fetid air of airplanes and dingy hotel rooms on the way up here.

By this time Buster had a staff of twenty-five people at camp, and because they knew us, we were mobbed by the cooks, Buster's younger sons, and almost everybody else at camp. They mobbed us because they remembered that we brought goodies on our last visit and so we distributed newspapers, magazines, comic books, candy bars, honey-coated peanuts, cassette tapes of music, some fresh Tourond garden produce and 'hot from the hen' eggs. Papaluk shyly asked if I'd brought my fiddle. Buster made a solemn speech welcoming us, but at the same time he demanded to inspect my duffel bag. He informed the camp staff that on my last visit I had violated an important rule, in that I had secretly brought home-made personal goodies for my late night tea time, and that I didn't share with him. He was referring of course to Sheila's 'giant butter tart' and oatmeal cookies.

We had arrived just before suppertime, so we sat down to a massive feast of local seafood, noodles, roast caribou and baked Arctic char. The single side band HF camp radio was on loud, with chatter on the 'game channel' from many Inuit communities and hunting expeditions being discussed. Some of this was in Inuktutuk, so Papaluk or Maurice Kukkiak would translate for us ignorant *kabloona*. Buster and Cathy did know some of their language. We ate in silence so we could hear the gossip, and were amused to hear that three new scientists had come into Saq camp, with a lot of jokes about what they were doing there. We also learned who had a new baby, the prices for new 20 hp outboard motors, fuel cache locations, and that there were three polar bears in the vicinity of Saqvacjuac Inlet. This last news was taken seriously around the kitchen table, several .308 rifles were brought out for oiling and boxes of ammunition were put on top of the radio cabinet beside the door. A spotting station on top of a ridge behind the camp was established and staff took turns watching with field glasses. The worry was justified, as two bears were spotted on the other side of the inlet a few days later, but they never came close to camp while we were there.

Saq Inlet is an estuary, mostly seawater, as freshwater inflows are small. But it is stratified, with lower salinity and warmer water on top. The Inuit name for this large Hudson Bay estuarine inlet is Saqvaalut, whereas the freshwater river rapids is Saqvaqjuac. The purpose of this trip was to collect water, mud, plankton and benthos, terrestrial soils, plants and animals to measure the concentrations of the uranium series elements for predictive estimation of impact of uranium mining near Baker Lake, not too far west of our field camp.

It took us several days to unpack a small mountain of komatiq boxes that contained all of our sampling equipment and supplies. And of course some of the essential materials were missing, including all of our strong acids and bases. We later found that these were still sitting at the airport in Churchill because we had puritanically labelled these boxes as 'corrosive liquids', as required by the Canadian transport industry. This meant a special flight was needed, which only happened when human traffic was near zero. Over the next few weeks we converted the diesel powered landing barge (the *Sea Cow*) to a floating radiochemical lab, with nine 200 litre steel barrels on the front deck.

These barrels were lined with heavy food grade clean plastic sacks. Water was pumped from surface and deep water from Saq Inlet into these barrels and this water was processed to obtain the uranium series nuclides. The water was acidified with concentrated hydrochloric acid to a pH of about 2, a known amount of radioactive isotopes similar to our target nuclides was added (so that we could determine how efficient our sampling method worked), some acidic iron chloride, the acidic solution was stirred mightily with a special clean paddle and the barrels were allowed to stew (equilibrate) overnight. The next day we added strong base (ammonium hydroxide), which caused the pH of the 200 litre solution to rise to about 10 and all the iron that we added precipitated out as rust-colored flakes. These iron rich flocculent particles attracted most of the uranium series elements in the water sample and these flakes settled to the bottom of the barrels in a day or so. We siphoned the top water out of the barrels with a 'J' shaped copper tube and retained only the flocculate iron hydroxide that had settled to the bottom of the barrel. This sample was bottled and shipped back to Winnipeg, where Paul Wilkinson's lab dissolved the sample in acid and separated the fractions for radiochemical measurements of the alpha and gamma emitting nuclides.

Of course things did not proceed smoothly. Scott Welch and Guy Amarok took the *Sea Cow* and us outside of Saq Inlet to sample Hudson Bay coastal water and we nearly lost the whole works in rough weather. We got blown up the coast into areas where the navigation charts were blank, which made Guy nervous. We took shelter inside a U-shaped island to see if the wind would decrease. I walked around the rocky island and built an *inukshuk* on top of the highest elevation on the island, maybe five or six metres above sea level. An *inukshuk* is a human shaped pile of rocks on top of a hill, sometimes with a pointing hand to give directions to a hunting party. Guy chuckled at this effort, saying it wasn't bad for a *kabloonak* but it was not useful, as the winter ice floes would completely cover the island in winter. *Kabloonak* is their word for a white southerner, them being generally ignorant of the basic fundamentals of Inuit social, ecological and survival skills. We didn't get back to camp until 1 a.m., cold and wet, but with a dozen precious 200 litre water samples.

We were grateful to be back in the calm waters of Saq Inlet for the rest of our work. We carried our barrels, acids and bases up to the perched lake basins above the estuarine inlet to sample the freshwater lakes and some of the local rivers. One of the little nearby estuarine inlets (Hawk Inlet) had heavy salty (over twice the salinity of Hudson Bay water) deep water which did not mix with the dilute surface water and there was no oxygen and lots of hydrogen sulphide, iron and methane in the deep water. Our flocculation methods did not work for this deep water, even after aeration and excessive acidification. We tried to strip the hydrogen sulphide gas in the 200 litre barrels of this water by bubbling with air, which generated mountains of foul black foam on the *Sea Cow* deck, and two of our helpers got sick. This little inlet had lots of small jellyfish blooping through the surface water, which we caught with dip nets. They had pockets of bright red cells within their bodies and we suspected that these were sulphur bacteria that live near the top of the deep salty water layer. The camp Inuit called this stinking deep water the 'devil's milk'. We tried to dig trenches in the soil so we could sample the soil layers for the uranium series elements, but this was difficult, due to the rocky soil and permafrost. Every day we would dig out a few centimetres and then let the warm July air thaw the soil until the next day's digging. This sampling allowed us to measure the century-scale accumulation of radioactive lead, which comes partly from atmospheric deposition, and the rate of emanation of its many radon grand-daughters.

Under Buster's guidance we tried to sample the terrestrial, freshwater and estuarine food chains of this region, and to determine the accumulation of these nuclides up the food chain. This allowed Buster to go hunting with the local Inuit during the previous summer and winter and he had collected bones, teeth and flesh of polar bear, fox, sik sik (a big Arctic squirrel), ducks and ptarmigan, seals, walrus, freshwater and estuarine fish. We brought a large otter trawl with us to capture some of the estuarine bottom animals (crabs, clams, worms). Buster and Martin Bergman used SCUBA to collect kelp, sea squirts, sponges, sculpins and some shrimp-like crustaceans in the shallow bays. A capelin run for a few days allowed us to collect tubs full of them, both for deep frying and for scientific analyses. Ray Hesslein and Mike Capel spent some happy evenings fishing for Greenland cod, with great success, but they were nearly devoured by mosquitoes. The local Inuit were granted special permits to kill a few beluga, bowheads, for our work.

And as you might guess, once we had subsampled all these creatures for our radiochemical analyses there was a lot of good meat left to eat. We had fresh cod, deep-fried capelin and char frequently. One of the best feasts was from 3-5 gallon buckets of frozen walrus stomach contents, which Buster had collected on one of his winter hunting trips with the Inuit. After we had subsampled this brew for radiochemical measurements, the remainder was washed in seawater and sieved and hundreds of long-necked clams were pulled out, to fill a giant pot on the kitchen stove. With some milk, onions and chilli powder, this made a wonderful chowder for the whole camp. The necks of the clam were already partially digested and they would melt in your mouth with wonderful flavour. The walrus apparently swim along the bottom of the coastal sea, snipping off the long necks of the clams. There were a few camp tricksters around. Buster had an Egyptian graduate student who was doing his thesis on Greenland cod life cycles and he was always asking for chocolate pudding deserts. The gelatinous mud from Hawk Inlet looked very much like chocolate pudding, so we sprayed some Dream Whip on top of a bowl of this mud and left it in the fridge for his late night treat. With one bite he knew he had been tricked, and he immediately pointed at me.

As previously mentioned, the camp SSB radio was left on the 'game channel' most of the time, to get the local news from the Inuit in the region. There was a lot of discussion about the proposed uranium mine and they knew that there were some scientists at Saqvaqjuac that were trying to study this situation. Rankin Inlet radio personality Andre Tautu of the Keewatin Inuit Association flew into our camp to interview us as uranium experts. With his tape recorder on the kitchen table, next to a pile of pilot biscuits, coffee cups and the inevitable mosquito coil burning, Ray Hesslein and I tried to explain the radioactive decay of uranium to its many daughters, and how some of these natural nuclides were taken up into animals with their food and water. He got a bit excited when we said that all of these radioactive elements were naturally present in all of their food and water. At that time we had few data for the Saq region, and we could only summarise what was known about radionuclide bioconcentration in food animals from studies in the USA, Europe and Russia. We later heard that Tautu's Inuktutuk radio broadcast gave a very good summary of what we had said, and that he talked about the 'mother uranium and its evil daughters'. He used the word *Ayurngnarngmat* (meaning 'it can't be helped') several times, to indicate that they were already eating small quantities of this stuff, even before the mine development began. The problem with the uranium mine is that some of the plants and animals downstream from the mine would have higher concentrations of these radioactive nuclides in their tissues.

The Sacvaqjuac Symphony Orchestra was back in business, with Mike Capel on dobro guitar, Ray Hesslein on harmonica and clarinet and myself on fiddle. We were joined by Papaluk on her button accordion, after we had all re-tuned to her pitch. Papaluk played her version of 'Jingle Bell Jig' and 'You Are My Sunshine' as a Scottish strathspey. The camp stereo often played Stan Rogers ballads, Silly Wizard, De Dannan, Gordon Bok and Jean Redpath. Several years later, Papaluk's husband traded her button accordion for a team of sled dogs. *Ayurngnarngmat* ('It can't be helped'). It took me a year to find a similar model, but I bought a new button accordion for her and I could just imagine her bright smile when she opened the case.

While all of this science, gluttony and music was happening, the camp pet was wandering through the labs, bedrooms, and kitchen. This was Ichabod, a very young sandhill crane, whose mother had been shot and eaten by the Inuit. Camp staff fed this squawking gangly bird (about the size of a skinny laying hen) on chopped fish and crab meat and it followed me around a lot, sleeping beside my bed. In my absence on *Sea Cow* expeditions, camp staff, and especially Kathy Martin, decided that I should take Ichabod with me back to our Tourond farm, as they knew my wife Sheila was good at taking care of wild birds. The camp was to be closed for the winter when the lakes were covered with ice in a month or so and this bird was obviously too tame to survive on its own. Kathy Martin tearfully made a cardboard box with a special lidded hole for its long neck to stick out, and lined the box with seaweed and moss. Kathy prepared many small whirlpac sacks of frozen chopped fish for his trip to the south. Sandhill cranes in the wild fly south in the fall to warmer climates and wetlands in Manitoba and the Midwestern USA, but Ichabod would be flying economy class with me. I was a little concerned about the reaction of the airline stewardesses to a sandhill crane passenger, so we used the radio to contact the airline bosses in Churchill to get special mercy dispensation for this trip.

Our departure from Saqvaqjuac was an emotional affair, at least partly due to Ichabod going south. The barge *Sea Cow* was heavily loaded with a tonne of our equipment, frozen samples and komatiq boxes of Buster's samples for the Winnipeg labs. There were the normal hugs and kisses, waving and hollering, the camp horn blaring as the *Sea Cow* left the jetty. I pulled out my fiddle and played a tune as we drove down the estuary, Guy Amorak at the wheel, his deckhand Medard Naitok calmly taking snuff in his lower lip. The strong tide was coming in and the Volvo diesel engines were almost defeated at the race, a narrow passage leading out to Hudson Bay. Out in the rough waters of Hudson Bay the sea spray was cold, and Ichabod snuggled inside my rain jacket. The awkward landing barge *Sea Cow* had a hard time banging through the two metre swells.

We arrived in Chesterfield Inlet at 11 pm in pitch dark, and I'll never know how Guy Amarok could find and gently slide the *Sea Cow* into the little cove gravel beach beside the community centre. We were stiff with cold and wet with sea spray. It took us two hours to unload our gear and haul it in wheelbarrows up to the 'Blue House', where transient Saq people used to await aircraft transport. Our flight was the next morning, so we got out our sleeping bags for the night, amidst two empty cupboards, two chairs, one table and an oil stove. Ichabod unfortunately left a lot of partially digested fish on the floor and the next morning I had to scrub the floor with soap and water, under the disapproving scornful eyes of Guy and Medard. I imagined them muttering in their own language about how white fellows were soft in the head. The aircraft was delayed of course but we made the trip down the coast, through

Rankin Inlet, Eskimo Point, to Churchill. At each stop I managed to get some ice cubes to put into Ichabod's cardboard box, as he was getting too hot in the aircraft. And at each stop we would search for a nearby thaw pond so Ichabod could play in the cold water. The stewardess fussed over Ichabod, even if he was a forbidden passenger. I had the normal troubles with our frozen freight in Churchill, getting it into safe freezers, as it would come down to Winnipeg on separate flights. By this time my lap was getting rather ripe from Ichabod's defecatory efforts. Mike Capel was busy writing the ballad of Ichabod the Crane and I tried to find a melody line on the fiddle.

We arrived at the Winnipeg airport dead tired at one a.m. in early August, with my lap covered with Ichabod effluent. We hustled our gear into two pickup trucks to deliver the perishable goods to the institute freezers and labs and to get some fresh ice to cool down Ichabod, who was wilting. I drove south, a half hour out of the city, to deliver Ichabod to drowsy Sheila in our Tourond farm home. Sheila flopped down on the kitchen floor at the sight of Ichabod, saying 'Who are you? What do you eat?' and he promptly puked up a load of fish on the floor. Sheila put him in the bath tub so he could clean himself. We played with him till about 3 a.m. The next morning was a big shock, seeing the lush green fields, large yellow sunflowers, lilacs in bloom and chokecherries ready to be picked.

Over the next few years, this trip was repeated several times. One trip in the winter was for coring mud, to obtain estimates of removal rates of the uranium series elements. It was easier to guide the coring operations from a solid platform of ice, rather than a bobbing wind-twisted *Sea Cow* floating on the water. In the Winnipeg lab, Paul Wilkinson and Eva Slavicek had troubles dissolving the oily meat of the sea mammals and birds, and we had to learn how to grind bones and teeth to get the samples in an appropriate form for radiochemical measurements.

We found that uranium and radium were relatively mobile in the water but were not bioconcentrated by the biota. Most of the thorium, polonium and lead isotopes were firmly particle bound and were removed from water quickly, to be buried in the bottom mud. The inert radioactive gas radon was found in high concentrations in lake water under winter ice and in ice-covered soils. However, accumulation of naturally produced polonium in liver and kidneys of fish and some invertebrate excretory organs was very high, but these parts were rarely eaten. This information was about the natural uranium series elements being cycled through the freshwater and estuarine ecosystem at Saqvaqjuac. If large amounts of uranium and daughters were added to Baker Lake from mine wastes, it would require centuries for uranium and radium to be flushed to outflow to Hudson Bay. Thorium, polonium and lead isotopes would be removed to sediments rapidly, but could affect bottom animals living in the surface mud. We reported that natural and mine enhanced radon gas could be a health hazard in ice-covered water and in areas of melting permafrost.

Some of these results were considered confidential by the Department of Indian and Northern Affairs, who gave us the project money. Most of this work was never published, consequently, except for a few restricted government technical reports.

And so here we go again, on the CBC media assisted tours of the western Hudson Bay communities, giving workshops on our results of our study. And again, Buster and I tried to explain the uranium series elements and radioactive decay chains to Inuit people of western Hudson Bay...'Uranium is a bad mother with daughters that have other daughters, some like to live in the water, some like to live in the mud.' We used salt and pepper shakers and a glass of water to illustrate how some stuff (salt) dissolves in water and some stuff (pepper) does not, with particles remaining. Most of the older Inuit were more rational about our information than the southern white people, who got irrationally excited about pollution with radioactive elements. The older Inuit explained that there were some old taboos against eating certain organs of the sea mammals but few of them ever ate crabs.

And again, if I played some tunes on my fiddle in my hotel room, the older people would come to listen and tell stories and legends about the old days. The younger Inuit had decent high school level education and knew some chemistry and biology. They would coach us on ways to explain complicated science to their parents and grandparents and would translate some of this directly into their own language. Buster had some of the Inuit language, which impressed the audiences and greatly disturbed our CBC and DINA government minders, who didn't have a clue what he was saying. Although Saqvaqjuac scientists were *kabloonak*, we were more respected by the local Inuit because we had worked for several years in their lakes and coastal seas. We usually got a different reception from the Inuit compared to the distrust shown to most white people from the south, who visited only for a few weeks.

During all of this Hudson Bay work our home life was progressing in interesting and sad ways. Sheila's father died in April 1980 while having an evening nap. We flew down for the funeral and our two young sons were introduced to the somber meaning of death in the family. We spent Christmas with Sheila's mother and her six siblings at the Ontario family home, with sadness in the winter air. Sheila and I played all of her father's favourite songs ('Buffalo Gals', 'Redwing') and I cut and split a lot of firewood for the fireplace. And then Mount St. Helens blew up in Washington State, killing fifty-seven people and spreading ash over much of western USA and Canada.

My piano teacher grandmother England had a stroke in early June 1980 and was sadly confined to a bed in the Rapid City hospital. We drove down to South Dakota to see her and also to attend the 75th Anniversary Jubilee of my home town in June 1981. And here was another small prairie town celebration, with a rodeo in the mud, where a bull rider lost his thumb in the chute gate. There was a variety show where Sheila and I performed some hammered dulcimer and fiddle tunes and I mentioned to the audience that the fiddle came from my grandmother England's family. My mother was horrified that we sang a song about Copenhagen snuff but the men in the audience loved it. There were pit slow-cooked beef suppers for the whole town, main street parades, dances and lots of home-made ice cream with strawberries on top. My high school class of 1959 had a reunion in the Methodist Church basement.

Grandmother England died in the Rapid City hospital in early August 1981. We drove down for the funeral, our small sons now paying more attention to the extent of our wide spread extended family. Our nine year old son Ian had played the piano for his great-grandmother on previous trips and Jesse remembered her oven-warm banana bread, so there was a real connection. And here it was, the prairie people gathered for the passing of a very important pillar of the community, bringing plates of food, pies and cakes, tubs of thick cream, flowers, old photographs, and we had house visitors for three days. Her Butterfield relatives came from Minneapolis, George and his florist son Ned, with a giant pot of flowers with Chopin sheet music and 'Home on the Range' music embedded. Many local and distant folks arrived to sit around the table with coffee and cake, telling stories about my grandparents in the dirty thirties and during the war, helping other people, feeding hoboes. Shorty, one of grandfather England's 1920-30s road-building mule skinners showed up, big felt hat and rodeo belt buckle, he could hardly speak from sadness but he told stories of my grandmother pulling a dozen apple pies out of the oven during the road camp days. My mother gave Sheila a sack full of Grandma England's sewing and darning tools, an old mixing bowl and lots of beeswax for smoothing sewing thread. She gave to me my grandmother's old sheet music cabinet, a crudely built set of shelves made from old packing crates. One of the England family (Bruce van Bockern) gave Ian MacGreggor a catcher's mitt and a baseball. Ian and Jesse got many rides in the pastures on Old Joe, one of many offspring of Sugar, who was the best horse for kids. By now all of my three brothers were married and had children, so there were lots of kids in the house.

Sheila had ordered a new hammered dulcimer from Harvey Prinz in Kansas and it was delivered to the big house west of Murdo. Ian and Jesse found a brightly coloured western milk snake and they were determined to bring him home with us for an aquarium pet. We drove down to the local Don Hight buffalo ranch to get some very thick sun-bleached rib bones from behind his barn. This was for Sheila's brother Alistair, who played the bones as a percussion instrument in folk music groups. Our mushroom van was stuffed with gifts as we got ready to go back to Canada. There were the usual teary goodbyes from the women and shuffling of boots in the dirt from the menfolk. Both Sheila and I had wet eyes as we drove northward through the rolling prairie, where combines were working on the wheat fields, haying crews were moving bales into their barns, ducks and cormorants were busy on the wetlands and stockdams. And as usual, Sheila was terrified of the customs agents at the US/Canadian border, as we had concealed in the van a new hammered dulcimer, a pet snake, a magpie and a turtle named Iversen.

Life on our little Tourond hobby farm went on as usual. Our neighbouring farmers were harvesting their wheat and sugar beets near our home. We had our normal complement of laying hens, ducks and geese, a few big turkeys, six pigs and our ever-expanding vegetable garden. We were selling excess eggs at my institute, 'Hot from the Hen' brand, for $1/dozen. Our dressed and frozen poultry cost us 60c/lb to produce, the pig meat worked out to 50c/lb, mainly due to free grain sweepings. We traded dressed pigs to Buster Welch for one of his home-made butternut sofas and a loom bench. The pig heads and skins were saved for winter treats for the dogs. A large pile of firewood was cut and split for our wood stove. By now Jesse had his own puppy, a 28 lb Newfoundlander named Tifty, who rapidly grew to 120 lbs in less than a year. Ian was performing on piano at local music festivals, playing extremely well. Our folk music band was actually being paid for gigs now, with the addition of our philosopher friend Hans Hansen. Our luthier/musician friend John Weier made a viola for me, which added to the rich texture of some of Sheila's Celtic ballads. We took the train to Regina and Thunder Bay for folk festival gigs, with two sons and fourteen pieces of baggage. Our fellow musicians were amused to see Sheila using an ordinary ironing board to support her hammered dulcimer on stage. We were asked to play traditional dance tunes for the Royal Scottish Country Dancing Society Annual Ball in a Fort Garry hotel ballroom, where it felt very much like a Hudson's Bay Company affair in the early 1800s. The Gentlemen and Ladies were dressed in old-time costumes, full gowns, kilts, high white stockings and gillies. We also played for the English Country Dancers, with our friend Al Thomson on Northumbrian bagpipes. And as usual, most of Sheila's family gathered to help operate the Winnipeg Folk Festival at Bird's Hill Park.

Being one of those rare eggheads in our small farming community, I was invited to be a judge at the Niverville Collegiate science fair, where Manitoba farm boys and girls tried to win the prize while their parents glowered at the judges. Ian MacGreggor was in fifth grade in the fall of 1981, and little Jess was in third grade, bright and shiny boys, curious and obstreperous. Sheila and I were now involved in PTA, band boosters and the community hockey club. Ian was a goalie and got sore legs from doing the splits to stop the puck. Jesse was taking fiddle lessons from classical violin teacher Emanuel Horsch in Winnipeg and I tried to learn a little from the sidelines. Ichabod crane had become a family pet of course, but he had some growth problems in the late fall and died in our barn as the sandhill cranes were flying south. Old Jenny, our Gordon Setter bitch that gave Sheila several batches of puppies, died in a blizzard in early January '82. She was survived by her offspring Nancy Whiskey and Banjo.

Meanwhile, other government scientific duties continued with relentless and bizarre irony. Because of my work on the Mackenzie River I was invited to a series of meetings to consider a hydroelectric dam on the Liard River to provide power to British Columbia industry. We spent many months making models of the various impacts of changes in water flow on Nahanni Mountain ecosystems. The dam engineers gradually lost their enthusiasm for the project after I had calculated the life of the dam being rather short, based upon the huge erosion rate and river sediment trapping in the dam. I estimated that the dam would fill with river mud in about twenty or thirty years. Buster and I were involved in a series of workshops in 1982 to create a large study on Hudson Bay and Arctic seas, but most of this effort was wasted as none of our ambitious proposals were funded. In September 1982 I was invited to an oceanography meeting in Fairbanks, Alaska, for a special session on the Beaufort Sea and the influence of the Mackenzie River inputs. At that time, my group probably had the best data for physical and chemical erosion rates by rivers and delivery of these erosion products to the Arctic Ocean.

As if I didn't have enough to do, my Winnipeg bosses delegated me to participate in writing a cross-Canada review of radioactive nuclides in the atmosphere, water, soils and continental seas. My research group had been accumulating scraps of radiochemical information in central and Arctic Canada and this was to be added to the information from the experts at the Atomic Energy of Canada (AECL) laboratories in Pinawa and Chalk River, Ontario, as well as some involvement of experts from the UK and the Oak Ridge National Laboratory in Tennessee. I spent way too much time (1980-82) in boardrooms at the AECL labs, reading old unpublished manuscript accounts of post-war time research on nuclear reactors to produce radioactive nuclides for medical and nuclear research projects. But I also found that peaceful Canada had worked hard to provide 250 kg of plutonium for the USA nuclear arsenal from 1955-76: the bomb dropped on Nagasaki used about 6.4 kg of plutonium. I also discovered that in 1952 a power surge caused a loss of coolant for the Chalk River reactor rods, resulting in a melt-down and hydrogen explosions. Some 4500 tonnes of radioactive water was dumped into ditches around the plant on the banks of the Ottawa River, a release of some 370 TBq (370×10^{12} disintegrations per second). The USA sent a Navy team to help with the clean-up, and one of them was named Jimmy Carter, later to become President of the USA. In 1956 uranium fuel rods caught fire, melted, and contaminated several buildings: nuclear scientists and maintenance men in protective suits carried buckets of sand into the burning building to extinguish the nuclear fire.

None of my input to this report was nearly so exciting. I contributed soil and water budgets for natural uranium series elements and global thermonuclear bomb test fallout (mainly ^{137}Cs) in the watershed and sediments of Lake Winnipeg and in some lakes at Saqvaqjuac on the West coast of Hudson Bay. And of course, being a government funded study, there were big banquets after our meetings and I got to know some interesting radioecologists from Oak Ridge, Lowestoff (UK), and the nuclear power consultant army. Sadly, before our study was published, our group leader Ivan Ophel died suddenly.

In February 1982 David Schindler and I escaped the cold winter of Winnipeg and went to San Antonio, Texas, for a large American Geophysical Union meeting. And here we met again our scientific colleagues from around the world and the Lone Star beer, Hunan Chinese food, hot chili beef, jalapeño jelly and hot pralines were wonderful. Warm and sunny San Antonio, the Alamo, Tex-Mex music and palm trees were a big change from the Canadian Arctic. I met with my Augustana College trumpet teacher Jim Ode at Trinity College there and he took me home to his wife Edna, for a good South Dakota supper of well-cooked roast beef and roast potatoes. These AGU meetings were huge affairs, with perhaps five thousand scientists involved in oceanography, limnology, climate and geochemical research. My hotel was beside the Trinity River and I spotted a long barbed object on a sandbar, which looked like the snout of a sawfish. I waded out into the water to retrieve it, hauling the prickly bugger into my hotel room, washing it in the bathtub, planning to take it home as a trophy to hang on our wall. I showed it to a native Texan who laughed, telling me it was only a palm frond and that he would not expect a Canadian to know what it was.

While at the AGU meeting, a phone call from Sheila came to my hotel room late at night, to let me know that the Dennett family Loretto house had burned down on 17 Feb '82, her mother and brother Bill were ok but homeless. Most of the family mathoms, books and clothing were lost. Sheila flew to Toronto to rescue her mother and bring her back to live with us. Sheila's mother was shy and reserved at entering her daughter's house and she didn't want to be referred to as 'grandma'. But she rapidly got involved in the spring garden planting, fall butchering and sausage making and educating our sons in a wide variety of exotic subjects. She coached our son Ian MacGreggor for a speech on Edmonton Oiler hockey deity Wayne Gretzky for a 4H event and Ian won 1st Prize. In April 1982 Sheila's older sister Mary arrived in Winnipeg to take a job as a City Librarian, specialising in children's literature. She was to have a large influence on our two sons, who were saturated with good library books for the following years. Now Sheila had a brother and two sisters living in Winnipeg, which meant many grand family feasts around our kitchen table. Sheila's brother Alistair introduced our sons to rocketry and explosives. Sheila's sister Heather's young son Gordon became infected with a desire to play the fiddle, from watching music sessions in our house. Several extended family canoe trips explored the ELA lakes and we made fresh-picked blueberry pancakes in the rain at one of our camp sites. That spring, cruel blizzards blew the snow off the ploughed wheat fields around our home and black snow drifts were piled high around our barn. Sheila and I continued to be enthusiastic supporters of the Winnipeg Folk Festival and I was the money bagman for a casino fund-raiser in the city. At the folk festival Ken Hamm sang a song about 'The Daughters of Radon', which was partly inspired by our work on the uranium series elements in coastal Hudson Bay.

I was to attend a big science meeting in Lyon, France in 1983, so Sheila and her sister Mary embarked on rapid learning session to try to get me to speak passable French. This was gradually dismissed as being nearly impossible, with the minor exceptions of ordering food, beer and the location of the toilets. My version of 'lippity day jenay' brought raucous laughter and tears to their eyes. But I soon discovered that Lyon was a nice place, in spite of being French, as it was an agricultural region rich in good food. I met all my old buddies from around the world and Jack Vallentyne, Giuliano Bonomi (Italy), Han Golterman, Sikko Parma and Jaap Kalff (Dutch) and Heinz Loffler (Austrian) managed to get me to drink some wine. I was amused to listen to wild stories about the great party feast that these Europeans had attended at our Tourond hobby farm at the 1974 SIL meeting in Winnipeg...this outdoor smoke-grilled caribou feast had become a legend in their minds. We toured a special chicken farm near the town of Brest, where chickens were allowed to go outdoors and eat grass and bugs. All the Europeans were greatly impressed by this but I was a bit puzzled as to why this should be unusual.

On the way home I took the train from London up to the Lake District to visit old friends and take photographs of some key locations that occurred in an Archie Fisher ballad 'The Witch of the Westmereland', which I was trying to sing. And then on to Edinburgh for the Fringe Festival, where the whole city was running berserk with music, drama and art, and I got to play some fiddle tunes with a band of very good minstrels on Princess Street.

Later that year I was in New Orleans for a big AGU meeting in the Superdome. This was one of the largest gatherings of geophysical scientists in the world and it was exhausting to give my paper (on Saqvaqjuac bioconcentration of uranium series nuclides), attend the science sessions all day and party most of the night in the French Quarter. And there it all was, the magical home of Louis Armstrong, the birthplace of jazz, Preservation Hall, the wonderful food at K. Paul Prudhomme's Louisiana Kitchen, where they served me Cajun blackened catfish and a wonderful creamed spicy seafood stuffed eggplant baked in batter. You could walk down Basin Street, Canal Street and Beagle Street and hear old-time blues, bebop and

Dixieland jazz, Cajun fiddles playing Cajun creole French ballads, big swing bands, rhythm and blues singers, electric blues with electric blue fiddles in the shape of Louisiana, amidst the pan handlers, strip joints and drug peddlers. I had chicory coffee and beignets with a castor sugar dusting down at the Café du Monde, right beside the Jacks Brewery.

Over the years of my attendance at New Orleans AGU meetings I became addicted to this city. I visited several parishes during Mardi Gras, all of them different in cultural and racial background, good-looking bare-breasted girls throwing necklaces of beads and candies from the parade floats, parish and school marching bands, Creole and Indian costumes, the Coochie Man, the Flag Man. Sheila and I had recordings of Ramblin' Jack Elliott singing a talking blues song about a party at 910 Toulouse Street, so of course I had to spend some time trying to find this place. And as Ramblin' Jack says in the song, there was no public entrance to that address as it was an enclosed courtyard. But I could easily imagine musicians playing all night, living in New Orleans, never seeing the light of day. I watched a group of street Dixieland and bluegrass musicians doing rowdy goodbye songs for friends boarding a train (in a grubby dark train station): 'This Train is Bound for Glory...', and 'Freight Train, Freight Train, Goin' so Fast'. There were tears rolling down my cheeks and I was shivering with goose bumps. Flying out of New Orleans after a long week of heavy science, bleary-eyed and sleepy, I watched Lake Pontchartrain vanish under the clouds and the lyrics of the ballad 'Do You Know What it Means To Miss New Orleans' floated through my head.

By 1984-86 our sons were teenagers. Ian was wearing Black Sabbath and Iron Maiden T-shirts and started a rock band that rehearsed in our garage. Our flock of Pekin ducks would waddle out to watch and hear them. Jesse was more interested in motorcycles and hunting. We kept taking them on canoe trips with their Scout groups, and down to special Camporees in the Black Hills of South Dakota. Our visits to my parents' ranch near Murdo allowed our boys to learn about feeding livestock, riding horses, rabbit and deer hunting and an increasingly large group of cousins, the sons and daughters of my brothers. My mother taught Jesse how to make cinnamon buns and Ian MacGreggor got to drive the four wheel drive Bronco in the cattle pasture. Sheila found a large part of an ichthyosaur fossil in the White River bluffs and we spent two days digging up gizzard stones, fin digits and backbone vertebrae. Ian and Jesse were prize winners at the school science fairs, where Ian had a project on the effect of acidification on aquatic amphipods and Jesse excelled with a project on the effect of good and bad food on rat reproduction, as well as showing how an oscilloscope revealed the complex overtones of fiddles and trumpets. Sheila and the boys would come with me to ELA for radiochemical sampling of the lakes, and they liked to ride on the four wheel bush bikes and motorboats. Ian was lucky to be selected to attend a summer music camp at the International Peace Garden in Rugby, North Dakota. Here the circle came around again, as I had been at the very first Peace Garden Music Camp in 1956. Our present for our sons at Christmas 1985 was a computer, which of course meant that all sorts of computer games were demanded. Their uncle Alistair was living in Winnipeg, taking a course in electronics and computers, and he became the computer guru for our family. But the Mennonite game crokinole was still popular at family gatherings.

Chapter 20. Chernobyl Fallout and Viennese Opera

As the spring of 1986 came slowly, melting four metre high snowdrifts in our yard and windbreak trees, Sheila and I got the barn 'cold frame' ready for our garden seedlings. This was a shed on the entrance to the barn where we had removed the roof cedar shingles and replaced them with old glass windows from our house. Most of the garden vegetables were sown in peat pots, germinated in our kitchen and put into the barn 'greenhouse' when the snow was gone. Some nights we had to put an electric heater into the greenhouse to keep it above freezing. Parent-Teacher meetings, Winnipeg Folk Festival Board meetings, Meadowlark rehearsals and performances and lots of house parties.

At the institute Paul Wilkinson and I began a program of collecting snow and rainfall for radiochemical measurements of natural and bomb isotope fallout. Willy Burton of the workshop constructed a beautifully smooth and clean 2 m diameter steel tank on legs, with a sloping funnel bottom outlet where we could collect hundreds of litres of melted snow and/or rainfall. These precipitation samples were chemically processed and concentrated down to a few tens of millilitres before extraction of the nuclides. We were trying to determine the annual cycle of fallout of natural ^{210}Pb, ^{210}Po and bomb fallout ^{137}Cs, so we could better use these particle-reactive nuclides as tracers of time and sediment mixing in our cores of lake sediments. Similar tanks were put into operation at ELA and Saqvaqjuac and supplemented with data from many soil profiles in those locations.

We had also installed large air filtration stations at ELA, initially to measure organic contaminants (DDT, dieldrin, toxophene, etc), but we also took some of these air filters to measure our ecological tracer nuclides, to help with understanding wet and dry fallout events. These air filtration stations pumped many cubic metres of air through fine pore filters for weeks to get enough sample to detect anything. All of this was pretty boring humdrum work, as we expected to measure these nuclides for several years before we could trust the data.

But exciting trouble seemed to find us. During the last week of April and early May 1986 our ELA rain samples and air filters suddenly appeared to be contaminated by relatively high activities of dozens of short-lived nuclides. At the same time our TV news was showing a developing disaster in the Ukraine and western Europe, as a large nuclear power plant at Chernobyl in the Ukraine had exploded. At 1:23 a.m. on 26 April 1986 reactor number four exploded at the Chernobyl plant in Ukraine, then part of the Soviet Union, spreading radioactive fallout across Europe. It seems Deputy Chief Engineer Anatoly Stepanovich Dyatlov wanted to do a risky safety test, to determine what would happen during an electrical power failure. Sure enough, they could not control the steam heat, the graphite tips of the control rods melted and contributed to production of explosive gases and the reactor exploded, blowing the 1000 tonne concrete roof high into the sky. The technicians in the control room all died within a week of radiation exposure but Deputy Chief Engineer Dyatlov had already left the site. He died of a heart attack in 1995. It wasn't until three days later, when the fallout was detected in Sweden, that the Soviet authorities would admit there had been an accident. The irony was that the reactor operators were involved in a safety test (in the event of an electrical power failure) when everything went wrong.

There was mass confusion for a few weeks but it eventually appeared that the explosion had spewed radioactive fallout over much of Europe, the Mediterranean countries and Scandinavia. Some of the reactor dust cloud must have gone into the higher atmosphere and was transported by winds around the temperate zone of the northern hemisphere. Within a month fresh vegetables and other foodstuffs from Chinese and Mediterranean ships in Vancouver and Toronto harbour were blocked, as their produce had measureable activities of reactor nuclides. Within a month Chernobyl nuclide fallout was detected across Canada and USA and also in the Canadian Arctic at Saqvaqjuac. Our data was amongst the best in Canada on this problem and our lab quickly became an important part of a North American network of reporting stations.

Because of our work on radioactive nuclides expected from the NWT fallout path of the nuclear powered Russian satellite *Cosmos 954* and preparations for Skylab impact, we had some of the literature that allowed us to calculate what nuclides we should expect from the Chernobyl nuclear reactor explosion. But we didn't have the appropriate calibration standards for the dozens of short-lived reactor fission products so we had to rapidly import small samples of these nuclides from Canadian and USA Atomic Energy labs. Many of these nuclides were activation products from irradiation of the Chernobyl power plant metals and thus we were detecting gamma-ray emitting radioactive isotopes of iron, nickel, cobalt, caesium, zinc, scandium, beryllium and dozens of other elements that normally would not occur in earthly nature in radioactive form.

Our measurements indicated that there was negligible ecological or human harm to be expected from this dilute cloud of reactor products in Canada, but it was enough to scare the dairy and vegetable crop industry, as radiostrontium accumulates in milk and cheese products. In Europe, Scandinavia and the Mediterranean countries, much of their agriculture was shut down, as the fallout was high enough to be dangerous for human and livestock health. The Chernobyl explosion killed more than fifty people in the vicinity of the nuclear power plant and over the next decades, perhaps four thousand people developed cancer that was attributed to the Chernobyl explosion. The reactor was covered in concrete and a large area of surrounding land was fenced to keep people away, because of high levels of radioactive contamination of the soil.

Speaking quietly now, many of my Canadian and USA scientific colleagues were not horrified that we now had a low level fresh sprinkling of ecological tracers of variable half-lives onto our Canadian soil, rivers and lakes. This was equivalent to the common hospital procedure of dosing your body with radioactive barium, to determine where it accumulates, if your organs are working properly, and how fast it is eliminated from your body. Only in this case we have dozens of differently behaving nuclides being quickly added to our soil, lakes, rivers and oceans. Some were soluble in water and some quickly stuck to particles and were not mobile in water. This was a gigantic whole northern hemispheric spike experiment, with ecosystems being labelled with dozens of nuclides with half-lives from several days to several years. For example, we already had one isotope of radioactive caesium (^{137}Cs, 30 year half-life) in our soil, water and lake mud, from nuclear bomb tests in 1950-70. Now we have another two year half-life ^{134}Cs isotope as a tracer for physical, chemical and biological processes. Cs behaves similar to potassium (K) in chemical and biological processes and can also be used as a marker of time of deposition of lake mud.

Out at our super clean bush camp ELA, David Schindler, Ray Hesslein and others had been doing research on industrial acid rain and industrial metal pollution for several years. They did this by adding carefully measured amounts of reagent grade acids of various kinds and trace amounts of radioactive metals to some lakes, and compared what happened to these metals in control and experimentally spiked and acidified lakes. They used tiny amounts of radioactive metals in these experiments, to avoid outright pollution of the lakes with high (toxic) loads of stable metal elements. This was possible because high quality gamma radiation detectors were now commercially available, being linked to computers to sort out the large amount of complex data from these liquid nitrogen-cooled germanium detectors. These new detectors could measure the activity of gamma-emitting nuclides at activities far below drinking water hazard levels. These experiments were designed to quantitatively test ideas about the impact of sulphur dioxide, nitrous oxide and metal emissions from mining, metal refineries (such as Sudbury and most of NE Canada and USA) and coal burning power stations. These industrial emissions were converted to sulphuric and nitric acid in rainfall and the increased acidity made the atmospheric metals more soluble and mobile in the environment.

Although the ELA lakes were far from any major polluting industry, metal pollution was world-wide, as some of our sediment core analyses from ELA, Saq and other Arctic sites had shown. Other international research groups were showing that many industrial metals were 2-10X enhanced in alpine glaciers, peat bogs, remote high mountain lakes and Arctic and Antarctic ice cores, in some cases beginning during the Bronze and Iron Age cultures of pre-history. And so now we had a new batch of Chernobyl isotopic tracers being added daily to our experimental and control lakes.

Like many other research groups in the northern hemisphere the ELA research team quickly scrambled to monitor the pathways of these Chernobyl nuclides through the forest soil and lake food chain, which ones remained dissolved in the water and how fast they stuck to particles and accumulated in the bottom mud. The Chernobyl atmospheric fallout continued for about a year and then diminished to undetectable levels. Most of these nuclides were rapidly bound to soil particles or accumulated in the bottom mud of lakes. With a few exceptions, Chernobyl nuclide activities were below the natural level of soil and lake mud radioactivity, such as natural radioactive potassium and the uranium series elements. We could however detect some of these Chernobyl nuclides for the next decade in soil and lake mud. We were repeatedly called to Ottawa to report to Canadian government agencies, who were usually puzzled at our enthusiasm for measuring tiny activities of these Chernobyl nuclides at levels way below any human health concern. Once they learned that our dose of Chernobyl nuclide fallout was not going to kill the fish or make fish unsafe for human consumption, they were not interested in extra funding for this work. Government officials were however very concerned about how this information was to be presented to the public, and there was more than one gag order issued to restrict unauthorised leaks to the media.

For example, I was going to report on Chernobyl nuclide fallout at ELA and Saq for a big AGU Ocean Sciences meeting in San Francisco in early December 1986. This was kiboshed from on high and I retreated to a presentation on our LRTAP work at ELA-Saq.

So here we go again, leaving Winnipeg in cold snow on 8 Dec and arriving in the green grass of San Francisco where winos were asleep in alleys (some of them in US military uniforms), Jamaican and Puerto Rican hookers of both sexes on the street corners, Vietnamese push-cart vendors calling out their goods. But also here at my hotel can be found lots of scruffy bearded scientists and grad students with briefcases labelled as being from NASA, WHOI, Scripps, LDGO, French, German and Australian institutes. A group of unruly grad students and I attended the smoker on the sixth floor with plastic sacks in our brief cases. The plastic sacks were for taking large quantities of melon slices, cheeses and finger foods from the food tables, to take back to our hotel rooms. Good science papers were presented all day and by night we explored the harbourside restaurants (fresh squid, octopus, abalone, conch, crab, oysters, Anchor Steam beer, mud pie and espresso), the Ghirardelli Chocolate Factory, mariachi bands, Dixieland jazz, a gay men's chorus singing Xmas carols, drug peddlers and gospel preachers, Salvation Army brass bands, porno video shops, skateboard daredevils and Chinatown.

One night we attended a party in honour of Wally Broecker, who had just survived an operation to remove a cancerous part of his jaw bone. Hans Suess, Minzie Stuiver, Han Oeasager, Phil Abelson and Tsung Hung Peng were there, all major players in climate change research, and Wally was in full oratory about his latest passion for a Geosecs of the Atmosphere. Several sessions were held on the difficult problem of the coupling between atmospheric, terrestrial and freshwater and oceanic cycles of the natural and contaminant elements and how do we discriminate a level at which things are no longer healthy for ecosystems and humans... is there a 'tipping point' where our present ecosystems all fall apart? Examples were provided from India (Bhopal), Russia (Chernobyl), a Swiss factory fire, tsunamis and earthquakes, river floods, Amazon deforestation, greenhouse gas warming of the planet, submarine vent fluxes of methane and the threat of nuclear winter.

Packing up to go home, with several unsuccessful attempts at not eating a box of Ghirardelli chocolates for my dear wife, I go to the airport at six am on 14 Dec, where the place is crowded with International Rodeo Association members in boots and big hats and shiny belt buckles, drawling and hooting, the National Rodeo Finals having finished in Las Vegas. Over the Sierra Nevadas to Calgary, where my suitcase is lost, and on to Winnipeg where it is white and cold but Sheila is there, smiling and warm.

Sheila and I and Scout Leader Gerald Toews led two patrols of local Boy Scouts up into the empty Precambrian Shield lake country east of Lake Winnipeg in January 1987. Gerald Toews had a cabin on Lake Wanipigow and we quickly augmented the wood pile with standing dead timber pulled to the cabin with skidoos, split and covered with tarps. Everybody took turns cooking for fourteen people on the big old wood stove and the food was good. There was a lot of fresh fish, sausages and bacon, canned corn and bannock with various kinds of dried fruit cooked in it. Not everybody was interested in eating fresh pike caviar, which was surprisingly good. There was a crokinole board stashed in the attic which was used for tournaments during a brief blizzard and after supper. We had a large collection of snowshoes, cross country skis and freight sleds for the skidoos, so the young men could explore the abandoned mine shafts and mining camps, and some wanted to learn how to track moose. We also found some tracks of the woodland caribou but never saw them.

In February 1987 David Schindler and I attended a big freshwater science meeting in Hamilton, New Zealand. David gave a presentation on the ELA metal spike experiments and I told the story of the fate of the Chernobyl nuclides in ELA soil and lakes. And here was that international gathering of colleagues, JRV, Han Golterman and wife Nell, Giuliano Bonomi (The Tiger of Pallanza), Dieter Imboden and Joerg Bloesch from Switzerland, David Culver, Rob Peters, Charles Goldman and some curious NZ military blokes with swagger sticks under their arms.

I had never been to the southern hemisphere before and I met a lot of rowdy New Zealanders and Australians who had a similar disdain for the formal plenary sessions and fancy banquets. Also they make good breakfasts here, eggs, bacon and sausage. I joined a group for a white-water rafting trip down a sequence of rapids in the Rangitaiki River. David, Gabe Tessier and I hired a car so we could explore this little island, and we tried to drive on the wrong side of the road up to the Coromandel Peninsula. The car broke down and David marched off angrily down the road to find a farm house with a telephone. I stayed with the car on the road margin and was playing my fiddle when a small boy came up through the road ditch with wonder in his eyes. He listened quietly and explained that his grandfather used to play the fiddle but he had died last year. He wanted to know how hard it was to play the fiddle. I had him playing 'Bye Bye Black Sheep, Have You Any Wool' in about fifteen minutes and there were tears rolling down his cheeks. He said that his grandfather's fiddle was at home but his parents would not let him touch it. He said now that he had a lesson maybe they would let him play the fiddle, and he marched off through the road ditch weeds.

The car rental agency vehicle appeared with a tow truck and another vehicle for us and we had a good time on the beaches of the Coromandel Peninsula. But the highlight of that trip was the fiddle lesson in the road ditch.

Later that summer, back home in Canada, an Australian named Bernie Chapman wandered into our institute asking for a visit to ELA. The Australians had heard David and I speak at the NZ conference and wanted to build a similar freshwater experimental facility in their country. I got the duty of taking him to ELA. The last two hours of the drive to our ELA camp was on a rough logging truck road, sometimes driving through flowing creeks and around fallen boulders and logs. I think Bernie might have been a city person, as he was apprehensive about the casual way I was driving on this rough road. But he was positively alarmed when I rolled a cigarette, put a harmonica rack on my shoulders and began to perform the ballad of 'The Man

From Snowy River', while roaring down this narrow road. But he was able to prompt me with the words when I couldn't remember the next line. Many years later I was to hear the story of this drive to ELA from Australians who had heard the story from Bernie Chapman.

Several of our Freshwater Institute gang were asked to attend a major government environmental planning conference in early October 1987 at the Canadian Centre for Inland Waters in Burlington, Ontario. In my opinion our CEO picked the wrong people for this meeting, as we were not very polite or diplomatic or connected to government logic. Our contributions were much too scientific and long-term for the politicians and there was a deafening silence after our input to the discussion.

Sheila came with me, as most of her family were living in the Toronto area. Sheila's mother was unhappily at the Blue Mountain Manor, a hospice that had none of the pleasures of the country. We borrowed a guitar, banjo and fiddle and played some old family tunes for her and she said she wished she still had a piano to join with us. We visited Sheila's brothers Ian, Alistair and Bill, during which some good feasts were endured late into the night. Ian drove us up to the north shore of Georgian Bay, Thornbury, where Sheila's sister Noddy and husband Ted lived, where there were nearby warehouses full of apples that smelled wonderful. Noddy was a social worker at a home for disturbed and disabled people, where Sheila and I performed some of our music. These people responded to us, coming close to watch Sheila's fingers on the guitar, trying to clap but out of rhythm, some had tears rolling down their cheeks. This was a difficult place to perform happy music and the institutional disinfectant smell disturbed me.

My colleague David Schindler had been invited to present a plenary address for an ecological society meeting in Siena, Italy. For some reason I can't remember, he decided he did not want to go and asked me to do this. So here we go again, Winnipeg, Toronto, Montreal, Amsterdam, Rome and at 0500 on 20 October I was starving for a good breakfast, having seen the Alps with creamy snow. The Rome airport had extra military security with machine guns, sniffer dogs and the usual chaotic madness of nationalities and languages. And then there was the usual bus and train ride through back streets, bums and pukers in slums, grand rich homes, black leather jacket punks with many earrings and bikes, horse drawn carriages, farmers in their fields with tractors discing, ancient brick walls, medieval towns, red tiled roofs, loud distorted ghetto blasters. I had a scary confusion to find train station to Chiusi. Dudes on the train had t-shirts 'Born in USA' and 'NY Dudes' with ghetto blasters, teenage girls in tight tank tops and jeans. The conductor of the train enjoyed watching me roll a cigarette and called me an 'American cowboy'. I arrived in Siena at 1600 on 20 Oct, the bus to Hotel Athena was a great adventure...the bus driver had to consult the police to find it...ah yes it is on the university grounds.

The Tuscan city of Siena has at its core a medieval city within high stone walls, many museums and galleries of ancient and modern art, tapestries and banners, endless diptychs and triptychs of Madonnas, Etruscan and Tuscan dukes, pagan skull/demons/snakes/ghouls/wolves/vulture symbols at the six floors of the Pinacoteca.

The city square is famous for 'Palio' horse races in the medieval square and the ladies wear scarf-banners for their riders. You can buy 14th century copy silk banners in the curio shops and Giuliano Bonomi bought one for Sheila. The university in Sienna was founded in 1240. The meetings were held in an ancient university hall, I was thankful that Rob Peters was here on sabbatical and he rescued me in a sea of Italian and French and German language presentations...Rob was married to a Pallanza lady and has his kids in Italian schools. We suffered through three hour lunches of guinea fowl and pasta and cheeses and gallons of Chianti wine. Some of the lunch and supper food lists in my journal are hard to believe: in one sitting we ate prosciutto and melon with fresh hot bread, scampi in rice, steak in butter sauce, cappuccino with fruits, melon, mango, strawberry, pineapple, cognac and cream. Rob explained to the waiter that I was a cowboy, then he beamed and the next day brought me a lumpy chunk of wonderful rare beef, seared on the outside, rare inside...speciality of Sienna he says...also soufflés, lark breasts, melons and golden tomatoes. Well, in spite of all this, it was difficult to get the Italians to make me a proper breakfast of bacon and eggs.

My talk on the ^{210}Pb mass balance for ELA and Saq lakes went over like a lead balloon... mostly biological ecologists here, not interested in this, they wanted to hear DWS talk about ELA. Many papers about ecological models of diversity, botanical and terrestrial, almost nothing about lakes. Very boring science.

My roommate was Daniel Simberloff, a Harvard-trained Florida State University ecologist who is a bluegrass/old time music purist, plays mandolin. We played tunes, me with my fiddle, sitting on our little veranda, attracting a small crowd of students, some of them dancing. He gave a very good lecture on the evolution of insect biodiversity in disturbed habitats. Cannelloni with calf brains deep fried in marrow batter with Giuliano was delicious. I explored the Rossi Academy for Ancient Music, attended several recitals of Renaissance motet singers and got lost in the rain, wandering the wet streets trying to find my hotel.

I left Sienna at 1500 on 25 October, the train neat and clean this time, good food, Chiusi, then Rome chaos at 1800, taxi to posh hotel but restaurant closed, starving...seek open restaurant amidst hookers and pan handlers. Italian TV full of airport scenes...waitress explains that there was an airport worker strike. Overnight, up at 0530 for taxi to Fiumicino airport where too many 'ugly Americans' are found, cursing and speaking loud...vast cavern open to outdoors, roof struts full of pigeons, droppings on the benches, flies, baby puke, barking dogs, all benches full, people sleeping on the floor, a variety of military guards and police walking through the crowd, fully armed. I'm waiting for Al Italia (later KLM) with twenty-six members of the Northern Italian Chamber Orchestra personnel and we took our time playing fiddle tunes amidst all this chaos. The next bit of history was told many times at supper parties.

The Rome airport story

I was invited to Siena in Italy to give a paper on Canadian ELA radionuclide mass balances for a European Ecology Congress in the mid-1980s. It was a wonderful culinary experience, but the science meeting was terrible. I enjoyed the food and visiting the ancient museums and art galleries, some of them featuring art work from medieval times. The old city has horse races inside the central square on fair days, and banners were flying for the different local country districts that sponsored the horses. Chianti wine comes from this area. Rob Peters of the University of Toronto was also there and he could speak Italian well, a good fellow wanderer. He helped me select some wine to bring home to Sheila. On the way back home I travelled by train from Siena to Rome, and by taxi to the airport. Upon arriving at the airport I was informed that Al Italia was on strike, and baggage handlers and control tower employees were not working. I was told to wait for a while, as German and Swiss airlines were planning to visit later in the day and alternate flight arrangements could be made.

The airport waiting room was open to the air, it was sticky hot, pigeons were roosting on the rafters, Italian ladies were breast-feeding babies, kids were playing games, old men were playing card games. I was trying to read on a bench, with my luggage close by, when a group of young people came up to speak in hesitant English, asking if I played the violin. I had my fiddle with me and I explained the difference between violin players and fiddlers. It turns out they were a string quartet from the University of Padua and they wanted to hear US fiddle tunes. So I played them a few rounds of 'Arkansas Traveller', 'Turkey In the Straw', 'Whiskey before Breakfast' etc. and they got their instruments out to play along. We were having a good old time waiting for the airplanes to fly.

But then two large Italian Airport Police in top coats with two German Shepherd dogs came to stand beside me. The music stopped abruptly. They spoke in Italian and motioned to me to come with them, with my luggage. They took me into an empty room, where the policemen took off their top coats to reveal machine guns strapped to their shoulders and pistols in belt holsters. I was informed that the dogs were trained to smell drugs and explosives and they identified me as having that smell. I was asked to put my luggage in one corner and to walk across the room. The dogs followed me. They asked me to undress, which I did. The dogs were sniffing at my boots. The light dawned in my brain, but not in the brains of the policemen.

I asked for an English speaking stewardess to help me explain the situation. She came bouncing into the room...the Al Italia stewardess was stunningly gorgeous in a very short skirt, a sister of Sophia Loren I suspect, and spoke good English while chewing gum. And there I was, standing in my underwear, grinning. With her help I explained that I came from wild savage Canada and that my son had shot a bear, from which I had rendered the bear fat into grease for my boots. I suggested that the dogs were attracted by the bear grease on my boots. These boots are high-heeled South Dakota cowboy boots, tall tops nearly up to the knee. The stewardess began to giggle and said that the policemen would not believe this story, as they didn't know about bears and she couldn't think of the Italian word for bear (*orsa*). Later I found out that bears had vanished from Italy about 25,000 years ago. Somehow she convinced the police not to pry the heels off my boots and demonstrated that the dogs were licking the boots, not giving the appropriate signal for explosives or drugs. They did open my luggage and asked for me to empty my brief case, but they eventually let me go.

The University of Padua string quartet had vanished from the airport lobby and the Lufthansa desk had a long line of anxious ticket holders. I must have got nearly the last seat and left Rome on KLM in the late evening, hungry. We all arrived at the Schiphol Amsterdam, where there were neat and clean toilets, helpful and fast staff and lots of help for the stranded Al Italia crowd. A special Dutch bus was provided for Al Italia crowd and I travelled with the chamber orchestra (they are booked for two concerts in Toronto tomorrow evening), we got special treatment for transport to the Ibis Hotel. What a difference, a feeling of order and safety and people that respond rationally.

The next morning it was a good Dutch bacon and eggs breakfast, oh so gratefully consumed, and then a bus to airport, three security checks, guard dogs, armed security on tarmac...off we go in mid-afternoon, Toronto, following the sun again, Winnipeg at 1830, and there is bright and shining Sheila who hauls me off to a Meadowlark rehearsal at Edigers' until midnight, no mercy for the weary international traveller.

Sheila's older brother Ian sold Sheila's family home in Loretto, after being fixed up due to the damage from the fire. This was of course a painful step for the Dennett family.

At this time I was president of the Canadian Folk Music Society and at our Annual General Meeting in Quebec City we tried to resolve some of the academic conflict between Quebec French and English traditional music words, tried to make a constitution and revitalise our publications. Edith Folke, Jay Rahn and John Leeder are editors of our publications, but these were not widely distributed. The Quebec fiddlers in this bunch taught me a few fiddle tunes not often heard in western Canada.

Once the Canadian human health threat from the Chernobyl reactor explosion was shown to be negligible, the excitement died down quickly. A year after the incident, however, an invitation appeared on my desk for a UN sponsored meeting at the International Institute for Applied Systems Analysis (IIASA) in Vienna, Austria, to discuss the ramifications of the Chernobyl fallout on ecosystems, agriculture and human health. I knew nothing about this organisation and was certainly not an expert in radiation health, but I had some friends in Vienna that would be fun to see. So here we go again, flying from Winnipeg to Montreal to London to Paris to Vienna in early November 1987, a sleepy tour of too many airports, bad food and worse airline coffee. And then the Vienna IIASA staff rescued me at the airport at midnight, brought me to a luxurious hotel where they fed me good coffee and giant pastries lathered with whipped cream. I got to sleep for most of the next day, wandering out in the evening to find good food.

And then they hauled me out to the Laxenburg Palace just outside of town, where the science meetings were to be held. I was immediately made to feel like a North American wild savage in a polite European court. I was wearing my usual Niverville thrift shop blazer jacket, Levis and South Dakota cowboy boots, and here I was in one of the Imperial Hapsburg empire residences, built in the mid-1700s, a lavish baroque palace fit for the Empress Maria Theresa and her servants arriving in horse-drawn carriages. There were several formal gardens of sculptured trees and hedges, with neat pathways past fountains and ponds on the way to other royal buildings and accommodations for IIASA staff and visiting scientists. There was a very good kitchen staff, who served full meals as well as morning and afternoon teas with amazing pastries. And in the usual old-fashioned European way, it took a full day for everyone to be introduced properly, to inspect the library, computer and communication rooms, the artworks and architecture. The meeting room was a large amphitheatre with headphones for translations, very clever projection facilities for multiple display of data slides, videos and recordings of our discussions on view.

By the second day, this amphitheatre became self-sorted into three groups of people: 1) radiation ecologists and human medical radiation biologists, 2) the industrial food chemists, and 3) the government regulators and administrators. In addition to the complexity of hearing over a dozen European and Asian languages being translated into English in my headphones, it was also apparent that these three groups of people also had very different priorities and objectives. The ecological scientists were gleefully showing how the Chernobyl nuclides were being incorporated into soil, plants, rivers and lakes and the ocean, fish and agricultural foodstuffs around the world. The dour human radiation health scientists were fussing about the historical mortality data from the World War II Hiroshima and Nagasaki bombs, relative to measured dose rates from several dozen radioactive nuclides. They wanted to predict the cancer and death rate for humans and livestock in the path of the Chernobyl fallout in Europe and Asia.

The industrial food chemists were busy showing that the current regulations for radioactive contaminants in nuts, meat, milk and potatoes were unnecessarily restrictive. These were representatives from Nestle, Procter and Gamble, Lever Brothers, the megacorporations that manufacture food products for the world, and whose economy was threatened by Chernobyl fallout contamination. And of course the government regulators just wanted the whole problem to be easy to understand and/or go away. The Russians silently showed air photographs of the steam and dust plumes rising into the atmosphere from the melted reactor and the radiation exposure meters of the reactor operators who died at their duty posts.

In contrast to our low impact scenario for direct Chernobyl fallout on Canada and USA, the Europeans and Asians did have some real radiation hazards in some food products. I learned that within a month of the Chernobyl explosion, Hershey and other candy makers could no longer use nuts from Turkey and other Mediterranean countries. Cheese and other milk product manufacturers had to find other sources of raw milk, as most milk from Scandinavia, UK, western Europe and southern USSR had too much radiostrontium activity. Sheep in Britain and Spain and reindeer in Finland had too much radioactive iodine in their thyroid glands and too much radiostrontium in their bones. Commercial vegetable farmers in Greece and Italy could no longer sell their produce. Depending upon the agricultural product and fish species and the half-life of the nuclides incorporated into the food items, some of these food contamination problems were predicted to continue for several years. The food industry wanted the radiation dose regulations relaxed, as many European foodstuffs for the international market were above acceptable levels for certain nuclides and commercial trade had been disrupted. The arguments went back and forth over the week, tempers rose, there was some compromise and disagreements, and the food safety regulators were very unhappy with the scientists.

This meeting was also concerned about other industrial accidents involving food safety, medical drugs and contaminated livestock feed. For example, the group considered several cases of radiation poisoning from medical facilities. In Sept 1987 two men entered an abandoned medical clinic in Goiania, Brazil, and dismantled a steel container, which they wanted to sell for scrap metal. The scrap dealer kept part of the container in his living room, because the blue powder within glowed in the dark. Some of his family smeared the blue powder on their bodies, laughing. The container held a large quantity of cobalt-60, a gamma emitting radioisotope used to kill cancer cells in medical practice. Five people died and 249 people were badly contaminated from this episode.

Outcomes from this meeting were published by IIASA as referenced in the section on further reading.. Most of these reports were buried in the 'grey literature' that is rarely seen by the public, probably because it would alarm many people to know how their food safety regulations are influenced by commerce, profit margins and the power of the international megacorporations. We also were treated to long lectures on the radiochemical impact of Windscale (UK 1957) and Three Mile Island (USA 1979) nuclear power plant failures.

But during the evenings we were all friends. Well, sort of, as the European social class system put the big executives from the food industry at the head tables of the lavish banquets. I usually ended up at the rear banquet tables with disrespectful and rowdy Americans, Scots and Australians, where we drank good Austrian beer instead of fine wines. The food was very fine but the desserts were killers. Even by my wife Sheila's high standards, these desserts were too large, too rich, and gorgeous to look at. The Austrians are famous for their cakes and pastries, usually '*alles mit schlag*', or with copious amounts of brandy-laced whipped cream.

After one such feast we were offered the chance to attend an opera on the following evening. A small group of us held up our hands and gave our names to the secretaries. Little did we know what the consequences would be! These determined grandmotherly secretaries promptly took our body measurements so that proper evening wear could be rented for us, at IIASA expense. Apparently we would never be allowed to enter the *Staatsoper* in our inappropriate North American clothing and we were fitted with tuxes with tails, pleated white shirt fronts and bow ties. These determined ladies were very upset when I appeared in this outfit with my cowboy boots but they sighed, threw up their fluttering hands and let me go.

Richard Strauss's *Der Rosenkavalier* (The Knight of the Rose) was performed in the Viennese State Opera House that night, where the full Austrian nobility royally dressed attended in this vast baroque building. Here was a very different world, where the grandeur and historical weight of the Hapsburg Empire was very heavy on my naive prairie boy shoulders. The pit orchestra had a completely different sound from North American orchestras, partly because some of their instrumentation were period instruments such as rotary valve trumpets. A smouldering : Mechthild Gessendorf sang the role of the Marschallin (the Princess Marie Therese) and I must admit to having shivers up my spine and tears running down my cheeks. I realised that the audience was listening to a musical history of their own nation, during a time when it was much greater and stronger in world affairs. During the intermissions tables of good food appeared, with wine and beer. Being a skinny hungry prairie boy, I ate large amounts of this finger food. The opera concluded at one a.m. and I was astonished to find that we were expected to go to a good restaurant nearby for our supper. So here in the wee hours of the morning we ate roasts of poultry and pork, sautéed leeks and parsnips, vinegary tomatoes, mountains of whipped cream on top of filo pastries filled with fruits and nut pastes, sugared plums and Sacher torte. We were next door to the Hotel Sacher.

We had several days off from the meeting, when we were supposed to write contributions to the synopsis of the conference. I had already written my Canadian data input previously, so these days became free to explore Vienna. I was startled to find a large Turkish and Islamic influence in some of the architecture and history of this city, as it had been within the Ottoman Empire in earlier centuries. In the Museum of Natural History I found a whole floor of Cretaceous dinosaur skeletons, *Baculites* and ammonites, many of them from the South Dakota Badlands, near where I grew up. This museum also had many of the Greek, Cretan and Egyptian artefacts discovered by Heinrich Schliemann in the late 1800s, which I had read about from my mother's library when I was a young boy. And then there were dozens of sarcophagi with real mummies inside. My piano teacher grandmother England would nod knowingly about all the statues of Mozart, Strauss, Goethe, Schubert and Liszt on many street corners.

Having done some hunting, I was curious about the Hunting Arts Museum, in the lobby of which were gigantic fossil skeletons of European elks recovered from various bogs in northern Europe and Scotland. I had regarded hunting as a way to get some meat but the Austrians apparently considered this activity to be some sort of communion with the forest and wild animals. Endless floors of displays of snares, pit-fall traps, Stone Age spears, early rifles, bird nets, bait traps, antique fish traps and lures and stuffed game of every sort, including wild pigs. The walls were covered with large paintings of hunters, fishermen, birds and other wild game. I was puzzled that such a tiny country full of cities and towns would have any wild game left but perhaps up in the Austrian Alps there might be some game for hunting.

Despite so much culture, the Australian contingent insisted that I accompany them to explore the seedy porn district, where Hungarian hookers, rock and roll bands, gambling dens and drug peddlers were found. I left Vienna on 8 November, flying over the Alps with glaciers like rivers of whipped cream, Frankfurt, Toronto, Winnipeg at midnight, dear Sheila bright and shining there to meet me, she being in the midst of her 'extended birthday'.

The human error Chernobyl accident ruined the reputation and public acceptance of nuclear power for years to come. In 1979 the Three Mile Island incident at Harrisburg, Pennsylvania also scared the public witless, despite there being little radiochemical hazard from this break-down of their cooling system. At present, the next door twin of the failed Three Mile Island reactor is still in operation and is surrounded by occupied homes. But within the scientific community the growing concern about global warming and energy security gave nuclear power an important option for future energy supply. It is not widely known that after the smouldering reactor was covered with concrete by a French contractor, the three remaining reactors at Chernobyl were restarted to provide electricity for this region, seven months after the explosion, and thousands of people returned to work at this site. The nearby dormitory town of Pripyat had been evacuated and two decades later it was decomposing into the ground.

Over the next decades the Russian agricultural and soil scientists, ecologists and wildlife biologists studied the distribution and impact of the gradient of radiation exposure in the 40,000 km^2 exclusion zone (called the 'zone of alienation') around the entombed nuclear power plant in the Ukraine and Belarus. Radioecologists determined that most of the plutonium did not travel far in the atmospheric plume and most of this Pu was found in the soil profile nearby the plant. Some other more volatile isotopes travelled around the world. Various schemes were proposed to decrease the activities of the nuclides in soil, including deep ploughing and planting crops that accumulate the most dangerous nuclides, and methods for capturing these nuclides in the ash of these plants. An Irish company wanted to grow sugar beets on this contaminated land to produce ethanolic or diesel biofuels, the radioactive portions of which would be captured in the non-fuel residues. Twenty-five years after the nuclear explosion, the European Bank and the French construction company Novarka have plans to build a better $2.1 billion containment structure over reactor 4, to be completed in 2014. The trouble at the Chernobyl reactors was not over, as within the plant a sudden hot neutron event occurred in June 1990 and too much snow on the roof of the concrete sarcophagus caused a partial collapse of the containment structure in February 2013.

The Chernobyl reactor 4 malfunction caused an explosion that spewed 14×10^{18} Bq over the southern USSR and Europe. Some genetic non-lethal mutants appeared in local biota within a decade of the accident. Despite the higher levels of radiation in the plants and soil of the land area around Chernobyl, closure of this region from human use for several decades resulted in development of high plant and animal biodiversity, and productivity. Wolves, moose, bison, wild pigs, lynx, raccoon dogs, fish, Przewalski horses, eagles, butterflies, reptiles and amphibians are abundant in healthy populations. There may have been radiation damage to these populations of wildlife but if so, most of the damaged organisms perished early and were not found by ecologists. Nearby farms and small towns have tractors, hayracks, cars and trucks, ferris wheels, dodgem car rides, school houses sitting abandoned, with lush vegetation growing around them. This was due to the exclusion of people and agriculture. Curiously, the demilitarized zone between North and South Korea reacted the same way, with the narrow forbidden land area becoming a nature reserve with teeming populations of geese, deer, cranes and bears. Nowadays, in 2012 tourists and journalists visit the site and they stay at the House for Foreign Specialists (originally built for scientists and engineers), which is now called the Chernobyl Hotel. From this horrible accidental experiment we might conclude that the greatest ecological stress came from human use of the land and lakes, not the radiation stress.

After all this polite social and acrimonious scientific excitement, it was a relief to come home to our quiet Tourond farm home to get ready for the winter season. Our fourteen year old son Jesse came with us on a moose hunt that fall, proudly carrying his own 30.06 Remington rifle. We had Christmas and New Year's at home, with most of Sheila's family attending gigantic feasts of home-grown ham, moose roasts and large fat geese. Some of my Augustana College Renaissance choral music books appeared somehow, and the whole family took to singing four part motets in Latin, with uncle Al on piano and/or pump organ. This soon created the need to humorously educate me once again in the language arts, as most of the family were more liberal in language skills than I was. Amongst the kid presents this Christmas were two Rubik's Cubes (invented by Hungarian sculptor and Professor of Architecture Ernő Rubik). And of course the adults stole these from the kids and had speed contests around the extended supper table.

While all of this was happening, the solar system persisted in spinning around and world events got more and more bizarre. We visited my parents in South Dakota later that year, and the Skelly fuel station and coffee shop conversation continued with my father's friends...

'Sorry, we don't have any sour cream pie today.' Said the waitress, waiting for our order.

'Apple pie would be fine, especially if you have some cinnamon sugar to sprinkle on it, and a cup of tea please', says I.

'Tea? What's going on, you've gone British on us!' says Wayne Sanderson, grinning.

'I hear you've been over in Austria? What's it like over there, behind the Iron Curtain?' says Ken Foster.

'Austria is not behind the Iron Curtain, at least not now' says I. 'I've been in Vienna, and there were no Commie guards at the airport.'

'Ah, Austria is where that paper-hanger Hitler came from', says Helge.

Ken Foster grinned, looked at my father, and said he'd bet $10 that Austria was behind the Iron Curtain. Apparently my father had already told them that I'd been in Vienna and they had already argued about this point of geography and history.

'So how can we settle this bet? Maybe we should look at an atlas?' says Ken.

So Ken Foster came out to our house on the hill, we dragged out my mother's giant world atlas, which had a jagged line on the map for the Iron Curtain countries. Austria was indicated to be with the rest of free Europe. I won $10 right then and there.

Chapter 21. Mother's Milk, Ethyl, and Oomingmak's Adventures in Search of Poisonous POP

Civilisations, ancient and modern, mess around too much with poisons. They invent new poisons to kill other people, plants and animals they don't like. The trouble is that these poisons don't stay where we use them, but travel around the world for decades and centuries and kill innocent targets. Historical records, sediment cores of peat bogs and lakes, cave deposits and tree rings tell the story of about five thousand years of local pollution from Bronze Age metal mining and processing, mainly lead, copper, silver and mercury. These ecological and historical records come from South America, China, India, Egypt and the Greek/Roman civilizations of the Mediterranean coast. Some of these archaeological records also tell the story of accumulation of these toxic elements in local plants, animals and human bones and teeth, with consequent changes in ecosystems, metal poisoning and human mortality. Not content with these dangerous activities, we continue to invent new poisons in modern times.

The powerful German chemical industries began exploring the properties of organochlorine and organophosphorus gases and powders in the interval between World Wars I and II. These were intended for military purposes, as bioweapons to kill soldiers from enemy nations. Both the Allied Forces and the German military used these poisons in human experiments and battle. After the war, European and North American chemists decided these poisons were useful to kill mosquitoes and agricultural pests. With the enthusiastic blessings of the various departments of agriculture, health and welfare, we poisoned the soil and atmosphere for several decades. By the 1950s when I was working with my father and grandfathers, they were spraying DDT, dieldrin and other organochlorine/phosphate pesticides and herbicides for weed and cattle insect control. Rachel Carson recognised this mistake and wrote her book *Silent Spring* in 1962. She was initially ridiculed and persecuted by the agricultural industry and government regulatory officials. It took the discovery that the eggs of the symbolic American bald eagle (and many other birds) were weakly shelled and broke in the nest before people woke up to this self-inflicted tragedy. Some of these poisons were banned in the 1980s but many are still used today.

Residues of these long-lived organic poisons are now found in biota in the deep ocean floor and in polar bear fat. Most of these poisons only worked on the target species for about a decade and then they developed resistance to the pesticide/herbicide. DDT usage began in the 1940s and by 1946, houseflies were resistant to it. By 1966 over 165 species of insects were DDT resistant, including the malaria-carrying mosquito *Anopheles gambiae*, but species of other animals were damaged by this poison. The story is similar for herbicides, where some of the worst agricultural weeds became resistant to the organophosphate poisons within decades of first usage. Other industrial organic compounds, such as the polychlorinated biphenyls (PCBs) and dioxins, were used and discarded to waste for several decades and bioaccumulated in ecosystems. Between 1947 and 1976, 500 million kilograms of PCBs were released from two factories on the Hudson River and much of this toxic load is still in the river mud. Three decades later, it was discovered that Atlantic tomcod in the Hudson River had developed mutations that resulted in resistance to PCBs and dioxins, although fish loaded with these poisons are not safe for humans to eat.

The USA agricultural industry and the military happily sprayed thousands of tonnes of Agent Orange over local weedy crops and most of South Vietnam, Cambodia and Laos (Operation Ranch Hand) during 1961-71. This stuff (2,4-dichlorophenoxyacetic acid (2,4-D) and 2,4,5-trichlorophenoxyacetic acid (2,4,5-T), in iso-octyl ester form) was contaminated with a dioxin (2,3,7,8-tetrachlorodibenzodioxin (TCDD)), and the dioxin was causing many of the previously unexplained adverse health effects which were correlated with Agent Orange exposure. This dioxin has been described as perhaps the most toxic group of molecules ever synthesised by man, causes neurological cancers and birth deformities, and this dioxin family of toxic compounds is now widely distributed over most of the earth.

An even odder story comes from the rapidly emerging and energetic oil and automobile industry. Early petrol fuels for internal combustion engines in the 1920s would 'ping' or 'knock', due to the premature and inefficient combustion of the fuel/air mixture in the engine piston cylinders. Du Pont industrial chemists provided a long list of 'octane' additives to fuels to prevent this, including ethyl alcohol, iodine, iron, selenium, tellurium, and some organic lead reagents. All were corrosive and expensive, except for ethanol, which worked very well and was industrially produced in Europe for reasonable prices. Henry Ford's early Model T cars ran very well on pure ethanol. The only trouble with ethanol was that it could not be patented by the fuel and auto industry. So they settled their patent target on tetraethyl lead additives to auto fuels in the mid-1920s, even though they knew that this was poisonous to humans and would be emitted to the atmosphere in the exhaust of engines. Tetraethyl lead (TEL) was initially synthesised by a German chemist in 1854, who considered it to be a deadly poison, causing hallucinations, breathing problems, madness, spasms, palsies, asphyxiation and death in a series of laboratory technicians. Du Pont, General Motors and Standard Oil established several production plants for this TEL additive in the mid-1920s and promptly tens of USA production plant employees died from contact with this poison. Several university and health officials tried to stop this TEL industry, citing well-known cases of lead poisoning, especially in children, but they were ignored by the industrial chemists and government health officials. Strangely, the fuel with TEL additive was commercially sold by the name of 'ethyl' fuel. By the 1950s most US automobile and truck fuel was ethyl and I remember my grandparents claiming that engines worked better with TEL, largely due to advertising from the fuel industry. For a hot blooded version of this story, see the reports by Kitman (2000) and Pearce (2017).

Now, we all knew that industrial cities were cesspits of contaminated air, soil and water. I was naïve enough to think that the northern boreal forests, taiga and tundra of the Canadian north would be relatively free of most of these industrial contaminants. Our naïve views were shaken by newspaper reports (based upon a science journal article quoted below) that Ungava Quebec sub-Arctic Inuit mother's milk contained measureable levels of industrial PCBs and our clean Arctic environment became defiled overnight.

In the late 1980s the Canadian government started another scientific task force to determine the extent of the problem, the past history of contaminant delivery to the industry-free Arctic and the mysterious sources of these contaminants. The name created for this research program was 'Long Range Transport of Atmospheric Pollutants', abbreviated to LRTAP. Over a decade of research the cold trap hypothesis was tested, as most of these temperate zone industrial contaminants are volatile in warm climates but tend to stay in the ground or water/ice in cold climates. This suggests that decades of agricultural and industrialised warm temperate zone use of DDT, PCBs, other pesticides and some metals would slowly evaporate in tropical and temperate climate zones and migrate toward the polar climates, where they would be thermally freeze-trapped. As anthropogenic greenhouse gas warming of our planet progresses, these 1950-1980 metallic and organic contaminants in polar and alpine glacial snow and ice are being released from their cold trap and are entering the food chain again in the following century, even though many of these poisons have been banned for use in western nations (Schuster et al. 2011, Arellano et al. 2011). Even worse, Breivik et al. (2011) suggest that recent increases in PCBs and halogenated flame retardants in tropical meso-America, Africa and SE Asia (where these contaminants are not produced or used extensively) are the result of illicit export of organic contaminated wastes from the rich western countries.

Over the 1980s my little crew had taken and analysed sediment cores from many lakes at ELA, Lake Winnipeg, sub-Arctic and Arctic lakes for a variety of purposes. Some of these radiochemically dated historical sediment records had been analysed in Derek Muir and Lyle Lockhart's labs for DDT and several other pesticides and their degradation products, PCBs, polycyclic aromatic hydrocarbons (PAHs), and Mike Stainton measured a variety of heavy metals associated with industrial activities. We expected to see high levels of contamination in lake sediments near the Canada/USA border and declining contamination as we went northward. The data were irregular and inconsistent, possibly from our own contamination of field equipment and handling, or maybe the samples were too small.

With new money coming in from the LRTAP project we began to create new sampling methods, equipment and transport containers to allow super clean collection of sedimentation history of these contaminants. It was a difficult situation, as clean sampling of the organic contaminants in the field required the avoidance of most plastic materials, fuels, oils and grease, insect repellents and foodstuffs. Clean sampling for the heavy metals required the avoidance of lead, copper, cadmium, zinc, mercury, which were all common in dust, fingerprints and vehicle exhaust. Willy Burton and the workshop crew built us a small 40 x 40 x 60 cm box corer and mud sampling tools from high quality stainless steel, with a battery powered winch on a lead-free painted metal tripod for use on the ice. We had a special graphite grease gun for cold weather lubrication. He made us a heavier gravity corer with a stainless steel weighted head. All of this could be washed clean with organic solvents and strong acids, kept in cleaned plastic sheets and stored in sealed transport boxes suitable for rough aircraft and skidoo sled handling. Special shipping containers were built to allow our gear and mud samples to be protected from organic and metal contamination during travel and field operations. We had a small gas heater to warm water in the field tent, so we could wash our hands with soap before handing the equipment and mud sampling tools.

All of this was done on the smell of next year's money, as our project was already overspent for the year. We were anticipating big money to arrive from Ottawa for the LRTAP program. We took three van loads of equipment out to the Experimental Lakes Area to test our equipment on the late winter 1988 ice. Sheila and our sons came along for the adventure. Sheila and Ian helped with large volume water and sediment sampling, while Jesse took off on the camp skidoos and 4-wheeler ATV. We almost lost our big new box corer twice, and we learned what modifications were needed for further work in remote locations. Paul Wilkinson (radiochemist) and Brian Billeck (organic contaminant analyst) kept me from using WD40 on rusty gears, and they made notes on how to make our core slicing tent analytically clean and more comfortable in cold weather.

Winter coring at ELA, 1988

ELA Box coring, Brian Billeck and Paul Wilkinson, 1989

And here we go again, another Arctic expedition, packing our gear for shipment to Saqvaqjuac on the northwest coast of Hudson Bay. We shipped our gear two months before our travel tickets because we knew of the inevitable delays for Arctic freight. Buster and Cathy Welch and field crew had over-wintered at Saq, so we had a warm and comfortable camp for our field headquarters.

In late April 1988 Paul, Brian and I took the small plane trip from Winnipeg through The Pas, Thompson, Churchill, Whale Cove and finally Chesterfield Inlet, where an inukshuk on a rock ridge near the airport greeted us. Maurice Kukkiak, Peter Amarualik and Buster's son Scott loaded our personal gear on three komatiqs, fuelled the skidoos and took off over the rough Hudson Bay ice northward to Saqvaqjuac. It was -24°C and blowing wind, and we had heavy goose down parkas with caribou skins wrapped around us. We would stop the caravan behind rocky islands or ice ridges to pee and tighten the load ropes. We arrived at Saq camp mid-afternoon, where our thirty-two freight boxes were stacked beside the dock and the cook had hot soup and sandwiches ready for us. Supper was caribou steaks and rice, with fresh baked bread.

Saqvaqjuac Research Camp

We spent two days unpacking our gear, cleaning and testing everything, and repacking for komatiq travel to our work sites. The komatiqs were attached to the skidoo by a long stretchy rope, which worked fine on level ice, but we had to learn how to handle these rigs on hills and sharp corners when the komatiq would have a mind of its own. In one case I rode one of the komatiqs like a bucking bronco over the rough ice and snow, hanging on to a rope tie, loudly singing 'The Strawberry Roan'. Buster and Maurice would bring the big Bombardier snowcat to our work sites, with a 318 cc Chrysler engine powered 50 cm diameter Sigurdsson ice augur attached to the rear bumper to our work sites, to drill several holes through the 2-2.5 m thick ice. We had to ban Buster from the work site since he insisted on oiling the ice augur shaft. Like all good Arctic experts, he had discovered that electrical transformer oil worked better in the cold. I'm sure it did, but transformer oil has high concentrations of PCBs, one of the contaminants we were searching for in the sediment cores.

Since we were 2-6 hours away from Saq camp, we took enough food, first aid and cooking gear to overnight if something went wrong or the weather worsened. We icebar chipped shallow holes in the ice to bury anchors for our work tent and filled the holes with fresh water (not seawater, it turns to rubber) to freeze in the anchors. Once we had carefully lowered the corer into the sediment and raised it for inspection, it could be taken inside the heated tent for clean sampling, centimetre by centimetre down the mud layer history book. In most cases, the history of mud and contaminant accumulation for the last several centuries could be determined from 50-100 cm of a sediment core from the deepest spot in the lake. We had a tiny propane stove so we could have hot water to wash our hands and sampling tools between each sediment layer in the core (and for tea breaks and thawing out our sandwiches).

We knew that most of the organic contaminants were man-made and that at some point in the mud history book we should reach zero concentration somewhere deep in the mud core. Since all of our clothing, food, freight boxes and equipment were made in a big city down south, we had to assume that all surfaces were contaminated with the compounds and elements we wanted to measure. We had previously sampled the dust inside the airplane freight containers, skidoo exhaust, dirt from near the Saq camp, rusty metal from the ice augur and grime from the back window of the Bombardier, so we could know the sources of sample contamination.

One of our sampling sites was Hawk Inlet, a small basin inside a steep walled valley, almost pinched off from the Saqvaqjuac Inlet seawater tides. The surface water was fresh but the deep water was seawater, and charged with carbon dioxide, methane and hydrogen sulphide. The Inuit helpers did not like the smell of deep water and called it 'the devil's breath'. One of the young Inuit helpers was Lukie, who had an epileptic fit during one of our delicate core slicing operations. We wrapped him in sleeping bags and laid him on a komatiq for rapid skidoo transit back to main camp. Some of the Inuit said that the devil had hurt Lukie, but he recovered within a few hours and was ready for work the next day.

One of our tasks was to collect fresh clean livers, bile, kidneys and flesh of Arctic Char from each of our sites. These samples were immediately frozen in liquid nitrogen and analysed for enzymes and proteins that acted to sequester or degrade some of the pollutants we were looking for. Brian Billeck was the main man for this job and he used an under-ice 'jigger' to stretch out the sixty metre gill net between two augured ice holes, with the help of Peter Amarualik. Northern fishermen are used to pulling fish out of nets in -30 °C blowing snow weather but my hands quickly became stiff and useless. You might imagine the puzzled concern showed by our Inuit helpers, who watched us open a large Dewar flask full of liquid nitrogen (at -90 °C), when it was plenty cold to freeze fish by laying them on the ice.

Work tent on lake ice at Sacvaqjuac 1988

We had mostly good sunny weather but when the wind came up and visibility was reduced, we gratefully stayed in camp. The warmth, good food, the sputtering radio and Buster and Cathy Welch's crew were wonderful, and some good parties were held after supper. Papaluk Kukkiak would shyly play old time dance tunes on her button accordion, and I could sometimes join her on fiddle. Peter and Maurice Kukkiak would hum and go into a trance dancing. Maurice, out of breath and giddy, tells everybody that they've given me an Inuk name: 'Umingmak', meaning 'the bearded one' because of my scrawny whiskers, also meaning the musk ox.

Papaluk with komatik sleds and sampling gear, Saqvaqjuac, 1988

During this time I interviewed and recorded Papaluk's music for a folk music (Canadian Traditional Music Society) journal article. Buster told the story of how Papaluk's husband, some years ago, traded her button accordion for some new sled dogs. Some of my photos show wind-burned red faces around a kerosene lamp on the kitchen table and a big jug of Martin Bergman's home-made yeasty cider.

After one of these fierce wind storms Buster noticed that some of the snowdrifts had a brown colour in certain pockets. We collected this brown snow, melted it in clean plastic sacks in the kitchen and carefully funnelled the tiny residue of dirty brown stuff into clean vials. This, and all the other samples, were analysed by a team of metal and organic chemists back in the Winnipeg laboratories, led by Derek Muir, Lyle Lockhart, Mike Stainton and John Armstrong. The brown snow was very interesting, as we later discovered that the wind pattern for that storm came over the North Pole from China. This dirt contained agricultural organic pesticide contaminants that were manufactured by Asian methods different from those in Europe and North America and could be shown to be a globally transported contaminant. And here it was: Long Range Transport of Atmospheric Pollutants (LRTAP).

After three weeks of sampling, we packed up our samples and equipment and began the perilous trip home. Well, it seemed perilous, with a variety of aircraft engine and fuel problems, weather difficulties and two days of living in cold airport sheds and hot stuffy aircraft cabins with strange food. We ate a lot of honey roasted peanuts and airplane cheese and crackers over the two day travel time from Chesterfield Inlet to Rankin Inlet to Whale Cove and back to Rankin Inlet, then to Baker Lake, Thompson, Churchill and finally Winnipeg, where my bright-eyed Sheila was there to meet me. Of course we were dressed for Arctic conditions, long johns and wool sweaters, but it was warm spring weather and green grass in Winnipeg! Samples in liquid nitrogen and dry ice were delivered to the lab.

Culture shock in more ways than one. Sheila had our mushroom van full of musical instruments and we immediately drove to her brother Alistair's home in Winnipeg, where he was waiting with new recording equipment in his basement. I barely got time to shower and change clothes when a fiddle was thrust into my hands. Meadowlark taped stringband music all day and into the night for the next few days.

The next week our son Ian performed in the percussion section of the Manitoba Honour Band and the Winnipeg Youth Orchestra. Younger son Jesse was off (with his new 30.06 semi-automatic Remington rifle) on a bear hunting trip with Gerald Toews and came home with a full hide, loin and hams of a big cinnamon bear from the wild country east of Lake Winnipeg. The bear fat was rendered down for use in waterproofing our boots. During this spring and summer Sheila and I were involved in running a folk music concert series, a Folk Arts Council, the Niverville Country Fair, several other small community fairs and the big time Winnipeg Folk Festival, at which our band Meadowlark performed. And all at the same time we managed to plant our big garden, start up our poultry and pig operation, pick some chokecherries and have lots of family feasts.

In May 1988 one of my all-time favourite jazz trumpet players, Chet Baker, was found dead and full of cocaine and heroin on the sidewalk beside his Amsterdam hotel room. So I guess my grandmother was right, these jazz musicians were dangerous druggies.

Even worse, Iran Air Flight 655, a commercial civilian Airbus A300 B2-203, en route from Tehran to Dubai, was shot down over Iranian air space by a guided missile from the US Navy cruiser *USS Vincennes* on 3 July 1988. All 290 people on board were killed, including 66 children: President Reagan's government never admitted guilt, but eight years later the USA paid US$61.8 million in compensation to the families of those killed. Nobody seems to know why the USA got involved in this Iraq/Iran war between Shiite and Sunni Muslims from 1980 to 1988, longer that the two world wars. Nobody won and it poisoned the Middle East for the next few decades. The financial and human life cost of this war was large: US$452 billion for Iraq, US$645 billion to Iran, 180,000 dead for Iraq and 500,000 dead for Iran. Ain't religion wonderful?

With some wonder and a little fear we watched our sons growing up and becoming more competent and independent. Ian and three of his buddies took our van and two canoes and spent over a week on a canoe trip to a series of lakes and rivers in northeastern Manitoba. He was also gathering and editing music tapes of his performances for an application for funding to participate in a July 1988 European tour with the International Youth Band (the band played concerts in Germany, France, Switzerland, and the Netherlands). Jesse was rebuilding a motorcycle, tuning our little Ford tractor, helping his Scout troop build a canoe and making tapes of his ability at playing the trumpet. By now both sons were Chief Scouts. Our little family visited the South Dakota folks in August, with turtle catching, prairie dog hunting, got some rattlesnake skins and saw all my brothers once again. Sheila and I dug up more of the dinosaur skeleton we had previously found up in the Pierre/Fox blue shales of the White River bluffs and we brought home some highly polished and greasy gizzard stones and a few of the larger vertebrae. On the way back home I bought some new cowboy boots at Tim Tibbs saddle shop in Fort Pierre, Tim being the nephew of the world famous bronc rider Casey Tibbs.

The biggest house party we ever hosted was a reunion of all of Sheila's six siblings (and their families) in mid August 1988. We borrowed six trailer houses to accommodate all the people, built extra outdoor feast tables, expanded the outdoor grill and woodpile and doubled the barn fridge capacity. Sheila made several 120 litre batches of ginger beer and John Bull dark beer and lots of cakes and puddings in advance. We were able to present meals with mostly home grown sweet corn, melons, beans, lettuce, onions, strawberries and raspberries, as well as many kid-cranked batches of home-made ice cream with chopped jelly snakes added. In addition, we could grill whole rib sections of moose and deer, bake hindquarters of Jesse's bear and home grown pigs, chickens, ducks and geese. We had 22-25 people at the table during this week and many softball games, music sessions, fireworks and rocket sessions in the evening. This was probably the last time all of the Dennett siblings were together.

The end of the year was a blur of government task force science planning and funding meetings about LRTAP in Toronto, Burlington and Halifax, each of them requiring a twenty slide presentation. Similar research efforts were being made in USA, Europe and SE Asia, and we tried to link to these people. In Halifax I stood on the docks and looked wistfully at the seagoing oceanographic research vessels and found folk music shops with piles of old dusty fiddle tune books. The Toronto meeting happened the week before the annual general meeting of the Canadian Traditional Music Society, where university scholars gave their research papers on how folk music persists in the age of rock and roll. We had music parties and dances and general feasting and drinking during the evenings. I had my fiddle with me and I joined the orchestra for the dances.

An early October blizzard at home caught us off guard and we had to butcher our six pigs in very cold weather. Jesse and I spent a lot of energy tracking moose up around Wanipigow Lake but came home with only track soup. Our family went down to my parents South Dakota ranch for Christmas, where we had some exciting blizzards, deer hunting with my brothers, lots of feasting, a town dance, New Year's oyster soup, cinnamon buns hot from the oven, sour cream currant pie and a lot of uncles and aunts and teen-age nieces and nephews (fourteen) running around. I spent too much of the Christmas holidays writing a Polar Continental Shelf Project proposal for more LRTAP field sampling on Cornwallis Island, which is the subject of the next chapter.

Well, while we were safe and warm down south Buster and his Arctic crew were having exciting times. I won't try to rewrite Buster's email message, here it is...

From: Buster Welch
To: Gregg Brunskill
Subject: Bear at Saq

Norman Fredet, old RCMP friend, came up from Rankin in March-April 1988 with his special assistant, Lucien Taparti (Maurice Kukkiak's uncle), a game officer, and Lucien's son. Next a.m. they and I and Dan Pike (grad student) and Kukkiak headed for Cape Fullerton to replace a damaged grave stone for an RCMP member who had died there decades ago. Weather was a bit blowy and the 'trail' we follow up the coast, through the land and then onto the sea, was very tricky. We replaced the stone, then Fredet and party went back south by Pequulik (Depot Island, south of the Lorillard), taking their time because Lucien had a bear tag. Kukkiak, Dan and I went NE to Yellow Bluff where I wanted to look for kasigiak (harbour seal) which Kukkiak said overwintered there.

We proceeded south along the floe edge, then came upon bear tracks coming towards us (north), blood in them. We surmised the bear had seen us and jumped into the water. I spotted it holding onto a pan of ice. We were surprised because there were no hunters from Chesterfield Inlet looking for bear at that time. We waited by the floe edge, glassing the bear, which climbed out onto the ice once. We could see a wound in the top of its shoulder. After several hours the tide brought the ice closer. Kukkiak ran out onto the thin ice. The bear jumped up onto it and charged. Kukkiak emptied his .243 into its neck. I ran out and together we quickly skinned the bear and cut off the head; the ice was so thin we were ankle deep in water. Dan stayed by the second sled, where I had an aluminum skiff lashed on, but we didn't need it. The bear had in fact been shot through the top of both shoulders.

We headed for camp, arriving late that afternoon. Tim Siferd and I forget who else was there ran out and said 'Lucien was mauled by a bear'. To which I said, 'Aha! We shot the bear.' Near Pequulik they saw a big bear. Lucien's gun sight was loose or something so he borrowed Normand's .308. The sights were set such that it shot a bit high at close range. Lucien dismounted and went ahead, climbed a hillock of ice, and shot the bear. The bear turned on him, he slipped on the ice and slid down toward the bear, the gun jammed into the snow, and by that time the bear was on Lucien. It tore his scalp and ears up like a patchwork quilt. This bear had about the biggest skull of any bear I've seen, and Lucien was lucky his head wasn't crushed. The bear bit down on his forearm, hitting his watch and leaving a huge deep bruise. Further back Normand saw Lucien shoot and disappear, but then the top of the bear showed up over the hillock and he knew something was wrong. He drove up on the skiddoo, scaring off the bear. They did what they could for Lucien, and then they had to make their way back to camp. Unfortunately their guide couldn't guide. Lucien's 17 year old son, however, had paid careful attention to the way they had come out and was able to take them back to camp, arriving that night. Tim got a medivac plane to Chesterfield Inlet, and thence out to Winnipeg. Cathy went to visit him the next day. His head looked like a patchwork quilt but he was up and about and chipper.

The wounded bear had made its way north along the floe edge, where we found it the next day. It was a mean bugger.

Buster

Chapter 22. Even Farther North for More LRTAP Adventures ..., Cornwallis Island, Lake Hazen, Axel Heiberg, and Carbonised Tree Stumps in the High Arctic

In January 1989 there were lots of blizzards in southern Manitoba. With the help of our neighbour's big John Deere tractor snowblower, there was only a small canyon through 3-5 meter high snowdrifts in our yard, and getting down our driveway was an adventure in snow shovel artistry. Tifty was happy, gorging on frozen pig butchering left-overs in the cosy barn. Jesse kept our little Ford tractor working at -40 deg., which was useful to pull stuck vehicles from our driveway. Our house calendar was black with scout meetings, high school basketball games, concerts and school meetings. But it was difficult to get to town, and easy to stay home. Buster taught our boy scouts how to build igloos in our front yard, using good hard wind-packed snow blocks cut to shape with hand saw. Inside the igloos was a sleeping platform and a hearth for the seal oil lamp (*kudlik*, usually carved from greenstone).

When I could get to work, my institute had no money, nearly all sections were broke and we couldn't even get ready for next planned Arctic trip. I attended the big New Orleans AGU Ocean Sciences meeting in January '89 and proudly gave presentations on our adventures in the high Canadian Arctic to collect sediment cores for contaminant inventories. This was for sessions that featured people from around the world trying to figure out the source and dispersal pathways of these pesticides, herbicides and heavy metals. I got a good dose of Dixieland jazz and Cajun fiddle music and K-Paul's Louisiana Kitchen food. I brought home some beignets and beignet flour for Jesse, who now considered himself to be an expert on these things. He was pretty damned good at making his grandmother Brunskill's cinnamon buns!

In March '89 we were urgently funded at the last minute to attend another meeting in the Baker Lake Community Centre, where various Inuit organizations demanded assurance from the government that they would not die in their sleep from radiation poisoning from Urangesellschaft operations in uranium mining. There was a lot of hype for this mining operation but no real action had begun. And here were radio, TV, press/media gangs from the south, mixing with old wind-burned Inuit hunters with caribou parkas and seal skin mukluks, most of their teeth gone, eating pilot biscuits and Coke/Pepsi in the hallway. There were babies playing on floor, ladies in the Men's Room, NO SMOKING signs everywhere, but Inuit ignore this, most smoking cigarettes and pipes. There were over a dozen skidoos roaring up and down the Community Centre alleyway with komatiqs attached, parked pickups idling and emitting ice fog and two bulldozers idling outside after enlarging the parking lot.

Canadian Atomic Energy Limited (AECL) officials, French, German, Japanese and Korean industrial agents were also present in suits and ties and there was a weird language flux, several translations going on during the presentations. The Inuit complain that pamphlets and literature reports are not in their language, nor are any of the presentations. Some of us try to explain that much of the scientific information would be difficult to translate into Inuktituk and that we would have to create new words. Our Saqvaqjuac group presentation was one of the few that the local Inuit trusted and that had useful data on U series elements in water, soil, rocks, mud, and biota.

Meals at local restaurants sorted people into interest groups: nuclear industry, engineers, environmental NGO groups, Indian and Northern Affairs, AECL officials and the Keewatin Inuit Association (KIA) and their scaremongering white consultants. The meeting ended poorly, with people angry and nothing resolved. I was ashamed to be part of the government group, as we had not made a good effort to explain ourselves to the local people. At the airport I met some of the CBC employed translators, who explained the difficulties of making up Inuit words for the uranium series nuclides, chemistry and irradiation impacts on biota. The word for uranium is not the substance but rather the emanation, and all the daughters are part of the family of emanations.

And so the trip home was a series of feasts of honey coated peanuts, plastic frozen cheese and crackers, with stops in Rankin Inlet, Eskimo Point, Churchill, Gillam and finally Winnipeg, where bright and shining Sheila was there to meet me with a cappuccino...home to the driveway in snowdrifts, Tifty in the barn chewing on frozen pig heads.

We were in despair, about ready to give up our LRTAP research program due to lack of interest and money in Ottawa. As a last gasp attempt to fill in a south to north transect of data we drove a caravan of trucks and vans with trailers to Bob Hecky's South Indian Lake camp in northern Manitoba in the cold winds of March. South Indian Lake had been flooded with water from a dam on the Churchill River, the overflow being directed into the Nelson River, to increase capacity for hydroelectric turbines on the Nelson River. This flooded huge areas of muskeg and taiga forest, resulting in the termination of the Indian fishery due to mercury contamination and creating a lake full of standing dead trees. One little bay of this lake was supposed to have laminated sediments, and we wanted to use these chronological deposits as indicators of the history of deposition of atmospheric contaminants. We tried our new box corer, KB corer, Bob Hecky's Livingston corer and a freeze corer. I was designated 'El Caballero', the corer cable man for the Livingston corer. As usual we had trouble with bombardiers and the coring tent blowing down in strong winds. I bought a Cree beaver hat with warm earflaps. Carl Hrenchuk was camp manager and also made beautiful big canoes, with curled up stern and bow, closer to the design of the early trapper freight canoes (he had three of them gleaming on a rack in his shed).

This was essentially the end of the SIL research operations. The local Cree were resigned to having their land and lake flooded but they got money and material goods in compensation from Manitoba Hydro. Everybody knew this was a sad and soul destroying kind of compensation.

Back in my institute office I was furiously writing and re-writing proposals for LRTAP expeditions to the Arctic, being refused, cut back, or even worse, being ignored. Now we have email, this stuff can go from my Wpg office to Ottawa, get evaluated and be back on my desk in two days. In early April there was a ^{109}Cd spike experiment going on at ELA and we needed cores of natural and anthropogenic and spike Cd deposition history. Since we had most of the coring equipment and experience, my crew went to ELA to do this work. Our big box corer failed but our new freeze corer worked well, freezing mud laminations to the side of an aluminium letterbox full of dry ice and ethylene glycol. This frozen sediment core provided the chemical information that anthropogenic cadmium was being added to these remote lakes shortly after World War II, and that current Cd deposition rates were probably three times higher than natural background.

Back home, by the end of April 1989 the strawberry plants were beginning to poke through the winter straw cover and the snowdrifts were reduced to pools of water and mud. It was a dry spring and the windy days were dusty. By early May I was cleaning out the pig and poultry pens, dumping the manure on the garden and tilling. Potatoes and onions were planted by the end of May. Our barn glasshouse was thick with tomato plants, which we dare not put into the garden yet for fear of frost.

And then a small amount of money was put in the empty budget for our LRTAP research and Polar Continental Shelf Project provided accommodations and aircraft services out of Resolute Bay. So we packed all our gear for Cornwallis Island LRTAP operations at six coring sites. Paul Wilkinson and Brian Billeck, the experienced crew, took off early to handle our equipment arrival. Buster has a temporary camp there but we have arranged accommodation at PCSP, Distant Early Warning (DEW) station at Eureka, and the joint USA/Canada weather station. And so in late May 1989 I left my family in Tourond with green grass and planted garden and flew from Winnipeg (I can see farm tractors going round and round in black soil fields, raising dust) to Edmonton, and then Yellowknife and Resolute Bay at -15 °C.

Some might call Res a desolate wind-swept gravel pit with variable amounts of snow cover...but it is home to many Inuit and Polar Continental Shelf Project (PCSP) staff, perhaps the best logistic support service I was ever to work with. There is little evidence for Inuit living in the high Arctic, as they were smart enough to go where there was good hunting for food, mainly around the coast of Hudson Bay and along the north coast of mainland Canada. High Arctic settlements were done by the Canadian government after World War II, for sovereignty, research and military purposes and many Inuit were hauled north to justify the claim of the land.

During my working time in this difficult region PCSP was a wonderful support service that made our work possible. They provided a variety of air, skidoo and bombardier transport assistance, hardware for remote field camps, accommodations, meals and rough medivac facilities. For any research institution interested in the high Arctic, this was a regular hang-out for us. It was no surprise that colleagues were already there, waiting for the weather to change to get to their research sites. Mike Retelle and crew from the University of Massachusetts and Bates College, Lewiston, Maine were there, intending to core laminated sediments of Sophia Lake (one of our target sites, on the eastern coast of Cornwallis Island). Retelle was a student of Stuart Ludlam, an old colleague from my Cornell University graduate school days. Also here was Dennis Gregor and crew from the Department of Environment in Regina, Saskatchewan; his crew were waiting to go onto the Agassiz Glacier for snow and ice cores to estimate contaminant loads to this region. They maintained a small warehouse full to the roof of equipment and supplies for their research. Bedford Institute of Oceanography Bayfield Lab had their own warehouse and small oceanography lab there and they let us pile our 1.5 tonnes of ragged and dented gear crates in their shed. Buster Welch and DFO also had a temporary camp at Resolute Bay and they provided some gear for our use, including experienced Inuit field man Peter Amarualik, who had SCUBA training and was familiar with most science activities.

Hurry up and wait, a mantra that would drive you crazy. To get ready to go we need to get our long john thermal underwear on, full insulated body suit, unwashed wool sweater, felt insoled boots, heavy goose down parka and usually my beaver fur ear-lap hat.

On 25 May we tried to fly a packed Twin Otter (Pilot Duncan Grant) into Sophia Lake, but we couldn't land due to a pea soup fog. We returned to PCSP, unloaded, reset, took off our warm clothing. The gloomy grey skies persisted, so we sat in comfort at PCSP watching big screen TV and Calgary won the Stanley Cup. It was cold but lighted outdoors all night. Hurry up and wait...

Finally three days later Duncan flew us into Sophia Lake (75°06'N; 93°31'W) with a bumpy landing. We unloaded our gear, set up our tent camp for 2-3 days work and erected our radio mast, PCSP Resolute scheduled radio contact all ok at 1700. Retelle's group was already here and they had a separate tent camp down at the other end of the lake. They are geologists, interested in varve (annual layers of lake sediment) formation and they plan to stay here into July to monitor inflowing streams. The Retelle camp was on a little raised terrace at the opposite end of the lake and we were invited for supper several times (corned beef hash and pilot biscuits), as long as we promised not to use the 'S' word (= shower).

The annually laminated mud at the bottom of Sophia Lake was important to us, as this meant that there was no mixing of the mud by physical or biological mechanisms and the sediment core should give us a true historical sequence of organic and metal contaminants as they fell out of the sky, into the lake and into the lake mud.

LRTAP Crew on Sophia Lake. Ice augur on komatik sled, 1989

We used a Raytheon Sonar depth sounder on a sled to find a deep section of this long estuarine lake. Sonar waves would penetrate the snow and ice if we applied a puddle of ethylene glycol (car radiator anti-freeze) between the bottom of the transducer and the lake ice. Peter and Brian drilled six holes through 1 m snow and 2.5 m ice for sediment coring and fish netting work. This is not a quick process and it involved putting on successive

1 meter long sections of augur onto the petrol powered engine shaft. When the third section of augur goes through to liquid water, the water spurts up into your face due to the pressure of the snow and ice on top of the lake water. This means that there was usually a 10 cm deep mess of slush around our working ice holes.

We were lucky the next day and got two near perfect KB cores of 40-60 cm of laminated silty mud. Retelle was very jealous of our coring gear, so we spent the next day getting two more good cores for his use. We found musk ox tracks on icy snow. Peter Amaraulik and Brian did the fishing under the ice with a jigaloo. They got mostly spiny sculpin and only three big (3-4 kg) char. Peter got two ptarmigan for our lunch with the .410 shotgun and he ate the heart, liver and intestine as a delicacy.

We found that surface water was clear and fresh, good for tea, but deep water below 12 m depth is seawater, 30-40 ‰ salinity. After three days of this work we returned to Resolute and PCSP comfort in Duncan Grant's Twin Otter. Both pilots were due for an inspection test, so the last time we took off from Sophia Lake, loaded to the roof, all the emergency bells and red lights and warning signs came on as we flew through a narrow valley. The pilots calmly went through their emergency procedure list but nobody had warned us and there were some alarmed faces and white knuckles amongst my tired and wind-burned crew, crammed into the back end of the cargo plane.

The pilots were alarmed to see no lights on the landing strip at Resolute Bay and we had arrived during a PCSP power failure. The big diesel generator shed was silent, the generators down for twelve hours, but it was light enough to read at midnight. Brian put out a cigarette fire in a waste basket in the pilots waiting room at midnight, big excitement all around, but unfortunately I slept through it.

Brian Billeck's racehorse won at some race track in Winnipeg and Brian had bet $7k on him. Paul Wilkinson was perfectly happy in Twin Otter aircraft but he was terrified of helicopters that we planned to use on our next trip. Hurry up and wait, up at 0700 ready to go, but bad weather keeps us in camp...so I play fiddle tunes, read books, videos in the lounge, Satellite TV basketball games, chopper transmission schematic on walls, good food. Our Quebecois cook demands to know if we want our eggs 'ovair or easie'. Wind and sun burned faces were common for the field operators but the inside service staff were pasty white skinned.

A Quebec lady reporter wanted an interview with us but she was unhappy that none of us spoke fluent French. She flew into Sophia Lake on a helicopter to film us coring and showing off the laminated sediment in the core tube. She was disgusted by my coring costume: leather top-rubber bottom boots, thick wool shirt, very dirty Carhartt zoot suit, beaver fur earlap cap and big polarised sun glasses.

With Peter Amaraulik's help, we skidood into Amituk Lake (to the east of the settlement of Resolute Bay, on Cornwallis Island) with two komatiq full of our gear, set up camp, built snow shelter and windbreak, put up radio tower, flattened emergency helicopter pad, did depth survey, drilled holes, set fish nets, got good KB cores but box corer failed. Over a few days we got about thirty char 40-70 cm, pink spots, red flesh. After we had sampled the liver, kidney, heart, bone, the bulk of the good fish (about 40 kg) was given to the PCSP kitchen, and what a feast it was! Our team got toasts and bravos and I played fiddle tunes for the dancing cooks.

Midday is quite warm from the sun's glare off the snow. Here we are in early June and there is some mushy liquid snow on top of the ice. Up on the ridges, where the snow has melted away, tiny flowers are already there. Musk oxen were nearby and Peter skidoos close to them...they circle their young, pawing the ground...Peter looks at me, strokes his bare chin grinning...Umingmak, my nick-name. Inuit hunters along the coast have found some big bearded seals (Ugjuk) along the terraced coast line. Lots of meat, good hides.

While we were at Amituk Lake there was a big party at the Ministry of Transport building in Resolute...there were some knife fights, lots of people were in jail and two cooks missing. The pilots and my crew help do dishes in the nearly empty PCSP kitchen galley. Our Pilot Duncan was a very good fellow, telling endless stories in the pilot's card-game room, amidst weather maps on walls and the chatter and static of the radio. He told about stupid Ottawa bureaucrats who demand to go on pleasure trips when the weather is bad, they would hunt game just to kill, not to eat. He tells of Ellesmere Island and Hazen Lake area, where Walter England and wife regard this area as their private kingdom, a garden paradise, where strange plants grow in protected sunny nooks in canyons along the north shore. He described in great detail a cabin on the south shore of Lake Hazen, maintained by Borek Air and Parks Branch, and the need to ship in diesel and propane fuel for the furnace and stove and generator. We began to arrange a schedule a visit next May on Lake Hazen, but we knew it would be a miracle if DFO funding could be found for one more trip.

Duncan told stories of Arctic flying disasters (there are few old pilots in the north), dumb tricks by newspaper reporters, naïve explorers. Most pilots that we worked with spoke of the de Havilland Twin Otter with reverence and respect, a reliable aircraft in dangerous situations. The next favourite airplane used in the Arctic was the old World War II DC3. Dunc told of a Japanese rich man and his motorcycle...he arrived at Resolute Bay in a pinstripe suit and wanted to ride around in a circle on the North Pole. PCSP was paid big bucks to take him and a photographer there, on a good clear sunny day.

But PCSP does a lot of goodwill work, to host artists, writers, musicians and other non-scientists, who want to get a taste of the Arctic in a safe way. While we are here a Twin Otter crashed near Norman Wells on the Mackenzie River, and six RCMP died...according to the radio...the pilot's room goes silent. PCSP hires local Inuit elders to give survival training for dumb white visitors and Canadian soldiers, teaching them how to build an igloo, primitive hunting and fishing skills, navigation w/o compass, skidoo repair skills with tinfoil, pliers and duct tape. Pilot Duncan told stories with great relish, about some trouble with polar bears knocking down igloos and chasing terrified soldiers.

As the weather was not improving we used skidoos to visit Char Lake, just up the hill from the settlement of Resolute Bay. This lake had been well studied in the early 1970s by biologist Buster Welch and colleagues for the International Biological Program but few sediment cores had ever been taken to gauge the input of atmospheric contaminants. We also cored Meretta Lake, where all the sewage from the Resolute Bay settlement went. As you might guess, the winter's accumulation of sewage generated a lot of bad gases, which erupted when we drilled a hole in the ice. After handling this rank smelly job we were all grateful to be able to take hot showers at PCSP camp.

We were tired, sunburnt, out of reading material and tobacco, so on 7 June we packed up 3,000 lbs of our samples and gear in 31 pieces for shipment back to Winnipeg for repairs. Some of our gear was left in storage PCSP shed for next year's trip to Lake Hazen. I phoned Sheila to warn of our return...she says they had an early June frost which damaged some of the tomato and melon sets. And so we spend another day living in airplanes. We left Resolute Bay on Canadian Air 695 at 1835, fly to Cambridge Bay, Yellowknife (Great Slave Lake still ice covered with moat), Edmonton, several suppers along the way. Our noses are peeling and our wind burned faces feel tight and sore. Overnight in an Edmonton airport hotel we watch TV news of Tiananmen Square massacre of students.

The next morning we are startled by the colour of green grass everywhere, it looks like they painted the ground green. And then we flew back to Winnipeg, freight cargo to the Institute loading docks and then home to my Tourond family, the garden, the poultry pens need fixing, we need to get four weaner pigs and half tonne of feed grain. I am getting accustomed to this culture shock but I needed a lot of sleep to get my head back to normal.

Back in my institute office things are falling apart. We had overspent on our Arctic coring trip operations and I needed to find a way to pay for this. There was far more concern about the budget than the scientific accomplishments in our LRTAP operations in the difficult high Arctic environments.

My colleague David Schindler resigned in late June 1989, sick and tired of low funding and no opportunities for exciting new experimental work. He took a position as the Killam Memorial Chair at the University of Alberta in Edmonton.

Sheila and I hosted a big farewell party for David and his wife Suzanne at our hobby farm near Tourond. We fed pit BBQ caribou hind quarter/black bear/home grown ham and 100 litre deldrums of home brew beer to forty-six people, sang songs, read bad poems about David's career at ELA and thanked him for his good leadership in establishing ELA as one of the world's best field research camps. We also made several batches of home-made ice cream for dessert.

Years later David wrote about his observations of the 1960-73 excellence of the scientifically led Fisheries Research Board and his frustration and bitterness at the following decline of Canadian civil service science. And so, one more of the original gang of ELA pioneers left the institute, Wally Johnson, Bob Newbury, John Cherry, Jack Vallentyne and now David Schindler.

After Schindler left the institute to go to the University of Alberta in Edmonton I was involved with an effort to test the time value of the laminations in ELA Lake 227 sediments. We had been working on Lake 227 for many years and had taken hundreds of sediment samples, but no one had used the bottom sediment laminations (light and dark layers in the soft mud) until we tried a dry ice/ethylene glycol freeze corer (first done by Ph.D. students Jeff Davidson, Jeff Curtis and John Crusius). Brent Woolfe (M.Sc. student with Jim Teller, Earth Sciences, University of Manitoba, and myself), Hedy Kling, and I tried to use lamina counts, lamina Pb-210 and Cs-137, algal microfossils and chemical data to demonstrate that the deep (pre 1968) laminae were likely annual, but surficial (post 1970) laminations were the results of multiple algal blooms caused by our additions of nitrogen and phosphorus fertiliser.

Back on the home front, our son Ian MacGreggor graduated from Niverville Collegiate High School on 26 June 1989. This electrified his family. His mother bleached her garden-browned feet, Jesse demanded one of my neck ties and Ian was decorated with cummerbund and bow tie, corsage and boutonnière. The ceremony took place in in a church, hot and sweaty and full of perfume, forty-seven grads looking earnest in mortarboards and gowns. Many groups of students pose for photos by their proud parents beaming. Ian and a buddy and their dates took off to Winnipeg for dancing, swimming pool, club scene and came home 0800 the next morning.

Younger son Jesse (sixteen) took a summer job as a fishing guide at Kasba Lake in northern Sask/NWT border with Colin (Crash) Welch (son of Buster Welch of Saq fame). In July, for no good reason, we gained two more dogs, Buffalo Gal and the black Wombat, chow and Newfie cross. Son Ian was now working 0500-0800 and 1500-1800 at the Oak Island Goat Dairy north of Ile des Chenes, milking, feeding and cleaning goats. He was also part of a very loud rock and roll band, drums, two guitars, one singer and they rehearse in our winter garage shed. This band and Meadowlark are headliner acts at the Niverville Summer Fair on 12 August. The rock band had a more enthusiastic crowd than did the old time stringband Meadowlark!

In early March Jesse's Boy Scout Venture group were at Gerald Toews' shed building a cedar strip/fibre glass canoe, under Buster's supervision. The result was a work of art, inlaid fancy wood and compass, a thing of beauty to be auctioned/raffled off to gain money to support the cost of a moose hunt airplane trip.

And how the circle turns again...I'm remembering how my father drove me down to Sioux Falls and Augustana College in our grain truck in fall 1959. In late March '89 I drove Ian and friend David Redekop west to Calgary for registration and enrolment for their first year of university. We stopped for the night with Peabody and Al in Regina, where we got butterless tarts and a huge breakfast the next morning. We arrived at the University of Calgary and promptly got lost, wandering the tunnels of campus. We finally found the correct dorm room, registered, got groceries, paid bills and checked into the campus computer system. We attended a far out drama (Medieval crusaders, lesbians, Muslims, science vs the church, lots of profanity, strange electronic music) which should have offended half the audience, but our son Ian was coming alive at all this. He found his percussion room and buddies and he was in heaven. I got to meet his teacher Glenn Price, who greatly influenced Ian to become a real musician. Ian is in his heaven here, he auditioned for a place in the music program and was successful. Ian's development from a rock and roll drummer into a university educated musician all came about over coffee and pie with Ian's music teacher at Niverville Collegiate, who suggested that Ian take some lessons with Glenn Price. We were a little startled to see our son settle into university life so easily and quickly! With a big lump in my throat I headed home, driving past melting snowdrifts, geese flying over, cattle and haystacks, alkali flats, antelope and deer, country music on the Celica radio. There was probably no need, but I wished I had given my son the advice below...

Take your time and do it right.
Leave yourself some room to fall.
Sometimes you got to give it up,
If you want to have it all.

Chris Whiteley,
Toronto blues and jug band.

In late July 1989 graduate student Bill Buhay arrived and I took him to ELA to find big trees for coring for stable isotope climate change signals. Sheila and I took the time to canoe Lake 240 again, one of the sites of our romance nineteen years ago. In August we had home garden sweet corn by the wheelbarrow full, excellent melons, squash and pumpkins to store for the winter.

In mid-August I went to an international limnological society (SIL) Congress in Munich, southern Germany. Here we go again, life in the fuselage of an airplane: Leave Wpg 13 August, to Toronto, my seatmate a German returning to Karlsruhe, a professional pilot who also plays twelve string guitar and was a big fan of Leo Kottke. They still had smoking sections on the aircraft, so en route to Frankfurt we had a good time with roast chicken, beer and wine, Drambuie, cigars and unfortunately some loud drunk Germans were also going home. We arrived in Frankfurt 0845 on 14 Aug, my suitcase is missing already, SIL people appear on the flight to Menschen and we flew over lots of grain fields, corn, channelised streams and rivers arrow-straight. The train to the SIL meeting place and hotel was full of my colleagues, JRV, Sikko, Giuliano, Jack Cornett, Ross Jaffree, John Robbins, Jaap Kalff, Dave Culver, Han Golterman, Jon Overpeck, Curt Olsen...all part of the conspiracy of science, together for lunch of bratwurst, hot rolls and dark beer. Registration was done amidst more wine and finger food and then we all shared a giant supper of cheese-stuffed chicken and dumplings, tomatoes and onions, hot strudel in a bowl full of whipped cream and thick rich coffee.

Schindler was session chair for a three day series of papers on eutrophication. My journals record that I was already checking out Australians for job options, consulting with Paddy Osborne (who was working on lakes in Papua New Guinea) and Richard Pearson (of James Cook University in Townsville, north Queensland). This consultation was done in a beer garden, with litre mugs of good beer, and a polka band that had a long whip snapper percussion. Things get bizarre quickly at these science meetings.

I gave a paper on our LRTAP work in the Canadian Arctic, in a session with similar reports of work in Africa, Europe, Scandinavia and SE Asia. We were finding that most of the man-made organic contaminants (PCBs, toxaphene, dieldrin, DDT, etc.) were everywhere, in biota at the bottom of the ocean, in glacier snow, in the Arctic and Tropical regions.

The meeting organisers put us on a bus tour up into the mountains to a violin maker's cottage and museum. It was a nice instrument building shop and they had some old 1600 violins on display, viola de gambas and medieval viols, zithers and autoharps. Back to Munich for a beer garden party, more brass polka band with whips, beer maids handling 4-6 one litre steins, lots of pigs feet and dumplings. Again I sat with Australian Richard Pearson, who tells me I should come to James Cook University in north Queensland, where he works. The Germans certainly know how to put on a good breakfast here, big fat sausages, bacon and eggs, thick toast with four kinds of jam, thick coffee and piles of fresh fruit.

I was impressed with a big session on climate change impact on freshwater resources, greenhouse warming effects on fish and lake biota. Jonathan Overpeck reported on some Canadian fjord lakes, Herb Wright reported on Minnesota lakes and Dan Livingstone provided data showing declining trends in river flow in Africa. Overpeck wanted to know if Lake Hazen has varves...I dunno, but Sophia Lake does. I promised to report on Hazen Lake next year.

I was chair for a session on radiochemical dating of lake sediments, with good friends John Robbins, Curt Olsen and Jack Cornett. As usual, such a simple thing requires complicated efforts to quantify how fast mud accumulates at the bottom of lakes.

After the science sessions we would wander the city, watching the Marienplatz puppets in old city tower, street performers playing Bach on lutes and viols, pantomimes by silver painted actors, chocolate and wood carving shops, jugglers and tafelmusik from balconies. There were also gypsy singers, beggars, drunks and panhandlers. I bought a Turkish drum for Ian and a German hunter cap (with boar bristles) for Jesse. I spent a long time in the Museum of Hunting and Fishing, this being a very serious activity around here. There were endless rooms full of hunting and fishing gear, paintings, books, guns, powderhorns, creels and fish traps, stuffed birds and fish and boars, canoes, boats, knives and a giant Irish elk skeleton in entrance room. I also visited the 'Alte Pinokotek', four floors of old European paintings from Germany, Austria, Holland, lots of de Breughel, Hieronymus Bosch, Rubens...stuff from 1600 to 1800...and then I discovered that there was a 'Neue Pinokotek' for art work 1800 to 1920s. All this tramping around gave me sore feet. My hotel room has a free shoe shine machine, and it keeps my cowboy boots gleaming.

Most of my gang left Munich 21 Aug '89, British Air flight to London. German friends warned us to be careful for 'dippers'...pickpocket thieves at the Munich airport. London airport ladies allowed me to snooze on big sofas in the departure lounge, a wonderful luxury. Air Canada flight to Halifax, then Toronto by 0330, arriving in Winnipeg 0550, I'm stuffed with lots of airline breakfasts, lunches and suppers. I suppose if you keep flying west you could eat suppers around the world? My brain gets a bit fried on these long trips...and then my dear Sheila is there smiling and beaming to meet me and I am brought delightfully back to earth, literally, straight to our big garden at the Tourond farm. Some melons were ripe.

We collected Jesse at the Winnipeg train station, him returning from being a fishing guide at Kasba Lake with 'Crash' Welch. He regaled us with very funny stories about how he handled incompetent but wealthy Chicago fishermen with expensive fishing rods and clothing, letting them catch 30 lb lake trout and then releasing them. He would take them up the Kazan River for grayling fly fishing but the river was so shallow they had to get the boat going fast enough to lift out of the water and tip the boat to keep the propeller from hitting the rocks. He would take these rich folks up on a river bank on warm sunny days and cook their catch on a campfire for lunch.

With barely time to catch my breath, we took a big family trip to Calgary in early September to deliver Ian to the University of Calgary School of Music. He got his dorm room, meal ticket, bank account, a good bike, a key for percussion room and we met some of his future buddies in the world of music performance. Ian was anxious to vanish into the world of percussion and he looks secure and happy in a university environment, not at all uncertain.

Jesse bullies us the next morning to drive further west into cattle ranch country and the Rocky Mountains, Banff, Sulphur Mountain Hot Springs (36°C) with tufa and steam, gondola further up the mountain to ice and snow, bighorn sheep. Sunset in the Rockies...this is Ian Tyson country...'when the Rockies turn to rose'. We drove back home through Medicine Hat, Swift Current, cowboy country, where the deer and the antelope play, sunny and clean. We rolled into Regina in time for supper with Sheila's sister Peabody and Al's family and the next day we gratefully arrived home in Tourond to hogs, puppies, garden and poultry. Jesse immediately took over Ian's room and set up his ghetto blaster and motorcycle repair manuals.

On 24 September we plucked, gutted, cleaned and froze forty-seven big Cornish Giant chickens, average weight 7 lbs, 330 lbs total in our freezer. Sheila's brother Alistair and sister Mary came down to help us with the big all-day job. Our summer kitchen wood stove was fired up so we could dunk the chickens in wash tub full of hot soapy water to make the plucking easier.

Quickly shifting gears, on 26 September 1989 I was a leader/organiser for a DFO meeting in Quebec, Institute Maurice Lamontaigne (Sainte Flavie), Mont Joli/Rimouski/Auberge, and I gave a paper on LRTAP field sampling decontamination methods. The mood of the meeting was sombre, as little funding supported our grand plans and we were being left behind by our European, Asian, African and South American colleagues. The country here looks European, with small farms, small village culture along the Gulf of St Lawrence. On the way home I visited Sheila's brother and wife (Ian and Angie) in Montreal and I had a big potato gift for their son Timothy in my suitcase.

Early October, back home we helped Jesse pack for a fly-in moose hunt on the Bloodvein River with Conrad and Buster...two canoes, tents, grub, rifles and meat handling equipment. Jess got a spike bull, dressed and boned in the field and brought home in plastic sacks, a weary and tired boy. Meanwhile, Ian is in Calgary performing for King Hussein and Queen Nad. By mid-October it was time to slaughter our pigs (affectionately named Nils Bohr, Henri Becquerel, Rutherford and Schrödinger). We gutted and split them on a singletree in the barn and let them hang for three days. We brought in some 4X8' sheets of plywood to cover our big kitchen table and cut meat for a full day. And then we made some very good sausage from pigs and Jesse's moose meat scraps. On 28 Oct Jesse got the Chief Scout Award at an impressive ceremony at the Manitoba Legislative Building in Winnipeg. Somehow our family saw no conflict or contradiction between playing music for a king, moose hunting and butchering hogs.

I was still an adjunct professor at the University of Manitoba Earth Sciences Department, giving lectures this fall in geolimnology and biogeochemical evolution in the Microbiology Department. The AGM of the Canadian Folk Traditional Music Society was held in Winnipeg, which I helped organise (at this time I was President) with help from Sheila's family and the Edigers. This meant that the CFTM board of directors assembled from all provinces of Canada to present scholarly papers and give a series of concerts. And of course there was a banquet at a good downtown restaurant. My colleague Arnold Nauwerck was visiting from the Austrian Academy of Science, on a scientific tour with Russians, Poles, Germans, and we managed to entertain them in grand style. Meadowlark played several concerts around town and my journals are thick with ideas for new ballads, tune lists and band jokes.

Sheila got the news that her mother died 7 November 1989. She had been in poor health and had fallen. So I hauled Sheila, her brother Alistair (Wump) and sisters Mary and Heather (Peabody, bulging eight month pregnant) to the airport for the funeral trip. Sheila's mother, Pamela Germaine Trechmann Dennett, was born in 1922 at 'Kiora', a house at Norton (Stockton)-on-Tees, Co Durham, England. During World War II she was in the Women's Auxiliary to the Air Force in London, where she operated the radar screens that tracked British pilots over the North Sea. This is where she met radar engineer Canadian Squadron Leader Lionel John Dennett and they proceeded to have seven children over the next several decades.

Boarding the airplane at the Winnipeg airport Sheila had excess baggage, a large duffle bag full of frozen chickens, ham, pork and sweet corn. Jesse and I stayed home batchin' and we ate lots of moose steak, home-grown potatoes and onions. I phoned Ian in Calgary to inform him of the death of his grandmother and he was upset as expected. Jesse was uncertain how to act with family death. Jess and I hauled in five wagon loads of standing dead timber to cut and split for winter firewood and stacked it neatly in our barn. Sheila and her siblings return on 12 Nov., weary, pale, sad and sorry looking. In December 1989 Jesse (sixteen yrs. old) got two deer down south of us, one on my tag. They were hung in our barn, butchered, packaged and froze. Our family ate quite a lot of meat. And now Jesse has his driver's licence, heaven help us!

Buster and I and others flew into Rankin Inlet (4-6 Dec '89) for another information session concerning uranium mining impacts. Like so many of these government briefing meetings, it was held in posh Siniktavik hotel, with good food, spotless table cloths and standard rooms. We met Paul Wolf and several Inuk media men and some Inuk lawyers. The meeting began by singing the National Anthem. Anti-Kiggavik buttons were prominent amongst the older Inuit. Each of us gave a summary of our expertise and experience in the Baker Lake region (where the uranium mine was planned). There were few questions and the local crowd was sullen. Many of the Ottawa government organisers of this meeting were alarmed and frightened, which tickled us pink, as we knew some of the local people. Outdoors it was pitch dark for most of the day, but Inuit kids were roaring around on skidoos and three wheelers well past midnight. And then we flew home to Wpg where it is -35 °C and Sheila had extinguished a chimney fire from our wood stove. She used up two fire extinguishers.

After the usual preparations for the Christmas holiday season the extreme cold weather and south wind froze our upstairs bathroom water pipes, bursting the copper pipes, and the water dripped down through the ceiling onto our hot wood stove and Xmas tree in the living room. This in the midst of Sheila's preparations for a family Christmas party, the steam rising from the water dripping onto the stove. Luckily, we got a miraculous fix by a local plumber, while Sheila and sister Mary polished brass candle holders, silver tea service and silver goblets. Meanwhile Jesse and I went outdoors to split firewood. Ian was supposed to arrive from the University of Calgary before Xmas, but bad weather had slowed him down. He phoned on Christmas Day from classmate Brent Van Dusen's hog farm house in Pilot Mound, Manitoba. So I drove to pick him up in our cherry red four wheel drive Subaru over snow, ice and blowing wind. New Year's oyster soup very good and Ian went back to Calgary on 5 January with Brent. Jesse's Niverville Collegiate high school basketball team was doing very well and I was acting as an assistant to the school coach. Meadowlark was busy with rehearsals and performances throughout the winter, with many late night trips home in blinding blizzards. We sponsored several mini-festivals and concert series, and made a lot of friends with local and far away musicians.

Fiddler and friend Don Zeuff's house burned down in early Dec. So we had a benefit concert for his family at West End Cultural Centre in early January 1990, a dozen bands performing including Meadowlark.

Meadowlark 1989: Walter, GJB, David, Mary, Sheila

In late January I travelled to Vancouver, Victoria and the Sydney Institute of Ocean Science (IOS) for DFO/LRTAP meetings. This was a desperation meeting to plan a minimal program of atmospheric contaminant dispersal research in Canada, as funding was not on the horizon. Several good papers were presented on PCBs and sooty charcoally PAHs in the high Arctic, with some evidence for origins from Asia over the poles. After the meetings were over we sat in front of the institute windows and watched sea eagles fly over this beautiful coastal scenery. I was offered a job at IOS in sediment geochemistry, but their funding situation seemed to be equally dismal to my institute in Winnipeg. Back in my institute office I struggled with DFO and DIAND applications for PCSP help for our next LRTAP trip to Lake Hazen in the high eastern Arctic. I had to provide endless lists of gear, weight and volume of our freight, safety procedures, survival training and lots of requests for illustrations of how this work was going to benefit commercial progress for the nation of Canada. It was getting harder to keep my crew committed to another difficult Arctic LRTAP trip due to no money and little Ottawa administrative enthusiasm.

In early Feb 1990 I was again off to New Orleans for the big AGU/ASLO/AAAS meetings. Nelson Mandela was released from jail, according to news screens in the airports. The Minneapolis airport has my distant relative Ned Butterfield's flower shop, and I phoned my mother's cousin Toady for a quick talk. They still had boot shine stands and my boots were polished several times. My goodness, the fresh pastries were good in Cincinnati, where they served cinnamon buns fresh from the oven. The good smell would drive you nuts, even from the most remote concourse. I arrived in New Orleans to breathe the moist warm air and had a proper New Orleans supper, 'po-boy,' fried bread with raw oysters and hot sauce and Dixie Beer.

These AGU meetings are science big time, six thousand registered scientists, twelve concurrent sessions, most of my colleagues are here...what a blast to see enthusiasm and warm camaraderie. Wally Broecker and the LDGO crew are here, pushing for funding of a GEOSECS of the atmosphere. My good friend Swiss Peter Santschi appeared and we had supper at K-Paul's Louisiana Kitchen, seafood stuffed deep fried eggplant, then a walk down to the Café du Monde for coffee and beignets. Dozens of my science colleagues are sitting around outdoor tables drinking chicory spiced coffee and the river boats are honking nearby. We walk the French Quarter, Basin Street, Canal Street, Louis Armstrong Park, such a warm comfortable part of the city with good book and record stores, music, mimes, poetry and panhandlers on the street. Cajun, Dixieland, pop, R&B, delta blues, reggae/voodoo, rapping gymnasts, mimes painted silver, smells of cigars, horse manure, smoke from street vendor carts, fishy slime and river smells. I bought melons and tropical fruits for my hotel room and staggered back there to sleep.

Susan Kobany and Bob Hermmeler were here, one of the pioneer internet providers of the Omnet service. Lot of papers were given on laminated sediments, varves, in the geological record and in recent freshwater and marine basins. I was a leader in a SIL working group on varved sediments and we were planning a series of sessions for the next congress in Barcelona. A bunch of us piled into a cab and went to Jefferson Parish to see the Mardi Gras Parade, many bizarre floats, garish costumes of some leaders, painted naked ladies and men throwing baubles, beads, panties to the crowd, ROTC marine band, school bands, jazz dancers, decorated horses. The sidewalks featured smoky vendor carts with seafood, hot pecan pralines, fresh bread rolls stuffed with spicy seafood. I phoned dear Sheila back home, where she had a sore back from painting walls and it is -35 deg C and howling wind. Yes, I felt guilty.

Leave New Orleans 0840 on 17 Feb after three security checks (some bad people are in the airport with guns), we took off over Lake Pontchartrain and brown swamps with straight channels, fishermen setting nets. I'm quietly singing the long ballad about Lake Pontchartrain as the airplane lifts above the clouds and I can't see the Mississippi River Delta anymore. And then we visit Memphis, Minneapolis (snowing, cold) and finally Winnipeg by 1435, Sheila and Jesse there to meet me and it is almost twilight, blowing snow and cold again.

In March 1990 everything seemed to fall apart. Our farm home porch door broke off its hinges and our home water well went dry. We had to hire a driller and plumber to put new pipe down 157 feet for good sweet water. Then the furnace quit running and I moved Sheila to the

Edigers' house. At the same time we were sponsoring a Home Town Folk Music Festival and we coordinated the billeting of the out of town musicians, kept the sound technicians happy, managed the stage performances, as well as taking our turn as performers.

Our detailed plans for a high Arctic Hazen Lake LRTAP trip had attracted no money or support and we were ready to abandon the entire project. Big meetings were held in Winnipeg to discuss the future experiments at ELA and there were serious proposals from Ottawa bureaucrats to close the research camp. David Schindler had come to these meetings from Alberta, to help with the presentation and promotion of ELA. The meeting ended with steaming anger and distrust between the working scientists and the administrative bosses (who controlled the purse strings). With great irony, in the following week I was a panel discussion leader at a UNESCO meeting at the Canada Centre for Inland Waters in Burlington, Ontario, about global LRTAP efforts and coordination. Here I was, coordinating Arctic research efforts from Sweden, Finland, Norway, Alaska and Russia, while my own research program had been abandoned.

Sheila came along on this trip, as her youngest brother Bill was getting married to school teacher Carla on 17 March. All of Sheila's family were there amidst a flutter of bridesmaids' dresses, men in tuxes, bowling parties and the standard wedding tricks. At the wedding reception Bill's jazz band played Duke Ellington, Glen Miller, Count Basie tunes, and I got to dance with all of the young nieces and Sheila.

In March '90 after several Meadowlark concerts in small Manitoba town halls, we drove down to visit my South Dakota folks. We drove through the dry dusty prairie with no snow, alkali flats and dust devils rising in the cold wind. As usual we arrive late in the evening, but my mother had the cinnamon buns and coffee ready to go on the ranch kitchen table. Over the next week I helped my father to pull calves in the warm barn at midnight, coming into the kitchen at 0200 to warm up, snacking on cinnamon buns, sour cream raisin pie, chokecherry jelly on toast and tea. Due to it being a warm winter a lot of the wildlife was very visible: antelope, hawks, white tail and mule deer, buffalo berries, yucca pods, badger, eagles, ducks, geese, deer and snowy owls. We drove to Philip, where my brother's daughter Donna Sue was teaching high school science. I gave a slide show about our Mackenzie River pipeline study and Arctic LRTAP research. During this time an Indian fellow tried to get into the liquor store by crawling down the chimney. He got stuck, was there all night, and the next day two squaws tried to pull him out with ropes in broad daylight. Jesse, now a handsome Canadian fellow, went to a Murdo dance and he apparently got tangled up in a girl war. When he woke up the next afternoon he said 'South Dakota girls are so much better than Manitoba girls...they drive big pickup trucks and chew tobacco and drink beer and cuss a lot.' Jesse slept most of the way back home, through dry prairie, lots of ducks, swans, geese, pheasants and grouse.

And the culture shock wars didn't end there. Sheila's uncle Freddie and Aunt Ester arrived from England in April. They came down to our Tourond farm home for big roast chicken feasts at our round table. With Sheila's father and mother now gone, I was able to get more details of Sheila's English family history for our genealogical treatise. Uncle Freddie was born in India, had been a British government administrator in Uganda and the Seychelles and was familiar with the introduction of Nile Perch to Lake Victoria and the devastation of endemic species that followed. He was also very knowledgeable about tropical fish. We argued late into the night about the philosophy of science and social/cultural/religious impediments to intelligent behaviour, part of which later got published in two short books. To their great amusement we cooked a haunch of venison for a feast. Our Tourond yard was now green grass and the garden was about half planted.

Back in my Winnipeg office I rewrote, one more time, applications for government funding for our Hazen Lake trip, as we needed air travel fare and charter Twin Otter funding for Polar Continental Shelf for scientists from the east and west coasts, as well as my crew. Despair had set in and we were expected to abandon our project. In late April I went to Chalk River AECL for a review committee job with a Belgian health physics professor, a radiation ecologist fellow from Alabama, a Battelle engineer from Washington State and an Oak Ridge radioecologist. When I returned to my office on 2 May there was an email message indicating that we had the appropriate funding and we should proceed with the expedition. Talk about quick response, on 3 May 1990 Paul and Brian left our Winnipeg institute early to prepare gear in the Resolute Bay PCSP warehouse. Margaret Koshinsky was hired to cook for us at the Hazen Lake cabin.

And so here we go again, flying from Winnipeg to Edmonton, Yellowknife and Resolute Bay on Cornwallis Island, where the weather was clean crisp cold sunny and the snow still crusty. The pilots had been having a bad time as two planes had damaged ski struts, there had been a four day blizzard earlier on, everybody was on edge and needed a few days rest. So we back off our plans to allow peace to develop...and in any case our Montreal food shipment had not arrived. This meant that we enjoyed PCSP movies, pool table (Brian Billeck is a hustling pool shark), cards, shuffle board. Most of the deck roustabouts here were from Newfoundland and they spoke in a near foreign language, a thick maritime dialect. There was foggy grey sky, no horizon, white-out conditions feared by pilots and Inuit hunters. Dennis Gregor (an Environment Canada scientist from Saskatoon) flew into Agassiz Glacier for ice coring but he and his plane were trapped there by bad weather for four days. PCSP staff were busy trying to rescue a party of Norweigan dog sledders up near the north pole but weather is bad, so we all wait for clear skies. Rum and peach liquor, NBA basketball on the big screen and walking the moat around Resolute Bay were our distractions.

PCSP was also asked to rescue some Korean walking explorers at 86°N who have run out of food. On 5 May some unwelcome VIPs arrived from Ottawa and overseas to sample the pleasurable life in the Arctic. Barry the PCSP Operations Chief was not amused, makes them wait with everybody else.

On 6 May RCMP and NWT wildlife officers and myself (a witness) flew in bad weather to inspect and photograph an illegal polar bear kill site south of Eureka, 80° N, perhaps done by wealthy Swiss adventurers. We overnighted in Eureka, where musk ox herds were nearby, and wolves were howling at the dump. Eureka is a Distant Early Warning (DEW) site, partly military staffed, for early detection of Russian missiles. So again we have ping pong tables, fully equipped exercise room, good food, showers, laundry room, we were not suffering. We observed a radiosonde balloon launch for sampling atmospheric metals and persistent organic contaminants upwards of 40 km high, as part of the LRTAP research program that I was involved in. Marcus Bischoff, a Yellowknife dog sled adventurer, was camped nearby, begging for dog food and airplane rides. The RCMP confiscated his rifle, him being a little unstable. Some of our equipment, two skidoos and two komatiq sleds, was here so we loaded them in a Twin Otter for transport to Borek cabin at Lake Hazen, plus 200 litres of diesel fuel for cabin stoves, all courtesy of a Twin Otter chartered by the RCMP. Everybody helps each other up here in the north.

And finally on 8 May good clear weather arrived and pilot Duncan is roaring to go. He remembered us from previous years' work and he considers us moderately useful humans, about as high a compliment as we could expect from him. Paul, Brian, Margaret and Peter Amaraulik flew with Duncan's Twin Otter from Resolute to Eureka with 2200 lbs of two skidoos and sleds, our scientific gear and food. We then flew three Twin Otter loads from Eureka into the Borek Cabin on the SE shore of Lake Hazen near the Ruggles River outlet. The Twin Otters were on skis and they drove the planes up and down the landing location to pack the snow. The skidoos and sleds were unloaded first, so we could ferry the equipment into the cabin. Duncan enthusiastically helped us start the furnace and stoves and explained how to operate the cabin services.

Dunc demanded coffee and sat back to tell us stories of Weldy Phipps and Gordon Kowtow, who built the first cabin at Lake Hazen, mainly as a safe hide-out for pilots caught in bad weather. Also told of small patches of valley on north shore of Hazen that trap heat from the sun and the ground gets up to 30°C in summer with flowers and green grass, but we are not sure if he was pulling our legs. Dunc tried to leave but weather turned bad, so he stayed for supper and sleep. We were able to monitor radio messages between Cape Columbia and Grise Fjord ice coring parties (Derek Muir, Ray Koerner, Dennis Gregor) where some major equipment failures were happening.

We settled in the Hazen cabin, hung up our cold weather clothing and parkas, Margaret got the stove working and cooked up a mountain of food. The next day we skidood sleds with a sonar sensor in anti-freeze on the lake ice to find maximum depth (maybe 255 m). We came back to the cabin, red-faced in blinding glare light at nine pm, for warm supper of pork chops and rice, coffee and cherry pie. By 10 May we had a rough map of deep parts of the lake. We used the old Defence Research Board camp site (1958) for our lunch shelter. We wandered around this abandoned scientific research camp, looking at old sleds, antique skidoos, fuel barrels scattered around, while musk ox herds above us on mountain slopes watched us carefully. After having determined the maximum depth location on the lake ice we set up our big tent for the coring operation at this location, about 8 km from Borek Cabin. We had radio communications in our coring tent camp, which was called 'Faulty Towers' and Margaret always giggled when we identified ourselves on the radio with this tag. Brian and Paul set nets under the ice between Johns Island and DRB camp and they got fourteen skinny white-fleshed char. We opened our battered and dented coring equipment shipping crates to wash everything with acids and organic solvents. We needed a super clean core sediment slicing bench, washing area, vial/sample storage bins, fish dissection bench with a dry ice Dewar, emergency supplies and first aid kits. Paul and I set up our coring frame tower with a battery powered winch. After half a day of augur work on a big ice hole we winched the big corer down slowly and we could watch it descend through the cold water on the sonar screen. For unknown reasons the winch motor got hot and we cooled it with plastic sacks full of snow. The first two coring attempts failed because the mud was very stiff.

But the warm cabin and Margaret's good cooking cheered us on. We had sour cream and currant pie, pork chops, baked potatoes, fresh baked biscuits and more cherry pie for supper that night. Margaret was excited about her photos of nine big wolves circling the cabin while we were gone, sniffing at old waste barrels. Most ladies might not find this amusing. We did burn rubbish and waste food every day, and the wolves soon lost interest in the cabin. But they did gather around the open water channel of the Ruggles River, where we got our drinking and cooking water. We kept kettles of water on the stoves all day, useful for washing up and to humidify the dry air. We were all getting cracked dry lips and hands. My fancy black skidoo suit was too hot for working, so I reverted to my old ragged dirty South Dakota Carhartt zoot suit.

By 12 May we changed our coring method by cutting the heavy plastic core tubes to 25 cm length, hand filed a sharp bevel, and we finally got good cores of 15-20 cm of sticky laminated clay mud. Still trying two days later for deeper cores (and longer environmental history), we tried our big brass box corer but we got no deep penetration into this sticky clay mud. But we were successful, a big deal, this is what we came here for! Our sediment cores probably represented at least four hundred years of sedimentation history.

Every night Margaret stuffed us with good food. We had baked fresh caught char and roast pork, fresh baked bread, cinnamon buns, apple pie and cherry tarts. Paul and Brian were unhappy to have wolves around the cabin and the water hole and they were ready to go home. My left knee was swollen, red and stiff, probably due to kneeling on wet ice to insert the plug-in bottom of the core tube. On 15 May we attempted to sample a glacier ice core for Pb-210, our radiochemical method of dating the lake sediment core. Glaciers of ice and snow flow from the mountains on the north shore of the lake. Doing science in remote locations requires a certain mania...one of our magical coring mantras was 'Skitter skatter, let's get atter.' We also developed a ritual prayerful bowing to Mount Umingmuk looming above us, with musk ox frequently watching us from above. Peter Amaraulik did a 250 m zooplankton tow with skidoo on the line. Everybody went fishing at the lake outlet on the way back to the cabin so we could have more baked char for supper.

By 16 May we began packing our gear, took down the coring tent in the middle of the lake and hauled three sled loads of gear back to the cabin. After making arrangements for Twin Otter transport to Eureka and then back to Resolute Bay and PCSP, we took down our radio tower. By this time my knee was very painful, swollen, ballooned up bright red. I got served blueberry tarts and coffee in bed and Paul, Brian and Peter refused to let me help assemble the freight pile near the Twin Otter landing area. Wolves were howling throughout the night, as if to say goodbye to us. Wildlife was abundant here: snowy owls, ptarmigan, fox, snow bunting, musk ox, gulls at the dump, flies inside the cabin.

Markus Bishoff (a Swiss adventurer) arrived with his sled dogs, having travelled seven days from Eureka, and he begs for our kitchen scraps to feed his dogs. He was a very odd person...he lived in a tent near Yellowknife and he says he has a wife and kids in Tunisia. Peter pours diesel fuel on our waste pile and human excrement barrel, and it burns through the night (daylight) with a sour bad smell, leaving black specks on clean snow.

Radio communications with Eureka and Resolute Bay were still possible from a short antenna on the Borek Cabin. Our scientific colleague Dennis Gregor was on Agassiz Glacier, to the north of us. Dennis was in a very dark mood, as he was three days behind schedule, having had two blizzards in a row, and his 11 m deep sampling pit filled up with snow. His work complemented our work, as he was essentially trying to get a history of industrial contaminant deposition in snow. Our Hazen cabin radio sked had Dunc flying in with Dennis Gregor's crew, who want to sample Hazen Lake snow and water. Hazen Lake water below 200 m was at 3.3 °C. Dunc demanded coffee and told us the story of flying a US Navy Admiral to coordinates far out on Arctic Ocean ice, carrying two brief cases. He drilled a narrow 1 m hole in the ice, put some dynamite down the hole, using electronic ignition he blew a huge hole in 3 m thick ice. Then he put a hydrophone in the water and proceeded to talk to a US submarine waiting below.

We had arranged PCSP Twin Otter pickup on 17 May, a cloudy warm foggy day. The air craft was at Eureka and the radio ominously revealed that other field crews were stranded due to bad weather. Our Twin Otter arrived at 10 am, we quickly loaded and moved our gear to Eureka in three trips. And there we find a crowd of twenty-seven new military fellows at Eureka camp for microwave tower maintenance (Operation Hurricane). Here was culture shock again: soap operas on large screen TV, microwaved food, pool table bets, card games going on 24 hrs/day and more pilot stories late into the night. A Canadian military electronics technician Leo Carrierre was from St. Pierre, Manitoba, (just south of our Tourond farm) and all of his family play old time Metis fiddle tunes. The Eureka military nurse says my sore knee is aggravated arthritis, due to kneeling on wet ice. Peter Amaraulik loftily told me that the Inuk know enough to put bear hide on ice to protect their knees. We left Eureka that same evening for Resolute Bay, all of us bedraggled, dirty, tired, four Twin Otter plane loads of gear to our PCSP warehouse by 0100...shower and bed, but it was full daylight outdoors.

On 18 May 1990 it was -12 deg C in Resolute and I slept till noon. My face was wind burned, my swollen knee was tensor bandaged and I was using a cane to walk slowly to the galley for food. At lunch we talked with Hans Webber, a veteran of Coast Guard Survey seismic/gravity surveys in Ellesmere region. He wanted to know how well the Borek Cabin cut the wind. He knew many of the early researchers that had studied the fish, musk oxen and wolves in Hazen Lake region, and he had many long tales of disasters in field camps over the past forty years. Our science colleagues drifted back to Resolute Bay from their field camps, some successful, some had met disaster and accomplished nothing useful. Fritz Koerner, Mike Retelle, Dennis Gregor and I met in the pool/billiard room to organise a special AGU session on varves (annually layered sediment) and environmental history in Arctic lakes, fjords and estuaries. Our science colleagues and veteran pilots told us about the unusual character of Buchanan Inlet on Axel Heiberg Island, just to the east of Ellesmere Island. Apparently the Inuit people are afraid of this fjord basin because it smells bad and there are dark spirits that blow strong winds from the nearby hills. I took the time to look up the air photos in the library at PCSP and checked the maps for this location for next year's expedition. Brian and Paul were packing and labelling our frozen samples and equipment to be shipped back to Winnipeg and some gear was left in the warehouse for next year's work from PCSP.

Peter Amaraulik's wife Nancy, kids and grandmother came over to visit us from Buster's Department of Fisheries and Oceans camp. Nancy was proud to have got a bear, assisted by her twelve year old son Mark. Peter took our excess camp food and Buster's DFO equipment back to the DFO lab camp. Peter and Nancy took us on tour of 'Crystal City,' North and South Camp, Signal Hill, Char Lake and showed us their fossil and artefact collection. They have a snug home in the Inuit part of town, with TV, VCR, stereo, good furniture, but there was a bear skull in the corner, seal carcass on kitchen table, haunch of caribou in freezer, bear hide draped over a komatiq outside and hard wind-packed snow almost to the roof eves.

I phoned dear Sheila, who was having a rough time back home on the Tourond farm. The house was cold, as the furnace was not working again. The van driveshaft fell off, our Subaru muffler broke, it had been snowing, a cold spring and many of our garden plants were killed. In addition, the telephone was working irregularly. Son Ian and his buddies were preparing for enlistment with the Ceremonial Guard band in Ottawa. They had to go through boot camp, which certainly had nothing to do with music. Jesse and buddies were trying to fix our van at Stott's garage. And finally, our Braun neighbours have a horse that needs riding out...a little ornery. And of course Sheila read off a long list of Meadowlark performance bookings for the next few months.

Finally our scheduled commercial airline arrived and off to the airport at 1600 with Tom the chief cook, him going out for a three week break in Edmonton. Leave Resolute 1540, arrive Cambridge Bay, snowing and poor visibility, arrive Yellowknife 1950, we find civilisation here as they confiscate my hunting knife and 30.06 shells in my side pack. Leave YK 2040, Paul and Brian drunk on airplane cognac, arrived in Edmonton 2205 and the colour green is shocking, tulips and Spirea blooming. Meet up with Canadian Wildlife Service scientists Ian Stirling and Wendy Calvert, coming down from Polar Bear survey at Radstock Bay (east of Resolute Bay, 70 miles). They had been trying to get age estimates for female bears and they found it was ~ 20 years, less for males. Wendy tells the distasteful story of Mech and Brandenburg, who wrote a fancy coffee table book about Ellesmere wolves...Wendy says the whole thing was staged, all their wolf photos came from the Eureka dump and they used skidoos to herd musk oxen into the wolf pack so they could film a wolf killing a musk oxen calf. This book has a lot of very bad biology.

We had to overnight in Edmonton so I phoned Dave Schindler at his new home. He sounded happy to be free of government institute bosses, had more intellectual freedom at the University of Alberta and was planning some adventurous expeditions to sample dioxins and other POP in high altitude lakes and glaciers of the Rocky Mountains. He bought two llamas for his Wildwood rural home and was building some fences and a garden. On 20 May

Sunday we left Edmonton at 1020, Margaret still had some of her Borek Cabin blueberry tarts for our airplane coffee treats on the last leg of our trip. Arriving in Winnipeg, Sheila, Jesse and Rosalyn were there to meet us. It was strangely odd to see colours other than white and dark.

Our frozen mud and fish tissue samples in dry ice coolers went quickly to our institute cold room. Over the next several months Paul Wilkinson would determine the activities of natural radioactive lead and nuclear bomb fallout nuclides in the mud slices of our cores, to determine the accumulation rate and rough ages of each mud sediment slice. Brian Billeck, and the labs of Mike Stainton, Lyle Lockhart and Derek Muir would determine in each core slice the concentrations of many heavy metals commonly emitted by industry, some of the sooty products of coal/oil combustion (the polycyclic aromatic hydrocarbons, PAHs), and a selection of the pesticides used in agriculture around the world. The core depth distribution of the natural and human industrial contaminants would tell us the time history of deposition of these contaminants and in some cases, suggest the source of the contaminants.

But back home on our hobby farm, our livestock and garden duties increased with the summer air temperature. By late May 1990 our son Ian MacGreggor had survived Canadian army boot camp in Ontario. Sitting at our kitchen table later that summer, he described crawling though muddy swamps with a 90 lb pack, gas mask drills for poisonous gas warfare, breaking down, cleaning and firing rifles and pistols, nuclear war drills and camouflage camping. He polished his boots and wore the high bearskin hat and red uniform for the ceremonial military band. Our seventeen year old son Jesse was observed putting a hot pad on his head to accelerate learning of the French language so he could pass his end of year exams. Jesse went to northern Saskatchewan/NWT Lake Kasba as a summer fishing guide again, this time an experienced old hand. Due to several late frosts we had to plant our garden all over again. We got our little piglets from a nearby piggery and Sheila promptly named them Hamadeus Mozart, Ribski Korsokoff, Chopin and Blossom. Warning to other hobby farmers...you should never name animals you are going to eat. We now have 54 meat chickens, 20 ducks, 2 dozen laying hens, and 4 pigs. With an empty nest Sheila and I had to do all the chores, but we enjoyed this work.

Because I was the coring and radiochemical expert at our institute I was asked to lead a coring expedition in NW Ontario for other institute scientists in early August. They wanted to know the carbon, nitrogen and phosphorus burial rates and mercury contaminant deposition history in these lake sediments. Sheila and I were flown into Trout Lake camp where we used big freight canoes and Lake Winnipeg yawls. We used simple manual coring methods and sliced the cores in the boat immediately after raising the core. We were lucky to have good weather and we took the time to explore old graveyards and abandoned settlements on Green, Orange, Linge and Trout lakes. We viewed caribou, eagles, osprey, loons, gulls and had just enough wind and rain to make it interesting. Sheila also accompanied me to ELA to sample 200 litres of lake water from lakes that had been spiked with cadmium. We took some excess melons and sweet corn with us to treat the field camp staff. We also performed some of our Meadowlark music at the field camp variety night, a big blast for everybody.

GJB Sampling at ELA, 1990

In September 1990, nineteen year old Ian returned from the Ottawa military ceremonial band and promptly burned his uniform and boots with due ceremony. He was not enamoured with military logic and life. Shortly thereafter we hauled him to Calgary for another year of studying music at the University of Calgary. We were involved with more Meadowlark concerts in small towns of the region, attended parent-teachers meetings and I helped coach the school basketball team.

In October I helped to organise and lead a meeting about Arctic climate change in Yellowknife. It was very odd to be considered an 'old veteran' in Western Arctic studies, and many presentations were given using our 1970-75 Mackenzie River Valley research data. My early working partner Herb Wood was now a Water Survey boss in Inuvik and Dave Brackett, who worked with me in Twin Otter water sampling for several years, was a senior man at NWT Natural Resources. Political and policy wonks droned on about the wonderful opportunities for a warmer Arctic, but the old timers were worried about melting permafrost, increased erosion, soil desiccation, fires and wildlife breeding seasons. Some of our long-time Yellowknife friends told us the story of Marcus.

Dear reader, remember our coring expedition to Hazen Lake and the Swiss adventurer on skis who visited us with his dog sled? His name was Marcus Bischoff and he had been arrested by the Eureka RCMP for stealing waste barrels behind the kitchen. Apparently he also stole a parka from PCSP in Resolute Bay and was arrested again. This fellow travelled with his dogs across the northern land and sea and returned to Yellowknife a few months ago. He kept his dogs on an island in Yellowknife Bay (now surrounded by water) and he would take his canoe across the bay to feed them with meat and bone scraps from town restaurants. He was scheduled to appear in Yellowknife court to hear charges for his bad behaviour in Eureka and Resolute, but he did not show up. The RCMP found his empty tent outside of town and later found him dead, eaten by his sled dogs on the island. It appeared that he had tipped over his canoe, got too cold swimming to the island, and the dogs ate him. They found only his head, the rest of him had been dismembered and chewed up by his dogs.

On the way home from this science meeting in Yellowknife I stopped in Calgary to visit Ian and his buddies in the percussion room. These really were drum maniacs, their whole world was composed of drums and percussion instruments of all nations. I got to meet Dr. Price, Ian's teacher, who was curious about our Canadian Folk Music Society, Sheila's hammer dulcimer and techniques for spoons, bones and bodhran. I got to listen to a jazz band sight reading exercise and I could sympathise with the trumpet section. Ian liked his astronomy course, the only science he was ever interested in.

Later in October in Calgary with Sheila we got to visit Ian again. We were both officers in the Canadian Folk Music Society and this annual general meeting featured us on fiddle, hammered dulcimer, banjo, Celtic ballad and cowboy work song workshops, as well as contra-dance, round dance and square dance music. Edith Fowke, Jay Rahn and John Leeder started a war about the society publications, whether we should be academic or popular in our audience. On the way home we visited Sheila's sister Peabody and Al who were cleaning their Regina house to sell, as they were planning to move to British Columbia for a better job.

In November and December I was again writing proposals for another Arctic LRTAP expedition to Buchanan Inlet, another expensive and difficult freshwater/seawater stratified estuarine fjord to visit. I was also working with graduate student Brent Wolfe, who did a very nice thesis on ELA Lake 227 mud laminations. I was also on the Ph.D. thesis committee for Jeff Curtis who had worked on iron and some trace element cycles in experimental tubes at ELA. At this time both Omnet and Bitnet (precursors to standard internet service) were working well, which was fortunate since Brent was in Ottawa, Jeff was in Calgary, Schindler was in Edmonton and I was in Winnipeg.

We drove down to South Dakota to have Christmas with my family. This was probably the last time all the cousins were together, and it was a fearsome view to see over twenty cousins wanting to make homemade ice cream, toboggan sledding on the dam grade, go rabbit and deer hunting, drive the Bronco, ride horses and eat cinnamon buns. My father was then feeding two hundred head of calves and Ian and Jesse got up early to help him feed ground oats/corn in the big wooden troughs in the corral by the barn. Ian was also upstairs with his xylophone practicing *Flight of the Bumblebee* over and over again, in preparation for a concert later in the year. We got to watch my high school's basketball team play and there was the usual Christmas/New Year's dance, with a live country music band. My father's 74th birthday was on 1 Jan '91 and we had the standard homemade ice cream, roast beef and oyster soup with saltine crackers. We drove back to Canada in January 1991, drifting snow blowing across the road, with our Subaru 4WD station wagon loaded with Ian's xylophone, an antelope skull, deer meat and hide. Sheila was terrified that customs would arrest us but no worries, we made it back home in fine time. Ian went back to Calgary to continue his studies. And then our furnace stopped working again, and it was – 30 °C outside. Al Podjan replaced the furnace for $2800, now 20 kW electric, and we no longer needed a fuel oil tank beside the house.

With a struggle to get to the airport through blizzard conditions, I was off to Burlington, Ontario and the Canada Centre for Inland Waters for a 6-10 Jan '91 meeting on experimental design for organic contaminant (PCBs) spikes in ELA lakes. This was more complicated and difficult than our previous nutrient additions, metal and radionuclide additions to ELA lakes. The whole ELA program was also suffering from the loss of leadership by David Schindler. A group of good institute scientists (Hecky, Hesslein, Rudd, Muir, Klaverkamp, and Malley) were now trying to lead research programs at ELA, in coordination with US and European research groups. These new experimental studies were hopefully to be linked to our remote field studies on persistent organic pollutants (POP) and metal contaminant distributions in high Arctic lakes and estuaries. My role was to have been to measure the particle sinking rates and PCB removal rates in the experimental lake, and to do this we needed sediment traps and sediment cores before, during and after the addition of the organic contaminants.

I was going crazy trying to get the measurements and analysis of our Hazen Lake samples graphed and ready for publication, and I didn't need another demanding project. In addition, it was getting harder to work with little support from management, funding had declined, and there was no hope for increased technical support for these new ELA experiments. I was getting disillusioned and bored with my bosses. Later in January I applied for jobs at Sapelo Island field camp in the US state of Georgia, a Mexican marine research institute and an Australian marine institute, just to see what would happen.

I visited my good colleague John Robbins in Ann Arbor, Michigan (NOAA lab) over 13-19 January, to learn about his new computer models of 'conveyor belt mixing' in sediments. I had been interpreting our sediment core chronological profiles of our radioactive sedimentation tracers in a simplistic manner, but I knew this was over-simplification of a complex process. Robbins' models incorporated the process of variable sediment mixing by mud burrowing bottom animals. The effect of sediment mixing on the time history of deposition of our radioactive tracers and the industrial contaminant was to smear the signal forward and backward in history. For example, thermonuclear bomb tests in began in 1952 and practically stopped in 1975, but we could measure some of the bomb fallout isotopes in the top slice of our cores, two decades after fallout had ceased. This was probably caused by bottom animals burrowing into the bottom mud for food and pulling deeper mud up into the surface layers of mud, as well as erosion of watershed soil into the lakes. John had his lab populated by an exciting mixture of students, post-docs and ex-students, new gadgets, big computers, simple and complex models of radiochemical tracers of variable half-lives and contaminants that do and don't decay. At this time a 486 personal computer was the top of the line and the bigger VAX was used for Omnet email and spreadsheet data storage. We had produced good sediment core data for ^{7}Be, ^{137}Cs, ^{234}Th, ^{210}Pb in our Arctic lake and estuary expeditions, and from this data we wanted to know what the total load of metal and organic contaminant had been over the past fifty years or so. We were also anticipating how to measure and interpret experimental lake spikes of ^{109}Cd and others isotopes at ELA. John was helping me interpret the known tracer and unknown contaminant profiles for the ELA, Saq, Cornwallis, and Hazen Lake cores, and I was delighted to be his student.

John's wife Renee was a piano teacher maniac and John was a good pianist too. It was nice to be around somebody who liked both science and music and the arts, as well as good food. We were visited by an ex-student of John's, Lt. Mike Brakke of USAF Intelligence Unit from Augsburg, Germany. Over coffee in the Robbins living room he predicted war in Iraq in twenty-four hrs. And so the TV went nuts on 17 January with invasion of Iraq...and the Fertile Crescent goes up in smoke and flames one more time.

I hurried back to Canada for a 25-27 February meeting in Vancouver about climate change, melting permafrost and changes in water cycle in the Arctic. Out there beside the warm Pacific Ocean we had misty weather, birds were chirping, fog horns at docks, unlike Winnipeg at – 30 °C. I helped a group of good scientists write proposals for over $6 million worth of research but we all knew none of it would be likely to be funded. Steve Zoltai and I wrote a list of scientific and political criteria for judging these proposals and sent them off to Ottawa for decisions. We stayed a few extra days for meetings to write proposals begging for funding of our group LRTAP expeditions to Arctic, mainly for helicopter and Twin Otter costs.

Back in cold Winnipeg we immediately had a miserable expedition to ELA with Brent Wolfe and Paul Wilkinson for more freeze coring of Lake 227 and Lake 375. We had skidoo trouble and had to walk back to camp several time over an icy crust on top of a metre of snow at – 28 °C. The cadmium spike in Lake 375 had begun and we had to dig out a large ice hole to find our sediment trap float, which was our tool to collect and measure the downward flux of radioisotope-labelled particles to the sediments under winter ice cover. Everybody got wet from slushy snow and hidden blowholes and we were much colder and more miserable than our experience in the high Arctic (where the snow and air is very dry). The only salvation was the hot potato chip and gravy wagon in Kenora on the return trip back to Winnipeg.

In March 1991 our Red River Folk Connection ran a Hometown Festival at the West End Cultural Centre. In addition to months of preparations and financial posturings, we had a good crowd for some wonderful local and outside Canadian musicians and we could pay all the bills. Immediately afterward Sheila flew out to British Columbia to visit her sister Peabody and family, leaving Jesse and I behind to batch, again we ate a lot of moose and bear meat and almost nothing else.

The US Environmental Protection Service were part of our international LRTAP team and they had asked me to train one of their fellows in cold weather operations methods, so they could work in Alaska. My technical support team was busy with other work so Jesse came with me to help. Being rather lavishly funded from the USA, we took Chad Gubala (from relatively warm Portland, Oregon) on a coring trip 25-29 March, to fly into Vermillion Bay and Red Lake, Ontario, where we got rooms at the Red Dog Inn. We were coring a group of lakes that were under study by Everett Fee and Bob Hecky from my institute. Chad was not accustomed to working at – 30 °C., so we were teaching basic cold weather survival techniques. We had instructed Chad to buy two sets of insulated overalls, wool sweaters and socks, waterproof boots, a good goose down parka and thermal underwear, with the strong hint that he would be getting wet and miserable. We all dubbin greased our boots in preparation for sloppy slush under the snow around the ice hole. We loaded our single engine Otter with coring gear and flew into crisp snow of a snow covered Ontario lake. Jesse was good with machines and became an artist with the chainsaw and ice augur in making big holes in the ice. We promptly got good cores with laminated sediment, set up our double wall tent for sediment core slicing and had tea, hot chocolate and shortbreads and smoked oysters for lunch.

Jesse in tent coring through the ice on Red Lake, March 1991

Jesse, the expert ice augur artist, discovered the pool room, Snickers and tea back at the hotel. Chad got puking sick and stayed in bed for two days while we cored several other lakes. Jesse was a top notch field man and in addition he was in charge of buying lunch groceries for our coring tent, which meant we had some weird lunches of tinned sardines, tuna, corned beef, fried bread and tea. On most of the lakes we had 30 cm powder snow on top of 1 m of ice. We were comfortable at -35 °C in our tent, with our Coleman stove for heat and tea. We had the right equipment, an A-frame and car battery powered winch over the ice hole. In the tent we had our core slicing platform, radio antenna and radio set, and survival packs in case something went wrong. We usually got good 50-80 cm long cores, maybe 2000 years of lake history, and we sliced while blowing snow came up in the mid-afternoon. This afternoon cold wind often delayed pickup from our aircraft, so we had cocktail hour with lumpfish caviar on crackers, smoked oysters, Camembert cheese and half-frozen fruit juice.

Jesse was ice fishing when the Otter arrived just at dusk, scary windy blowing snow and poor visibility. Finally, the pilots revolted and wanted to give up on this project, as the radio weather report predicted blizzards in Kenora and Winnipeg. So we packed up our gear for later freight haulage and Chad, Jesse and I returned home to Tourond for rest and recuperation with Sheila's puppies, Tifty the Newfoundland dog, Fang the budgie and Max the Brazilian Matamata side-necked turtle in the downstairs bathtub. I would have liked to have heard Chad Gubala's report to his bosses on this trip.

And at the end of March, waiting on my Winnipeg office desk was a note indicating that our funding proposal for LRTAP sampling of high eastern Arctic Buchanan Inlet and working out of Resolute Bay and Eureka was fully approved.

Dearly beloved Sheila returned from visiting her sister Peabody in BC, with her pockets full of sand dollars and muddy rocks, just like a kid exploring the ocean shore. And immediately the house felt better, and the food improved immensely.

By 6 April all snow has melted and our home yard was a lake of muddy water. The Red River ice was moving, robins and thrushes were already here, so I fixed up our barn cold frame/greenhouse for garden seedling plants. I got my first small laptop (my payback for coring Rudd and Fee's Ontario lake sequence). It was geared up for linkage to our phone line, so I can do Omnet and Bitnet email correspondence at home. In mid-April I was off to the University of Waterloo for Bill Buhay's Ph.D. thesis final exam. Bill was mainly trying to sort out the usefulness of stable isotopes of C, N, S and O in different extractions of tree ring cellulose.

On my return, Sheila tells me of a phone call from Australia, asking for an interview for my job application to the Australian Institute of Marine Science. Wow.

Shock, puzzlement, do we really want to do this??? To add to the feeling of uncertainty and family future decision, it was a very rough spring. It got almost too warm in mid-April, but by early May we had snowfalls for two weeks. The prairie wind came up, blew a lot of dirt around and blew the porch door and window off. We lost most of our greenhouse seedlings but by 11 May the irises and daffodils and tulips were flowering. Son Jesse was preparing for his high school class graduation ceremony and a big party afterward at our home.

By 11 May we were off for the Buchanan Inlet LRTAP expedition. Buchanan Inlet was an estuarine fjord-lake on Axel Heiberg Island, just to the west of Ellesmere Island, where we were last spring. We suspected that the surface water would be fresh, the deep water would be diluted seawater, and there would be no oxygen or bottom animals to disturb the mud deposition history in the deepest locations. This was a gamble, as there was little known about this region of the Arctic. All gear had been shipped two weeks ago and we had left some of our equipment in the PCSP Resolute Bay warehouse last year. If this is sounding a little repetitious, beware, as things were changing, both for our family and for my scientific career. It wasn't that this was the only thing we knew how to do, but that this was the only well-funded research project for our group at our Fisheries and Oceans institute. In addition, our work was linked to an international study of the global distribution of contaminants, so that the local bosses were reluctant to discontinue our work.

Paul, Brian and I left Winnipeg 0730, to Regina, Edmonton, Yellowknife (Great Slave Lake ice roads were still in use), to Resolute Bay at 1600, where it was -14 °C and sunny for twenty hours/day. And then the same old mantra come into play when the weather turned bad. Hurry up and wait. Sleep and eat. Sleep and eat.

Peter Amaraulik shows up with a bottle of Captain Morgan rum in his parka pocket, wanting to come to Eureka with us, with his wife and kids. PCSP bosses said no, as Peter was at that time not an employee of DFO. Peter was upset and angry at us. So we radio-phoned the Winnipeg head office to organise a temporary hire of Peter. Our purpose on this trip is to take sediment cores and sample fish from Buchanan Inlet, another fjord-like basin linked to the sea, the surface waters being mostly fresh, and the deep water more salty, perhaps fairly old water isolated from the atmosphere. Anaerobic deep water would protect the bottom mud from mixing by bottom organisms, so we would have a better chance of seeing the history book of lake mud and contaminant accumulation.

It was comfortable to sit in the pilot's radio room for two days with endless supplies of tea and pilot biscuits, listening to the radio chatter. We listened to our pilot Duncan Grant tell stories about Lt. Buchanan surveying Axel Heiberg Island in 1928-30. Some of his gear is still there on landslide on the south shore and there is a sled and steel food cache near the outlet. He said 1 m long char come up blowholes in summer ice and swim around in melt water on top of the ice. He warns us of strong wind gusts that keep the estuarine lake ice relatively clean of snow. Finally on 14 May '91 the weather came clear and we were up at 0645 for urgent packing of Dunc's Twin Otter for transport to Eureka.

We couldn't find Peter Amaraulik and his wife and father say he's 'gone away'. He was supposed to help us. We found out he was in jail for drunken fights and brandishing a gun at people, so we paid his bail, helped him pack (cigarettes, rum, sleeping bag, snow knife), and ran back to the loaded plane. The pilot Dunc was furious with us and he retracted his statement that we were 'moderately useful humans'. He had a big day scheduled. He was to transport us to Eureka, then fly to Agassiz Glacier and to the ice island camp, delivering mail and supplies. He had scheduled to leave Resolute for a month off-duty in Yellowknife. We noted that for this trip Dunc's total survival gear was a parka, a thermos of coffee and two cartons of cigarettes.

But we gotter done. Dunc flew 2.5 tonnes of our gear over North Hamilton and Baille Islands, Devon Island, over the Axel Heiberg Island mountains to Eureka, our base camp for the next week. We could see seals in the spring ice leads and polar bears being successful at hunting them. We were bunked in an old Quonset hut with an oil stove in the middle of the room. Eureka Sound has brown black dirty ice due to 60-90 knot winds and little snow cover on the land. Some of the glacier ice feeding into the sound was white, like rivers of cream. Here we wait again, as we are to be ferried into Buchanan Inlet by helicopter, as there is no good place for a Twin Otter to land. No worries, as we were in the lap of luxurious DEW facilities, big screen TV, pool tables, ping pong, enormous piles of food at meal times, lots of ice cream and pies any time of day. One of the cooks drinks too much and does not appear to sleep, as he is cooking 24 hrs/day for military and pilot operations. We could get fresh made pizza at 0200, but sadly no beer.

The local trouble is in the form of two young Spanish fellows from Madrid, adventurers who had travelled by dog sled from Greenland to Eureka in three months, intending to go to the North Pole. Eight of their twelve dogs died from starvation on the crossing, both men were sick (physically and mentally), they were without dog food, money or support and they spoke very poor English. Peter and I offer to shoot some seals for the dogs but Eureka Officer in Charge (OIC) Bob McInnis wanted to shoot the dogs and ship the two men back to Spain. I don't really know what happened next but the two Spaniards and their dogs vanished, so I assumed McInnis did what he had threatened.

The next day was cloudy and foggy and still no chopper, so I walked across Slidre Fjord to view icebergs looming over the coastal sea cliffs. Tracks of wolves, fox, bear, musk ox were abundant and there were lots of ptarmigan droppings. This was certainly not a biological cold desert! We had wonderfully huge rib steaks and fatty pork chops with shrimp for our late supper, followed by Canadian NHL hockey and US NBA basketball games on the giant TV screens, lots of betting, playing cards. We were embarrassingly comfortable in this far north outpost of civilisation, with our home families and institute friends probably thinking we were suffering in the cold, eating our boots, etc. But then our chopper arrived later that evening, heavily loaded with survival gear, spare parts, oil/greases, propane engine heaters and cargo nets.

Later that night I was summoned to the HQ military radiotelephone room, which was normally off-limits to civilians. It was my dear wife Sheila on the phone, giggling amongst the static and squeal of the radiophone, saying that the Australian Institute of Marine Science had phoned, wanting to negotiate on salary. Sheila had already researched the value of the Australian dollar vs the Canadian dollar and AIMS was in the right ballpark for money. Wow, we knew nothing about tropical AIMS and what am I doing up here in the Arctic? I must have been very focussed on our Arctic research logistics and all the problems with equipment, as my journals say nothing more about this important (and eventually life-changing) message. I must have tucked it away in the back of my head and moved on to the next Eureka problem.

For example, I also took a phone call from Nancy, Peter Amaraulik's wife, who just had another baby (named Star) in Resolute Bay. Now Peter wants to return to Resolute to see his wife and baby but we need him for our work in Buchanan Inlet. A little more turmoil and conflict in our crew happened then.

On 16 May we were up at 0400 to load the first belly net cargo for helicopter pilot Vic Cobb. Peter and I hooked the loaded cargo netting links underneath the chopper and rode it to Buchanan Inlet, where the pilot dumped the cargo net and us. The chopper returned with two more loads, all dumped in a grand pile at the landward end of an inlet in a rocky landslide terrace. Finally, Brian and Paul were brought in with two skidoos and sleds, with 2-200 litre fuel barrels and two insulated tents. Brian got lunch together with Coleman stove tea. Peter and I did skidoo sonar soundings along the central axis of the basin, mostly 45 m deep, with two holes 56 and 82 m deep. We skidoo and sled hauled our camp gear to the maximum water depth station and ice bar chipped holes for 'deadman' anchor ropes for tents. We noticed coal lumps all over inlets and outlets, blocks of columnar basalts, volcanic lavas and organic shales in the exposed valley walls. Up on the northern ridge was a line of black vertical stumps, which we eventually discovered were carbonised Eocene tree stumps. We used our radio connection to Eureka to call for the chopper to pick us up, as we hadn't got our tent camp built yet. We returned to Eureka too late for supper (nuked spaghetti and meatballs, big loaves of fresh baked bread) and then collapsed in our bunks after a frantic long day of work.

The next day, 17 May Friday, 0645 and –8 deg C, sunny, the birds were singing, and a strong wind was blowing dirt and snow across the runway. We return to Buchanan Inlet by 0900, where we finished hauling our cargo net dump of gear to the central lake tent camp using two skidoos and sleds. The little Honda generator fuel regulator got smashed, so Peter jury rigged a repair. Our radio mast and radio set checked out and our emergency gear for a week of overnights was carefully stashed at the back of the tent. A very strong wind was scouring the ice surface, polishing the ice with blown sand. Skidoos would sit stationary on the slippery smooth ice with the track running and we had to give it a push to make it move. Brian and Peter augured three holes in 2 m of ice to set 200 m of gill net with an under ice jigger. The underside of the ice was irregular and spiky, causing troubles with under ice snags and tangles of the gill net. Electrical conductivity (a measure of water salt content) and temperature profiles of the deepest hole showed no deep sea water, but there was no oxygen in deep water and the bottom half of the lake had high concentrations of hydrogen sulphide. This meant that the deep mud of the lake would not have been mixed or disturbed by biological mixing, and it was more likely that we could trust the sequential chemical history of our sediment cores.

We were helicoptered back to Eureka by late evening, dead tired, with headaches from trying to sample deep water hydrogen sulphide. A Hercules with twenty-four US Army blokes arrived, crowding mess hall, toilets and entertainment rooms. One of the US military mechanics fixed our generator fuel regulator and our Coleman stove propane hose, which had been damaged during our helicopter cargo net release at Buchanan Inlet.

Finally, on 18 May, Saturday, we organised and set up our coring tripod and accessory gear over a large ice hole, with our acid and organic solvent cleaned slicing table and tools in the tent. Our first core had tipped sideways and was discarded and our second core was better at 1400. We were slicing in the warm tent with heater full blast, bare handed, but our feet were cold on the slippery ice floor. Brian got three skinny char from the gill nets and he performed the fish dissection with acid and organic solvent clean table and tools like a surgeon in a hospital. Some of the fish had small lumps of charcoal in their stomachs; the poor things must not have many good things to eat. We later noticed that the bottom mud of this estuarine fjord had the same charcoal lumps. This of course caused problems, as part of our target contaminants were the sooty PAHs from industrial smoke stacks of North America, Europe and Asia. Much later we discovered that the charcoal, coal and amber in Buchanan Inlet mud came from much warmer greenhouse Eocene times (55 Ma ago), when trees, ferns, crocodiles and camels roamed this land. These woody carbonised plant fossils on the uplands above Buchanan Inlet were being eroded by ice and glaciers into the lake.

We returned to Eureka for repairs to our gear and there was a serious ripping wind that made it difficult for all four of us to tie down the chopper while the pilot held it steady with the rotor. We were happy to see lots of big fatty pork chops, BBQ ribs and apple/cherry strudel in the mess hall. We were tired and wind-burned, irritated by too many klutz Army fellows at this base camp, most of them helpless loud-mouthed clipboard clerks doing some sort of inventory in day and night shifts.

The next day, 19 May, Sunday, we were up at 0500 ready to go but outside it was almost zero visibility and a roaring forty knot wind. Chopper pilot Vic says 'no go', so we all went back to bed gratefully and I slept till noon.

At 1330 Vic says maybe and we cautiously fly over the white landscape under white overcast sky to Buchanan Inlet, where we can barely see the ice for blowing snow and dirt. There were dust devils about 30 m high above the outflow gravel bar. We had doubts if we were in the right place as there was no tent, no skidoos, and no fuel cache where we left it in the middle of the inlet. We landed on a gravel spit and scanned the place with field glasses, finally finding our skidoos, komatiqs and fuel barrels half way up the rocky slope of the inlet valley wall. With great difficulty we walked/slid in the ripping wind, with dirt in our teeth, to our tent and ice hole location, to find broken tent ropes and steel cable wires (tent grommets still at the end) still frozen in the ice. There was no sign of the tent.

We later found U-shaped tent poles, our battered food supply box, propane tanks, radio antenna, stove, fish dissection tool boxes, frozen apples and oranges, tins of coffee and corned beef and our precious coring gear crate blown along the south side of the basin up against rocks at shore. We were lucky that some of our equipment was inside sturdy shipping boxes so that even if they were blown halfway up a rocky cliff, the gear was protected. The weather was terrible and we couldn't do much, so we flew back to Eureka to let the wind die down. There was strong wind and blowing dirt storm all the way back to Eureka and I think the pilot was somewhat scared too.

Eureka engineers fixed our bent radio antenna, battered generator and Coleman stove hose. We had a wonderful roast chicken supper while the wind was rattling the tin roof and wind-blown sand rattled the windows. PCSP seemed to anticipate our difficulties and they had replacement gear (tents, fuel barrels, skidoos and komatiqs) in a warehouse here. Several pilots tried to explain katabatic wind gusts to us...dense cold Arctic air sliding down mountain valleys onto the inlet ice surface, apparently at hurricane force wind speed. We were lucky to be absent when it happened.

Finally, on 20 May, Monday, it was sunny and clear, fifteen knot winds, so we flew again to Buchannan Inlet, ready to rebuild our camp. We gathered our skidoos down from the rocky slopes, straightened the handle bars, attached new skis, but the engines were still working fine. Most of the fuel barrels were still full and our gear boxes were mostly intact. We set up a new tent, radio antenna, stove and the sampling gear was organised again. Peter and Brian were fighting the irregular underside of the ice with the jigger to set a net for char. We took big net samples of phytoplankton (algae) and zooplankton for colleagues Hedy Kling and Kaz Patalas back at our institute. *Voice of America* was on our fish finder radio, reporting sixty-four dead in India, while we dissected three miserable looking char. These skinny fish had charcoal and amber in their stomachs. Brian took samples of liver, gonads, bile and muscle, for measurements of contaminants and physiological enzymes/proteins involved in biological sequestering and excretion of dangerous contaminants. We got two good Kajac Brinkhurst cores, revealing that the sediments were laminated as we had hoped. We got 35 cm of stiff mud sliced into clean containers, representing maybe 500-1000 years of history.

The helicopter picked us up at Buchanan Inlet at 7 pm and the ride back to Eureka gave us beautiful light on mountains, glaciers and Eureka Sound. There were seals basking in the evening sun and twelve musk ox on parade. We gratefully ate big steaks and we were hungry for fat and sugar. We had several late teas and many toast sandwiches with honey and jam. In the shower our faces were sore and red from wind burn. Our sleeping bags were cold and damp from blowing snow on the chopper cargo rack, from our ride to and from Inlet.

21 May, Tuesday, -6 deg C, clear and calm. Chopper fuel line ruptured en route to Buchanan Inlet, spewing fuel all over our gear on the cargo rack, including our sleeping bags and parkas. But we set up the big KB corer and tripod again and got an even longer core from

the deep hole, 46 cm long, it took all day to slice cleanly and wash up. Brian was dissecting fish again and went hunting *Saxifrage spp.* (a small flowering plant) in the evening. Peter and I fixed the battered and dented skidoos and sleds so we could take a long ride up the valleys to the top land, where we found parts of our tent and other gear, carbonised tree stumps, musk oxen bone piles, mainly skulls and ribs, big flat stones in rings (Peter told me not to touch them, as they were magic. These flat stones were not likely recent Holocene Inuit, as no Inuit were native this high in the Arctic). We were searching for a flat safe place for the helicopter to pick up our equipment in cargo nets. We were now finished with our sediment coring work and all of us had a desire to become a little more familiar with this place. We skidooed down to the outlet gravel bar, where the remains of a pre-World War II emergency supply cache had been packaged and put into sealed steel barrels. We packed up our gear for our final retreat, ready for multiple helicopter cargo net haulage.

We flew back to Eureka in 10 pm twilight, viewing musk oxen herds with calves, whipped cream glaciers draped over mountain slopes, the ice flowing down valleys into the sea. We had the culture shock of eating small buckets of jalapeno chicken wings, veal cutlets, ham, three kinds of cheesecake and espresso for supper, sports on big screen TV, pool table hustlers with ten dollar bills on the table. I made the radio call to Resolute PCSP to arrange for a Twin Otter to carry our gear and us back to Resolute Bay in a few days from now.

The next two days were cloudy whiteout poor visibility, -5 deg C., 25-35 knot wind, blowing snow and sand, marginal conditions for the chopper. But we did make it into Buchanan Inlet, which was now strangely protected from the wind. Brian and Paul were still jigger gill net fishing, while Peter and I moved most of our freight boxed equipment on sleds to suitable high level ground (800+ m above sea level), so that the chopper could pick up our gear in cargo nets. Above the estuarine lake valley the rising wind made it increasingly dangerous for us to stand underneath the chopper to hook the net. The chopper tried to land but the wind blew it sideways across the frozen icy gravel and almost tipped over. So we selected a different place for cargo loading, at the inland end of the lake on big rock landslide, where it was more protected from the wind. Brian's Bistro supplied hot coffee and hot corned beef sandwiches at lunch back at Inlet tent camp, where Brian and Paul had caught six big char and were dissecting them, nice char fillets for snacks. We tested our survival .22/.410 rifle by target shooting. Peter and I skidoo sled hauled the rest of our gear to the cargo loading spot for the next day and we all few back to Eureka for a hot supper of spicy ribs, egg rolls and an endless supply of banana cream pie.

We were back to Buchanan Inlet the next day for another seven char in our gill nets set overnight. The fish were dissected and the stomach contents, organs, bone and flesh were frozen in liquid nitrogen. It was time to take down our tent and radio antenna (Faulty Towers) and clean up the site. We had 1.5 tonnes of gear piled up for several chopper loads in cargo nets. We were sweating at this work, stripped down to wool shirts, as it was relatively warm, just below freezing. In spite of the weather we got two loads back to Eureka, feeling good, having accomplished what we wanted to do, tired and cold. Radio communications to Resolute Bay indicated even worse conditions there, fog and wet snow, and we were advised to relax for a day or two. The PCSP Twin Otters were busy with traffic to Taconite Inlet, Agassiz glacier and several other research camps. So we settled down to a ham and pineapple supper, Detroit TV, movies and Ghandi's funeral pyre of sandalwood. Our crazy Spaniards reappeared in the foggy windy night, having been floundering around on the sea ice. After a good supper, they departed again with only four dogs and one big sled, intending to travel to Grice Fjord. Peter tried to teach them how to hunt seals from snow blinds but they wouldn't listen. I think this was partly a language problem.

Over the next two days we hauled our gear from Buchanan Inlet to Eureka by chopper cargo nets. We then prepared two Twin Otter loads for flights to Resolute Bay. Now our work was done and we can afford to relax, get lazy, tired and sleep. The adrenalin of adventure leaked out of Brian and Paul quickly and they wanted to go home. They were no longer amused with multiple pool tables, ping pong or TV. But still we hurry up and wait.

Ed Darlington of the University of Washington in Seattle arrived with radar and computer gear for measurements of glacier ice thickness (100-800 m) at Mer de Glas. He says the Chinese glaciologists found lots of black carbon (soot, charcoal) on their glaciers, and this sooty black stuff was causing increased glacier ice melt rate. His crew were also wiring sensors in icebergs but the polar bears kept digging them up. So he hired Inuit hunters who took out seven bears in two days. He had a great little super-fast Toshiba laptop, which he used in the field inside a plastic cooler full of hot water bottles, on a komatiq behind a skidoo. This little computer was recording glacier ice thickness from sonar and mini-seismic shock explosions, and they could see under ice mountain ranges and valleys. They drilled 100s of meters into the glaciers and lowered a microphone into the drill hole to listen to flowing liquid water underneath the glacier ice.

Peter Amaraulik was puzzled by the pictures in the book I'm reading about Australian aboriginal culture...they look like Inuk but are naked, don't they get cold? He was also puzzled by all the violent movies and TV shows here...poor language and no respect, is all of USA and Canada like Rambo and Schwarzenegger?? Why was everybody so loud and rude? Finally Dunc and our Twin Otter arrived and we returned to Resolute Bay late 24 May. Peter was met by his wife and new daughter Star.

The next day at Resolute Bay PCSP we sorted our gear, some of it to stay here for future work, some belongs to Buster Welch's DFO camp, and some to be shipped to our home lab in Winnipeg. Fish tissue samples were in dry ice or a Dewar of liquid nitrogen, and would travel with us. Sliced sediment cores were relabelled and packed secure inside shipping crates. The *Voice of America* radio was blaring Cajun music and Inuktitut country music, broadcast over the huge warehouse sheds where we were working. And then I could feel my strength draining away, my adrenalin declining, muscles and brain relaxing. I no longer had to push myself, the crew and the pilots.

Peter Amaraulik came to PCSP to say goodbye. He was oddly quiet and serious, perhaps sensing that we might not be coming back. He gave me a little piece of paper with the Inuk symbols for 'Star', in English letters it is 'Uluriak', the name of his new baby girl. I finally beat him at a game of snooker at the pool table but then, maybe he let me win? He asked if we would return again next spring and I said I couldn't promise this, as funding was getting harder to find. I hadn't mentioned the job offer in tropical Australia, as this was just too bizarre to even think about here. Peter was a well-educated, technologically competent Inuit, but he hadn't quite abandoned the old ways. He talked of Sedna, the legendary spirit in the sea, and that Sedna would keep calling us back to the north. In previous discussions he said that some shamans could fly, inhabit the souls of wildlife, and knew more about the world that the southern newspapers. Some years later I found this in a blog about Neil Armstrong walking on the moon:

> 'We in Nunavut know, of course, that that is not true, and that the accolades extended to Neil Armstrong (for being first to walk on the moon) are entirely unwarranted. No, I am not a conspiracy theorist who believes that Armstrong's moon landing was faked at some mysterious location in the Nevada desert. Armstrong reached the moon. But his accolades are undeserved because he was not first. All right-thinking Nunavummiut know this, because we know that Inuit regularly visited the moon for centuries.'

David Iqaqrialu said as much in a heated exchange in the Nunavut legislature on May 6, 2002. We know it was heated because he prefaced his remarks by telling the Speaker, 'I am starting to get hot under the collar...' He then went on to say, as reported in *Hansard*, '...it is not really related to the question that I posed, but this is background material. Inuit had reached the moon quite some time ago during the shamanistic ages, prior to the Americans reaching it with their machines and finding out it wasn't what they thought it was.' In fact, Inuit pre-Christian beliefs held that shamans often travelled through the air to visit the spirit of the moon man, who was always willing to give men good hunting. Many Inuit continue to believe that shamans had this ability in the past.

We left Resolute Bay at 1645 on 25 May, all of us sleepy, groggy, wind burned. We tried to read new copies of the *Globe and Mail* and the YK *News of the North*, sipping Drambuie with honey roasted peanuts.

A small sadness crept into my brain...if I go to Australia, I won't see this country again. And here was Cambridge Bay, and then Yellowknife by 10 pm. Getting off the plane we were snow blind, our whole bodies felt stiff and sore. Great Slave Lake was still solid ice covered, but some of the smaller inland lakes were ice free already. Airport security worried about my jackknife and half a dozen .410 shells in my briefcase. We arrived in Edmonton at midnight, where it was + 22 deg C and there was green grass under airport taxi lights. We overnighted here and in the hotel I went to sleep with hot towels on my face.

My journals show lists of home and office gear I need to get ready for a move to Australia, a long list of questions for the negotiation phone call and queries about biogeochemistry of coral reefs and tropical mangroves. I should have been scared of this proposed move, as I actually knew almost nothing about this Australian institute or tropical oceanography.

On 26 May we were up at 0500, to the Edmonton airport by 0630, where there were green leaves on trees. I was groggy, stiff ankles and feet, face red and sore and my brain was in culture shock. So here is Saskatoon and then Winnipeg by 0940 and there is my Sheila bright and shining and brown already, driving a borrowed car (our van broke down). There was not a moment's rest for the weary, as she hauled me off to a Meadowlark rehearsal at David Ediger's home that afternoon. Finally we got down to our green peaceful Tourond hobby farm, to find that our (Jesse's) eleven year old Newfoundland dog Tifty had died peacefully, curled up on a pile of straw in our barn. It was almost as if Tifty knew we were leaving before we knew it, and she could not come with us to Australia.

This has been a very long chapter about our Long Range Transport of Arctic Pollutants (LRTAP) work. Before I go on to the next chapter of my life, I need to take some time to explain our results and our role in this international project. I have described mostly the outdoor field work but of course there was a small intellectual army of biologists, chemists and administrative support staff back in the Winnipeg home institute, as well as collaborators around the world that helped us do this work. As I've described before, I got a delicious vibration/tickle/emotional swelling when our international team (mostly Canadians, Swedes and Norwegians, Russians, Finlanders and Alaskans) would informally gather for supper at AGU scientific meetings in New Orleans, San Francisco, San Antonio, Munich and Lyon. We would argue loudly about how to measure low-level contaminants, new methods of sampling, tales of mistakes and field adventures that ended badly, and how to operate on small budgets. Here was exciting worthwhile collaborative science in operation.

Unfortunately not so back in my office in Winnipeg, where oddly parochial and administrative matters were dominant in our bosses minds. As evidence of the decay of my institute to managerial efficiency, cost cutting and allegiance to Ottawa priorities, our excellent metal and electronic workshop was closed, expert machinist Willy Burton, general handyman Cliff Jones and others were fired and their shop work space was converted to administrative offices. Willy had built all of our custom designed coring apparatus and our stainless steel atmospheric contaminant collectors, and Cliff was our main man for building ELA from a bare rocky sandy lake shore.

We were in the early stages of providing quantitative historical evidence for global dispersal of organic and metallic contaminants, and the impact that these organic poisons and extra supplies of dangerous metals might have on ecosystems and humans. Rachel Carson ignited the public about this concern in 1964, and it has taken several decades to show that she was correct in her brave assertions based upon limited data. [Of course, the denial industry has been at work on Carson's book over several decades too, trying to rescue the profits of the big pesticide/herbicide/poison manufacturers...see the book by Morriss et al. cited in the references.]

After I left Canada our lab and field crews continued this work and published science journal papers and book chapters on our results. Much of our work and that of our international colleagues is summarised and available from the Arctic Monitoring and Assessment Programme website. AMAP is an international organisation established in 1991 to implement components of the Arctic Environmental Protection Strategy (AEPS). Now a programme group of the Arctic Council, AMAP's current objective is 'providing reliable and sufficient information on the status of, and threats to, the Arctic environment, and providing scientific advice on actions to be taken in order to support Arctic governments in their efforts to take remedial and preventive actions relating to contaminants.'

As I began this chapter our funding and research interests were stimulated by the startling finding of PCBs in Inuit mothers' breast milk in Ungava, Quebec. Since then, similar measurements have been done around the world, in Asia, Africa and South America, and are still being done as of 2012. Some of the toxic pesticides have been banned (DDT) or reformulated, but new agricultural and industrial organic poisons are being added to the atmosphere and ecosystems almost yearly, such as the dioxins, fluorocarbons, bromocarbons and organophosphorus toxins. Concentrations of organic contaminants in Canadian Inuit mothers' milk have declined in the last few decades but this is at least partly due to changes in sources and types of food (less traditional wild meat, more imported groceries). Since our work, research has shown that top wildlife predator (foxes, sled dogs, polar bears) ingestion of low levels of organic and metal contaminants from the food chain cause reductions in reproductive success and physiological functions. Inuit from northern Quebec and Faroe Islands eat a lot of seal and whale meat, their levels of body contaminants are high, and subtle negative impacts on foetus and child development and learning have been found. Hudson Bay Inuit have been excreting high concentrations of PCBs since the 1960s, and it was still high in 2010 (see Hermanson et al.). A 2014 review (see Malisch and Kotz in the references) indicated that dioxins and PCBs declined slightly between 2000 and 2010 in European food (mainly meat, fish and dairy products).

There has been a global decline in lead (Pb) emissions from banning tetraethyl lead in automotive fuels in 1976, despite fierce opposition from the coal, oil, automotive and electrical energy industries. This resulted in measureable declines in oceanic seawater Pb concentrations, coral annual growth bands, human blood levels and wildlife health improvements (see Wu et al. 2011; Wong et al. 2011; Sherrell et al. 1992; Shen and Boyle, 1987). But even with this tenfold decline in human blood lead concentrations, this was still a factor of 1000 higher than pre-industrial North American Indian bones. Twenty-five years after our work, modern researchers have found that various methods of cooking fish and other seafood do not diminish the concentrations of PCBs or dioxins...isn't this wonderful (see paper by Rawn et al.). Three decades after our work on mercury in fish and sediment cores, atmospheric deposition rates have declined, due to better pollution control in coal fired electricity generation plants. However, mercury in fish muscle in northern Ontario lakes has not declined at all (see Tang et al.).

From the review of Matsuyama et al (2011), Hg pollution of Minamata Bay in Japan began in 1941, where a factory was making acetaldehyde from acetylene and carbide, using mercury sulphate as a catalyst. After discovery of mercury sickness in the local people, the discharge of Hg was stopped in 1965 but the fish and sediments were already loaded with 25 ppm Hg. The Japanese actually dredged the bay, removing most of the surface contaminated sediments, and put it on 60 ha of land for reclamation. In 2011 the fish still have 1 ppm Hg and are not fit to eat. But we did help accomplish better regulations on mercury emissions (Driscoll et al. shows a big decline in Atlantic Ocean surface water), and PAHs (Slater et al.) Some of the organochlorine and organophosphate pesticides, estrogenic and antibiotic products, perfume musks, caffeine and nicotine, recreational drugs in sewage (cocaine, marijuana, meth, heroin, etc.), and medical/agricultural pharmaceutical products have been blamed for all sorts of human and wildlife problems such as the international demographic increase in diabetes, autism, sperm counts in Danish men and changes in sex ratios of various fish and birds. And so it goes on, maybe we are our own worst enemy?

Our hypothesis that a lot of the metal and organic contaminants came from over the poles from China and India (the Buster Welch 'Brown Snow' paper) is still valid, but further research has shown the pathways vary for each contaminant, and with the seasons and extreme weather events. Our research on the enhancement of mercury in sediment cores has been challenged (see Outridge et al., Stern et al., and Kirk et al.), claiming this enhancement is due to rapid climate warming and increased organic matter production and sequestration of mercury from the land and dust. Other studies on mercury indicate that over half of the atmospheric deposition in industrialised nations comes from burning coal (see Wang et al.). The danger of ecosystem contamination from carcinogenic industrial polycyclic aromatic hydrocarbons (PAHs) remains confused because of our finding abundant PAHs from natural erosion of geological deposits of coal and amber near some of our Arctic study sites, and we found coal particles in the stomach contents of fish in Hazen and Buchanan Inlet. Our hypothesis of the cold finger 'freeze-out' of organic contaminants in the Arctic is probably still valid, as lots of models and measurements show the 'degassing' of PCBs and DDT from the St Lawrence Great Lakes and tropical zones (see Wong et al), and the likely 'grasshopper' migration of these volatile contaminants toward the cold earth poles for relatively permanent storage in ice and bottom sediments (see Semeena and Lammel, 2005; Jurado and Dachs 2008; Blais et al.).

Ironically, global greenhouse gas warming of our planet appears to be melting glacier and polar ice and thus liberating about fifty years of excess metal and organic contaminants to Arctic ecosystems (Blais et al., Donald et al.). This kind of pollution from the past has the ironic title of 'legacy contaminant supply'. It was/still is both rewarding and also frightening to see our work followed up on different continents, often finding similar results, to add to the world map of distribution of these manufactured industrial poisons in humans, wildlife, soil, lake mud and polar ice. And so it is that there are no pristine environments left on the surface of our planet, from polar ice fields to the deep sea (see Stemmler and Lammer, 2013).

But during this time positive action was attempted. In 1987, the Montreal Protocol was established for reducing chlorofluorocarbon gases in industry appliances and aerosol spray cans. Many of these chlorofluorocarbon (CFC) gases were implicated in changes in ozone chemistry of the atmosphere, leading to the 'Ozone Hole' in polar regions. These CFC gases dissolved in the world oceans and became useful tracers of ocean currents, vertical mixing and age of ocean waters. This was of concern because reductions in atmospheric ozone allowed more solar UV-B radiation to reach the ecosystems of the earth, where deleterious impacts on DNA was feared. These industrial CFC gases were banned but new industrial products were created and dispersed that were even worse, more persistent, greater greenhouse impact, such as the various fluorocarbon fire retardants, dioxins and furans that will be with us over the next century (such as the flat screen TV gas NF_3, a much stronger greenhouse gas than CO_2, see Prather and Hsu, 2008).

Of course, as fast as new insecticides were produced, mosquitoes evolved resistance to these poisons. *Culex pipiens* evolved resistance to organophosphate insecticides within a decade by replicating several copies of one gene that allowed the production of an enzyme that could decompose the organophosphate insecticide. This new mutation is now spread world-wide. However, the old persistent poisons continued to work very well for many other organisms that were not the target, many of them beneficial, such as bees.

About the time I was to leave Canada, the Stockholm Convention on Persistent Organic Compounds began its work. They established the dirty dozen, the worst contaminants of the lot, such as DDT, PCB and dioxins that needed to be controlled. This was an international environmental treaty which began signing in 2001 and became effective in 2004. The dirty dozen = Aldrin, chlordane, dieldrin, endrin, heptachlor, hexachlorobenzene, mirex, toxaphene, polychlorinated biphenyls (PCBs), DDT, Polychlorinated dibenzo-p-dioxins and furans. By 2010 the list was changed, adding lindane, pentachlorobenzene and the bromodiphenyl and perflurooctanesulfonic acids (fire retardants). The list is updated as new industrial chemical products are created in the lab, tested by the market, dispersed and sometimes banned because of some obscure impact on ecosystems. And of course, the whole idea of government regulations was despised by conservative/Republican politicans and most industrial lobbies and they did everything they could to discredit or confuse the message.

The unreasonably long list of references for this chapter indicates that we had only scratched the surface of the problem of 'long range transport of atmospheric pollutants', and that research labs around the world continue to quantify the tragic story of industrial poison dispersal into all compartments of global ecosystems.

But the sad news was that there was no more a 'natural ecosystem' out there in the remote parts of the world. Pure pristine clean nature has been lost to industrial pollution, and we are swimming, breathing, eating, growing and dying in our own wastes. From the frigid poles to the equatorial Pacific and the bottom of the Marianas Trench, plastics, pesticides and excess metals are now influencing the health of biota and humans.

Chapter 23. A Tipping Point, Change of Pace, Major Upheaval, Jumping Off a Cliff, the Summer of 1991

Within a week of my return from the high Arctic, late May 1991, my mother phoned from South Dakota wanting to know if I'd been eaten by polar bears. She was giving me the news of lots of rain and grass on the South Dakota prairie, the comings and goings of nephews and cousins and wanting to know if we are coming down for the big town celebration in July.

Ian MacGreggor phoned from Ottawa, having been excited by a Pierre Boulez concert. He was planning on a NYC trip, and would attend the Montreal Jazz Festival. He was now studying at the University of Toronto, with mentors from the percussion group NEXUS. Sheila and I were both amazed and comforted that our young son had found his life passion and career.

Younger son Jesse was totally buffaloed by the Australian job offer; this did not compute in his brain. Jesse said he would not go to Australia and wanted to stay with his Niverville school friends. Sheila and I sat at our big round kitchen table after supper, alternately afraid and excited to discuss the option of selling our Tourond home for $50K and going to Australia. We were startled at our decision to pull up stakes, leave a nice home, family friends, our folk music band and a safe well-paying job.

On 2 June, Sunday, we had a long phone call from Dr. Alistar Robertson of the Australian Institute of Marine Science near Townsville in north Queensland, Australia, who explained the deal with the Australian Commonwealth employment situation and promised a technician (offsider). They wanted someone who could measure and interpret river mud accumulation in mangroves, deltas and river fed coastal embayments of the Great Barrier Reef inner shelf. I startled myself by saying yes, I'd come, even though it was a slight reduction in salary. I put down the phone and immediately Sheila was bubbling with plans, lists of jobs proliferate, maps of Australia appear on the kitchen table and we located the small city of Townsville on the central Great Barrier Reef coast of Queensland.

Within a week Alistar phoned again, saying that the AIMS CEO and Canberra had approved of hiring me. I now had a technician and an office there, and he was sending me new business cards. I was given a moving allowance (US$4K) which wasn't much, some benefits, pension and a superannuation plan. This all seemed to be happening too fast and I kept thinking that I really didn't know much about tropical seas, mangroves or northern Australia. But here it was again, serendipity offering a completely new adventure. Dear Sheila was shaking her head at the kitchen table, saying she was sure I wouldn't leave our comfortable (fully paid for) Tourond home, and was afraid to say that she did want to move to Australia.

Wow: Here we go!

We set about to make the yard, barn, and house neat and tidy, mow the grass, burn rubbish, trying to make it presentable to anyone who would buy it. We invited a city real estate agent to visit, a lady in a white silk pant suit and high heels come down to look us over...that was a mistake, a city person should not have been involved in this, as many of the features they wanted in their photos were not feasible. But our neighbouring farmers and small town folks began to visit and make offers. And of course we knew most of the old time farmers in our area referred to our home place as the P. G. S. Friesen place, about to return to its Mennonite heritage.

We were a little surprised at most of our musical and scientific friends, who were stunned that we would leave our home and friends. Some of them were upset and in tears. Sheila was hurt by some of this and wished they would be happy for us, to have a new adventure in a far distant land.

And of course I had to tell my colleagues at the Freshwater Institute about this, shut down my involvement in research program responsibilities, make arrangements to 'retire' from the Department of Fisheries and Oceans, arrange for severance pay and pension transfer to the Commonwealth government of Australia. And it was becoming apparent that we were leaving behind twenty-four years of friends, and it was good to have been in the comfortable association with high achiever colleagues. There was also some frantic re-arrangement of visits to relatives in North America and England. I had previously committed to a 2-10 July Murdo Class of 1959 High School Reunion in South Dakota, a 22-29 August visit to Sheila's relatives in Toronto, Montreal and Ottawa, and 7-25 September trip to England and Scotland, paid by my institute for my attendance at a big conference on long distance transport of contaminant metals in Norwich, England, and a conference on radionuclides in Arctic marine environments in Edinburgh. My little notebook journal was full of priority lists of things to do.

In late June, Jesse's high school graduation and class parties happened and we were amazed at our handsome young man in graduation gown with his good looking date, Joanne Friesen. Joanne had a laugh you could hear half a mile away. This is one of the rare times that Sheila washed her feet in bleach, to get the blackness off her heels, and to dress up in a proper gown. We had a good time helping Jesse get into his tuxedo and cummerbund. We went to the usual Niverville school and church graduation ceremony, the church basement dinner with speeches. I gave a goofy speech with my South Dakota cowboy hat and moose hide leather vest. We set up our yard and barn for the class party, big bonfires, some fireworks, Xmas tree lights everywhere, huge coffee urn, ice cream, cake and finger food tables. During the evening thirty-five students showed up, and we had to relieve a few drunks of their car keys. We were lucky to have our friend the St. Pierre RCMP Cliff Russel to help out with law and order. We made breakfast for them all the following morning at 0600, as most of them slept on the grass by the bonfire ashes.

This is yet another key milestone in our family, our second son graduating from high school, making it possible for us to change our lives, pull up stakes and consider going to Australia...a turning point, a tipping point, a pivot was now possible.

In July 1991 we drove down to my parents' home in South Dakota, where all my brothers and classmates were there for the big town reunion. North and South Dakota were unusually wet, water in road ditches, lots of sweet clover, tall grass, prairie flowers, hawks, ducks and geese, this being unusual in the arid high plains. All the usual comfortable family stuff happened in rapid succession, making homemade ice cream and fireworks at the fairgrounds by the town fire wardens. Sheila couldn't tolerate the town water, which gave us a good excuse to get tea water from a White River floodplain sand point well. Jesse and I went swimming in the muddy White River, set some lines for catfish, but had no luck. My father and youngest brother John were making hay, the big 800 lb bales being stacked behind the barn. Gentle horses Dusty and Terrible Tom were brought out for all the kids to ride in the corral. The regional cowboys put on a good town rodeo, and nearly all the competitors were local people. At the town talent show Sheila and I did Guy Clark's 'Homegrown Tomatoes' and Robert Earl Keen's 'Copenhagen Song', and my mother afterwards was upset about the Copenhagen song. You need to understand that most of the good old boys in this region used Copenhagen brand snus under the lip and would politely spit tobacco juice onto the ground or into brass spittoons. And of course all the good women hated chewing tobacco. All of my brothers had a little circular can of Copenhagen in their shirt pockets, and my father kept extra cartons of snus in the freezer to keep it fresh. There was the usual Country Western Dance in the high school auditorium, all decorated with old class pictures back to the 1930s. There was a lot of cowboy boot stomping on the floor and flinging the young ladies around the room. There were hot air balloon rides from town cemetery. Jesse and his cousins gleefully went off in a pickup truck for rabbit and prairie dog hunting late into the night.

I am remembering sitting on the back porch, visiting with my brothers, telling stories about old timers who were now dead and gone. This was all done with a droll dry humour often found in desolate corners of the cattle country. Brother John tells the story of one of his steers that went through the dam winter ice and drowned. It surfaced in the spring, bloated, surrounded with a lot of big happy turtles eating it. John tells of despair at beef prices...'Too many calves, when it comes time to feed them; not enough calves, when it comes time to sell them.' Sheila and I went fossil hunting in the White River bluffs and visited our dinosaur skeleton still partially buried. Mother and her granddaughters were busy in the kitchen, as we had catfish rolled in corn meal, a fried chicken feast, four roast pheasants, watermelon/muskmelon, sour cream raisin pie and rhubarb cream pie.

And finally my family was stunned quiet at our announcement that we are migrating to Australia. Again, this simply did not compute in South Dakota brains. Down at Skelly's gas station over coffee and heavy cinnamoned apple pie with thick real cream, the conversation was...'why on earth go to Australia??? That place is full of sheep! Bad enough to be up there in Canada'. And then my father and Aunt Margie had to go down to Sioux Falls for the funeral of Elida Gregg, one of Grandma Brunskill's family.

It was good to meet with all my high school classmates of the graduating class of 1959 in the Methodist Church basement. Many of my classmates had done well, gone off to become bankers, doctors, nurses and successful in business or ranching. Several of our teachers showed up, including Mrs. Kuhrt, who taught me well in English literature and world history, and insisted that boys take typing. In my career I have thanked her many times for being so feisty and demanding of us prairie kids. My class was remarkably lucky to have good teachers and some bright students that pushed the rest of us along. *UVA UVAM VIVENDO VARIA FIT.*

On 10 July we packed up to go home. My mother stood on the porch waving to us, crying, saying she probably would never see us again. She was right, as she died of cancer within a year. The men folk tried to show no emotion, pushing their boots through the dirt, heads down, covered with big hats. And then we drove away through the unusually green Dakota land, the big bluestem grass waving and blue and reddish glinting in the sun. We all had big lumps in our throats, realising that we won't see all the folks and this country very much in the future. A full day's drive to our wheat straw coloured home on the Tourond farm in Canada, where it was raining, our dogs were muddy and delirious to greet us, the yard was wet tall grass and the vegetable garden was lush with weeds.

We were quickly immersed in Meadowlark rehearsals in town and we performed old time stringband folk music at dozens of small town summer fairs. At the same time we were emptying our house, packing our precious books and photo albums, deciding what to leave behind or sell and getting ready for a yard sale. Burton Lysecki (a Winnipeg bookshop owner) came to view my book collection with a big panel truck and book trolley. He paid $2000 for my Arctic exploration book collection, but they were probably worth ten times that.

We found out from the real estate company that they could not sell our house, as it had urea-formaldehyde/urethane foam insulation, which had been installed a decade ago with a provincial grant to rural homes. Apparently it was recently discovered that this insulation slowly leaks formaldehyde gas into the house, which is not good for you. Ah, but the local people valued our little property. Farm worker Dave Hiebert arrived at our door, all spiffed up, sat at our kitchen table and awkwardly offered $40K cash for our home and land. By 17-18 July we met with a real estate agent and lawyers to seal the deal, even explaining that the urea-formaldehyde insulation was bad for his health. Dave had taken care of our house and farm animals previously when we were away, and he said he slept very well in our house. After the horrid meeting with lawyers and real estate people, Dave and I went to the Montcalm pub for beers and watched some strippers in front of loud music, a very bizarre situation, and Dave was thrilled to own our home by 31 August.

On a hot and sunny 19 July, Jesse and I helped Sheila set up fifteen tables from our local Community Club on the neatly cut green grass of our front yard. We put out all our possessions to sell: 110V barn tools, dishes, garden implements, Ford tractor, disk and harrow, pitch forks, hoes, shovels, John Deere mower, crocks, dressers and antique stuff that Sheila had collected. We found that our Mennonite cupboard was surprisingly valuable as an antique: we had been using it in the barn for shelving and it had to be cleaned up. All of our goods were priced and sat neatly on table cloths. We ate supper and went to bed feeling uncomfortable and miserable. It began to rain at 0200 the next morning and we had to scurry around covering our tables with tarps and plastic sheets, amidst hordes of mosquitoes, wading through muddy puddles, feeling even more miserable.

But the next day we were up at 0500 to see clear sky, our yard was wet and muddy, some of the price labels were smudged and we were about to cancel the yard sale. But no, wait, it was hot and sunny so we went ahead, removed the tarps, fixed the price labels, wiped off moisture. The Winnipeg antique collectors arrived at 0745 to buy all the old ceramic crocks and the old Mennonite cupboard. By 10 am the yard was full of local people, and our remaining books and LP records sold quickly. Jesse's girlfriend Joanne Friesen bought Jesse's childhood baby highchair. So goods accumulated over twenty-four years of our life vanished before our eyes, and we pulled in $2000 for it all. Sheila still calls it 'the yard sale from Hell.' Dave Hiebert wanted all the stuff that didn't sell, especially some silver wine glasses. We protected our small residual belongings and hauled them to Gerald Toews, a local carpenter friend and co-Scout Leader, who built a padded shipping box for our eighteen foot moose master canoe, musical instruments, precious photo albums, non-electrical shop and garden tools, some clothing, bedding, camping gear, Jesse's bearskin, my hunting rifles, my South Dakota handyman jack, log chains and crowbars. Cliff Jones and the Freshwater Institute workshop men made some wonderful sturdy shipping boxes for my office books and mathoms, but I had to give away my office foetal moose barometer (from D. S. Rawson's shed beside Great Slave Lake, in Hay River) and a very large jade plant that lived on my office window frame.

On 24 July '91 Sheila and I went to the spruce/pine scented ELA in a large institute van to collect six 120 litre deldrum samples of Lake 302S water for radiochemical and trace element analyses. We also performed some of our music at the famous ELA Variety Night stage show and said our goodbye to lots of old friends that knew us from our romantic 1970 days. Sheila had tears in eyes when she returned her pet turtle (Adleberry du Pres) to Lake 373. We picked blueberries, watched a moose feeding in the wetlands, a bobcat sniffing at a beaver lodge and of course the loons were calling at night. We had had lots of good times here, winter and summer, hot and sweaty, cold and miserable. And again we had big lumps in the back of our throats as we drove our heavily loaded van through the twisting lumber camp road to Kenora and back to Winnipeg, a four hour trip. We dumped the deldrums at the institute, along with notebooks on the field measurements, and then went home to find Jesse with his high school buddies, sitting on bales of hay drinking beer around anti-mosquito smudge fires in the yard.

Lake 302 Filtering Crew, 1991

All sorts of unpleasant duties awaited us. We visited a Financial Advisor to get professional advice concerning my Canadian commonwealth severance pay, pension, house sale money and something called a registered retirement savings plan (RRSP) which we didn't know we had. We were worrying about having enough money to buy a house in Queensland. Australian immigration required an AIDS test, so we went down to St. Pierre Hospital for blood tests and police interviews, finger prints and new passport photos for the medical report form.

Jesse packed for two months at Kasba Lake as a fishing guide, and we asked him to seriously think about coming with us to Australia. On 27 July we hauled Jesse to the Winnipeg airport for his flight to Saskatoon, and then an old military DC3 food supply flight to Kasba Lake Lodge on the NWT-Saskatchewan border, where he returned to his fishing guide job. And then we rushed home to greet Dr. Birkes, who wants to buy our old piano for $1000 after having a piano tuner inspect it, having called it good value. I hoed and watered the garden, feeling uncomfortable in abandoning it, such good rich black soil. Our tomatoes were big, fat, red, almost begging us to stay home and eat them. Our house was now near empty, except for a few pots and a frying pan and our airplane luggage to take with us overseas. Sheila made one more blueberry pie and we ate more of our home grown chickens from the freezer.

The one thing that was very hard to shut down was our folk music band (Meadowlark), as we had been booked around central Canada to perform at small town festivals, carnivals and celebrations. On 3 August Meadowlark was again at rehearsals at the Ediger house, after a good supper from Barbara, and her Dean family was there demanding all the old cowboy songs. Meadowlark, the band that plays on its stomach. On 11 August we assembled at the West End Cultural Centre with Cecil Anderson the sound man and played twenty-five tunes straight through, recorded direct to cassette. I still don't know what happened to that recording. Our musical instruments were already packed for shipment to Australia, so we used borrowed instruments.

Why do we get so attached to our pets? We felt bad giving up Gladys the Goose to a neighbour. Sheila sold Wombat (her black dog) to a neighbour, but Wombat was back in our yard with chewed leash by evening. It was very hard for me to push happy bright Buffalo Gal into a dog cage and haul her to town. But Sheila's sister Beek had arranged to send her to Mission BC to live with younger sister Peabody and her kids, so she stayed in the family.

On 17 August Meadowlark was on our last international tour down to Grand Forks, N. D. for the Prairie Heritage Festival. We played four 45 minute sets out of an amazingly large repertoire. It was also good to see old kerosene engines, oil burning tractors, saw mill tools and shingle making demonstrations. It was a good hot and sunny prairie day. We went to a good restaurant for our 'Last Supper' with Meadowlark. The next day we came home to our sad, nearly empty home, fridge and piano gone. Folkie friend Tom Janzen's mother bought our pump organ for $500, a bargain, as it was an antique, having come from a pre-World War II whorehouse in Winnipeg. Edigers came down to take tubs of sweet corn, tomatoes and strawberries from our garden, and they bought our washer and dryer.

Ian and Jesse's grownup babysitter come down to say goodbye and she was in tears too. We were still packing up leftover stuff, giving some away. Neighbours bought my No. 9 John Deere sickle mower, wheelbarrows and the old horse wagon that my father had fixed up for hauling wood. Sheila's brother Wump and sisters Mary and Peabody came down to our nearly empty house, sadly collecting stuff Sheila wanted to give to them. Mary got our big old wooden study desk and Peabody and Al took the Hada sleds and goose sleigh. We all had one last feast on home grown chicken, and the rest of the frozen chickens were taken by Peabody and Al.

On 21 Aug I was back at my government research institute, which was abuzz with gossip and stories about my departure. I had to hose down a dispute with one of the directors, who was concerned about me running off with my shipping boxes (good solid wooden boxes, and some old red komatiq boxes from Resolute Bay) being used for my personal office stuff. I walked through the inventory of all my accumulated scientific equipment, almost half million dollars of science gear, which was now left in the hands of Paul Wilkinson and Ray Hesslein. Many good friends there still could not believe I was leaving, because I was one of the pioneers in this institute. I wrote my letter of retirement, acknowledging my golden handshake deal: 72 more days of accumulated vacation time allow me to retire 1 January 1992, with a severance pay of $30K, and my pension is apparently worth $73K. All this was tucked away in some account by Gary Senft who worked for a local finance company. We arranged to sell our bright cheery red Subaru to Mike Stainton for $7K, same as our purchase price. The Winnipeg IMAX theatre obliged our upheaval by showing a feature on the Great Barrier Reef.

On a sunny and warm day in late August we flew down to muggy, foggy Toronto and Montreal to say good bye to the rest of Sheila's family. I was deep inside Robert Hughes' good book *The Fatal Shore*, about the history of Australia. Sheila's older brother Ian and wife Angie, daughter Rachel and son Tim were at Dorval Airport to meet us, and they took us out to their home in Beaconsfield on the west island of Montreal. Sheila's younger sister Noddy and husband Ted arrived, and we had some interesting adventures on uncle Ian's sailboat.

During this summer our son Ian MacGreggor was in the Canadian military Ceremonial Guard band performing the 'Final Mount' in front of Parliament Building in Ottawa. We took the bus from Montreal to Ottawa to watch the band perform. Ian and his buddies were soaked in sweat, wearing heavy wool bright red braided uniforms and bearskin hats. We visited the Ottawa museums, art galleries, visited with colleague Jaap Kalff, listened to

David Essig's music, had an Abyssinian African supper with Ian, and then slept on the bus trip back to Montreal.

My brother John phoned from South Dakota, saying that my mother was in the hospital having an operation for a gall bladder problem in Rapid City. A day later brother Gerald in Arizona phoned to say that mother was ok but she had a malignant tumour in her liver and would stay in the hospital for at least a week. This put an additional damper on our spirits.

Now here again there were strong emotions saying goodbye to relatives. Rachel (Grade 3) and Timothy (Grade 1) went off to their first day of school, uniformed and bright smiling. Ian MacGreggor was now finished with his military service and he packed his percussion stuff, polished boots and ghetto blaster and came back to Winnipeg.

Since our house was empty, we were camping at Walter Lysak's house, as Walt and Bean were in Virginia on holidays and folk festival touring. Ian MacGreggor came with us down to our empty farm home and selected a big pile of jazz records and books to take with him to his Calgary rental house. Our garden was too beautiful and it was breaking our hearts to leave it. There were fifteen melons ripe, abundant ripe sweet corn and big fat red tomatoes. I was hoping that we could have this kind of wealth in our new home in Australia.

I drove Ian to Brent Van Dusen's home near Pilot Mound, where Ian and Brent will drive all night to Calgary, for their last year at the University. This was goodbye to our oldest son, who wanted to stay in Canada and study at the University of Toronto with NEXUS. In early September son Jess phoned from Saskatoon, where he will have some party time with other guides, very tired due to riding down from Kasba in a DC3 full of fuel barrels and propane tanks. Jesse finally decided to come with us to Australia to help out for two or three months getting us settled, and then he wanted to return to Niverville to be with his classmates.

On 3 Sept our good farmer neighbour Peter Klippenstein came to our Tourond home in his grain truck to haul the remainder of our boxed goods to Gerald Toews for the shipment crate. This included our disassembled butternut sofa, Sheila's Quebec maple loom, my grandmother's music cabinet and way too many books and records. It took ten helpers to lift our crated goods into the grain truck, which took it to international freight movers Ross and Day [and shipping Agent A. D. Rutherford and Co. Ltd] in Winnipeg. We paid $5300 to move our gear from Winnipeg to Toronto to Halifax to Singapore to Townsville, Queensland, ETA 1 November 1991. We visited a lawyer to oversee our Niverville Credit Union account to make sure the movers, freight and travel all got paid promptly. We also bought our airplane tickets to London and Edinburgh for science meetings.

On 4 September we went to the Winnipeg airport to collect our fishing guide son Jesse, with several heavy army surplus duffle bags of his clobber.

And then we had another big party in town, as niece Rosalyn Jane and my 50th birthday are coming up soon. There were all sorts of goofy tropical Australia related presents, such as gaudy shorts and muscle shirts. Podiatrist and excellent Celtic bouzouki player Martin College demanded that I take my cowboy boots off to exercise my Achilles tendon, as I would hereafter probably be living in flip flops. We had one last late night music party with Martin and Nan, Walt and Barb, David and Barbara, Joe Kinsella, Wump and Susan. Ian MacGreggor phoned, saying he was happy in his rented house in Calgary, which was a 45 minute walk from the University of Calgary campus.

On a warm and sunny 7 September Sheila and I left Winnipeg for London, to visit with her Auntie Ester and uncle Freddie in the Cotswolds, and I was to attend science meetings in Norwich and Edinburgh. Sheila brought a large bottle of Canadian maple syrup as a present for Ester and Freddie. We arrived in London at 0945 to see unusually clear sky over Buckingham Palace and London Bridge. Ester and Freddie were there to meet us, whisked us away in a tiny Hyundai at 100 km/hr on M40, somewhat terrifying. Sheila needed a toilet, so Freddie madly dove off into spaghetti of tiny lanes and we promptly got lost in Banbury Cross. There were police everywhere, smoke bombs, parachutes and sky divers. Completely lost, Freddie mutters 'It's a small island, we'll come to the edge soon.' We finally arrived at their home in Blockley, an old house which used to be the sheriff's office and prison cells, thick stone walls. We were cosseted with tea and little cakes, peace and quiet, but the air smells like oil on hot metal. We are being coddled in the Cotswolds.

Our lunch was Freddie's recently caught trout with homemade mayonnaise, new potatoes from France, two salads and Pimms champagne. We walked the alleys of the little town, stone bridges over streams with ducks floating by, stone fences with styles, stately homes, ancient Norman churches with high bell towers and graveyards with stones to the 1700s. I had to cook my steak and onions in a wok, as Ester appeared to be afraid of South Dakotan eating habits. I stayed up late arguing philosophy and religion with Freddie, who was writing essays about this. The next day Ester demanded attention to family photo albums, giving us a history lesson of the Dennett and Trechmann families. There were English and Scots ancestors that had served in government positions in South Africa, Uganda, Seychelles and India. Sheila and I had to walk for a while to get an appetite for our supper of roast Welsh lamb, mushroom gravy and ice cream with delicate English strawberry puree. We reviewed train schedules to Norwich, booked a room at a bed and breakfast in Norwich, planned everything in detail as it is a small island and all shops along the way could be visited.

On 10 Sept 1991 Sheila said I was fifty years old. Surely this can't be true. Off to Moreton train station, £37 RT ticket, Brit Rail terrible coffee and hard fruit cake, to Paddington Station in London, underground to Liverpool Street Station, to Gate 9 and Brit Rail to Norwich, everything I hate about big cities. On the train there was a glass windowed emergency box containing a hammer, to use to break the glass for an emergency stop or fire. So how do you get the hammer out? We arrived in Norwich and took a taxi to Fuchsias Guest House, which was locked. So we wait an hour for a lady who comes up riding a bike, shows us a little stuffy room with windows that will not open.

We walked two miles through the old city to the University of East Anglia halls for registration for the International Symposium on Radionuclides in the Study of Marine Processes. We were told to go to the Norwich Castle for greetings from the Lord Mayor, finger food and wine, and here were all my buddies again...Curt Olsen from LDGO, Santschi from Texas A&M, J. David Smith from Melbourne, Murdoch Baxter of Glasgow (now Monaco Lab Director)... the international biogeochemical mafia coming together again. Here again was that comfortable tingle of being with a group of international colleagues, a consortium of expertise that floats around the world, independent of nationality and institutional allegiance. Aussie David Smith had visited AIMS, where I was going, he told stories of wallabies learning to approach electric doors opening, they go inside, get scared and tip over tables in the library. His hobby was building oboes from Australian acacia wood, which he said was as good as or better than the black African acacia normally used. Sheila and I wandered the back streets to find a small restaurant for a good supper of Stilton cheese soup, Scots grouse, scallops in shells, cheesecake, all the while we had a purring cat under our table.

Over the week of 11-14 September hundreds of presentations were given in multiple sessions, all about natural and man-made radionuclides in ecosystems of the world. I gave my paper on the variation of the flux of excess ^{210}Pb (from the uranium series decay chain) from the temperate zone to the high Arctic in Canada. This isotope of lead arises from the decay of radium in soil, to the noble gas radon, some of which escapes to the atmosphere, where it decays with a half-life of 3.2 days to ^{210}Pb. In porous soils of the temperate midcontinent of North America lots of radon can escape the soil, to form lots of ^{210}Pb in the air. In the high Arctic you would expect that the ice and snow covered frozen soil would not allow much radon to escape, but surprisingly, quite a lot must escape because ^{210}Pb fluxes from the atmosphere to the land/sea are not greatly diminished.

During this time Sheila visited Norwich Castle, where dungeons and prisons held poor people that were shipped to Australia in the early 1800s. She also visited Colman's mustard factory, lots of tiny shops, narrow cobbled streets and Yarmouth town, famous in folk songs. We took a boat ride on 'The Broads' which apparently was the peat mined residual lake land of the river floodplain. The muddy water had lots of geese, pigeons, swans, herons, moorhens and coots. We had some interesting breakfasts at Fuchsia Guest House, usually curried eggs, big fatty sausages with hot chutney and ceramic pots of lukewarm tea.

And then we had the usual big bash for the science meeting wind-up at Black Friar's Hall, where a dance band in the 1950s style of music performed in a high arched vaulted ceiling room, some brave elegant dancers swooped under paintings of church fathers and a huge organ that does not work anymore. I was proud to dance with my elegant Sheila in her ballroom gown. There were endless tea and pastries served, and some small but delicious Eccles Cakes!

On 14 Sept we went back to the train station on a hot and sunny day, amidst diesel fumes and human sweat on a train to London, Oxford. It was a complex path back to Ester and Freddie in Moreton-in-Marsh, where we had to eat even more tea and cakes. We spent the next day in Freddie's English garden writing postcards and updating our journals. Freddie hauled us to Cotswold Falconry, where they had Indian falcons, eagle-owls, an Australian boobook owl and a buzzard. And then we had some pub lessons all about properties of English bitter, lager, ale and cider. These lessons were kindly given as if to an ignorant child who had no culture back home. Then it was back to the sheriff's house for a curry chicken, lentils, pappadums, lime pickle, three chutneys, fruit salad and little wafers, tea and more cakes. We were stuffed, bloated, saturated and moaning in good agony. And then Freddie whips out a video of Papua New Guinea dancers with percussion (*kundu*).

No rest for the scientifically wicked, so on 16 September, in the rain and fog, we are off to Moreton-in-Marsh train station again, £46 RT to Edinburgh for a big science meeting on contaminant trace metals all over the world. Ester was fussing that we didn't pack a lunch and we are going up there to the savage wilds of Scotland. The train had only three passenger cars and it went slowly, sleepily past fields of Brussel sprouts, cabbages, fava beans, leeks and sheep. We had to change trains at Worcester (where there was a big factory for Lea and Perrins sauce), and we continued northward through grubby industrial towns, tunnels, switches and finally neat clean fields of vegetables and berries, canals with long boats. And then the Birmingham switch, more hot sweaty crowded train cars with warm brown ale and soggy chips. But we survived and arrived in Edinburgh at four pm in the rain, Edinburgh Castle glowering down at us from the cliff. My science conference was at the Royal British Hotel (I was told that no Scot would ever stay there), and registration was followed by the usual smoker/banquet. We walked Princess Street to view 'Buffalo Girls' fashion displays in department stores, western fringe shirts, belts with big rodeo buckles. What's going on here?? We later discovered that western US clothes and country and western music were common here, due to lots of Texans in the offshore oil industry being here.

Despite the hotel concierge insisting that I take a taxi, I walked to the science meetings, inhaling the curious mixture of medieval and modern city features. And all of a sudden there were all my familiar science colleagues again, many of the scientists who had been storm bound in Resolute Bay, or on Agassiz Glacier (Dennis Gregor), Rod Allen, Jerome Nriagu, Brazilian mangrove scientists, Italian mountain lake mud chemists, all of them with stories about their latest field trip disasters. I gave my LRTAP summary to an audience of friends who were trying to do the same kinds of measurements in their country. I told of how we measured the century scale history and annual fluxes of metals (lead, cadmium, mercury) in Canadian Temperate Zone and high Arctic lakes and estuaries. The gossip had got out that I was moving to Australia. Several Australian coral coring geochemists warned me about life in the 'deep north' of Queensland...similar to the deep south of the USA. They told hideous stories about the corrupt and primitive conservative Premier of Queensland, Sir Johannes Bjelke-Petersen, who had recently left office in disgrace.

Despite my admiration of her formal gown, Sheila complained about wearing it to the Lord Provost reception, which was boring as expected, but we did get hear some highland pipers, sample haggis and some neat single malt whiskey. During the rest of the week Sheila toured Edinburgh Castle, Perthshire and Glen Turret distillery in Sterling, where she sampled even more exotic smoky Scots single malt whiskey. Peter and Kathleen Maitland (friends of my institute workshop champion instrument builder Willy Burton, back in Winnipeg) drove us fifteen km out of Edinburgh for a fine Atlantic salmon supper and some traditional Scots fiddle music. They told stories about Willy being a featherweight boxing champion and making claymore swords, all of which was easy to believe. Willy had often come to my office in Winnipeg to show off his home-made replica Colt .45 Peacemaker pistol, which looked like it had come out of a John Wayne film.

Sheila and I abandoned science meetings to tour the Royal Museum of Scotland, where we got an overdose on the history of engines, golf, ships, looms, fishing gear, scientific instrumentation, some of the *Challenger* Expedition oceanographic equipment and notebooks, Sverdrup and Nansen's Arctic exploration gear, whale skeletons...too much, it hurt our overloaded heads. And then we had another fancy conference dinner in Signet Library (roast lamb, skink, raspberry with cream and porridge) with a few of my irreverent (but well dressed) scientific colleagues trying to enliven the dour serious affair with football songs and dirty stories. Being just about fed up with science, Sheila and I visited the Royal Botanical Gardens, where we could see bamboo forests, jackfruit, pineapple, huge lily pads in ponds and Aussie eucalyptus trees. And strangely, there was South Dakota Black Hills gold jewellery on sale in shops on Princess Street.

Rather gratefully, we escaped our stuffy Fuchsias cottage rooms and on 21 September we walked back to Waverly train station. I suppose this must be a famous place in the UK, where they have a 'superloo' and it costs 10 p and features sofas, TV, showers, shaving place, comfortable toilet and reading material beside the throne! Wow. And then on the train at 0930, off to the Borders, Carlisle change again, and we watch the soft rolling hills with stone fences and riffle/pool streams, into the English midlands, industrial Birmingham, and the air smells like oil on hot metal again. We had one of those true English rail lunches of Stilton cheese, beetroot sandwiches and coffee. And then we were back to Moreton-in-Marsh and the quiet jail house cottage of Ester and Freddie. They had been worried about us up there in wild Scotland, where anything could have happened. We had to suffer another steak and kidney pie, strawberries and cream.

Freddie drove us around the Cotswolds, telling the local history and legends, pheasants were crowing, we visited some stone circles, observed highland cattle and sheep in stone fenced pastures. The Cotswold region is famous for their 'Hunter horses' and hounds used for the fox hunts. Some tiny pastures had signs saying these horses are Chargers, bred for knights in armour. They looked bigger than Clydesdales. And then back to the prison cottage for smoked trout and salad, elderflower champagne, parsnip wine, four different homemade chutneys, thick cream on hot scones with currant and crab apple jelly. TV news was showing a civil war in Yugoslavia. Especially for Sheila, we visited a bird park where lots of tropical parrots, falcons, hawks and eagles sit on perches, and there were enclosures with cranes, emu and storks. Gypsies were camped beside Rollright Stone Circle, charging admission, and uncle Freddie says this is wrong, and he got a little hot under his calm English collar.

On 24 September 1991 Sheila and Ester went off early to shop for cheese and antiques. I was grateful they left me behind and I looked forward to snoozing in the back yard. But no, Freddie was very anxious that I hurry my breakfast, as he wanted to take me to his fishing lake (Top Lake, a dam, stocked with numbered trout, mowed grassed path, shade trees with benches, list of rules on shed door), where members only fly fish. He got two rainbow trout and insisted that I try the fly rod...I got one, but I didn't play it long enough, lacking sportsmanship, I just dragged the poor fish into the landing net. Freddie paid £195 for June-Sept fishing, fifty-four fish allowed for the year. Sponsor for the fish stocking program was General Electric of America. Members of this fishing club arrived in suits, vests and polished shoes, but they have a little closet in the shed where they can put their suit coat on a hanger and put on a fishing jacket, rubber boots, a fishing hat, dip net, creel and side pack with a selection of flies. Freddie had six metal boxes of flies, leaders, snaps and swivels, and a 'priest' for killing the fish. We fished for four hours in the damp misty rain, and SMB and Ester finally showed up with a thermos of tea and BDHS fruit cake. Uncle Freddie was upset at this, as I suspect the fishing grounds were for men members only. Home to the jail house, Ester cleaned the four fish, put two in the smoker, and we had them for supper with pork chops, spicy new potatoes and almond cakes with whipped cream and five kinds of cheese. Sheila and I did our separate English/Scots and Cowboy versions of the classic ballads 'Barbara Allen', and 'Come Boat Me O'er'. Freddie was determined to talk about his treatise on science and religion, as he was in the midst of writing about the demise of religious faith, the established churches, the rise of science and technology and the loss of morals and spiritual values.

By 25 September we were all packed up again, off to the Banbury Bus Station at 0655 to struggle through traffic jams, car accidents in the road ditches and finally Terminal 3 at Heathrow Airport. We left London at 1245 and we got to sit together, had a decent lunch of beef and noodles, snooze, read and neck with my giggling Sheila. Arrive Toronto 2030 London time, 1530 Toronto time, Customs no problem, as for once we were not smuggling anything. Left Toronto 1720, getting hungry, more noodles and chicken in little plastic trays with weak lukewarm tea. Arrived in Winnipeg 1845 local time and it had been raining a lot, mosquitoes and standing water all over the place. Conrad Reimer was there to meet us and he hauled us off to find Jesse at work at the University of Manitoba Campus kitchen. And then we were off to our musical friends Edigers for tea, toast and sleep.

Over the next few days I cleaned out my institute office for the last time. Lawyer Alex Cudney got our house sale money deposited at our account at the Niverville Credit Union. I packed up my office books, journals, file boxes, slide boxes and reprints in seventeen heavy wooden komatiq boxes built by Cliff Jones in the institute workshop, some 1772 lbs of books. Leo Yana, the finance officer at my institute, was very nice to us and he helped with the 'retirement' game. We visited Great West Life finance man Guy Nolan, who will take care of our severance pay and RRSP investments. We paid $5652 for three one-way air tickets to Townsville, Queensland. We were a little miffed that the travel agent would not accept a personal check for these one-way tickets! And then we had to fuss about our Australian Visa, the Ottawa tax office records and our Canadian medical insurance.

There seemed to be endless goodbye parties. I borrowed a banjo from Mitch and a fiddle from Susan Israel for Barb Ediger's birthday party. On 28 September we were taken out to Jack Klaverkamp's country home near Anola for an Institute farewell party, about sixty people there, steaks, melons and sour cream raisin pie (made by Margaret Koshinsky). Brian Billeck, Buster Welch and Paul Wilkinson gave a eulogy speech and presented me with a Mary Valentine painting of our skidoos and sled on the ice of Buchanan Inlet. Mary had travelled the Arctic with PCSP and knew Buster and Cathy in Resolute.

On 29 September we gave our last full Meadowlark concert at the West End Cultural Centre. Hans Hansen, our good philosopher friend and Meadowlark pioneer, showed up from Ontario. Sheila's sister Mary propped a South Dakota buffalo skull on a chair in front of the band. Most of the audience were good friends and we had a hard time shutting down the performance, as everybody was calling for their favourite tunes, Celtic ballads, cowboy songs, fiddle medleys, a rowdy audience indeed, and a good one.

And out by Tourond there were two rural community farewell parties for us at the Tourond Superette Store and in a farm yard. We were given a big cake, speeches, Sheila and I sang our homemade Tourond Store song, sitting on picnic benches beside fuel pumps. Mostly farmers in pickup trucks and feed mill caps sat around, their wives watching from the passenger seat, neither of them wanting to show any emotion, but the farmers came up to shake our hands.

It was now early October and the leaves were falling, fall colours were out, golden and crisp weather, birds flocking. Beek invited us over to her place for supper, tears and a gift of a telephoto zoom lens, so she could see good photos of Australia. She demanded a full account of Ester and Freddie's lives and house, and then we had tea, biscuits and Stilton and Oka cheese. And finally we packed up to leave Canada for Australia.

There was a last phone call to my South Dakota folks. My mother was sick from liver cancer chemotherapy and her grand-daughters Marcy and Donna were taking care of her. My father was hauling big hay bales to the barn and brother John was stacking them in the evening. The big news was about arrival of water pipes from the Oahe Dam (on the Missouri, beside Pierre), so there was no more cistern water from local stock dams, and our house had a new washer and dryer. Despite her illness my mother was determinedly making a baby quilt for Sheila's brother Bill and Carla. After a final supper with Edigers we had teary goodbye champagne toasts. We loaded Sheila's brother Wump's rusty Volvo for the trip to the airport, and Walt, Barb and others were there to say goodbye. Conrad and Dave Stott were there to bid farewell to Jesse with rum and coke. I had a great swelling of emotion during all this farewelling. We were lucky to have so many friends who care about us, links through community schools, scouts, music, science, relatives. I'm sure lots of people leave town and nobody notices. Talk about jumping off a cliff! *UVA UVAM VIVENDO VARIA FIT.*

We left Winnipeg 1730 on 9 October 1991 and the Manitoba prairie was gold, brown and black squares and rectangles in the setting sun. What are we doing here...is this the right move? Arrived in Vancouver, where Al and Sheila's sister Peabody were there in a station wagon to greet us, and off we go to their home in Mission Beach. Buffalo Gal on a chain in their yard. Over the next four days I built a fence for Pb and Al, to keep school kids out of their yard and Buffalo Gal inside. They had borrowed a post hole digger, some big corner posts with brace wire and deadman, it was good to do some physical work. Jesse and Ian removed hedges of thorny blackberry brush with a chain saw. The boys and I took Buffalo Gal for long walks and I felt guilty leaving her behind. Neighbour Fred brought over two fiddles so we could play waltzes and reels together. My godson Gordon was playing his wooden spoon fiddle, watching us carefully.

I remember sitting quietly with Buffalo Gal, watching the clouds and thinking that I'd miss one of the signature sounds of Canada...the high pitched 'baby crying' of the loon.

On 15 October my brother Gerald phoned from Arizona to say that our mother is in poor health in Rapid City hospital with liver cancer. Gerald was *en route* to Montreal and northern Quebec for BHP business and will visit mother on the way home. I had one last walk with Buffalo Gal, who seems to know I'm leaving, wants to cuddle. I don't know how dogs know this. It is even hard to write this.

And so we were off to Vancouver airport for a 2010 flight, leaving North America. Next to me on the airplane is a New Zealand rare book librarian returning home. Honolulu fuelling stop at 2245 in the rain and dark, and we had some beer and pizza. Lv HNL in big CP 747, good food, efficient service, customs forms, folk music on head phones, Eric Bogle, Banjo Paterson, Slim Dusty, Bushwacker fiddle tune stringbands at 35,000 feet and 700+ km/hr, 9 hrs and 20 minutes to Sydney, Australia. Robert Hughes' book tells of all the poor misfits, orphans, criminal that initially populated white Australia. Sleep. Well, I slept, but Sheila was nervous and couldn't sleep.

16 October, Wednesday, did not happen, it vanished into the maw of the International Date Line.

17 October, Thursday, Sheila woke me up to see ruby red/wine red/roseate/blue black dawn over the Pacific Ocean. Then there were cool towels and spray cans to kill bugs and Northern hemispheric germs. Oz news on the radio about footy, the rugby world cup, Oz Wallabies beat Wales, and a lot of hysterical horse race mania. There was a good Aussie breakfast of spicy omelette, bacon and sausage, pawpaw and oranges. I nearly lost my South Dakota cowboy boots underneath the seat in front of me. I was wearing corduroy long pants, cowboy boots, a Fort Pierre Stetson hat, big rattlesnake skin belt buckle, and it occurred to me that customs might not approve of the rattlesnake skin buckle. Ah, well, we had truly jumped off the cliff...no way to turn back now.

Chapter 24. Dinky Di Septic Tanks, True Blue Aussies, Bugs and Prawns in the Daintree Mangroves for the Melbourne Cup Horse Race, and Our New Alligator Creek Home

We arrived in Sydney, New South Wales, 0630 on 17 October 1991, and for sure 3.5 million people live here, all upside down. Sheila wanted to confirm that the water went down the sink in opposite Coriolis rotation. The famously fussy Aussie customs inspection was surprisingly fast, and they let both my cowboy hat with a South Dakota pheasant feather in the hatband and my rattlesnake skin belt buckle go by. They had intelligent breakfasts here and Jesse and I inhaled a large plate of fried eggs, bacon, sausage and ham, even though we'd had breakfast on the airplane before landing. This was a bright and clean airport and we took a bus to the domestic airport to find our flights to Brisbane and Townsville, way up north in Queensland.

I was happily surprised to notice that the land north and inland of Brisbane was looking more and more like the Dakotas, with patches of trees, grasslands, dry sandy rivers, herds of cattle raising the dust as they walk toward windmills and stock tanks of water. But this was all up against the Coral Sea, salt pans, mangrove-lined creeks and not very many towns. This country just might be to my liking. I was no longer potamodromous.

We arrived mid-afternoon in Townsville, a small sleepy tropical town in the middle of the Great Barrier Reef coast. Dr. Alistar Robertson of the Australian Institute of Marine Science was there to meet us. Our host scientist was recovering from dengue fever (bonebreak) as we arrived, and had been off work for six months. He sure didn't look like a distinguished scientist, dressed in shorts, T-shirt and with a long brown beard nearly down to his belly button. It took us a while to get used to the far northern Aussie accent, and we were startled to hear Alistar say (in his deep booming voice) that he was going to take us to the 'Spanish Lice' motel to recover from our long trip from Canada.

We later found the motel to be called 'Spanish Lace' and of course Sheila and Jesse walked around the motel grounds to view tropical flowers, palm trees, geckoes, several kinds of parrots and the inevitable mynah birds. For some reason the town, cars, shops, and appearances made me feel I had gone back in time to the 1950s and everything was starting all over again.

The very next day I was hauled out to AIMS, a federal marine science research institution on a promontory jutting into the Coral Sea, about thirty minutes out of town. We drove casually through scrub forest fires along the base of the coastal mountains, this being the normal state of affairs in late winter apparently, the burn-off season. And here it was again, a piece of paper to be signed, asking me to swear allegiance to the Queen of England, this being another Commonwealth country. Again I refused, and again they found another form that I could indicate I was not a terrorist, illegal immigrant or a criminal, and I promised not to embarrass the Australian Commonwealth.

I was introduced to Irena Zagorskis, my 'off-sider' or technical support officer, who led me to a wonderful large corner office with a balcony facing the beach and Coral Sea. I had arrived with Wordstar and Lotus 123 on my laptop, and the institute here was mostly Microsoft mainframe supported terminals. They had very good computer helpers (Information Technology, IT) here, and Peter Hewitt had me running MS Word, Excel and Vax email within a week.

There were about 160 scientists and technical support staff here in this wonderful hide-away location. Maybe I'd died and gone to heaven? Now I had to figure out how to be a saltwater cowboy, to learn the geography, culture, language and habits of tropical Australia, nearby Papua New Guinea and Indonesia.

We were given an AIMS visiting scientist house to live in for a couple of months, while we searched for a home that we could afford. In a very short time we got a bank account, a tax number and Medicare card, purchased a cheap old rusty car, learned how to drive on the left side of the road and started viewing properties for sale in the newspaper. The CEO of AIMS invited us to supper at the Caltex Roadhouse near Alligator Creek Road, and inspected our skin to determine the likely development of skin cancers...this location apparently being the melanoma capital of the world. From our temporary AIMS accommodation we could walk to the beach and jump into the warm Coral Sea water, and we would leave our clothes, towels and personal stuff in a pile on the beach. We discovered the keys to our AIMS accommodation had vanished from this clothing pile, but we hadn't locked the doors anyway, no problem.

But we later listened to stories of how bower birds steal colourful items to decorate their bowers and attract females. 'Aha!' The light bulb went on, we searched the AIMS beach for bowers, and found our house keys (with a bright blue tab) in front of the nest.

Sheila wisely pushed our son Jesse into SCUBA lessons and he promptly was offered volunteer diving jobs on the AIMS ships on Great Barrier Reef expeditions. While visiting the local pub Jesse met many of the local sugar cane farmers and prawn trawler/commercial fishermen. He shortly had a job driving cane trucks, hauling the harvested cane to the railway siding.

We inspected several houses in the nearby rural community of Alligator Creek, and put in an offer on one with cobwebs in the rafters, doors that wouldn't close and badly scratched wooden floors. But I liked the view from the toilet seat, where you could look down on the tops of a steep slope of gum trees and view two local mountains in Mount Elliot National Park. The owner was sad to sell it, but showed me the water supply system needing repair, and showed me how to kill snakes with a short piece of heavy chain. He had tried to grow citrus in the creek valley, but the trees hadn't been watered or fertilised in a long time and were near dead. The house needed a lot of work to clean it up, but we put in an offer and hoped for the best.

Almost before we could catch our family breath, in late October Alistar invited me to accompany his northern rivers research cruise along the Queensland Cape York coast to measure mangrove tree zonation, core the trees and bring back some logs for wood research. I abandoned my family at the wharf, Sheila smiling and waving at me, shaking her head at the idea of a South Dakota cowboy on a research ship in shorts with no shirt on.

RV Harry Messel departing for Pascoe River, October 1991

This was the *Harry Messel* (named after a crocodile researcher), 107 tonnes, 20 m long, 4000 gallon diesel tanks, radar, two sonars, GPS colour video maps, gyro compass, 4 radios, and a cruise speed of 10 knots in decent weather. The ship Master was Jim Wardell, a deep voiced spinner of tales and sly jokes, veteran of the Aussie Navy in Indonesian seas, who could recite paragraphs of Joseph Conrad's commentary on mangrove jungle creeks. He also could recite Robert Service poetry about the Klondike gold rush, and he wanted to know about the Athabasca tar sands. He provided a case of good Aussie wine for meal times. Two large cool lockers in the aft lab were stuffed with 24 bottle cases of beer (these are called 'slabs' here). Master Jim liked to use the word 'defungerate', and he often departed with alacrity.

So here we go, I'm trying to become a mangrove mud scientist, and I was reading a book on mangroves as we steamed northward up the coast for fifty-seven hours. Upon arriving into the Cape York mangrove lined small rivers, we would put on white boiler suits and lots of mosquito repellent, and use small inflatable rubber (zodiac) boats to get up into the shallow intertidal salinity gradient between seawater and freshwater, where the mangroves like to grow. All of a sudden I had to learn about spring and neap tides, and the duration of low tide. There were about two dozen tree species for me to learn, with help from Alistar, Irena and Paul Dixon. I helped map the species zones, measured tree chest height diameters and cut down a few big trees for tree ring studies, wood chemistry and silvicultural experiments. Some of the big trees were probably over two hundres years old, but nobody had been able to prove that their tree rings were annual. It was bloody hot and near 100% humidity, with sticky mud all over your body. Sand flies would bite your ankles and wrists, the mangrove mud would get inside your boots and rub your feet raw, and your skin was covered with a salt crust. I had to learn how to roll a cigarette with wet muddy hands in the wind on a rolling deck.

At the end of the day, a cool shower and the ship air conditioning was gratefully acknowledged, and you could scrub the sand fly bite scabs off your ankles and wrists. These tidal creeks were everything that Joseph Conrad promised in his book *Lord Jim*, about 10-15 m wide, lined with over-hanging vines, *Nypa* palms and mangroves rich with birds, snakes, clicking crabs, sucking mud and hard sand banks. Some of the *Rhizophora* sp. trees had tangled buttress roots higher than a man's head, and the out-going tide often carried a neat row of hibiscus and other mangrove flowers, coconuts and mangrove seeds.

Somewhere up in the Pascoe River, we nosed our zodiac boat up beside a clump of mangrove roots and I gazed up into the eyes of a cassowary. His big blue helmeted head and beady eyes gave me the impression that he didn't really appreciate our visit, and that we should go somewhere else. Alistar later told me that few people get to see a cassowary in the wild. There were also jabirus (a kind of stork), spoonbills, several varieties of egrets, huge sea eagles,

Torres Strait pigeons, wampoo pigeons, large palm cockatoos, wild pigs, lots of crocodile slides in the creek mud, and sea turtles and dugongs near the river mouth.

On the Daintree River on 5 November (Sheila's birthday) Alistar stopped work in early afternoon, spread a red and white checked table cloth over our gear boxes in the middle of the muddy zodiac and brought out some cold prawns, bugs (a kind of Coral Sea lobster), quiche, hot peppers, cheese, pineapple, a loaf of French bread and cold beer. Everybody else knew what was going on except me, but then he turned his little transistor radio on loud, and the Melbourne Cup horse race was on at exactly two pm. And there we were, in the shade of some big trees and palms, delicately shelling cold prawns and drinking beer while the race was on.

We did similar work up the Claudie, McIvor and Lockhart Rivers. Some days the wind was blowing hard, the little *Harry Messel* pitch and rolled, and half the science crew were sea sick. This means that we got extra desserts at supper time, as I was not affected by sea sickness in rough weather. Yup, this was good, learning a new science down under with these crazy people.

On the way back home we anchored overnight in Cairns harbour and I phoned Reg Watson, who used to work for the Canadian DFO Institute in Winnipeg. He picked me up, gave me a tour of town and his DPI fisheries lab and introduced his wife Jackie, who worked for the Qld. Parks and Wildlife Service, fighting off developers who want to flatten the rain forest. She grew up in Kenora, Ontario...small world. My scientific leader Alistar Robertson announced that I was Chief Scientist for the rest of the cruise, and he defungerated to the airport to go to a science meeting in San Francisco.

It was nearly a two day run back south to Townsville port, and on the way the crew set mackerel lines off the aft deck. These lines had bait fish (a kind of gar) on large hooks decorated with feathers and plumes of garish fibres on a heavy steel leader. When a fish was hooked, the ship stopped quickly, to recover the fish and reset the line. They would throw away fish larger than 1 m length, as they might have ciguatera. There was a special gutting table rigged for the aft gunnel, and soon there were several dozen big mackerel fillets and steaks in the freezer. We also had some fresh mackerel, bonito, mud crabs and squids for our galley meals. The ship cook was very very good at his job. Yup, I could get used to this job...quite a bit different from exploring Hazen Lake on Ellesmere Island at 80 deg N. latitude. We steamed past the port of Townsville and into the AIMS Jetty at 0300 on 8 November, and unloaded 2.5 tonnes of mangrove logs with the jetty crane when daylight came.

Very brown and very happy Sheila and Jesse collected me at the jetty, explaining that our bid for the 84 Alligator Creek rural ridge top house and five acres of land was accepted. This meant we had to get dressed up for a flurry of visits to real estate lawyers and the bank before anybody changed their minds. We had a carpenter/builder visit the house to vouch for the timber frame and cyclone standards for the roof. This was a Queenslander style house on posts, 1-1.5 m of free space underneath the floorboards, exposed plumbing and wiring, to allow for free flow of air underneath the house and avoiding termites getting into the wooden frame. We briefly remembered that there was to be no winter to freeze the exposed water pipes, and the steep driveway was not going to be covered with black ice in the winter. We had to get used to the idea that there was no furnace, no chimney, no triple pane storm windows and no house insulation. Fire had burnt our creek valley clean of brush and grass, which was probably good for me to start my orchard plantation. We could see that the previous owner had a few citrus trees down in the creek valley, and a lot of the irrigation system had been melted by the fire.

Brother Gerald phoned on 16 November 1991, to inform me of the death of my mother the previous day. She died of liver cancer, had refused chemotherapy, and my father and John hauled her from the big city hospital to the small town Philip hospital to die close to her family. Brother Obie said all she asked for in the Philip hospital was for some home-grown garden tomatoes. She was born 18 Aug 1917 in Murdo, a graduate of Murdo High School, and had two years at the University of Minnesota nursing school. Mother had four sons, thirteen grandchildren, and seven step-grandchildren who loved her dearly. She was responsible for all of the wonderful scrapbooks and photo albums which her family regarded highly. My classmate Linda Haugland Brost, student of my grandmother's piano classes, was the funeral organist. The Methodist Church Choir sang her favourite hymn 'There is a Balm in Gilead', and they played a recording of Dvorak's 'Going Home' symphony at the reception. Something important slipped away here, much of the source of my artistic and musical abilities has gone. I sadly wrote a ballad for my mother, thinking of how the town would be bringing baked goods to our house before the funeral, and how awkward my father and brothers would be at this sad but expected event.

Our son Jesse was eighteen years old on 27 November, so we took him to a Malay restaurant in town. This age allowed drinking in Australia, so he ordered five exotic cocktail drinks that he had learned about while bartending at Melville's Pub. We had a good feed, and drove down the highway to our Alligator Creek home. There is a bridge over the tidal Alligator Creek, near the turnoff to the Alligator Creek community. I had unwisely described night croc spotting and hunting from a book I was reading (Tom Cole's *Hell West and Crooked*). Jess decided we should go night spotting for crocodiles from the railway bridge. So we parked our car underneath the bridge, Sheila firmly refused to participate in this unwise adventure, sitting with her arms crossed and not approving of this at all. Jesse and I walked out on the railway bridge with spotlights, and indeed there were glowing red eyeballs in the muddy banks seaward of the bridge. And then the railway began to jiggle and shake, a train whistle came through the humid night, followed by the headlight piercing the dark. Jesse ran down the railroad ties to jump off into prickly bushes near the shore. I was less mobile in my cowboy boots and sat down on one of the side platforms in the middle of the bridge. A long sugar/molasses train went thundering by, and then it was silent in the night. Sheila was sure she had lost her husband and son. But no, we hadn't perished, and we sheepishly walked back to the car to reassure her that all was ok, and that it was a good birthday adventure. Dear Sheila was not amused.

We had some trouble with Australian Customs concerning our shipment of household goods now in the harbour. We had to show proof that Jesse's cinnamon bear hide had been properly tanned, our moose antler cribbage board had been bleached, our guns were legally owned, and pay duty on our new bicycles. On 29 November 1991 our shipment of household and office goods arrived in the yard of our new home, and we carefully unpacked our musical instruments, the moose master canoe, hunting rifles and shotguns, butternut sofa and chairs, kitchen wares, mathoms, shop and garden tools, our home-made wooden ice cream maker bucket, crank and canister, and heavy komatiq boxes of books and music recordings. Now we were feeling a little more settled.

It was also the date of our bank 'settlement day', when we paid AUD $95,000 for our new home. Jesse borrowed a small John Deere tractor with front end weed tree puller, to eradicate china apple forests on our creek valley. This resulted in a lot of embedded prickles in our hands and giant bonfires of this terrible pest tree. We bought a whippersnipper to clear weeds and 1.5 m tall guinea grass in our creek valley and used the cut grass for mulch, as we began planting a fruit tree orchard. What exactly did we know about tropical fruit trees? Bugger all, but we planted eight varieties of mangoes, two varieties of lychees, figs, cashew, macadamia, rambutan, soursop, jackfruit, breadfruit, pumelo, pineapple, coconut, and all kinds of citrus, avocadoes and quite a few tropical fruit trees that we didn't know (longan, Malay apple, mabolo, some of which we later found we didn't like). Our house water supply came from a bore (well) in the creek valley, which also allowed an orchard and vegetable garden irrigation system. We built a garage/shed/workshop near the house and Jesse made a good long work bench and tool hanger board.

We took our moose master canoe out into the Coral Sea from the AIMS beach. Jesse was standing on the gunnels, playing around to swamp it in the warm 30 deg seawater and bring it right up again in the waves. Up until now this canoe had experienced only icy cold Canadian water. What a difference!

In early December the local mangoes were ripe and several neighbours invited us to pick as many as we wished from their trees. We made jackfruit and mango ice cream with our ice cream crank machine. Our neighbour Lorna Beecher had house-raised a possum and had turned it loose. This meant we got a tame possum that would come in our house and sit on the sofa with me to watch the late night CNN news. If I gave it a slice of melon or pawpaw, it would take it outdoors to eat. In our yard there were kookaburras, cockatoos and wallabies begging for handouts, which delighted Sheila.

Our first Christmas in Australia...the mid December AIMS Engineering Workshop Xmas party featured most of the big burly hairy engineers dressed in drag, high heels and thick garish makeup. A blues band was playing loud in the basement and hot sweaty jitterbug dancing was going on as if it were the 1950s all over again. Coral coring technician Bruce Parker did a Madonna sexy dance routine with a ship-board cannon, dressed in fishnet stockings and huge breasts made from grapefruits with chocolate nipples and licorice lipstick on oversized clown lips. There were two stock tanks filled with iced salty water and good Aussie beer and several grills cooking prawns, snags and buffalo wings liberally doused with hot sauces. This was all about the right level of crazy behaviour for a professional science institution staff party.

At home, Sheila used a small palm in a pot for our first Xmas tree and hung heirloom decorations from both our families on the spindly palm fronds. A few of these precious glass decorations were broken during the night, probably by possums climbing the palm. Of course, this is the hottest part of the Australian year, the beginning of the wet season, so Bing Crosby's 'White Christmas' didn't sound quite right, nor did 'Frosty the Snowman'. But you would hear these songs being piped into the shopping malls in Townsville! With a small amount of research, we found a lot of unique and poorly known tropical Australian Christmas carols. One of our favourites was 'Carol of the Birds', about the brolgas, a wetland crane that is common along our coast.

We were invited to a Boxing Day feast with ex-Canadians Rick and Bette Willis, who lived nearby and so were introduced to the idea of 'prawns on the barbie' during a hot weather Christmas. You need to understand that this time of year in Australia is a wet hot summer season, so the idea of Santa in his sleigh gliding over the snow didn't fit at all. Our photo album shows us sitting around our palm Xmas tree in shorts, with all the doors open to the outdoor sunshine and heat. We went to a New Year's party at Alligator Creek's own Melville's Pub, where the Sultans of Swag bush band and square and round dancing was going on, grandma and kids sleeping at the tables, BBQ ribs and snags, lots of beer and sweating dancers. In true Aussie fashion, Melville's Pub was open to the outside on three sides, with dogs and ponies wandering around in the dark just outside of the dancing feet.

It is amazing how fast we adapted to life in the tropics. I didn't wear my South Dakota cowboy boots very long before my feet wanted freedom. The poor white naked things preferred flip flops (called thongs here). My Levi jeans and long-sleeved snap-button shirts were shelved and I got a lot of shorts to wear. In our neatly indoor repainted, newly sanded and floor-varnished house, we have geckoes all over the walls and rafters, green tree frogs and mud wasp nests under the sofa and in the toilet tank and pythons in the laundry basket. We educated ourselves on all the pleasures and dangers of the tropics. Where we now live you can die from a taipan or eastern brown snake bite in three minutes. Box and Irukandji jellyfish will kill you swimming at the beach. In addition to death, these jelly fish cause priapism (prolonged erections in male victims), of interest to the drug companies seeking the next Viagra. Bats can scratch you, giving you lyssavirus, a form of rabies that is often deadly to horses and humans. The big fruit bats were strong enough to carry away ripe mangoes and drop them on our corrugated metal roof at night. Pythons and harmless tree snakes were around the house, feeding on mice and lorikeets. There were lots of big spiders and poisonous thorny bushes that we quickly learned to avoid. Digitoxin (a lethal heart stimulant) was excreted from ear glands of the abundant pest cane toad. Town car wash employees complained about roo and wild pig hunters coming to town to wash down their bloody utes (ute = Aussie for utility truck, a pickup). I got a fungal infection in my feet, probably from walking in the mangrove mud up on the Pascoe River. I got a note from the doctor, saying I should not wear closed shoes, which meant I could wear flip-flops in the AIMS HQ building, where everybody else had to wear steel-toed shoes. The Health and Safety Officer was not amused.

Huge barramundi could be caught in tidal creeks and coastal muddy water, some over 1 m long. The local blokes would cut off the fillets and then proudly nail the bony carcass (frame) to a nearby fence post. This was some sort of macho symbol. I hadn't realized how good mud crabs were to eat till Jesse and a friend trapped a few of them in the local mangrove creek. Sheila could get fresh prawns and seafood from the ships in the harbour. Where we live, we don't have winter or summer. We have a wet season (December to March) and a dry season. The Alligator Creek Fire Brigade came to our rescue several times when grass fires came down our valley. You can work up a sweat at breakfast just by drinking a cup of coffee, even with no shirt on. I never appreciated the word or the phenomenon of zephyrs, tiny breezes that come and go, which were wonderful on hot sweaty days.

I was learning the ropes of surviving in the AIMS HQ, finding all the good people who knew how to do things. An AIMS government car came to pick me up at our post box every morning at 0740. I had my Stanley thermos full of strong coffee, and I could sit on my office balcony reading about the Coral Sea and SE Asian seas. I made a five year research plan, with the necessary equipment to accomplish this work, which was largely accepted by the AIMS administration, much to my surprise.

This grand plan was especially good for the engineers in the workshop, who happily began to build several large sediment coring devices for use on the AIMS ships. I got expert advice on designs for Kasten and gravity corers from Charles Nittrouer of the State University of New York in Stony Brook, whom I had met at an AGU meeting in New Orleans.

I was getting office supplies from the AIMS stores one day, run by an old military veteran fellow. Some young guy had called me a Septic and I had to ask for an explanation. The old fellow said it was cockney slang for Yank (Septic Tank), but that the younger generation didn't appreciate how the USA came to the rescue of Australia during the Japanese attacks during World War II and the bombing of Townsville and Darwin. Most of the Australian Army was off fighting for Churchill in Europe, but Douglas McArthur and some Australian/New Zealander troops chased the Japanese out of Australia, PNG, Indonesia and Malaysia. The old fellow said most older Aussies honour the USA for this wartime support, which was not forthcoming from the UK.

In December 1991 I was at the Australian Nuclear Science and Technology Organisation (ANSTO) in Sydney for an International Atomic Energy Agency (IAEA) sponsored workshop to plan a teaching session for SE Asian scientists about the use of coral annual growth bands as a chronological tool in ecosystem history and pollution. IAEA had been pushing expert radiochemical workshops in many SE Asian countries for a decade, and many SE Asian institutions had been outfitted with gamma and alpha detectors and stable isotope mass spectrometers for research. New research was showing that the stable isotopic ratios of carbon and oxygen, and ratios of strontium, magnesium, manganese, cadmium and calcium in the coral carbonate skeletons gave information about a few centuries of sea water temperature and salinity. Often the SE Asian institutions didn't know what to do with this fancy complicated equipment. This coral workshop had been proposed by AIMS CEO Joe Baker earlier, and I was selected as the AIMS representative, but there were representatives from James Cook University, ANSTO, Australian National University and IAEA (Monaco) at this meeting. We had to find funding to support Asian participants and the cost of presentation, food, entertainment and housing at AIMS. So I came home trying to schedule and estimate costs of a gathering of fifty or so people at AIMS for a week. This was sort of like planning a Boy Scout Camporee.

In February 1992, after an EPA meeting in Seattle to review Arctic contaminant research, I flew back to visit my South Dakota folks, my father, Obie and John's families, uncle Gerald and Joyce down on the Ranch, Uncle Bernard and Aunt Margie over on Prosperity Flats.

Ian MacGreggor came down from Calgary, looking like a long haired hippie musician. My father had moved out of the big house, to allow John's family to move in, and was living in a trailer house out beside a stockdam, alone and sad. He'd been in the hospital with a bad heart and was looking thin and gaunt and wasted. He cried when I came to visit him, because things were not so neat and clean with mother gone. On doctor's orders he was eating hamburger, frozen vegetables and tinned grapefruit, which he really didn't like very much. He was curious about what our sons were doing with their lives and wanted to know about the cattle in north Queensland. I told him about the lop eared Brahma 'Drought Master' cattle standing in the salt pan beside the AIMS road. Dad had two garden seed catalogues on his supper table, marked up with a red pen, thinking about planting a garden in February. February in South Dakota is windy and cold, gloomy, icy snow, grey skies, very different from our new home in tropical Queensland.

I visited my mother's grave in the Murdo Cemetery, where all the Brunskills and Englands were buried. Brother John asked me to help sort a basement full of our mother's file cabinets and cardboard boxes of photos and family history. Wartime letters and photos from my uncle Raymie were given to his daughters Janet and Gail. Raymie also sent letters and photos from Fort St. John in British Columbia, where he was working with a construction crew on the Canol Pipeline, a wartime project. Mother had carefully labelled boxes for each son and some of the nieces and nephews. Several nieces had taken mother's antique doll and doll clothing collection and Aunt Margie took care of an extensive fabric collection. A lot of my letters to my parents from 1960-1992 Augustana, Cornell, Winnipeg, European travels and Australia were bundled up in ribbons, along with Hot Crotch, a Russian tea cosy we had given to her in 1971.

Obie's daughters took me to the Rapid City airport and I flew to Boston, rented a car and drove through blowing snow and icy roads to the Woods Hole Oceanographic Institution on Cape Cod. I was seeking advice on building sediment coring equipment, how to approach a coastal tropical marine sedimentology, and visiting some of the best radiochemical laboratories in North America. I was planning to build a new radiochemical lab at AIMS, but the new varieties of gamma spectrometers were different from my gear in Winnipeg. I also visited another ELA student (now professor at the University of Connecticut) Tom Torgersen, who was greatly amused to find that I was trying to be a chemical oceanographer.

And then I flew back to Seattle to visit my colleagues Glen Shen (coral core history), and Steve Emerson (early ELA radium spike student, now professor at the University of Washington) and continued up to Eugene and Corvallis, Oregon. Here much of the JOIDES oceanographic ship equipment that I needed for my work in coastal Queensland was stored, such as the big box corers, Kasten cores, and large piston corers. Allan Mix gave me advice on tropical sediment core slicing and processing, and yet again instructions on how to be super clean on a dirty ship with diesel exhaust everywhere.

Next I flew down to Albuquerque, and drove a rental car to Santa Fe, New Mexico where I was scheduled to give two papers for the American Society of Limnology and Oceanography and attend a climate change panel discussion at the Santa Fe Institute, a highly respected think tank. Here were many of my colleagues, David Schindler and Susanne Bayley from the University of Alberta, Wally Broecker from Lamont, Gene Likens of Hubbard Brook fame, Dave Culver, who helped me with my Cornell Ph.D. thesis work, and John Robbins from Michigan, who taught me all I know about sediment core ^{210}Pb tracer modelling. We all had a good time at evening suppers in wonderful Mexican style restaurants. I spent a whole day walking through an abandoned Indian cliff archaeological site, just outside of town.

I then went to San Manuel, Arizona to visit my brother Gerald's family, up in the San Pedro River valley. Gerald was a supervising mill engineer for the San Manuel Magma Copper mine, which had been recently bought by Australian megacorporation BHP. He gave me a tour of the refinery, where the copper ore was crushed, extracted, flotation cells, and the final copper ingots dripping on racks. Gerald took me on a tour of the country, past the little abandoned mining and cowboy towns...this is the country of many of the famous cowboy ballad writers, Power River Jack, Gail Gardner ('Work's All Done this Fall'), the Sierry Petes were to the north, Mongollons, cougars and pumas of Badger Clark's poems and Larry McMurtry's novels were all here. This was the land of the Hopi, Navajo, Comanche, Apache, the last refuge of Geronimo and Cochise. Gerald pulled his banjo out from under his bed so I could play for all his kids.

And then another long long flight across the Pacific Ocean to Australia, and Valentine's Day vanished into the maw of the International Date Line. *En route* home, I spent three days in Canberra at a science meeting at the Australian National University, where I met some very useful biogeochemists to help me get set up at AIMS, and walked through the National Museum. From there I visited Sydney and the Australian Nuclear Science and Technology Organisation, where I found more new colleagues anxious to join me in research in the warm tropics. Finally a sleepy airplane ride took me to Brisbane and Townsville, where a very brown and bubbling Sheila and Jesse were there to meet me.

I was bothered by leaving my South Dakota family in a period of stress, me being the oldest brother, and realising that my home was now halfway around the world away. As it turned out, my feelings of family upheaval were justified. My father died 6 March 1992, having had a heart attack in the parking lot of the Murdo Supervalu, with a load of groceries in his arm. He didn't get to plant his garden. I suspect he didn't like much living without his wife. John took him to the Pierre Hospital in his pickup, but he was dead when he got there.

Dad was born in Murdo, 1 Jan 1917, and grew up on the White River ranch. He was a good shot with his .22 Winchester and .300 Savage rifles, and he knew how to set a White River fish trap. He graduated from Murdo High School, and his 1935 class photo was hanging on the wall in the typing room when I went there. His basketball team went to the State Tournament, and he followed school sports even after his sons graduated. He was a quiet but very smart man, a University of Iowa graduate mechanical engineer, and he could solve my high school algebra and physics problems in his head.

I didn't go to the funeral, but asked my brothers to make a bouquet of White River *Phragmites, Spartina* and Yucca seed pods. Gail England van Bockern, Donna, Marcy and Annie did the final assembly. The weather went bad the day before his funeral, strong wind and snow, and it must have been miserable at the cemetery. Brother John was saddled with most of the legal problems of the lack of a will and the dispersal of our father's estate. Apparently, the ownership of the big England house and land had not been transferred from Grandpa and Grandma England to my father, and the lawyers had to scramble to fix this. Nor was my father's share of the White River Brunskill ranch easy to understand. The land, livestock, tractors, trucks and Bronco were subdivided amongst the four sons, but John remained to run the England ranch and cattle. I sold my quarter share of the home ranch land to my youngest brother John, and this money was used to pay off our Alligator Creek mortgage. This was also a signal to my brothers that I was not coming home to South Dakota. All of this was deeply unsettling to me, so far away. Both Sheila and I had now lost our parents and we became the old folks, with our sons, nieces and nephews now grown up to adulthood.

During this hot summer of 1992 we were invited to pick mangoes and lychees at neighbour Marg and John Scorse's place. We enjoyed our neighbours Mike and Lorna Beecher's swimming pool when they were away on holidays. Their daughter had a riding horse down in the creek valley, which Jesse and I got to know fairly well, even if we had to ride in an English pancake saddle. I slowly planted, irrigated, pruned and fertilised my orchard, after cleaning out the chinee apples and weeds. Meanwhile, Jesse appears in our photo album on many different ships, holding many different kinds of fish.

Sheila and I accompanied AIMS fishery researchers Sharon Kramer and Steve Newman on the research vessel *Lady Basten* out to the Myrmidon Reef edge to explore the feasibility of hand line fishing for the big red snappers, worth a bundle in Japan. We were catching big red snappers (*Lutjanidae*) at 400 m water depth on gunnel reels (Alvey Reef Queen Handline Reels), with many of the fish chewed in half during ascent by sharks. Captain Dave Beard caught and gutted a 40 kg mackerel on aft A-Frame wire. We drove inland to the Mingela Rodeo in May and it was not too different from a small town South Dakota Rodeo, except for no saddle horns, more drunks, and the massive mountains of empty beer cans behind the grandstand. Jesse worked on a prawn trawler for a while and learned barefoot waterskiing. He found work with local blokes Robbie Lyons and Peter Connolly, because he knew how to drive tractors and trucks.

Over the next years Sheila populated the house and yard with a variety of pets and semi-tame wild creatures, orchids and vines and built concrete ponds in the yard for the wildlife. A white van appeared in our yard and someone we didn't know from Adam provides us with Charles the Cockatoo, unannounced, free with cage. We should have known better, but we were suckers for the parrots. Sheila raised some magpie geese, some Australian magpies, a fruit dove, an emu (see below), all of which wandered through our house night and day. She calmly named a medium sized python in our laundry basket Pythagoras, and I took the basket out to the back yard for release. Pythons are good to have around, as they eat rats and mice, similar to bull snakes in South Dakota barns. A Children's python was observed on the kitchen rafters, and a big spider caught and ate a little bat above our fridge. At night I had to be careful walking around, as the big rhinoceros beetles would bite my toes. There were big green tree frogs sitting beside my late night reading chair, gobbling insects attracted to my reading light.

GJB and Charles, the sulphur-crested cockatoo, 1992

In most years, from January to March it rained. And rained and rained, and the humidity was about 100%. The eucalyptus trees released the odour of mentholatum (eucalyptol) when they were wet, and frangipani flowers gave off a perfumy fragrance and would glow in the dark. Our local white gum and ironwood trees are more transparent, compared to North American opaque trees. You could see right through the ridge top forest trees when the western setting sun went down in a salmon-belly rose colour. Sheila was delighted to view over two hundred species of birds from her back veranda, as evidence of the expected tropical biodiversity. Inland, flocks of grey and pink galahs would fly over the red earth, like plumes of smoke. Compared to the Dakota and Manitoba rich gumbo soils, our north Queensland soils were terrible, useless, pathetic, unless greatly augmented with mulch, manure and inorganic fertilisers. Our first garden near the house was pitiful, and what came up was quickly eaten by the wallabies, possums and parrots.

Chapter 25. Salt Flats and Mangrove Mud Adventures in the Fly and Herbert River Deltas

One of the long term worries in north Queensland was the conflict between the inland cattle industry, the coastal sugar cane and banana plantations and the busy small town harbours loading mineral ores to ocean-going ships, up against the concerns about protection of the environmental qualities of the World Heritage Area Great Barrier Reef (GBR), and the associated tourism industry and seafood harvesters.

The agricultural scientists in CSIRO (the Commonwealth Scientific and Industrial Research Organisation) were anxious to measure and limit the impact of soil erosion and river transport of contaminants to the estuaries and inner shelf of the GBR. Much of this river sediment was trapped by extensive mangrove forests in the estuaries, sort of a kidney filtering system for the clean water of the mid and outer shelf reefs. CSIRO began a research program, from which Alistar Robertson had already obtained funds for his work on mangrove biology. Shortly afterward, the national government started a coordinated research centre (CRC) at James Cook University for land/GBR research and Alistar and I wrote proposals for major funding for a decade of work in the Herbert River estuary and Hinchinbrook Channel.

In May and August 1992, continuing to find my niche in this part of the world, I assisted Alistar Robertson and his crew of Irena Zagorskis and Paul Dixon in measuring growth rates, mortality, species distribution, leaf litter fall, root density and nutrient chemistry in the Herbert River estuary mangrove trees. I slowly learned the species of the mangrove trees and how to walk and work in this hot and humid environment. I guess most people would avoid walking through mangroves at low tide, when the smell of sulphur and rotting vegetation, the popping of the fiddler crabs, abundant sandflies and mosquitoes and the very abrasive bark on the tree roots makes life difficult. But I learned from tall gangly Alistar, who seemed to be especially adapted to striding through the tangled mangroves with a beaming smile on his face.

Mangrove is a word used to describe the habitat of emergent vegetation that can tolerate being alternately inundated with tidal seawater and then with freshwater from adjacent rivers. It also refers to a group of species of trees, shrubs and ferns that have variable preferences for low and high salinity regions of estuaries, and can tolerate highly saline rocky, sandy mud with low oxygen concentrations around their roots. In our region, from Hinchinbrook Channel (Herbert River) up to the Papua New Guinea coastline, there are 30-40 species in this habitat group (amongst the highest biodiversity in the world). In some locations this ecosystem is highly productive, producing organic material that is food for microbes, crabs and fish. The zone of interlocking mangrove subsurface and emergent roots is also an efficient sediment trap, where much of the terrestrial sediment carried by rivers is deposited.

Along the Great Barrier Reef coastline of north Queensland this ecosystem function is very important, as it keeps the mid and outer continental shelf relatively clear and free of riverine mud, so that coral reef communities can grow down to depths of 20-30 m. I was lucky to be in a good group of AIMS scientists that worked on mangrove ecology, Alistar, Dan Alongi (microbiology), Norm Duke (taxonomy), Barry Clough (tree physiology and productivity) and some very good technical support people.

We did our work out of an inflatable rubber boat (Zodiac) with a little outboard motor, zooming up the shallow mangrove channels at high tide and trying to get back out before low low tide. There were egrets, sea eagles, jabiru, pelicans, kingfishers, spoonbills, cormorants, ibises, mud crabs, crocs, dugongs, polychaetes that glow in the dark and foul sulphide brine in mangrove mud pore water. We spent a whole afternoon penetrating up the Herbert River delta trying to find a reputed patch of the *Nypa* palm, probably the southernmost population of this lush tropical species. We tried to get permission from sugar cane farmers ('cane cockies') to enter the mangrove forest from the freshwater end. Some of them did not like academics messing around on their property and they asked us to leave and not come back!

I was paying the price of being a 'newbie', as the mosquitoes, sand-flies and no-see-ums feasted on my flesh. Insect repellent and sweat would pour into my eyes, and I developed an infected toe splinter. At night I would soak my foot in disinfectant underneath the table. On these Zodiac trips we stayed at Lee's Hotel, reputed to be the famous location of Slim Dusty's song 'A Pub with No Beer'. It took me several months to build up resistance to the mangrove sand flies and mosquitoes. Back home, my skin itched for months.

In June, Sheila and I played an afternoon forty-five minute set (with fiddle, banjo, guitar, hammered dulcimer and autoharp) at the Pangola Park Rainforest folk festival south of our home. The audience appeared to be mostly hippies camped out in the park, with only minor interests in music. That evening we drove to the north side of Townsville to perform for Jennifer Hultgren's NQ Bluegrass Society Festival, which was set up at her home back yard, with a good stage and sound system. We did a twenty minute old timey music set for an appreciative audience and were happy to hear lots of other good performers.

It was NQ winter now, so it was chilly in the evening and we needed woolly sweaters or jackets. It was very dry, with smoke in the air from bush fires, most of them intentional 'burning off' for property protection or from cane fires. We harvested our own paw paws for the first time!

Sheila and Jesse volunteered for an AIMS ship expedition to Lizard Island. The wind was up, there was rough water, and Sheila was seasick most of the time. When the work was done she decided to come home on the bus from Cooktown, arriving in Townsville at dawn, very brown and tired but happy.

Michael Rosen was a Perth scientist and bluegrass musician. He had heard about us from folk musical circles. We sat and played tunes for each other for several evenings, and in the daytime we explored the local salt flats near AIMS.

Mike was interested in geochemical and hydrological aspects of sabkhas, or intertidal evaporative lagoons. As if we didn't have enough work to do, we started a collaborative study of the local Cocoa Creek salt flat, with the purpose of figuring out the composition and age of the brine contained in the sediment underneath the salt crust. This salt crust would be stiff enough for a vehicle to drive on it during neap tides and would be completely dry and windblown, often glistening white sea salt. During high spring tides, the salt flat would be covered with 50-100 cm of seawater, often pink coloured due to hypersaline algal growth. It would get very hot while over the shallow area, and evaporation would almost double the salinity of the tidal water flowing back into Cleveland Bay.

The tidal creeks were rich in fish, mud crabs and birds of many species, and local people would set up camp on the creek banks for a weekend of fishing and crabbing. We discovered aboriginal rock art in cliff caves on the Cocoa Creek trail beside Mount Cleveland. The aboriginal people knew about these caves and they wanted to make sure I would not photograph the art or advertise the location of the caves, and would treat the salt flats with respect. This whole region was a safe hunting ground for the early aboriginal people, especially during the dry season, when the billabongs, rivers and wetlands had dried up.

Neither Mike Rosen nor I had the manpower and equipment for a study on the Cocoa Creek salt flat (sabkha) hydrology and geochemistry. So again we gathered interesting colleagues to help us. Paul Givney and Ken Woolfe of the JCU Geology Department helped us drill holes for the piezometers, which were basically slotted plastic pipes from which we could pump brine from different depths below the salt flat surface. The big diesel powered drilling rig was tenderly backed out onto salt pan, the dry surface crust undulating over jelly-like saline mud underneath. We had several anchored safety cables on nearby chenier ridges, in case we needed to pull the whole works out of the mud.

GJB and Jesse sampling salt flat piezometers, 1992

This reminded me of flexible seawater ice between leads in the high Arctic, where the Inuit knew how to judge the safety of walking on this thin ice.

We drilled down 4-12 m deep to find different kinds of brine, sandy, gravel, shelly, muddy layers, down to 18,000 year Pleistocene soil. We also drilled on the nearby terrestrial vegetated chenier ridges, which we found to act like channels for freshwater runoff moving toward the tidal creek, avoiding the salt flat because of the high salinity/density of the water there. Sheila and Irena helped with installing and pumping the piezometers and Peter Ridd of JCU Marine Geophysics contributed several graduate students who were involved in thesis work over the next several years. By several lines of evidence, we discovered that the deep brines were many centuries old and did not flush to the sea very often. Several valuable trace elements (Li, Sr, U, Th) were enriched in both the deep brine and briny mud. The layers of different kinds of sediment in the sediment cores appeared to indicate different times and speeds of sea level rise.

We also discovered a recently acetylene torch-opened safe in the bushes, which we reported to the police. This caused me all sorts of trouble, as I had to go downtown to the police station to sign photographs and maps of the location and got fingerprinted.

Paul Givney was a stickler for a 10 am and 3 pm smoko campfire for brewing tea out there in the middle of the salt flats. He taught us that smoko was a serious requirement for coring crews, but he provided dreadful stale biscuits with it.

In August I was asked to contribute to a review of the Townsville Port Authority Dredge Spoil disposal plan. This was to become a recurring irritation, as the port had somehow convinced the state and national government that they should be allowed to dump dredged harbour and ship channel mud out in the Great Barrier Reef. This should have been a clear violation of the Great Barrier Reef Marine Park Authority provisions, as well as the World Heritage Area status. But, no—jobs, money and industry overpowered the environmental advice.

Colleagues back in Canada refused to let me go. I was a Ph.D. thesis reader for two of Jaap Kalff's students at McGill University in Montreal and was working on proofs of two science journal manuscripts about our Canadian high Arctic contaminant work.

I had to do several major revisions of my proposal for the Herbert River estuary study, as my Aussie colleagues told me I was a little too aggressive in explaining the outcomes of my work before I had even begun the study.

In September my fifty-first birthday rolled around, and I got a wheelbarrow and candles in a pawpaw for my presents.

BHP and Ok Tedi Mining Ltd supported Fly River Delta research on mangroves proposed by Alistar before I arrived here. Although I knew nothing about Ok Tedi Mining Ltd (OTML) at this time, I eventually learned that it was a giant gold, silver and copper mine up on top of Mount Fubilan in the Star Mountains of western Papua New Guinea, in the headwaters of the Fly River. For such a relatively small island, PNG has a lot of world class big rivers, due to 6-12 m/yr of rainfall up in the interior rainforest mountains at 4-5 km elevation. The Fly River delivers an average of 200 km^3 of muddy water to the top of the delta every year and some of this mud goes into the Gulf of Papua at the northern end of the Coral Sea. OTML was one of many international mining projects of BHP (Broken Hill Proprietary Company Limited, HQ in Melbourne, one of the biggest and oldest mining companies derived from a 1885 New South Wales silver and lead mine, but now diversified into other metals, oil and steel production).

The little stream named Ok Tedi was a small mountain valley tributary to the Fly River, where the waste rock and tailings from the mine were dumped. The mining company had tried to build a tailings pond, but landslides and high rainfall made this completely impractical in these high mountain valleys. BHP and OTML were very wealthy and powerful forces in a country with little or no environmental control, and beginning in 1984 they hugely polluted the entire Fly River, one of the world's major tropical rivers, with mine tailings and especially copper. Downstream villagers (some 50,000 people in 120 very remote villages) were complaining that their floodplain gardens, the river and oxbow lake fishery and the floodplain forest were being damaged by OTML mine tailings. BHP and OTML hired some outside scientists (including several scientists at AIMS) to assess the damage. OTML also maintained an environmental laboratory and scientific staff in Tabubil for their own environmental research capability. To their credit, BHP and OTML assisted, funded and published scientific research on the Fly River and Gulf of Papua for over thirty years.

Alistar Robertson had funding from OTML for mangrove tree ecology studies but he was away in San Francisco for a science meeting when his next expedition was scheduled in September 1992. He asked me to ramrod his technical helpers, as they knew what to do, and so Sheila and Irena and Paul Dixon and I gathered our gear. I was very much acting as an assistant to Alistar in this work, which was mainly related to mangrove tree biology and productivity. But Alistar needed to know the burial (loss, storage) rate of nutrients and organic matter in the mangrove intertidal mud, which was my job. The AIMS workshop had built a manual coring device for me, along with a portable light jungle working table and mud sled. OTML flew us from Cairns up to Daru in a Twin Otter, where we were met by Bradley Tulloch from Topaz, Queensland, Master of the OTML ship *Western Venturer*, with a crew of PNG nationals.

OTML ship Western Venturer, *1992*

The PNG customs officials took their time looking us over. They searched our luggage and confiscated my coucal hat feather. We walked down to Daru jetty, where a Western Tug and Barge company vessel was unloading essential supplies for the village: Pepsi, chips, rice, candy and SP lager, plus a few big plastic septic tanks for the government offices. We walked the town market, an open dirt space under the shade of big sea almond trees. The ground was saturated with blood-red betel nut spit and there were smoky fires where fish was being cooked in banana leaves and turtle meat was being roasted on spits. Mangoes, betel nuts, crabs and clams were all for sale or trade. This was 17 September 1992, the second day of their PNG Independence Celebration week. The town was in party mode and there were soccer games, stalls selling Jesus weavings, beadwork, crochet and big hand crocheted doilies for sale.

Our science equipment shipment was delayed and we had to wait for fuel and freshwater supplies for the ship, so the Master took the *Western Venturer* to Missionary Passage for two days anchorage, to keep his crew sober and out of jail. We finally got our scientific gear and ship fuel by incoming barge and we were able to steam for the western channel of the Fly River Delta in choppy sea in the dusk. Of course things did not go well, as we hit a sand bank, lost steering power, a hydraulic line to the rudder had exploded and the crew tried to fix it at an awkward anchor position in full wind in the dark, thumping up against the sandbar.

The *Western Venturer* was a catamaran design and should have been shallow draft, but we kept hitting sandbars for a week. Meal time was a character-building experience: we had Donald Duck napkins, chlorine flavoured cold tinned yellow beans, nearly raw steak tough as nails, tea and shortbread fingers, brought to the table joyously by cook Leo Warakan. Leo also made lunches for us when we were off in the mangroves in the Zodiac: bologna, cheese and tomato, corned beef with cabbage and mayo, every day identical. We had an endless supply of New Zealand apples and warm Solo lemon crush drink. We were taking lariam tablets once a week, to reduce the chances of malaria.

Some of us were a bit concerned at an official-looking notice taped below the horn button in the *Western Venturer* wheel house which said:

Six Lic Lic Bello
1 Big pela Bello
All pela jump.

This is PNG pidgin meaning 'six short bells, one long bell, all fellows jump', or the signal to abandon ship!

We took Zodiacs into the mangrove creeks at locations where Alistar's crew had previously tagged and measured tree chest height diameter (CHD) and tree density in 10 m^2 quadrants. We measured salinity of high tide water around *Nypa* at the freshwater end of the delta as 4.2‰, with mostly *Rhizophora* species in high salinity (25-30‰) water. On Purutu and Aibinio Island we collected *Nypa* seeds to estimate fertility/mortality of the seeds.

The AIMS workshop had built a portable aluminium tripod with a hook for a block and tackle about 2 m high, to pull out our manually driving plastic core tube. I didn't appreciate the force needed to pull the 2+ m full core tube out of the plastic liquid mud, as the weight and suction almost defeated us. We later learned to drive a small plastic pipe down beside the big core tube, to pump air (using the Zodiac air pump) underneath the core tube as we were trying to lift it. Our first good core came from a Purutu Island *Heritiera* sp forest in the tidal mud. It took us five hours of hot sweaty work to get a 1.3 m core. We built a working platform of Nypa and palm fronds on top of the soft mud, to allow good footing while we worked to take the mud cores, and to set up our core slicing table. If we stepped off this palm frond platform we would sink in the soft mud up to our hips.

Irena Zagorskis, Paul Dixon and GJB taking sediment core of mangrove mud, Fly River delta, 1992 (photo: Sheila Brunskill)

We also cut several big sections from the biggest mangroves we could find, mainly the cannonball mangrove (*Xylocarpus granatum*), for tree ring radiochemical studies back at AIMS. We wanted to find a way to determine whether or not the mangrove tree rings were annual, as in most northern hemisphere temperate zone terrestrial trees. These were 60-100 cm in diameter, very heavy (they would sink in water), and we had a special mud sled to slide them out to the deeper creek water, so we could load these 70-100 kg wood sections onto the ship with a crane assist.

Just to keep us amused and off balance, we found a telephone floating in a mangrove creek. It rained almost every day for a few hours, so we were soaking wet and hot for most of our working time in the delta mangrove forest.

The native PNG *Western Venturer* crew took us into some of the Fly River Delta villages on Purutu and Aibinio Islands, over a muddy footpath and single log bridges over tidal creeks. The people lived in *Nypa* palm-thatched longhouses perched 1-2 m high up on big mangrove tree stumps. The people wore little or no clothing, as expected in the hot muggy tropics, but some kids wore baseball caps backwards and Michael Jackson T-shirts. There were big dugout canoes tied to the house ladders, some with sails and outriggers, some with 15 HP outboard motors, and some with platforms to carry cargo to Daru for trading. Great piles of corned beef tins on the outside of the village indicated some connection to grocery stores in Daru, but we also walked past carefully tended gardens for sweet potato, pawpaw, mango, banana, cassava, sago and coconut palms, each marked by the owner. Pigs, kids and dogs wandered about in complete freedom and pet parrots, eagles, cassowaries and crocodiles were kept in crude cages or pens. Hoop fish traps and gill nets were neatly stashed in the village centre. Wood smoke came from cooking fires on small clay-covered wooden platforms up on the little porch of the elevated long houses, where fish, crabs, coconut, breadfruit and sweet potato were being cooked. Pith from the sago palm was made into flour, packed into palm leaves or bamboo cylinders, and brought to Daru for trade.

After we had finished most of our work, Master Bradley insisted on visiting the NE Fly Delta community of Kaviopo, near the freshwater end, where they had a village craft centre and a grocery/supply store. Master Bradley wanted to buy toilet paper, biscuits and fruit, but none was for sale here.

Kaviopo Village, Fly River Delta, 1992

We then understood why we had been pushed up here, as the villagers were anxious to sell their craft products, all carefully laid out in piles under a *Nypa*-roofed open-walled display area. A lot of it was intended for the tourist trade, but at prices one-tenth of that in Port Moresby or Cairns. Large and small, plain and decorated *kundu* drums, war shields, arrows, garden sticks, grass skirts, woven baskets, carved hooks, carved crocs and birds, cassowary and parrot head dresses, wooden spoons, tame parrots in woven sapling cages were all for sale. Sheila and I bought 50 Kina worth of kundu drums and wooden wall hangings, but it was difficult to find the right person to pay. Two fully dressed native ladies in wool suits carrying bibles walked through the market, clearly not approving of us buying pagan symbols.

Our mud skipper Bradley took the *Western Venturer* around the corner for peace and quiet overnight, and the next morning we steamed down the delta channels for Daru. We hit several sandbars, lost the steering power twice, but arrived in Daru just before midnight in the pouring rain. Despite our best efforts and the efforts of good shipping agents at AIMS, no provision for shipping our scientific equipment and samples was ready.

This meant starting from scratch with the local barge company, who claimed to have a four month backlog of shipments waiting in their warehouses. 'Hurry up and wait'- - have you heard this before? Master Bradley phoned his bosses at Ok Tedi (Ian Wood, Murray Eagle) in Tabubil, who must have then phoned the barge company (and/or paid the necessary bribe) to tell them that we had priority. Within a few hours our freight was labelled, put on pallets, loaded with a crane and the barge lines were cast off the jetty, a little tug pushing with engines whining at full bore. This meant we could leave Daru and Master Bradley hauled us to the airport, through hot smoky dirty streets with giant pot holes, fish guts, upside down turtle shells strewn everywhere, piles of Pepsi and corned beef tins, lots of mosquitoes and mud.

Daru Customs and Migration did not like us, scowled and prodded our luggage, but we were in the midst of Ok Tedi employees from Tabubil going to Cairns, and they knew how to intimidate the Customs agents. We left Daru at noon on 28 September 1992, flying over the gorgeous Torres Strait coral reefs and islands, fires over Cape York cattle land, arriving in the clean air conditioned Cairns airport, where the Australian Customs Agents sighed and smirked when they saw cassowary feathers sticking out of my luggage and bundles of wooden *kundu* drums with untanned lizard skin drum heads. Most of this stuff had to be fumigated, but we eventually got it all back. Our 1.5 tonnes of scientific equipment and samples were temporarily confiscated, as we had no import permit. We fell into the Cairns airport food services with glee, having some hot meat pies, cold beers, good black coffee and mango ice cream.

Sheila and I had been away from our Alligator Creek home for over two weeks, while Jesse was working for the cane farmers down on the Burdekin River region. It hadn't rained and our orchard was parched, hot wind sucking the water out of the fruit trees. Many of the young sapling leaves were chewed off by grasshoppers and bugs, but they came back well after several days of irrigation from our bore.

Something important seeped into my addled brain, as the adrenalin from working in exotic locales wore off: I felt a link, a kinship, a personal heart-murmur for that big muddy Fly River. As a young boy, I had grown up on the floodplain of the White River, and got to know the annual cycle of drought and flood in South Dakota. I had spent six years working on the huge sub-Arctic Mackenzie River and tributaries in the NWT of Canada, and now I was being lured into a relationship with an even bigger tropical river in Papua New Guinea, the people and villages, and the industry associated with it.

In November I was sent down to Brisbane to attend an AIDAB meeting, Asian Development Bank, World Bank, Australian Department of Foreign Affairs, and other agencies interested in collaborations and investment in SE Asia. I was probably sent to this meeting because I was the front man for AIMS research projects in Papua New Guinea and possibly Indonesia. Alistar and the AIMS CEO Russ Reichelt had given me a half dozen names of good people to seek out, to support our proposals for work in the Gulf of Papua.

I was really really out of place here, in the midst of pinstripe suits, bankers, investors and slick talking industrial hucksters. I did meet some people concerned about environmental problems related to developments, mines, oil and gas exploration in countries with little or no environmental protection laws. Many of the lectures were trying to educate Australians on how not to insult SE Asian cultures in business meetings, how to handle fee and bribes and leverage, and the double meaning of hand gestures and facial/body language clues. None of this appeared to be pertinent to my recent muddy experiences in the Fly River Delta villages. Well, ok, the hotel was very posh, the meals were pricey and good, smoko featured wonderful pastries with cream, jams and fresh fruits...but I had forgotten how much I disliked the smell of cities, diesel fumes and all, and was very happy to return to the sweet warm fragrant tropical air of Townsville and our quiet peaceful home at Alligator Creek.

I was only dimly aware of OTML Murray Eagle and Ian Wood's plans for environmental studies on the Fly River. They wanted to fund Alistar and my research, and invited us to a big meeting at the BHP/OTML mining company HQ in Tabubil, up in the Star Mountain rainforest of the Western Province of Papua New Guinea. And so in late November Alistar and I left Townsville at 0600 and rode the airplanes to Brisbane, Port Moresby, arriving in Tabubil at 1640 in the pouring monsoonal rain. We were travelling with Graeme Batley and Simon Apte from a good environmental chemistry CSIRO lab in Sydney, as they had been working on contaminant copper chemistry in the Fly River. This is a 2000 m elevation mining town, built completely for OTML employees and their families and the mine administration facilities.

We were briefed at the OTML Environmental Lab by Ian Wood and Murray Eagle. Just outside the veranda was a riot of orchids, torch gingers and heliconia. The forest was close by, dripping with water, moss and vines covering all tree trunks and rocks. There was a large mesh enclosure nearby with black cockatoos, green parrots, tree kangaroos and a couple of very large carpet pythons. There were endless rows of joined ATCO trailers for miner accommodations, kitchens, laundry and playgrounds for kids, but they put us up at the posh Cloudland Hotel with satellite TV, a swimming pool and a bar fridge. We had a lavish supper of lobster, prawns and barramundi with Wim Salomons, a Dutch geochemist, while the rain pounded loudly on the tin roof.

Over the next few days, a dozen OTML contracted researchers gave reports on their research activities on the river. Alistar's presentation on mangrove biology went very well, and my naive presentation at least gave them the impression that I knew how to core and measure mud accumulation rate. Wim Salomons, Graeme Batley, Simon Apte and several BHP chemists gave good presentations on mine contaminants in the upper Fly River, just below the mine effluent. Gary Parker (University of Minnesota) and Bill Dietrich (University of California) spoke on their research on river channel hydrology and geomorphology. The mine had hugely increased the suspended material and dissolved copper in the river, and the fish and other biota had greatly declined. Australian biologist Jim Tait spoke about his fishery research with many of the downstream villagers, who relied on barramundi, catfish and the famous Papuan black bass for their protein. The increased sediment load into the upper Fly River also increased the elevation of the river channel bottom and resulted in flooding of many of the sweet potato and fruit tree gardens in downstream villages.

During these sessions, our noon lunch was with the miners at the Poon Brothers dining hall. [I need to point out that the Poon Brothers were a Western Australian company that provided services to remote mining communities in PNG, Indonesia and Malaysia. 'Poon' is also a derogatory term for aboriginal people.] There were two lines for the food tables, one for PNG people (a lot of sweet potato and fish in coconut milk) and another for the Indonesians and white people (more red meat, potatoes, spicy rice dishes, salads and fruit).

At the end of the week our host gave us some free time and we visited the craft shop, run by wives of the miners. There were lots of finely carved and decorated canoe paddles, kundu drums, lap laps and a great variety of penis gourds for sale. We were also invited on a helicopter trip down the Fly River, to view all of the different zones of impact from the Ok Tedi Mine up in the headwaters, down to the lower Fly River near the apex of the delta. Just below the foot of the mountains at the town of Kiunga, the big copper ore carrying ships were loaded but anchored. They were stranded upstream, as there was a drought and there wasn't enough water to get down to the port at Daru, where the ocean-going freighter ships were waiting. Even further downstream, we could view the floodplain lakes and wetlands, endless miles of tall grass, with flocks of magpie geese, Sarus cranes, jabirus, little chital and rusa deer bounding along trails splashing through shallow water. Many of the connecting channels between the main river and the wetland lakes were dry, which apparently did not happen very often, according to old folks that lived there. The channels erode through rose-red ochre dirt (a palaeosol), with river bank ledges of harder material being used as docks by the floodplain villager dugout canoes.

Jim Tait knew some of the mine people and I got invited to a musical party, where guitars, mandolins, fiddles, a very good Gibson banjo, didgeridoo and bodhran appeared, and a strange mixture of Slim Dusty, Nashville country and western, pop music and traditional Celtic fiddle tunes were played long into the rainy night. It turned out that Jim Tait was born in PNG, his father had a tourist resort fishing business there, but Jim now lived on a commune north of Townsville. He worked as a photographer and occasional fish ecology contract work, and his family were now living down on the Burdekin Delta Sheep Station Creek wetland lagoons.

After we had been 'pooned' for breakfast at 0600 on 28 November, we were herded to the little Tabubil airport for a two engine Beechcraft flight to Mount Hagen (also high in the mountains, at 1630 m elevation), where even more native PNG miners from Porgera joined the push for seats, often with their possessions in fibreglass feed sacks, and some very worried anxious missionaries carrying bibles. From there we flew Air Niugini to Port Moresby and Cairns, and another inspection by worried Australian customs. They scowled at my collection of penis gourds, put them in a fumigation room briefly, and waved me on by.

Back in Townsville dear bright and shining Sheila was there to meet me and Alistar's wife Charlotte and son George were there to meet him. Sheila rushed me home to help fight a grass fire burning close to our creek valley. After all this excitement I had a nice shower and demonstrated my PNG lap lap, much to the alarm of Sheila. A lap lap is a simple piece of cloth wrapped around the hips down to the knees, which is about all what many New Guinea people wear.

Sheila's family were good friends and high school classmates with Nick Fog's family in Ontario in the 1960s. We were surprised to find Nick and his partner Brigitte living in Port Moresby, PNG, where Nick was a CUSO social worker and Brigitte worked for the UN/

World Bank on economic development. Shortly after my return from PNG, Nick phoned, urgently requesting that we collect Brigitte from the Townsville airport. Brigitte had been home alone in Port Moresby and their apartment was repeatedly attacked by hoons (called rascols in PNG). It was too stressful to stay any longer. Brigitte needed some decompression and recovery time, and Nick arrived a week later, having sold their apartment and put their PNG affairs in order. They stayed in our granny flat into late December and showed us how to make jackfruit curry and mango liqueur. They eventually found jobs and a safer home in Brisbane.

By now I was trying to form two consortia of research scientists, one for some work on the Herbert River and another for the Fly River Delta and Gulf of Papua. I didn't yet have much field sampling gear, a sediment chemical lab or radiochemical tools, and I needed expert collaborators with good analytical laboratories. I had been in contact with scientific colleagues Chuck Nittrouer and Bob Aller of the State University of New York at Stony Brook, who were just winding up their monumental work on the Amazon River estuary and coastal oceanography. I got the blueprints for Chuck and Steve Kuehl's Kasten Corer, and got the AIMS steel workshop engineers (George McNaughton, shop boss, and Moroccan-French steel craftsman Henri Le Donne) to build me one that would fit on the AIMS research vessel *Lady Basten*. By email, Nittrouer and I had tentatively planned a research operation in the Gulf of Papua called 'Flamaz', for Fly and Amazon. AIMS management was hesitant to get involved in such an operation, and there were some worried CEO level consultations between AIMS, the Australian Department of Foreign Affairs (Ruth Blunden), and the Papua New Guinea Department of Environment (Guykovina Kula, Patricia Pepina). But something must have clicked, because I got the go ahead to execute this large international consortium and enough money to build the Kasten corer.

By a series of strange coincidences, AIMS was invited to participate in a CSIRO sponsored study of riverine estuary impacts on the coastal zone, mainly the Herbert River and Hinchinbrook Channel and the continental shelf coral reefs to the seaward, out to the Coral Sea. The basic problem was to demonstrate whether or not the upstream cattle properties and the river floodplain sugar cane industry were causing watershed soil erosion and pollution in the Great Barrier Reef inner shelf and lagoon. My work was to be monitoring of the suspended sediment concentrations in the river, and measuring the locations and rates of sediment accumulation in the estuary and inner shelf. Funding for this work allowed the beginnings of assembly of my radiochemical laboratory at AIMS and to build our big ship mounted Kasten corer. Other AIMS scientists were studying the physical and biological oceanography of this region, which had some important tourist sites that needed protection.

I went down to Sydney in early December '92 for a big UN meeting on isotopes in coral growth bands at ANSTO, RSES ANU and CSIRO's Centre for Advanced Analytical Chemistry. Here was a scientific bonanza of high powered instrumentation and scientific talent for me to exploit. My ace bargaining card was the work of AIMS scientists Peter Isdale and David Barnes, who had amassed the world's largest collection of coral cores and sawn slabs over the past two decades and had created models for their mode of annual growth variations. Other AIMS scientists had decades of experience in coral biology. I burrowed into the knowledge and expertise of the ANSTO radiochemical labs, and found how to obtain World War II gun barrel steel and pre-1940 French cathedral lead, to use for shielding on my (yet to be purchased) gamma spectrometer detectors. Most metal manufactured after 1960 contained trace activities of radioactive elements from global thermonuclear bomb explosions.

Here I met ANSTO CEO Helen Garnett and scientists Ron Szymczak, Steve Short, Peter Airie and Wally Zuk. Part of the purpose of the meeting was to organize an IAEA/UN radiochemical workshop for SE Asian institutions, to educate them on how to use coral growth band chemistry as indicators of pollution and climate change history. It was agreed that myself at AIMS, John Brodie at GBRMPA and Ron Szymczak from ANSTO would be the hosts and organisers for a four day August '93 workshop at AIMS, with a demonstration of seawater and sediment core sampling in nearby Bowling Green Bay.

Back home in Alligator Creek, the wet season had truly arrived and it rained and rained and rained for most of December and January. Our little creek valley was briefly flooded, and all the vegetation grew like blazes into a jungle. Our second hot and humid Christmas in Australia was again marked by Sheila decorating a potted palm with our family heirloom decorations. Sheila and Jesse must have been very very good, because Santa brought them new snorkels, fins and face masks for reef diving, and I got books on fruit tree horticulture and lots of shorts and singlets. As is the custom in north Queensland, neighbours showed up for a lunch of cold meats, prawns on the grill, champagne and orange juice, cold beer and wild stories about the wet seasons of the past. You must understand, we are considered to be 'newbies' or 'new chums', and we have to be educated in Aussie slang, jokes and shaggy dog stories. We phoned our son Ian MacGreggor in Medicine Hat, Alberta, where he was being coddled by girlfriend Gemira's family for Christmas. There was another New Year's Bush Dance at Melville's Bar and Grill, where Jesse was now a bartender.

With all the rain, the soil in our creek valley was soft, so we planted a lot of fruit trees, Buddha belly bamboo, paw paws, pineapples, eggplants, tomatoes and improved the irrigation hose water distribution system. Grasshoppers promptly ate all of the tomato plants and wallabies ate the paw paws, so we decided we needed to have a herd of chickens down within the orchard.

Several leaks in our roof were discovered during all this monsoonal rain, so I spent some time on the roof with a silicone/putty gun, trying to plug the holes in the corrugated metal roof sheeting. It was very humid, and we discovered how to remove green mould from the walls, my leather boots and belts. Mainly this is done by wiping with a dilute solution of clove oil.

But there was no rest for the good mud doctor. On 2 January 1993, Alistar, Dan Alongi and I (with the help of Irena and Paul Dixon) were loading the *Lady Basten* with four truckloads of deck and lab gear, including my new Kasten corer. This was another trip funded by OTML/BHP to the Gulf of Papua. The AIMS ship crew were a little concerned about the handling ability of my 1 tonne weighted corer on a railroad track down the middle of the aft deck. My goals were to determine century scale sediment mass accumulation rates, and to search for geochemical changes in sediment composition after 1980s, which could be attributed to OTML mining pollution. Dan had his big box corer for the collection of benthic biological fauna, and Alistar's task was mainly zodiac work in the outer deltaic island mangrove forests. The ship crew were excited about the opportunity to pre-order cheap cigarettes, CDs and booze from the duty free shop in Cairns.

The first version of my Kasten Corer on the aft deck of an AIMS ship. We would put about a tonne of lead weight collars on the square top of this rig.

Here we go again. As I had learned in the Canadian Arctic, and I would learn again on AIMS ships in tropical seas, it is a game of hurry up and wait. The R/V *Lady Basten* travels at 10-11 knots in good weather, a little less if the waves are huge, and it pitches and rolls badly in rough weather. It was about a 1200 km (600 nautical miles) trip up to Daru, 18° up to 8° S., near the Fly River Delta, so we all settled down for a long ride. We had lunch rocking and rolling northward, past the Palm Islands, where blue green cyanobacterial booms were seen in patches to the seaward. The mackerel rigs were streaming out behind the ship, and the ship would pause to allow recovery of the big mackerel and reset the lines, while the gulls would gather to feast on the guts pitched over the gunwales. For the science crew, there was nothing to do but eat, read, watch the coast or TV, lots of very bad films, study the charts and tell stories. Most of the AIMS ship cooks were excellent, and the meals and snacks were supremely good. During these early years, the ship master usually brought a case of good wine, and we all brought a case or two of beer with us. We got a fine view of Hinchinbrook Island as we steamed past, a site of my work for the next decade.

RV Lady Basten

We arrived in Cairns at 0130 on 4 January 1993. The duty free shop was wide open and brightly lit. We all bought more beer and cigarettes, and I bought a small bottle of Drambuie and a box of cigars. We had a peaceful sleep at anchor in the harbour and the next morning we steamed north again. Bare breasted suntanned girls waved at us from incoming sailboats (there was a scramble for field glasses amongst the crew), and the green coastal mountains floated by in the mist. We passed the Daintree River in the dusk, and the aft deck was a beehive of activity from the crew, building crab pots, floater traps, rope ladders and splicing buoy lines. We ate coral trout for supper as we passed Cape Flattery in flat calm seas, watched movies and played Trivial Pursuit.

This was my second voyage up the coast, and I was able to identify many of the coastal landmarks in my field glasses, including some interesting glistening white silica sand dunes near Cape Grenville. I learned that Werner Krall, the engineer, loved avocados and ate them smeared thickly with red Tabasco sauce and salt on rye crisps for smoko, with espresso. This confounded me, as I thought only ladies in their South Dakota bridge club meetings ate avocados in salads and in jello. The cook Graham was from England, and he knew about Kent and the little suburb of Sidcup, where Sheila was born. Tough life, this coastal oceanography cruising.

We reached the Horn Island diesel fuel depot at noon on 6 January, and moved north to Thursday Island in the pouring rain, through the clear clean turquoise water of Torres Strait. As is common in many remote ports, Thursday Island coastal region was littered with abandoned half sunk ships, rusty barges and scrap metal piles, around which big sea turtles, sea snakes and dugongs were playing. The Island hills were bare of trees, covered with low grass. Australian Customs wanted to see our passports and food supplies, mostly to protect us from PNG Customs inspectors farther north in Daru. They advised our AIMS ship Master Brian McCarthy to wait a day or two, to get some agreement with the Daru Customs people, who were upset about something. The pile of passports from our ship was interesting, as there were a few Australians, a Norwegian (Stein), two British, a New Zealander, an Austrian (Werner Krall), and my USA passport in the pile.

The little town of Thursday Island had about an equal number of churches and pubs, and lots of early war memorials, including Crimean War cannons and World War I plaques listing names of Torres Strait islanders who had lost their lives fighting for Australia. We all migrated toward the veranda of the Grand Hotel, drinking Kent Brown Ale, feasting on grilled seafood, waving off flies and telling lies. Big fat mamas and kids were fishing on the jetty, laughing and playing in the soft evening light. When the sun went down, the town went dark, except for a few swinging kerosene lanterns, and the lights from the luggers and freighter ships in the harbour.

After radio communications with our ship operations manager John Hardman back at AIMS, who had smoothed the situation with the Daru Customs Agents, we cast off our harbour lines the next morning and slowly made our way through the tangle of islands and reefs northward to Daru. Many of the Torres Strait islands were of granitic or basaltic bedrock, surrounded by white coral sand/gravel beaches. There were many big ship wrecks along the way, attesting to the difficult navigation in this region in early days. A giant mother ship for prawn trawlers was aground on Pierce Cay, ripe for stripping. Alistar and Ray McAllister knew people who lived on these islands, school teachers, social workers and some AIMS drop-out hippies.

These are picture book tropical islands, rich in solitude, wind-blasted leaning coconut palms, clean white beaches, seafood, poor in freshwater and garden soil. The wind came up suddenly and everything was difficult all of a sudden, while the cooks of course came up with lots of greasy smelly food, cabbage and fatty pork chops, cold prawns and garlic cream soup. But we made it into Daru harbour, again in the pouring rain, at midnight.

While we waited for the PNG Customs inspectors, we had radio contact with AIMS, where the purchasing office was upset at my requisition for a Packard-Canberra gamma spectrometer which had been quoted for $47K, but the new cost was $57K. The AIMS CEO Joe Baker himself authorised the purchase go-ahead, even though I didn't have enough money in my budget to pay for this increase. AIMS also had to authorise payment of ship clearance fees to enter PNG waters.

Back here in Daru, by noon the various Harbour and Health, Customs and Migration officials had arrived, all seven of them in neat uniforms and white knee socks, with big air force sunglasses, briefcases, permits, forms, clip boards, check lists and cameras with flash bulbs to take pictures of the engine room, bilge and underneath our bunks. Some of this was hilarious, sort of like a Peter Sellers film, as the officers interviewed each of us, wanting to know if we had any bad diseases, what we were going to do in PNG, how much money we would spend, what we would buy or sell, and what we were going to bring back to Australia. They confiscated several *Playboy* magazines from the crew, as this was considered pornography. They were genuinely puzzled at our answers, that we were going to bring back mud, water and mangrove leaves for scientific analyses. I would learn from many field trips to the Gulf of Papua in the future to expect this routine. After surviving this pompous inspection, the cooks and some helpers went into the market to buy bananas, mangoes, sweet potato, coconut meat, venison and pig meat.

By mid-afternoon on 9 January 1993 we were cleared by PNG Customs, and we steamed south of the 'Merry England Shoals' into the tidal mixing zone of the muddy Fly River water with the clear Torres Strait waters. Big 3-8 m diameter eddies swirled in front of us, clear turquoise water surrounded by muddy water. AIMS colleague Eric Wolanski had already shown that most of the muddy Fly River water turned east into the Gulf of Papua, and did not enter Torres Strait.

The skies were grey, and there was lightning, thunder and rain squalls. We were off the Fly River Delta front in choppy waves and Dan Alongi put his box corer down several times, to get gravel and sand at 50-120 m water depths. It wasn't until we had steamed eastward to the vicinity of the big Kumul oil loading platform that we found soft mud, and then the real work began. Alistar and Paul Dixon were measuring algal productivity, respiration, nutrients and chlorophyll in water samples, Dan and Paul Kristofferson were screening box core samples for benthic animals and squeezing the mud to get pore water for chemical analyses. Dan was also measuring bacterial growth rates by spiking samples with radioactive thymidine, a substrate used in microbial growth. We deployed a CTD (an instrument on the ship winch wire to measure profiles of conductivity, temperature and depth) to determine the stratification of coastal river water of low density/salinity overlying deeper seawater of higher density/salinity. Paul Dixon deployed large zooplankton nets. Dan had two sediment traps, big settling tubes which were deployed by Ray McAllister with a float above, and attached to half tonne railway wheels, so they could sit near the bottom and capture settling particles while we went off to do other things*.

We were 50-100 nautical miles offshore from the mangrove coast, but even this far away there were twice daily tidal streams of muddy water carrying mats of aquatic macrophytic vegetation, coconut rafts, whole *Nypa* palms, entire mangrove trees with attached birds and snakes, and little islands of rooted vegetation. On either side of this salinity 6 muddy river stream, there was clear salinity 31 Coral Sea water, and as our ship moved through the muddy water, we could see clear water underneath. There were the occasional buckets or plastic jugs or ribbons tied to the floating trees, suggesting that they came from near one of the coastal villages. The Kumul oil platform was a city of light in the night, far offshore in the Gulf of Papua, a pipeline pumping station to load oil from the PNG interior Kotubu Lake area onto ocean going ships.

When it was flat calm, we would anchor out in 20-30 m water depths, and I could play my fiddle of the aft deck, where First Mate Stein and the crew would drink rum and tell stories, and the flying fish would flop on the deck at our feet. Both Werner the engineer and Irish Brian Hyslop the deckie were in Vietnam in 1967-68 on fuel tanker ships, and they had many wild stories to tell. The kind cooks would bring out fresh baked pastries and strong coffee, and this was a good life indeed.

Well, the main *'other things' were getting my brand new Kasten corer apparatus installed on the aft deck. This meant bolting my railway tracks to the deck, winching the roller cradle into place and mounting about a tonne of lead weights on top of the coring apparatus. This gadget was moved by clever arrangements of shackles with the big winch wire, lifted vertical, and liberated into the water on the winch wire, lowered slowly until it entered the mud. A critical and goofy part of this whole apparatus was a 'core catcher flap' at the bottom, held open by a thin wooden satay stick and a rubber band, which was pushed out of the way when the core tube entered the mud. If we had done all this right, the 1-4 meter long 15 cm square steel coring tube would be driven vertically into the mud, collecting several centuries of deposition history, and could be brought back on deck, man-handled onto a special core tube table. Here it would be opened on one side and cut into 1-4 cm slices (pages of the history book), some packed by Irena into vials which were frozen for chemical analyses back at AIMS, and the remainder going into a plastic sack for radiochemical measurements of radioactive lead, radium, beryllium, thorium, and bomb fallout nuclides caesium and americium.

All of this Kasten coring operation was a learning experience for me and the ship crew, who had to learn how to delicately operate the winch and the ship position. We were worried about the 'pull-out force' needed, and the strength of the A-Frame and wire block, but we had no problems with this. After the first few successful cores over the next week, there were smiles on the science and ship crew faces as we learned how to handle this dangerous monster. Down in the galley at smoko one afternoon, Alistar quietly told me that I was being called 'mate' by the crew, and that this meant I had passed muster, qualified as being a decent ship scientist, and was no longer an untrustworthy newbie or a Septic. My nickname was now 'Flash', as I was the operator of the aft deck Death Ray Machine...the Kasten corer sticking out horizontally from the aft deck looked a bit like a cannon. At smoko the ship crew joked about backing the ship into Daru Harbour and taking it over for Australia. But it was bloody hot, even at night. Irena, Paul and I could carefully slice a 3 m Kasten Core in about six hours on the aft deck, and we rigged a tarp shade over the slicing table, or did some of the slicing at night. On advice of doctors, we were taking anti-malarial tablets (lariam) and salt tablets.

By 15 January, we steamed into the Fly River Delta channels to anchor at Umuda Island, where Dan and Alistar had previously worked. Most of the Delta channels had sandy or gravel floors, the box corer often failed, and my Kasten corer could take only 1 m cores, much to the relief of the sediment slicing crew. These delta channels are very large, sometimes 4-5 miles wide, and big swells came in from the sea, and our ship had to hide from the SW wind. So were the copper ore freighter ships hiding here, waiting for better weather to slip out to their loading platform for OTML ore on ocean-going ships anchored out in the Gulf of Papua. There were also two huge barges loaded with giant rainforest logs, being tended by a group of little tugs operated by Malay fellows that spoke little English. The local PNG villagers took advantage of this, coming out in dug-out canoes to sell bananas, mud crabs, melons, cucumbers and unknown slabs of meat covered in flies. They really wanted to trade for petrol and cigarettes.

We steamed up the delta to Sumogi Island and almost freshwater for the next few days, as radio contact with big ships in the Gulf of Papua said there were 3-4 m waves outside the delta in the Gulf. This allowed Alistar and Dan to do some mangrove work for the next few days, tagging and measuring half a dozen species.

Deploying Grab Sampler in Fly River Delta, 1994

That evening the cooks went berserk with good food, as they figured we would be in bad weather on the return voyage...we had breaded prawns, bugs, roast beef, roast duck, roast potatoes, steamed spiced cabbage and sweet corn. We had bought two large bunches of green bananas. Werner Krall anxiously waited a week for them to ripen. He actually got a little angry at these bananas, hanging there on the A-Frame, hard and green, so he sealed them up in a big plastic rubbish sack, pumped in some acetylene from the welding equipment tank and made a fuse leading to the sack out of waxed string. The entertainment that night was to watch the bananas blow up on the aft deck, and there were bits of hard green bananas all over the ship.

By 22 January we were getting tired. Master Brian's leg was infected and swelling, he was out of antibiotics, and the cooks were out of fruit and vegetables. So we cancelled a visit to Bramble Cay, ignored Daru customs (AIMS Field Operations Officer Frank Hardman arranged this due to medical emergency), and steamed for Thursday Island (TI), where there was a hospital and Australian medical services. We washed up our deck equipment, chained everything to deck bolt rings and generally prepared for a rough ride. The *Lady Basten* plunged through the big waves with a pitch and roll movement that made most of us woozy all night.

We arrived in TI a day later at 0900 in the pouring rain, and deckie Jonah took Master Brian to the hospital for treatment. Alistar, Dan, Paul, Irena and I booked air travel back to

Townsville, as the ship would stay here for several days while Master Brian was taken care of. We had to wait a few days in TI, so Paul and I walked to the top of the hill, where World War II 150 mm cannons in concrete bunkers were still pointed at the harbour, intended to blow up invading Japanese ships (which never came close to TI). We also visited the Anglican Church, which was a museum for the *Quetta* disaster (a British Royal Mail Ship that sank off the coast of northern Cape York in 1890. It ran into a big rock in the middle of the Adolphus Channel, wasn't marked down on the chart, and 134 people drowned). I got interviewed by a Torres Strait radio station jockey, and we spent a lot of time at the Grand Hotel veranda eating, drinking beer and watching the black cockatoos rip branches off the sea almond trees by the beach. The Mills Sisters were performing at the Federal Hotel. They were very good singers and they took not the slightest notice of several fights going on at the back of the room. On 25 January we took the ferry to Horn Island, where we boarded an airplane for Cairns and Townsville, arriving at 2100, with bright and shining Sheila there to meet me.

By now, Sheila and I were beginning to recognise the pattern and rhythm of this oceanography job. I would come home to Sheila, my skin roasted brown from the wind and sun, dead tired, sore swollen hands, and the house would be rising and falling underneath my feet. There must be some habituation of ear and brain balance to the ship's deck movements, and after several weeks of living on a ship, the ground is still heaving and pitching. Sheila was spending more time at the local Billabong Sanctuary, a place to view Australian animals. Jesse was becoming a bloke, driving cane trucks and getting to know most of the local farmers in our region.

Chapter 26. Big Ships, the NW Pink Coast of Australia, and the Foundations of my Research for the Next Decades

The year 1993 started innocently enough.

The first of many Canadian and USA visitors arrived in early January to taste the warm Australian sun. Hedy Kling and her children Joey and Tyler visited us. Hedy is a phytoplankton taxonomist, one of Sheila's colleagues who invaded ELA during the summer of 1970. Jess took them mud crabbing and they returned pink from the sun, covered with sandfly bites, a normal introduction to north Queensland. They are the first of a long string of Canadian visitors that Sheila took to the Billabong, Mount Elliot Park and Melville's strange animal farm and pub. The kids knew us from our Tourond farm days, and they demanded home-made ice cream with chopped snakes. So we dragged the old wooden bucket out of the shed, and a possum was sleeping in it. The wood had dried out, so we had to soak it in a tub for two days before it would hold the saltwater ice mixture. Jesse was now driving a cane truck, hauling cut cane to the sugar mill, becoming an expert at backing up truck trailers to a narrow unloading chute. We harvested our first bunch of bananas and some paw paws. We picked mangoes from our neighbor's trees, as our orchard was still struggling to grow.

There was a brief two months of peace in my office, and I was trying to write journal papers on our work in the Fly River delta and Gulf of Papua. I was also writing plans, schedules, budgets and invitations for a workshop on nuclides in coral cores in May, and another FLAMAZ workshop in June, and requisitioning gamma spectrometers, computers, large liquid nitrogen dewars and lead/steel/copper shielding for the detectors.

But then all Hell broke loose...a baptism by fire! For reasons never explained to me, Dan Alongi and Alistar Robertson had proposed and were successful in obtaining the *R/V Franklin* for shelf and deep water work on the Gulf of Papua, partly with OTML funding. Dan was sick and Alistar was away on leave in California, so AIMS asked me to be Chief Scientist for this cruise...with about one month of preparation time. Whoa back here...what does a South Dakota prairie boy know about organising and bossing a big ship oceanography voyage? Well, not too much actually, but it isn't too different from organising a Boy Scout camporee or an expedition to Lake Hazen in the Canadian Arctic.

The *RV Franklin* was the supreme oceanographic vessel in Australia, the big time. The *Franklin* had been operated by CSIRO Marine Research as a National Facility since 1985. This ship was on its last legs, as the steel hull had rusty patches that paint could not hide, and the aft deck had been repaired and rewelded many times. But to a prairie boy, it was a bewilderingly big ship with three winches and bunks for thirty. My journals are filled with lists of ship deck plans, ways of operating a limited crew for different tasks on a 24 hour/day cycle, and how to accomplish Alistar and Dan's tasks. The AIMS Assistant Director Michel Pichon wanted us to deploy a giant beam trawl for his research on deep sea creatures. The science crew was Otto, Irena, Paul Dixon, Paul Christopherson, Ray MacAllister (Ray Mac), Ron Szymzcak, Ken Woolfe and myself.

CSIRO Research Vessel RV Franklin

And so on 2 April 1993 our semi-trailer truck brought our AIMS equipment (8 tonnes) to the Cairns harbour, where the *Franklin* had a comfortable berth (which was not available at Townsville Port). There was frantic unloading by crane of our gear from trucks to wharf to the ship all afternoon, very hot and sweaty work till dark, and then the ship Master Paddy stopped all work. We were grateful to use the ship showers, and off we went to a seafood restaurant and cold beer. The ship went out for a sea trial, as it had some new engine parts, some trouble was found, and they didn't get back to the wharf till the next morning.

With Ray Mac and Henri Le Donne's help, we welded/bolted down the Kasten Corer railway tracks and slicing table on the large *Franklin* aft deck, and figured out how to mount the winch wire through the blocks and shackles to operate our coring railroad. All this took place while the ship is steaming inside the reef northward to the Gulf of Papua. The ship crew deckies and engineers were impressed when I opened my tool box, where I had large crescent wrenches (spanners) suitable for huge bolts and nuts. Ray Mac assembled floats and anchors (rail car wheels) and wire for the sediment trap and current meter buoy deployments. The ship was actually a mess, rusty patches on the deck and hull, grease and oil leaking, untidy gear not secured well. But we had very good cooks, a nice TV and a library room. I was in unaccustomed luxury in the Chief Scientist's cabin, with my own fridge, TV, computer terminal, phone intercom, file cabinets, map table and shower.

My crew were not accustomed to operating 24 hrs/day, but we split our small scientific crew to keep the ship busy. The ship crew operated on eight hour shifts, so the kitchen was open for meals 24 hrs/day as well. The science crew night shift (2100 to 0600) involved Ron and his Kraken trace metal sampling system most nights, pumping large (3 m^3) volumes of water from surface to depth for thorium-234 and trace element measurements. Ken was logging CTD and a chlorophyll/fluorometer, running a Smith-Mac grab sampler and box coring operations at night.

The rest of us worked the day shift, and I focused on Kasten coring and writing the next day's schedule of operations in the computer and map room. Ray Mac was the master deckie, and he deployed Alongi's sediment traps with current meters attached to the wire. Ken used the ship sonar to map part of the cliff edge at the continental slope, where detached pinnacles appeared in water depths of 1-2 km.

I quickly learned how to be a chief scientist, the person who schedules and controls the scientific affairs on the ship 24 hrs/day, in consultation with the ship Master Paddy Lorraine (a snazzy dresser, who wore a necktie in the wheelhouse), 1st and 2nd Mates Dick Dougal (a Nat King Cole fan) and Brice Bathe (who wants to live in Indonesia), and the aft deck bosun Stormin' Norman Marsh (his wife was from Saskatchewan, and he used to bosun Canadian freight ships from Montreal to St. Johns and Halifax to Churchill). The ship was like a small city far away from the shore, the engines throbbing, ship crews working day and night.

Some of aft deck workers seemed unfamiliar with our kind of work, and I had to draw pictures of some A-frame, block and winch wire movements to obtain safe corer deployments. Our grab sampler and Kasten corer required careful and slow winch wire movements, and it took several days of bad operations for the winch driver to get the message, especially when their eight hr shift was over and we had to train yet another winch operator. One day a deckie moved the A-Frame out while the wire was tight on the Kasten corer, and it lifted the corer and welded railway track right off the deck, scaring the bejeezus out of Ray and myself, who were standing beside it.

Back deck of RV Franklin, *Gulf of Papua, 1993*

Much of the time we had 2-3 m swells with good wind, often pouring rain. Sometimes the aft deck was covered calf deep in green sea foam if we tried to back into the wind for a coring position. We were trying to map sediment properties and accumulation rates from shallow (10 m) to deep (1600 m) water along this PNG coast. Sometimes the Master of the ship shut us down due to dangerous conditions, 4 m swells and ripping wind and rain...and our science crew gratefully went to their bunks to rest, our faces wind and sun burned, hands raw from operating metal equipment in sea water.

We deployed Michel Pichon's huge 4 m wide trawl with a big cod net behind it, towed it for half an hour at 900-1200 m depth and recovered a lot of *Nypa* seeds, coconuts, mangrove logs and leaves still green, exotic colored crabs and prawns, sponges, scallops, starfish and brittle stars, startlingly blue urchins, scaphopods, sipunculoids and strange snake-like fish. Paul and Irena sorted, weighed and preserved all this for Michel, Dan and Alistar's work back at AIMS.

Max Cameron, the big burly 1st Engineer, came to our rescue several times, welding broken gadgets and helping instruct the winch driver. He also jigged for squid most evenings, and we ate a lot of stuffed squid on this cruise. He kept bedraggled terns and shore birds that landed on the ship far out to sea, gave them freshwater and apple bits in the tool room. He actually welded a special perch for them above the work bench on the aft deck workshop. We spotted whale fins and tails breaching while scanning with field glasses for our sediment trap floats rising after a week's deployment. Some of the sediment traps came up from the deep onto the deck upside down. I drew a better design for big ship sediment traps (which eventually were built at AIMS later on). We accomplished all that was planned by Alistar and Dan, and as well completed a dozen transects of sediment sampling for my own work. Over the next decade, we were to keep improving these transects, making it possible to draw maps of the river sand, mud, marine carbonates, shell and hard coral bottoms and oolitic carbonate sand.

The cooks were very good, and we had melting butter on hot cross buns for Easter smoko. Irena had brought along chocolate Easter eggs for everybody, and she placed them beside everyone's cabin door. The cooks would bring into the galley large trays of rosemary slathered pork loin and lamb roasts, and the hungry crew would applaud heartily. Several times the cooks provided an endless supply of fresh made cream puffs with dark chocolate drizzled over them. The Chief Cook Gary Hall had fun providing a menu for the day, including Gutters of Calcutta (soup), Jockey Whips (shoestring potatoes), Murphies (baked potatoes), Porcupines (roasted whole potatoes with cuts along the top), Train Smash (stew), and Dead Man's Eyes (fried eggs).

17-18 April was spent unbolting our aft deck gear and lab bench equipment in preparations for the turn-around, as we were steaming toward Thursday Island. We brought Peter Harris's gear up from the ship hold, onto the deck with the ship crane, and put our gear down in the hold. Peter Harris was a scientist with the Australian Geological Survey Organization in Canberra, and he was scheduled to use the ship in this region for the next three weeks. We tarped down our pallets with about two tonnes of carefully sliced and packaged mud, ready for the forklift at the Cairns wharf. We were lucky to have dead calm weather, but it was very hot on deck and in the diesel fume-filled hold.

We arrived at Thursday Island at 0545 on 19 April, tired and sore. We left most of our gear and samples on the ship for recovery later in Cairns, and took our personal luggage to the ferry for Horn Island to fly to Boigu, Cairns, and Townsville. The airplane served cold beer, something we all appreciated, as the *Franklin* was a dry ship. And there at the Townsville airport gate was Sheila, brown and beaming, my ground truth. The floor of the airport was heaving like the deck of a ship in rough weather.

And now for something completely different: Back at Alligator Creek, Jesse had organised for us to buy half a pig. Alligator Creek cane farmers had purchased a few 250 lb pigs from the local agricultural school down on the Burdekin River, and several people bought shares in these pigs. So we bought a little freezer, cleaned my meat saw, skinning and boning knives, and drove down to the shed of our neighbour Peter Connolly. There were six nice pigs hanging, debristled and frozen, with a large industrial band saw cleaned up for meat cutting. The question was 'You want roasts, chops or bacon?' Two of the entire pigs had been smoked and salt cured and four were fresh. All meat cutting was done zip zip on the band saw, and there was no need for my hand saw or boning knives. That was the fastest butchering session I have ever seen.

The rest of late April and early May was a blizzard of planning sessions for our IAEA coral history workshop in mid-May, formation of a Coordinated Research Centre (CRC), my FLAMAZ workshop for Gulf of Papua work with Indonesia and the USA, and a new program of research on the NW coast of Australia. Having just got off the R/V *Franklin*, I was now asked to write proposals for coastal research offshore from Exmouth Gulf and NW Cape using this ship, as well as complete a voyage report for our April research cruise in the Gulf of Papua.

In addition, I was still gathering equipment for my radiochemical and geochemical laboratory. I emphasise again that I was on a very steep learning curve here: what did a South Dakota prairie boy know about centuries of coral annual growth bands and the trace elements and isotopes incorporated into the calcium carbonate skeletons? It was like being in graduate school again, only with more pressure and unknown consequences. I spent a lot of time in the library catching up on thirty years of research in this region, and also keeping my mouth shut in some of these workshops. I was lucky to be surrounded by scientists who had spent their life career in this region.

The IAEA steam-roller organisation moved into AIMS visitor offices in mid May '93, and our workshop on radioisotopes in tropical seas and coral reefs began. All sorts of funding had appeared from magical sources, and AIMS became a scientific resort for a week-long series of lectures and field demonstrations. There were accommodations for twenty-five visitors at AIMS, and some of the lecturers had accommodations in town hotels. AIMS had a nice canteen that would seat all of our visitors for meals, and I helped grill chicken for Hindus, Muslims, Jews, Christians and almost certainly a few pagans. Rather large quantities of good Australian wine and beer were consumed, and a few international songs were sung in merriment.

We organised an evening of entertainment at Alligator Creek's own Melville's Pub, where a bush band played square dance music, PNG dancers were featured, and tastes of crocodile, emu and kangaroo meat were offered. Two young ladies (actually sisters) from Thailand, with very long names (Suchin Udomsomporn and Boonsom Porntepkasemsan) became attached to our colleague Ron Szymczak. For decades later, Ron was kidded about his affair with the Porn Sisters. Note well...in the Thai language 'porn' means 'wish', and has nothing to do with the ancient profession.

We had radiochemical and ecological lecturers from Sweden, Monaco, USA, France, Israel, Holland and Australia, speaking to about twenty-five SE Asians from Sri Lanka, Malaysia, Indonesia, Vietnam, Thailand, Papua New Guinea, India and China. They gave lectures on radionuclide contamination of the Arctic Kara Sea by Russian ships dumping nuclear wastes, Persian Gulf war oil spills, plutonium contamination of the seafloor near Thule, Greenland, rare earth elements in Sahara desert dust and the surface Atlantic Ocean and Caribbean Sea failure of coral ecosystems due to pollution and greenhouse warming. There were several major disagreements between the coral biologists, who studied the physiological formation of the carbonate skeleton growth bands, and the trace element and isotope geochemists, who had discovered oscillations in trace element and isotopic ratios across decades and centuries of coral growth bands. These oscillations in trace element and isotopic ratios had been shown to be directly related to changes in seawater temperature and salinity, and were thus valuable for an account of recent changes in climate and oceanographic currents.

Peter Isdale at AIMS had discovered that fluorescent organic matter from river floods (and maybe terrestrial groundwater) was measureable in coastal coral cores of centuries of growth band history, showing small and large flood events back to the early 1800s. This fluorescent banding in coral cores did not appear in corals from mid ocean atolls, where there was no significant runoff from terrestrial vegetation and soil organic matter. There was a demand for an international coral core repository and coral carbonate reference material. AIMS had the world's largest collection of coral cores, largely due to the work of Peter Isdale, David Barnes and Janice Lough. I got to know Glen Shen of the University of Washington in Seattle at this workshop, and we developed plans for collaborations on coral growth band geochemical histories in the Torres Straits and adjacent to the muddy plume of the Fly River in PNG.

CEO Joe Baker had wrangled more funding from Canberra to launch a Western Australian branch of AIMS, and he wanted this $ 4 million spent pronto. Mangrove physiologist Barry Clough and I spent two weeks in early June '93 flying in a 165 Cessna over the North Western coast of Australia, from Broome in the east to Exmouth in the west. Karratha hosted a little AIMS research station where a temporary HQ was run by Andrew Heyward (Smiles). We were trying to become familiar with the huge pink tidal salt pans inland from the thin ribbon of miserable mangroves trying to grow in very saline trapped seawater. Braided rivers of dry rose-red sand made small deltaic structures 10-20 km inland from the sea, debouching into the landward side of the huge salt pan. The airport landing strip in Onslow was simply road grader-bladed salt pan on top of huge *Porites* coral bommies shaved flat by the road grader. We flew across Exmouth Gulf where prawn trawlers were busy and the giant radar dishes of the US Naval and Air Force Communication towers at Lone Pine were jutting out from NW Cape.

We rented a 4WD Toyota Cruiser to visit mangrove sites from Onslow to Broome, a very long dusty rough ride in very empty country, where the hotel rooms and road house cafes showed every sign of rough miner occupation. This country has some very wild history that was evident from photos on the Balla Balla roadhouse café walls...big cyclone storms, sharks, crocs, dingos, aboriginal camps, old mine disasters, lots of outlaws, murders, prospectors, union organisers massacred, camel depots, big cattle stations and usually several aviaries in the shade with dozens of different kinds of parrots. I was educated in tidal ecology around Broome, where the diurnal tide oscillation was about eight metres, and the steaming hot mud flats seem to go on forever.

We got desert and salt pan botanical education from several WA Conservation and Land Management fellows who worked with Dampier Salt evaporation ponds in Karratha, and with Morton Salt Company in Port Hedland. Inland from this saline coastal world was very much like South Dakota, rolling hills with sparse grass and little water, a few cattle and a lot of very empty country. Up by Port Hedland even the sheep look pink, as they eat some of the salt tolerant samphire weeds on the edges of the pink coloured salt pan. The pink colour is from salt tolerant algae that live in the brine. Some of the mangroves along this coast had huge basal stumps, but only a few thin spindly branches. Barry explained that this coast has lots of severe cyclones, and a direct hit would have ripped the leaves and branches off of the mangrove trees. This is a tough and salty place to live.

We flew back to Perth, on the Indian Ocean coast of Australia, to meet some CSIRO scientists who had expertise on north coast oceanography and mangrove ecology. From the Herdsman Hotel we could see black swans on the river and lots of different parrots as we lounged on the cool veranda with cold beer. We met several CSIRO scientists who explained the history of the post-war UK nuclear bomb tests on Montebello Island during 1952-56. This was of great interest to me, as we could use known inputs of bomb nuclides into ecosystems of this region, such as the growth bands of mangroves and corals, and in sediment cores. They gave me reprints and references that showed the efforts of UK and Australian military to gather up the bits of tower metal and ships that were highly radioactive, after the bomb tests, and cemented them into a cairn on Montebello Island. So much for the unspoiled pristine NW Coast of Australia.

We flew across the desert country from Perth to Cairns on a red-eye flight, refuelling in Alice Springs in the early morning. The roof of the airport building in Alice Springs was also desert sand rose-red...and as the airplane landed, the whole roof rose up and few away. It was a large flock of Major Mitchell cockatoos. The Alice Springs lounge was full of shops selling didgeridoos, clapsticks, aboriginal bark art and books about the Australian red interior. We arrived in Cairns to be startled by the colour green everywhere, lush and verdant. I was reminded of the same sensation when travelling from the Canadian Arctic black and white land of ice and snow, to temperate green Winnipeg.

I knew we couldn't do large scale multi-disciplinary coastal oceanography with AIMS staff and funding alone, so I began stringing together the elements of an international consortium of people interested in the tropical islands of Papua New Guinea and Indonesia and their coastal seas. I had won a pile of money from the Australian Academy of Science to fund a workshop to set up such a consortium. Our AIMS CEO Joe Baker was very supportive, but he was also aware of how much it was going to cost AIMS in ship time and staff.

Charles Nittrouer (University of Washington in Seattle) was the leader of a big project on the Amazon River input to the tropical Atlantic (AMASEDS), which was just winding down. A half dozen of his USA colleagues wanted to work on the big rivers of the island of New Guinea and their continental shelves. We contacted Anugerah Nontji and Gani Ilahude from LIPI in Indonesia, Guykovina Kula from the PNG Department of Environment and Conservation, Murray Eagle and Ian Wood from the Ok Tedi Mine environmental laboratories and Hugh Davies of the University of PNG, so that we had representation from the countries involved. All this came together for a 17-26 June '93 workshop at AIMS, with lecture series on the Amazon work, Australian work on the Fly and Sepik Rivers, the Gulf of Papua and the Bismarck Sea. We informally called this the FLAMAZ workshop, for Fly and Amazon.

Chuck Nittrouer, his wife Debbie, son Jeffrey and daughter Christine arrived two days early, so we could plan the schedule and entertainment. Alistar Robertson, Irena Zagorskis and Dan Alongi at AIMS were also key organisers in this affair. So here we go again, about twenty people arrived, we had extra cooks in the AIMS canteen, all of the visiting scientist accomodations were filled. Once again I was at the AIMS outdoor grill with buckets of prawns, bugs and squids, and liberal amounts of Australian wine and beer were available at lunch and supper time.

We heard two days of lectures about the Amazon and two days of presentations by Australians and Indonesian oceanographers. A full day of national expressions of interest, funding options and ship commitments was discussed, and of course the diplomatic arrangements for the cooperation of national research institutions of these three countries.

It was obvious that we needed the blessing of each of the governments of these three countries, and that would take nearly a year to accomplish. In addition, the Indonesians requested that Nittrouer and I lead a tour of Indonesian oceanography institutions, and to promote collaborative coastal ocean research, to loosen the purse strings in Indonesia. We also agreed that we would need another workshop next year, to further these scientific, diplomatic, ship time and funding plans.

Towards the end of the workshop, we took the participants to Alligator Creek's own Melville's Pub on the evening, where PNG dancers and a bush band entertained until the wee hours. The PNG people also cooked several whole sides of pork along with sweet potato, sweet corn, bananas, plantains and whole pineapples in a pit BBQ with hot coals and stones underneath, and banana leaves and sand on top. For the final two days, Alistar and Dan took the participants on a tour of Hinchinbrook Channel mangroves, and I led an excursion to the local salt flats and the delta of the Burdekin River, the largest input of water and sediment to the central Great Barrier Reef. We hauled all tired bleary-eyed participants to the airport by 1400 on 26 June, and then went home and slept for two days. Over the next month, Nittrouer and I wrote a summary of this workshop for the AGU Eos Newsletter.

Unfortunately our good CEO Joe Baker could not shield us from the shifting winds of politics, and some new policies appeared that irritated everybody. This scared me, as it was going in the same direction as the Freshwater Institute of DFO back in Winnipeg, which I had just left. Some minister in Canberra wanted to amalgamate AIMS with CSIRO Marine Science and the Fisheries Department. They also wanted budget cuts, quarterly project evaluation and changes in management of the AIMS research ship crews. We wasted almost two months in staff meetings, trying to find a way to keep our research institution working well with less money and with more bureaucratic drag on staff.

Our CRC organization for coastal research was in the process of being born, and in anticipation of funding and fame AIMS sponsored several research expeditions to this sector of the central region of the Great Barrier Reef. Most of the work was focused on offshore coral reefs, coral diseases and aspects of species diversity. Alistar Robertson, Dan Alongi and I focused on the estuary of the Herbert River and Hinchinbrook Channel.

My job was to make maps of the river mud deposition sites and try to measure river mud accumulation rates, with hopes of doing a comparison of river annual sediment load supply and the summed rates of accumulation in the mangroves and estuarine regions. You need to know that most of the sediment on the mid and outer continental shelf was made of skeletal carbonate minerals from marine organisms, so it was not difficult to map these to very different sediment compositions.

In late August 1993 Irena, JCU geology grad student Michaela Toohey and I took the little *R/V Sirius* and the flat-bottomed barge *Hercules* up to the Herbert River delta and Hinchinbrook Channel, to try to map mud chemistry and take a few shallow water cores for measurement of accumulation rate. The smaller the vessel, the more the rock and roll, and it was a rough trip up the coast with Captain Peter Eden, Mate Simmo De Veaux and John Sharp, Master of the *Hercules*. The *Sirius* was about to be decommissioned. It stank of diesel fumes and mould, and the bunks were damp and clammy even in good weather.

We sure did explore the back stretches of the mangrove creeks and got stuck several times, having to wait for the tide to come in to lift us off the mud or sand banks. Sharpie had a CD of Eric Clapton's *Unplugged,* which he played over the *Hercules* loudspeaker as we were slicing sediment cores. We were using our hand-driven corer, in locations where we could stand on the mud bank. All went well until the bottom half of our corer snapped off deep in the mud, bearing our fancy metal core catcher, of which we only had one. I marked the location with an anchor line and some floats, as the tide was coming in. The next day Peter and Simmo recovered the buried core tube by some sort of Pacific Island magic.

Simmo (from Noumea) was also a very good cook, and we feasted on rosemary smothered lamb roasts, fresh mud crabs, barramundi, pineapple/mango pie and fresh baked bread. John Sharp introduced me to 'pepsies', small beach clams which were abundant from Ramsay Beach, boiled in sea water, pretty damned good, but it was hard to get a good mouthful of the tiny buggers. We watched sea eagles fly toward each other in an aerial courtship dance, lots of jabirus, egrets, pelicans, sea snakes, huge sea turtles, dugong and of course the damned sand flies. After two weeks of this work we ran out of coffee and eggs, the electrical system was faulty on the *Sirius*, and everybody was tired, so we went home. This was the beginning of almost a decade of Herbert River coastal mud mapping by my crew.

As mentioned above, many of my Canadian and USA colleagues seemed to use me for an excuse to come down to visit Australia. Colleague Jack and wife Janice Cornett from the Chalk River Atomic Energy Commission Ltd. arrived in mid-September. He gave a seminar at AIMS and had several good feasts with us. He was very helpful in providing details of the availability of new kinds of gamma spectrometers and radiochemical yield tracers, neither of which were commercial or openly available to the public. I was still putting the pieces of my radiochemical laboratory together, bit by bit.

As part of the bureaucratic re-organisation and shuffling of research responsibilities, AIMS and the marine science part of the huge CSIRO were supposed to 'make nice' at a workshop in late September '93. I was one of those selected to attend, and flew down to Melbourne and Hobart in pouring rain and thunderstorms. I didn't have a good raincoat so

I bought a DrizaBone coat, intended for saddle horse riding, with fittings to strap the legs inside the flanks of the coat, and a warm flannel lining. Now surely I was a real Australian!

Hobart, Tasmania, looks and feels like England or Scotland, chilly and damp, ships at docks everywhere, and all the hotels had UK names. The workshop was a painful waste of time, as we spent too much time discussing 'linkage and collaboration', with anguish over ship time costs. There was some conflict between John Finnigan's CSIRO Coastal Zone Project and the new CRC being created at JCU, which did not include CSIRO. The Australian Geological Survey Organization (AGSO) was also running environmental/pollution surveys from its ship the *Rig Seismic*. We toured the big research vessels tied up there, the *R/V Southern Surveyor*, the *R/V Aurora* (used in the Antarctic Southern Ocean), and the *R/V Franklin*. On the way home, I got the 'Gentleman of Distinction' sticker from Qantas stewardesses in Brisbane, because I helped a lady with two babies get off the plane.

Our long lost Canadian musician son Ian MacGreggor appeared at the Townsville airport on 1 October 1993, laden with hockey bags full of drums, metal percussion instruments and cymbals. He was strangely white-skinned, long haired, looking very scholarly and intellectual, a bit out of place in North Queensland, full of beach bums and hardy weathered fishermen. He was immediately immersed into the Alligator Creek style of living, and Charles the cockatoo took a liking to him. Ian immediately began tuning up his drums, tabla and other hand drums.

So our little family was all together for Christmas 1993. It was bloody hot, over 40°C, so we spent most of Christmas day in our neighbours' swimming pool, as they were away for the holidays. As traditionally required, I cooked a stunning Xmas breakfast of macadamia/cinnamon/black currant pancakes, sausage, bacon, French egg bread, three kinds of fruit juices, fresh mangoes, rambutan, lychees and strong coffee.

In late December Sheila, Ian and I ran off to see the Maleny Folk Festival, just north of Brisbane. Ontario friends Nick and Brigitte Fog were finding their home in Brisbane, and they took care of our lodging. The Festival was very good, with lots of Aussie, Celtic, Canadian and USA material, and a big poster for the Winnipeg Folk Festival! The opening ceremony featured aboriginal dancers and Sirocco. Later, there were some very good bush bands playing music intended for dancing, such as Wongawilli, which gave Sheila one of her favourite Aussie ballads 'And When They Dance'. I was taken by Ted Egan and his beer box drum, singing ballads about Torres Strait Japanese pearl divers, North End cattle drovers and mistreatment of aboriginal people during the 'White Australia' period of history. There was a wonderful tribute to Pete Seeger, complete with historical film, songs, poetry and lectures. We listened to several Scots singers, some family bands (one with a pet fruit bat named 'Charcoal'), some good *a capella* choirs, but Eric Bogle blew us away. To Ian's delight, there were North Indian dancing instructors and a small band with tabla and sarangi, mridangam and viola. Can you imagine two hundred music—inebriated hippies trying to follow the Indian instructor demonstrating 27 foot positions in 5/4, 7/4, and 9/4 time signatures, 27 face expressions, and many arm and hand positions?

The whole thing ended New Year's Eve with a glorious fire ceremony, processions of white sheet people with candles in colored paper lantern holders on bamboo poles, drums and didgeridoos throbbing, burning of statues and figurines, some fireworks, massed choirs and different bands, recitations, spidery stork figures on stilts appearing to leap into flames and reappear, jugglers with flaming batons, spinning wheels burning, very primeval symbols of giving up the old year and welcoming the new year. I wished Mitch Podolak could have seen this ceremony. It was emotionally and musically important for Sheila and me to see and hear this good festival, as we had been involved in planning, executing and performing in folk music festivals for almost two decades in Canada, and we hadn't yet found musical friends here in Queensland.

We were certainly in our own world, down here in the southern hemisphere where tropical fruit grows and the tides are diurnal.

Chapter 27. Exercises in Creating a Research Consortium: TROPICS. Rivers, Coastal Mud and Diplomacy in Hawaii, Indonesia, and Papua New Guinea, 1994-95

I was assembling an international team for research on runoff from the Island of New Guinea, and how this impacted the coastal seas. I badly underestimated the diplomatic and administrative skills needed for this operation. From here on, I seemed to be plagued by administrative troubles that delayed research or made things more difficult.

For six months I had been planning a *Lady Basten* plus the *Hercules* landing barge coral drilling cruise with Glen Shen and Joanne Ellison. The week before our departure, all sorts of trouble erupted. The previous year I had made collaborative arrangements with Shen (U. Washington, Seattle) for a coral coring expedition to Torres Straits and on the margins of the Fly River muddy plume, to see if we could get century scale information on Fly River discharge and influence on coral growth. Joanna Ellison had recently arrived as a Post Doc from Stanford, wanting to work on mangroves and sea level rise.

But then, as of 26 January 1994, our PNG ship clearance was blocked, both by Australian Workmen's Compensation not being valid outside Australian waters, and by PNG Immigration, as our Research Visa was not approved because it did not include the ship crew. So the voyage was cancelled.

Not to be denied my adventure, I rewrote the cruise plan in about six hours, letting the AIMS ships work in Australian territory for coral coring and mangrove work in Cape York (but not in PNG), while OTML Murray Eagle allowed us to use the *Western Venturer* for Fly River salinity gradient work in PNG.

My collaborator Glen Shen had arrived 25 January 1994, with his diving gear and coring drilling gear. We left Townsville harbour on 27 January, after replacing fuel injectors in the *Lady Basten* Cummings diesel engines. Dave Beard was our R/V Master, and we rigged a towing bridle for the *Hercules*, meant as a diving and core drilling platform in shallow water Torres Straits. This towing bridle didn't work very well, as the rope and/or steel cable broke several times during the night *en route* north, in 1 m choppy water. We spent some miserable dark hours searching with field glasses and radar for the *Hercules*. Between Cape York and Horn Island it was very rough and even the cook was sick, the galley a mess.

We arrived in T. I. 0915, where the *Western Venturer* double hull ship was waiting for us, with Master Ron Clark and Engineer Robbie Kellie fresh and ready to go. We were woozy, tired and needed rest. The Cook Malachi told the story of the previous cook Leo's death, as a big tree rolled under the ship and flipped the aft deck platform up, while Leo was peeing at night. They couldn't find him in the dark. So we transferred our chemical equipment for a Fly River Trace Element and salinity gradient study with Glen Shen and Ken Woolfe, Irena and myself. Joanna Ellison and her helper took the *Basten* back to Cape York estuaries for mangrove coring work.

We went through PNG Customs in Daru in the pouring rain on 31 Jan 94, and the Daru agents found something wrong with Glen's visa. Our OTML agent grabbed the passport/visa and a fistful of our kina, and vanished into the muddy streets of Daru for a few hours, returning with it all fixed. To work in places like this you need to know how to find the right people to fix problems. The muddy smoky roasting fish smell of Daru was unmistakable, and the ground was blood-red from the spitting of betel nut juice. Despite the heavy rain, salinity of Daru water was 36 ‰. We steamed full throttle into the protection of the west channel of the Fly River delta and Umuda Island for the night, grateful for some quiet weather sleep, and some time to organise our lab and aft deck field gear.

Glen Shen was one of those maniacal trace element chemists that try to avoid all chances of contamination of the water sample from the ship, winch wire and Niskin bottle metal parts. He had a special wooden pole wrapped in clean plastic, with a little gripper at the end where he could put an acid washed 250 ml plastic bottle in the gripper, repeatedly fill and dump the bottle about 1.5 m away from the rubber zodiac boat well upstream from the *Western Venturer*. My job was to hold the zodiac boat steady and pointing upstream in the strong current. Glen wore plastic gloves to handle the pole and bottle, and the sample went inside an acid washed plastic bag immediately after retrieval.

Meanwhile, Irena and Ken Woolfe collected water for nutrients and major elements using an ordinary clean plastic bucket, and sediment samples, along this same salinity transect, from zero salinity (170 μmhos/cm) up in the river, to 36 ‰ a hundred miles downriver through the delta and estuary into the Gulf of Papua. During the hot sunny days, the aft deck was so hot we had to spray seawater over the deck from the ship's pump continuously, and the water would steam up our sun glasses. Even with safety boots on, you could feel the heat. Irena and Glen would not allow us dirty sediment people inside the lab, where careful clean handling and filtration of the water samples was being done.

Glen was trying to determine the variations of the concentrations of dissolved and particulate trace elements Fe, Mn, Sr, Cu, Cd, Pb along this salinity gradient. It turns out that most of these elements are relatively high in dissolved concentration in the river water, and these concentrations rapidly decline in low salinity water up inside the deltaic estuary. The dissolved concentrations decline as the deltaic water flows toward the sea, but these elements were then sorbed to the abundant fine particles held in suspension in the estuary. And then most of the fine particles settle into the sediments along the outer continental shelf of the Gulf of Papua. It was to take me about a decade to complete the budget for some of these trace elements in the huge Fly River Delta and Gulf of Papua. In addition, Glen wanted to know how much of these dissolved trace elements survive to reach the coral reefs of the clear water Torres Straits. Most of the suspended mud in the deltaic Fly River water appeared to turn east, and move along the muddy coast.

Due to mechanical problems, we had a break for two days while the ship engines, steering and rudder were repaired. Luckily this happened at the apex of the delta, beside Sturt Island, where the water was fresh at low tide. We had time to walk around the island, feed the mosquitoes, and visit the saw mill operated by Dickie Randolf (Scungie). He paid local PNG people in big dugout canoes with two 20 HP Evinrude outboard motors on reinforced transoms to snag and pull big logs into the nearshore above his sawmill. You need to appreciate that this is a large river, at this point it was about four miles wide. He had a giant winch cabled to big mangrove trees (and steel cables attached to buried dead man anchors far inland) that was used to drag the drift logs into his little holding bay. Some of these logs, when sawn and dressed nicely, were almost 2 m square. Some of this was rosewood, merbau, and other precious tropical wood species. The whole yard was thickly covered with sawdust, and the houses and tool sheds were on piles about 2 m above the sawdust ground. The saws were antique belt-driven circular saws about 3 m in diameter, powered by antique diesel engines with many home-made repairs visible. Scungie was an ex-Kiap, a pre-1975 Australian patrolman who helped PNG become an independent country, and he had been in PNG for over thirty years.

Ken Woolfe tried to map the parts of the deltaic islands that were accreting sediment, growing in area due to mud and sand accumulation, and those parts of the islands that were eroding due to the current. We also had maps of this river delta made by early explorers and from post-World War II air photos. As best we could determine, there was no large net accretion or erosion of the islands, but they appeared to migrate somewhat, due to growth on one side and erosion on the other side. According to OTML *Western Venturer* ship masters, the locations of the sandbars and deep channel appeared to change only very slowly. OTML needed to know all this, as their copper ore ships travelled these channels to reach the ocean-going vessels anchored in deep water off the delta front.

Half way down the delta, we visited some of the communities to buy bananas, fish and mud crabs. The head man of Domori Village at Mile 50 was a national MP for the Western Province who spoke a little English, and he gave us a proud tour of his village. They had a *puk puk* pen (crocodile) to raise little ones up to eating size, some penned cassowaries, ducks and chickens, lots of pawpaw and coconut piled about, a sweet potato garden and houses on stilts with *Nypa* palm thatch roofs. We also bought some very good yellow and red meated watermelons, which I enjoyed. They had their own church and Samoan missionary, a First Aid House, a rainwater collection basin and reservoir with spigot, and kids running everywhere with Michael Jackson and Motorhead T-shirts.

We left this village in the evening and promptly got stuck on a sandbar in the pitch dark, the engines screaming in reverse, but no luck, so we anchor till the tide come up the river to lift us off. The sandbar was composed of shiny black sand, probably magnetite, and the mosquitoes and sand flies were unbelievably thick, clogging the air intake to the engines. So we watched ridiculous Arnold Schwarzenegger films until 0230 on 3 February, when we could feel the tide moving in to us. Meanwhile, Glen and I were ready to give up, and we joked about discovering the new rare earth elements *Rhizopterium* and *Newguineium*. You get a little goofy in this business.

Finally by 4 February we were out on the calm clear water of the Gulf of Papua, with frigate birds flying over us. This completed our salinity gradient work, and we steamed to Daru and the smell of roasting fish, wood smoke, rotting seaweed and sewage. We tied up at the wharf, where the kids and mommas were all lined up fishing for their suppers, quietly talking and singing. Customs was nowhere to be found, so we waited, tea and toast with jam, bad movies. The next day Customs showed up at 9 am neat and tidy with sunglasses, brief cases and clip boards, regarding us as pretty strange characters with bottles of water and mud.

We were in radio contact with the AIMS R/V *Lady Basten*, and Joanna had done well to get lots of mangrove sediment cores from Jacky Jacky Creek in Cape York, and had also worked on Sassie Island and Saibai Island mangroves. We joined them on 5 February and went to the Bramble Cay to explore for big bommies of *Porites* for Glen's work on trace elements in coral growth bands. Glen, Irena and Ray Gray were diving, and they found several huge bommies, maybe 5 meters in diameter at the base and over 7 meters high. Sharpie ran the *Hercules* over to the coring sites, where the *Basten* was afraid to go, and we spent two days drilling this huge coral history book. I got to do some snorkel work to deliver tools and to watch them work.

Peter Isdale of AIMS had warned me about this place, as it was a fisherman gutting station pre-World War II, and the sharks were reputed to be abundant. We didn't see any big sharks, but on the sea floor there were lots of stinging ctenophores, rays, skates, sea cucumbers, old anchors, piles of rope and coal, and fragments of wrecked ships. Green footed boobies were nesting on this sandy island, only about 2 m above mean sea level.

With three long drill cores of these coral bommies, Glen was happy, and we steamed for Cairns. Glen left the ship with his precious coral cores and superclean water samples and flew to Townsville to have Monty Devereux help him cut the coral cores into quarters lengthways with a special clean bandsaw. We steamed onto to Rockingham Bay, just north of the Herbert River delta, where Ken Woolfe, Irena and I got eight good sediment cores. By 16 February we were back at the AIMS Wharf, tired, sun and wind-burned, but happy with our work.

Back at my office, the administrative problems multiplied in March. We had a new AIMS CEO Meryl Williams, a good lady who immediately recognised me as having pet parrots at home. Puzzled, I asked her how she knew this and she grinned, pointing to my scarred arms, where Charles Byrd liked to sit at breakfast. But then she took off for a month for work in the Philippines and Acting Director Peter Moran learned how nasty the job can be.

Due to insurance and labour disputes, the AIMS ship crews were laid off for the months of May-July, and all scheduled research cruises were cancelled. Well, this sort of cuts the heart out of marine research activities! Several important cruises to the northwestern coast had been funded by outside agencies but we couldn't run them, and we had to return the funds. We were all told to write journal papers and stay in our offices, or take leave.

Ok, so we all relaxed. Ian MacGreggor was playing in several bands. One group played in the local pubs, and another was called The Flaming Stars, which probably could not be classified by normal rock and roll categories. Sheila and I took the time to explore the Tablelands and inland cattle country of North Queensland, watched platypuses in streams, collected seeds and bedding plants for our orchard. By July, Ian's girlfriend Gemira arrived from Calgary, with a beautiful trained singing voice. The Flaming Stars went on a North Queensland road trip, two station wagons crammed with sound equipment, guitars and drums, three blokes and two sheilas. Jesse was busy driving sugar cane trucks in the Giru area, going to work at 0400, and working late burning cane before the harvester cut the next lot of cane.

Yet my brain was seething with plans for an international consortium of researchers for PNG work, and the email and fax traffic was fierce on this topic. Chuck Nittrouer and I had secured conference rooms in Honolulu for 15-18 July, just before the Oceanographic Society meetings. It was essential for us to have official government lab representation from Indonesia and Papua New Guinea, and we had to courier deliver the cash to pay their travel and other costs. Several key scientists cancelled, replacements were found, visas were obtained at the last minute.

And so on 15 July, Ken Woolfe of JCU, Ron Szymczak of ANSTO, Ian Wood of OTML, Gaikovina Kula of PNG DEC, Eric Wolanski, Dan Alongi and I flew to Honolulu with grand hopes to create a research program much larger than we could do ourselves. We had a wonderful meeting room in the University of Hawaii East-West Centre, complete with contemplative Japanese style meditation garden with koi fish ponds, small gongs and a tea room. Nontji of Indonesia and Kula of PNG gave opening addresses, and the strong USA contingent took over for most the first day. The USA group were all veterans of the Amazon research group, many of whom I knew from past AGU meetings: Bob Aller, Don Wright and Ed Scholkovitz. We created our name (TROPICS = Tropical River Ocean Processes in Coastal Settings), a set of accomplishable objectives, a rough schedule of ships and technical staff availability, and the beginnings of a Prospectus to use in administrative and government funding offices. Another proposed name for our groups was 'SPAM', partly due to common consumption of this pork product in PNG, but also as an acronym for 'Sediment Plumes And Mud'.

The Honolulu organizing group for Project TROPICS, in 1994.

Both Kula and Nontji warned us that it would be difficult to control their Customs and Immigration Agents, ship clearance, and the import and export of scientific equipment and samples. PNG and Indonesia were wary of fast-moving US/European/Australian researchers taking away valuable intellectual property, oceanographic data, plant and animal samples that could be used for money-making biomedical products in their home countries. Part of the price of their agreement to TROPICS participation was to have data repositories in their countries, copies of all publications, and some of our ship space for their people to learn how to do our kind of science. PNG had almost zero oceanographic capacity, but Indonesia had eight big ships with fancy equipment that roamed their seas every year. We had to learn how to attract these busy oceanographers.

The following Oceanographic Society meetings in Honolulu were very good, with major presentations by the scientists of the International Geological and Biological Program and their Joint Global Ocean Flux Study (JGOFS) program in the western Pacific. We met top scientists like Jim Murray and David Karl, who had been working in the equatorial Pacific for much of their lives. They were interested in our TROPICS plans, and urged us to consider adding a strong climate change/greenhouse gas/CO_2 exchange component to our Prospectus. They also suspected that a lot of the micronutrient iron in the western and central Pacific equatorial waters came from PNG rivers and coastal sediments, via the equatorial undercurrent, which flows along the north coast of PNG, does a U-turn, and then flows east toward South America. And here was the international science community, helping each other with ideas and passing on experience to younger folks.

When there was free time, I walked the crowded beaches to smell the coconut oil suntan lotion and watch the buxom babes proudly strutting their stuff. I visited Telu Li, Keith Chave and Fred Mackenzie at the University of Hawaii Oceanography group, good colleagues since my summer course at the 1964 Bermuda Biological Station, and my 1977 sabbatical days at Lamont Doherty Geological Observatory. Nearby was the University horticultural laboratories and gardens, where I bought lots of books on tropical fruit and viewed a lot of tropical plants I didn't know. There were some good music stores too, many of them featuring very good ukuleles and slack key guitars. There were some bombastic salvation preachers standing on park benches yelling out warnings about the end of times being near. I visited the Honolulu Zoo, and I could have spent a week there...lots of Australian parrots, SE Asian Golden Tamarins, lemurs, hornbills, gibbons, orang-utans, gorillas, several kinds of apes, Sumatran tigers, a large aviary of PNG birds of paradise, South American parrots, African mammals and a huge Reptile House with a pond full of matamata turtles. One evening the TROPICS group toured the University of Hawaii Oceanographic ship *Moana Wave*, a sleek and well-designed ocean research vessel.

And now here is one of those flash-back events that don't happen very often. My parents often spoke of their Murdo high school friend John Norris, usually in a derogatory tone due to his left-wing jazz-man hippie newsletters he sent around. I knew he lived in Honolulu, so I looked him up in the phone book. His good wife Lois answered the phone, and at hearing my name, without any further information, she said...'You must be Virginia's boy. You come on over right now and I'll feed you.'

And so on 20 July, after the day's science presentation were over, John picked me up in his bright red van full of musicians' tuxedoes and two cellos, and he hauled me to the Hale Koa, apparently a military hotel full of on-duty and retired officers, with an old time Dixieland jazz band in full swing on stage. The singer almost immediately called John out of the audience, asking him to 'get your ass up here right now'. John had been carrying a velvet sack, sort of looked like a Chivas Regal whiskey bottle sack, but I now saw that it contained an old cornet. And wow, for an old fart, he sure knew how to play that music, for the next two sets, no sweat, comfortable and easy. Between half hour sets, he brought members of the band to my table, and we talked about the history of jazz in Hawaii.

John had been in the US Navy during World War II, and was shipped to Hawaii just after the Pearl Harbour bombing. He spent most of his Navy service time playing seven nights a week in jazz joints in the Waikiki hotels for military personnel, during the time of swing big bands, the popularisation of Hawaiian slack key guitar blues, and the evolution of bebop and cool jazz in the 1950s. John then toured the world with US Navy bands from 1943 to 1960, playing with some jazz greats like Jack Teagarden and Trummy Young, but he was based in Honolulu. This hotel where I was visiting was apparently the last hold-out for old time jazz, and was thus haunted by a lot of old military geezers, as all the other hotels now played tapes of mushy elevator music and karaoke bars. Most of the beachside hotels now catered to wealthy suntanned tourists and the feel of gambling chips. Several of us gawked in awe at buxom young ladies coming in with surfboards under their arms, with only body paint for a swimming suit. John retired from the Navy in the 1960s, took up the cello, and held a chair in the Honolulu Symphony Orchestra for many years.

The next late afternoon, John picked me up again and took me to his big home on a jungly mountain slope, well back from the city and beach culture. And then his wife Lois came to the door, took my arm as if I were a long lost son, and led me into the dining room, where the table was set for a true South Dakota feast. Lois had a disturbing habit of clucking, cooing and pushing the food toward me, just like my mother. There was a giant well-cooked roast beef, roast potatoes, sweet corn dunked in melted butter, onions and mushrooms, home-made chutney and home-made beer. And one of those god-awful green jello/shredded carrot/

olive salads that are still common in western South Dakota kitchens. And then a macadamia-chocolate coffee cake, which led to a vicious argument about the origins of macadamia nuts (they originally came from Queensland, not Hawaii). Lois also gave a lecture on the beneficial and healthy properties of macadamia and cashew oil in salads and cooking. Many of John and Lois's family and adopted family were here, perhaps a dozen people round the table, and they all treated me like I was part of their family. They were all health freaks, would not eat meat, and some wouldn't even eat eggs, butter or cheese. Their yard had pomegranate and passionfruit vines, pomelo, mango, lychee and cherry trees.

John told of his Murdo high school days in the 1930s, when he arrived in town as a new preacher's son, and how my Grandpa and Grandma England treated him nicely. John told of wanting to play the cello when he was a sophomore in high school, and his grandfather got him one, mail order from Minneapolis, arrived in a cardboard box. But nobody in town knew how to play it. John had this theory that many Midwestern kids were successful in the outside world because they haven't the slightest clue how badly the cards are stacked against them, and so they go ahead and do it. [That is sort of like me trying to play the high trumpet parts in Bach Cantatas with a B flat trumpet! These parts were written in the 1700s for trumpets in the key of F.].

Lois must have been a special friend of my mother in high school. She spoke of her in a quiet glowing tone, with insight to my mother's artistic and musical abilities. According to Lois, she and my mother went to a girl's camp on Long Lake, Minnesota one summer, staying in a tent together. They got into some trouble by going swimming after lights-out and walking into town to get candy. She said my father was the most handsome and smartest kid in his class, and all the girls were after him, even if he was a pretty rough bow-legged cowboy from the White River bottomlands. This was all mind-blowing, a lesson in my family history.

After all this, John took me to a local inland country pub where farmers and cattlemen were singing around a piano, most people at the tables were singing, sometimes in harmony, call and answer songs. And then he took me to the airport at 2330 for my late flight back to Australia. That is 8700 km to Australia, about ten hours of flying time at 10,000 m altitude, -38°C over Noumea, and it was winter when we landed in Sydney, chilly and cold rain at the airport, I was grateful to get home to Sheila and the tropical warm sun six hours later in North Queensland.

After the next two months of mind-numbing administrative wrangling and horse trading between CSIRO, AIMS, the feds and state governments, I was again startled to have another blast from the past. A very good blast it was.

In early October 1994, with minimal warning, Jack Vallentyne appeared at the Townsville airport with his globe on his back, light bulbs blinking inside it. Dr. Vallentyne was my Ph.D. thesis supervisor and mentor, now 70+ years old, long retired from the Canadian Department of Fisheries/Environment, and was touring the world speaking to young people, engineers, town planners and government bureaucrats about the need for environmental conservation and ecological education. He had been in Perth and Melbourne for speaking engagements, and he stopped to visit us. Sheila and I fed him prawns, bugs, croc meat and some good Australian wine. We were startled to hear that he was learning to play the guitar, was taking singing lessons and ear training, so that 'Johnny Biosphere' (his stage name) could entertain kids with Woody Guthrie ballads. He knew that Sheila and I had been involved in organising the Winnipeg Folk Festival, and had performed this kind of music for over twenty-five years, and he wanted to hear some of our repertoire. We had long walks in my creek valley orchard, viewed wallabies, possums, parrots and dingoes, and then he went to sleep in a chair on our veranda in the warm sun. He gave lectures at AIMS and JCU and went with Professor Richard Pearson's limnology class to some rivers north of us. Sitting at our pergola feast table, JRV reminisced about his early days of growing up in Ontario, being a sharp-shooter in the Canadian Army, stories about George Evelyn Hutchinson at Yale University, the last days of Lamont Cole (ecologists) and John Wells (marine invertebrate palaeontologist) at Cornell University, some aspects of early ELA and Ottawa history that I knew nothing about, and the importance of Frank Rigler and Rob Peters to Canadian science.

I had this very warm glow in my chest, listening to this seventy year old man reviewing his life in science, and reaching out to others for help in new projects. I felt deeply honoured that he took time to visit us. He said he wanted to live his life out at his Albert Beach cabin on Lake Winnipeg, but his good wife Anne needed access to her cello, the symphony orchestra, and string ensembles in the City.

And then Sheila's brother Alistair (Wump) arrived with his wife Susan and two kids. Both Alistair and Susan had been members of our band Meadowlark when we lived in Canada. Their arrival meant that we scheduled a few concerts and house parties to try to remember all of our old repertoire. And of course there were many visits to the beach, the Billabong Sanctuary, and major feasts of seafood.

The rest of October was wasted in a series of agonising psychobabble, self-flagellation and group power-building exercises in Perth, Sydney and Hobart amongst the major marine science institutions in Australia. AIMS sent me to these meetings because I was outspoken and was pushing for major international research consortia, both for our Herbert River estuary and shelf work and our TROPICS research plan for New Guinea. We had nice rooms in expensive hotels with good restaurant banquet meals and heart-warming presentations on how we should all collaborate. I managed to gather scientific and ship time support for our Herbert Estuary and TROPICS work, so I guess it wasn't a complete waste of time. In Perth, I walked out of the meetings in the afternoon to visit the Fremantle harbour, where ships from all over were tied up at the wharves and Portuguese restaurants served good seafood. Huge wool sheds nearby smelled of lanolin and the wood had that wonderful varnish from many decades of use.

A major improvement in my prospects for using radiochemistry in my research was the arrival of John Pfitzner on 21 October 1994. John had been working as a radiochemist in Jabiru and Darwin, Northern Territory, and we offered him a job at AIMS to operate my newly purchased gamma spectrometers and associated accessories. He arrived in a very dusty 4WD ute with a trailer full of his worldly possessions and two dogs. We gave him a good supper and a bed for overnight, and Sheila helped him house hunt in town.

Over more than the next decade, John improved, expanded and operated the best marine radiochemistry lab in northern Australia, mainly focussing on ^{210}Pb and bomb fallout chronology in sediment cores, uranium series elements in water, radon and radium families of nuclides for tracing water mass origins.

Due to the raging success of our IAEA Marine Radiochemistry Workshop on Coral Reefs last year, I was then saddled with the opportunity to organise and host another IAEA Marine Training workshop 31 October to 5 November 1994. Scott Fowler of IAEA Monaco and Ron Szymczak of ANSTO showed up three days early to help me organise this affair, and over the weekend sixteen SE Asian participants (lab technicians, lab directors and a few experienced radioecologists) arrived and were neatly accommodated in the AIMS visiting scientist houses. We had a dozen seasoned lecturers (from Sweden, Monaco, USA and Australia) who had done a lot of marine radiochemistry. As before, the AIMS canteen went into overdrive, and interesting meals (suitable for Hindus, Muslims, Jews, Christians and a few pagans) were provided for the week. Participants came from China, India, Bangladesh, Pakistan, Korea, Thailand, Vietnam, Malaysia, Myanmar and Indonesia, and we had several translators helping us with information transfer. We had the usual fun evening *hungi* banquets and bush dance at Melville's Pub, a visit to our local Billabong Wildlife Sanctuary, and bird-watching in our Cocoa Creek Salt Flats.

We also provided several demonstrations of how to take short sediment cores, and quantitatively collect some radionuclides from seawater, in cubic meter water samples in large plastic tanks, using acids, radiochemical yield tracers, strong base, a canoe paddle, and a 'J' shaped copper pipe siphon tube. Most of the SE Asians watched from a distance, all dressed in suits and ties, looking worried about being close to radiochemistry in action. We tried to explain that the natural ^{4}K strong gamma emitting isotope was about 100 times more abundant that the 1960 bomb nuclides we were seeking, but this had little impact.

We also offered to take workshop participants out on Bowling Green Bay, nearby the AIMS ship wharf, to show them how to pump sea water from various depths, through filters and sorbent-coated acrylic fibres, for the measurement of radium isotopes. We used the AIMS vessel *Hercules* for this operation, this being a World War II landing barge with a big open, shaded deck. Only a dozen of the SE Asians were brave enough to come along, and those showed up in suits and ties. To me, it was a relatively calm day, and the flat-bottomed *Hercules* rocked gently as we pumped a cubic meter of water through a series of filters and sorbent cartridges. But our SE Asian students were puking their guts out over the rails, paying little attention as Ron Szymczak, Terry Hamilton and I ran the hoses and pumps and filters. Upon arrival back at the AIMS dock, the SE Asians leaped off, grabbed the bollards on the wharf and sat down, faces white and scared. We had to help them into vehicles to take them back to their accommodations, and very few of them showed up for supper. Field Operations Manager John Hardman allowed us to use his brand new Toyota Troop Carrier to move these people from the wharf to the accommodation houses, and they puked all over the back end of this shiny new vehicle. I judged this operation to have been a miserable failure, but the other organisers were highly amused at this affair.

The second part of this marine radiochemistry training course was given at ANSTO in Sydney, and I gave a long illustrated lecture on mangrove ecology and sedimentation. Again, we were in fancy hotels with good food, and I could escape evenings to explore music and book stores.

While all of this lollygagging around was happening, Sheila at home had no water supply in early November, as our water system had failed. The house switch had burned out, the bore pump appeared to be dead, and the bore hole no longer produced water. So we got a driller to put a new bore down to 36 meters, with good water flow at 24 m depth. We got household water through a garden hose from our neighbour, until we could get a new Grundfos pump installed, and the electrician fixed the bore pump and house switching system.

Just about run ragged, Sheila packed my clothes and remembered where my passport was, hauled me to the airport on 21 November for a big UNESCO science meeting on sustainable development and environmental problems. On the way there, I was reading Adrian Vickers' book about the post war history of Bali, warning me of the surprising abundance of European and North American artists, writers and musicians on that island. I had also struggled through Ricklefs' book in preparation for this trip. I can do no better than quote my letter to my son

Ian, who had anxiously asked to be on the lookout for opportunities to purchase gongs for his collection of percussion instruments.

23 November 1994

Sahid Bali Hotel

Kuta, Bali

Indonesia

Ian MacGreggor (& Gemmy Lou too):

Oh Ian, Oh dear me, you should be here. I'm feeling guilty about being here and having to describe it to you. Yes, Yes, I've made inquiries about a gong, and I've walked the back streets amongst the street-side vendors and grills and fruit stands and hucksters...This whole place reminds me of some novel I read a long time ago about the Spice Islands being magic, and how white people sort of forget to go home and die here, pickled in gin and tonic and spices. Bali is a very small, very hot and damp island mostly occupied by volcanic mountains, dense jungle, with a coastal margin packed tight with cities and people.

I got here a few days ago, via Cairns, Gove and Darwin in torrential rain. Actually, I got kicked out of the Cairns Qantas Club because I was wearing flip-flops. The Denpasar airport was nearly blacked out on our arrival, no people, one bus for a 737 load, mass chaos and shouting and whistles and dogs barking and a long wait at midnight till 0130 to get to our hotel. At 2 am the hotel is full of music (western and Balinese), bar crowded, sarongs and lap-laps galore, you can hear the language of Balinese, Javanese, Malay, Japan, Arabs, Russians, Japanese, Brits and the Dutch floating in the soft air, thatched pagoda style cottages by the sea. Kuta is the section of Bali that Indonesians visit, not *boulies* (whites), so it is not too bad a tourist trap, and most things are disgustingly cheap (9 course meal for A$6 = 10,000 Rupiah). The hotel is Balinese, not an international chain, and the menus and signs are Bahasa and Balinese, not English. There is some sort of a large bamboo xylophone set up in the registration area that is played casually by hotel staff when they get bored. Ornately carved sandalwood everywhere, stone and wood statues and wall bas-relief carving on all surfaces. Servants and hotel staff everywhere waiting to do what you ask them...they come in your room and rummage through your clothes to take dirty stuff to the laundry.

Most people in Bali are Hindu, so we can have non-beef meats, seafood and beer, in contrast to Java and Sumatra which is Muslim and is more strict about alcohol and meat and dress codes...Most people here chain smoke clove-spiked cigarettes, so the air is heavy with clove smell.

Ok, Ok, what about music...last night after a full days scientific meetings in very hot very humid rooms full of drowsy audiences, we were hauled off in buses to the Governor's Palace in Denpasar, where I ended up sitting with some government zombies from Washington D.C. and a Papua New Guinean Princess (?) at a table on the grass in the courtyard being served with soursop cocktails and whole pickled baby squids for appetizers, to be dunked in a variety of sauces, most of which were lethally hot. And then we get to tour the groaning tables full of seafood and satays and weird things of unknown sources, eating our way with fingers and fingerbowls to rinse...all this is in the heat and humidity that would melt your average Bedouin. I was later told by a Bahasa savvy Englishman that amongst the curries on the table were spicy rat, spicy dog, and spicy bat. Rat meat here is good, it comes from a big water rat, bigger than a South Dakota muskrat. And while I was prowling the fruit table trying to find out the name for things, the gamelan orchestra started up behind a curtain somewhere. With my fruit plate piled high with savak, mangosteen, rambutan, white melon, durian, jakfruit, I clambered across the flower garden to watch the gamelan outfit.

They were all dressed in white suits with white, gold, and red cloth wrapped around their heads, six fellows on small flat steel bar xylophones pounding the Hell out of them with geology hammers, two fellows with pot gongs cross-legged on the floor on one side, and then another six fellows on larger pot gongs with bigger sticks on the other side, two fellows on larger very deep sonorous steel bar xylophones behind them, and the master gong player with one enormous (maybe 1.3 meter diameter) gong, one smaller (maybe 0.8 meters diameter), and one 0.3 m gong on the floor. The big gongs were suspended in ornate wooded frames carved everywhere, suspended on ropes with rags tied inside the gong on the rope. At the back and between these two groups were two drummers, each with three drums, a flute/whistle player, and a curious fellow who fondled a gold metal Komodo dragon frame that held a variety of small finger cymbals. All of these people had expressions on their faces of being very spaced out, madly pounding the bejesus out of their chosen instrument. They were sitting in a small highly decorated pavilion amongst the frangipani trees full of flowers falling onto the stage, beautiful glittering garishly coloured cloth banners and hangings on the pillars of the pavilion, candles and torches giving flickering light, no electricity, the air smells of cloves and sandalwood smoke and torches and sweat and the rotting seaweed and sewage in the canal nearby.

Meanwhile, the rest of the crowd finishes their supper and wanders over to find out what's causing all this noise (it is very loud). The Governor of Bali marches up and bellows out a welcome, demanding everyone to sit and watch his dancers. Then the gamelan orchestra went into overdrive, completely amok (that word comes from Bali, refers to the mental state of princely warriors going into battle expecting death) for a half hour, after which about a dozen brilliantly costumed dancers appear suddenly, with their heads moving sideways and eyes very large, hands making motions that seemed unnatural. They did four dances (a modern welcome dance, a bird of paradise mating dance, a dance about a prince going into battle, and then an old classic dance on a courtship theme) which brought both the dancers and musicians to dripping sweat.

This took from about 9:30 till about 1 am, soursop and citrus drinks being served, Ambonese boys waving large fans over the governor's position, everybody else drenched in sweat and mesmerized by the incredible racket and splendorous setting. Some less appreciative souls of western origin snuck off to visit less cultural activities (whores and strippers, and Michael Jackson videos are projected against the backs of souvenir stalls). To tell the truth, your recorded gamelan music required a lot of effort for me to listen to...but this outfit was fascinating...the resonance and sustains and rhythmic and tonal conflicts were more obvious, more sonorous, and the pot gongers had some neat ways of getting a sock-cymbal like doooonngg-chuck doooonnngg-chuck rhythm. And then the local United Nations rep gets up and thanks the governor, and a rep from all the countries gets up and thanks the governor, and there's lots of bowing and holding the hands in prayer form in front of the noses, and its all over.

After they finished, I hung around the pavilion as they packed up, and I got one of my Jakarta friends to ask the gamelan master if it was possible to purchase a gong, and how much they cost. The big gong would cost over US$800 and couldn't be obtained quickly, but could be arranged in several months. The smaller gongs and 'pot gongs' could be had for $50-100 US. All of these costs were quoted in millions of Rupiahs, but realize that R10,000 = A$6.50. I made an arrangement to look at some gongs at his house on Friday, and he'd sell one to me suitable for a musician. He is Wayan Suweca (Home address: Jalan Noja Saraswati No. 9, Kesiman Petilan, Denpasar, Bali, Indonesia) a teacher at the State School of Art in Denpasar, phone # 227 316. The name 'Wayan' means first son, I think.

This orchestra is the Governor's private group that plays for state visits and important functions, and it goes on international tours. Wayan stressed that this is Balinese gamelan, not Javan or Sumatra...he says Balinese gamelan is more dynamic and active, whereas the Javan form is more static, formal, restrained. It sounds like each of the Indonesian islands has its own culture and music. The local abbreviation for this school is SS Kar or STSI. The small gong I played with sounded about like your mother's heavy aluminium bowl hit with a wooden spoon, but it had a longer sustain and it cost a lot more. The big gong was truly awesome, made your liver shiver. Wayan said his group had just returned from touring Germany, and they were about to take off again for another European tour. He taught gamelan at the University of Montreal with Jose Evangelista, and he has taught at several US universities. He had heard of Glenn Price, but didn't know him personally. He speaks English well, has a wife and a small son, and they were very curious about my conference and oceanography and where we live at Alligator Creek. He invites you to visit him, and he said he would help with learning about gongs and drums. Our conference has been in the local press and TV a lot in SE Asia. He came to my hotel for coffee and to show me several smaller gongs. I ended up paying R170,000 (= A$111.18) for a small gong about 35 cm in diameter, with an appropriate padded wooden striker, and I rejected the fancy carved wooden holder. Did I get ripped off? Everyone here says to bargain, and pay only 50-75% of the asking price, but he seemed straightforward and honest.

Anyway, after the gamelan music and dance, I took a taxi back to my hotel, took a long shower, drank two beers, and wrote my talk for the next day, the din of the gamelan still in my head. The meetings are for the Intergovernmental Oceanographic Commission of the UN, trying to figure out a way to understand tropical oceanography and stop the dreadful pollution of the Indonesian Archipelago, and China, Thailand, Viet Nam, Cambodia, India, PNG, Sarawak, Australia are all here with their scientists. During the sessions, a bell goes off at 11:30 am everyday and all the Muslims get up and go to the mosque for prayers and don't come back till 1400. Oh yes, and after the officials did all their welcomes and little political speeches for the press, the meetings were opened with a wallop on a big gong. We get to go up to the top of a nearby volcano on Saturday, one with gases and cinders still flying up out of the cone. The newspapers reported today that one other volcano erupted near here, killing 30 odd people. This country is very crowded in the lowlands, and empty in the mountains, and so full of ancient culture that it hurts your head to think about these people having over 2000 years of civilization. It would have been even more interesting if the damned missionaries and Dutch spice traders hadn't screwed up their culture. I am told that Bahasa is easy to learn and well worth the effort. But you have to realize that Bahasa is only the official language of modern Indonesia (post Sukarno), originated in Malaysia, and each island has several or dozens of local languages of great antiquity. The local Balinese people can be divided into 4 social castes, each of which has their own language distinctions, with their occupation and caste status being included in their name. And then you add the complications of having Buddhists, Hindu, Moslem, Christian, Military, and Jewish regions of the city. My Jakarta (Java) friends tell me that the Balinese are fiercely independent, were savage warriors, and have recently begun re-establishing their ancient pagan animist religion/culture. People from Jakarta tread lightly here, considering this to be a foreign place to them. And the whole place smells like cloves. And oh yes, the beach is beautiful, surfing waves 3 meters high, and the city traffic is dreadful, and each coconut tree has a number on it for the owner, and they have Dunkin Donuts, McDonald's and Coke.

25 Nov...I walked down some rat hole streets behind my hotel a few nights ago, trying to find some nice Balinese cloth for your mother, and a fancy shirt for Jesse the Birthday boy (you remembered, didn't you?). I saw a lot of junk, and some nice leather, cloth, and carved wooden objects, carved bone, porno movies, poverty and filth. The pimps drive small motorcycles with their fancy dressed whores on back, they drive right up beside me walking and offer 'What is your name? You want boy or girl? Cheap cheap, you come with me'. The more persistent ones grab your penis, but their other hand is trying to find your billfold. They want to know your name, because your name should reveal your status, caste, and this determines how they are to treat you. The street

entertainment joints all have little bands, keyboard, bass, guitar, big speakers and boxes with glowing lights, fancy dressed singer and cordless microphone, but the whole thing is faked, air band, there is a tape playing the song, the most god-awful stuff...'take me home, West Virginia, mountain mama...', the same pretend band will do C&W, rock, rap, Tony Bennett with big band. Street vendors are everywhere, selling satay, sweet corn, hot candy, sweet black rice confections, every fruit imaginable, soft drinks and beer, rice wine, clove cigarettes, cigars, dope of various unknown sources and names, and bottles with Jim Beam or Johnnie Walker labels but selling for less than the equivalent of A$1. Kuta Beach is miles long, dark volcanic sand with white coral fragments and shells, all the Asians are fully clothed with big straw hats, all the western people are naked or in the tiniest of swimming suits, lots of Australians, Scandinavians, Arabs, surfers out on the breakers even at night, most of them Americans or Europeans. The water looks clean and clear, and there is no garbage, as Balinese keep picking it up with wooden spears and carrying it off in handmade woven baskets.

26 Nov...Today was excursion day, the formal meetings are over, Suharto gave a little speech at the closing ceremony, and they hit the gong again to close the congress. We were all herded onto big buses and whisked away with police and military guards and escorts on motorbikes. I guess security is maximum here at the moment, because there are several other big meetings going on, such as OPEC, and some economic conference attended by Chretien and Keating amongst others. The bus guide with the howling microphone was completely hopeless in English, trying to give us lectures on Hindu culture, cremation ceremony, mangroves and science. We drove past miles of mangroves cut down for prawn ponds in the mud, a horrible mess, grubby little grass thatched huts on the margins. We went through an art colony where they worked silver, gold, wood, leather etc and paintings for the international market...most of the items would not be used by Balinese themselves. The painters work on wood slabs, some modern and natural history subjects, but most of it was inspired by ancient Indian legends and religious sagas, done in ancient styles with modern acrylic paints, incredibly detailed and packed with information. The wood carvers had more of a sense of humour, carving fat mommas fondling their huge breasts, lovers in copulation, roosters fighting, Buddhas, Beethoven, Michael Jackson. The buses went speeding down the tiny narrow roads, sirens and horns going full blast, little old ladies with loads of firewood on their heads scuttling for the ditch, men on bicycles with crates of chickens or fruit went flying off the road into irrigation ditches. The cattle here are bentang, an Asian breed which is older and different from European cattle, they have a half moon crescent on their hind quarters, the horns grow to the rear. The little plots of land for each home have a shelter for one or two cows, they have little wooden bells, and there's lots of shrines in the rice paddies and back yards, with Hindu basket offerings in the notch of shrine gates. They took us up to the volcano top for lunch, where you can see the smoke and fumes rising from some of the cones. The restaurant was right in the caldera, and we had some good satays and beer, and I bought a kg of mangosteen for about $4 and ate them all the way

back to the hotel. Your mother will have a fit when she hears this, as she loves them, and they cost $3 each in Townsville. In Balinese, they are called *mangis*. On the way back we stopped at a batik and ikat factory, where they were looming cotton cloth, and applying the wax to the pencilled patterns. I bought some bolts of fabric for your Mom, but it seemed awful expensive to me. Silk was out of sight, over US$300 for 3 meters, but I suppose they import it from Japan. Most of the shirts I saw were ugly, metallic looking patterns, probably destined for some big city somewhere. One of the shop clerks liked my shirt and asked if I had bought the shirt I was wearing in Kuta, but I told her that my wife made it from material in Townsville.

It is extremely hot and humid, and I'm afraid to admit that I long to go back to my clean, air-conditioned hotel room. This island has many conflicting symbols and images that shouldn't co-exist. Pizza Hut and Hindu shrines, Mercedes wealth and poverty, gamelan and fake rock and roll, crowded cities and empty mountain jungles with no roads, some of the world's most diverse and unique reef ecosystems being used as sewers and blasted with dynamite for coral blocks for building houses, barongs vs Big Bird. A barong is a mythological creature of similar stature to the Canadian Plains Indian 'Wisakedjak', a mischievous spirit that is not evil, but plays tricks on you. The postcards and statues of barongs here show a creature that looks like a giant Pekinese dog, only garishly coloured. TV here is all in Bahasa, loaded with political propaganda, and based upon US style advertising of processed cheese, Coke, Dunkin Donuts, hashish shops, fancy cars, and big business, with soft rock TV videos in Bahasa but with all the Hollywood schmaltz and posturing and backdrops. The soft, gentle Balinese have their own language and style, and consider the 'Indonesian' culture to be something new and difficult from Java. The old art is sensual, rich, reverent, and meaningful...the new stuff is cold and brittle, lacking in humour and depth, and is largely copied from European and US models of 20 years ago.

Well, I finally got back to the hotel to take a shower and snooze, and when I woke up, the Oz TV station had a program on about a jazz festival at Bond University (Sydney?). One of the bands had a percussionist who beat the Hell out of garbage can lids with railroad spikes, and he used a bowl (just like your mother's shiny large stainless steel bowl) which he filled with water, rolled the water around as he hit it from underneath, the whole thing held up close to the microphone. I'm going to sleep till noon tomorrow, walk the beach in the afternoon, and then catch a midnight plane to return to Alligator Creek.

Thus endeth the testament from your father in Bali.

I can't improve much on that.

There were NO DURIAN signs all over the city, major hotels and government offices. So of course I had to have some durian, which was supplied to me in the hotel restaurant as a tart, a flaky crust with a dimple of custard-like substance in the middle, sort of tasted like nut paste. Ah, but some Indonesian friends heard me talking about this, and during one of our excursions to volcanoes, we stopped at a road side stall selling durian. The Indonesians warned me that the smell of fresh picked durian was bad, but I didn't expect the aggressive smell of rotten fish/skunk/two month old dead horse. Despite the bad smell, if you take the two nuggets of fruit flesh out of the spiny outer capsule, put in a plastic sack and put it in the fridge, it doesn't smell bad. Later on in North Queensland, I had some in a pie, which looked like a lemon pie, but tasted more like cashew paste in jelly form. These vile fruits are extremely valued in rural SE Asia, people go upstream in canoes to find them in the jungles, where they ripen on very tall trees. The ones I saw were about the size of a small watermelon, and apparently there are a surprising number of deaths attributed to durian falling on your head. I later found durian cakes and durian soft drinks in a food market, along with dozens of other fruits I had never encountered, such as rambutan and sarak.

The hotel I was given was very posh, and the staff were very bossy. There was a gong outside my door, which a servant would hit loudly to signal that I was to bring out my laundry, and they wanted to polish my shoes (Ha, I was wearing flip-flops). A full 1.5 litre bottle of Glenfiddich was on the table beside my bed, and the little fridge was packed with Bintang, the local beer. I had a few sips of the whiskey, and it tasted like real single malt scotch. There was the usual worry about finding a useful breakfast, in a country where rice or noodles were usually eaten for the morning meal. But I'll be damned if they didn't have a completely separate cook for us honkeys in the dining hall. He would cook eggs in any manner, but I'm not sure what the bacon and sausage was made from. You couldn't get plain hot black coffee, it was all spiced, sugared and lukewarm. Part of the restaurant fronted onto the beach, and I could have chilli mud crab and chicken satay with Bintang for my supper, sitting on a bamboo-wicker chair on the beach sand, the liquid languid air making me drowsy.

My presentation on the relative environmental impacts of mining, urban development and agriculture was appropriate for this meeting. And here in the audience were Indonesians, dressed to the hilt in dark suits, vests, white shirts and neckties, whereas most of the Europeans and North Americans were decked out in gaudy Bali beach shirts and dress shorts. Part of my talk was a promotion for our Project TROPICS, and this was well received. Anugerah Nontji and Gani Ilahude (from LIPI, one of the main oceanographic institutions) were my scientific partners in promoting our TROPICS research program, and they guided me to the offices of important scientific and political leaders that had to be stroked in order to gain traction in Indonesia. There was the usual conflict between the environmental consultants (who usually worked for mining companies), and the oceanographic research scientists, who despised the dumping of urban, industrial, and mining wastes into the clear tropical seas hosting some of the world's most diverse marine biota. And there were the usual distrust between the nations:

the Koreans hate the Japanese, the Javanese hate the Chinese, the Balinese don't trust the Java people, and all the SE Asians consider the whiteys (US and Australian) to be gullible rich bumpkins lacking in social graces and wisdom.

I had a 27 November midnight flight from Denpasar to Sydney, 10,000 m high at 944 km/hr, outside air temperature - 39°C, and I'm writing a ballad about Bali. There had been a hotel lobby air band with three fellows with white suits, white guitar and bass, keyboard, singing US country western songs. They didn't even hide the tape machine and speakers, so it was obvious that they were not making the music live. But they were very good actors. I arrived in Sydney the next morning at 0830, drowsy and tired, but the ballad was finished. And on to Brisbane and Townsville and my darling Sheila there to meet me.

By mid-December, bore pump experts and electricians had got our bore running at 600 gallons/hr, so I could finally irrigate all our young fruit trees. Some of our trees were producing fruit, soursops, carambola and paw paws. A close friend of Jesse's was electrocuted while picking mangoes with a long pole, which touched a power line. Son Jesse took the Australian SAT test for admission to James Cook University. He did very well in quantitative skills and terrible in verbal skills. Sheila now has a new sewing machine and is making Bali shirts, tote bags, skirts, and a side pack for my coffee thermos and office homework books.

In late December 1994 I went down to Canberra to present the TROPICS package to the Australian Academy of Science. The AAS building is a very nice dome with an amphitheatre inside for big presentations, running multiple slide projectors on three big screens, pretty posh outfit. We were bunked at the University House, a rather posh English 'Oxford' style lodging and restaurant. Most of my Australian colleagues were there, as well as the major international research committees (IGBP, LOICZ, IOC) with reps from UK, USA and the European Union. The official Australian representative for LOICZ was Colin Woodroffe, who was in the hospital, having fallen from a window of a fourth floor parking lot. I must have done well, as some of the crusty oldtimer scientists came to greet me afterward, including legendary veteran PNG researcher John Chapell. So now I had officially pushed our Project TROPICS into the big leagues.

AIMS executive assistant Sue English later reported that Project TROPICS was in the Indonesian 'Blue Book', the list of approved government projects for the next decade. I had a hard time getting home, as there were several mechanical failures of the airplanes, a bomb scare, and some very bad weather. But it was a relief to see Sheila at the airport, taking me home to view her little palm Christmas tree, and I could stop traveling and hustling for a few months.

The parade of North American visitors to our home continued. In March 1995 John and Patricia Flannagan and their daughter Kirsten visited, taking command of our verandas and the warm tropical air, knowing that the Winnipeg snow was deep and the air temperature was probably -40°C. John was from Scotland, one of the first scientists at FWI who helped with the survey for ELA, and an international expert on larval benthic and aerial adult mayflies.

Also in March I hosted the first international Salt Flat Conference at AIMS. This is tongue in cheek, as it was simply a gathering of everybody that had worked with me on the Cocoa Creek salt flats over the past years, mostly JCU and AIMS scientists and their graduate students. Renagi Sam of Papua New Guinea got his Ph.D. working with JCU physicist Peter Ridd, designing an electromagnetic device that measured ground water salinity while walking on the crust of the salt flat. And of course we had a good feast and music party afterward at our home, Michael (guitar) and Laura (fiddle) Rosen being in especially good form.

We still had to attend to some diplomatic and scientific obligations to Indonesia and Papua New Guinea for our Project TROPICS. Nontji of Indonesia and Guy Kula of PNG wanted the USA and Australian scientists to tour their countries, and inform their science agencies of the proposed activities. So now we had to invent a schedule for a decade of research, committing ships, technical and scientific staff, institutional and national support for our program. This we did, with a lot of imagination and experienced guesswork. The US contingent already had NSF money for this project, and the Australians had to beg and borrow for travel money. But in mid-April 1995, half a dozen Australians and a lot of Americans travelled to Jakarta to begin our TROPICS 15-21 April 1995 road show. There were definitely some elements of an entertainment road show about our group, as we carried our own sound system, slide projectors, computers, bundles of prepared texts of speeches, pamphlets and prospectus documents, and official signed letters of intent from our own governments.

And so here we go again, leaving quiet Townsville airport in the early dawn, converting \$Aus150 to 1,000,000 Indonesian Rupiah, flying over Cocoa Creek Salt Flat, Cleveland Bay, and Hinchinbrook Channel (sites of my own research) in the pink early morning sun. On to Singapore, where the airport seemed to be full of Arabs in white robes and SE Asians in full dark suits. Our Australian contingent (Ron Szymczak, George Cresswell, Ken Woolfe, Steve Blaber and myself) were definitely outclassed, as we were in Aussie shorts, tropical shirts and flip-flops. We arrived in Jakarta with the airport and city in darkness, heat and smog, probably some problem with electrical power. Nontji and Gani Ilahude had sent two vehicles with LIPI drivers to collect us and put us into an Ibis hotel suitable for western honkeys.

Ron and I escaped our Indonesian guides the next day, 16 April Easter, and walked the Pasar (market part of the city), where vendors, stalls, carts, and cardboard boxes of stuff on bicycles for sale was loudly promoted. The open canals were rivers of sewage, feathers and tin cans, the air was richly scented with frangipani, the smell of decay, cooking, exhaust fumes, sweat and urine. Ron and I had lunch at the 'Press Club', a haven for expat reporters up two floors above the markets, where we ate some of largest prawns and mud crabs I'd ever seen.

Returning to our hotel by some miracle of navigation, we found Chuck and Debbie Nittrouer, Dave McMaster, Rocky Geyer, Bob and Josie Aller and a representative of their US funding agency, the National Science Foundation. Chuck had a sack full of TROPICS T-shirts, with our logo and 'Ski PNG' on the front. So our team was now together, and we had a short meeting to review our speaking tour plan. Amongst other things, there were complex financial, safety, and intellectual/scientific instrumentation property agreements needed by the US and Australian governments that we already knew were not appreciated by the Indonesians.

Within the next few days, we attended tribute sessions to high ranking Indonesian officials in oceanography (Aprilani), visited the department of fisheries down by the harbour, toured and spoke at two universities, toured a Marine Geological Institute and geotechnical lab in the mountains at Bogor, drove past tea plantations, durian and mangis roadside stalls, spoke at formal presentations at the military/Navy BPPT headquarters in Jakarta, and strutted our stuff at LIPI, the major oceanographic institution (which had sixteen affiliate institutions on all the Indonesian islands, including West Irian (Irian Jaya, or West Papua in Dutch times). Anugerah Nontji and Abduhl Gani Ilahude presented the 'IndoTROPICS' research plan for work on the Digul and Mamberamo River estuaries in Irian Jaya. Anugerah and Gani were also our protectors and advisors, and guided us in several official receptions to make sure we stroked the highest officials properly, and showed the right degree of respect. Gani was worried that we were all too young and lacking in status, as most US and Australian negotiators for research projects were senior officials ranking high in the government. Some of these meeting rooms were very posh, padded chairs and multiple slide projectors, with microphones and translation services, frequent coffee and sweet rice cakes on the balcony, the air heavy with clove cigarette smoke. Indonesian oceanographers were spoiled rotten, with eleven Republic of Indonesia Navy ships outfitted for oceanographic research. Their newest ship was the *Baruna Jaya 8*, built and outfitted for oceanography in Norway, larger and better equipped than any vessel in Australia. But we noticed that some of the multi-channel sonar and shallow seismic sounders were still in their shipping crates, as nobody knew how to use them yet.

The US and Australian oceanographers wanted to bring their research vessels into the Arafura and Banda Sea, Indonesian waters, but the officials at BPPT flatly rejected this, saying that we could use their vessels. Later, CSIRO physical oceanographer George Cresswell told of the difficulties of bringing Australian ships into Indonesian waters, as all data and scientific equipment was then claimed by Indonesian police. And then there was a long diplomatic and legal discussion about the performance of past fifty years of joint research agreements (COSTAI, LIPI-CSIRO, Indo-NOAA, ARLINDO, Snellius, most of them under the thumb of the Indonesian military). Some of the senior Indonesian officials were very well informed about us, they knew of AIMS research on genetic and medical uses of marine organisms in commercial enterprise, and they were determined to keep possession of intellectual property and ownership of discoveries in Indonesian waters that might lead to commercial or medical benefits.

Throughout this extravaganza road show, Suharto's Revenge (Bogor belly, stomach troubles) occurred mostly with the US contingent, and we were all exhausted by 21 April, for our final session in Gani Ilahude's office, where we agreed to exchange research personnel between Indonesian and US/Australian ship in the Port of Darwin. We also reviewed our performance and experiences in promoting the TROPICS Project, as we were going to do this all over again in Papua New Guinea in May. Djoko hauled us all in the big troop carriers to the darkened airport in the rain that night, and by dawn the next day we were home again.

It took me about a week to decompress from this Indonesian diplomatic expedition. I usually was invigorated after returning from research voyages, during which I would work long hours on sometimes dangerous activities on the rolling aft deck of ships. But this greeting and soothing the anxious concerns of Indonesian officials was not my normal task, and I found it frustrating and tiring, partly from lack of Bahasa Indonesian languages and partly from unfamiliar protocol.

But our little Alligator Creek property was an ideal place to deflate and rest, and our mango, pawpaw and lychee trees were beginning to produce fruit. Son Jesse returned from working with Burdekin River cane farmers and took up residence in our guest room. He was now enrolled in engineering studies at James Cook University, and was struggling with calculus, physics, chemistry and design drawing.

Meadowlark was stimulated by the AIMS arrival of post-doctoral fellow Stewart Walker to work with me. Stewart was a good violin, viola, and Hardanger fiddle player of Scots and Irish tunes, and we performed several public concerts with him. He had a disastrous year floating at JCU, without finding a good research topic. He finally came out to AIMS and took up the problem of the difficult measurement of low concentrations of mercury in several of our dated coastal sediment cores between the Burdekin River delta and Hinchinbrook Channel.

A week is about all I had to recover, as we packed up our TROPICS tour package prospectus, laptops, projectors, printers and maps in tubes for a similar diplomatic tour of Papua New Guinea. And so on 7 May we were up at 0430 to drive to the Townsville airport in the rain, exchanging Aus$800 for PNG Kina 715. We met a very tired USA contingent

(Nittrouer, Wright, Aller) and our Australian members (Ken Woolfe, Ron Szymczak, David Smith) in the Brisbane airport, and flew Air Niugini to Port Moresby. We were met by two technicians from the PNG Department of Environment and Conservation (DEC) who hauled us directly to the Lemana Hotel, behind locked gates, high barbed wire fence and four armed guards. There had been some violence between police and rascals over the past week, which was normal for Moresby. Our sponsors in PNG were Gaikovina Kula of DEC, Hugh Davies of the University of Papua New Guinea Geology Department, as well as the National Research Institute (NRI). Gai was especially good to us, but his mother had just died, his secretary had gone 'walkabout', and a big environmental report was due on some other project. Murray Eagle of Ok Tedi Mines had already given his blessing to TROPICS, and this was important influence on the PNG government.

Over the next few days, we gave our series of lectures for the various PNG government departments (Fisheries, Mines and Petroleum, Forestry, Environment, Wildlife, NRI, Foreign Affairs, Immigration, Labour, Customs and Quarantine) and an open lecture series at the University of Papua New Guinea. Hugh Davies at UPNG took us to his office for tea and shortbread biscuits, and he surprised me by indicating that his daughter Josie knew my son Ian in JCU music circles. Don Niles of NRI was our key contact for visas and ship clearances, and he wanted space on our ships for ethnomusicological researchers. PNG Immigration were worried about Indonesian participation, as they were essentially at war with them on the western border line. Several government and university people wanted to know how they could participate in our research, and how our work would benefit PNG. Since PNG had very little expertise in marine science, this was difficult for us to answer. Every evening the DEC technicians would haul us to a different interesting restaurant, Sri Lankan, Indonesian, Chinese, South Pacific (SP) beer, but no PNG food. They explained that PNG food was boring, mostly sweet potato, sago and fish.

We flew north over the Owen Stanley Mountain Ranges to the port of Lae, on the Markham River and Solomon Sea. We were hosted by Walt Benko of the PNG National Analytical Laboratory, who had a good clean modern analytical chemistry lab with competent technicians and instrumentation. But the lab was routinely shaken by earthquakes every week, and most of their work was on food safety and mine waste inorganic chemistry. The bed of the Markham River was huge boulders, and apparently these boulders moved down the steep slope into deep water during floods. These mobile boulders cut the telecommunications cables over and over again.

Our host had rented a near new 4WD Pajero for our excursion into the mountains, explaining that the word 'pajero' meant masturbation in Mexican Spanish. We drove through

rough roads towards the Ramu Sugar Plantation and Madang, fording shallow rivers, lots of hawks, eagles, cattle and horses, and roadside stalls selling fruit and sweet potatoes. We had two flat tires and the steering wheel became loose on this trip, and we had to pay extra when we returned this very dirty vehicle to the rental agency. We blamed it all on Ken Woolfe, who did most of the driving.

Back in Port Moresby there were more lectures, and a final feast with PNG friends at Tony Chin's Golden Bowl Restaurant, good spicy Hunan food. During this feast we got a bit silly and began discussion to alter our TROPICS logo to include the western Papua penis gourds, kundu, birds of paradise, and snow-covered peaks on the western border. We ate and drank until midnight, and were preparing to walk to our hotel. Tony Chin himself said that he forbad this, and he drove us to our hotel in his own troop carrier landcruiser with two shotguns in scabbards on the door. Chin was a Vietnam veteran helicopter pilot. We gave him a lot of our TROPICS T-shirts and prospectus papers for his future guests.

Our last official meeting was with Neal Jensen of Highlands Gold, who was planning a nickel/cobalt mine 80 km SW of Madang, and they wanted to dump the refinery tailings into Astrolabe Bay through a pipeline down to 120 m water depth. They had already done some CTD profiles to estimate the thickness of the warm surface water, and they proposed to slide the refinery tailings underneath into the colder deep water. As you will hear in future chapters, I was to get involved in the environmental impact of this project over the next decade. We all left Port Moresby on 13 May, tired but satisfied with our work in this difficult and beautiful country.

And so we had set up an international study group of coastal oceanographers over 1992-95, with informed government agencies and universities from Indonesia, PNG, Australia and USA. Now we were ready to organise the ships and begin the work...what could possibly go wrong?

Chapter 28. Busy 1995. Mercury and Cadmium, TROPICS Full Blast

Meadowlark now had Stewart Walker on fiddle, a real Scotsman. We performed at the Thuringowa Library for a good crowd in May 1995. This was our first performance of our original ballad 'Captain Jim Wardell's Story' and 'Oh, Rosella'. Jim Wardell was the Master of the *Harry Messel*, an AIMS research vessel, and he was in the audience when I sang it in public for the first time. He told me the story over tea in the wheelhouse of this little ship, as we were headed home in choppy water. All I did was put his words to rhyme and rhythm in ballad form. Sheila and I composed 'Oh, Rosella' for a tape of childrens' music. It is mostly an Aussie nonsense song. My home chores now included watering and trimming fruit trees in our creek valley, and by now we had homegrown soursops, lots of bananas, pawpaws, passionfruit, sweet corn, melons and pineapples.

Stewart Walker was also an analytical chemist who wanted to work on mercury in the Great Barrier Reef lagoon estuaries. Yasvir Tesiram was a JCU Honours student working with Michael Ridd and myself, and his thesis work was on cadmium in GBR sediment cores. We knew that mercury had been used to extract gold from crushed ore in the Charters Towers area, and that the guano and rock phosphate fertiliser used on the sugar cane fields contain some cadmium. Both of these elements are undesirable in excess in ecosystems and food products. So I organised a special sediment coring trip on the *R/V Harry Messel* for Stewart and Yas to get super clean samples, so that very low level contaminant and background Hg and Cd could be measured. Irena and I provided logistic services and operated the coring gear. We defungerated 3 June '95 for Hinchinbrook Channel, and Master Wardell was in full glory driving the ship into the mangrove creeks, quoting Joseph Conrad in his deep loud voice. Walker and Tesiram were funded in this Burdekin/Herbert River estuary research by the big new CRC, which wanted to identify threats to the GBR coral reef ecosystem.

During a lot of this special river mud research, the media was roasting the cattlemen and cane cockies for increasing erosion by bad land management. While this may have been true for selected properties, a lot of the agricultural people were deeply resentful that they were being accused of mistreating their land and moving soil/mud into the Great Barrier Reef.

I was invited to speak at several Herbert and Burdekin River Catchment Committees, where I tried to explain how we measured river mud input to the GBR, and how farmers and cattlemen could protect their land from erosion. Because I came from a rural upbringing, I was perhaps better able to reach these crusty old cattlemen and banana farmers. Several of the sugar cane farmers had already constructed a series of wetlands and dams downslope from their fields that slowed muddy water *en route* to the estuary. As is usual in these affairs, the ladies provided, and we had nice pastries, sausage rolls, coffee and cold beer afterward. I got to know several of the officers in the river drainage organisations, and contributed to their library.

On one of these speaking trips, a very large man walked up to me and introduced himself: He was Andrew Wood, who worked at a Sugar Cane Research Station in Herbert River delta. He was an ex-Kiap, a very quiet and strong person, and was very knowledgeable about the Western Province of PNG. He had a musical voice that fascinated me, and he sat down and wrote good advice in my notebook about how to behave in remote villages in PNG, and many names of obscure retired experts on the Fly River region of PNG.

In June 1995 AIMS administration was all in a flutter, reorganising the science program, appointing Charlie Veron as Chief Scientist and pushing for more involvement from JCU and CSIRO Marine. Charlie asked me to be a project leader, a third tier management role, which I tried to refuse, but he wrote down my name anyway. I was already an adjunct Senior Lecturer at the JCU Department of Chemistry, and I had about a dozen CSIRO, USA and Indonesian scientists involved with TROPICS. So I was held up as an example for what the AIMS Administration wanted to happen. I had submitted the AIMS portion of Project TROPICS to the management committee, which disturbed them greatly, as it was far too ambitious and expensive. But they approved research voyages with the AIMS research vessels to the Gulf of Papua and the north coast PNG Sepik River estuary. My colleagues down in Hobart CSIRO Marine jumped the gun and submitted the TROPICS proposal for *R/V Franklin* ship time for 1997-2000. I was very surprised to hear about two months later that this proposal was tentatively approved, subject to funding from the participating institutions. At this time, the subsidised fee for victuals, insurance and the use of the National Facility research ship was $30K per 2-3 week voyage, and we had asked for six voyages. This didn't cover the cost of transport of personnel and equipment to the ship, nor any of the post voyage analytical costs.

AIMS was justifiably worried about this expense, but my colleagues in USA, ANSTO, ANU and CSIRO came to the financial rescue before I got beheaded.

Sheila's older sister Mary visited us for a month in July/August 1995. As with all family visitors, there were lots of beach and Billabong visits, feasts and music parties on our pergola. Jesse surprised us all by getting top grades in his first term civil engineering at JCU. Once again, we realised that motivation was the key. In his spare time he was also driving cane trucks for the Alligator Creek harvester crews. It was unusually chilly and wet in August, with a big 110 mm rain event that flooded parts of town.

Reichelt, Chalker, Williams, Veron, Alongi and myself travelled to Perth in early September for largely political science policy discussions with Western Australian EPA and CSIRO staff, partly to bolster our planned AIMS substation in Karratha.

This Western Australian substation was a project started by our previous CEO Joe Baker, and it was finally coming online several years after his retirement. We flew up to Karratha for a very weird opening ceremony for WAIMS, with an audience of Hammersley miners, Dampier Salt and Woodside Petroleum and WA Conservation and Wildlife staff in their working clothes and hard hats. There was lots of wine and beer, fancy finger food, caviar, satay pork and fish and huge baked barramundi at the reception.

To get home from all this, we had to go through Perth, Sydney, Brisbane and finally Townsville, spending many sleepy hours inside airplanes and airports. I was reading a big book of Bob Dylan lyrics, which acted as a drug on my drowsy imagination. My notebooks show a complete lack of attention to science policy, and the beginning of several ballads about ghosts in the mangroves and Australian variations on the 'Sad-Eyed Lady of the Lowlands' and 'Desolation Row'.

And so 8 September 1995 was our 25th wedding anniversary, and my bright and shining Sheila was radiant. We never had a proper honeymoon, so we tried to declare the next few weeks a holiday. Two days later was my birthday, when I tried to insist that I was only twenty-nine years old. And so to continue my explosive life, the French exploded another nuclear bomb inside a cavern in Mururoa Atoll, French Polynesia. But our honeymoon plans didn't work right away, as I had to participate in a series of science popularisation entertainments in Townsville, something I liked to do. By the end of September there was a gap in institute activities, so Sheila and I took off in our little 1980 4WD Subaru station wagon, leaving son Jesse to take care of our house, parrots and orchard.

Busman's holiday, ok. We drove westward into the dry cattle country around the Burdekin Dam and Charters Towers. I was surprised to see that most of this dry country still had shrubs and trees, at least where it hadn't been cleared for agriculture or grazing.

Where the land had been cleared (with two caterpillar tractors with a chain between them), the soil was blowing away, there was little grass, cattle were gaunt and listless, and there was evidence of soil erosion and deep gully formation. The town of Charters Towers had been the nineteenth century hub of north Queensland riches and development, a centre for miners, cattlemen and the railroad. We visited a relic of one of the gold mining operations, the big Venus stamping mill (where the gold ore was crushed to sand size), the sand sorting/shaking table where quicksilver (mercury) was used to extract the gold, huge centrifuges and medieval-looking retorts for heating the gold-mercury amalgam to recover the mercury. Nearby were large pits where cyanide was used to extract the gold from deeper ore that was sulfidic.

Stewart Walker had given us a tip to visit the Inspector of Mines, Ross J. Thomas, in the Mine Registrar Office in Charters Towers. We walked into Ross Thomas's office without appointment or warning, and he kindly gave us the afternoon packed with precious information about mining history and the use of mercury in gold ore extractions. He explained that there were about 103 gold fields (shallow and a few deeper mining pits) in this region over the past century, and all of them used quicksilver to extract the gold from 1874 to 1892. Ross estimated that there was probably 10-15% loss of Hg in the retort evaporator, and maybe 75% recovery of the gold in the ore. After that time, most of the mine pits were deepened and had encountered sulfidic ores, for which quicksilver didn't work well, and the cyanide process (invented by two Scots brothers) was superior. However, some mercury was used up to the 1970 in small operations. The mercury was shipped into this region from Almaden, Spain, in 70 lb sealed iron flasks, many of which now litter the empty landscape of the upper Burdekin River catchment. Ross Thomas estimated that many hundreds of tonnes of quicksilver were imported for north Queensland gold mining, but the exact details are now lost due to the 1974 flooding of the Brisbane Mining Registrar Archives. We were startled to listen to a short lecture from Ross on the history of Australian tunnelling miners in World War I, and the origins of the egalitarian term 'Diggers' for most Australian troops. Years later we heard that Ross Thomas was an important impetus and technical advisor for a major film *Beneath Hill 60* about the 1st Australian Tunnelling Company's effort to dig underneath German bunkers and plant explosives, which they did repeatedly.

This information was valuable to Stewart Walker and myself, as Stewart had analysed a 3.6 meter long sediment core from Bowling Green Bay, near the Burdekin River estuary. Stewart found a large increase in mercury concentrations in sediment from 230 to 300 cm core depth. Mercury concentrations increased from a core depth 360 cm background concentration of about 20 ng/g dry weight, to over 400 ng/g at about 250 cm core depth. We had tentatively used radioactive tracers to date this core depth interval as 1870 to 1920, which was now neatly supported by the history of quicksilver usage in many of the gold mines in the upper drainage basin of the Burdekin River. There were some large spikes in Hg concentration after 1920, probably from flood events that eroded old Hg contaminated sediment along the river bed.

I can hear you now, muttering '...some honeymoon!' Hang on, it does get more romantic!

So Sheila and I drove further west to Hughenden, where the streets are lined with dust-covered utes with tool boxes and dogs in the tray, and the telephone wires were heavy with flocks of pink and grey galahs. This place felt good to me, like small town South Dakota. They had a good bakery, and we ate a wonderful meal of camembert cheese stuffed chicken, scallops Mornay, and custard apple pie at the Royal Hotel Motel.

The next day we drove past the Porcupine Creek Gorge on a dusty rough gravel road, through shrubby grasslands and alkali flats...one creek was labelled 'Poison Creek'. We had lunch in the big empty with red-winged parrots, galahs, bustards, emus and lots of hawks/kites. We stumbled into The Lynd, a very remote truck stop with bunks and showers, the world's smallest bar. Jenny Toohay was the boss lady here, and she wanted to know where we had come from, and how long we'd been on the road. After telling her we came from Hughenden, she shook her head and announced to the other truckers that...'These folks are in no hurry.' She served a limited menu of T-bone steak with a fried egg on top, chips, and pawpaw for dessert. If you didn't eat all the chips, a pet potoroo out back got them. There was a row of accommodation units, none of them locked, and Jenny said 'take your pick'.

We drove on northward in this dry dusty cattle and emu country, where wildlife was abundant around windmills, stock tanks and billabongs. Mount Garnet had a bakery too, and we filled our thermos with coffee and took off down the Gunnawarra Track, an old mining wagon road heading back east toward the coast, through the drainage basin of the Herbert River, not especially recommended for ordinary vehicle traffic. As we drove east, the trees became bigger, the grass was thicker and taller, there was more water in the creeks, and more birds and wildlife. As we came into the coastal mountains, we visited the upper Herbert River and the Blencoe Falls, a stupendous view that very few people see. It was good we were in a high suspension four wheel drive vehicle! The road became a twisting mountain road completely covered by arching rainforest trees, dark and gloomy in mid-afternoon, and I was sure we were lost forever. We came out of the rainforest on the coast to see Hinchinbrook Island in the misty evening distance, and I knew immediately where we were. We stayed overnight in the Hinchinbrook Hotel in Ingham, a little sugar town on the banks of the Herbert River. Our room was straight out of an early 1900s cane cutter history book, a creaky bed, a small fan, clapboard walls and cupboard with a white enamelled water pitcher and wash basin. But the hotel food and beer was good, and the next morning we got back home, where Jesse was busy vacuuming the kitchen floor and the parrots demanded out attention right away.

Back in the office in early October, trying to get into science mode again, I gave two lectures at JCU on the use of radiochemical methods in oceanography. But Jesse collected me after the second lecture, saying there was a big fire approaching our creek valley orchard and we skedaddled home to find Wayne Krogh and his local Alligator Creek fire brigade already back burning just up the hill from our home. We were very grateful to have this fire protection support, and we made sure to contribute funding to the brigade over the next decades. Jesse borrowed a big tractor from neighbour Rob Lyons, with a front end loader and backhoe on it, so we could massacre and burn the big china apple trees down beside our creek.

In late October Meadowlark was back in operation with Stewart Walker on fiddle. We were booked to perform at a Mary Lou Schonfeldt Townsville Community Music concert series, a Lucinda Queensland Sugar Conference, the Townsville Folk Club, and the Alligator Creek Bowls Club Annual Dance. Our gig in Lucinda allowed us to stay on for some more holiday explorations of the Herbert River catchment. We drove up a steep curvy road to see Wallaman Falls, some 300 m vertical drop. We explored this Stoney Creek country, seeing cassowaries and emus, and overnighted at the Forrest Beach Hotel, where they served wonderful plates of coral trout and pork chops with blue cheese melted all over it. It turns out they had a fancy chef from Melbourne, who like to take his holidays in the Forrest Beach kitchens. We visited Helen Sturmey and Brian McCarthy at their Abergowrie horse ranch on the upper Herbert River, where saltwater crocs had been known to eat a small horse. We got lost exploring the Bularoo Swamp along the coast, SE of Ingham. When the road got very bad, the roadside sign said 'The Orient, any further at your own risk.' But of course we went on further, through some salt flats, mangroves and silky oak beach ridges. The beach went on forever, empty, no cabins anywhere, so we went skinny dipping in the warm water, snoozed on the hot beach sand, had a fire going for our supper and slept on foamies in the Subaru under mosquito nets. Coming back out of this swamp, we got very lost, twice stuck in mangrove mud, and arrived again at a beach (a sign said 'Rex's Water Hole') where a very large fat naked man was throwing a bait net for prawns. His wife was back in the trees, screaming at him to get back to the shack and put some clothes on. We then drove generally SW, had lunch at the Trebonne Pub, where the cook was on holidays, so we cooked our own food at the back of the pub. We drove southward past Mount Fox on rough washboard roads, sparse ironbark and gum trees, this country being badly over-grazed in this long dry season. This was now part of the Burdekin River headwaters, and there were lots of gullies, washout roads and terrible soil erosion due to over-grazing. But there were some deer in this badlands country, heaven knows what they eat. We saw herds of peacocks, apostle birds, bony gaunt drought-master cattle, and a lot of dry sand in the creek beds. We got into Charters Towers late, and had a midnight 'two eggs up on Whiskey Toast' and cold beer supper. We were grateful to our good son Jesse for taking care of our house, parrots and orchard during these 'bumbles'.

By late November 1995 Jesse was doing very well in his engineering classes, learning CAD, big and little electronics, chemistry and computer science. He also became a master at touch powder, which alarmed his mother the first time it was demonstrated. This is a homemade explosive liquid (iodine and ammonia cleaning solution) which can be squirted onto a smooth surface, hit with a stick, and it will pop loudly. It was very hot and dry, and the first lychees were available from the roadside stands. It was peaceful and warm, and I was reluctant to pack my bags one more time for meetings at the State University of New York on Long Island, and a big sustainable development meeting in Washington, D. C. And since I was in North America, I would also visit relatives in South Dakota and Ontario. Sheila didn't want to come, but she provided gifts for her family and demands for stuff to bring back to our Queensland home. She also left little notes in my journal and suitcase clothing, asking that I get someone to iron my white shirts before my presentations, and to avoid carrying my jackknife when boarding airplanes.

Leaving tropical Townsville at 0650 on 23 November, tired already at Cairns, Honolulu, smog covered Los Angeles, overnight in Salt Lake City (the hotel restaurant had lime jello on the dessert menu and the TV was fixed permanently on 24 hrs of Clint Eastwood films). The next morning we flew over the muddy brown Great Salt Lake and Rocky Mountains to Rapid City, South Dakota, where my brother Obie and his daughters Marcy, Donna and Julie were there to meet me. Stepping out of the door of the airport, I was blasted with the smell of prairie grass, now in winter brown mode, but still tall and waving and fragrant. It was the smell of a different planet. And so we go across the open rolling prairie to Obie and Annie's ranch house, north of Philip, 'where the deer and the antelope roam.' And there was a huge roast turkey with stuffing, and pumpkin pie with heavy whipped cream for desert. It was deer hunting season, so all the young fellows were out in their pickup trucks hunting the cottonwood and willow creeks. We could watch pronghorn antelope parading along the ridges south of Obie's house, but they were too clever to let us get close enough for a shot.

The next day Obie took me down the Interstate Highway east to Murdo (with a minor stop to get some Nibs, sadly lacking in Australia), where my brother John was living in the England family home. Here were a headspinning collection of relatives and old friends. John's house is large, but it was full to the brim with his four kids, some of Obie's kids, Gerald's family from Arizona, my mother's cousin Toadie from Minneapolis, my uncle

Bernard and aunt Marjorie and their three kids, and uncle Gerald and Joyce from down at the ranch on the White River. Everybody wanted to pick my brains about life in Australia. Sheila had wisely provided two small photo albums with photos of Ian and Jesse, our house and orchard, and us on the AIMS beach. Toadie desperately wanted Sheila's lemon curd recipe and some black opal; she was about ninety years old and very bossy, but everybody loved her, a remnant of the Butterfield family of my grandmother England. She declared that I was a dead ringer for Elihu James England, my mother's grandfather. My uncle Gerald was old and thin, but he was well read on climate change and Australian agriculture. He felt sorry for the gaunt drought master Brahma cattle standing on the AIMS salt flat in my photo album. He was a civil engineer himself, and he was delighted to hear that Jesse was *en route* to a civil engineering degree. Gerald confirmed that Sheila's plesiosaur was still safe up in the foothills of the White River. He had dug back a bit, hoping to find the skull, but all he found were vertebrae. Gerald and Margie had some new information about our ancestor Simon Brunskill's parents in Darlington, Wisconsin, probably having come from farming/mining country of the upper Yorkshire dales in England.

All of this took place while a giant roast beef lunch with hot fresh baked buns was provided, and then homemade vanilla ice cream with crushed garden strawberries, and then more cake, tea, coffee. We all trooped outdoors to walk around, visit the renovated barn and grainery, and feed grain to the calves. The house has town water now, new foundations and a modern furnace, and the basement had been cleaned up for bedrooms and a TV room. It had been a long while since all of my brothers had been together, so there was a long photo session with John, Gerald, Obie and myself trying to look intelligent.

It was sobering to go down into the basement to sort out my father's personal stuff. I hadn't been home for my mother or father's funeral, so my brothers had just waited for me to come to divide up the personal effects: My father's high school sweater with sports letter, a framed University of Iowa Mechanical Engineering degree, fraternity/society pins, watch chain and Mason's fob, a complete set of his sons' basketball schedules and tournaments, his old Iowa engineering textbooks, and vet/agricultural/animal husbandry books. I wanted to have his spurs, and an old pair of batwing chaps that was hanging in the barn.

My mother had anticipated this very event. She had all her stuff in labelled cardboard boxes down in back dark part of the basement, with notes on who should have each item. Some boxes were also labelled for the families of her brother Raymie England, and Gail England got these boxes of photos, old World War II letters and family artefacts. Mother specified some of the books, kitchen doodads, and photo albums were for our sons Jesse and Ian, and she had saved every letter I ever wrote to her from Augustana College, Cornell University, overseas trips, and our Canadian and Australian days. There were book shelves to the ceiling, much of it old hymnals and song books, and school library books that she had rescued from small town prairie schools that were closed for lack of students. There were little sacks of sewing supplies, sock knobs, beeswax cakes, dolls, rolls of fabric, buckets of buttons, rolling pins, labelled copies of photos of distant ancestors, pioneer housewife gadgets and heirloom kitchenware from Grandma England pre-war days. Can you imagine my mother's four sons and grandchildren, sitting around in a circle, passing these boxes around, tears rolling down their cheeks?

Sheila and I had given her a 'Hot Crotch' tea cosy from our time in Leningrad, and I recovered this for Sheila. I sorted out a suitcase full of items for Ian, Jesse, and Sheila, and gave Meghan $120 to mail it to Sheila in Australia. After another giant supper, they hauled me off to the Murdo Auditorium where D. D. and the Fayrohs were playing country western music for a town dance, the cowboy boots galumping across the cornmeal strewn basketball floor. It was a chilly windy evening and I wore my father's wool dress coat. I had never seen him wear it.

The next day, John's sons Bill and Dan got two deer before lunch, now hanging in the granary ready for skinning. The weather was foggy, with sleet, rain, and glare ice on the roads. After a typical South Dakota lunch of roast beef, roast ham, baked potatoes and fresh baked buns, Gerald gathered his family and myself to drive to the Rapid City airport in blowing snow. Tears and sad goodbyes all around, even the tough deer hunters. Toadie drove herself back to Minneapolis, and Marcy and Donna were thinking of driving to Alaska to get higher paying teaching jobs. At the airport 0500 on 27 November, Gerald and I enjoyed cigars, admired the big spittoons in the café, and had big slabs of apple pie with whipped cream and coffee, because the Delta Airlines computers were down, all flights delayed two hours. The airplanes were being repeatedly de-iced, and it was white-out conditions all morning. I had to go back to Salt Lake to fly to NYC for another Project TROPICS organisational meeting, to keep the US science contingent and funding agencies happy.

I arrived at JFK airport at four pm and immediately got lost trying to find the luggage area. A man loomed out of the chaos with a big sign saying 'Brunskill' on it, and I was rescued by a limo driver, frizzy haired and in a t-shirt. In true NYC fashion he steals a luggage cart from the porters and whips me down noisy tunnels to his shiny black Lincoln, illegally parked, and drove through terrifying six lane traffic for two hours to get to Stony Brook on Long Island. In the gloomy dark we found my hotel, where Aussies George Creswell, Dennis Mackey and Ken Woolfe were waiting for me to order supper (good crab cakes and Samuel Adams ale, apple crisp).

The next day was for jet lag recovery, Ken Woolfe and I toured the book shops and coffee houses and watched the Long Island fishermen work up the salt marsh creeks. It was a chilly New England early winter wind and my father's long wool coat felt very good, smelled of cigars and prairie grass. Chuck Nittrouer invited the whole group to his house for supper. He and Debbie had tickets for a Red Clay Ramblers concert and a performance of *Stomp* in Greenwich Village. Over the next three days, at Chuck and Bob Aller's Marine Science Division of the State University of New York, we tore TROPICS apart and put it back together, planning ship tracks, personnel, equipment, sampling locations, sample and data sharing, funding limitations, and even label codes for sample archival storage. The purpose of most of this was to parade our collective expertise and determination in front of the various funding agencies, who had representatives at this meeting. On the advice of US funding agencies, we almost gave up on sharing ships and personnel with the Indonesians, as there had been trouble in previous agreements. We would work until nine pm, then assemble at our hotel for excellent lobster pie and beer and tell outrageous stories about research ships in trouble.

Our son Ian MacGreggor was at the University of Toronto, post-graduate, studying with Russell Hartenberger of the legendary percussion ensemble NEXUS. I had a few days free until my next meeting in Washington, so I flew up to be with him on 30 November. It was much colder, snow and ice everywhere, and Ian was there to meet me with a heavy backpack, tabla and other percussion gear. He got us onto a bus for a hotel on Jarvis Street, apparently selected by Sheila's sister Noddy and her husband Ted. Ian and I had fun at Thai restaurants and a visit to Pimblett's Pub, where cats and dogs sleep beside the fireplaces, Elvis ceramics and velvet painting on the shelves and walls. Ian of course wanted to hear all the news about his South Dakota cousins, and was practicing tabla at the same time. We visited Allan Gardens Conservatory, close to Ian's old apartment, where prostitutes, druggies, drunks, panhandlers and abortion clinics abound. Ian knew where to get cheap CDs of my and Sheila's favourite musicians. I got to visit with Russell Hartenberger, Ian's teacher and hero. We also visited Ravi Naimpally, Ian's tabla teacher and mentor, for a long session with keyboard and tabla, *masala chai* (Indian tea with cardamom, fennel, ginger and maple syrup) and *gulab jamun*.

Sheila's sister Noddy and husband Ted, and brother Bill and wife Carla and their kids arrived at our hotel in the adjoining rooms on 2 December, and I think this was a blow-out party for them. They brought some rural road stand cider and sacks of local apples from north of Georgian Bay, and the room suddenly smelled much better. Noddy's son Peter, who was large and computer savvy, quickly fixed my hotel room internet connection to my little laptop so I could email Sheila. Noddy's daughter Ana was quiet and shy, but she sidled up to me, asking about our parrots back in Queensland. Bill and Carla's boys were there. James was four years old and demanded to be thrown up in the air, and to walk on the ceiling. Little Thomas was only ten months old, but he took a liking to my whiskers. Noddy was very bright and happy, and wanted to hear about her sister Sheila's tropical house and activities. Part of the reason they had specified this hotel was that it was near the huge Eaton Centre, which they enjoyed visiting. The Mr Greenjeans restaurant was terrible, noisy, poor service, and Ian and I wanted to leave, walking through sex joints, hookers and tattoo parlours to our hotel. The next morning we had a very odd Eaton Centre breakfast with everyone, snapped a photo of all of us in front of a hotel Xmas tree with a penguin, and they left to return to their homes, half a day's drive north and NW of Toronto.

Ian and I then toured the art galleries and several music stores with exotic instruments from Asia, mainly Balinese gamelan orchestra stuff. We had coffee and croissants as I was waiting for my airport limo, and Ian was in tears, trembling as we hugged and said goodbye. I had given him my father's long wool coat to keep him warm in this Canadian winter. And off he went, with his tabla case, jugs of cider and a small sack of Georgian Bay apples. Ian needed reinforcement and intellectual support for his chosen career path. His whole world was music, percussion, creative expression, experimental music, and he had no interest in money, city shopping or the normal comforts of life.

And so here I am in the Toronto airport again, hot and sweaty from hurrying, only to wait in long lines, hungry and tired. Christmas carols were being played over the announcements for gate numbers, departing and arriving flights. Back to the real world I guess. I got a small cold airplane with no food to fly to Washington, D.C., arriving at nine pm in the foggy moist darkness. Limo to my hotel, and I'll be damned if the hotel restaurant wasn't serving some of the best damned southern deep-fried chicken and hot baking-powder biscuits with gravy I've ever eaten.

I was in Arlington, Virginia 4-6 December 1995 for a conference on Sustainable Development and Global Change. I had been outspoken at AIMS and several Australian government planning meetings about the lax attitude toward some idea of being able to feed and water ourselves, and how this was going to change with accelerating greenhouse gas warming of the atmosphere and oceans. So they sent me to this meeting, and I had a thirty minute lecture on some Vallentynian parameters for tropical, temperate and polar regions (agricultural land area vs continental shelf area, human and livestock population vs continental shelf area, and coal/mineral/petroleum mining output/continental shelf area). My analysis of UN data showed that countries with large areas of agricultural land and high human populations had the worst problems with continental shelf pollution, fisheries, tourist industries and greenhouse gas emissions. In other words, the human population exerted a stress or pressure on coastal seas, and this was going to get worse, not better.

There were many top level speakers at this conference. Bert Bolin, the co-founder (with Gro Bruntland) of the first 1988 Intergovernmental Panel on Climate Change, was the lead speaker, assuring us that CO_2, CH_4, CO, N_2O had been rising in the atmosphere and oceans, most rapidly since World War II. Soot, acid sulphate particles, biomass burning, water vapour and ozone all acted to confuse the issue in model atmospheres and oceans. He showed that it was now warmer than at any time in the past two thousand years, and that the cost of no action (to control greenhouse gas emissions) was much greater than emission reductions now, and early mitigation and adaptation for the next several centuries. Most scientists believed this brave prediction, but the media and political creatures in Congress ignored it. Peter Tans of NASA tried to explain the impact of greenhouse gas emission on the carbon cycle, but it was too early in the game to see these results in crop productivity, the hydrological cycle and food security.

Several panel discussions were rather heated, because they had officials from government, agricultural and food industries, the manufacturing and automobile industry, and the coal/oil/ electricity generating industry. One speaker from General Motors claimed that 'what's good for GM is good for the USA'. Several agricultural speakers claimed it would be possible to feed nine billion humans in 2050, if only we had more fertiliser, pesticides, and no regulations on their usage. Some of the big energy consultants (MIT, Batelle) gave serious papers on carbon sequestration, recycling of carbon, and other technologies that would allow continued use of coal and oil, without once mentioning solar, wind, geothermal, nuclear or fusion energy production.

It appeared that wealthy countries could use technology to modify climate change impacts, but poor countries, especially in the tropics, would have little chance to avoid reductions in food supply, increased diseases and rampant population growth. It was shown that draconian restrictions on coal and petroleum fuel combustion would be necessary to avoid CO_2 concentrations rising above 400 ppm, a level they hoped would never occur. [At that time, atmospheric CO_2 concentrations were 380 ppm, already about 100 ppm higher that pre-industrial atmosphere.] Several speakers gave accounts of how a Chicago carbon credit scheme, or carbon tax (US$5-125/tonne carbon emissions) might work, but there were shrieks of outrage from the industrial participants, claiming most countries would go bankrupt under a carbon tax. It appeared that the success of the Industrial Western Nations since about 1750 was dependent upon cheap coal and oil, and the industry was not about to take responsibility for dumping its gaseous wastes into the atmosphere, like some giant rubbish can. Some industrial consultants claimed it would be cheaper to reduce emissions in 2010, but not now. Remember, this meeting was in December 1995. There was a dreadful silence after some of these simple facts were displayed on the big screens.

We were treated to two days of satellite technology displays showing planetary zones of productivity, water deficits and excess, some of the factors promoting the spreading of human, livestock and crop diseases, and the sudden melting of temperate and tropical glaciers. We heard reports on Kansas prairie and New Hampshire hardwood forest experiments, where additional heat, more and/or less precipitation was varied on small watershed plots. Forestry scientists reported that eastern USA and northern Europe were experiencing rapid changes in forest species and growth rates. There appeared to be an increase in the area and frequency of coastal ocean 'dead zones', red tides, and malaria/cholera/tuberculosis around the world, almost all of these ills associated with large human populations and dense agricultural land use. Many SE Asian, Gulf of Mexico, and Mediterranean coral reefs were crumbling away, bleaching, shedding zooxanthellae, and were infested with new diseases not seen previously. We were presented with detailed accounts of the consequences of 'weak sustainability' (technological attempts to maintain food and water supply, such as concreting wetlands for parking lots, and engineering wetlands somewhere else), or 'strong sustainability' (action in favour of future generations and natural biodiversity). Good presentations on the problems of accelerating populations in the megacities, most of them on the coast, and the food and water footprint necessary to support them. This was a weak solution to the population growth problems: stack them up in vertical skyscrapers close to jobs, where they are easier to feed and water, and keep them away from fertile soils of the agricultural zones. The beginnings of geoengineering proposals were discussed here, suggesting fertilisation of the oceans with iron filings, shooting sulphur into the atmosphere, or giant reflective screens in space, for the purpose of temporary control of atmospheric greenhouse gases and temperature. IPCC members reported on similar meeting results in Madrid, Beijing and Rio. Many global science program leaders reported from Nairobi, Costa Rica, Pacific Island nations, Brussels, Paris, Moscow and Stockholm.

At one of banquets I sat beside an Indian geochemical scientist, Haroon Kheshgi, originally from New Dehli, but now working for Exxon as a climate change modeller. He wanted to know how much organic and inorganic carbon was annually deposited and buried on continental shelves of the world. Wow, good question, and right in my field of expertise. Furthermore, he wanted to know why industrial carbon dioxide could not be pumped into the deep sea, where it would stay forever in the cold water under high pressure. I took both of these questions with me, and I eventually answered them via email over the next year.

He apparently had asked the same questions of dozens of people at that meeting. The amount of organic matter buried per year on planetary ocean continental shelves is about 0.68 Petagrams C/yr (Sarmiento and Gruber, Broecker). Carbonate carbon, mostly as biogenic skeletal debris, piles up at a rate of perhaps twice that of organic carbon (Inglesia and Milliman) during interglacial times. Deep water in the Pacific, Atlantic and Indian Ocean circulates on an approximate time scale of about 1500 years...that is, sinking water in the polar North Atlantic Ocean and polar Southern Ocean migrates around the world, and eventually rises to the surface in the North Pacific, there it releases most of its accumulated carbon dioxide. Therefore, pumping carbonic acid in the deep ocean only postpones the consequences of rising greenhouse gases in the atmosphere for 1000-2000 years, an unwelcome gift to future civilisations, if we last that long.

Throughout this very good science policy meeting there were at least three major camps of people: 1) the industrial, coal, oil, economic and financial consultants wore dark suits with white shirts, dark neckties and polished shoes, and they carried immaculate leather briefcases with brass locks, 2) The academic scientists, NGO greenies, and wild-eyed climate changers were dressed in running shoes, open collar shirts, occasionally with sports jackets with pockets full of pencils and note paper, and little backpacks slung over their shoulder, and 3) the social and political scientists, an intermediate force observing the policy battle. There were some very useful headlong confrontations between the greenies and the industrialists, where it became clear that the greenies were interested in ethical and moral choices, alternate pathways to good lifestyles, and avoidance of war and pestilence. The industrial camp appeared to be concerned about financial, economic, and political stability, from which they asserted all life comforts and benefits are derived.

I had to admit that there was a grain of truth in both sides of the conflict. The social scientists were trying to teach us elementary language and logic gates for communication with the general public. Much of our language was opaque, not useful to the media, and didn't have the 'hook' needed to catch the fifteen second blurb on TV news. They explained that most non-scientists did not know how flower or tomato greenhouses work in capturing solar energy/heat, couldn't appreciate the importance of extinction of obscure species of frogs or bees, and did not associate their household electricity with burning of dirty coal someplace else. They also explained that most voters did not know the climate change position of their congressmen/women. I was thrilled to be in the presence of so much high level expertise on these difficult subjects, and also in despair because there seemed to be no simple scientific pathway to change the behaviour of governments, voters and industry. My own research institute had no policy or action that related to these issues for the coastal ocean.

After three days of this high level thinking, my head was spinning and in need of relief. I had been in internet correspondence with a Murdo High School classmate, Patricia Peck, who was now Dr. Pat Gossel, Science Curator in the Smithsonian Institute

(American History). She was a specialist in the history of microbiology, but had written major articles on the birth control pill, the ozone hole and climate change. We met one evening in a local restaurant, and she was matronly, white-haired, scholarly, not at all like the classical pianist (student of my grandmother England) and jitter-bug dancer I knew in high school. Her mother Elsa was one of my mother's best friends when I was young, and her father Harold went to the University of Iowa with my father in the 1930s.

The next day I took the train to the Smithsonian for a guided tour of Patricia's empire. There were rooms full of early microbiological lab equipment from the days of Salk, an iron lung, a history of dietician's recommendations on what's good and bad to eat, a Los Alamos/Hanford nuclear reactor control panel, part of Enrico Fermi's Chicago football field subterranean reactor pile, several 1950s bomb shelters, research balloons for atmospheric work, George Washington's uniform, a display of Duke Ellington scores and tapes, Bill Clinton's tenor sax, Dizzy Gillespie's trumpet with the vertical bell, a history of computers from Enivac to the laptop, a history of black people and women in science, and a massive collection of folklore and folk music. Patricia took me to lunch at 'The Castle', an old administrative tower building of 'The Gentleman Scientist Dr. Smithson of Oxford'. I visited with some of Patricia's colleagues, some of them from Cornell University, one from North Dakota. Over a good supper, she told me that she had uterine cancer, and not much time left. Indeed, within a year, she was dead.

In the taxi back to my hotel, I went past homeless people sleeping on cardboard mats over building hot air vents, t-shirt booths, panhandlers, hookers and AIDS syringe depots. In the hotel lobby was a magnificently polished grand piano, madly playing Christmas carols with nobody at the keyboard, the keys being pressed by some electronic madman. There must have been a security scare, because there were armed guards in the elevator and in the corridor of my fourth floor room.

My next hurdle was to promote Project TROPICS at a National Science Foundation 'Future of Marine Science' meeting here in Washington. It was in my hotel, and I was a bit startled to hear that NSF actually owned a large interest in this hotel. Our meeting room had large framed photos of NSF oceanographic ships, ice island camps and maps of locations of current ocean science programs. For the next two days, we poured it on thick in front of a special group of NSF advisors, who would probably have a vote on how much money we might get for our research program in the coastal shelf of far distant New Guinea. Our small audience was composed of NSF officers responsible for ships and logistics, biological, chemical and physical oceanography, earth modelling science, and several international affairs specialists. Nittrouer, Aller, Sternberg, Kineke,

Geyer and Kuehl all gave their spiel, partly going from their huge Amazon experience toward PNG.

Our TROPICS program was designed to illustrate the century to millennial scale ecological and geochemical processes that move organic matter, the chemical elements, and sediment from the mountains to the sea. I was the only one present with muddy hands and photos of the PNG mangroves, jungles, giant rivers and the coastal seas, and my Hobart colleague George Cresswell was there for physical oceanography of the Coral Sea. We were especially grilled about our previous experience with Indonesian oceanographers, and whether or not we had access to their fleet of Navy operated oceanographic vessels, and whether or not USA and Australian vessels had permission to work in Indonesian waters. Cresswell knew more about this, and in general he said it was better to use their vessels, and avoid bringing our equipment and ships into Indonesian waters.

And then suddenly it was over, we all shook hands, packed up our laptops and suitcases, and headed for the airport. A group of us had a really good supper of bratwurst and beer, followed by excellent espresso and croissants with dark chocolate filling, and that wonderful tingle of international science conversation.

And away we go on Delta Airlines, with good views of the Capitol building, the Lincoln Memorial and Washington Monument, and I slept through the evening as we crossed the Appalachian Mountains, to Cincinnati, St. Louis, Kansas City, Denver and the Rocky Mountains, Las Vegas, and finally Los Angeles, where I was stupidly tired and desperately in need of a beer. At this time the LA Qantas Club had two levels: Ordinary Snob and SuperSnob with gold credit cards. Neither were very useful, as what I needed was a shower, some fruit, pastries and tea, a smoking room, and a nice sofa so I could take a nap. But no, here we go again, on 7 December 1995 the 2150 Qantas 12 flight took thirteen hours to Sydney, there were howling babies, a good jazz channel on the headphones, I slept for most of the time and the eighth of December vanished. At 0730 in Sydney, there was the usual hassle with customs and my laptop, but I rechecked my luggage to the domestic terminal for flights to Brisbane and Townsville, arriving 1215, and there was my brown smiling Sheila, and the warm moist perfumed air of the tropics. She stuffed a peeled lychee in my mouth and took off my necktie.

And so we drove home in the hot sun and went directly into our neighbour's swimming pool, where the *Poinciana* trees and *Bougainvillea* vines were in brilliant orange and red flower. The contrast with the grey chilly skies of New York and Washington was exploding in my head. Sheila grilled me about my meeting with her sister Noddy and brother Bill, Ian MacGreggor and my Murdo family. We picked several big sacks of mangoes, and then more rain came down in buckets. In mid-December the famous AIMS Christmas Party was held in the workshop, on the theme of nuclear disaster. For reasons unexplained, I got a book award called 'Magic Pudding', which refers to one of the iconic Aussie pieces of literature.

Our CEO Russ Reichelt asked Kathy Burns and myself to review Canberra deliberations on approvals for a mining company to dump 30,000 tonnes/month waste into the sea east of Tasmania. This mean reading a 20 cm high pile of EPA and CSIRO reports, and consultants' documents for Pasminco. The mine and refinery waste was mostly iron oxides (jarosite) spiced with a suite of interesting heavy metals, not likely to be beneficial to fish or bottom animals. We flew down to Melbourne to meet with the proponents, the regulators and a few competent scientists who had done some shallow seismic sounding of the seafloor. The main reason for dumping waste in the sea was the easy cheapness and the escape from responsibility of taking care of the waste in tailings ponds on land. Although there was a good fishery in this region and benthic biology was diverse, Pasminco provided jobs for the region, and could not be denied the right to pollute the coastal seas. I left this meeting with a very bad taste in my mouth.

Ah, but back home it was very hot (44°C in the afternoons) and humid, and I took some time off to replant part of our orchard that had been burned in a grass fire in November. We planted more Bowen mangoes, macadamia, blood oranges, cashew and some sour Spanish oranges for marmalade. We also liberated a lipstick palm from the nursery, and it got named 'Pedro the Madagascar Redneck Palm' by Jesse the Christmas Elf. Pedro was hung with South Dakota and Ontario heirloom Christmas ornaments, in hopes of St. Nicholas visiting. Jesse sewed a Christmas stocking for Charles the Cockatoo, with big parrot claws instead of toes. This stocking got filled with parrot toys, pencils and erasers, all of which he would destroy within the day. The day before Christmas, Sheila and I went to the AIMS beach for skinny dipping in the big waves, and Sheila almost lost her glasses. Christmas morning breakfast meant Jesse and I had to cook French toast for Sheila, and we had mango/banana pancakes, bacon, sausage, eggs, toast and coffee. After opening our presents, we proceeded to the neighbours' swimming pool to rest on Aquaduck in the pouring warm rain.

We had a good house party for my AIMS Marine Geochemistry Group, with roast duck, macadamia ice cream, macadamia liqueur and some Meadowlark music. On New Year's Eve Meadowlark was hired to provide a short segment of a stage show at the big Civic Theatre. For once we had proper dressing rooms with makeup artists, shower and racks of costumes. Sheila went wild and put on some lipstick. I was startled at this. I hadn't seen my wife with lipstick before. But that was ok, when we went back home to Alligator Creek she put the Russian Tea Cosy on her head and we danced to a CD of Bob Wills and the Texas Playboys. Later, we watched the midnight New Year's fireworks from nearby Melville's Bull Ride from our veranda, while sipping tea and eating mince tarts.

As my dear wife keeps saying, we led a protected life.

The outside world went on its merry way, providing bizarre improvements and deathly terrors.

Chapter 29. Sadness In The Family, TROPICS Bismarck Sea Cruise, and Herbert River Expeditions, Cromarty Wetlands and the Pat Meehan Family, NW Shelf Cruise on the Franklin

The year 1996 didn't start off very well. Sheila's brother Alistair phoned from Winnipeg with the sad news that their older sister Mary had terminal cancer. Sheila and Jesse got air travel tickets to fly to Winnipeg two weeks later. At this time, son Ian was in Toronto, getting ready to return to Calgary. He planned to change his schedule so he could be with his mother and Jesse in Winnipeg.

I stayed home to take care of Charles the Cockatoo and my orchard, which was partly under water. I was scheduled to lead environmental discussion groups with the sugar cane industry in Ingham. We were having a good wet season, over 300 mm had fallen in just a few weeks, our little tributary creek rose above the banks and flooded my orchard, washing away my carefully accumulated mulch. But many of our fruit trees were obviously happy with all this water, and soursops, mangoes, paw paws, bananas, sour oranges and lemons were ready for picking all at once.

Mary died in the Winnipeg hospital on 3 February 1996, with all of her brothers and sisters around her. It was -45°C in Winnipeg, grey and dismal, much like the faces of her sisters and brothers. Jesse went down to Niverville to visit with his old high school classmates and found most of them as usual in the Chicken Chef coffee shop; not much had changed down there. *En route* home they visited with Ian in Calgary and Banff, with some spectacular winter scenery of the Rocky Mountains. I picked up Sheila and Jesse at the Townsville airport on 16 February. Both of them were pale, tired, quiet, and murmured about how nice it was to be warm. They quickly changed Canadian clothing to shorts and flipflops, and demanded mango fizz when we got home in the pouring rain. Despite the rain, Sheila quietly began watering her yard plants, even though we had received over 500 mm of rain in the last month. We all noticed that we were now calling Alligator Creek our home.

Andy and Betty Hamilton visited in March 1996, after attending a POP meeting in Manila, Philippines. Andy had been with us for the very early days of creation of the experimental lakes area in Canada during '67-'68. Andy was a midge fly (chironomid) taxonomist and ecologist during that time, but had long since moved to Ottawa where he was involved with the International Joint Commission and other Canadian government environmental agencies, trying to interpret ecological logic to politicians and lawyers.

My office was a busy place: John Pfitzner and I had advised Clodagh Curran for a Master of Science thesis on soil integrals of nuclear bomb fallout ^{137}Cs, and her thesis defence was in March. Post-doctoral student Stewart Walker and I helped organise a sugar cane technology conference 15-17 March, where Stewart gave a paper on trace element impurities in cane trash after harvesting, compared to the ash from burning the sugar cane trash.

Sheila and I took a bumble on 30 March down to Cromarty Wetlands for birdwatching. We met Pat Meehan, daughter Majella and son Bill moving cattle along the road. They invited us for tea and honey on scones at their cattle shed. Mark Stoneman, Queensland State Government Parliamentary Secretary, has a fancy home here and wants to make these wetlands into a conservation area. Mark spoke highly of Gavin Blackman, who wrote a book on the vegetation of these wetlands and the birds that eat it all. Gavin's son Mark Blackman was one of our son Ian's band buddies. It is odd how things twist together.

On 8 April '96, my motley crew loaded the *R/V Lady Basten* at the AIMS Jetty: Irena, Tenshi Ayukai's hot ^{14}C productivity gear, Ken Woolfe's kayak, Sue Codi's supersucker (for Kathy Burns hydrocarbons), John Jaeger from SUNY at Stony Brook, Gail Kineke (from Boston College, Massachusetts) with transmissometer and tripod cage for Sepik Canyon and deckie Troy Shepard. This was one of our international science collaboration efforts in Project TROPICS. It took three days for the *Basten* to get to Madang. Meanwhile we were having passport problems for some of the USA partners with PNG Embassy in Brisbane. The AIMS science crew flew out of Townsville on 12 April at 0820, hitting Cairns, Port Moresby, arriving in Madang, PNG, by 3 pm. There were betel palm nuts on sale at airport just outside the luggage racks. We rode in the back of an open truck to the very nice Madang Hotel, where immigration fellows awaited us with stamped passports for Gail, John and Troy.

We wandered the hotel grounds and craft booths. Thatched bird enclosures housed toucans, birds of paradise (*kumul*), tree kangaroos looking miserable, and red and green parrots. Near town are the coconut plantations made by the pre WWII Germans. Most of them are 50-100 years old, but now they are owned by the German Lutheran Church. They also have a copra factory here, where oil from the coconut husk is expressed, for soap, cooking oil and livestock supplement. We visited Jais Aben Resort and the California-supported Christensen Institute, just outside of town. Most of the Christensen Institute scientists were off doing rainforest field work involving relationships between fig trees and insects. We had a very nice supper of chili mud crabs, with good tropical blues singers in our hotel restaurant. Irena and Sue got up to dance to the music, so enthusiastically that Sue injured her ankle.

I sat with Tenshi, a very quiet private Japanese person. His father was a temple-keeper, but quit that to become a school teacher. Tenshi is now obligated to learn the job of a temple-keeper. He has samurai warrior ancestors, but now he loves potato chips with ketchup and Coke. Our ship the *R/V Lady Basten* arrived 11 am on 13 April, with crew dead tired from driving steady, dodging big logs at night along the fast currents on north coast. We boarded the ship and felt right at home. Cook Peter Underwood made a big supper of roast lamb and cheesecake with strawberries. The ship crew (Peter Gair, Master, Jonah Johnson Mate, Ron Bacci Engineer, Troy and Gary deckies) ran off with Irena and Sue for another session on the dance floor. Gossip from the bar indicated that they had a good wet season this year, and there were lots of logs floating in the Sepik River plume in the Bismarck Sea. US colleagues Gail and John were still stuck in Moresby without passport visas and did not arrive in Madang as planned, so we told them to rebook to arrive in Wewak, farther west along the coast, so we could use some ship time at sea moving west.

Finally on 14 April we take the ship pilot to move the *R/V Lady Basten* through the Madang reefs, and then we are away seaward, trying to map the near coastal water and seafloor distribution of the mighty Sepik River turbid water and bottom fine mud. Most of the seawater farther north is clean and clear, and the bottom sediment is volcanic rubble and ash.

Spicy chicken wings and cold beer for lunch, volcanic islands of Karkar, Bagabag, Koil and Manam to the seaward and coastal fire smoke along the shore. The ship fishermen were complaining about the pesky tuna hitting the mackerel lines. We were mapping the shelf and slope but our sonar blew fuses at 5 pm and we lost some control over computerised navigation map ship control, and had to handle the ship's wheel manually. There were dolphins swimming joyfully just ahead of our ship's bow and sailfish jumping just outside of the Sepik muddy plume. In the evenings, flying fish would land on the aft deck and wriggle back off the deck to the water. Such a tough life. There was a gorgeous sunset approaching Manam Island, and we spent the evening trying to find Sepik River mud at 300-1200 m water depth between Manam and Laing Island.

On the evening of 15 April we steamed into Wewak and took zodiacs into shore, but had to wade hip deep to find Gail and John sitting in the restaurant veranda watching us. I wandered around this disorganised town in search of Francis Sumanop, the local Chairman of the East Sepik River Province Research Committee. It turned out he was a Canberra educated lawyer, who was very amused to see our TROPICS prospectus and AIMS brochure. Gail and John were very happy to finally board the *Basten* with us, bringing some additional research equipment for mapping water turbidity and key parts for our box corer. Gail had previous experience measuring submarine mud suspensions sliding across the continental shelf of the Amazon River. She and Tenshi built a continuous flow system to the CTD and transmissometer, so we could measure temperature, salinity and turbidity of surface water while the ship was underway.

Just offshore of Wewak were some big freighter ships at anchor. Local people were in big dugout canoes with kerosene lanterns hanging from tripods at the bow, fishing all night for wahoo and skipjack tuna. You could hear the soft voices murmuring throughout the night and the click click of the 'fish callers', making soft clicks with several white dry coconut shells hitting together.

After several days of working in split shifts 24 hrs a day, partly because all the water was over a kilometre deep in our working area, offering no chance of anchoring, we decided to go get some rest by anchoring near one of the little offshore volcanic islands. As we approached Koil Island, about twenty dugout canoes came out to meet us, kids waving their hands madly, ladies offering bunches of bananas and slabs of smoked dried tuna. There was a nice little protected bay for our anchorage and all the islanders applauded when the anchor chain went out. Within minutes the aft deck of our ship was full of kids and our ship cooks were madly handing out apples and ice cream cones. In exchange, the local people offered canoe loads of their local fruit and fifteen kg slabs of smoked tuna. We had to quickly cover some of our equipment with tarps, as there were 'no touch' gadgets and sharp edges on some of our gear, and the kids were swarming all over everything.

Irena, Gail and Ken went to the school to distribute the TROPICS prospectus and to give an explanation of what we were doing in their Bismarck Sea. They didn't call it the 'Bismarck Sea' of course, and they looked puzzled when we used this phrase. We walked through three villages to find the school teacher Joe Gubio (sick with malaria) and the local chief, who really did not want to talk to us. Underneath the chief's big house was a large screen TV set but as far as we could see, there was no electrical power on the island. Next to the TV set was a very large *garuda,* a highly decorated and painted log drum used for signalling to all the villages and the nearby canoes fishing at sea.

We walked through the neat villages along coral gravel pathways raked and smooth, past palm frond thatched houses on stilts, with lots of kids explaining that we should walk to their house. Koil Island is a cone-shaped volcano rising out of the sea, with a narrow flank of coral and basaltic rubble where the people lived. There were big dugout canoes pulled up all around the island in little protected coves adjacent to each little village. There were flower gardens everywhere, breadfruit, oranges and lemon trees, kumquats, palau and *gallup* (candlenuts), melons, bananas, plantain, white and red yams, taro, *kau kau* (sweet potato), sago flour wrapped in palm fronds, coconut and passionfruit, along with lots of dogs and pigs. I don't know how they kept the pigs from eating all this good produce. The local people have large grills to smoke, cook and dry huge slabs of several kinds of tuna, red emperor, mackerel, kingfish, rainbow runners and coral trout. This apparently goes on constantly, for locals to eat and for sale in Wewak, and the smoky smell of roasting fat fish was wonderful. Kids would walk by the fish grills and pinch off a handful to stuff in their mouths, grinning shyly at us. The old men tell us that the tuna are mostly gone now, the big Taiwanese, Korean and American ships have netted them all. Two of the big ships in Wewak harbour were Taiwanese ships. The old men say that when they were young, they could catch a few tuna, smoke it, and sell it in Wewak or Madang for enough money to buy a 200 litre barrel of outboard fuel good for a whole year.

This tiny island had been sadly infected by missionaries, and had a Seventh Day Adventist mission and a Roman Catholic church (built by the Germans before World War II). The church had a corrugated metal roof which collected rain water into a huge metal tank for the people to use. It had only one wall, behind the pulpit, and they had large artistic panels showing a black bare-chested Jesus in a *lap lap* and several very attractive bare-chested Mary Magdalene portraits nearby. They even had a sleeping house for visitors, called the Tabu Guest House, available for K15/night with food provided. Our ship crew were near mutiny, they all asked to stay an extra day here and planned to visit on their holidays.

We were herded toward an open shed where several old men were offering carvings and artefacts for sale. Koil Islanders were apparently famous for their small *kundu* with fish skin drum heads, decorated paddles, canoes, ancestral masks, bowls, fish spear-throwers, head-rests and three legged stools. Ok, so yes, we were being treated as tourists, but they must not get too many visitors out there. I bought a fire-blackened mask made from a coconut shell. The ship's cook bought a canoe with several long paddles and an out-rigger, and a bundle of fish spears for K60, which he put on top of the wheelhouse for transport.

Here and on other islands we got stories from the old people of Japanese hidden relics of World War II along this coast. The Japanese had carved or detonated explosives to make deep caves in the limestone cliffs and shoreline, in which their troops and ships could be hidden from aircraft attack. The old men said that rusty hulls were still floating in some of these caves. Many of these north coast islands were occupied by Japanese troops, and they held and tortured Australian prisoners on Mushu and Kairiru Islands, off the coast from Wewak. Several members of the 'Z Unit' of the Australian Special Forces were beheaded and buried there, and have recently been recovered for full military honours and burial at the Lae War Cemetery.

We reluctantly left our new friends on Koil Island on 17 April to continue our efforts to map the extent of the Sepik River muddy water surface plume and bottom sediments. Koil and Wei Islands were only rarely reached by the largest Sepik River flood waters, and the bottom sediments were black glassy andesitic sand, pebbles and gravel from volcanic explosions. Closer to the mainland, the surface two metre layer of water was turbid grey and salinity 32, but only ten metres deep. Deeper water was clear and salinity 34, increasing to salinity 35 in the water 20-1200m depth. Physical oceanographers told us that this cold deep (500-1200 m) water came from the Antarctic Southern Ocean, had moved across the Coral Sea deep basin, and flowed northward into the Solomon and Bismarck Sea. Close to the continental slope of PNG we could find some of the fine soft mud from the Sepik River, but of course our bottom grab sampler trip wire broke, and it took half a day to repair it from the ship's stores.

Bismarck Sea, RV Lady Basten, 1996

Several days later we tried to map the continental shelf submarine canyon produced by the descending liquid mud slurry from the Sepik River. This was Gail's expertise, and she had two laptops recording our transect of salinity, temperature, turbidity from offshore, up the submarine canyon, and into the river. There was a shallow sandbar at the river mouth, but the ship snuck over it with inches to spare. We drove up the river to find full freshwater and huge wetland swamps on both shores. Obviously these wetland swamps were good fishing locations, as there were big canoes fishing there day and night. The local villages were built on stilts over the variable water depth of the wetlands, with floating canoes tied to the steps of the houses.

We also tried to monitor all these parameters in the Sepik River submarine canyon at 200-1000 m depth, but the currents were very strong, whipping our winch wire around such that we got a very confused signal. Our depth sounders (sonar) did map the canyon fairly well, but George Cresswell did a better job on a later cruise on the bigger ship *R/V Franklin*. We were also taking 100 litre barrel water samples for large suspended sediment samples, and for radium and radon isotopic measurements. After several days of this, we were tired and needed an anchorage so we could all rest.

We sought anchorage in Hansa Bay, near Laing Island, to the SE of the mouth of the Sepik River. We could watch the steam and lava being puffed out of the volcano on seaward Manum Island. It was wonderful to anchor, shut the engines down, to have silence, without the big diesel engines and the mind-numbing howl of the winch going down to 1200 m mud.

During some of our shipboard activities, the ship crew and I often had greasy or muddy hands, and we designated one person to be the 'clean hands person' for doing chemical sampling of water or sediment (usually Irena). Box coring champion John was known to start cold mud fights when things got boring on hot days, and Ken amused us all with his long shaggy dog stories about trout fishing in mountain streams in New Zealand. Tenshi's job was to measure algal productivity using ^{14}C photosynthetic uptake rates in light and dark bottles suspended on a long wire underneath a big free-floating buoy. To recover the bottle string, the ship backed up onto the buoy so we could snag the wire, but there was always a danger of entangling the wire onto the ship's aft propeller. During birthday festivities Ron Bacci, the first engineer, dropped a big brown hydrocarbon water sample bottle on Sue's good foot. So now she was injured in both feet, and she was not much good for anything except maybe croc bait.

We hadn't realised that there were people on this little Laing Island. A big canoe visited us the next morning, and we were politely invited to tea and pastries by a young Belgian fellow. On the other side of the island was a large bamboo and palm thatched building up on stilts, near a bamboo and bridge plank dock with several dugout canoes and metal boats with outboard engines (tinnies). We were politely informed in an officious European manner that this was the King Leopold III Biological Station, and our hosts were Dr. Jean-Marc Ouin and his girlfriend Barbara Rennie (from Edinburgh, Scotland), the managers of this estate. There was no one else there, but they had bunk room accommodation for fifteen people, long lab benches with many aquaria, 240V 7 kWa generators, several computers somehow linked to towers for internet connections, and a good Channel 82 VHF radio communications room. Marc was an expert in the genetics of the big trumpet shells (*Olividae, Triton* sp.), and Barbara was working on crinoid reproduction. The big trumpet shells were apparently abundant on the flanks of the volcanic cone islands offshore, but were rare elsewhere along the coast. They had careful bathymetric maps of Hansa Bay and some of the bays and slopes of the offshore volcanic islands. There were three trumpet shells on a special platform on the veranda, alongside of a big *garamut*. The trumpet shells had carefully drilled holes in the inner whorls, so that there was a high, medium and low pitched sound from the shell when vigorously blown. The local PNG people in the nearby mainland village of Nubia had taught them the various signals from the log drum and trumpet shells. Marc and Barbara served us tea, toast and jam at a long table on this veranda, the tea coming from a giant European porcelain hot tea urn. We hadn't really anticipated so much social life on this research cruise. As we were leaving, Marc gave me a big trumpet shell, with a black mark where the hole should be drilled if we wanted to blow it. We invited them aboard the ship for a supper of roast pork, roasted potatoes and cheesecake with coconut and rum liqueur. They brought us a welcome gift of a burlap sack of SP beer in one litre bottles.

We reluctantly left Hansa Bay the next morning to continue our efforts to find the dispersal fan of Sepik River mud in offshore 1000 m deep water, and to get some good cores of the mud, avoiding the volcanic gravel bottom which would damage our core catchers. Much of the soft mud in Hansa Bay was from the smaller Ramu River. Most of the mud from the much larger Sepik River apparently slid down the submarine canyon as a dense slurry to 1000 m deep water, as there was almost no continental shelf here. At the base of the canyon we found mostly sand and gravel, so most of the fine mud must have been moved by currents to the NW along the coast past Wewak. A few of our deep sediment cores came up frothing at the surface, probably due to lots of methane and CO_2 gas in the sediment. Later big ship multibeam sonar work by TROPICS partners George Cresswell, Steve Kuehl and John Milliman showed that some soft mud spurs along this coast were actually old coral reefs that were now covered with river mud. Only a small fraction of the mud in the Sepik River water floated on the surface, probably the colloidal and organic fraction, forming a twisting plume 20-50 nautical miles northward in the Bismarck Sea. This surface muddy plume, far out to sea, also had floating islands of mangrove vegetation, big rainforest logs complete with green leaves, perched birds and lizards, coconuts, some plastic bottles and styrofoam. This surface muddy plume was only a metre or so thick, as when we steamed through it with our ship, the prop wash showed clear water underneath the turbid water.

We anchored close to Karkar and Manum Islands on some evenings, and we could see smoke, steam and red-hot coals coming out of the volcanic calderas. Lightning seemed to be attracted to these volcanic emissions. Local fishermen worked at night with lanterns, catching tuna, sharks and barracuda. From far out at sea we could see the village fires along the coast, and we could almost smell the roasting fish.

As we steamed back to the port at Madang, we cleaned up our equipment, freshwater washed and oiled our tools, washed down the deck, stored our gear in the hold, and prepared to depart the ship. We spent a lot of time itemising our electronic equipment by serial numbers, and our water/mud/biological samples in the freezers, because we knew the PNG Customs and Immigration officials in 'white shirts with aviator sunglasses' would be looking for guns, drugs, girly magazines, booze and other bad cargo to confiscate. On 23 April 1996 the 'white shirts' visited the ship in the rain and dark, ate supper with us, drank two bottles of rum, took all the fruit, and left us at midnight without even looking in the hold.

I had been deeply affected by the helpless sadness of some of the Bismarck Sea islanders' description of their life being upset by the big factory fish freezer ships from Korea, Japan, Taiwan, Spain and the USA. They used to be fierce warriors, head-hunters, and feasted on ceremonial tuna. Their tuna catch had been considerably reduced, the missionaries had damned much of their culture, and they felt abandoned in their paradise. Now they carved masks and fish spears for the tourists and bought twenty kg sacks of rice and tinned corned beef in Wewak.

The next day was still, hot and very humid, and our ship left Madang harbour for the long trip home. The science crew got to wander the harbour and muddy streets of Madang, visiting some good craft stalls and aviaries full of exotic birds. The harbour was busy loading woodchips onto big freight ships and there were the usual huge cranes moving railway containers from ship to shore, and shore to ship. At noon we rode in the back of a truck to the airport and flew south via Mount Hagan (where I bought three wooden flutes and two jawharps), Port Moresby, overnight at the JCU apartments in Cairns, and home to Sheila in Townsville by 0800 on 25 April. Gail Kineke stayed with us for a day, visited Eric Wolanski at AIMS and flew back to the USA the next day. It would take seven days for the *R/V Lady Basten* to arrive in Townsville for the usual Australian Quarantine Service to inspect our ship's hold full of mud, canoes, fish spears and arrows.

It was good to spend some time at home, mainly whipper snipping pathways through my orchard in the creek valley. We had a good wet season and by May the grass was about 1.5 m tall.

This peaceful time was rudely interrupted by a phone call from my ANSTO colleague Ron Szymczak, who wanted my carcass at a thing-tank session at Lucas Heights Sydney about the environmental consequences of energy production. So I spent the next week in posh Sydney hotels and ANSTO meeting rooms discussing what research should be done in the next decade. The Australian Nuclear Science and Technology Organisation was a major laboratory in this country, with 300+ people employed in research and development, and a lot of international treaty obligations. They really really did not like my contribution, as I tried to push them into an aggressive uranium mining and nuclear power program as a means to lower greenhouse gas emissions. The Australian government has been allergic to nuclear power since Hiroshima and they didn't even want to discuss it.

It was pretty well a complete waste of my time and I flew home on 7 May to be greeted at the airport by Jesse in his old Toyota ute. We hit Baskin-Robbins for a jamocha ice cream sugar cone, and then home to Sheila in Alligator Creek, where everything made good sense again.

My AIMS office was crowded with upset people in the next week, as the local CRC organisation decided not to renew the next year's research contract for several of our post-doctoral fellows, including our good friend Stewart. The old mainframe Vax computer had finally been shut down, and a new intranet system and in-house website designed by Peter Hewitt was begun. My personal office computer had been upgraded, and our old email platform (Eudora) was replaced by Microsoft Outlook. Of course this meant a massive blockage of electronic traffic when it started working again. It took two days to clear my email windows. I much preferred to be on a ship on the coast of Papua New Guinea somewhere out of reach of the bureaucrats and computers. Ah, but I must admit that both Sheila and I were addicted to email communications with our relatives and friends in North America and UK.

I escaped this madhouse, as I was invited to give a paper at the Royal Australian Chemical Institute at the University of Queensland campus in Mackay, about 500 km southward on the Queensland coast. Sheila and I took our time driving down the coast, visiting some beaches and good restaurants. A Queensland custom in roadside motels here was the delivery of breakfast on a big tray through a trap door, complete with fresh flowers and seasonal fruit trays. We drove up to the Finch-Hatton rainforest and walked the hiking trails, had tea and croissants and watched platypuses in the river with light rain at dusk. We hiked back to our hotel where a fireplace dried us out and we could pick off our leeches.

We drove back home through the coal mining town of Collinsville, and to Bowen and Edgecombe Bay. There was lots of good seafood to eat along the way. Much to the delight of Charles Byrd and Jesse, we arrived back home on 14 May and I could start picking a good harvest of grapefruits and lemons.

Our AIMS CEO flew me down to a marine science forum at the Academy of Sciences dome in Canberra, the goal being to upgrade marine research efforts in the face of climate change, population growth, and increased pressure on estuarine and marine fisheries. I had been giving a lecture to outback cattlemen and coastal sugar cane and banana farmers, showing how landscape activities affect the quality of the coastal ocean. I was using the analogy of anthropogenic pressure (agricultural land area/natural land area, fertiliser and pesticide usage rate, livestock and human populations, deforestation, fish harvest, tourism investment and income) on the coastal zone, showing how these parameters varied in Europe, North and South America, Asia and Africa. In many cases, countries with high human populations, big industry and little natural landscape also had polluted coastal seas and over-exploited seafood harvest, whereas Australia was relatively lucky to have low population and large areas of natural land. We had the Great Barrier Reef, one of the wonders of the world. However, increasing human and livestock populations, increasing cultivated and fertilised agricultural land, coal mining and shipping, and uncontrolled industrial expansion will exert deleterious pressure on this relatively clean coastal zone. The CSIRO scientists did not like my implications that agriculture was the dominant land use change and contaminant source in river catchments along the sensitive Great Barrier Reef lagoon.

In early June, Sheila and I performed a set of our Meadowlark music for Jennifer Hultgren's Queensland Bluegrass Festival in her back yard, with maybe a hundred people under the white gum trees. Sheila was a big hit with her blues 'Nobody Knows You'. We met a good fiddler, Kevin O'Connor, a Sydney human rights lawyer, who we jammed with for an unplanned improvised set of tunes. The next day, we drove down the Bruce Highway to the Palm Creek Folk Festival, where it was raining buckets. We did our thirty minute set with a sound system that was shorting out from the rain. The ferals (hippies, unwashed musos) were abundant with earrings and peace signs, amidst complete chaos. Later in July, we performed a set in the local art gallery, the first performance of my ballad 'The Battle of Bali'. Meanwhile, Jesse was in JCU Engineering second year examinations, fussing about Laplace Transforms and steam engine thermodynamics. Between exams, he was hauling sugar cane molasses to local livestock operators.

As part of my international hustling of science institutions in a dozen countries, I had fought for a budget to support travel and accommodations for Indonesian and PNG people to come to an AIMS/TROPICS workshop, as they were often not well supported by their home institutions. I was also struggling to assemble enough funding to manage our proposed CRC

(Coordinated Research Centre for Coral Reefs) ship cruises and lab analyses. My institution also wanted to begin research on the NW Coast of Australia, into the western Indian Ocean seas between NW Cape and Sumatra. I was saddled with an *R/V Lady Franklin* cruise that had been planned and budgeted during the previous CEOs term...but now we didn't have enough money to do this cruise. All of these funding efforts failed, and I was in despair for these good research programs. So we had a 22-27 June 1996 TROPICS workshop to organise ship cruise timing, personnel and equipment for the next five years. Scientists from Japan, USA, Germany and Australia attended, but no Indonesians or PNG people were there due to lack of travel funding. But even without the necessary funding, good metal workers in the AIMS workshop were building my large sediment traps, six big trap arrays on gimbaled hooks, each array with six three metre long tubes, to be deployed from the *R/V Lady Franklin* on the NW shelf.

With a strange combination of funding sources (CRC, KEPCO-Japan) a previously scheduled *R/V Lady Basten* research cruise went ahead with inadequate funding in late July to early August 1996, with New Zealander Mike Rosen, Ph.D. student Joanne Cavanaugh, JCU/CSIRO/Geology students Paul Leeming and Gavin Dunbar, UK University of Manchester Professor Barbara Maher with technician Andy Perkins, AIMS technical support Irena Zagorskis and Paul Dixon, and myself. This was a Kasten coring trip, to locate the distribution and accumulation rate of river mud in the Herbert River estuary, Hinchinbrook Channel and Rockingham Bay. Peter Gair was our ship Master, with first mate Jonah Johnson, the crew working hard to bring the ship to the long list of planned coring sites.

We spent three weeks doing this, taking 3-4 cores/day of 1-3 metres of mud, slicing them cleanly into 2-4 cm samples, washing up and getting ready for the next day's work. Some of these cores were for Joanne Cavanaugh's Ph.D. thesis on agricultural organochlorine pesticides, so everything had to be clean of oil and grease. Dr. Maher wanted to collect samples for magnetic mineral analysis, and especially the microbes that form pyrite spherules.

After two weeks of this exhausting work, we took the ship into the little town of Cardwell to pick up four Australian Broadcasting Corporation news crew from Melbourne, who wanted to film and interview us doing this work. We took them into the deep mangrove channels, where we got some good hydrogen sulphide smelling mud, and later out into clear water where they could film jumping flying fish, sea turtles, dolphins, sailfish, minke whales and dugongs in the seagrass. The ship crew were being their usual irreverent larrikins, teasing the ABC film crew with stories of koala stew for supper, having just butchered some fat platypuses and dugongs for lunch. They filmed our best core (2.9 m) at Hecate Point, which turned out to be a key bit of evidence for all of us. It was rich in microbes that form pyrite (FeS_2), had a very rapid accumulation rate, was useful for tracking organochlorine pesticides, and most exciting was the presence of an 1870-1900 peak in mercury concentrations, which almost certainly came from the Burdekin River catchment gold mining process around Charters Towers. Barbara, a UK scholar of the classics as well as mud, informed us that Hecate was the Greek goddess of darkness and terror and the crossroads, who roamed the earth on moonless nights with ghostly hounds. Hecate was worshiped by witches and magicians who sacrificed black dogs and sheep to her. This story came to us in slow bits as we sliced this core into the wee hours of the night, with some tea and pastry breaks.

Having performed well with international colleagues, it was a letdown to return to AIMS, where the atmosphere was one of defeat and retreat from ambitious science.

Part of the explanation for this slowdown in research funding was due to the new budget of the current conservative government of John Howard, who was not exactly a supporter of environmental and research activities. On 20 August, our CEO Russ Reichelt gathered AIMS employees in the corner of the library to announce twenty-three forced redundancies to reduce the salary bill, so there would be some money for ships and research activities. So some very good scientific and technical support staff went away to hunt for new jobs, and the rest of us at AIMS were stunned and shocked. 'Gutted' was the north Queensland phrase used. Two of the main workers in the AIMS workshop left, including George McNaughton and Henri Le Donne, who had built my Kasten corer and other nice marine steel gadgets for my shipboard work.

I spent most of September and October speaking at conferences, workshops and advisory committees. I was involved in several outback land care and agricultural based river catchment management committees for the Burdekin and Herbert Rivers. I enjoyed talking with cattlemen and remote small town people.

Towards the end of 1996, we were all involved in hosting and presenting reports for our Herbert River project for two funding agencies, the James Cook University CRC Reef and the Japanese energy industry KEPCO/KEEC. It was curious to have a big industrial energy company sponsoring research in a mangrove-lined estuary, but I guess this was some sort of guilt money. We had an inside track on Japanese funding through our colleague Tenshi Ayukai, who cautioned us on polite behaviour in front of senior Japanese industrialists. A graduate student, Paul Leeming, helped me create a slide show that made it easier to see the impact of century scale land use change, contaminant history, and coastal mud accumulation amongst the estuarine mangroves and wind protected bays. This was based upon an older IBP idea from Frank Rigler called the 'Pressure-Impact-Response' model.

An earlier AIMS CEO wanted to expand the operations of AIMS to the Northwest Shelf of Australia. AIMS had been traditionally limited to the Great Barrier Reef shelf and the Coral

Sea. This was plenty big for one small marine research institution, but CEO Joe Baker was determined to put AIMS ships into the Arafura and Timor Seas, the Gulf of Carpentaria, and the Australian shelf and oceanic water of the Indian Ocean between NW Cape and Sumatra. As part of this administrative thrust into new territory, which began four years ago, I had been naïvely led to propose a big ship *R/V Lady Franklin* research cruise on the NW shelf out past the Barrow and Montebello Islands onto the Exmouth Plateau.

And now the time was up, and we were expected to mount an expedition over 30 October to 23 November 1996. Our institute was woefully unprepared for this; not many scientists wanted to participate, we had little background information, and it would cost a bundle to get our equipment and technical staff far across the country to the Indian Ocean. Nevertheless, we loaded a long semi-trailer truck with our shipboard equipment in two railway containers in mid-October, and it drove across the dusty red interior of Australia to Fremantle, where we were going to load the ship. Our purpose on this expedition was to measure how fast the planktonic algae were growing, to sample for hydrocarbons (natural and oil platform contaminants) and the microbes that degrade the oil, and to measure the removal rate of carbon and nutrients by means of sediment traps, sediment cores, and profiles of natural (Th-234) radionuclide that stick to water column particles. I was also interested to seek evidence for the 1952-58 British nuclear bomb tests done on the Montebello Islands, and I expected to see some of the longer-lived nuclides in our sediment samples.

Sheila drove me to the Townsville airport in the cool early morning of 29 October, where Tenshi Ayukai (algal primary production), Sue Codi (hydrocarbon chemist), Irena Zagorskis (nutrient chemist), John Pfitzner (radiochemist), John Soles and Cary McLean (aft deck riggers), Russ Hill (microbiologist), Jo Johnston (microbiologist) and Jo Cavanagh (organic chemist) joined me to fly to Cairns, over Ayers Rock (very hot and lots of flies, $100 shirts in the airport) and Perth, with a van transport to the Fremantle harbour, where most of our red dust coated equipment was being loaded on the ship by cranes and forklift tractors. I had a planning meeting with the ship Master Neil Cheshire, who was a bit nervous about the Australian/Indonesian boundary offshore from the Exmouth Plateau. The ship crew quit work at dark and we all went to a good Portuguese sidewalk seafood restaurant near the ship docks, and found a good ice cream shop for dessert. I had been earlier warned that the Chief Scientist's cabin had an internet connection, a new stereo radio and CD player, as well as a better lighted map table and a fridge, and I brought along a few CDs of jazz and fiddle tunes.

The next morning it was hot sweaty work with hard-hats and steel toed boots on, guiding the cranes and forklifts to sort our equipment to the proper places on the ship, to bolt down my Kasten coring apparatus, and assemble the six large sediment trap arrays. Inside the ship Irena, Jo and Jo, Russ and John created a chemistry and computer data logging laboratory. After a wonderful ship lunch of rare roast beef, Yorkshire puddings, strawberries and real cream, the ship cast off the lines at 1330, and we steamed north for two days at thirteen knots with a twenty knot tail wind. During this time we had lots of safety and fire drills, instructions on ship capabilities, practiced rough weather procedures, and tried to get our deck and lab equipment to work properly on a rolling ship with fluctuating electrical power supply.

By 2 November we were in Exmouth Gulf pumping cubic meters of seawater into deck barrels for determination of radionuclides and oily residues. In the little coastal town of Exmouth we picked up ANU marine geologist Bradley Opdyke, who would help us interpret sediment core logs and be the night time Chief Scientist, as this was a 24 hour operation ship with no down time. The ship's cooks were providing us with lamb brains scrambled with eggs for breakfast, with lots of tabasco sauce. We got smoko at ten am and three pm, so we were certainly well fed. Some of the ship crew were the same as those on our earlier Gulf of Papua cruise, so there was some recognition that we were half way competent on the aft deck. Stormin' Norman Marsh, the bosun on the aft deck, became a good friend.

Over the next two and a half weeks we rarely had the chance to settle into a routine, due to equipment breakage, failure of some deployment or recoveries, and rough weather. But this was like a floating small town, a culture far from land, where stuff got fixed one way or another, and the engineers were very clever. We were trying to run a series of ten sampling stations from interior Exmouth Gulf northward to the Indonesian border, northwest of Barrow and the Montebello Islands. Our first task was to deploy the six big sediment trap arrays, mostly in deep (500 to 2000 m) water, and this went well in the experienced hands of riggers John and Cary, and the ship crew winch and A-Frame operators. This was a complex operation that required intense focus for the deck riggers, with three different winches going, hand signals flying, and a lot of dangerous heavy stuff hanging on the wire. These gimbaled traps were suspended on a steel wire anchored by railway wheels (attached with a radio-activated release link) and kept vertical by large subsurface bundles of floating buoys. These moored traps also had current meters below on the wire. These were to collect sinking particles for about two weeks, and then we would recover them by activating the acoustic release link, if all went well. From these samples, we could determine the removal rate of organic matter produced by algae, carbon, nitrogen, phosphorus, petroleum hydrocarbons and all the other chemical elements, in units of mass/square metre per day. Kathy and Sue would later show by gas chromatographic analysis of the particles collected by these traps that the hydrocarbons were different from the local drilling platform oil.

After completing this job, the routine was supposed to be 1) CTD profiling of temperature and salinity of the water column, light wave length spectrum and fluorometer profiling of chlorophyll, 2) depth profile sampling of the big Niskin bottles for nutrients and hydrocarbon microbes, 3) incubation of water samples with ^{14}C spike to determine algal photosynthetic growth rates and respiration rates, 4) deployment of Kathy Burns and Sue Codi's SeaStar large volume *in situ* free floating filtering rig for hydrocarbon particles, left to filter for 12-14 hours, and then recovered, 5) pump one or two cubic meters of sea water from surface and deep water into deck containers, for radionuclide determinations, 6) pump more sea water through MnOOH coated acrylic fibre columns to measure the radium isotopes, 7) take a grab sample of the seafloor sediment, to determine if the mud was good for coring, and 8) deploy the Kasten corer, to obtain several centuries of sediment accumulation history and Montebello bomb fallout nuclide history. Of course, most of the time at least one of these activities encountered trouble, especially in forty knot winds and a rolling deck; but some days were flat calm, the deck was steaming hot, and everything worked according to planned schedules.

Most calm evenings we assembled on the foredeck to watch for the 'green flare' as the sun sank into the Indian Ocean. We saw breeching whales and dolphins at the ship bow, and the cooks jigged for squid. They continued to astonish us with piles of wonderful food, day and night. Often we were slicing sediment cores late at night, finishing and washing the core slicing table just before midnight. The acoustic release control box that released the sediment traps from their anchor chain malfunctioned, so we emailed AIMS to get another one flown to Exmouth. This meant we spent half a day steaming back to Exmouth Gulf from far out at sea. Greaser Lindsay got sick, so we took him into Dampier so he could fly to the Perth hospital.

By mid-November, we raised the sediment traps at the four deep stations, getting plenty of material in the trap bottles for radiochemical, geochemical and hydrocarbon analyses. The trap array farthest north came up fine for the 300 m trap, but the deep trap on the wire at 1200 m came up tangled and upside down. This was a dangerous big mess on the deck, with taut steel wire wrapped around the big trap frames and broken floats, all covered in algal and gooseneck barnacle slime, getting slippery and stinking in the hot sun. Then the wind came up and angry green sea foam came came rolling up on the aft deck, so we tied the whole mess down to deck bolt rings and went to have supper. After supper, the aft deck lights came on and it took us four hours to untangle the mess.

Having accomplished well our goals for the NW Shelf transect, we steamed east into the Timor Sea, visiting Mermaid and Scott Reef to recover current meter and CTD moorings for colleagues who had been here two years ago. This region is populated with lots of oil drilling platforms, which appear to be a small brightly lit city at night. At some of the reef locations we could see the rusty hulks of wrecked ships that were not paying attention to the charts. Brightly painted wooden Indonesian *prahas* with two sail masts were seen fishing in Australian waters.

At Pee Shoals, our microbiological crews sampled sediment from a region of oil spills in the previous year. At 0830 on 22 November we steamed into the port of Darwin, where harbour stevedores were waiting with cranes and forklifts to unload our gear from the ship and put it all back on a semi-trailer truck for shipment back to AIMS. We had to keep some of our samples in large insulated boxes frozen for chemical analyses, so we had to search Darwin to find some big freezers before we could eat supper and get some sleep in a hotel. The next morning we left Darwin with lots of excess frozen baggage, flying over lots of smoke from bush fires, to Brisbane, and then to Townsville, where Sheila met us beaming and brown.

These big ship research cruises were a lot of work to plan and execute, and as Chief Scientist I rarely got much sleep on the ship. I would be up late at night with the navigational charts, planning and writing out the next day's ship track and science activities, and posting them in the mess hall and labs. So when I got home, I needed a few days of sleep to regain normal brain function, and the floor of our house seemed to move like the deck of a ship for several days. In rapid succession, I was involved in two workshops on our Herbert River research, to explain it all to government bureaucrats and the local cane farmers and cattle station operators.

The wet season came in late December, and Christmas happened again. Jesse made special mango pancakes for his mother for Christmas breakfast.

Meanwhile the earth continued to circle the sun no matter what happened on our ships in the Bismarck or Timor Seas.

Chapter 30. Pythons, Cyclones, Minihasan Wedding Music, Spicy Dog/ Snake/Bat Curry. 1997

We were beginning to get accustomed to the year's cycle here, so we welcomed the wet season of 1997. We had over 500 mm of rainfall in February and the consequent abundance of mosquitoes, flying ants and carpet pythons. One of our usual yard pythons was named Pythagoras, and he was comfortable in the head level notch of a big fig tree in our front yard, happily digesting a young wallaby, his stomach swollen hugely. We were on the margins of a few cyclones, and strong winds laid down some big trees in Townsville and Alligator Creek. Jesse had purchased a tinny (North Queensland slang for a small metal motor boat), and he was trying to park it beside our shed with the boat trailer attached to the Subaru bull bar. Of course, the boat and boat trailer took off down the steep hill, pulling the bull bar off the car. Several expletives coloured the air. Due to the good wet season, there were some cases of Ross River fever, dengue and meliodosis reported from our area, one of the prices we pay for living in the tropics.

Good friends Barbara and David Ediger (and sons Ian and Jaremy) from Winnipeg, Canada visited us in February, in the midst of the heavy rain. During brief periods of sunshine, Sheila led them to visit all of the parks and wildlife sanctuaries within driving distance. Their young son Ian was heard to remark 'I'm sick of seeing more waterfalls and beaches.' David played guitar, fiddle and mandolin in the Canadian version of Meadowlark, so we put on a few concerts in town. Rainfall intensity was great enough to cut off the highway to the south and north of us.

I had become a project leader at work, which meant trying to keep fifteen people happy and funded, reporting to the CEO, and remaining focused on some useful/ interesting research activities. Most of us were involved in Project Herbert, a CRC supported CSIRO/AIMS study on the hydrology, agriculture, soil erosion and Herbert River sediment transport to the estuary, and the marine biology of the estuary and Hinchinbrook Channel. In addition to the scientific value of this work, there was a significant effort to involve and consult with the Herbert River catchment cattlemen, sugar cane and banana farmers, and in general they responded positively. The worry was that expanding agriculture, with associated fertiliser and pesticide use, was causing increased soil erosion leading to increased mud and contaminant transport into the clean clear waters of the Great Barrier Reef lagoon. In my work I found that most of the river mud was trapped in the mangroves and estuarine channels, with very little terrestrial mud escaping to the inner shelf coral reefs. Our radionuclide profile chronology of sediment cores provided no evidence for huge increases in river sediment deposition in the estuary or the Hinchinbrook Island region.

Our project staff helped with public environmental education booths in Townsville, Ingham and Ayr. As usual, we were in the precarious financial position of trying to do too much with too little money. Big meetings in Canberra in February 1997 provided better support from CSIRO and the CRC, and Tenshi Ayukai began negotiations with Japanese industrialists for financial support for our work. Our work in Papua New Guinea, under the banner of TROPICS, was hopefully being supported by hustling money from the US National Research Foundation, in-kind support from Ok Tedi Mining, AIMS, CSIRO, and several UN agencies. We had an important TROPICS meeting in Sydney in early February 1997, with representatives from Indonesia, PNG, USA and Japan, where we had to demonstrate we had enough external funding options to justify being awarded sequential Australian National Marine Facility *R/V Franklin* cruises over four years, the total ship cost being over two million dollars.

These kinds of meetings were uncomfortable for me, as we were asked to provide the practical benefits, useful outcomes and commercial gain as a result of our proposed scientific activities. Would tourism benefit? Would prawn and tuna catches improve? Would the marine seafood industry benefit? Would the impact of climate change be revealed?

We had to conceal the true fact that 'If we knew what it was we were doing, it would not be called research, would it?' Albert Einstein said that.

On 16 March 1997 Sheila drove me to the airport for a major trip to a coastal oceanography meeting in Indonesia. And off we go flying again, Townsville, Cairns, Gove (a BHP bauxite and alumina mining town), Darwin in the pouring rain, Denpasar on the island of Bali by midnight. We can see the golden arches of McDonalds as we land. A Qantas guide hauls us to the Dynasty Bali Hotel, with rooms much too nice for us, featuring hand carved tropical wooden furniture, two TVs, lavish large bathroom with shower and tub and cool granite slabs on the floor. The whole place smells like cloves, cinnamon and citrus flowers.

I had to be up by 0430 to join my group on the bus to the airport, where Ian Wood (ex-Environment Chief at Ok Tedi) led us into a posh club lounge where we had our breakfast of hot biscuits, coffee with spicy fish paste sandwiches. Off we go again at 0640 to fly to Surabaya in east Java. I was sitting beside the CEO of an Indonesian food company working for a subdivision of McDonalds, who wanted to talk about tilapia ponds, fish/oyster farming and prawn ponds. Their Indonesian outlets supply fish burgers and spicy oyster stews to Indonesian restaurants. He advised me not to eat the boxed lunch provided by Garuda.

We sat on the tarmac in the rain in Surabaya for too long, but the stewardess finally brought us some hot coffee and mushroom omelettes, while the Indonesians all had cold *nasi goreng* (rice) for their lunch. The aircraft cabin was full of cigar, tobacco and marijuana cigarette smoke, and very spicy clove smoke.

By 0930 we were in Balikpapan in eastern Kalimantan, where my friend Ian Wood works with a small mining company. And then farther north on a Sympati Air Fokker 100 prop plane, eating our way onward, fish balls, spring rolls and *sambal asli*, washed down with warm Sprite, arriving at noon in hot rain at our destination Manado on the north coast, the capital city of Sulawesi. From the air, the region around the city is like lime-green velvet grass, similar to the treeless coastal regions of PNG, where all the trees were harvested repeatedly. Away from the city was rainforest, thick and deep green, interrupted by coconut plantations in orderly rows. Behind the city were two raw looking volcanic cones.

My presentation at this meeting was sponsored by some Australian/Indonesian agreement related to our Project TROPICS, and the Novohotel hosts greeted us by name at the airport. We were hauled to our hotel which again was embarrassingly lavish, with ornate wood carvings and shell/ivory inlaid furniture, frosty air conditioning, a patio opening onto a raked beach, marble fittings in the bathroom, four large bowls of good fruit, CNN, ABC and USA TV stations available.

I slept away the afternoon and woke up just in time for a shower and a fresh white shirt and necktie, for our banquet at 1900. Here were drummers, *tok tok* dancers, all sorts of curry and skewered meat dishes, sculptured mounds of white and black rice, a massive fruit table (most of which I could not identify), and lukewarm sugared coffee. And of course we had to be respectful and listen to speeches by the Agricultural Assistant to Minister Farid, two presentations by Italians running SCUBA diving tourism, a talk by the second richest lady in the country, and two more ministers and the mayor welcoming us, most of this in a mixture of Baha Indonesia, the local Minahasan language, and an English patois.

Modern Indonesia is a conglomeration of thousands of islands large and small, many of them with their own cultures and languages, and not very much allegiance to the Sukarno nation. This region had been exploited by the Arabs, Portuguese, Spanish, Dutch and Japanese for several centuries, but since World War II all the islands were united in the Republic of Indonesia.

The Minahasan people were not easily subjugated, and there was fierce armed resistance to Dutch and Indonesian rule. The National Government in Jakarta had to bomb Manado to bring them into the Republic. Sulawesi, and Manado in particular, had been Muslim under the rule of the Sultan of Ternate, and presented a distinct Minahasan culture, music, language and flair for traditional knowledge of marine biology. The Bunaken National Marine Park just offshore is one of the star features of the 'Coral Triangle', a magical eastern Indonesian/ Papua New Guinea region of the planet's highest marine biodiversity. This part of Sulawesi gets about three meters of rain per year, the air temperature was rarely below 30° C., and the coastal vegetation was lush. The famous English evolutionary biologist Alfred Russell Wallace was in Manado in 1859, and he praised this region for their rainforest and coastal coral reef communities. They were very proud of their local Sam Ratulangi University, which was the host for this coastal oceanography meeting.

The next morning, I was able to check the sequence of my 35 mm slides while sitting on the balcony, overlooking the beach and islands far out at sea. They brought to my room a 'Christian breakfast' of bacon, omelette and large bowls of fruit. This part of Indonesia is mostly Christian, whereas most other islands are either Hindu or Islamic.

The various lectures were held in university classrooms, with creaky wooden floors, slowly spinning ceiling fans and gorgeous carved woodwork everywhere. The science presentations were similar to science meetings anywhere in the world, but the setting was exotic, hot and humid, multilingual and smoky, as many of the audience smoked clove cigarettes and cigars.

I gave a paper on how agriculture and mining activities put an anthropogenic pressure on estuaries and the coastal zone in North Queensland. The local media wanted an interview from me, and I soon discovered that there was a gold mine just up the coast from Manado, which had sparked controversy about polluting coastal water with arsenic and mercury. This is a delicate business, as I was an Australian scientist who was not supposed to get involved in international disputes without going through a briefing from the Australian Department of Foreign Affairs.

We were served a wonderful lunch of *laksa, manis, dukak*, hot peppers in noodles, with Bintang, the most common Indonesian beer (probably Dutch in origin).

In the evening we had another long ceremonial banquet with local politicians and university officials welcoming us yet again. Many of the Minahasans and SE Asian speakers were formally dressed during the science meetings and banquets, some with academic robes and little puffy hats, the local business men and politicians in dark suits with white shirts, neckties and black polished shoes, the ladies in full gowns carrying little parasols. Meanwhile, the Australians, Americans and Europeans were casually dressed, usually without ties, wearing flip-flops on our feet. There were 170 scientists at this symposium, most of them Indonesian, Malay, Thai, but also a few invited speakers from Japan, Australia, USA, UK and Norway.

The Minahasans were 'putting on the dog' for this science meeting! There were mounds of flowers on the speakers table, place name cards, four large screens for simultaneous projections, six computers and video cameras operating to capture video and audio, large portraits of Sukarno and children on the hallways, frangipani flower decorated chandeliers, many welcome posters and assorted flag displays. The speaker's table had rows of bottled spring water, one litre bottles of Bintang, packets of clove cigarettes and cigars, huge ash trays and a big gong nearby to signal the next speaker. There was a morning and afternoon break in the conference, provided with tables of interesting (unknown) finger food, goblets full of flowers and rice wine, a very lavish tea and coffee service.

A few of us were overwhelmed by this lavish affair and we escaped to wander the streets. I found a food market and bought a kilogram sack of *mangis* (mangosteen) for 1700 R, a bargain, as they are Aus $2 each in Townsville. The hotel provided a swimming pool, where we would retreat after the science meetings, drink beer and tell stories about our research expeditions that went bad.

The hotel servant boys would come to find us at the pool and politely request that we join the others on the second floor lounge for the farewell banquet, where more exotic food was served, dark chocolate balls with *salak*, fish curry in coconut milk, fiery spiced spit-roasted chicken, spicy dog, spicy rat, spicy snake (with chili-banana paste), and spicy bat, mountains of rice, and endless bottles of wine and litre bottles of Bintang. The spicy rat was actually a large water rodent the size of a muskrat. They had a shiny black grand piano in the lounge with a singer who performed Paul Anka's 'My Way', complete with gestures and fedora hat, and a lady in a ball gown who struggled with Barbra Streisand songs.

Representatives from each country represented at this meeting went to the microphone to thank everybody, usually in several languages. I sat at a table with a UK and USA group that lived in Manado, and they told of shipping their daughters via USA MEDEX to Singapore for medical services, usually related to malaria fevers and dengue.

We had made friends with Chris and Cristina, a French/Italian couple that ran a dive shop and dive expedition resort just east down the coast from Manado. So after a wonderful 'American breakfast' of sausage, bacon and eggs, they picked up four of us at the hotel in their Toyota troop carrier and drove through the city and coastal towns on pot-holed twisty roads through coconut and banana plantations, nutmeg, cloves, mangis, durian, almond and cashew groves. The dive hostel was on a low-lying promontory in the South China Sea, with mangroves everywhere. We had nice cabins with Indonesian showers (a porcelain pitcher of water with a cup), put on our swimming togs and went out to a nearby reef to snorkel for the afternoon. Later that evening they showed us the little tarsiers with huge brown eyes and long fingers climbing around in the strangler figs and beach almond trees. Chris and Cristina had been here for twenty years, but in the last five years tourism exploded for them. Cristina explained why tropical vegetables were so tasteless (compared to her native Italian vegetables). She said it rained too much here and all the flavour was washed out of the vegetables, and that is why they spice everything with hot peppers.

We gratefully slept in the next morning and had a wonderful brunch of grilled reef fish with hot strong Italian espresso and good fresh-baked bread. They took us farther out to the Bunaken Reef, where we could use SCUBA to see huge fish, electric blue *Diadema* starfish, miles of hat corals, *Porites* and *Acropora* corals, giant potato cods, eels and sea snakes.

There were also native fishermen treading water with spearguns in hand, no boats nearby, but floating baskets for their fish catch. As we were leaving, a large dugout canoe with sails appeared silently with a cooking fire on a clay mat deck platform. It was operated by big mommas and lots of kids, coming to get their men. I guessed this must be an example of a 'multiple use national park'.

We headed back to our dive shacks on shore for a snooze, then sat around in the late afternoon smoking clove cigarillos under the nutmeg and clove trees. This was a real tough life. We gorged on spicy roast chicken, mackerel and rice, and they tried to serve us Coke, thinking this was a big treat! We asked for Bintang beer. Several of the dive guests were going to Ningaloo in Western Australia next. I gave Cristina my Meadowlark business card, hoping for a paying job playing music here.

One of the resort Minahasan cooks asked about the music we played and then proceeded to explain some of the traditional Minahasan music. He shyly told of 'wedding music', which originally had nothing to do with weddings, having originated after MacArthur's US Marines chased the Japanese out of Manado in World War II. The US troops apparently had a marching band with trumpets, flutes, trombones, tubas and drums, and they marched through the liberated town before going on to the next battle. The Minahasans were impressed, so they constructed trumpets, trombones, flutes, marimbas and tubas out of bamboo or roofing corrugated metal sheets. These instruments were made to sound a bit like their US counterparts, but of course they were not in tune and did not obey any note intervals. They also made tubas out of brass, tin cans and kerosene cans. No matter what they looked like, they played with glee and fierce determination, 'Stars and Stripes Forever', 'Bridge over the River Kwai', 'Beer Barrel Polka' and other US Marine marching band tunes. Nowadays you can buy this 'wedding music' on cassettes in most shops. It's not really being used for weddings, but most roadside stands had this music blaring from speakers on top of their roof. The Minahasans had their own versions of gamelan orchestras, but this was a folk music of the old times, not so popular anymore.

The next day, Cristina drove our small party of Europeans, US and Australian guests to the airport in the red Toyota Troop Carrier, which smelled of cloves and sandalwood smoke. We drove through coconut plantations, where brahma cattle were tethered to the trees. We passed several spring-fed pools, and the sulphurous vapours were in competition with the smell of cattle dung, drying seaweed and wood fires roasting fish on sticks. There was field corn drying on mats beside the road, mattresses drying on a shack roof, washed clothing drying spread out on shrubs, stacks of coconut husks ready for loading to the copra market and fruit stands at every corner, many of them featuring *mangis* (mangosteen), for which I developed a craving. Brightly dressed school kids were coming home from school, their mothers holding umbrellas over the little girls so prim and proper, the boys being rambunctious in the road ditches. Cristina explained that the mothers wanted their girls to have white skin, but with the boys it didn't matter.

We got to the airport within about ten minutes of departure time, but of course the Garuda airplane didn't leave for another two hours in late afternoon. After paying our departure tax twice (one of them was probably a rip-off bandit with a pistol on his hip), we flew right over one of the huge volcanic cones inland from Manado. My seat mate was journalist Phil Erari from Irian Jaya, who wrote about environmental and political issues of the western part of the island of New Guinea. He had attended university in Basel, Switzerland to study ecology and oceanography, a rather odd place to study oceanography. I told him about my efforts to get research ships into the Mamberamo River estuary and Centriwashi Bay (on the north coast of Irian Jaya), and he explained that this region was off limits to most people, being a pet development project on the twenty-five year government plan out of Jakarta. He said the only research ships that would be permitted in that region would be the LIPI ships. I gave him three of our TROPICS Prospectus, and asked him to write to me about a useful pathway into the political situation of the Indonesian Navy and the LIPI ships and oceanographers.

We landed in Ujung Pandang, the ancient capital of southern Sulawesi, what used to be called the Celebes. From the air, the coastal plain was almost totally covered with fish and prawn ponds, complete with walking paths on berms and small cottages for the fishermen.

The Portuguese and Dutch had colonised this region, and a very European 'Fort Rotterdam' was in the centre of the city, the walls of this 16-17th century fort being shaped like a giant turtle. I was startled to see these ancient fort walls from the airplane, as they appeared similar to the English Hudson Bay Fort Prince of Wales built in the 17th century across the river from the present town of Churchill, Manitoba.

This Sulawesi city is on the SW coast of the island, facing the Makassar Strait, the main channel for north Pacific Ocean water to flow through the Indonesian Archipelago into the Timor Sea and Indian Ocean. The Makassar Strait is also the dividing line between Alfred Russell Wallace's Asian (Sunda, Orientalis, Sumatra, Malaysia, the islands of Borneo and Java) and Australian (Sahul, Wallacea, Sulawesi, the Moluccas, Halmahera, Ceram, Lombok, Timor and New Guinea/Australia) ecological and faunal zones, drawn in 1859 while he was collecting bird skins. To the north and west into SE Asia, you will find monkeys, tigers, elephants, and woodpecker kinds of birds, whereas to the south and east of Wallace's Line you will find marsupial mammals (kangaroos, wombats, koalas) and many parrots not found anywhere else.

The huge airport was neat and clean, with many shops and food stalls, some of which were filled with Dutch goods: the Makassar people were famous for their silver and gold smiths, wood and ivory carvings, penis gourds with leather thongs, garishly coloured tropical shirts with shiny metallic threads, long silk shirts with fancy embroidery (I bought one) and long distance sea trade and piracy. The Makassar Bajau and Buginese people built elegant ocean-going schooners long before European invasions, and they traded goods to the Aru Islands off Irian Jaya and the north coast of Australia. The Kimberley coast and the Gulf of Carpentaria in Australia were one of their sources of *trepang* (dried sea cucumber) and mother of pearl (mollusc shell) for knife handles, and they probably brought their dogs and tamarind tree seeds. I bought a Makassar-crafted silver hair clip for Sheila, wondering if it was really silver. This part of Sulawesi was Islamic, with many women with their faces covered, the men in little fez hats, and the call to prayer sung over the airport loudspeakers.

As many have discovered before, travel provides a kind of education that is rarely found in schools. Leaving Ujung Pandang on Garuda Air, I found my seat mate to be a German (Landau) missionary fellow who was trying to coordinate development and education for the native people of the Mamberamo River delta region of the north coast of West Papua. I told him of our hopes for coastal oceanographic research in this region via Project TROPICS and our involvement with LIPI in Jakarta. In typical Germanic style, he then proceeded to give me a two hour lecture on the social problems of northern Irian Jaya.

Most of the native people in Irian Jaya are closely related to the Papuan cultures, and have little allegiance to the Malay/Indonesian culture and government. They had been loosely governed by the Dutch from 1602 to 1945, until the World War II invasion by Japan. After the defeat of the Japanese, Sukarno captured the country for Indonesia. Part of the present day Hababie government twenty-five year plan was to export millions of people from crowded Java to Irian Jaya, the *transmigrassi* project. This project also exported millions of Javanese to drain the rainforest swamps of Halamahera and establish coastal cities. These projects were meant to grow more rice to feed people back in crowded Java, and to build cities and industry for steel, aluminium and timber production. Apprently many of these transplanted people were politically undesirable radicals and intellectuals. The Javanese and Koreans were logging and burning the rainforests of the Mamberamo floodplain, cutting down the precious sago palms, bulldozing the sweet potato gardens and family coconut plantations belonging to the native people. As you might guess, the rainforest swamps had poor peaty soils that once cleared and drained, were not suitable for rice paddy operations, so nobody benefited from this savage transformation of the land and human culture. Much of the estuary had been polluted with attempts to grow rice, build industry, and the land was covered with plastic soft drink bottles, Coca-Cola and beer tin cans, plastic litter, and abandoned corrugated metal sheds that nobody wanted to live in. The Papuan people did not trust the Indonesians, refused to go to their schools or speak their language, and most of them migrated up the river to find a new home.

My German missionary seat mate was well educated, was fluent in most of the Indonesian and Irianese languages, and he shuttled between the Mamberamo River valley, Jayapura, Jakarta, Basil (Switzerland) and Heidelberg. He also explained to me that most of this airplane was full of a Manado soccer team en route to Denpasar for an Indonesian wide tournament. They were singing football songs and some Christian hymns, and the Garuda stewardess appeared to be leading the songs. Coming into Bali, we flew right over the top of several volcanoes, where wisps of smoke and steam could be seen down inside the caldera. The rowdy soccer team applauded and cheered the pilot's soft landing in Denpasar, where the black volcanic sand beaches were waiting for their pleasure.

Gani Ilahude and LIPI were partners in hosting the Manado coastal science meeting and supporting the costs of some of the external speakers, including me. LIPI sent a posh black Toyota Land Cruiser with darkened windows, air conditioning and a loud radio, to pick me up at the Denpasar airport in Bali, with Joko from LIPI as a driver and guide. With much honking of the horn, dodging pedestrians, bicycles and potholes, he delivered me back to my disgustingly posh Dynasty Hotel on Kuta Beach. Air conditioning. Almost too cold in my room to take a shower! I walked the beach to see the sunset. My hotel supper was on the patio by the beach, a buffet featuring Italian food, pasta with chili, seafood in giant woks, pickled veggies, and mango cheesecake with espresso. A four piece band (two nylon string guitars, conga/bongo drums, a white painted bass) was wandering around playing pop/jazzy tunes. They were good musicians and they performed credible versions of Beatles and Harry Belafonte songs. It just about blew my mind when they started singing 'Do You Know What it Means, to Miss New Orleans?'.

The backs of my legs were sunburnt from snorkelling on Bunakan Reef, so I reeked of clove oil balm. Having slept on the airplanes, I couldn't sleep in this hotel, so I watched bad movies (*Blade Runner)* and ate *salik* and *mangis* in my room.

I had a full day layover here, so I slept in till 10:30 and had a wonderful 'American breakfast' with rich black hot coffee spiced with cloves and cinnamon. It was blazing hot outdoors, but the hotel pool and beach was thick with Dutch, English and Australian tourists, most of them pink with sunburn, many of them loud and rude to the quiet gentle Balinese trying to serve them with drinks and finger food. Motorised *praha* pushed up on the beach to sell trinkets and 'antique' carvings. These peddlers knew their business, as they took lots of money from these tourists.

I had my evening meal in a beach longhouse that had a stage for performers. While I feasted on nutmeg and clove spiced lamb, sate of unknown meats with many hot sauces, pickled vegetables, exotic fruits and cheesecake, a full gamelan orchestra began setting up their instruments. After a short musical introduction, the big gongs announced the entry of the dancers, all in full costume with painted faces. They did four long historical story dances, including one that was called the 'Frighten Enemy Dance' that probably incorporated the ancestry of the word 'berserk'. Then there was the inevitable lounge piano and singer, which included the blues 'Nobody Knows You'. The tables around me were occupied by obviously wealthy Dutch, German, Norwegian and Danish families drinking port. The men were smoking big cigars. The ladies were smoking little cigars. The Dutch family demanded and got a bottle of Advocaat (egg brandy), which seemed a bit out of place in Bali, but then this hotel was certainly set up for Europeans and Americans.

I checked out of my room and sleepily waited in the lobby till 10 pm, when Joko honked in the driveway to take me to the airport. It took the full two hours to get to the airport in sluggish traffic, nobody seemed to be in a hurry. The airport duty free shop had clove cigarettes, cigars and macadamia liqueur (made in Australia). It was full of sunburnt Aussie tourists loading up on the way home. We finally left Denpasar at 0100, and they served us another good supper of cold Bintang and spicy chicken in rice.

We arrived in Darwin at 0435, all groggy, trapped in the airport waiting room, as there was no customs inspection until we got to Cairns. It was supposed to be a short layover, but Cyclone Justin was on top of our destination in Cairns on the Queensland coast. They had no electricity and the airport was closed. So there we were, two 737 loads of passengers jammed in a small waiting room in Darwin. They finally opened the bar and coffee shop to us. I re-booked to go to Brisbane, departing at noon, and in the meantime I phoned Sheila to revise arrival time.

Here it was again, the international collegium of science, with young and old scientists from UK, USA, Germany, Holland and Australia, telling stories of adventures at sea in SE Asia. My good friend Andrew Wood of CSIRO, a retired PNG Kiap and full time agricultural scientist was there, as well as Wally Zuk and Ross Jaffrey of ANSTO, going back to their labs from a visit to the Rum Jungle uranium mines in the Northern Territories. The Aussies delighted in telling wild stories to the North Americans, who were definitely out of their comfort zone. Eventually the airplanes flew us to Brisbane and I got to Townsville by 9 pm, where we were on the margins of the cyclone up north. There was my bright and shining Sheila with son Jesse. They drove me through the driving rain to the Baskin-Robbins ice cream shop on the way home.

By 24 March AIMS and JCU were closed, many of the highways and town streets were flooded from rain and high tides/storm surge, and everybody was preparing for the cyclone coming south from Cairns and Innisfail and Tully, where the banana plantations were flattened and a few roofs were ripped away by the wind. We cleaned up our yard, closed up our shed and battened down our windows and doors. Our house is on a high ridge, so we had no fear of flooding, but some of our creek valley fruit trees and banana patches were laid down on the ground by strong winds. A boy died from electrocution in town, walking over live lines from downed power poles, and a lady was killed in a landslide on a nearby coastal mountain road.

Two weeks after my trip to Manado, the AIMS workshop electrical engineer Eric Gill invited us to a party at his house in Townsville. His wife Rose was a Minahasan from Manado, and they had invited all the Indonesians at JCU. Rose cooked up a wonderful feast of reef fish, sate, and spicy pork, brushed with lemon grass basting brooms. We brought our ice cream freezer to make home-made ice cream, which was a big hit. We had Manado people singing songs about their Banakan reef, an Ambonese fellow singing about the big groupers, and a girl from Java who did a little dance while clapping hands and snapping fingers. Rose explained that all Indonesians went to schools that had the same music instruction books, so they all knew folk music from all of the many many islands of this archipelago, as well as some European, UK and USA popular songs. Sheila sang some Canadian songs, and the audience was very quiet to hear such a strong singer. Eric quietly explained that his wife Rose had been one of the leaders of the Resistance Army in Manado during various revolts and changes in ruling governments, and she knew how to operate a machine gun in the jungle. In addition to the Manado scientists, some of my Jakarta LIPI scientists were with us, including some high level government lab leaders, such as Anugerah Nontji and Gani Ilahude. Nontji was to become an important Indonesian representative for TROPICS, and this tall quiet man led the Indonesians in singing a song about cockatoos.

Chapter 31. Dancing With Radon Again, TROPICS Madness On Ships, and Princess Diana In The Jungle 1997

CSIRO radiochemist Gary Hancock, John Pfitzner, Tenshi Ayukai and I took the *R/V Harry Messel* up to the Herbert River and Hinchinbrook Channel mangroves in early April 1997 to try to use the radium/radon isotopes to estimate the hydrological and geochemical flushing of mangrove forest tidal creek sediments.

We could measure the very low activities of ^{223}Ra (half-life, 11.4 days), ^{224}Ra (3.7 days), ^{228}Ra (5.8 years), ^{226}Ra (1600 years), and some of their very short lived radon decay products in incoming tidal seawater, and then measure very high activities of these isotopes in low tide draining of the water from the crab holes and sloppy mud into the creek. The very short half-lives of these isotopes gave us a chemical clock to model the rate and source of supply of salts, organic matter and nutrients from the open ocean and from river mud trapped in the mangroves. This process also sucks nutrients out of the tidal forest floor to fertilise the nearby coastal water. Tenshi was measuring the growth rate of algae in the coastal water during high and low tides, to find that low tide water draining from the mangrove creeks was richer in most nutrients. We wanted to perform transects from the inshore estuarine mangrove creeks out to the shelf edge but 40-50 knot winds prevented this, and we had a very rough ride home.

The morning of 16 April 1997 Sheila went into a complete tizzy, washed her feet, washed my armpits and applied deodorant, ironed my shirt and pants, even got us into some shoes... SHOES?...we hadn't worn shoes since we arrived in Australia. This was all in preparation for our Australian citizenship ceremony at Townsville City Hall with Mayor Tony Mooney. Jesse escaped from his university classes to be with us as we recited the pledge (without God and without the Queen version). And then Tony clapped his hands, we were all citizens, and Tony gave us a tree in a little plant pot. Sheila actually held citizenship in the UK, Canada, and Australia, and I was a dual citizen of USA and Australia.

I, Chuck Nittrouer (University of Washington in Seattle), Anugerah Nonji, Gani Ilahude (LIPI, Jakarta), Gaikovina Kula, Hugh Davies, and Patricia Pepena (Environment, Marine Research, PNG Government), Bernhard Peucker-Ehrenbrink (Germany), Jim Veness (Environment, Ok Tedi Mining Ltd.), and Hodaka Kawahata (Geological Survey of Japan), along with quite a few Australian and USA coastal oceanographers, had been working hard for several years to plan and equip a six year research consortium for work on the big rivers, estuaries and continental shelves of the island of New Guinea. On a global scale, the little island of New Guinea delivered a huge amount of water and sediment to its coastal seas. Although I had been on ships into New Guinea coastal shelves several times, the formal Project TROPICS was now beginning.

The National Marine Research Vessel *Lady Franklin* was tied up in the port of Townsville on 6 May 1997, loading equipment for TROPICS Legs 1 and 2, chief scientists being my friends physical oceanographer George Cresswell (CSIRO) and sedimentologist Ken Woolfe (JCU Earth Sciences). Over the next weeks, George mapped the submarine canyon of the Sepik River, increasing our knowledge of the New Guinea Coastal Current and the NG Undercurrents in the Bismarck Sea. His super-technicians deployed my big sediment trap array at the shelf edge and deep water Coral Sea. Ken Woolfe began a sediment coring program to map the distribution of river mud on the Gulf of Papua continental shelf.

Deck rigger John Soles, Chuck Nittrouer and I joined the *R/V Franklin* in Port Moresby for our special expertise in Gulf of Papua sediment coring over 3-11 June. Ken's JCU crew were interested in long Kjullenberg cores to see millennial scale sedimentation history of 10,000 years of the Holocene. Chuck and I were interested in century scale sedimentation processes and physical/chemical changes over short periods of time.

After clearing PNG customs and immigration, we were hauled to our hotel behind locked gates and high barbed wire fences. Apparently there had been several incidents of kidnapping of white tourists by the 'rascals', and we were prime targets. We were rejected at the hotel restaurant because we were wearing thongs/flip flops and shorts. So we went back to our rooms to get our grubby safety boots, the only proper shoes we had with us. Chuck had brought several jars of especially strong Seattle horseradish for Sheila, but the jars had broken inside his boots. So he appeared at the entrance of the restaurant with a good fragrance.

The next morning the *R/V Franklin* arrived from the north coast work, our colleague John Milliman and his technical crew got off, and we got their bunks. We opened the hold to bring up and assemble our Kasten corer and Chuck's box corer and X-ray apparatus. We spent the next seven days working aound the clock on the aft deck, taking Kasten and box cores and processing the samples. With John Soles' help, we successfully used the deck box pinger to locate and activate the acoustic release of six moorings of our big sediment trap arrays, which had been deployed by George Cresswell earlier this year. And low and behold, up they came, their bright yellow floats leaping into the air, to be met by our grapple and dumpy hooks. We took samples from the trap cone bottle and re-deployed them to continue their work collecting the rain of particles *en route* to the seafloor. We steamed home to Townsville in 40 kt winds, arriving 0800 for Customs to inspect all our mud. It took two AIMS flat bed trucks and two utes to get all our gear off the ship and back to AIMS, and then we collapsed to sleep for a few days.

But not very much rest. In mid June, Sheila and I flew to Cloncurry and Mount Isa to perform two 45 minute sets of music at their folk club. Sheila gave a hammered dulcimer workshop and I gave a frailing banjo workshop. Most of the banjo players there used Earl Scruggs bluegrass style with metal finger picks. We met some real Isa characters out there, Butchie the goat, pig and camel hunter, Alan Shaw, who had a Pete Seeger long neck banjo, and Bazza Rogers, a miner, who sang ballads of bad times in Vietnam. Bazza also gave us a tour of the deep lead and zinc mine, Sheila and I being dressed up in overalls, hard hat with lamp, safety glasses, harness belt and big boots. We tootled around very deep underground in the hot dripping wet tunnels in a Komari truck, where many people worked their lives away in the dark. The little country Flight West airlines didn't bat an eyelash at our luggage (guitar, hammered dulcimer, autoharp, banjo, fiddle) and they crammed it all in the forward bathroom. We even got complimentary champagne on the way home, over the dusty red ridges and braided dry river beds to the coast.

All this was interrupted in July by an extensive workshop series of safety, SCUBA, first aid, fires in the ship engine room, radio communications, distress signalling, medivac procedures, helicopter safety, survival on the water and on a coral island, and ship rescue training sessions for two weeks. Not too surprisingly, nearly all of this training was useful in the next decade. We got trained in how to launch the lifeboats of various designs, some of which were not seaworthy.

I got selected for jury duty. It was only for the trial of a local fellow who had chopped up his wife with a machete over an argument about the TV remote control, buried her in the back yard, and the neighbour's dog dug her up and dragged an arm into the street. This was made even more bizarre as the court house was close to the Perc Tucker art gallery which was displaying a collection of Albrecht Durer's woodcuts from 1512: Some highlights were *The Prodigal Son amidst the Swine, The Torture of St. John the Evangelist in Boiling Oil, The Whore of*

Babylon, The Circumcision of Christ and *The Beheading of St. John the Baptist*. This was certainly an alternate education! Also on display were Michael Wolgemut's 1493 depictions of *The Anti-Christ, The Beheading of Faith, Hope, Charity, and Wisdom*, and a very sensual and voluptuous *Crucifixion of St. Regina*. I'm not sure how this viewing of medieval art influenced my actions in the jury duty.

On 22 August 1997 the *R/V Franklin* returned to the Townsville wharf from four successive Project TROPICS research voyages on the north and south coasts of Papua New Guinea, involving seventy scientists from six countries (Legs 1-4). Amongst other things, my big sediment traps had been deployed and recovered with a month's particle rain collection. All of this equipment was unloaded from the ship to AIMS trucks on the wharf and hauled to AIMS for temporary storage until the next *Franklin* cruises in following years.

At the same time, I was helping to load gear on the AIMS *R/V Harry Messel* for TROPICS Leg 5, with Ron Szymczak from ANSTO, J. P. Walsh from the State University of New York, Pete Swarzenski from the University of Louisiana (LUMCON), Peter Davies from Southern Cross University, Michael Holzapfel from ANU, Irena Zagorskis, John Pfitzner and Bruce Parker from AIMS. We were using the small *Harry Messel* because we wanted to go into shallow water of the mangrove estuaries of the south coast PNG Fly, Kikkori, and Purari Rivers to measure freshwater to seawater salinity gradient changes in phytoplankton, chlorophyll, particulate organic and inorganic carbon, nutrients, trace elements and the radium/radon radionuclides. The ship crew steamed north for two days, and the scientific crew flew up to meet them in Thursday Island (in the native language, Waiben) in the Torres Straits.

On arrival, we had a few free days in Thursday Island, waiting for our research ship to arrive and waiting for PNG ship clearance. So we wandered the streets, visited museums (Quetta disaster, fishing fleet disasters in Princess Charlotte Bay, ships lost in cyclones) and Crimean War/World War I cannons on the hill. We all bought gaudy islander T-shirts and dress shirts at Moma's Store. We had nice rooms at the Jardine Hotel where the fresh seafood was excellent. I visited the church where the white 'protectors' tried to lock in the native shell divers in the early 1900s and bully the local people into diving for pearl shell for low wages (see Nonie Sharp's book). The native people refused to deal, and jumped out the window, because the doors were locked. This is an example of defenestration. I had visited a museum in Prague which described 'The Defenestration of Prague' in 1618, when politicians were thrown out a window, one of the events that started the 'Thirty Years War'. Nowadays, defenestration refers to computer nerds abandoning MS Windows to use Linux.

The *R/V Harry Messel* arrived at Thursday Island just at dark and anchored in the bay away from the wharf. Our scientific party used a zodiac boat to board our ship, and immediately we went through a long series of safety and fire drills, survival boat launching, engine room fire drill and medical evacuation options. AIMS people had been through these training sessions repeatedly, but AIMS was nervous about having so many foreign scientists on board.

We were steaming past Bramble Cay on the morning of 26 August '97 to begin our transect from full strength Coral Sea water at salinity 34. Irena was chief data recorder, and managed the salinity meters and nutrient sampling. Ron was taking superclean water samples for trace element analyses (Cd, Cu, Pb, Fe, Mn) in his lab at ANSTO, samples for osmium isotopes for Sholkovitz at Woods Hole, Puecker-Erhenbrook in Germany, and Candace Martin in New Zealand. Ron also used the ship's zodiac to get away from diesel fuel and rust contamination from our ship, and took acid washed glass bottles and expensive GoFlo Teflon bottles to sample water for Kathy Burns's measurements of hydrocarbons in water and suspended matter. Ron also spent many late nights trying to get Kathy's 'supersuckers' working properly, as they did not like working in freshwater at all. Peter Davies was determining some of the nitrogen and phosphorus nutrients and the photosynthetic pigment families of chlorophyll in the ship lab, and preserving or freezing samples for further analyses back in the home lab. John managed the large volume sampling pumps and cartridge filtration systems for the measurement of the radium and radon isotopes in cubic meter sized samples. He also obtained large samples of suspended sediment for geochemical analyses back in the AIMS labs. Mike was responsible for shipboard titration of the carbonate system for Bradley Opdyke's estuary models.

My job as chief scientist was to coordinate science activities with the ship master and crew, keep all the scientists busy and happy, and to take regular bottom sediment samples with a big heavy dredge. A big problem was trying to avoid metal and hydrocarbon contamination from the ship's exhaust, the smoke from forest fires on land, and the normal grease and oils that are necessary to keep the ship's engine, winches and anchor line running well.

This routine continued up the Fly, Kikkori and Purari River estuaries, sampling every other salinity change to zero salinity fresh water. We could work several days and nights, and then we would all get tired and ask for a night's rest at some quiet anchorage behind an island in the estuary. I got so tired I could sleep standing up. When I closed my eyes, my cheek bones felt like they were dissolving. If I was asleep sitting on an aft deck bench with my back against the bulkhead, I would waken when the pitch and vibrations of the diesel engines went down, which meant we were approaching the next sampling station.

To some observers it might have looked like a tedious job, doing this over and over again along the salinity gradient inland, but all sorts of bad things happened to keep us on our toes.

It rained almost every afternoon for two hours, and then the sun came out to broil the aft deck. Even with the ship's air conditioner, the humidity was close to 100% most of the time and all our clothing, bedding and towels were wet with either sweat or the rain. We hit a few submerged sandbars, and had to wait for the tide to come in to lift us off the sandbar. Ron and I broke one of Kathy's special superclean glass sample bottles, so we spent a week inventing a long story about how a crocodile ate it. Two sea snakes somehow crawled into our zodiac when it was being towed behind our ship. Our rest nights usually were made difficult by masses of mosquitoes, sand flies, and enough beetles to clog the air intake of the ship's engine room. Night and day we heard bird songs from cockatoos and swallows in the mangroves and big frigate birds flying overhead. The ship food was wonderful, huge roasts, chops, curries and disgustingly good desserts. It was supplemented by visits from the local delta villages, people in canoes wanting to trade seafood, venison and fruit for fuel, cigarettes and ice cream. On rest night we would gather on the aft deck, smoke cigars, tell sea stories, then retire to the galley for tea and crumpets with bluegum honey. Sometimes we watched Hollywood films, the strangest one starring Bruce Willis, Brad Pitt, and Madeline Stowe in *Twelve Monkeys*.

It was an interesting life, and I often wondered how a South Dakota boy got tangled in such a situation.

We hadn't been inside the estuary of the Purari River before, so we slowly explored Port Romilly, an inlet bay leading to the Pie River and the Wame Channel of the Purari River delta. Huge mangroves, coconuts, branching *Pandanus*, tiered ranks of *Nypa* palm, red and yellow hibiscus trees greeted us, and it smelled different from the Fly River estuary.

We discovered the community of Baimaru, which wasn't marked down on our navigation chart. They had a wharf, some sheds and shops, a school, a first aid medical house, a sawmill, piles of logs, thatched houses on stilts, and an airport for Twin Otter airplanes and a helicopter. Laura Bada barges from Moresby were tied up at the wharf, full of supplies for the store. The helicopter was owned and operated by the grocery store owner, Philip Taudevin, who treated us to cold soft drinks. He could provide helicopter services, lodging and logistic support for research in this area. They proudly showed us the radio room, fax and telephone connections. He said this valley had over a thousand people spread out for mud crab and barra pond culture, and coconut, banana, rice, and tropical fruit farms. He was concerned about the Malaysian crews moving into this region to cut down most of the big rainforest trees, using Cats with bulldozers and truck tractors, leaving the land bare for erosion and weeds, and paying no attention to the local people. He gave us instructions for navigation of the Wame Channel, the westernmost channel of the Purari River Delta, it being navigable for 200 miles inland. He also gave us critical information about navigation into the Kikkori River, which had an oil pipeline running down the river bed.

So on 1 September, we confidently steamed up the Wame Channel past Arana and Kapuna villages, to find freshwater almost fifty miles up the channel. This was very muddy water, like heavily milked coffee. It was draining a different geological watershed and had different chemistry from the Fly River. This was better than we expected, and Ron and I took a zodiac up river to get clean freshwater samples.

Our solitude on the river was broken by a flotilla of canoes full of excited young people chattering, eating candy bars, dripping with mango juice and pointing at us, while we were desperately trying to be superclean. So we gave up, and one lady with a full suit of clothing stood up in her canoe and loudly informed us that 'Lady Princess Diana' was dead, in very correct London English. It turned out that she was the local school teacher, and she was horribly embarrassed that most of her happy smiling students in the canoes were totally naked. Ron and I had no idea what she meant by Lady Diana, maybe some local important person? It wasn't until we got back to our ship out in the Gulf of Papua, where radio links to our home lab worked well, did we find that indeed, in the UK Princess Diana, former wife of Prince Charles, was killed in a car accident near Paris, France, along with her boyfriend Dodi Fayed and her Mercedes-Benz driver Henri Paul. I guess the jungle telegraph worked pretty well.

We laid anchor at dusk in outer Port Romilly for roast lamb chops, sweet corn, roast porcupine potatoes and a good night's rest. The cook had traded a sack of flour to the local people for mud crabs and bush turkey eggs, which were excellent for breakfast the next morning.

The next morning we cautiously followed the Chevron/Kumul Oil Pipeline signs up the Nakari (Kikkori) River delta channel. It would be easy to get lost in this delta, as you needed to go through many large and small channels, past big and small mangrove islands, many islands with small villages with cheering kids on the river bank hooting at us.

We kept going inland until we found freshwater in the river, near the village of Kikkori. Here we encountered huge bumblebees, sweat bees and very large biting March flies to make us bloody and miserable. Shoreward were endless coconut plantations, big coconut husking operations and piles of copra, sago palm plots, and hibiscus, poinciana, *Nypa* forests protecting river bank caves. The freshwater here was again different from the Fly and Purari Rivers, as this river drained limestone country with volcanic basalt in the midst of it, and the river was nearly milky white in colour, much like the South Dakota White River where I grew up. After we collected several cubic meters of water, getting grounded on submerged black mud banks twice, we hightailed, yes skedaddled, even more so we defungerated it back to the deeper water of the Gulf for supper and rest.

We had finished most of our planned work and I was writing the cruise report and notes for a lecture I was expected to give on my return at JCU. We dissembled our aft deck gear for storage in the hold and brought out the vibra coring gear for J. P. Walsh. I also had to account for every water and mud sample taken in PNG, for the export records of the PNG Customs and Immigration. They were worried we were stealing their biodiversity and samples for medical research, which probably had happened before, but certainly not by us. We contacted both PNG and Australian Customs and Immigration to warn them of our approach the next day or two. Our administrators back at AIMS got us on the ship radio, informing me that it was AIMS policy not to allow foreign people on our research vessel. What a wonderful executive support team back at AIMS, inspiring in a negative sort of way. We had been organising TROPICS for five years, and they finally woke up.

Our USA and German colleagues were now waiting for us at Thursday Island, in Australian territory, for the next leg of this TROPICS research cruise. We contacted our colleagues to apply for PNG visas and gave them the AIMS edict, which we promptly ignored. We were in Daru the next evening for the PNG Quarantine Service inspection (all in neat white knee socks, pressed shorts, khaki shirts with epaulettes, and even Ray-Ban sunglasses, even though it was dark), they came aboard to crawl around in the hold inspecting our nally bins of neatly sacked mud, freezers near full of frozen filters and suspended sediments. The ship Master Matt had a few cartons of cigarettes, which was enough to send them away with signed forms on their clipboards.

We arrived in the bay of Thursday Island the night of 4 September, amidst lavender phosphorescent seas and jumping dolphins around our ship. Irena and Pete went ashore in the zodiac to hunt for the next science crew, J. P. Walsh, Peter Swarzenski and their technical helpers (from the USA) and Bruce Parker from AIMS. They had sent their passports to Brisbane via express bags for PNG visas, and were happily sitting on the veranda of the Jardine Hotel. After a long shower at this hotel, it was strange to notice the absence of the smell of sweat and mould, the ship's exhaust, and the endless sound and vibrations of the ship's engines.

My science crew flew out of Horn Island on 5 September, flying over smoke plumes from fires on Cape York, and the wonderfully clear blue water of the Great Barrier Reef. I need to insert here that Irena and John were destined to work for at least a year on the precious samples we had taken, using the AIMS laboratories and equipment. They would gradually report to me progress on the analyses for nutrient elements, major and trace elements, radioactive isotopes, and tabulated records of locations of sampling stations (longitude, latitude), salinity, temperature and local weather.

Jesse and girlfriend Sarah (in an iconic old Toyota ute) were at the Townsville airport to collect me with my tubs of frozen filter and mud samples for the freezer. Sarah had two new galahs, an Indian Ringneck parrot, and a very large python as pets at home.

And now I should explain about the appearance of Sarah Jane Hazzard in our family. As is probably usual, fathers rarely know what is going on until the critical bits are over and settled. Jesse had been working part time hauling sugar cane to the mill down on the Burdekin Delta and part time as a bartender at Melville's pub just down Alligator Creek Road from our home. As I learned much later, Sarah and a lady friend had decided to have a VW van trip around Australia. Well, she got as far as Melville's pub, less than a half day's drive south, when she met Jesse behind the bar. The connection clicked, Sarah abandoned her friend and VW van, and came to our home with Jesse. She fit in very well, as she was a wildlife nut, connected with wild creatures, was a good snake catcher, and eventually worked at our local Billabong Sanctuary.

GJB and Casper, Alligator Creek 1997

As usual when returning home from a few weeks on a ship, the floor of our house seemed to be rising and falling, and I had to balance as if I was on the ship. I got busy whippersnipping the grass and weeds in our growing orchard, and even harvesting some citrus fruits. We had our twenty-seventh wedding anniversary, quiet and peaceful, on 8 September, but then Ron Szymczak, Jesse and Sarah demanded that we go out to a good

Thai restaurant to celebrate. Sheila brought a big green tree frog in her pocket, to put down Jesse's neck at supper. I was fifty-six years old on 10 September, and I was wondering when I would have to grow up and behave.

John Pfitzner and I had to suffer through a three day radiation safety training class. Most of it was for high hazard materials that we would never touch. All of our gamma spectrometer work was on very low level activities of natural and 1960-70s bomb-fallout radionuclides. I was also helping Joanne Cavanaugh with her Ph.D. thesis, which was linked to our Herbert River Project, trying to describe the mass balance of organochlorine pesticides in sediment cores of the Herbert River Delta and Hinchinbrook Channel.

Meanwhile, all was not well with TROPICS Leg 5B on the *R/V Harry Messel* up in Torres Straits. When we left, all systems were go and the science crew was fired up for action, but the ship's cook Bob, engineer Al, and deckie Jason had fallen sick, badly enough to be in the Thursday Island hospital. Two of them were diagnosed with serious kidney infections and were air lifted to the Townsville hospital. Alarm bells went off at AIMS. The safety administrators wanted to know details of what we ate and what kinds of poisons were on the ship, and whether this sickness came from eating mud crabs, bananas and venison obtained by trading with native people in the river deltas. The ship was swabbed down for clues, then washed down, disinfected, the kitchen dissembled and cleaned, but no clear cause was found.

It was curious that none of the scientific crew got sick during our eleven day voyage, eating the same food as the ship crew. In any case, the USA scientists and their technicians sat in the Jardine Hotel for five days until AIMS sent a new ship crew to Thursday Island on 9 September, and then TROPICS Leg 5B research team went to work in the muddy water of inshore Gulf of Papua.

In late September I was invited to Innisfail for a Great Barrier Reef Marine Park Authority (GBRMPA) sponsored four day workshop on coastal wetlands, where I gave my expanded views on the agricultural, mining, and urban pressures on the estuaries and inner shelf of the GBR. I also presented our data on mercury pollution from sugar cane fungal inhibitor Shirtan (an organomercurial fungicide) in sediment cores from nearby Hinchinbrook Channel and Rockingham Bay. I talked about our research on the saltpans near Townsville and the relatively great age of the brine inside them. Coastal wetlands, swamps, saltpans, and mangrove estuaries are important for the health of the World Heritage Area GBR, as they tend to trap terrestrial erosion products and agricultural/industrial contaminants, and they are often important nursery areas for fish and other marine organisms.

Although they had no authority on land, GBRMPA was doing the right thing by trying to engineer various mechanisms to protect existing and newly constructed wetlands. This would include tax breaks for wetland preservation by cane and banana farmers, and restrictions on draining wetlands and landward mangrove forests. In the midst of the workshop, Anne Clarke, Sheridan Morris and Warren Lee Long took a few of us on a tour of the Eubanga Wetlands, supposedly a protected wetland, where there were sugar cane and bananas growing right up to the boundary. The wetland itself had 2 m deep drainage ditches on the seaward boundary, so that the original wetland could no longer hold back land runoff, and it was bone dry, with parts of it on fire. In some other locations, there were drainage ditches with one-way flow chutes inside the mangrove zone, to keep salt water out of the cane land and release freshwater to the sea.

It was an El Nino year, and it was very odd to have the nearby wetlands and rainforest dry, and forest fires in the region. There were international wetland experts from the Netherlands, France, and the USA, and a few RAMSAR officials concerned about conservation and preservation of wetland species. The whole workshop was an embarrassing jurisdictional and resource exploitation catfight amongst GBRMPA, Department of Natural Resources, Department of Environment (State and National), JCU, various sugar cane grower and miller organisations, the upland cattle industry, the North Queensland Conservation Council, several branches of CSIRO, the commercial and sport fishing industry, the tourist industry, several Rural Municipal Offices and Town Councils, and two aboriginal organisations.

Harry Bonanno of the Sugar Cane Growers Association took exception to my suggestion that their organomercurial fungicide could be polluting Hinchinbrook Channel biota. I had no data on plants and animals, so I kept my mouth shut. Unfortunately a reporter from the *Cairns Post* newspaper gave this story, causing some stir amongst the anti-environmental groups in attendance. Most of the state and national government departments were doing bugger-all for wetland preservation, and there appeared to be no regulations for conservation of wetlands, whereas there were many rules and regulations in the Great Barrier Reef marine environment and the fisheries industry. I met a few cane and banana farmers who had already built small wetland step ponds to trap runoff from their agricultural activities. Behind all this whinging, the quiet aboriginal people make their claim for most of the coastal wetlands.

The only useful and pleasurable result of this workshop was to meet Sheridan Morris, a Natural Resources officer, whose father had a tropical fruit tree farm inland from the little town of Babinda. Joe Zappala of Harvey Creek proudly walked us around his exotic fruit trees, many of which don't have English names, most of them from South America, SE Asia and China. Sheridan invited us to supper with her family, and we had durian pie for desert. Now, in its freshly harvested state in Indonesia and Malaysia, durian smells like a long dead animal. But the meat of this fruit had been whipped, frozen and topped with brandied whipped cream. Pretty damned good.

In October '97 we had yet another TROPICS meeting at AIMS with Australian, the USA, PNG and Indonesian government officials and scientists, mainly to sign a 'memorandum of understanding' between the respective governments. This was concerning exchange of personnel on ships, use of equipment, customs and immigration clearance for ships and science equipment, data management and intellectual property, and options for commercialisation of bioprospecting discoveries, insurance and medivac options. This was to cover a dozen separate research cruises over six years on five different research vessels in territorial waters of three countries (some of which had already taken place).

This was a diplomatic and logistical nightmare for the institution CEOs and their departments of foreign affairs. For example, there were fifteen Indonesian ministers involved in marine science and fisheries, and six PNG ministers involved. There were some delicate discussions of past collaborations that had gone sour, and how to avoid this in the future. The Indonesian government said that they could not guarantee our safety in some parts of West Papua (Irian Jaya), but we were probably safe at sea on one of their Navy research vessels. Most of the working scientists were bored with it all, and wanted to go to the pub to drink beer and tell stories of adventures on ships at sea. This workshop was held at AIMS, where we organised accommodations, meals and evening amusements for most of our overseas guests, that in itself being a major social, diplomatic and culinary operation.

I had now worked for federal government research organisations for all of my professional life. One of their strange characteristics is their tendency to disembowel themselves periodically, question their purposes, and make grand plans to reform themselves with 'strategic directions', strength and weakness lists, decadal research plans, and identification of stakeholders and clients.

My institution, the Australian Institute of Marine Science, had an annual budget of $A24 million, a staff of 170, two ships, a decade-old research plan, and reported directly to the Environment Minister in Canberra. [This was small potatoes compared to the much larger CSIRO, or most American or European research institutions.] We also had a Science Council of Directors, important people who assisted in defining and focusing our directions and science policies, and they all came to AIMS in November 1997 for a science conference. Many of these Directors were industrial scientists or CEOs, political and/or policy wonks from state and national government departments. And of course, the big divide amongst the Council was about whether it was enough to do work for the 'public good', or whether we ought to be providing economic and technological stimulation, innovation and measureable outcome products for the market. We were being urged to cooperate and communicate better with our clients and stakeholders, and even better, we should seek to get them to invest and pay for our work on their behalf. If you are bored by the words in this paragraph, your eyes are glazing over, then please imagine how this affected my fellow working scientists!

We all dutifully paraded our research activities with colourful slide shows, inserting comments about imagined clients and stakeholders, desperately trying to apply our critical input to policy decisions. I was proud of our Herbert Consortium and TROPICS organisation, and I brought in local sugar cane farmers, Indonesian, PNG, and USA scientific partners to explain our multinational efforts in describing new ways of viewing river inputs to the ocean and continental shelf dynamics. Our work in PNG ranked highly against international efforts to study the world's largest river inputs to the ocean, the Ganges, Amazon and Huange. But of course this did not impress the Council, who wanted to know how this all benefited the Australian economy.

I was reminded of similar questions in my previous life in Canada, when we had been working on atmospheric contaminants in the high Arctic for a decade...'Why should the Canadian government be spending all this money in the high Arctic, when all the people are close to the Canadian/US border?'. This was all very tiresome, and I desperately needed to escape.

For the above explained reasons, we took a week off work in early December to have a major 'bumble', that is, a trip with no clear destination or purpose.

Jesse had been customising his old Toyota ute, souping up the engine, installing two radio antennas and a big stereo with speakers in the doors. Jess and Sarah took off in this posh ute to deliver 'Stinky', an evil-mannered semi-tame rainbow lorikeet, to her parents in Tully, 200 km north of us. Sheila and I followed in our Subaru, also fixed up with a UHF radio for communication with Jesse's ute, so that Sheila and Jesse could consult on the next pit stop. We suffered along the way, stopping for ice cream at the Frosty Mango and lunch at the Mud Crab Café in Cardwell. We overnighted in a remote Tablelands resort called Possum Valley, a wonderful place where electricity came from solar panels and a twenty meter drop Pelham Pump. The resort owner, Paul Tredgett, also had a woodworking shop and built beautiful furniture from local rainforest tree stumps left over from logging a century ago. This region of North Queensland, the Tablelands, inland from Innisfail/Cairns, was a rich red soil high diversity rainforest over a century ago, but was extensively logged, and is now mostly grazing land for dairy cattle, with a few valleys and creeks that display fragments of the original forest. The Tablelands are several hundred meters above sea level, so the weather is cooler and suitable for some subtropical crops. We saw a lot of big Holstein milk cows standing at the barn gates with full udders.

Over the next few nights we went spotlighting, to search for all the eye gleam of the nocturnal possums, bats, Lumholtz tree kangaroos, boobooks and frogmouths, high up in the rainforest trees. Sheila and Sarah got up early for bird watching, to watch platypuses snootling around in the local creek, and spotted a Boyd's forest dragon too. In the early evenings the local cicadas started up their racket, like some demented group of castanet players. We visited too many waterfalls and rainforest shaded mineral springs with sulphur cyanobacterial mats. And of course there were many visits to the pubs for cold beer and a meat pie, or sometimes a bowl of ice cream with squashed lychees. At one place, pademelons lurked behind the pub, waiting for handouts of hot salted chips. Around Herberton, Ravenshoe and Mount Garnet, we talked with old timers who told how the tin miners (1930-63), copper and silver works (1880-1903) had ruined the local rivers and creeks by dredging and damming and poisoning the water, making it unsuitable for livestock. Over by Atherton was Margit's house, where Sheila and Sarah went totally gooey over a baby echidna (sort of like a dirty water balloon with spiky hair), spotted quolls, pademelons and a young tree kangaroo living in a kitchen cupboard. We found an Italian restaurant in Yungaburra where an accordion accompanied a tenor singing opera excerpts, and the food was good too.

Then it was Christmas again. Sheila decorated the Norfolk Island pine with family ornaments, and we made some home-made grapefruit/lime rind ice cream. We had a Christmas dinner that couldn't be beat. Jesse got an angle grinder, Sheila got a scanner, and I got a big globe from Santa. Sheila and I sleepily went out in the rainy night to wave sparklers around for New Years, and that was it for 1997.

If you think the above was a bewildering sequence of travel, research voyages, diplomatic workshops and meetings, you are right, but it all went by quickly.

Chapter 32. Visit to North America for TROPICS Workshop, Visits With Family, and AGU Ocean Science Meetings in San Diego

In January of 1998 I was writing our Exmouth Gulf paper. Our AIMS research work had been done four years ago on the NW coast of Australia, and all the data had finally come together.

At home in Townsville we had a big rainfall (half meter in three days, the 'Night of Noah' in the newspapers) from Cyclone Sid 10-12 January. Homes were damaged, perhaps millions of dollars worth of infrastructure lost, and some homes, cars, refrigerators, pets and cattle were washed into the Coral Sea. Access to pubs for beer and grocery stores was by boat or canoe. There were several days of 80 km/hr winds which brought down power lines and destroyed roof tops, roads and railroads were closed due to flooding, and some sections of asphalt highways were washed away. There was mildew and mould all over.

In late January Mick Roche (of Mineral Resources of Australia) tried to create the Cleveland Bay Consortium, including the users and abusers, the sewage plant, Townsville port, Yabulu (Queensland Nickel) and MIM Copper Refinery, environmentalists, tourism and fishermen. This was another brave attempt to integrate industry, domestic waste and science.

Some very odd stories came out during our workshop: The beach sand on the Townsville Strand (an important local and tourist attraction) was trucked down from Charters Towers Macrossan Bridge over the Burdekin River. Townsville beach sand had accumulated over the last 10,000 years, but was no longer being replenished, due to a dam upstream trapping most of the sediment. This was one unanticipated consequence of building a dam.

Metal concentrations (As, Cr, Ni) were revealed to be high in dugongs and sea turtles, and coral growth bands showed Pb and Cu increased in last century by 5-10X. Our local industrial

Korea Zinc (Sun Metals) produced 170,000 tonnes Zn ingots/yr and produced 30 ha of waste refinery liquor evaporation ponds close to the Cleveland Bay mangrove creeks. Queensland Nickel/Yabulu was fed by four ships per month at our port, delivering nickel, lead and zinc ore concentrate. MIM Cu Refinery shipped 200,000 tonnes of copper/yr, and they irrigated the nearby meat cutters' plant pasture with waste water enriched in Cu, Ni, As, Pb, Cd, Hg. They try to recover Ni, As, Pb, Se, and Te to ship overseas.

We did manage to produce a CD of our survey of Cleveland Bay environmental knowledge and industrial statistics, and then the whole subject vanished...according to my journals. I guess people didn't really want to know how polluted their harbor was.

On 27 Jan '98, after having been given written orders for North American goodies from Sheila and Jesse, I was off to New York for SUNY meeting to organise TROPICS with Nittrouer, Aller and others; Toronto to visit with son Ian; South Dakota for a visit with my family, Arizona to visit brother Gerald, and finally San Diego for our TROPICS special session in a huge AGU meeting. Leave Tvl 0600 in the hot rain, Brisbane, Sydney, Los Angeles (hot cinnamon buns) arriving 0500, an hour before I left home, an afternoon flight to NYC/JFK where it was a cold city grimy gloomy evening; a trench coated limo driver with 'Brunskill' sign hauled J. P. Walsh, Beth and I to Three Village Inn on Long Island, arriving 11 pm on the same day.

The next day the sun tried to shine but everything was grey, grimy and cold, New England Ivy, coastal marshes and bare tree branches brittle and dark. The purpose of this meeting was to help Chuck Nittrouer get NSF money for TROPICS support, organise our AGU presentations, organise a whole issue CSR journal on TROPICS research, write a 'TROPICS prospectus' to hand out to the press, and plan upcoming ship equipment and operations.

For the US scientists to participate, it was important to show that Indonesia, Australia, and PNG were contributing money, logistics, ships and government approval. Chuck and Bob Aller had assembled scientists who were already involved in tropical river/ocean research around the world, including some NSF officers who grilled us on why PNG was a good target for research. We heard experts from other disciplines explain why PNG was in a position to be a trigger location for ENSO events, climate change and important carbon sequestration on land and sea.

After an exhausting three days of intense negotiations and planning, Chuck and wife Deborah took me into NYC to see the musical *Stomp* off Broadway, preceded by pizza on the street in the cold air. This was a good lesson in how to make music with percussion only, using brooms, clapping, broomsticks on garbage cans, 20 and 200 litre steel barrels, plumber's helpers, paper bags, hub caps, kitchen sinks, suitcases, rubber hoses, spatulas and frying pans.

I wish my son Ian had been with us. We ate hot Indian curry after this, and got back to the Inn at 0100 in a cold wind. The next day, Chuck's son Jeff drove the Long Island Expressway (3-4 lanes of solid traffic, two buses disabled, three wrecks, one in flames, road rage) to get me to La Guardia airport for my flight to Toronto.

I arrived in the Toronto airport to find son Ian and Sheila's youngest brother Bill in the beer parlour, where we stayed for several hours looking at photos. And again in the cold dark, we wandered the airport trying to find the car park, once that was found, Bill and Ian couldn't remember where the car was parked, and then Bill couldn't find his ticket to get out of the six story car park. Bill finally talked his way out of a fine with the exit lady and we drove to Ian's apartment which he shared with Mark Duggan and Alan. This was a neat and clean two story apartment behind a grocery store, where all rooms were full of percussion equipment, marimba, electronic keyboard, thumb pianos in huge calabash gourds, a PNG kundu drum, Indonesian gongs, African rattles, xylophones, CDs stacked to the ceiling, piles of musical scores in big folders, incense burners, scented candles and one round table. There was a tiny clean kitchen and fridge, but it was easy to see that food was not a big deal with these musicians, except for espresso. Foamy slabs on the floor were our beds, and we gratefully slept late the next day.

My son Ian surprised me with a good bacon and eggs breakfast, espresso with toast and strawberry jam. And then Marcy (daughter of my Brother Obie) arrived, promptly demanding to hear some music in this house full of music. Marcy was living in Chicago at that time, studying to be a wealthy megacorporation business executive. She had a list of stuff other South Dakota relatives wanted from Ian, including a 'ribbit frog'. I had my fiddle with me, so with Mark on marimba, Ian on various drums, and Alan on keyboard, we played several fiddle tunes surprisingly well. There was lots of talk about Australia, Chicago and South Dakota news. We wandered the streets to find 'Obie's Restaurant', scared Marcy by having lunch in a dingy (but good) Indian restaurant, and ended up at a Bob Becker concert. Ian guided us to a late night croissant shop, the subway, and back to his apartment for tea and more music, and then we slept on the floor again.

Sheila's sister Noddy and husband Ted arrived the next day, determined to extract every last bit of news from me. They had rented a hotel room, so that there was room for everybody to visit. They had brought a big sack of Georgian Bay apples for Ian. Bill's wife Carla and the two sons James and Thomas were there. These young gremlins wanted to walk on the ceiling and play the marimba. Sheila had provided Aussie items (a wallaby scrotum coin purse, Meadowlark tapes) to trade for Neocitran, maple syrup and strong horseradish (which were not available where we lived). Marcy claims most of the Murdo relatives are still talking about our concert there a few years ago…'Well, I had a little chicken and it wouldn't lay an egg…', and of course the Copenhagen song.

Later on, Ian proudly led us through the University of Toronto percussion storage rooms and recital halls. That evening, after everybody had left, Ian wanted to know where I was going in Japan later in the year. He asked me to obtain 'temple bowls' for him, bronze bell bowls used in Buddhist and Shinto temples and in home reliquaries for ancestor worship. He gave me dimensions and even the frequency for the tones (C, D, E). Ian wanted to finish his studies at U of T, and then come to Australia. He said there were too many percussionists in North America. He also made an apple crumb pie (to his father's amazement), but wistfully wanted to know how to make pastry for the pie, and for strudel, Danish pastries and donuts. He was remembering his mother making donuts in the Tourond kitchen when he was a small boy.

The next day I was up at 0530 for a good breakfast and a taxi to the airport with Marcy. We were subjected to excessive security for some reason. Both Canadian and US Customs and Immigration were carefully looking at all luggage, we were X-rayed, frisked and interviewed by large armed men. Marcy was going back to Chicago, and I was aimed toward Minneapolis. My airplane was full of brightly dressed Koreans as we left Toronto behind in a dull grey yellow smog.

It was strangely warm in Minneapolis. The hockey rink was melting, there was no ice on the river, and it was raining on 2 February 1998. So here I am, in the land of a thousand lakes, where Paul Bunyan and his blue ox Babe wanders the terpenoid pine and spruce forests, with Russian tea cakes in my briefcase. My Northwestern Airlines flight across South Dakota, heading toward Rapid City, was like flying through turbulent cotton candy. I could not see the ground, and visibility at the Rapid City airport was less than a quarter mile. So we returned to Minneapolis, where I phoned my brother Obie, who should have been waiting to pick me up in Rapid City. He was home at the ranch north of Philip, not feeling well, said there had been icy rain for the last few days…not good for his cattle. His wife Annie had been at the airport waiting for me, but had returned home by now. He said a big pot roast and some grasshopper pie was waiting for me. He gave me the phone number for Toadie, my mother's cousin who lived in the suburbs of Minneapolis, with instructions to phone her. Toadie was in her 90s, but she sure didn't slow down much. She yelled at me about distant relatives I'd never heard of, and she wanted photographs of me with the Butterfield fiddle, wondered what on earth was I doing in Australia anyway, and gave me a long lecture about communicating better with family, mainly her. But she was aware of the brilliant green tree frogs in Australia, and wanted to have one. She had a frog fetish, as well as bad osteoporosis with steel plates in her backbone.

The adventure continues. After waiting in line to rebook for a late night flight to Rapid

City, I had supper in an airport restaurant, sitting at a table below a giant poster for the Kasba Lake Lodge in northern Saskatchewan, featuring a leering Colin 'Crash' Welch holding a giant Lake Trout on outstretched arms, son of my Canadian Arctic colleagues Buster and Cathy Welch. I found a shop selling fresh baked cinnamon buns just in time, as I had finished all my Toronto Russian tea cakes. My flight to Rapid City was delayed and then cancelled due to bad weather, so I took a Holiday Inn room for the night. The next morning, the airlines advised me to fly to Denver and wait for the weather to improve in Rapid City. The lounge waiting room for this Denver flight was filling up rapidly with men wearing Stetson hats, big belt buckles and cowboy boots, and the pattern of speech changed abruptly from city talk to cattle and horses. And so we fly across the checkerboard prairie, the James, Missouri, and Platte Rivers all winding their way to the Mississippi and on down to New Orleans, and I tried to write a ballad about this, my brain in a tired fog of travelling.

Denver airport is a good place for me. They have the Buffalo Chip Café, where you can get the food my father liked, such as a hot beef sandwiches with brown gravy, spicy beans on toast, chipped beef in white gravy with melted slabs of Velveeta cheese on toast, BBQ chicken wings, Coors beer, coffee ice cream, and endless refills on your coffee cup. Even better was the smoker's lounge, up on the roof of the main concourse, which was full of cigar and cigarette smoke, sawdust and peanut hulls on the floor, cowboy hats, boots on the bar rail, spittoons everywhere, snuff and plug chewing tobacco for sale and a huge humidor for exotic cigars. There was Michael Martin Murphy on the speakers singing old time cowboy songs (not C&W!), and the waitresses had big hair, tight Levis, and their cowboy shirts unbuttoned and tied above the bellybutton. You could buy one dozen Skol or Copenhagen snuff for $27, and most of the older stockmen here had that circular bulge in their left shirt pocket, along with their little notebook for keeping track of the cattle auction prices. There was a full wall window so you could watch the flat brown prairie turning white with snow. Down on the main concourse, a group of square dancers were twirling around to the music of a ghetto blaster in the midst of their circles, all dressed in proper western regalia. There was also a side room with computers hooked up to the internet, where I could send a message back to Sheila in Alligator Creek.

The flight to Rapid City was delayed several times, but by late afternoon they finally let us board a small propeller plane and we arrived in Rapid City, South Dakota in the gloomy dusk. There was Annie, patiently waiting for me, and outdoors it was above freezing, most of the snow had melted, very unusual for this time of year. Annie drove to her home ranch north of Philip, where Obie was groggily recovering from the flu. There was a batch of puppies in the coat room, and indeed a giant pot roast for late supper. There was grasshopper pie, a concoction that surely only prairie people would make...crushed Oreo cookie crust, a lime green filling made of marshmallows, whipped cream, lots of butter, crème de menthe and crème de cacao, served cold. Much of the conversation was about Annie's son John Paul, who was a bull rider but had smashed his knee, and now he wanted to join the military. The next morning we toured Obie's pastures, fed his fat cattle, and admired his horses. Obie drove me through the small towns toward Murdo, where my brother John had moved the old Grandpa England house to allow new concrete foundations to be poured. John now had a more modern home, with city water supply, a propane furnace, and big clean basement rooms fixed up nicely by his carpenter son Dan.

And then the deluge of relatives arrived. Brother John and Pat filled the kitchen table with warm roast beef and home cooked bread, and kept making coffee for all the visitors. My uncle Gerald and Aunt Joyce demanded a full account of my life in Australia. They told stories of some newcomer cattlemen (funded by Ted Turner of CNN) up north who bought 30,000 acres and stocked it with purebred Angus, but had no barns, haystacks or resident manager. They lost 800 head of cattle during the April '97 winter, and Uncle Gerald says...'What kind of a cattleman are they?' During this same blizzard, my brother John lost one cow out of 400, and Uncle Gerald had no losses down on the White River ranch.

Uncle Gerald also had discovered some new sources of information on our ancestors Joseph and Simon Brunskill in Wisconsin in the mid-1800s, recent immigrants from Yorkshire. Aunt Joyce gave me news about one of my favourite high school teachers being on her death bed in Huron, urging me to write her a letter right now. Mrs. Kuhrt taught me history and literature, and was an important influence on my life. My aunt Margie and uncle Bernard with most of their kids arrived, Bernard being interested in getting an electric car, which causes my brother John to snort loudly. Brother John is frustrated at the media for giving conflicting information about climate change, greenhouse gases coming from beef cattle and cutting back economic growth to allow sustainable developments. In addition, he thought that President Clinton was a crook, there was a $30 trillion deficit, and everybody he knows is going broke. I had to suppress a chuckle, as this sounded just like my grandpa George Brunskill about fifty years ago.

The next morning I had a wonderful big South Dakota breakfast of piles of bacon, ham, fried eggs, pancakes, whole wheat toast with chokecherry jam and good coffee. I helped feed hay to John's black cows, and then we drove three hundred miles west to Rapid City on the wide open rolling prairie as far as the eye could see. We visited John's oldest son Bill at Cenex just outside of Rapid City, where he drove a propane truck to remote cattle properties to the north. Obie's daughter Julie and kids arrived for lunch, along with Obie and Annie, John and

Pat. I had to complain about the lack of sour cream raisin pie. And then everybody waved goodbye as I boarded a flight for Minneapolis, and then to Phoenix, Arizona, by dark. There was my brother Gerald and son Josh to meet me. [If this list of brothers is getting boring, just move on to the next chapter. There are only four brothers, surely that is not too many?]

Gerald and Doreen's house was two and a half hours' drive from Phoenix to San Manuel, a mining town high on the San Pedro Mountains. Most of Gerald's kids (Sarah, Becky, Christopher and grandkids) were there to meet me, some hiding behind the sofa, curious about life in Australia upside down, demanding immediately to see the photo album I was carrying. Gerald was a metallurgical engineer working for the Australian giant mining company Broken Hill Pty (BHP), but his mine works was closed down and the huge tailings pile was being cleaned up by various reclamation schemes. About the only vegetation was big cactus, where cactus wrens and a tiny cactus owl lived. Gerald also had all the England and England road gang photos framed on his living room wall, as well as several Harvey Dunn, Charlie Russell and Remington cowboy paintings.

The next day all the grandkids were yelling at me to play Gerald's old banjo, hidden under his bed. Some strings were rusted or busted, but we managed to crank out 'Comin' Round the Mountain' with some screwy verses. I tried to sing our Australian song 'Oh Rosella', but it got interrupted by demands to see photos of our house parrots, kookaburras and wallabies. Son Christopher was a rubber gun maniac, but I got to tell him about our rubber gun wars when I was a kid in Murdo. Gerald took the whole family out to supper at a local Mexican restaurant. Everybody in the restaurant called Gerald by his first name, and we ate spicy pressed/pickled meat, guacamole with chillies and deep fried corn fritters.

After a chili rich breakfast the next day, we went back to Phoenix to buy a set of banjo strings, as the kids were clamouring for more banjo songs. We toured some of the dusty abandoned mine sites in this region and admired the desert saguaro and cholla cactus with wrens and owls. This is Apache country, where Cochise and Geronimo resisted the US Army, and we drove through the Tonto National Forest. A lot of my repertoire of old time cowboy songs came from around here, including Powder River Jack's 'Tying Knots in the Devil's Tail', Badger Clark's 'Spanish is a Lovin' Tongue' and Captain Jack Crawford's 'California Joe'. I had warped them all to fit South Dakota of course, so some of the original place names got changed. And then Gerald, Becky and Chris drove me to the airport on crowded four lane highways, and I was off at 8:30 on 7 February 1998 to fly to San Diego, California, for my AGU Ocean Science Meeting.

The last few paragraphs were about my family, where I was increasingly viewed as a prodigal son who wandered in error away from his home country. I had slipped into the upper

Midwestern form of English, with continual reference to horses, cattle, prairie grass and the weather. With me living on the opposite side of the world, many eyebrows lifted in a sceptical way, and a lot of my older relatives still expected me to come home one of these days.

But now, arriving in San Diego in the gloomy chilly dark, here was a different kind of family and a different, more international culture.

The airport shuttle bus to the Town and Country Inn had a dozen of my scientific colleagues hollering greetings and shaking hands. At the hotel registration desk there were more colleagues, most of whom were speakers for a special session on TROPICS that I had organised. I had also provided funding for Indonesians and Papua New Guineans to attend this meeting. The hotel lobby was an international feast of languages and scientific jargon, where you could hear German, French, Baha Indonesian, Japanese, Mandarin, Pidgin PNG, many varieties of English, and a few languages I did not recognise. I had a double room, the extra bed for my Indonesian LIPI friend Abdul Gani Ilahude, who arrived from Jakarta cold and shivering at midnight. He was dressed neatly, like a North American scientist, heavy wool black top coat, dark suit with white shirt and dark tie, polished black shoes and a shoulder-strap briefcase. This was not how he dressed for science meetings in Jakarta! While he was having a hot shower, I phoned the hotel concierge to determine the easterly direction in our room. Gani then spread his prayer rug and knelt down for his Islamic prayers to Allah.

We had a few free days before the big meetings began, and our TROPICS gang gathered for meals and informal meetings. We were preparing to strut our stuff in front of our peers and the representatives of oceanographic science funding agencies in USA. We had to impress them that we were capable of doing good work, that we were coordinated, and could store, manage, interpret, present and publish complex information that would be useful to world science. Nittrouer and I had prepared a glossy prospectus handout to attract admirers and collaborators. We had to organise the options for taking Indonesian and PNG people on our Australian and USA ships, as part of the deal with their governments. We also planned to have reporting workshops in Jakarta, Port Moresby, and Australia, with digital and print mechanisms for depositing records of our work in their government archives. We had already conducted five research voyages to the PNG coast, and we were now planning Legs 6-12, including Indonesian ships working the north coast of Irian Jaya. Most of the ship time I had already secured from AIMS and the CSIRO National Facility *R/V Franklin*, but there were options for US and Japanese research ships to assist.

Our Australian contingent was George Cresswell (CSIRO Hobart), Ron Szymczak (ANSTO), Kathy Burns and myself from AIMS, who were the main scientific and logistic hosts for the field work in coastal Papua New Guinea over the next few years (assuming we could raise the funding). We were expected to explain, repeatedly and patiently, to US and European scientists why it was important to go halfway around the world to study big tropical rivers pouring water into the coastal ocean.

It wasn't all science. John Robbins wanted a full account of our son Ian's music career. Chuck Nittrouer and Don Wright wanted to know about the health of our house parrots Sir Charles and Casper. Several dozen of these scientists had been treated to a feast by Sheila at our Alligator Creek home, and were disappointed that I had not brought Sheila with me to this meeting. We visited the famous San Diego zoo, where we could see orangutans, marmosets, exotic tropical monkeys, gibbons, chimps, gorillas, barking deer, tusked deer, hairy pigs, and many tropical birds, including the PNG bird of paradise.

The main plenary sessions for the AGU Oceans Meeting began on 8 February in a huge convention centre, with alarmingly good presentations by Jim Murray (JGOFS = Joint Global Ocean Flux Study) on the Equatorial Pacific, and Jorge Sarmiento on the impact of greenhouse gases on world ocean circulation, deep water ventilation, warming and carbonic acid titration of seawater carbonate alkalinity. Murray's presentation pointed to large importance of big river inputs from the north coast of PNG into the equatorial undercurrent, especially iron, which appears to limit algal growth in this region.

We held our all-day session on TROPICS the next day, in front of a huge, critical crowd. Many people in the audience were fighting for funding of their own research programs, and viewed our well-organised international mega-project with dismay. Our session featured twenty minute presentations on the Fly, Sepik and Mamberamo River catchments, physical, chemical, and biological oceanography, and geochemical work on the sediment budget, all of a magnitude approaching that of the Amazon. We created a good buzz in our audience, and all of us were busy late into the night at the bar, explaining what we hoped to do in the next five years.

The rest of the week went by in a blur of eight concurrent sessions every day, each of which featured dozens of good speakers with two or three slide screens going at once, sometimes videos of action, and handouts of interesting web sources of data and methods. These AGU Ocean meetings might be the largest gathering of oceanographers in the world, with attendance of over five thousand registered guests. Many colleagues I hadn't seen since graduate school showed up and demanded a beer session. My Canadian colleague John N. Smith gave a talk about sampling nuclear power plant radionuclides (from British Sellafield wastes, the largest nuclear power site in Europe) in seawater underneath Arctic ice in US Navy submarines. Several colleagues invited me to the 'Albatross Award' session, one of the rare times that joke honours were given to members of our exotic peer group. This time a physical oceanographer was awarded the old ratty stuffed albatross, for his attempt to deploy dozens of current meters in the Fram Strait during a snowstorm and hurricane. Colleague George Cresswell whispered in my ear that there was a similar tradition in Australia called the 'Galah Award'.

It was bewildering, tiring, and exciting to try to swallow all of this new information, like trying to drink from a fire hose, and to meet good people who have made incredible new measurements in unusual parts of the world ocean. After the late afternoon sessions we all went out to find interesting restaurants, where we argued science (and a few other things) late into the night. Recurrent themes were climate change, contaminants everywhere, and sustainable harvests of the world ocean energy, seafood, and perhaps mineral and freshwater via desalination. Population growth was making many scientists worried about water, food and energy supplies in the coming decades. The biologists were concerned that most of the whale species were down to less than 10% of their pre-whaling days' abundance, the big carnivorous fish (tuna) were gone in some parts of the ocean, and fish/prawn pond aquaculture was becoming a dominant source of seafood. In some parts of the Asian and African coastal ocean, the people were eating almost everything on the continental shelf, and harvesting edible creatures before they had a chance to reproduce.

Meanwhile, here we were, in one of the richest and best educated countries of the world, eating grain-fed beef and imported SE Asian seafood, with hot and cold running water, and no fear of starvation or war.

During our last night there, my Indonesian roommate Gani had the TV blaring loud about riots in the streets of Jakarta. He spread out his prayer mat, pointed himself to the East, and said his prayers to Allah. Then he proceeded to grill me about my absence of religious faith, why was the USA fighting with Saddam Hussein, did I believe in any of the cosmological models (such as the big bang), and did I believe in quarks and gluons. We talked until 3 am. The next morning we walked the San Diego malls, where consumerism was in full display, viewing goods that were neither useful nor necessary.

That evening I checked out and caught a taxi to the airport at 6:30 pm for the usual adventure of trying to get to my departure gate on time, past determined security guards, and long lines of US Marines going somewhere. A small propeller plane took us to Los Angeles, where we boarded a Qantas plane to Sydney. I ate a good chicken curry and gratefully slept the whole way across the wide Pacific Ocean.

Meanwhile, we crossed the International Date Line, and Valentine's Day vanished into the maw of planetary rotation. I did buy a card in San Diego for my sweetheart Sheila. The arrival in Brisbane was comforting, with warm sunny skies, cricket on the telly, and everybody is tanned brown. I was becoming a fair dinkum Australian! There in the Townsville airport was my bright and beaming Sheila, waiting for me with a hibiscus flower in her hair. The Alligator Creek air was sweet, fragrant, warm and moist...it felt very good to be home.

There was little time for rest. Swedish Ph.D. student Elsa Pienerud arrived in May 1998 for an extension of her Swedish estuary work on biogenic silica in fossil diatoms from sediment cores. I had to secure funding, accommodations, and laboratory access and methods for her. Irena provided subsamples from our sediment cores of Hinchinbrook Channel estuarine mud. Elsa was one of these good students, a self-starter who knew what she was doing, and proceeded with the work.

In May and June I was involved with a Technical Advisory Panel (TAP) on rivers and water supply problems for coastal towns and agriculture. This meant attending planning sessions with water engineers from the Department of Natural Resources and week-long tours of the interior headwaters of the Burdekin River, the Tablelands inland from Cairns, and some of the Tableland crop irrigation schemes that took water from rivers flowing to the Gulf of Carpentaria. I learned that in this semi-arid country irrigated lands often suffer excessive salt loads (and reduced plant growth) after decades of irrigation with bore water. We tried to moderate the enthusiasm for irrigation and river dams, but this was an unpopular stance with many frustrated farmers and stockmen.

Also in May and June, Christian Roth and Andrew Johnson of CSIRO in Townsville began a series of workshops for the purpose of writing a major review of coastal agriculture, water management, and impacts on the Great Barrier Reef. This project gathered steam over the next years, with input and funding from both state and national governments.

Chapter 33. A Perfect Storm in the Herbert River Mangroves, and a Search for Temple Bowls in Osaka

I was part of an academic, government, agricultural, fisheries and community consortium that organised a late March 1998 workshop on the social, cultural, economic and environmental aspects of the Herbert River catchment area, its delta and estuary, and the adjacent sector of the Great Barrier Lagoon continental shelf.

Our workshop group had been funded by CSIRO, Australian Research Council, the Coordinated Research Centre (CRC Reef, CRC Rainforest), James Cook University, Great Barrier Reef Marine Park Authority (GBRMPA), the Queensland Department of Natural Resources, the Australian Department of Environment, and a few other sources. We had previously created a Herbert River Information Centre (HRIC), which sponsored the workshop in a community hall on Ripple Creek, right in the lush Herbert River Delta. The cultural setting was as follows: environmental science and green NGOs claimed that agriculture was causing erosion, increased sediment and contaminant addition to the river and GBR Lagoon; national, state governments and the sugar industry wanted to expand and increase land use and production; the local towns wanted higher dike flood protection; the upland cattlemen didn't want anybody telling them what to do. We were naïve, we should have known better, and it was a perfect storm.

Several speakers from the State Natural Resources had plans and funding for the expansion of the sugar cane industry, mainly by increased drainage channels in the Herbert River floodplain and delta. The natural creeks and mangrove channels were to be widened, deepened and straightened, big tree snags to be removed and riparian vegetation to be 'trimmed' to allow faster runoff. The small floodplain sugar town of Ingham and delta villages wanted to decrease the danger of flooding their civic centres, and they viewed the wetlands and mangroves as slowing down wet season runoff. Some of the sugar industry agents explained that they wanted to extend sugar cane agriculture further into the rich soils of the delta mangroves, but that it took over twenty years to desalinise and neutralise acid sulphate soils in previously deforested and levelled deltaic land. Deep drainage channels had been excavated through the cane lands, with bermed culvert seaward outlets inside the mangrove forest having one way flow shutters to allow freshwater to flow out during floods, but to stop seawater entry during high tides. Many of the floodplain and delta cane cockies got 60-90 tonnes of sugar cane/ha per year, at a profit of $300-350/ha of cut cane, or a little over $1 billion/year, statewide. Flooding and drought reduced these profits. With the exception of marijuana, sugar cane was the most profitable crop in north Queensland. Most of the pro-development and pro-agriculture speakers also said they wanted to have a sustainable industry, to support natural biodiversity and local communities, and to avoid introduction of toxic wastes into the system.

Various agencies wanted to build dams on the headwater canyons of the river catchment area, with the intent to irrigate water intensive crops (mango, citrus) on land that was naturally arid and better suited for cattle. The cattle grazing industry was already worth $1.2 billion/year, and some old stockmen wanted the government to leave the grasslands alone. Some of the headland regions had subsurface groundwater that was more saline than seawater, and there was some danger in causing this salty water to rise under irrigation.

The Local Council of the Hinchinbrook Shire (the lower Herbert River basin and the large mangrove forested estuary) were trying to do their best to find a path of sustainable, environmental and economically successful development. They had already commissioned a study which recommended limiting agricultural encroachment onto mangrove saline soils, fertiliser and pesticide usage, protection of wetlands and riparian vegetation; but they also needed better flood response plans for the town of Ingham, perched on the Herbert River bank and floodplain, just upstream from the delta.

In great contrast, a Department of Environment speaker was concerned about aboriginal sacred lands and cultural heritage sites that had already been bulldozed and levelled for sugar cane production. The Department of Primary Industries were concerned about preservation of wetlands, freshwater fish habitat, mangrove and sea grass habitats for estuarine fish, dugongs and turtles. They were seeking solutions to problems of invasive fish and aquatic plant species, water pollution and acidification of runoff from oxidation of pyrite (FeS) in canelands on top of mangrove (drained marine) soils. GBRMPA and the North Queensland Conservation Council spoke strongly of the clear need for preservation of wetlands, riparian vegetation and the fetid mangrove swamps, all of which act as a giant filter for sediments, nutrients, and contaminants flowing to the sea. The problem of rising sea levels, increasing salinity of floodplain groundwater, and landward migration of mangroves and estuarine plants was unpopular with the cane cockies.

I gave a talk on the relative strengths of the rate of growth of human population and industrial pressure on the coastal zone in tropical SE Asia and Australia, and some of the expected consequences of 'business as usual'.

An illustration of the 'business as usual' scenario was our discovery of a two to five-fold increase above background mercury in some of our sediment cores from Hinchinbrook Channel in the Herbert River estuary. Some of this excess Hg probably came from the use of organomercurial fungicides (trade name Shirtan) on sugar cane setts over the past century. Our radiochemically dated sediment cores showed that some of this Hg may have come from 1870-1900 gold mining operations in the Burdekin River catchment, indicating that Hg contaminated sediment from the Burdekin River mouth had moved northward along the coast for 250 km, to be deposited in the mangroves of the Herbert River estuary and Hinchinbrook Channel. On the other hand, in contrast to previously assumed information, our sediment cores did not indicate large increases in terrestrial sediment delivery to the delta and estuary over the past two centuries. If there had been increased erosion in the upper Herbert River catchment, most of the sediment must have been trapped in the lower river floodplain and upper deltaic mangroves. Joanne Cavanagh was a post-doctoral student working with Kathy Burns and myself on the history of Herbert River sugar cane pesticide/herbicide usage, persistence and flux to the sea. She found that most of the persistent pesticides were apparently degraded in the cane land soils and upper delta mangrove mud, and very small fractions of historical pesticide usage reached in inner shelf of the GBR. This was good news.

The local District Catchment Committee chairman tried to sum up the results of this workshop and called for input on sustainable options and floodplain engineering plans for the next century. There wasn't any. Many of these different agencies had funding for their programs, and were in the process of expanding the cane lands, increasing cane and banana production with fertilisers and pesticides, building flood dikes, commercialising a local wetland for tourist bird watchers, and trying to build fish ladders over the fast flowing drainage ditch outflows. DPI and AIMS had research programs in the Herbert River estuary and Hinchinbrook Channel. Some of these programs appeared to be working against each other, and there was little coordination. We had been seeking a synthesis and improved coordination for the whole river catchment area, estuary, and continental shelf of the GBR Lagoon, but this did not happen.

So it was back to the office to try to write up our research on the Herbert River catchment, estuary and GBR lagoon. We had a contract with KEPCO (Kansai Electric Power Company of Japan) for research on the ecological functions of estuarine mangrove forests, funnelled to us by our Japanese colleague Tenshi Ayukai. We spent most of this money on research in the Herbert River delta, estuary, and Hinchinbrook Channel, which hosted one of the largest and least disturbed mangrove forests in Australia. This funding essentially tripled our ability to construct annual budgets for river inflow, estuarine chemistry and biology, carbon sequestration by the mangrove forest, and the burial rate of organic carbon in the mangrove mud.

We had a few 'show and tell' progress meetings with KEPCO officials at AIMS, and I sat beside one Japanese gentleman who asked about my career. Once he found that I was the son of a South Dakota cattleman, he demanded to see me in cowboy boots and Stetson hat, just like John Wayne and Clint Eastwood in the US movies. So the next day I appeared at AIMS with my big Fort Pierre Stetson hat and some high heeled boots to give my presentation for the Japanese officials, who clapped merrily because of my hat. My AIMS colleagues were puzzled by this.

I had a few weekends to cut the grass and weeds in our orchard and harvest the last of the mangoes. We had some Meadowlark rehearsals followed by good feasts, and then did two concerts in town.

I was scheduled to go to Osaka, Japan for another KEPCO meeting in late April 1998. Sheila had her order for several jars of Japanese hot green horseradish, and Ian MacGreggor wanted two temple bowls. Jesse and Sarah drove me to the airport in the rain. I was travelling with several AIMS, JCU and GBRMPA scientists, who loudly wanted to know why I was not wearing my cowboy hat. I found Tim Flannery's *Throwim Way Leg* in the airport bookshop, and I was set for a long plane ride to Tokyo, with good cappuccino and an apple Danish before I went to sleep.

We arrived at Nagoya airport at dusk, the city smothered in a grey-green smog, where we caught a van to travel the crowded streets to the bullet train, a very fast (300 km/hr), quiet, neat and clean passenger train to the ancient capital city of Osaka, the second largest in Japan. Our little Australian party was met in Osaka by a uniformed driver directing us to a polished black KEPCO limousine that took us to our Holiday Inn Hotel, obviously set up for western visitors. The restaurant however was very Japanese, and at midnight we dined on sushi, raw beef in sauces, seaweeds in raw egg batter, with good cold Japanese beer. We had CNN news on the TV and Japanese versions of Hollywood films. It took a while to figure out how to operate all the buttons, levers and knobs in the hotel room shower stall.

We were herded into a Japanese garden for breakfast, with little waterfalls, pools with fish, trimmed bushes and quiet benches for contemplation. Much of this was lost on our rowdy bunch of Australians, but we ate hardboiled purple yolked eggs with tiny fish sticks, croissants and tea.

Our colleague Eric Wolanski arrived from Tanzania, jet-lagged, where he had been a consultant for management of giant wetlands. We were led into several shiny black limousines that whisked us off to the lecture halls for KEPCO/KANSO, where thirty Japanese in immaculate black suits and ties awaited us with notebooks and clipboards.

In addition to our Australian group, there were also study groups from Thailand and Malaysia. All of these presenters were using printed posters, blackboard drawings, overhead projectors, 35 mm slides, PowerPoint, as well as several videos of life amongst the roots of mangrove trees. With three different language translators, it was an exhausting process to listen to the talks, and even worse to be a presenter, pausing so that the translators could catch up explaining our slides. The translators had laptop computers with several screens showing previously submitted texts, dictionaries, and alternate labels for each slide being shown. Here was technology gone berserk: everybody seemed to have cell phones and small hand-held computers, sometimes you couldn't figure out if they were talking to you, or to their phones.

So here we were, giving quantitative ecological information about tropical deltas and estuaries where the few remaining mangrove forests exist, that have survived agriculture, deforestation, prawn and fish aquaculture and urban development. Our group, and similar study groups from Malaysia, India, Indonesia and Thailand, were presenting information on the effect of deltaic deforestation, changes in river hydrology due to dams, mangrove tree species growth rates, rates of river mud accumulation trapped by the mangroves, rates of decomposition of tree wood and organic matter in mud, and other plant and animal species life cycles that live in the mangroves. Several presentations were about the impact of the absence of coastal and deltaic mangrove forests, where tsunami, cyclone, and storm wave energy was directly thrust upon coastal villages and businesses.

The elephant in the room was climate change and sea level rise, which was poorly discussed.

It was never clear why a large multinational Japanese industrial energy company would fund all these studies and host this conference. Perhaps part of their agenda was to find sustainable ways of living, but more likely it was guilt from burning enormous amounts of coal for electricity production, and consequent emission of greenhouse gases. They were attentive to the idea of finding commercially interesting countermeasures for the removal of CO_2 from the atmosphere, schemes to collect and bury industrial exhaust greenhouse gases, and plans for reforestation of mangroves and wetlands.

Barry Clough of AIMS gave a good presentation on our Hinchinbrook estuary mangroves, where their carbon biomass was about 600 tonnes/ha (much of this underground as roots and woody debris), and growth rates were 15-30 tonnes of carbon/ha.year. Some Malay and Thai mangrove forest remnants, closer to the equator, grew at 70 tonnes carbon/ha.yr. There were several other Japanese (and multinational corporations) in the room, all with unexplained acronyms, KEEC, KANSAI as well as KEPCO, all presumably industrial giants in SE Asia, seeking a corner of the problem for their businesses.

The president of KANSO, Ichiro Adachi, came up to me, shook my hand, and said that he was disappointed I was not wearing my cowboy hat. He was a fan of 1950-60s US cowboy films, especially John Wayne and Clint Eastwood, and wanted to hear about the American western culture. He led me up to the seventeenth floor where we were served an immaculate evening meal of jellyfish, sea urchin gonads, fish eggs, shark fin soup, lots of raw tuna, as well as rare roast beef, prawns and scallops, Camembert cheese and an apparently endless supply of beer. After the meal, it was customary for the guests to stand up with a glass of sake, to make a little speech of gratitude to our hosts. And then of course several important politicians and businessmen had to toast us in response.

The next day it poured rain and the Japanese provided us with large umbrellas. We heard presentations from Malay and Thai scientists on their protected mangrove forests. In sharp contrast to the previous feasts, we were served small triangle sandwiches of white bread with some sort of fish paste inside. Some big important Japanese executive attacked our leader Tenshi Ayuki that afternoon, claiming that we had not studied dugongs, fish, and sea grass, which was true. There was a long furious argument in Japanese during which the translators were silent, and we never did know what was happening. Tenshi later told us not to ask what it was all about, apparently something about proud Japanese wanting to be in control.

The next day was spent arguing about the next four years of the research contract, the level of funding, signing the contracts, during which the institutional directors, presidents, lawyers and other mysterious big shots did the work.

The next few days were open for visits to various Japanese research institutions. We also spent a lot of time walking the streets, sampling the street food vendors (whale, shark, tongue, heart, liver, dried fish, jellyfish, seaweed), visiting parks for peace and quiet, or listening to swing band jazz in the mall. Osaka was full of small parks, where families were enjoying lunch and watching their kids, bicycles with multiple seats for the children, immaculately trimmed trees, hedges and flower gardens, rows of yew trees, the whole works neat, clean and happy. This part of Osaka was efficient, clean, punctual, plugged in, GPS on bicycles, shrines everywhere, and the people were gracious and polite. Yet the young folks had Levi jackets with the collar turned up, backwards Michael Jordan caps with sunglasses, tight Levi pants faded and stressed with holes over their knees. The streets and sidewalks were swept, washed, disinfected and probably polished on weekends.

We found a small stall selling coffee and cinnamon buns for Y300, and stalls selling bamboo flutes and fiddles. It was side-splittingly funny to watch old *Ironsides*, *Gunsmoke*, *Bonanza*, and Clint Eastwood films on the hotel TV, all in guttural Japanese language. One evening Tenshi took us to the underground malls, where high fashion girls in platform boots pranced, fellows in hip-hop baggy trousers were dancing, with crying babies in backpacks, and old men bent over with huge cloth sacks on their shoulders, construction crews with loud electric tools, a frightening crowd with little body space, and the smells of street food everywhere. We stumbled into a seafood restaurant where the tables had a gas burner grill in the middle, and we could cook our own lobsters, prawns, scallops, squid, fugu, octopus, fish balls, mushrooms and cabbage, along with huge vats of a variety of noodles, eleven different sauces, and four litre bottles of cold beer.

I had asked Tenshi to help me find guidance to purchase the temple bowls for Ian. Tenshi introduced me to one of the executives of KEPCO, who listened carefully to my quest. An explosive 'Hai!' and he would bow, and then he took me to a higher executive's office on the fourteenth floor. 'Hai!' and he would bow, and took me to an office even higher in this building. And there, sitting serenely in a big office with expensive chairs and sofa, was a very tiny senior boss man, who again listened carefully, smiled gently in a condescending way, knowing that I was a savage Australian, wrote a note on a piece of KEPCO note paper, and put it into a company envelope. His English speaking minder told me to give the envelope to a taxi driver, and he would take me there and return me to my hotel. I was not to pay the taxi driver.

It took about three hours of driving through crowded noisy streets to get to the outskirts of Osaka, where the taxi driver beckoned to me to visit a large Buddhist shrine shop, and for me to give the envelope to the shop attendant. Without saying a word, the attendant took me into a back room full of ornately carved reliquary cabinets, each one with a series of drawers, inside of which were hundreds of temple bowls sitting on little rice filled cloth pouches. I could roll a small wooden stick around the top of the bowl edges, to make an ethereal clear musical tone. I had brought a small electronic tuner with me, so I could select bowls close to concert C and E. The attendant was somewhat disturbed at all this, but I paid Y7,000 for two of them, got back in the taxi, and returned to my hotel room, feeling victorious.

It was later explained to me that the shop I visited was a funeral service. The reliquaries and temple bowls were meant to worship your ancestors, and not for musical instruments.

I had little understanding of Japanese culture. I was impressed by the gracious and overly polite behaviour of business executives and their minders, scientific colleagues, technical staff, shop keepers and restaurant waiters; but I also detected a sense of superiority, an almost haughty but curious distain for our sometimes rowdy Australian and US people from western culture. I had felt the same polite superior reserve from senior Sioux Indian people, and the far northern Canadian Dene and Inuit people.

On the afternoon of Sunday 26 April we boarded the fast clean electric Shinkhausen to fly through the harbour district, over canals from Lake Biwa, irrigated fields of vegetables in the lowlands, to Kyoto and Nagoya. Coastal mountains loomed behind tiny patchworks of rice paddies and fruit orchards, all nestled between cities and residential areas. The taxi from the train station to the airport took one hour of absolute chaos, diesel fumes and soot, spaghetti cloverleafs, elevated highways ten metres above the traffic below. At the airport we ate bowls of rice with bits of chicken with cold beer. I was grateful to see for sale Canadian maple syrup, honey roasted cashews, macadamias, almonds, Cuban cigars, Gitanes and Senior Service cigarettes.

I was attracted to the loud voice of an American in one shop: He was about six foot four inches tall and maybe three hundred lbs, shouting at the tiny shop attendant, who had a limited understanding of English. Tenshi helped sort out the language problem, and I spoke to the American, to find that he was a Black Angus feedlot operator from just outside Lincoln, Nebraska; he had a brother named 'Rooster' and a nephew named 'Miller Lite'. He was here in Japan to find out if he could grow Wagyu beef in Nebraska. In the usual way that odd coincidences happen, I had read about Wagyu cattle in Australia, and I told him that there were two large Wagyu operations near Dubbo in New South Wales, Australia. He seemed a bit miffed about this information.

We boarded our Qantas airplane at 8:30 pm, ate beef teriyaki with cold beer, watched Jack Nicolson in *As Good as it Gets,* and then I slept all the way to Cairns. It was wonderful to get a double plate of bacon and eggs for an Aussie breakfast, and smell the clean rain and perfumed warm air outside the airport. I arrived in Townsville at 0750, and there was my beaming brown tropical Sheila to meet me. She whisked me away to Alligator Creek and peaceful sanity of our home.

Chapter 34. Another Family Visit to North America, and Further Adventures in the Mud of Tropical Australia and Papua New Guinea

We were going to visit Sheila's family in Canada and Sheila had been packing for two weeks. Housekeeping and orchard watering duties were given to Jesse and Sarah in written detail. We were up at the ungodly hour of 0500 on 16 July 1998, Sheila scowling at my Aussie outfit, definitely not suited for international travel (levi shorts, old blue work shirt, flip flops). Jesse and Sarah drove us to the airport, and off we go to Cairns (bacon and eggs breakfast), Narita Airport in Tokyo, Vancouver (Okanagan cider and Vancouver draft beer), and finally Toronto in the evening of the same day.

Our son Ian was there to meet us and he hauled us to his musician's hang-out, a lair for long-haired musical graduate student bandits in downtown Toronto the Good. This was a second floor flat, above a Portuguese butcher shop, reeking wonderfully of smoked meat, sides of pork hanging in the shop window, with cassettes of world music for sale on the sidewalk in front of the shop. And then there were tea and pastries, Sleeman lager, and finally the solemn ceremonial presentation to Ian of the temple bowls from Japan. There was reverent silence in the room as all the musicians got to play them.

The next day we toured Toronto music shops trying to find a good banjo for rent, and found only bad cheap ones, or $4,000 Gibson banjos for purchase only. Sheila borrowed a good guitar and a hammered dulcimer. We ate Indian food with Ian, met his tabla mentor Ravi Naimpally, and went to a classical Indian music concert featuring sarod, tanpura and tabla. Ian drove us north from Toronto to Mono Mills, where Sheila's brother Bill and wife Carla live, through fields of corn, onions, grapes, old barns and lots of pea green lakes.

An alarming number of Sheila's family arrived over the next two weeks, perhaps twenty or so (Sheila's brothers Ian, Alistair, Bill/Michael, and sisters Heather/Peabody and Lindsay/Noddy), our son Ian's aunts and uncles and all their families. Consequently, there began a number of feasts (wild strawberry-rhubarb crumble pie, wild blueberry cheesecakes, black currant jam on warm croissants!) followed by story telling and music sessions till way past midnight. The family rejoiced in exotic activities, like swimming in a gravel pit, a horse show, visiting an emu and ostrich farm, computer video games, antique shop explorations, formal figure ice skating routines, a mountain rope slide, water pistol fights, model diesel airplane flights, seriously big rocket launches, turtle food search expeditions and hikes in the woods.

Sheila's brother Bill is a good trumpet player, and he had my collection of old trumpet solo and exercise books, as well as his own collection of trumpets in the keys of B flat, C, D, E, and F, as well as an old cornet and a flugelhorn. He had also committed Sheila, Ian and me to perform at the Mono Mills pub, where his jazz group had been playing for the past week. The Dennett family were a good third of the audience, but the place was packed. A fellow from the Toronto music shop Long and McQuade showed up just in time, with a decent banjo for me to play. So these Ontario people got a dose of our old time cowboy music, along with Sheila's Celtic ballads and hammered dulcimer fiddle tunes.

One of the original Meadowlark musicians was Hans Hansen, who was a graduate student in philosophy in Winnipeg in the late 1970s. He played guitar, mandolin, and sang mournful Prince Edward Island songs about logging. By now he was a professor of philosophy at Windsor University, and he had a family cabin near Eugenia (Ontario) to escape the city in the summer. Sheila and I visited him in late July, and he was flustered to see us, showing off his three year old daughter Molly. Again, we told the old Winnipeg stories, and played a few tunes for old times' sake. Hans is of Danish origin and spent a lot of time in Copenhagen and Amsterdam teaching and doing research. We had raspberries, peaches and tea at his cabin.

Now it gets even more complicated. We flew from Toronto to Winnipeg, where we were swamped by more Dennett family and many musical friends. Various Meadowlark musicians showed up at David and Barbara Ediger's house for several long musical parties, including bass player Tom Janzen and good guitarist John Weier, Susan Israel on fiddle, Walt Lysak on many homemade string instruments, as well as Sheila's brother Alistair and kids. Feasting continued for several days with Saskatoon berry pie, butter tarts, chokecherry jelly, Lake Winnipeg goldeye on the grill, BBQ ribs and cheesecakes galore. Arlo Guthrie was performing at our old stomping ground, the West End Cultural Centre, and we were surprised to see a long white haired hippie singing 1960s folkie songs...including an updated version of 'Alice's Restaurant', one of our family code signatures. I guess it never occurred to us that we were getting old too.

We made the mistake of visiting our old Tourond home, half an hour south of Winnipeg. Our old house had burned down, leaving only a hole in the ground, with the rusted remnants of our steel wood stove sitting forlornly in the middle, and the yard with metre high weeds, the sheds and barn falling apart. Well, ok, that was another life some time ago, but it was sad to see. Son Ian took all his cousins down to St. Malo beach for swimming. While out in the Mennonite country, we visited John and Eleanor Kipe of the famous Tourond Superette Store, and good friends and Boy Scout leaders Gerald and Mary Toews, where we had to sing for our supper ('Tall Buildings', 'Copenhagen', 'Roseville Fair'). Sheila's 1970 ELA summer student friend Doreen Middleton appeared, having travelled from Edmonton to see Sheila.

Over the 1970-90 years, our folk music string band Meadowlark evolved through a dozen or more members. All these old friends demanded a concert, so we rehearsed several different band member configurations over a week, mostly playing our old tunes. Sheila and I played two forty-five minute sets at the Irish Club, with Peabody's son Gordon on bass, Ian and Alistair on percussion, Susan and David on fiddles and mandolins, Walt Lysak on Appalachian dulcimer and concertina, with Alistair's daughter Rosalyn Jane dancing to the fiddle tunes. It was heart-warming to look out at the audience to see nearly all close friends, fellow musicians, and some of my science colleagues from the Freshwater Institute and the University of Manitoba. This was the kind of rewarding and visceral connection that makes performance a lot of fun and very meaningful. Many in the audience demanded their favourite songs, so Sheila had to sing 'Roseville Fair', and I was asked for 'Buffalo Gals' and Stan Rogers' 'The Lock Keeper', as well as the inevitable Copenhagen song. Ian on tabla and I on banjo played an untraditional version of 'Waterbound'.

Throughout this visit to Canadian relatives, our son Jesse was at our Alligator Creek home in north Queensland, emailing us about the latest disasters along the coast (man found in belly of croc, our bore pump failed, hot and dry weather). On the last day with Al and Peabody in north Winnipeg, I helped their son Crowbar plant a garden, where I had buried two large salmon skeletons, with peas, carrots, potatoes, raspberries, the normal stuff. We had plum kuchen, beer and cider underneath the chokecherry tree, as many of our friends came to bid us goodbye, and I had the last Eccles cake with tea.

The next morning at a cruel 0500 on 22 August we packed and went to the airport, with several teary goodbyes from family. And then it was the airport circus, Winnipeg to Calgary to Vancouver (where we found T-John Beder postcards) to Narita/Tokyo (live orchids on the balconies) to Brisbane and Townsville, where the warm perfumed tropical air was a delight. Jesse and Sarah were there to haul our tired carcasses down to Alligator Creek, where the loud greeting of Sir Charles Cockatoo and Deputy Commissioner Caspar Galah welcomed us home.

It was very dry, an El Nino year, but the mango, peach and citrus trees had brave flowers all over. There were four large bunches of bananas hanging in the pergola. I took off my shirt and felt at home.

There was only one day of rest, during which I watered my orchard, cut bananas and collected horse manure from around our creek valley stock tank. Our neighbour to the west had horses in a poorly fenced pasture, and the manure was welcomed to improve the soil in my vegetable garden.

The next day I went to AIMS to help outfit the R/V *Harry Messel* for a sediment coring expedition to nearby Cleveland Bay, the Townsville harbor, a sewage effluent creek, and around Magnetic Island and Halifax Bay. This was part of a Ph.D. thesis project for Greg Doherty, who wanted to determine the impact of this small town on the inshore coastal region. The *Harry Messel* was a small vessel with a skegged propeller, suitable for going into shallow muddy water. All the usual excitement was experienced over 27 August to 2 September 1998, such as being up to our hips in sewage rich mangrove creek water with a rising tide and a stuck core barrel, and being caught in an eighty knot wind storm with 400 mm of rain in one day. But we 'got 'er done', and Greg wrote two good science journal papers showing that the inner harbour and port region mud had high concentrations of metals from industrial uses, but that outer Cleveland Bay mud was pretty clean, due to frequent flushing by tides and wind storms.

There followed a series of celebrations, the eighth of September was our 28th wedding anniversary, 10 September was my birthday (got a wheelbarrow!), and 18 September was the date that Sheila should have left ELA for home in 1970 [but rather she stayed with me, and we soon got married]. We took some time off, drove our four wheel drive Subaru up into the Paluma rainforest and bumbled around the hinterlands for birding.

I was at one of Peter Gilbey's TAP meeting in Charters Towers, where the CSIRO building was full of displays of all the bad/inappropriate/invasive/feral plant and animal introductions that were made by well-meaning CSIRO agricultural scientists. We were supposed to present our different versions of sustainable lifestyles and water management advice for sustainable agriculture that does not impact the GBR. My part was to suggest ways to reduce soil erosion and river transport of mud to the coastal ocean and Great Barrier Reef. My advice was to avoid over-grazing, destock during droughts, keep grass on the land, and keep riparian vegetation along the river banks. The parking lot was full of dusty old utes with dogs in shaded cages on the tray and a rifle rack in the cab back window, and our audience was dusty old farmers and stockmen looking very sceptically at us boffins with our 35 mm slide presentations. I felt good at these sessions, because the audience were people like my father and uncles, and I could sense when they wanted us to back off, or maybe needed a little more detail or examples. In addition, I didn't wear a necktie, I rolled my own cigarettes, and I wore Levi trousers and high-heeled boots.

Jesse was in the final agony of fourth year exams for his James Cook University degree in civil engineering in October-November 1998. He was assisted by a Queensland Main Roads scholarship, which obligated him to a job in Emerald, about five hundred miles inland. He passed all exams with high marks, was eventually awarded the Charles Barton Medal for the best fourth year thesis, and Sarah, Jesse, Sheila and I celebrated at a good restaurant in town, where we presented Jesse with a posh leather briefcase. By now, Sarah was more than a girlfriend, and she and Jesse were gathering goods for their house in Emerald [...does this sound a little like *The Wizard of Oz*?]

On 3 October, Sheila, Jennifer Hultgren and I played processional, ceremony, and reception music for the marriage of Jennifer and Ken's son Ross to Vicky, a classical violin player who worked at the Yabulu nickel refinery. The wedding was outdoors at the Townsville Dan Gleeson Park, with red carpets amidst the palm trees, and all the guests were in fancy gowns and tuxedoes. After the wedding, a tent party reception at Jennifer and Ken's home was set up for a hundred guests, where we played Morris dances and English country garden tunes.

All these good times were mixed with sad events, as usual. Dr. Sven Froiland, Professor of Biology at Augustana College in Sioux Falls, South Dakota, died on 4 October 1998. He was one of my mentors who pushed me into science, despite my initial attempts to be a music major. He taught me genetics, chordate anatomy, and several courses in evolution in 1959-63. Not too long afterward, I found out that my high school teacher niece Donna Sue Brunskill was a friend to Dr. Froiland's daughter Karen, who was a high school teacher in Custer, South Dakota.

Our washing machine quit working. Jesse took the motor apart and found frog soup inside; after cleaning the motor with rubbing alcohol, the motor worked just fine without the frog lubrication. Our home computer printer died. Jesse took it apart, to find the power supply coated with dead geckoes and gecko eggs.

One of the results of earlier workshops in Pentland, Charters Towers and Ingham (inland Queensland), about trying to provide environmental and hydrological information to cattlemen and farmers of coastal Queensland, was to establish a Resource Information Centre in Ingham. I had given special presentations for the Beef Cooee Club in Pentland, the District Managers Committee in Ingham, and the Grazier Club in Charters Towers. I was good at this, as I related well to inland cattlemen, but unfortunately AIMS administration didn't like my environmental missionary work, and asked me to stop it. Very curious.

Despite this, one outcome of the work was a Resource Centre, sponsored by agricultural and environmental scientists in CSIRO, myself, JCU's CRC Reef, GBRMPA, several state resource agencies, business and tourism groups, and some NGOs. In early October we had another Ingham workshop as an Open Day for the Centre, and once again we had a long line of dusty utes parked in front of the hall. We had three days of talks, with banquets at Lee's Hotel every evening, and I enjoyed sitting with the cane cockies and cattle station managers. Ladies provided the lunch, morning and afternoon smoko with tea and pastries, but it should also be noted that almost half the workshop speakers were women.

My organic geochemist colleague Kathy Burns and I were invited to lecture at the Central Queensland University in Rockhampton, and the Royal Australian Chemical Institute (RACI) in Gladstone, in November 1998. Dr. Vicky Vincente-Beckett was our host in Rockhampton, and she gave us a tour of the Fitzroy River mouth and delta, the 'barrage' (a sort of low gate above the mouth to restrict salt water progress upstream), and organised meetings with the Dean and Department Heads. The CQU people wanted AIMS to begin a study of the Fitzroy River estuary and the adjacent continental shelf. Vicky wanted more collaboration with AIMS marine scientists, and in particular, asked me to be on the M.Sc. thesis committee of a student working on Mount Morgan mine waste heavy metals in Fitzroy River estuary sediments.

I gave a lecture on the uses of radiochemical tracers in coastal oceanography, and Kathy talked about the chemistry of the thousands of components of petroleum in seawater, and how some of this oily stuff enters the food chain and is deleterious for marine organism growth and reproduction. Kathy and her colleague Norm Duke had done several experiments on the impact of oil spilled on enclosed sections of intertidal mangrove forests near the Gladstone Harbour nearby. Gladstone is the location of many chemical industries, ore refineries, coal and petroleum industries, and the harbour is probably the most polluted estuary along this coast. The RACI Conference in Gladstone was dominated by Stuart Oil Shale chemists, aluminium and coal chemists, and agricultural scientists concerned about the productivity of the beef industry and the chemical quality of the fat in cattle. They had a very fancy lecture room with triple slide projectors and screens, video facilities, a computer technology crew, and a separate balcony for the press gallery. It was obvious that the audience was a bit startled to have speakers talking about coastal oceanography and environmental science, when the main game in this town was industrial chemistry.

Partly at the urging by Kathy and myself, a group of JCU, AIMS, CSIRO marine scientists returned to CQU in Rockhampton in late November to discuss a joint research program on the Fitzroy River estuary and Keppel Bay. As usual, there was lots of enthusiastic talk, but nobody had any extra money. I took my research crew down there on the R/V *Lady Basten* for sediment sampling, but this effort was not rewarded with future funding. Eventually such a research program did get organised by CSIRO, but AIMS administration did not want to participate. We didn't know it at the time, but 1998 was a hot year, the Coral Sea surface water warmed more than usual, and it was the worst coral bleaching year so far recorded.

Jesse now was a proper civil engineer, graduated with honours, and he and Sarah were getting ready to move to Emerald in early January. We were grateful to have them with us for Christmas and New Year's Eve, and then they were gone for Jesse's first engineering job with Main Roads, Queensland. The 'empty nest syndrome' began in our house. Ian was still at the University of Toronto. Well, not quite empty nest. Sarah left her pet galah with us. This was Caspar, who liked to sit on my shoulder and eat gristly bits of beef heart at breakfast time. Sir Charles the sulphur crested cockatoo was much more refined and dignified, and would eat only small bits of my breakfast omelette.

Chapter 35. More Adventures in Papua New Guinea, Irian Jaya and Indonesia

As noted earlier, partly due to the severe 1997-98 El Nino and reduced precipitation in Australia and SE Asia, our TROPICS organisation called for more time on the smaller AIMS research vessels in the Gulf of Papua. Most of the rivers and estuaries of the Gulf of Papua were not accessible to the big research ships, because of reduced precipitation, river discharge and low water levels. The AIMS ships were built for inshore and estuary work, having protected propellers.

I helped my crew gather equipment from Woods Hole (Sholkovitz), ANU (Candace Martin), University of Washington in Seattle (J. P. Walsh), State University of New York (Bob and Josie Aller), ANSTO (Ron Szymczak), James Cook University (Jamie Bunt), Peter Davies of Southern Cross University in NSW, and added my own gravity corer, gantry and deck work table. This was all loaded onto the *R/V Harry Messel* on 5 January 1999, and the ship departed the AIMS Jetty at midnight to steam northward along the coast to Thursday Island in Torres Strait. The science crew stayed at AIMS for a further few days, to fly into Thursday Island when the ship arrived.

But in the meantime, JCU, GBRMPA and AIMS were dithering about the new Coordinated Research Centre (CRC) for Coral Reefs, fussing about where the money was going to come from, with some squabbles over leadership and directions of the research. We already had responsibilities to our Japanese funding sources (KEPCO) for work in the Herbert River estuary mangroves, led by Tenshi Ayukai, and they wanted us to work in Indonesia, Thailand and Malaysia. AIMS was not terribly excited that I was taking staff and ship time for TROPICS research in the coastal waters of Papua New Guinea, even if we did have big funding from the US National Science Foundation. I had already posted proposals for work on the Fitzroy River estuary down near Rockhampton, which would spread our personnel, ships, and equipment resources even more thinly.

Several high level meetings were held in early January, wherein it became obvious that we could not do all of these things on the money we had in our budgets. And underneath all this worry and conflict was a simple request from Kathy Burns, Tenshi, two technical support positions, and myself for an extension of our contract with AIMS. In addition, my right foot swelled and was sore, and the doctors failed to find any reason for this, other than perhaps some insect, spider or snake bite. I was using my grandmother England's cane to walk down to my orchard to pick mangoes, with my foot tightly wrapped in tensor bandages. I gratefully left all of this uncertainty behind to go play in the mud of the Gulf of Papua with my international colleagues.

I said goodbye to Jesse and Sarah at the Townsville airport on 8 January 1999, as their old Toyota ute was loaded to go to their new home in Emerald on 11 January.

My AIMS crew were grounded in Cairns because of torrential rain, but managed to get to Thursday Island in the Torres Straits the next day at 1:30 am, the rain still pouring down, with standing water on the Horn Island air strip. After the droughts of the 1997-98 El Nino, we now had La Nina torrential rain and flooding in the tropics: Land of feast and famine! After some sleep, we walked the town streets, had a good supper at the Jardine Hotel, with beer and cigars on the sheltered verandah. The AIMS ship Master Matt phoned to signal his arrival in the late evening, having had rain all the way and a very useful SE trade wind pushing the ship at ten knots. Bob Aller had been in the Jardine Hotel for four days, having arrived early to counter jet lag, he already knew the restaurant cooks and recommended feasts of Indonesian laksa with large loaves of hot fresh baked bread.

Many TROPICS research voyages began this way, with minor feasts in the outdoor verandahs of the Jardine and Grand Hotels, underneath sprawling frangipani and sea almond trees, where outrageous stories were told, some of which might even have been half true. J. P. Walsh told of wild fraternity parties at the University of Washington in Seattle, one result being a stripper suing the frat house into insolvency for 'impolite behavior', and the Latin motto emerged as... *Puella divestienta noli mordere*! Australian honour required a story to top the Yanks, and Ron Szymczak came to our rescue several times, with another motto...*'Ogrzewane kotletti nie som smaczne'*, a Polish proverb, roughly translated as 'Reheated rissoles never taste as good', meaning that old girl friends are not as exciting as new ones, or that new frontiers are always more interesting to explore. As you might guess, a few of these stories were not fit to print. Some of these mottos were chiseled into planks of wood and hung on the ship aft deck. This is yet another example of the international family of scientists that are linked by jokes and stories of adventure in odd places of the world.

Our good ship the R/V *Harry Messel* arrived with a tired crew, Master Matt, engineer Paul, Brett the cook, and Jason the aft deck hand. Australian Customs and Immigration inspected us in Thursday Island harbour, we all found our bunks, and the ship lifted anchor at nine am on 9 January 1999 to steam in the torrential rain to Daru for PNG Customs inspection.

This was Leg 7 of our TROPICS Consortium operation, the seventh expedition of a planned dozen. The weather improved, and our little crew explored the chemistry and biology of the estuaries of the Fly, Kikori and Purari Rivers and out to the edge of the continental shelf in the Coral Sea over 9-17 January. We steamed upriver to freshwater in each river, and then we slowly descended the salinity gradient (from fresh river water to sea water at salinity 35) to sample water and mud, for measurements of algal nutrients and chlorophyll (Peter Davies and Irena), trace metals and hydrocarbons (Szymczak), and variations in sediment chemistry (me). An important separation was maintained between the clean people (Irena, Davies, Szymczak, in the scrubbed clean, air filtered and air conditioned lab), and the muddy people (JP, Jamie, the Allers and myself). Irena Zagorskis was the comptroller of ship log, locations, CTD and sample numbers for each activity. JP Walsh and Jamie Bunt went off in two Zodiacs (big rubber tender boats) on a full day trip to vibracore the mangrove mud in the delta of the Kikori River, right behind Gairibari Island, where local people ate a Lutheran Missionary Society preacher in the 1920s.

This maze of deltaic mangrove islands had many villages, and I'm sure they were curious about these strange mud men. This region was also one of the research sites for Carleton Gajdusek (also known as Vin Zigas), who went to New Guinea in 1957 to discover that the people were dying of what is now called Mad Cow Disease (kuru, Creutzfeldt–Jakob disease) caused by eating the brains of your loved ones and enemies. He got the Nobel Prize for this work. In his old age he was mentally deranged and was eventually jailed for paedophilia at age seventy-two.

Bob Aller and I were interested in the time history of sediment accumulation and the depth variation in associated chemical components in the mud. Companion cores were sampled by Josie Aller for animals and microbes that live in the mud. The river and mangrove regions were choked with a huge supply of black sand, with occasional mud banks near the mangrove islands, but as we crossed the shelf to deeper water, the fine mud appeared, full of surprises to all of us. The rivers often had floating streams of coconuts, *Nypa* and other mangrove seeds, small floating islands of aquatic vegetation, complete with birds and lizards. We would occasionally encounter sediment that appeared to be finely ground wood, which we called 'coffee grounds', and sunken rafts of mangrove logs in odd locations of 100+ m water depth. Some mangrove wood has a density greater than water, obviously.

My journals are full of detailed times of deployments of various kinds of chemical, biological, and sedimentological equipment, with notes on failures and needs for repair of items that were damaged. I seemed to also keep notes on the good food served by Brett the cook, who also helped us slicing sediment cores late at night. JP Walsh kept us all entertained by singing the Frank Sinatra ballad 'My Way' on the aft deck while cleaning his vibracorer.

A few examples of Brett's fine meals were giant roasts of beef, crab salads, veal with mango mustard, barramundi with Jabiru Tingle Sauce, racks of lamb smothered in rosemary, half a dozen roast chickens, tubs of chili spiced battered onion rings, steamed mud crabs, bugs, squids, prawns in filo pastry, curried scallops, fresh baked whole wheat buns, cream puffs with strawberries inside and frozen figgy cream pies. There were some serious complaints about the hazards of eating with a crew that were cracking large mud crab claws with pliers: Bob got hit in the face twice with large fragments of crab claw shell, so he claimed compensation by asking for an extra large claw. Of course, there were immediate calls for Occupational Health and Safety glasses and face protection helmets. We were working hard for long hours in the tropical heat and rain, but we were not suffering for lack of good food.

Brett, GJB and Bob Aller retrieving gravity corer on RV Harry Messel*, Kikkori R Delta, 1999*

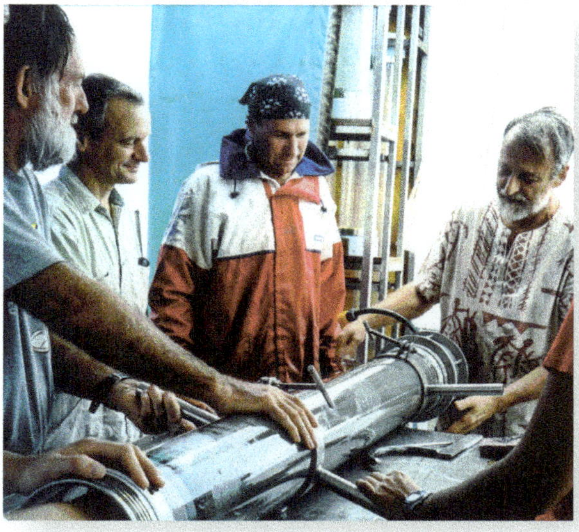

Paul Dixon, Bob Aller, Brett the cook GJB ready to open core tube, RV Harry Messel *1999*

Up in the Wame Channel of the Purari River delta, we encountered a fleet of canoes with fifteen or twenty native people from Araava Village, all very curious as to what we were doing. Some of them were from a missionary school and carried their Bibles with them. They understood fishing, but not collecting bottles of water and plastic sacks of mud. They wanted to trade a woven palm frond basket of mud crabs and bananas for cigarettes and outboard fuel. I let one young man named En Jack roll a cigarette from my Stockman's tobacco pouch, and he wanted to know how to get a job at AIMS. I gave him our AIMS brochure and a TROPICS pamphlet, and his eyes were wide with wonder.

We found that a lot of the fine black magnetite sand was coming from this delta channel, and it covered a long part of the continental shelf to seaward. Nearshore, at low tide, it seemed odd to see glistening waves of shiny black sand stretching seaward from the mangroves. Everybody thinks tropical beaches should be white sand, but not in this region. This sand meant that our coring equipment and radiochemical tracers would not work well. We could only take surface grab samples, but Josie found that there were lots of benthic organisms in the sand.

Farther west, in the upper reaches of the Fly River delta, we encountered big dugout canoes with outriggers and sails, loaded with neat palm leaf packages of sago flour and hind quarters of rus deer, heading downriver to Daru to sell their goods. Some of the canoes came to visit us, to trade baskets of live crabs for petrol. J. P. Walsh took a liking to the hand carved paddles, and he traded Peter Davies' shoes for one of them. The running shoes were hanging off an aft deck railing, to be dried out. Peter was asleep in his bunk at the time, and would not have approved of this transaction, but we told him the wind had blown them overboard. Bad JP!

By the morning of the 16th we had finished our planned sampling schedule and were cleaning up and packing our gear. The wind came up to forty knots with torrential rain, so we stayed at anchor protected in the lee of a Fly River Delta island, watching films, trying to eat up the rest of the food in the galley freezers. We could get satellite TV here, and it felt odd to watch CNN and ABC news, snug and secure as the storm raged outside.

We eventually tied everything down securely, passed PNG Customs in Daru, and steamed through two metre waves and strong wind southward into the Torres Strait. We passed Coconut Island at midnight, where we saw several police boats and tents on the beach under floodlights. We later found out that a small airplane had crashed into the beach, killing three women. We arrived at Thursday Island at 0700. Australian Customs and Immigration carried out their inspections at ten am while the ship crew were reverently watching the cricket. Our science crew abandoned ship to have sinful cheeseburgers, cold beer and garlic chips at the Grand Hotel. We took the ferry to Horn Island to board a Dash 8 aircraft to fly to Cairns (torrential rain) and then to Townsville by 8 pm, where dear Sheila collected all of us for delivery to AIMS accommodations and Alligator Creek. The *R/V Harry Messel* took another four days to steam south along the north Queensland coast, to dock at the AIMS Jetty on 21 January for unloading samples and gear into two trucks.

Now the hard analytical work began in the AIMS laboratories. Irena was supervising the geochemical measurements of major and trace elements, organic carbon, nitrogen, silica and phosphorus for all the water and mud samples. John Pfitzner had his own gamma spectrometer lab, where two or four cm slices from each core top to bottom were allowed to give off their gamma rays for the measurements of excess ^{210}Pb, ^{137}Cs, ^{7}Be, and ^{234}Th indicating the rate of deposition and the rough chronology for the core. The data were put into tables with estimates of analytical uncertainty, graphed, puzzled over, and eventually used for the next publication. The number of big metal five-drawer file cabinets was increasing in my office, and I eventually put several of these data files out on a little veranda where I ate my lunch and smoked my roll-your-own-cigarettes. Irena and John also had digital records of all this data, along with maps of the ship's track and coordinates of each sampling location.

As I've mentioned before, most research organisations have a habit of trying to re-invent themselves, to eviscerate and brainstorm themselves, to reveal strengths and weaknesses, and to get ready for new directions of research.

In late February 1999, AIMS, JCU, the CRC for Coral Reefs, CRC for Rainforests, GBRMPA, and several local industries gathered in a downtown hotel to explore what might happen in the next decade. Mick Roche of the Minerals Industry (BHP, mining) complained that our CRCs didn't have a charter or rules for procedures. Simon Woodley,

CRC Reef Chief, explained that we were in the fifth year of operations and needed to plan for the next five years of efforts in education, social engineering, economics, physical, chemical and biological oceanography. Where would the money come from? We definitely needed a 'sugar daddy', as the federal and state government were not likely to continue the current funding program. Port facilities from Gladstone Harbor to Cairns were discussed, with worries about coastal contaminants from port handling of ores, metal and oil refinery wastes, dredge spoil disposal, dust and air quality standards. Large amounts of coal, lead, silver, copper, and zinc ores were trucked or rail shipped into Townsville from the interior of Queensland. Yabulu Nickel Refinery, MIM Copper Refinery, and Sun Metals (mostly zinc) had large areas of liquid tailings waste evaporation ponds near the coast, which were in danger of overflowing or being swamped by tidal surges during cyclones. The energy companies were mainly concerned about rail access to coal to supply adequate electricity to these high use industries. They eagerly awaited the arrival of cheap natural gas from the Timor Sea and Papua New Guinea, as a pipeline was to be built shortly.

There were speakers promoting solar, wind, tidal, hydro, and sugar cane bagasse biofuels. JCU chemists and biologists had been measuring metals and organic contaminants in Cleveland Bay water, mud, and biota since the 1970s, and my Ph.D. student Greg Doherty gave a modern update on the locations of the worst contaminated locations. The tourist industry was disgusted with the dumping of port dredge spoil in Cleveland Bay, which provided lots of fine mud to cover their sandy resort beaches on Magnetic Island. The port administration and the mining industry kept asking for a clear statement of the legal jurisdiction over the coastal creeks, mangroves, mud banks, and inshore bays. Aboriginal lawyers claimed some of the land along the coast, and they have been ignored for over a century. The City of Townsville wanted to see a plan for the sustainable development of the region, with the ability to meet the demands for adequate water, electrical power, housing, food production and education for a growing population. This was a reasonable request, but not one that was easy to deliver!

Now, what do you think happened as a result of all this effort and information? Not much, mostly a lot of hot air, with hopeless efforts to summarise the wildly varying priorities of industry, city growth, port facilities, university research, GBRMPA regulations and environmental science. This little episode might be an example of the difficulties of urban planning for a sustainable style of living around the world.

After the drought of the previous few years, we were grateful to have a decent wet season, and our little creek was running strongly. Our orchard soil was soggy, the grass was over a metre high, and the mosquitoes were fierce. A newly planted group of passionfruit vines and grapes took off on a four post frame platform, and we had plenty of our own home grown mangoes. We feasted on red and yellow meated paw paws, limes, bananas, carambola, mabolo, figs, guavas, mandarins, grapefruits and lemons. We were fast becoming embedded in a tropical life style of living. We got word from my brother Obie that his daughter Marcy was to be married in the South Dakota Black Hills in June. I was busy on ships, so we sent our son Ian to represent us.

Chapter 36. Further Adventures in Japan

Because I was one of the leaders of the TROPICS consortium, I was invited to present a paper on our work in the Gulf of Papua at a workshop in Tsukuba, Japan, in honour of K. O. Emery.

Kenneth Emery was a Woods Hole Oceanographic Institution (WHOI) legend who was an innovative leader in continental shelf marine geological research from World War II to the 1980s. He and one of my Cornell thesis committee professors, John W. Wells, surveyed Bikini Atoll before and after the nuclear bomb tests in 1946-50. I had heard him lecture at Cornell University in the mid-1960s, read some of his papers, and had copied his diagrams of the variety of styles of continental shelf modes of formation. After his retirement in 1980, he lived at the head of Oyster Pond in Falmouth, Massachusetts, where he managed a productive fruit tree orchard and apiary, produced a bathymetric map of Oyster Pond from a row boat, produced a book about the Oceanography of Oyster Pond, and monitored the impacts of hurricanes on Oyster Pond. He died in April 1998 at the age of eighty-three, a man with thousands of friends, scientific colleagues, students, and admirers around the world. I knew nothing of his connection to Japan, but apparently he was one of the first academics to visit Japan after the nuclear bomb destruction of Nagasaki and Hiroshima, and helped Japanese universities and research institutions recover from the war.

As a consequence, the Japanese oceanographers held him in high regard and provided an international workshop in his honour.

I left Townsville at 0915 on 14 March 1999 in hot sunny weather, flying over the startlingly blue Great Barrier Reef lagoon, wearing my Niverville thrift shop corduroy suit with long pants, tie, shoes and socks, because it was only 9°C in Tokyo. I arrived at Narita at 8 pm, into the usual hassle of finding my luggage, customs, passport control, amidst a very crowded airport. A taxi took me to Tsukuba, a two hour drive through flashing neon signs, Kentucky Fried Chicken shops, used car lots, McDonalds, multi-layered concrete roads full of speeding cars with blaring horns, Mr Donut Shops, and finally to the Daiichi Hotel. One of the meeting organisers Yoshie Saito met me at the door, and we had another supper of fish, rice and Japanese beer while watching CNN (English) on a huge screen in the hotel restaurant.

The next day I found that my hotel could supply very good bacon and eggs breakfast, a wonderful surprise. Our group was kindly and gently herded through a geological museum on the Tsukuba campus. Right smack in the midst of it all were fossil *Baculites* and dinosaur bones from South Dakota and the Wyoming badlands, and photographs of the excavation site operations...I almost felt homesick.

There were also computer controlled models of the oceanic plates moving to form the Atlantic and Pacific ridges, submarine volcanoes and gas vents, and the genesis of tsunamis from earthquakes and landslides. I met an Indian geochemist from Goa at a table full of minerals of the elements boron and lithium: he was Shirodkar, who wanted me to write a paper on B and Li fluxes from the Fly River in PNG. In a few years, I actually did write such a paper!

In the afternoon we heard heavy presentations from Shirodkar about the erosion rates of the isotopes of boron and lithium by Indian rivers. Nittrouer and Kuehl spoke about their research on the Amazon River delta and continental shelf. We had Japanese and Korean guides to find us a good restaurant for supper, and nearly killed ourselves trying to eat all the good food as well as good beer and sake. It was cold and raining, and I bought an umbrella for 500 Yen, some Ceylon tea and a mix for hot chocolate for my late night snack in my hotel room.

The Japanese were very good at polite herding of the workshop people. After a day of plenary lectures by the bigwigs of continental shelf science, we were bussed to a fancy tea house for a massive seafood feast. There was a large smoked (not cooked) tuna stretched out on a table, snout to tail, with people digging in to one side of it. There were large pots of steamed sea creatures, some of which I could not identify. One room featured a small ensemble of folk musicians and dancers, drums and flutes with small finger cymbals, the dancers with masks. Some masks on men made them appear as garish beautiful ladies, some were fox faces, and some of the dances were about the medieval history of peasants and the emperor in this region. We were provided with way too many bottles of sake, palm wine, iced fruit bowls and cigars. Some important industrial and university people gave toasts to the workshop with long speeches in Japanese.

On the evening of the fourth day of presentations, there was a disrespectful party in Steve Kuehl's room, where the Americans had to unleash their spirits. Japanese society is very polite. Americans can't stand this for very long. Chuck Nittrouer's wife Deborah bought some green ribbon, and was making green bow ties for St. Patrick's Day. There was too much German dark beer and Korean vodka, and a lot of the stories about bad oceanographic ship crews, mistakes in navigation, and conflicts between professors and students. This was the camaraderie of international science.

On 18 March, I woke with a sore throat and a bad chest cold, probably from walking around in the cold rain hunting for good restaurants. I gave my paper, croaking and coughing, on sedimentation in the Gulf of Papua, along with Texan David Prior who spoke on the Markham River inputs to the Huon Sea of eastern PNG. Several papers came from Malay scientists on the coastal shelves of Borneo. Taro Takahashi from Lamont Doherty of Columbia University was there, an old friend. Nittrouer and his colleagues spoke on the Amazon River estuary and shelf and the Gulf of Alaska. Asahiko Taira, a tectonic physics guru, gave an amazing presentation of his work using side scan sonar following a submarine canyon for 700 km across the shelf, slope and into the abyss, where the submarine river of mud had levee banks and floodplains at 6000 m depth.

A fellow walked up to me and introduced himself as Richard Hiscott, a marine geologist from Memorial University, Newfoundland, Canada. I was flummoxed that he knew me from Meadowlark performances in Winnipeg, as he was the brother of button accordion player Jim Hiscott, who worked as a music producer for the Canadian Broadcasting Corporation. He was also a fiddler, and he was disappointed that I had not brought my fiddle with me on this trip. He described Sheila playing the hammered dulcimer with me on fiddle, on stage at the Winnipeg Folk Festival. It was one of those 'out of context' surprises.

That evening Richard and I sat together at our banquet in an old sake factory, where we ate plates of fermented rice with more curious pieces of seafood. The tables were low to the floor, we sat on the hard floor with our legs crossed and had hot pot fondue of big slabs of wagyu beef and mushrooms. Again, tributes were expected, as well as ribald stories about K. O. Emery, some sung ditties, songs from the old days on the oceanographic ships. I got up and sang Flanders and Swann's ditty from the 'Hippopotamus Song', highly appropriate for scientists who work with mud.

I was supposed to go to another symposium in Tokyo, but by 19 March I felt rotten, with a bad cold, congested chest, and Hirohito's revenge kept me close to my hotel room toilet. I slept till noon, had a huge US style breakfast, wrote some letters, and in the evening walked the mall. I had some nice hot fig rolls which I brought back to my hotel room to have with tea. I slept in the next day as well, groggy and weak, and had a very strange breakfast at two pm of a bacon and lychee pizza. I checked out and waited in a big soft chair in the hotel lobby for the airport bus. While dozing, my mind was viewing the paintings on the wall, odd Picasso modern style clowns playing fiddle and banjo with green birds perched on the fingerboards, Mount Fuji, rice paddies, cherry trees and cats. I got on the cold bus to the Narita Terminal 2 airport in the evening, immediately found a seat beside a hot air vent and went to sleep. There was a computer bar and a coffee lounge with Steve Goodman singing 'The City of New Orleans'. I must have boarded the airplane in a daze, slept across the Pacific, and woke up in Brisbane on 21 March. There followed the short flight north to Townsville, where dear beaming brown Sheila gathered me up to haul my carcass down to the peaceful safety of Alligator Creek..

I recovered quickly, and whippersnipped my way through the two metre tall guinea grass of our orchard. We had lots of rain, and my fruit trees were very happy. I helped Sheila haul big flat creek bed stones up the hill, for paving paths around the house. It was wonderfully peaceful, green and lush, and the birds, geckoes, possums and wallabies (and even a little Children's python) were part of our family.

Chapter 37. Getting Lost in the Irian Jaya Mangroves

As previously mentioned, funding for research was becoming limited and our ambitions for research projects greatly exceeded the money available from AIMS to support the travel, ship time, laboratory analyses and technical salaries.

We were encouraged to 'work with industry' and obtain externally funded consultancy work with high overhead fees to keep the AIMS lights on in the lab. My good friend Dr. Peter Isdale, a scientist gone over to the 'dark side' as a business manager for AIMS, helped me rather too much for the last few years. As a result of his hustle, I was entangled in US National Science Foundation supported AIMS and Franklin ship costs for our research activities in PNG, US NSF/Indonesian funded workshops for IndoTROPICS, Mineral Policy Institute (MPI) reviews of environmental impact studies (EIS) on nickel/cobalt mining and refining waste disposal into the Bismarck Sea (north coast of Papua New Guinea), a radiochemical analysis of several sediment cores from the Tamar River estuary in Tasmania, and a consultancy to explain the distribution of copper in big molluscs from a Freeport Indonesia polluted estuary on the west coast of Irian Jaya.

If this list makes you a little dizzy, just think what it did to me. My simple mind was not designed to worry about legal contract details, unforeseen personnel costs, insurance policies, ever-changing customs and immigration policies in three countries, reporting deadlines, and interactions with legal and government departments of Indonesia and Papua New Guinea.

Joanna Ellison came to my group via the Cayman Islands (British West Indies) and Stanford University, on a postdoctoral study of Queensland mangroves. She had helped me in earlier work in Exmouth Gulf mangroves in NW Australia, and had visited some of the mangrove forests on the south coast deltas of Papua New Guinea within the TROPICS research program. My little group was also enriched by the presence of JCU Earth Science

Ph.D. candidate Greg Doherty, who was studying the sediment and metal contaminant distribution in nearby Cleveland Bay harbor and port facilities. Joanne Cavanagh was doing a Ph.D. thesis (with Kathy Burns and myself) on agricultural pesticides in estuarine sediments from the Burdekin to the Herbert River. Jochen Mueller and his student Caroline Gaus appeared out of nowhere, wanting to sample my coastal sediment cores for the determination of dioxins, one of the more dangerous contaminants from the plastics industry. Yas Tesiram started a project on metal contaminants in sediments from nearby Bowling Green Bay and Hinchinbrook Channel. Dr. Jan Skjemsted of CSIRO Adelaide wanted to collaborate with us on charcoal abundance in our coastal sediment core archive. Irena, John and I increasingly became servants to this herd of good students and colleagues.

Joanna Ellison previously had a contract with Freeport Indonesia (PTFI, a subsidiary of the New Orleans, USA, parent company) to take some long sediment cores in the mangroves of the Ajkwa and Minijerwi River estuaries at the port of Timika, to determine the time history of trace element pollution. She discovered that the Timika Environmental Laboratories held a large database on copper (and other heavy metal concentrations) in the large coastal clam (*Geloina coaxans*), which was an important dietary component for the local Irianese people. She advised the leader of the Timika Lab (Kent Hortle) that we could analyse this data, as well as determine the radiochronological history of mine tailings metal deposition in the sediment cores.

Way too quickly, lawyers from Freeport in Jakarta and New Orleans, USA, and our AIMS business office had a contract signed, a big pile of money appeared, and I was to travel to Timika forthwith. Where is this? The contract also specified that I was to review the methods and reports from their Timika Environmental Lab, visit their coastal sampling sites and field sampling methods, recommend improvements in activities of the lab and field procedures of the Environmental Group, and recommend future environmental studies of the dispersal and accumulation of mine tailings in a dozen mangrove estuaries along this wild remote coast.

And so on 30 March 1999, Sheila delivered me to the Townsville airport for travel to Cairns and Darwin, arriving at 1630 in the pouring rain. The country was brilliant green, with standing water everywhere, and puddles coloured rosy red due to the rusty red nature of the soil.

Freeport Timika lab manager Kent Hortle (with his gorgeous Chinese/Malay wife and two kids begging for skateboards) met me with a bus. About half the passengers on my Darwin flight were going to Timika, and we all had large steaks at the Hog's Breath Cafe, Rydges Plaza Hotel in Darwin. This was part of the 'fly in, fly out' style of a big mining camp in remote Irian Jaya, also called West Papua, what used to be called Dutch New Guinea before Indonesia's Sukarno claimed it. Freeport Indonesia must have had a lot of muscle and money, as they chartered a Qantas airplane to move their employees back to camp. The aircraft was late and we left Darwin in the pitch dark, raining buckets at 2210 hrs, arriving in Timika, Irian Jaya, Indonesia just before midnight.

There were very few lights on the short single airstrip, and there was no sign of a town or roads. The Indonesian police operated a very quick and simple customs and immigration and quarantine office beside the air strip, using flashlights to inspect our passports. We were all piled into the back of several big trucks so they could haul the miners to their accommodations, and we were dumped at the verandah of a very posh hotel, solemnly called 'The Sheraton' by the Indonesian drivers. This hotel was obviously for important visitors, and it included a pub in full operation after midnight, a neat spic and span white table cloth restaurant and rooms with good beds, showers, and satellite TV with CNN in English and several Indonesian TV services. Not what I expected for a remote mining camp. Ah, but we only got to stay there on the first and last night. For the rest of the week we were given a miner's cabin near the port and science lab.

The next morning I was guided to a breakfast table with Dr. Fred Wells of the Western Australian Museum. He was hired to work with me for his expertise in marine ecology and especially the biology of molluscs. Fred had published papers on the impact of tributyltin on mollusk physiology, in particular how this contaminant changes the sex ratio of snails and clams. He also knew a lot about how the native people relied on mangrove and coastal seafood, including the giant mangrove clams, mud crabs, fish, and the boring toredo molluscs that eat dead mangrove wood.

So began our routine for the next week: We set up our office in the library, where we had long bookshelves of internal reports to evaluate and criticise. If it was not raining, we were hauled to the Topoeka Channel port, where Irianese men expertly handled the boats with long shaft outboard motors and canvas shades. Each day we toured the estuaries and mangrove channels of the major rivers, from the Kamoro in the north to the Otakwa in the south, the Irianese men being silent but showing us where we were on the map. These fellows certainly knew the dense jungley mangrove swamp channels, as they were all connected along the entire coast, and we only rarely glimpsed the Arafura Sea to the west. As the tide went out, Fred Wells could grab some of the big *Geloinia* molluscs and mud crabs for chemical analyses. I took surface sediment samples and short cores for heavy mineral mine tailings analyses. Most of the time it was very hot, humid and sunny and we were grateful to be inside the mangrove forest. It rained heavily every night.

Life in a mining camp certainly is a social and cultural experience. The Muslim call to prayer could be heard, and there was a Christian church as well. The Christian church was sort of evangelical, and their bell was a large nitrogen lab cylinder cut off at the bottom and struck with a heavy crowbar. Hosea Wally was the leader of this church, where a large bonfire meeting was held nightly. He assumed that Fred and I would want to attend his services, was horrified that we did not take up his invitation to worship, and mumbled something to the effect that he would pray for our souls. As I recall, Fred mumbled something back at him 'Fat lot of good that will do.'

We ate our meals at the Portside Cafeteria with everybody else, where there was an Indonesian food counter and a 'White Man' food counter.

The Indonesians ran the town camp and tailings pond operations, the Indonesian police looked for trouble, and the native Irianese had all the menial jobs. Most of the biological and chemical scientists and science assistants were Americans or Javanese educated at Bogor, with a few group leaders from the USA or Australia. We were given small backpacks with several plastic bottles of drinking water, labelled 'Clear Pure Spring Water Mississippi', bottled and sealed somewhere in New South Wales. We eventually found a hidden pool hall and beer parlor for white men only, with old *Penthouse* and *Playboy* posters pinned on the wall, a library of comic books, magazines and newspapers, soft drinks, chocolate bars, potato and durian chips, Dutch, Australian and German beer, and felt-topped tables for serious poker games. We could see the snow and glacier capped mountains from our cabin door.

The native Irianese people that inhabited the coastal mangrove swamps were well known for their carvings and their warfare. Perhaps the Asmat further south had contributed to the 1961 disappearance of Michael, one of Nelson Rockefeller's sons. They used large dugout canoes with outriggers and sometimes sails to move up and down the coast and up to the foothills of the mountains. As usual, I knew nothing about their culture, language, or ecological knowledge. I am sure they knew more than we did about the local tidal swamp ecology. The Indonesians asserted that they were primitive savages and should be ignored. But the Irianese had stories about the mangrove clams, mud crab caves, the weird hermaphroditic toredo shipworms in the decaying wood, the snow and glaciers onto of the interior Carstensz Mountains.

Dutch Captain Jan Carstensz sailed along this coast in 1623 and reported snow on the mountain tops, despite this being close to the equator. Of course, nobody in civilised Europe believed him. From early 1900s to post World War II, this region was called 'Dutch New Guinea', but it was a long way from Holland and the interior mountains were poorly explored and mapped. Several Dutch explorers tried to boat up the same rivers of our study, but most were not successful, and suffered greatly.

After World War II, in 1963 Indonesian leader Sukarno got the United Nations to agree to annexation of West Papua to the Republic of Indonesia, itself having won independence from the Netherlands in 1949. The wonderful drug induced explorations of Alfred Russell Wallace reached the northern part of Irian Jaya, and Ernst Mayr collected exotic species of birds for the Harvard Museum here in the 1930s. Jared Diamond and other anthropologists gave accounts of the native Irianese social and warfare arrangements. Zuckoff tells the story of World War II American airplane crash survivors in a jungle valley of the mountains of Irian Jaya in 1945. Both the Dani people and the Americans were somewhat startled.

Jean Jacques Dozy was a French member of the Dutch 1936 Colijn Expedition, which was the first to reach the snowcapped peaks around Puncak Jaya in the Carstensz Mountains. They used an oil company's Sikorsky S-38 to fly men and supplies into the base of the mountain ranges. Dozy described large black outcrops of rock in the high meadows at 3500 m above sea level, which he called *Ertsberg,* which is Dutch for Ore Mountain, and Grasberg. After half a century of hammer sampling, drilling, and geochemical measurements of the ore body, it was decided that this was perhaps the world's richest mine site for gold and copper.

US Freeport-McMoRan Sulfur executives from New Orleans and Phoenix arrived in Jakarta in 1966. They required about a decade of retraining in the language, culture, government and technology of Indonesia before a deal was sealed.

From 'Freeport in Indonesia: Filling in the holes'. By John McBeth. Jakarta, *Asia Times* - February 22, 2006:

> 'Freeport's chief executive, James 'Jim Bob' Moffett, enjoyed unprecedented access, flying into Jakarta in his private Boeing 757 jet on regular visits. A big, larger-than-life Texan, Moffett would leave invited Indonesian officials slack-jawed with his impromptu Elvis Presley impersonations at staff gatherings. But his roughshod style and a culturally taboo habit of loudly banging the table when he was upset also made him a host of enemies. One rancorous meeting with Moffett turned then-environment minister Sonny Keraf into a life-long foe. Now in his new role as a member of the opposition Indonesian Democratic Party for Struggle (PDI-P), Keraf also happens to be the vice chairman of the parliamentary commission on the environment.'

The rich deposits attracted investors, including Rio Tinto, and by 1973 a mill was operating at Tembagapura in the high misty cold mountains at a mill rate of five metric tonnes/day at 3% copper. It boggles the mind to wonder how they got all that heavy mining equipment up on the mountain tops, using very large military cargo helicopters. This mill rate (mine tailings production) increased from 1980 (20,000 tonnes/day) to 1995 (115,000 tonnes/day).

All the tailings waste from the mill grinding was dumped into the Akjwa and Minajerwi River headwaters. The waste rock and fine tailings were iron sulfide rich, and thus acid producing, some waste water in tailings had pH = 3, and the copper concentration was high enough that it would have been considered rich ore in other parts of the world. This sediment load was so large that the river bed got high enough for the water to escape the channel and establish new river systems through the rainforest. The ore concentrate slurry was piped down to the coastal port of Amamapare, where a dredged channel allowed ocean-going ships to load the copper/gold concentrate.

The main accommodation and service functions for the mining operations (and 8,000 employees, including 850 Indonesian military police) were developed halfway up the tidal Ajkwa River at the mining town of Timika, where we were camped. From Timika, Freeport had built a giant tailings pond (The Ajkwa Deposition Area, ADA) about 35 km long and 230 km^2 in area, vaguely along the course of the Ajkwa River, where the coarse sand tailings were supposed to accumulate. But of course the mine tailings spread out from the river channel across the rainforest, killing most of the trees, and the fine silt and clay sediment boiled over the tailings dam outlet into the mangrove swamps and the Arafura Sea.

Our analysis of the PTFI laboratory measurements of copper (and a few other heavy metals, such as zinc, arsenic and selenium) in estuarine mangrove molluscs (especially that large mangrove clam favoured by the native people), mud crabs, and shipworms (*Teredo* spp., also a mollusc that eats wood) were badly contaminated by the mine tailings, with copper concentrations 100 to 1000 times greater than allowable food levels (by either Indonesian, Australian or USA standards). This was the case for the Minajerwi, Ajkwa, Mawati, Otakwa, and Tipoeka River estuaries (in declining order of concentrations of Cu), with only the northern Kamora River estuary region being close to probable background Cu concentrations.

We recommended that PTFI institute medical studies on the local people, to determine whether or not this large dietary dose of copper had affected their health. In addition, the local Irianese should be warned not to eat these copper contaminated seafoods. The flesh of the mangrove clam (*Gelonia coaxans,* also called the mangrove cockle) is normally like that of an oyster, a silvery tan colour. These clams from the estuary of the Ajkwa and Minajerwi Rivers were a deep green colour, caused by high concentrations of copper.

As far as we could determine, this mine tailings contamination was present in the molluscs and mud of the major rivers in this area, draining into the Arafura Sea. From limited samples on the inner continental shelf, it was found that a large plume of mine tailings copper contamination occurred for 60 km along this coast, with highest mud concentrations in the estuarine mud below the Ajkwa River tailings dam, and was being pushed seaward at least 20 km offshore and southward from the mangroves. Surprisingly, the biota and coastal mud had near background concentrations of arsenic, cadmium, mercury, lead, selenium and zinc. Some of this inner shelf area was used for harvesting prawns, lobster, molluscs and fish by Indonesian vessels, and this seafood was likely to be contaminated by mine tailings rich in copper.

In addition to human health worries, these metal concentrations were high enough to endanger growth and reproduction of some marine organisms, and probably inhibited microbial processes (decomposition, nitrogen fixation, nutrient mineralisation) important in marine food chains.

In other parts of the world, copper sulfate is used as a molluscicide and microbial poison in harbors, ship paints, and orchards.

Fred Ward and I also reviewed their ship and boat facilities and field sampling procedures. We reported that their small boats for estuarine mangrove sampling were adequate for biological and surface sediment sampling, but not useful for coring operations. A larger V-hull boat with an aft deck A-frame was intended for offshore sampling in shallow water under calm conditions. These boats and sampling gear were kept at Timika and at Amamapare Port site, where massive amounts of atmospheric ore and tailings road dust and liquid tailings sediment was common. This meant that all boats and sampling gear would have to be washed with strong detergent for sampling activities in uncontaminated northern estuaries and offshore Arafura Sea.

We recommended that PTFI construct a clean room for an archive of biological and sediment samples, but acknowledged that there would be great difficulty in keeping mine tailings contamination away from this archive. The Timika Environmental Laboratories were found to be operated in a fully professional way, with excellent facilities and instrumentation, quality control, audit procedures and training programs in biological and chemical sciences. These staff however had little training or ability in the statistical analysis and publication of data from coastal, mangrove and Arafura Sea oceanography.

We further recommended that their biota and mud sampling program be extended farther north and south and seaward along this coast, as the mine tailings contamination appeared to be spreading toward the Lorentz Nature Reserve to the SE of the Otokwa River estuary. This 1.5 million ha protected area features some of the highest and most unusual terrestrial and estuarine biodiversity in the world.

On the evening of 5 April 1999 we were hauled from our Amamapare Portsite cabin to the luxury of the Timika 'Sheraton' hotel, where once more we had fancy white tablecloths and personal waiters to deliver our evening meal. Lab Manager Kent Hortle and several senior Freeport Mine managers came to discuss what we should do next. Amongst other ideas, I recommended that they entrain my Project TROPICS colleagues Anugera Nontji, Gani Ilahude and Meriweather from the LIPI Oceanographic Section in Jakarta.

The next morning we were up before sunrise with our duffel sacks bulging, and an old truck hauled us to the Timika Airport. As the sun was rising, we could see the snow and ice capped Carstenz Mountains to the west, and the glistening Arafura Sea to the east. The tiny airport was crammed with Indonesian miners' families with big padlocked chests and dilly bags bulging with fruit. Indonesian Air Fast Military police with pistols and shoulder-slung rifles with scopes stood ominously behind the counter, opening each suitcase, trunk, duffel and dilly bag. They didn't like my jackknife, which I had to give up yet again.

We were warned that the aircraft would execute three rolls after takeoff, apparently due to the proximity of the mountains and the air currents descending from the mountains. These abrupt rolls allowed us a good view of some of the river channels, the giant tailings pond, and the massive buildup of mine tailings along the river channels and mangrove tidal channels. The rivers loaded with mine tailings were muddy tan colored, but the rivers far to the north were tea colored, due to organic matter from the rainforest soils. Where the mine tailings entered the Arafura Sea, the muddy freshwater floated on top of the more dense seawater, and the prawn trawler ships would pass through the muddy plume, their engine propellers opening the view to clean clear blue water underneath.

By 11:30 we were in Ujung Pandang (Makassar) on the south coast of Sulawesi, where the airport was full of delicious smelling seafood vendors with charcoal burners on bicycle wheels, and mountains of fruit for sale. The whole airport smelled very good, that smoky food smell, and there was bedlam just outside where kids were playing soccer, amongst crying babies, barking dogs, whistles and sirens. The airport lobby was full of vendors selling gold and silver jewelry and colourful tropical shirts with shiny, glistening metallic fibers in the cloth. Departing Ujung Pandang, we could see fish and prawn ponds for miles up and down the naked coast, where the mangrove forests used to be.

At 1:00 we were in Denpasar airport on the Indonesian island of Bali, the airport being big enough for a South Dakota prairie boy to get lost. I was luckily rescued by alert Freeport Indonesia greeting staff with a big blackboard with my name on it, and they hauled me to a Holiday Inn near the black sand Kuta Beach, where I got a special US$75 room rate due to being linked to Freeport.

There in the hotel lobby was my mollusc ecologist colleague Dr. Fred Wells, who had missed his flight to Perth. After showers and a nap, we walked the gauntlet streets lined with crippled beggars and obstreperous street urchins selling dirt cheap TAG Heuer and Longines wrist watches, custom tailors ready to make a suit for you, snake venom parlors, leather goods, and all sorts of smoky street vendors of unnamed food items drowned in hot spicy sauces.

The culture of this part of Indonesia is polyglot, with Hindus dominating, followed by small minorities of Islam, Christianity, Buddhism and Confucianism. Little woven baskets with food or candles burning were positioned on the entrance steps to various shrines and temples. In a big department store called Mata Hari I bought a bag of mangosteens, some clove cigarettes, and two very large Snickers bars. We had our lunch at Kuta Beach watching the surf, with Bintang and mangis, till the torrential rain came and we walked back to our hotel soaked and dripping.

Fred got his 1945 flight to Perth, and I watched films on the hotel HBO TV while my clothes were drying on the balcony. I had a mediocre supper for R95,000 in this fancy hotel, where the band was playing pop tunes, 'Mrs. Robinson' and 'Bridge over Troubled Water' with outstanding American accents. [Note: rupiah 10,000 = Australian $2]. With no great effort, I ate 1 kg of mangosteen before my checkout time to go the airport that evening.

The bewildering Denpasar airport was a small city in itself, with endless shops selling clothing, booze, candy, ivory/silver/gold carvings, jewelry, electronic goods, all too cheap to be real, none of it was useful to me. Clove cigarette smoke hung over the malls. Sunburned Australian teenage girls and roughly dressed miners were hustling to their departure gates.

I boarded a Qantas airplane just before midnight, to wake up at 0545 on 7 April in Sydney, where it was bloody cold. I sought out the big windows facing the sun in the airport lounge, where it was warmer, and I could read Mealey's book about the Grasberg mine history. I had to wait till noon to get a flight to Brisbane and then Townsville at 1600 hrs where Sheila was there bright and shining in red, and the air was warm and perfumed.

For the next few months, the AIMS analytical lab determined a dozen trace elements in various parts of molluscs and mud crabs that we brought back from the Irian Jayan estuaries, confirming that the PTFI Environmental Lab was competent. We also determined radiochemical sediment tracers (^{210}Pb, ^{137}Cs) and the same trace elements on several mangrove sediment cores taken by Joanna Ellison. We found that the top 20-40 cm of mangrove mud carried huge concentrations of copper, and these mine tailings had been deposited since about 1980, the time of major expansion of the mining effort. With the statistical help of Greg Doherty and my Cornell University colleague Roger Green, we wrote up a report for PTFI and got our money.

In so many ways I was naive about dealing with big industry and legal contractual operations. I wanted to publish our Irian Jaya results in scientific journals, but of course all of our data and reports were owned by Freeport Indonesia. It took several years of legal scrutiny of my manuscripts, word changes and negotiations to get their USA lawyers to allow me to submit a paper to a science journal.

I was beginning to get a glimmer of the difficult relations between the Indonesian government, Freeport-McMoRan USA, the various conservation organisations (Conservation

International, Greenpeace, Rainbow Warrior, WAHLI), the 250 native Irianese tribes and their disputes over their land and rivers being polluted and no share of the wealth, the West Papua government in Jayapura, the gun runners and dope peddlers that financed the pro-independence from Indonesia movement for Irian Jaya, the Papuan Separatist Movement, the antipathy of all against the Indonesian Military Police, the Malaysian, Filipino, Korean and Chinese bandits who illegally took a lot of valuable forest trees, and the trawlers and refrigerated fish netter ships from half a dozen countries that were taking away many tonnes of seafood from the local people.

I began to get a glimpse of all this when Jane Perlez (*New York Times* correspondent in Jakarta) came to AIMS to interview me.

I should have known better than to participate in a *New York Times* interview. Jane was a very intelligent lady, spoke a dozen SE Asian languages, and was aware of all the political movements in Indonesia and Irian Jaya. She pestered me about details of the Freeport towns, the treatment of the Irianese and the pollution of rivers and aquatic foods. Of course much of the early chemical research on copper contamination of molluscs and mud crabs was private industrial information from Freeport Indonesia laboratories, and I was technically not allowed to say much on these matters. I did give her a draft copy of my science journal paper on our sediment core chronology and mine tailings heavy metal contaminantion, and I got into trouble with the Freeport lawyers and my own institution about this.

I didn't know that Jane was in cahoots with several NGOs that were trying to bring down the Freeport mining operations for aspects of slavery, murder, poisoning food supplies, deforestation of river floodplains, and inappropriate deals with politicians in Jakarta. Of course this concern was justified.

Jane also explained that investigating Freeport (and other Indonesian government mining and timber harvesting agreements) was bad for your health, and that her email and telephone were being monitored. She said it was good to be a little paranoid about these matters. The Indonesian Navy had seized thirty-two foreign ships in Arafura Sea Papuan waters since the beginning of 2000 for crimes ranging from illegal fishing to the transportation of illegally harvested timber and weapons. Thirty-two ships have been captured and detained during one month by patrols coordinated at the main naval base in Jayapura (the capital of Papua Province). Most of these ships were from Malaysia, the Philippines, China and Thailand.

At the Freeport's request, the Indonesian government stationed a battalion-sized task force to protect the mine, built around a strong core of well-disciplined Army Strategic Reserve (Kostrad) combat troops. Anxious to ensure they were well looked after, Freeport spent \$32 million on building new barracks and other amenities in late 1996, and also set aside an annual budget of $6 to $7 million for ongoing support. Although Freeport has never denied paying the Indonesian military, it only began disclosing the aggregate amounts in 2002-03 after the ambush deaths of two American schoolteachers and their Indonesian colleague on the road to the mine raised questions about the company's relationship with the military. Over the past nine years, 75-85% of that money has been for in-kind goods and services—housing, food, medical care and transportation. The soldiers use Freeport's dining facilities, with the company paying the food services contractor for an extra 850 people per sitting. The company similarly reimburses medical expenses for soldiers treated at two modern, well-equipped hospitals, something they would be unlikely to find in most other parts of the country.

The sore point is that the local Irianese people benefited little, and got nothing for the use of their land, rivers and sea coast. It is no secret that Freeport ignored environmental concerns for the first twenty years of their operations.

Although the company is portrayed in most media reports as a gold miner, sitting as it does on one of the largest gold reserves in the world, it is in fact a New York Stock Exchange-listed company that uses its rich gold grades to offset the production cost of copper. That enables Freeport to produce copper at 10 cents a pound, compared with the 50-60 cents for most other mines around the world. The Grasberg mine produced 16.1 billion pounds of copper and 23.3 million ounces of gold net since it began production in 1988. The mine's reserves are currently (2000) estimated at more than 40 billion pounds of copper and 46 million ounces of gold. Freeport wasn't about to abandon what is now a $12 billion investment. Nor does a government close a mine that has already brought in revenues of $33 billion and promises billions more over the next forty years. In the year 2000, a quarter of a decade after going into Papua, the company's revenues hit an all-time high of $4.1 billion, and it remains Indonesia's largest taxpayer and one of its biggest employers, with more than 18,000 workers.

From Jonathan Wootliff at the *Jakarta Post* :

> 'Their defense has not succeeded in placating a coalition of some of the world's biggest pension fund investors from Europe and the United States, which is campaigning to persuade the company to clean up its controversial pollution record by appointing an environmental expert to its board. From the other side of the globe, such reputable institutional investors as the New York City pension funds have joined forces with such reputable investors as the Dutch ABP pension fund for civil servants, the Swedish government pension funds and the General Board of Pension and Health Benefits of the United Methodist Church in the US to put the proposal on the ballot at Freeport's annual general meeting. The investors were going head-to-head with Freeport's board of directors, which has unanimously recommended that shareholders oppose the motion. But the company's environmental record has been a regular target of institutional

> investor discontent. In 2006, the huge Norwegian Government Pension Fund blacklisted the company over its disposal of waste residue into the local riverine system. Investors contest Grasberg's waste management practices are illegal in many other countries where Freeport operates. Earlier this year, the Norwegian fund also withdrew more than $600 million in investments from Rio Tinto, the London-listed international mining group, because of a joint venture with Freeport at the Grasberg mine. It is worth noting that the Indonesian government is a major shareholder in this enterprise.'

From 1998 to 2002, Freeport conducted a comprehensive Environment Risk Assessment (ERA) to analyse the potential risks associated with the Riverine Tailing Placement on human health, plants and wildlife and aquatic ecology. Our work was part of this ERA. The ERA took four years to complete. It included 96 studies by national and international experts, more than 10,000 samples and more than 200,000 data points. Parametrix Inc. said that it was probably the most comprehensive ERA done in the mining industry in the world. A multimillion-dollar 2002 study by an American consulting company, Parametrix, paid for by Freeport and its joint venture partner, Rio Tinto, and not previously made public, noted that the rivers upstream and the wetlands inundated with waste were now 'unsuitable for aquatic life.' The ERA process and report was reviewed and feedback provided by a team of nineteen independent experts from the government, academia and non-governmental organisations. The documents were submitted to the Indonesian State Ministry for the Environment and the Energy and Mineral Resources Ministry in December 2002.

And then of course all these reports vanished, except for pdf versions on the website of Parametrix Consultants at www.parametrix.com in Kirkland, Washington, USA.

For the next decade, I received Freeport newsletters, stock reports, and financial reports, as if I was an investor. I also got email newsletters and reports from Ed Colijn in the Netherlands, editor of The Indonesian Nature Conservation Database. Bruce Beehler of Conservation International Indonesia visited my office, much alarmed about the damage to exotic and endemic biodiversity of terrestrial and marine ecosystems of Irian Jaya. He was trying to publish a book on the Ecology of Papua, the last book in the Indonesian Ecology Series, which seems to have vanished from the book publishers lists. It appeared that few people get to visit Irian Jaya and the Freeport mining towns, and my brief contact with PTFI attracted a lot of people who wanted to know what was happening there. My education in SE Asian ecopolitics continued on for the next decades.

Chapter 38. Family Affairs, Acid in the Mangroves

Jesse and Sarah arrived at our home in mid-April 1999, having driven from Emerald in a ute crammed with cages and boxes containing a baby cockatiel, a bettong, several white rats, three rabbits and a baby possum being bottle fed. They were all tired and were grateful for the peace and quiet of Alligator Creek.

Jesse was here to participate in his official graduation from the Engineering Department of James Cook University, held in the Townsville Entertainment Centre basketball court which had been carpeted and spruced up with potted palm trees. There were the faculty all dressed up in robes, hats, a sceptre and the JCU Mace, while the 297 B.Sc, M.Sc., B.Eng and Ph.D. graduates in their mortar caps tried to look serious. Jesse had his Woody Woodpecker necktie on underneath his robe.

As usual with these affairs, Jesse got to visit and say goodbye to half a dozen of his close friends, who had helped each other survive calculus, drafting, and essays on concrete with steel reinforcing rods. We all went out to a good restaurant perched over the Coral Sea beach, Sarah's parents Doug and Dinah coming along to embarrass Sarah with stories of her early days with the Tully Girl Scout Group. We presented to Jesse a framed Mechanical Engineering Degree diploma earned by his grandfather Gregg at the State University of Iowa in the 1930s, a framed photograph of his great grandpa England's road building crew in Western South Dakota in the 1920s, as well as his Canadian Chief Scout and Gallantry Awards.

The next day, JCU presented Jesse with the C. H. Barton Medal and $250 for his prize winning thesis on steel rod reinforced concrete. Older son Ian phoned his congratulations from Toronto, where he was in the midst of rehearsing Igor Stravinsky's *Histoire du Soldat,* with Sheila's brother Bill sweating out the trumpet part.

After a few days of visiting friends, Jesse and Sarah packed their clobber into their old Toyota ute for the seven hour drive back to Emerald in the 'outback' of western Queensland. As they drove out of our yard, Charles the Cockatoo yelled 'Bye Bye', and there were tears in my Sheila's eyes.

Once again, there was a lump in my throat about all this, the cycle coming around yet again, a rememberance of my pride and bewilderment at my own university graduation, visiting my South Dakota parents after my first job in Winnipeg, driving out of their yard, with my mother in tears, my brothers standing around pushing the dirt with their boots, and my father walking toward the barn to avoid being seen as sentimental.

Here also was the recognition that the next generation was bravely charging off to find their place in the world. Jesse had earned a decent job with Queensland State Main Roads, which would lead to other good jobs in civil engineering.

Meanwhile, our TROPICS research project was in full swing. Kathy Burns and the AIMS *R/V Lady Basten* returned to Townsville after three weeks of work on the Sepik River estuary in the Bismarck Sea (Leg 8), on the north coast of Papua New Guinea.

I was down south in Rockhampton for a week's workshop on forming a steering committee for a Fitzroy River Basin and Estuary study. They already had graziers, cotton farmers, town businessmen and the mayor at the workshop, and they wanted a dominant role in management of this study, with the Canberra people being given the cold shoulder. Since I was from even farther north in the boonies of remote Queensland, they tolerated me, even if I was a 'Fed'. Most of the workshop was a bloody waste of time, a cat fight between the Canberra bureaucrats and the non-government local people. I had already connected with a dozen local friends, and we had good visits over tea and pikelets with whipped cream and strawberry jam at the smoko tables. Good friends Vicky Vicente-Beckett and Mary Ann Jones hauled me off to a good restaurant where we ate a large amount of roast duck.

I had a couple of weeks of peace to write some science papers in May. For several years I had been a lone wolf howling about the impacts of anthropogenic CO_2 on calcification of our coastal coral reefs, due to carbonic acid titration of the carbonate ion in seawater. The media were calling this impact 'acidification of the ocean', even though seawater was a strongly buffered alkaline solution. The institute bosses really didn't want to hear about this 'acidification problem'.

Because of my loud connections with this problem, the institute CEO sent me north to a largely political workshop in Cairns, a tourist hotspot in north Queensland. The 'acidification problem' there had no relation to anthropogenic CO_2, but rather was due to inappropriate commercial developments in the intertidal mangroves of Trinity Inlet, the main harbour for the city. Town developers wanted to expand the tourist facilities (hotels, restaurants, apartment complexes) along the estuary, so they cut down the mangroves and tried to drain the stinking black muddy sediment with deep excavated channels. This didn't work very well, because the tide kept coming in, the drying mud began to smell even worse, and the saline tidal creek waters turned very acid (pH = 3-4). They had exposed a few centuries worth of iron sulphide formation in the mangrove mud to oxygen in the air, resulting in the generation of hydrogen sulphide and sulphuric acid, which tends to kill all plant and animal species in this disturbed zone. The acid also liberates many natural and contaminant trace elements that had been sequestered in this black organic mud. Previous research estimated annual sulphuric acid production as 1270 tonnes/ha.yr. Now there were dead fish floating up and down the estuary with the tide, not good for the tourism industry! Since these strongly acidic waters would dissolve concrete, the development came to a halt and only then did they consult with marine scientists for an explanation.

In mid May 1999, the Mayor of Cairns, the Parliamentary Secretary for Science from Canberra, numerous state government MPs, several important representatives of local industry (tourism, sugar cane, mining companies, the port), and the commercial fishery lobby called out for scientists from the acronym salad of state and national government departments. CSIRO, DNR, QDOE, Geological Survey, DPI, GBRMPA, ANU were all there, and I was representing AIMS. They put us up in front of the grand hall of the Community Centre on a raised dais panel table with microphones and grilled us for three days.

Several of the CSIRO soil scientists (Rob Fitzpatrick, Greg Bowman) had been studying acid sulphate soils in New South Wales for decades, and they knew well what had happened in the Cairns estuary. A century of research in the Netherlands and Germany on saline mudflats resulted in alienation of the land for both agriculture and marine ecosystem productivity, and some disturbed sites never did recover or become useful for development purposes.

The local old-timers already knew most of this: two 70+ year old cane farmers described trying to drain 110 ha of inland mangroves for sugar cane plantations in the 1960s, and of course nothing would grow in this strong acid soil. The old timers also told of World War II (actually 1930-45) dredging of the Cairns and Port Douglas port access channels by the US Navy, with dredge spoil disposal in nearby mudflats and mangroves. They described fish diseases and fish kills being common in these early days. Of course, the town fathers immediately demanded remediation plans, or ways of elevating the alienated land above the reach of the seawater tide.

Most of the scientists recommended abandoning the project, planting mangroves, allowing the seawater tides to flow, with hopes that the ecosystem would heal itself within perhaps a century. This was not what the developers wanted to hear, and they started spouting the price of lime (AUD$90/tonne), building seawater barriers to isolate the disturbed site, and the cost of dumping sand and other waste soil/rock/gravel/concrete on top of the mangrove soils. None of this was likely to be effective, but of course several environmental rehabilitation engineering companies leaped on this opportunity for decades of work. Several of us tried to leave early, but Warren Entsch kept inviting us to good restaurants in the evening.

This was a part of my job that I enjoyed the least, but at least we were providing our expertise to the taxpayers. To my knowledge, nothing worthwhile was done and that patch of hugely disturbed mangrove intertidal zone was abandoned to a natural recovery process. It wasn't until 2008 that a Ph.D. thesis was produced on the East Trinity Inlet disaster, essentially recommending the 'rewilding' of the mangrove forest and tidal mud flats.

Because of my good relations with inland cattlemen, AIMS assigned me the job of being a marine representative on a Technical Advisory Panel (TAP) for planning and monitoring modifications to rivers and groundwater extraction in northern Queensland. The job was really to provide some caution to politicians and agricultural engineers who wanted to build dams everywhere and irrigate the dry dusty interior *Spinifex* savannah land.

In this region, west of where we lived, annual evaporation was 10-20 times annual precipitation, and the rivers were usually full of hot dry sand and gravel. But in this land of extreme events, when it did rain, the rivers flooded the land for miles, and many small inland towns moved their household goods upstairs. With input from the engineers, 50-100 years of water flow records, fisheries and wildlife biologists, geologists and hydrologists, soil and agricultural experts on dry land and irrigated crops, and consultants anxious for twenty years of water manipulation jobs, we tried to create a semi-quantitative model that would allow best guesses for eliminating water use disaster proposals, and provide suggestions for state approval of possible successful and sustainable plans for intelligent water use in agriculture.

All of this was cautioned by the observation that many dryland irrigation projects around the world usually failed within a decade, as the soil would become saline due to huge evaporation rates and accumulation of salt in the surface soil would kill most plants. For me, this also included the need to provide river water flow to the estuaries, natural flooding of the floodplains and estuarine mangroves, and erosion of river watershed soils into river-borne inputs to the estuary and Great Barrier Reef lagoon. As you might guess, engineers and agricultural scientists spoke different languages from the fish, wildlife, hydrological, and marine scientists. There was some difficulty in comparing the methods, units, and uncertainty of fairly precise (but highly irregular) river flow measurements, and the high uncertainty of most biological data. We had been struggling for several years with these problems at workshops in Charters Towers, Cairns, Ayr, Ingham, and the Atherton

Tablelands, adjacent to some of the main rivers (Burdekin, Herbert, Barron, the Gilbert and Mary Rivers).

In August 1999 our TAP group met again to rank various proposals on a scale of 1 to 5, rank 5 being catastrophic negative impact on communities and agricultural production. The Burdekin River was already partly dammed, and farmers downstream had options to buy water for irrigation of thirsty rice, sugar cane, fruit and vegetable crops during our dry season. Proposals for deep drilling for groundwater supply in the Burdekin floodplain were ranked 5, as many of the aquifers underneath the river bed contained saline water. Some proposals in the interior of the Herbert River and the Tablelands would rank 4 or 5, as the groundwater would likely be salty in a decade or so, crops wouldn't grow well, community water supplies would be ruined, and the soils were not suited for irrigation. Damming some northern rivers in Cape York with drainage to the Gulf of Carpentaria would seriously damage prawn and marine fisheries in the estuaries, and I would vote against these proposals, but this resulted in a contest between the value of irrigated agricultural produce and the value of seafood from the Gulf estuaries. The best option for both cattle and crops was the creation of off channel storage 'dugouts' which would be filled during floods and high water river flow by pumping. The best ones were roofed, to reduce evaporation losses.

Some of the proposals were to build dams for hydroelectricity production, but most of the dam sites were flatland and would suffer enormous water loss due to evaporation. There was a sincere desire from the politicians to make the northern part of Australia into a gigantic food production region for commerce with the huge populations of SE Asia. Few politicians had much understanding of the violent and severe oscillations in north coast hydrology, cyclones, flooding, droughts, tropical diseases, wildlife predators, and the importance of river inputs to estuarine/coastal dynamics.

The above paragraphs might be a little boring for you, reader, but they were more boring for me. I had to learn how to read engineering blueprints, fifty years of runoff measurements, agricultural productivity experiments in soil salinity, and the sensitivity of native dryland and introduced farm crops to variations in chloride and salinity of irrigation water. But I was paying my dues, contributing my expertise to the people on the land, the taxpayer, and the dreams of developers in the northern frontier.

In late August, Sheila and I packed up our old Subaru 4WD station wagon for a trip to visit Jesse and Sarah in their home in Emerald, west and south of our coastal home. We stopped in Charters Towers to visit musical instrument builder Taffy Evans, who agreed to fix my old black fiddle. He showed us his first bouzouki (a large Greek eight stringed mandolin sort of instrument), made from Australian woods, with a wonderful booming tone.

There had been some good rain, so the Gregory Highway roadside stock dams were being visited by deer, emu and camels. We arrived at dusk in Emerald, where the large Fairbairn Dam (1.4×10^9 m^3 storage capacity) provided irrigation water for citrus, cotton and melon crops. Sarah had a large garden, with rhubarb, cauliflower, beets, parsley and tomatoes. We brought paw paw seeds and banana plant sprouts for their yard. After supper, we were asked to play tunes for them: 'Do Re Me', 'Homegrown Tomatoes', 'Home on the Range', etc.

We drove down to Carnarvon Gorge, where there was an excellent display of aboriginal rock art on 600-1000 m high rock cliffs: hands, boomerangs, killing sticks, emu and wallaby tracks, vulvas, cross mesh, fish and possum/wallaby drawings. This is normally very dry country, but the Carnarvon Creek canyon must have had springs upstream, and there were 10' high *Zamia* trees, huge gum trees, tree ferns, and palms growing right in the bubbling clear creek water.

We visited a sapphire mining site, mainly being operated by individual lunatics sifting dirt behind their trailer houses and tin sheds. Some had their trucks hooked up to power take-off operated excavators, sifters and shakers, washing chutes, all in an effort to find the $100,000 stone.

Before we left, we had a giant roast beef feast and Jesse made cinnamon buns from his Grandma Virginia's South Dakota recipe. On our drive down the long highway home, there was the normal abundance of roadkill wallabies, emus, bandicoots and possums being eaten by eagles, kites and ravens.

We celebrated our twenty-ninth wedding anniversary on 8 September 1999, and my fifty-eighth birthday on 10 September. We were in serious Meadowlark rehearsals with Jennifer Hultgren, a bluegrass musician/school teacher, and we gave several good concerts in local galleries and concert halls.

I was also busy organising a local workshop about wetland protection, partly because I had become friends with local cattleman old Pat Meehan, who held the lease on the Cromarty Wetlands, the coastal sand and salt flats near AIMS. Colleagues at JCU were bringing in a new course about wetland productivity and evolution. We heard lectures on the history of development and loss of wetland habitats on the Burdekin Delta, mainly due to sugar cane and vegetable farming, and grazing from cattle. Coastal wetlands up and down the Queensland coast had been altered greatly over the past century. A lot of the damage was done intentionally by CSIRO agricultural scientists, who introduced invasive pasture grasses and weeds (Paragrass, *Brachiaria muticum*; *Hymenachne amplexicaullis*; water hyacinth, *Eichornia crassipes*; water lettuce, *Pistia* spp.; lily pads, *Nymphaea gigantea* and *Salvinia molesta).* These were all freshwater plants, which died when the wetlands were inundated with seawater during tidal surges and cyclonic storms, and their rotting biomass caused deoxygenation in the swampy basins.

Local politician Mark Stoneman was trying to declare the Cromarty Wetlands a private reserve, not under the control of Parks and Wildlife (because they didn't do a good job, in his opinion), and he wanted AIMS to be a partner in this wetland reserve. Stoneman brought together a group of investors who wanted to buy the Cromarty Wetlands, build a tourist hotel, conservation centre, a little tourist railway and boardwalk through these beautiful wetlands. This was in contrast to the conservationist view that the wetlands should be protected, fenced off from cattle grazing and human modifications. Stoneman and Meehan, cattlemen of course, thought that cattle grazing kept the introduced/feral/invasive water weeds, grasses and vines in check, and allowed better viewing of the wildlife. This was a good place to see black swans, jabiru, pelicans, several kinds of herons, egrets, ibis, ducks, geese, sea eagles, and the majestic brolgas dancing. Quietly behind these arguments was the aboriginal claim on much of this coastal wetland.

Using rented troop carriers, we hauled a lot of dignitaries and pompous politicians out to the Cocoa Creek Saltflat research site, and to nearby Pat Meehan's 1600 ha cattle paddock Cromarty Wetlands. My research group had drilled into the saltpan mud and sampled existing bore water all over this area, and I gave several presentations about how thin lenses of fresh groundwater sat on top of salty water of salinity three to five times greater than seawater. My main point was that these intertidal areas were unsuitable for long term agriculture or cattle that need freshwater, but they were appropriate for carefully designed prawn and fish farming.

GJB at Cromarty Wetlands with Bill and Majella Meehan, 1999

My friend and buddy of our son Ian, Jim Tait, had grown up on one of the lagoons in the Burdekin River Delta. He showed photos of beautiful clear water channels and lagoons with towering tea trees and river gums shading their old house in the 1950s. Now, decades later, this Burdekin Delta area had been laser levelled and planted to sugar cane, and the deep drainage ditches were leaking muddy saline water. Over a decade later, Stoneman's group did purchase the Cromarty Wetland property, and were still trying to find funding for their conservation and bird-watching tourist resort, to be called Wongaloo.

During this same time, AIMS was having one of their internal evisceration conferences, trying to look for new directions and new sources of money. It was very easy to see the sad and tragic difference between CEO administration views and scientific leadership. In my spare time, I was writing up the program for a TROPICS workshop in Jakarta in December and a TROPICS session speaker list for an AGU meeting next year in USA.

Chapter 39. Prodigal Son Returns, and More Indonesian Adventures

Our famous folkie band Meadowlark played a long concert at the Pinnacles Gallery in October 1999. And then there was a feast at Satay Mas afterward, the musicians being high on adrenalin. We had good rains in November, and our pomegranates, lychee, mulberry, passionfruit, wampi and mango trees all produced wonderful fruit, and of course the fruit bats and parrots feasted too. The air was damp and perfumed, you could almost eat the air!

Our older son Ian had been at the University of Toronto studying percussion with his Nexus mentors and living the student life with drum maniacs above a Portuguese butcher shop. He decided he'd had enough of university, packed all his gear into hockey bags, and flew to Sydney, Australia arriving 9 November. He tootled around Sydney, visited the zoo, botanical gardens and the Opera House, and slept on the grass with spiders and ants. He took the bus from Sydney to Townsville (that is a long long ride), and arrived with heavy bags full of strange percussive stuff (saw blades, gongs). This was before the high security luggage examinations. He moved into our 'granny flat', made peace with our parrots Sir Charles and Caspar, and was assimilated into the local music community here. It didn't take long for bands to be formed and paid performances were happening.

Our scientific and oceanographic consortium TROPICS had four major partners, Australia, Indonesia, Papua New Guinea and the USA. This was a 'grass roots' formed consortium, not administered by CEOs and bosses of the dozen institutions involved, but initiated and operated by participating scientists. This didn't sit well with the AIMS administration, as they had no control whatsoever of our research program. But to their credit, they provided staff, ships, and equipment for my contribution to the project.

PNG did not have a coastal oceanography capacity, but Hugh Davies of the University of PNG and Gaikovina Kula of the Department of Environment provided sponsorship and workshop space for our work. The Indonesians had a powerhouse of research vessels, mainly four large vessels of the Indonesian Navy, and half a dozen marine fisheries and research organisations. Anugerah Nontji, my good friend Gani Ilahude, and Muswerry Muchtar were our LIPI contacts, and our project was actually entered into the 'Blue Book' of approved Indonesian government activities. According to our TROPICS agreements, we were obligated to have workshops in each country and so IndoTROPICS embarked on a workshop in Jakarta.

I left home on 29 November 1999, loaded with a US $2000 check to support the workshop, several PowerPoint presentations, and orders from my son Ian to purchase a special kind of Javanese gamelan drum. Flying to Brisbane (raining, musical group in airport singing 'Brisbane Ladies Dance All Night', the ladies having tattooed midriffs and elevator shoes), Sydney by 7 pm. I had a meeting in a restaurant with TROPICS partners Ron Szymczak and Denis Mackay, trying to get our act together for presentations in Jakarta.

The next morning, we all bought Rupiahs at the airport, bought cigarettes, laptop charger cord adaptors and laser pointer batteries, and boarded an early flight to Jakarta, arriving late afternoon. Surprisingly, we were not greeted by our LIPI hosts, so we got stung R150,000 for a cab to our Sari Pan Pacific Hotel, but we found that our rooms had been prepaid by our BPPT sponsor. It was stinking city hot, so the cold towels, carbonated soursop juice, chilled fruit bowl, CNN on the TV and air conditioning were gratefully enjoyed.

The news from the USA was dreadful, street riots in Seattle related to the World Trade Organization meetings, tear gas, police with battle uniforms and clubs and the National Guard was called out to suppress the demonstration. It was easy to see how people in SE Asia have a fear of the USA.

I had previously emailed Gani Ilahude for assistance in finding a kendang sabet for son Ian. This is a double headed horizontal drum, one of the foundation parts of the gamelan orchestra. Gani's son was a manager for a chain of restaurants in the city-kingdom of Solo, to the east of Jakarta on the island of Java, which is a center of culture and literature, and where the best drum and gong makers were found. Gani organised the deal and I deposited R400,000 in his bank for the drum and stand, to be delivered to Gani in Jakarta. Gani would repackage it to ship to me. We knew there would be trouble at Australian customs, as the drum head was animal skin, which would need to be fumigated.

Packing up kendang for Ian, Jakarta 1999

The next day was free time. We had a long breakfast in our hotel of Indonesian and Japanese style food, tofu, rice cakes soaked in coconut milk, melon, passionfruit, cinnamon spice bread and strong coffee. We walked across the street to the giant high rise HQ of BPPT (Agency for the Assessment and Application of Technology). BPPT is a non-government agency, a department under the coordination of the Ministry of Research and Technology which has the task of carrying out government duties in the field of study and application of technology. This agency provides support for most Indonesian science.

On the 19th floor we found billboards advertising our ICOSTI/IndoTROPICS Workshop, with all our names listed, and here were IndoTROPICS offices, small and gigantic meeting rooms. Gani joined us for lunch (prawn lumpia) and proceeded to lecture us on a thousand years of Javanese history. Gani is not even Javanese; he grew up near Manado on the north coast of Sulawesi. Gani had just returned from working in the Arafura Sea on one of Kent Hortle's Freeport PTFI research ships. Ron, Denis and I tried to convince Gani to organise a joint IndoAustralian big ship cruise in the Arafura Sea, but it never seemed to happen, partly because the Indonesian government could not guarantee our safety in that unstable and sometimes war-like region. Gani introduced us to his protégé, Jorina M. Waworuntu, who had just earned her Ph.D. from the University of Florida. She did her thesis work on the physical oceanography of the Indonesian Throughflow, the massive current that flows from the Western Pacific through the archipelago of Indonesia, into the Timor Sea and Indian Ocean. Gani pushed us to find a job for her in Australia.

I walked the street shops nearby, bought five gaudy shirts for AUD46, and observed an advertisement for Metallica in concert with a Symphony Orchestra. It was very hot, stinking diesel fumes and garbage smell, busy and loud people, and I was hunting for mangosteens with no joy. Gani said that he always had a cough while in Jakarta, but when he was on a ship in the Arafura Sea, he had no cough. I had a good supper of mountains of seafood and two kegs of Bintang with Ron and Denis, cigars and brandy afterward. For extremely odd reasons, this hotel had some Spanish singers backed by guitars and electronic keyboard. Very odd music.

And so, with typical SE Asian pomp and ceremony, our little workshop opened on December 2 with the long singing of a huge gong. There were welcoming speeches in a variety of languages from half a dozen ministers of the Indonesian and Irian Jayan governments, The Australian Department of Foreign Affairs, the Aussie AUSTRADE booth, Dr. Arnold Gordon of the Lamont Doherty Geological Observatory, Gesang from Freeport Indonesia in Timika, a Norwegian research ship building company director, and a German CEO of a palm oil plantation. Ron, Denis and I gave short speeches to show our gratitude for being on their show. There were TV cameras, radio and newspaper interviews, and formal conference photographs. The nineteenth floor conference hall was not air conditioned, and it was very hot. Ceiling fans droned, and one whole side of the hall was open to a balcony and views of the city. Thank heaven we could get coffee, pastries and beer on the balcony. This welcoming took most of the day. Many of the speeches and formal presentations were in a variety of languages, not all Bahasa Indonesian, and we had access to little ear buds for English translations for some of the talks.

Ron and I found a giant department store where I could buy some mangosteen. There were also street shop displays of silk and leather goods, batik, hand carved silver sailboats (praha), gold ornaments, endless racks of garish tropical shirts with metallic threading, elegant Nehru shirts, shadow puppets, musical instruments (rain makers, anklung, hanging chimes, bamboo xylophones, bone flutes), and shops full of *parangs*, a kind of Asian machete. I was happy to take my sack of mangosteens back to my air conditioned hotel room, where I could snooze and watch CNN and 1950s Hollywood western films. The TV said a NASA lander was on the north pole of Mars, but it was not communicating.

After the workshop presentations, Ron had some BPPT contractual business for ANSTO. Denis and I took an open air cab to the old Batavia harbor where we could view hundreds of prahas tied up to old rotten wharfs, unloading giant logs of tropical hardwoods from Sumatra, large crates of coconuts for the copra trade, bales of hot peppers, crates of ducks and chickens, burlap sacks of rice and fruit and tubs of seafood on ice. Nearby we visited the old Dutch prison, where many disruptive Indonesians were held in leg irons and neck chains pre World War II. We had lunch at The Hollywood Café, overlooking tidal canals full of sewage, rubbish and European tourists in small sculling boats that dodged the big floating rubbish piles. This was the practical part of this city of eleven million, where side street shops had metal and wood working tools, motor bikes were being repaired, the smell of diesel, hot grease, oil, smoke and sawdust filled the air. I was grateful to return to my clean hotel room for a shower and CNN international news.

On 4 December, all foreigners were gathered into a fleet of minivans for a tour of the volcanic regions near Bandung, up in the mountains. We drove past rice fields with duck enclosures, a smelly soybean fermentation factory, Hindu temples and Muslim minaret towers, plantations of guavas, pawpaw, tea, durian, mango, mangosteen, rambutan, cocoa, coffee, tomato, chilies, rubber, bamboo, sweet potato, a giant sign saying 'Welcome to Kansas' (a US cigarette advertisement), and abandoned car factories.

The city of Bandung was crowded with bikes and carts full of durian and the smell would frost your eyeballs, but at least the high altitude meant that it was cooler. The city proudly featured McDonalds and KFC, and even a donut shop. I bought some chempiak, a mountain relative of jakfruit, smaller but somehow more pungent compared to my north Queensland orchard tree fruit. In the pouring rain, we visited Tangkuban Perahu Crater, a volcano which was still belching sulfurous gases, steam and the occasional shower of hot cinders. The hawkers were selling necklaces of carved volcanic tufa, some of which were rather pornographic. The nearby hot springs were crowded with mamas with infants, old men in sarongs, naked lovers, and the usual legion of Japanese with video cameras. It was the Sunday before Ramadan, which is a popular holiday time. Bandung also featured several craft shops, where you could buy penis whistles, slit log drums, several kinds of stringed instruments, anklung, bamboo marimba, rain tubes, exotic rattles and wooden cowbells. On the way back to our hotel in Jakarta I bought a sack of rambutans, mangosteens, and a small insulated sealed container of frozen durian meat.

For the next few days, back in Jakarta, the science workshop shifted from the eighteenth floor of BPPT to the LIPI building. We would start with a splendid English breakfast of soft boiled eggs, beef bacon, crumpets, croissants and strong clove-scented coffee served in a fancy furnished room with big stuffed chairs and ornate carved wooden tables. This was obviously reserved for visiting Yanks, Europeans and Australians. Our motley bunch of Australians felt a little guilty in this lavish situation.

This phase of the workshop was more oriented toward IndoTROPICS results from recent research voyages around coastal Irian Jaya, and what to do in the next five years. This was led by our friends and colleagues Anugerah Nontji, Gani Ilahude, and Muswerry Muchtar, and there was important logistic input from their ship builders from Norway, Danish equipment suppliers and the Indonesian Navy. The Indonesian group had mapped the bathymetry of the north coast of Irian Jaya, especially the coastal canyon eroded by the large discharge of the Mamberamo River. There were nineteen presentations on the physical, chemical and biological oceanography of the coastal plume of the Mamberamo, and the Bismarck Sea to the north. Most of this work represented the first exploration of the oceanography of this remote coast. From their work, we could see that the Mamberamo inputs influenced the chemistry and biology of water far to the west and north in the Bismarck Sea, and together with the Sepik River, was a dominant supplier of particles, nutrients, metals to the important Equatorial Undercurrent, which flows north and then eastward underneath the equatorial Pacific Ocean.

I gave a presentation on our decade of research on the carbon budget for the Gulf of Papua, Denis Mackey gave a presentation on the giant Sepik River inputs to the Bismarck Sea on the north coast of PNG, and Ron Szymczak talked about trace elements in river particles and dissolved in the water of both locations.

Some of the best collegial experiences happened in discussions during the evening meals in the back rooms of good restaurants. The Indonesians were fascinated by my origins in cowboy country in the USA. They knew all about New York and Los Angeles, but South Dakota was a great mystery to them. Over frog legs, roast duck, prawns, spicy noodles and rice, palm wine and Bintang (beer), Gani quizzed me about Mount Rushmore in South Dakota. He couldn't quite understand how they carved the faces on mountains of granite, and why anybody would want to do this. I later sent him a bundle of information about Gutzon Borglum's team of dynamite blasters.

On another evening feast in a Japanese restaurant with steamed lobster and crab omelettes, Islamic Gani wanted to know how come there were so many different varieties of Christians, because some of his Indonesian colleagues were Catholics, Lutherans and Evangelical Baptists. Each of these three faiths appeared to deny the legitimacy of each other. I surprised him by saying that this was also true of Sunni and Shiite Muslims. He was upset that I was not religious at all.

The Europeans and Australians from our hotel hired vans to haul us to the airport on 7 December. We were greeted at the metal detector security gates by Santa's helpers, tiny barefoot ladies in red velvet Santa suits with miniskirts, who guided us to the duty free shop. We left Jakarta at ten pm, flying over the twinkling light of this huge city, and I was furiously trying to eat the last of my mangosteens before our 8 December 0430 arrival in Sydney, Australia.

Quarantine Service people were very interested in the wood and skins of my kendang sabet and several rattles and wooden bells. I had to pay for gamma radiation and fumigation with some methylbromide poisons, and they promised to send it all to me within twenty-one days. We were also closely checked out by three different kinds of sniffer dogs, probably hunting for drugs. And then it was flying northward toward the tropics again, Brisbane for another breakfast, and Townsville by noon. What a wonderful thing to have clean, fragrant air...and there is my bright and shining Sheila to haul me home to Alligator Creek.

And here we are at another Christmas season! Our home grown mangoes were ripe. Ian MacGreggor and I made a desperate trip to town to search for Xmas presents for Sheila. We were invaded by Caroline Gaus and Jurgen Mueller from the University of Queensland in Brisbane, who sat in our pergola happily being entertained by our parrots Caspar and Charles. They managed to also discuss our next actions in a research program to figure out why there were dioxins in coastal Queensland soil, wetlands and marine sediments. Peter Ridd and Thomas Stieglitz from James Cook University appeared the next evening, demanding parrots, beer and discussions about our salt flat and wetland research plan. Throughout these weeks there were music parties in town, with several Christmas feasts that couldn't be beat.

Jesse and Sarah arrived 0200 on 25 December, adding to the pile of presents under the potted palm tree. Ian got his kendang, Jesse got a chainsaw, Sheila got a tree kangaroo book by Tim Flannery, and I got a bouzouki. The rain poured down night and day. Roads north and south of us were flooded. We were at a music party on New Years Eve, and we played fiddle tunes and Celtic ballads till three in the morning.

Many businesses and AIMS were worried about Y2K, what would happen to computer time codes when we entered a new century. Nothing bad happened on 1 January 2000.

Chapter 40. A New Century: More TROPICS Cruises, Rabaul Volcano, The Alamo

We all survived into the next century! After whippersnipping pathways through my growing orchard, I installed a new irrigation line for a row of avocadoes and a tropical peach tree. The peach tree died quickly and was replaced with another avocado tree. I got my new bouzouki tuned to suit my kind of music, and it was easy and fun to play. Sarah and Jesse packed up their clobber to return to Emerald and the Main Roads job. Ian MacGreggor went with them, to work in the cotton fields. Sarah cleverly left behind two baby magpies for Sheila to feed. For reasons that escape the logic of the cosmos, they were named 'Big Rat' and 'Little Rat.'

The political wind changed in Papua New Guinea, and a new PNG Marine Science Committee decided they needed to review the progress of our TROPICS research cruises, with some possibility of not approving our ship clearances and personnel visas. It appeared they were worried that we were stealing their biodiverse marine biota to harvest exotic compounds for commercial profit. The Washington DC Consulate of PNG was upset, holding back the visas for all USA participants in TROPICS Leg 9 and 10. I flew up to Port Moresby for a full day meeting to explain that none of us expected to make any profit from collecting plankton, benthos and mud. My life was becoming more complicated with bizarre political edges.

From previous TROPICS cruises we had accumulated at AIMS oceanographic equipment from several USA, New Zealand and Australian collaborators. Bruce Parker and I loaded several tonnes of awkward/fragile equipment onto a semi-trailer truck for shipment to Cairns, where the R/V *Franklin* was supposed to dock. We also drove a ute with a rubber zodiac boat on a trailer up to Cairns on 11 January, arriving after dark, happy to get a good supper of Greek style scallops and prawns.

The next morning we were at the wharf at 0700 sharp, only to find the ship piled high with rubbish and a rusty deck dirty with hydraulic oil and grease guns. The hold was occupied with strange rubbish. This was the result of a Pacific and Orient management dispute with the union and a crew dispute with the Master. The bosun had resigned and was somewhere in a Cairns pub. The CSIRO cruise manager Don McKenzie was badly sunburned and upset with the situation, and he told us to unload our gear on the wharf within reach of the ship's cranes, and then go away for a while.

At a nearby pub, hiding from the blistering heat and humidity, we found the other TROPICS scientists and technicians for the next cruise, Denis Mackey (Chief Scientist, CSIRO Hobart, studying algal pigments in the Sepik River plume and Bismarck Sea), Kathy Burns (AIMS hydrocarbon expert), Ron Szymczak (trace elements and radionuclides), Chuck Nittrouer (mud), J. P. Walsh (mud), Bob Aller (mud chemistry) and some others. The AIMS media office had alerted the press to our operations, and Kathy Burns gave a TV interview about her work on tracking the different kinds of hydrocarbons in the water and sediments. Meanwhile, our semi-trailer broke down near Ingham and would arrive later in the afternoon. We had torrential rain and the mosquitoes and sand flies were abundant. My life was becoming complicated again.

Just at dark, Ship Master Neil Cheshire came into the pub, demanded a beer, and invited us back to the wharf to load gear into the clean hold and clean deck in the pouring rain. Apparently the National Facility in Hobart had sent a different Master and crew to Cairns, and the ship was neat and tidy now. This dark wet loading included Ron's twenty tonne railway container clean lab, all sorts of superclean sampling gear for hydrocarbons, my Kasten corer and slicing table and equipment for our USA collaborators. We finished up around midnight, and most of the scientists went to their bunks on the ship for Leg 9 for the rest of January on the north coast of PNG.

Bruce and I were on the next leg of this ship cruise in February, so we drove home through the pouring rain, stopping in Ingham for egg and bacon McMuffins and two kilos of lychees. We made it home by 0300 the next day.

AIMS was in administrative crisis mode, and I tried to ignore the political bitterness and indecisive leadership. The finance office was upset about my accounting of my $2000 cash contribution to the TROPICS workshop in Jakarta, because they could not read the receipt (not in English). Irena was working on my presentation graphics for an upcoming AGU TROPICS session in San Antonio, Texas. I spent two days on the phone with the Brisbane Consulates of PNG, USA and Indonesia, trying to find assurance that medivac facilities were available to international personnel on our ships. Ph.D. students Thomas Stieglitz (groundwater fluxes to the GBR), Joanna Cavanagh (organochlorine pesticides in north Queensland estuaries) and Greg Doherty (sedimentary geochemistry of Cleveland Bay) were near ready with their theses. Does this sound a little hectic? You betcha.

The next two months were going to be hectic, so I took a week off work to stay home and catch up with my chores. With so much wet season rain, my orchard needed whipper snipping and I picked pomegranates, passionfruit, three varieties of mango, soursop and some big bunches of bananas. We have become fructivores! Sarah's magpies were now part of the household, following us around and pecking at our toes to beg for treats. They would perch on the shower head as I was taking a shower and follow me to the garage to move my tools around on the shop bench. But of course Sheila was their main target, as she was the source of all good things in their life. We were called to the North Queensland Customs Service in town, to collect crates of freight: this was our son Ian's collection of books and musical gear which had been shipped from Toronto, Ontario.

AIMS paid my way to San Antonio for a big AGU meeting, where we had a special session to report on our TROPICS Project. I paid a small additional fee so I could visit my Brother Gerald in San Manuel, Arizona. Life in airplanes and airports: left Townsville 0600 in the sleepy dawn of 18 January, hot ham and cheese croissant in Sydney, leaving Sydney at four pm, sunset and dawn comes up over the Pacific, arriving Los Angeles ten am on 18 January 2000. After a while, you get used to the International Date Line, arriving in Los Angeles before we left Sydney!

I was carrying my US passport, and proceeded outdoors to diesel fumes, smog, greased dreadlock hair, hip-hop clothing and rose coloured peeper sunglasses. Everybody had cell phones glued to their ears and there was a strong smell of sour boiled coffee, French's mustard, cheeseburgers and fries. I wandered this huge airport to find the domestic terminal and a one hour flight over dry mountains and desert to Phoenix, where Gerald and daughter Rebecca were there to meet me.

It took two hours to drive to San Manuel, a mining town associated with Magma (now BHP) Copper Mines. There emerged a whole lot of Gerald's family, with no intention of giving me any rest. Following a big family feast of pork chops and mashed potatoes they wanted me to play Gerald's banjo (rescued dusty from underneath his bed) but the high G string was broken. The next day I woke up to the house full of rampaging kids being day-cared for by Gerald's wife Doreen. Gerald had to drive me into Tucson to find banjo strings, and of course for Sheila's shopping list...horseradish, NeoCitran, a strap for my bouzouki, red neckties, and adapter plugs for US to Aus electrical cords (for my laptop charging)...stuff that was hard to find in Australia.

The next day Gerald took me to nearby Biosphere 2, a giant engineering and scientific experiment in how to operate an enclosed sustainable ecosystem with humans in it. Biosphere 1 was of course our planet Earth, which we were busy screwing up with contaminants and industrial greenhouse gases. It had the feel of what would be needed to build a human colony on Mars or one of the moons of Jupiter. Several kinds of ecosystems were enclosed in air and water tight multiple domes photosynthetically powered by sunlight, with the intent to measure how plants, animals and humans could live well and recycle wastes. My colleague Chris Langdon and his student Francesca Marubini gave me a behind the scenes tour of this giant bubble ecosystem, the labs, the water pumps, the recycling plant, a mangrove forest, a coral reef, grasslands, a rainforest, a foggy desert and the human quarters. Chris was a Lamont-Doherty scientist particularly interested in the response of coral reefs to increases in anthropogenic carbon dioxide in seawater. He and his technical staff could manipulate the rate of growth and calcification of corals by changing the concentrations of CO_2 in the atmosphere; at atmospheric CO_2 not much higher than at present, they could shut down coral growth completely. Associated with these giant experimental bubbles was a hotel, restaurant, library, book store and an astrophysical observatory. At that time, the cost of operation of the multiple isolated ecosystems was about US$ ten million/yr. Several of my colleagues from Cornell University and Lamont-Doherty Geological Observatory in New York were there, and we all had lunch together, with talk of coastal oceanography, ELA, David Schindler, Wally Broecker, and our upcoming TROPICS special session in San Antonio. Because of the presence of lots of international scientists there, I was able to use my AIMS credit card to get US$1000 cash to support my Indonesian colleagues Gani Ilahude and Sutomo at the TROPICS session.

What I didn't appreciate until later was that many of the greenhouse gases in the different compartments of Biosphere 2 fluctuated wildly in the early years of this experiment, pCO_2 sometimes reaching 4000 ppmv, with wild excursions of methane and nitrous oxide. This caused high temperatures and created the need for some kind of air conditioning (engineered lungs) to allow the humans to survive. Oxygen concentrations in air declined, partly due to oxygen consumption by degradation of organic matter in agricultural soils, and partly due to the formation of $CaCO_3$ films on the building foundation concrete. This caused discomfort to the humans, and more oxygen had to be supplied from outside of the biome. In some ways, the different ecosystems were roughly in balance, with plant growth being matched by decomposition rates of organic matter. The production and consumption of atmospheric gases was out of whack, and deleterious to parts of the ecosystem. The experimental humans experienced good health and weight loss, probably due to the food being mostly plants, but they were unhappy with high CO_2 and low O_2 in air.

Gerald drove me down to a Tucson National Park to visit the Desert Museum and to observe the wildlife in large enclosures of land protected from vandals, hunters and other predators. We saw lots of Mexican parrots, puma, jaguaramundi, ocelots, hummingbird nests, dozens of species of rattlesnakes and lizards, bears, wolves, javelin, long-tailed blue magpies, vultures, eagles, otters, beaver, bighorn sheep, and a rock/mineral/fossil display. There was also a wonderful botanical garden, native crafts for sale and a book about Bill Pickett (I had a song about him, a black Oklahoma cowboy).

Gerald's large family gathered for supper, mainly to grill me about their Australian cousins Ian and Jesse, demanding banjo tunes and songs about South Dakota. Sarah, Rebecca and Josh had met our sons during visits to South Dakota years ago. Josh was spell-bound by Sheila's little photo album and wanted to come visit to see pythons and wallabies.

On 22 January Gerald, Josh and Becky drove me to the Phoenix airport, where my Murdo High School classmate Jorgina Anderson Rederick was found sitting outside my departure gate. She was one of the brainiacs of my class of 1959, an A student, and now she had sons that were computer nerds. Then it was onto another airplane, to go to San Antonio for my science meeting. Flying out of Phoenix allowed a view of the Gila River valley, mostly a patchwork quilt of irrigated crops, and then fading rapidly to desert rocks and sand as we approached Texas.

For some reason, San Antonio always agreed with me, even though I dislike cities in general. I arrived just at dark and took a bus to the Davy Crockett Holiday Inn, which was across the street from The Alamo and beside a Haagen-Dazs ice cream shop. Coulda been a lot worse. I walked down to the Trinity River promenade to find my supper at a Hunan Chinese restaurant (scallops and beef in a spicy oyster sauce). I could watch the river boats putt-putting along, some loaded with loud cheerleaders and rowdy athletics teams. The streets here are named Crockett, Bowie, Sam Houston, Travis, Bonham, Santa Ana, all heroes of Texas/Mexican history. My hotel room information packet told me that Jim Bowie played the fiddle and was a great storyteller, often creating new legends as he spoke. Davy Crockett was also a ballad singer, a fiddler and played the banjo, with a repertoire of over a hundred songs After a good espresso and a cappuccino almond mocha ice cream, there was time to watch a Clint Eastwood movie as I fell asleep. Hard times on the road.

It was a chilly free day, Sunday, 23 January 2000, as I sleepily staggered across the patio to an outdoor hotel restaurant for a wonderful Texas breakfast of baking powder biscuits, cornbread, very thick bacon, four soft fried eggs, pepper sausage, pork chops and Texas Ruby Grapefruit...the only trouble was the dishwater coffee.

Here was my other family! The place was thick with oceanographers from the USA, Europe and some Asian countries, all scanning the book of abstracts for the meeting that begins the next day. There were unscientific hoots and hollers as each new guest arrived, friends that we only see at science meetings. I got stuck with a bunch of Woods Hole marine chemists (Ed Sholkovitz, Michael Bacon, Ken Beusseler), and we wandered under the live oaks, toured the Alamo, inspected old cannons built in Scotland and Spain, and reviewed the history of this area from Spanish missions in 1500 to the establishment of the Republic of Texas in 1836. It was also clear that most Texans did not give up on the Republic of Texas, and considered the rest of the USA as a backward third world country being propped up by Texan wealth.

Our TROPICS organisation had sponsored attendance by several SE Asian oceanographers who could not normally afford to travel to the USA. I found my Indonesian colleagues Gani Ilahude and Sutomo at the AGU registration desk, where I paid their registration fees and hotel room costs for the week. Gani immediately wanted to know where to find some pizza, as they had been traveling for thirty-six hours. After they had a nap and a shower, I took them to Landry's Seafood Place (not too far from the stadium of the San Antonio Spurs, one of my favourite basketball teams), where we had spicy crab cakes, prawn quesadillas, chocolate chili ice cream, cold Lone Star Beer and loud Tex/Mex music. Gani was always curious about Americans, and he wanted to know why he couldn't understand the Texan drawl language of the waitress ('...what language is she speaking?'). I had previously sent information to him about Mount Rushmore in South Dakota, which intrigued him for some obscure reason. He had brought some Rushmore photos along with him, so I could explain some aspects of this curious superpatriotic behavior of Americans.

On Monday the science presentations began, running 0800 to 1700 for the week, some eight session rooms running simultaneously for the different oceanographic disciplines. Most rooms had two or three screens being used for PowerPoint and video illustrations of enthusiastic reports from around the world.

I can't emphasise enough the curious combination of reverence, competition, respect and joy of scientific communication with our peers, almost like a ceremonial event (without the bloody sacrifices), but this allowed questions and criticisms after each talk. I was grateful to attend sessions on changes in Pacific Ocean physical and chemical oceanography, sediment trap results, global ocean carbon, nitrogen, and phosphorus budgets, impacts of mine wastes dumped into the sea, consequences of anthropogenic climate change in various parts of the world ocean, and where did all the mud from all the rivers go into the ocean.

Like our TROPICS organisation, there were dozens of other international scientific consortia such as JGOFS (Joint Global Ocean Flux Study), presenting their results and smooth-talking their funding agencies. The AIMS workshop engineers and I had built some giant sediment traps arrays for our work on the NW Shelf and Gulf of Papua, and I was happy to meet with Wilf Gardner, who had published good papers on the best design for sediment traps. The reason for such attention to catching the falling particles in the ocean had to do with the rate of production and removal of organic matter in the surface 100 m of the ocean. Our results, and those of others, indicated that the oceans contained maybe half of the anthropogenic carbon dioxide emitted by industry over the past fifty years. Much of this greenhouse CO_2 in the ocean was converted to algal organic matter by photosynthesis, consumed by herbivorous small marine animals, which are then consumed by the big fish, marine mammals and birds. Catching the annual rain rate of organic matter (and associated minerals, like $CaCO_3$ and diatom silica) in various parts of the world ocean indicated that only 1-10% of the surface ocean productivity reached the deep ocean (1-4 km), and much of this was decomposed in the mud and returned to the ocean as CO_2.

Part of my work was to quantify the organic carbon removal rate (with sediment traps) and the century scale burial rate of organic matter (using radionuclide dated sediment cores). Some of our measurements suggested that the tropical coastal and shelf ocean was burning off (respiring, decomposing) more organic matter than was annually produced in surface water, and was emitting CO_2 from the ocean to the atmosphere. Some of these sites were additionally incinerating organic matter brought to the ocean by rivers, in my case, the giant rivers from Papua New Guinea. And even more curiously, most of the organic matter carried by the big tropical rivers to the seas, to be buried on the continental shelves, was dated to be thousands of years old by radiocarbon. This suggested that most recent terrestrial and marine organic matter was rapidly decomposed, and the only organic matter left to be buried was very old soil and shelf lignins and celluloses that were resistant to decomposition by microbes.

There were fancy receptions at the Texas Culture Museum (I'm not kidding, they really do claim to have culture in Texas!) with wonderful Mexican finger food, chicken mole, blackened catfish, some smoky spicy barbecue ribs and barrels of Lone Star beer.

Our special session on TROPICS research went very well, with several hundred people in our audience. It felt like a Meadowlark concert, because we knew everybody in the audience. I was one of the conveners, and I could field questions to the speaker by calling out their names in the audience. Curiously, a lot of Canadians showed up, even Quebecers, hooting and hollering at me for having escaped Canada. At the end of our session we had a logistic meeting with everybody that would be on future TROPICS cruises, mainly about visas, cruise dates, sharing equipment, and trying to plan for bad weather and equipment failure. As usual, there were some political problems, such as PNG not wanting to approve the visa for a graduate student from Chile, for reasons we never understood.

These AGU meetings also featured a large display warehouse full of oceanographic and lab equipment suppliers, ship machinery, book and journal publishers, and examples of new electronic gear for deep sea rovers, submarine communication cables, and instrumented atmospheric and marine buoys.

As usual, some of the best social and scientific arguments happened over lunch and supper in interesting restaurants with big pitchers of beer and TexMex food. Every night it was like a different gang would gather around a big table, with no need for introductions, and most of us had nicknames from some ridiculous embarrassing event on a ship or a conference. And Texas was a weird place to find gentle Swiss geochemist Peter Santschi, whom I had known as a post-doc at Lamont-Doherty, but was now was a professor at Texas A&M University. People who knew me from the USA or Canada kidded me about the absence of my cowboy boots and leather vest, as I was now a tropical guy. I could honestly reply that I was wearing a Canadian thrift shop corduroy suit with flared trouser legs that was at least forty years old. French colleagues pretended to be upset by so many of the speakers not wearing neckties and complained bitterly about the lack of choices from the wine shelf. Two Belgian colleagues said they hadn't been able to find any place that sold fresh waffles, and were alarmed that so many people were tanned like the Marlboro Man. The Germans wanted to see some *Playboy* foldouts appear on the speaker's stage. The Russians quietly ate their food, and then wanted to go clubbing along the river. The Tennessee *faulschlamm* (German for stinking mud) David Culver showed up one night, now a professor at Ohio State University. I had worked with him in graduate school at Cornell University.

Some of topics of discussion shouldn't be repeated in polite company, but the more amusing stories were about disasters on ships at sea, equipment failures and how to fix them, strange food and stranger cooks on ships, and adventures in ports of call. You will understand how dangerous it might be for a young graduate student in oceanography from Nebraska or Vermont to get off his or her research ship to wander around in Rio, Calcutta or Amsterdam. An alarming thread emerged from these supper table discussions, that I was being regarded as a 'senior scientist', which startled me, until I recalled that some of my stories came from activities thirty years ago. Some of these people had been at Sheila's feasts in Canada and our Alligator Creek home, and they enquired about our parrots Sir Charles and Deputy Caspar, as well as our orchard productivity.

It really was winter, and even in Texas it was chilly. Thursday was cold and raining, and my Indonesian colleagues had no warm clothing. I bought them good insulated jackets, but Gani decided to stay in bed all day, coming out only for supper. Sutomo gave his poster, and I helped translate questions and answers. I also nearly froze to death in my Carhartt jacket. Gani begged for a private supper with me and Sutomo, so he could grill me about many things. He was worried about the progress of democracy in Indonesia, with many of the remote island people still living in pagan tribal societies where disease and pestilence were considered due to witchcraft. Not too far from where he grew up in western Sulawesi, leprosy, malaria, cholera and typhus were common. Gani was Muslim, Sutomo was Catholic, and they wanted to know how I could claim to be a druid, an infidel, a heretic. Gani's ideas about food were odd. He liked anchovies on ice cream, and he would put honey on a hamburger! They were curious about the big bang theory, Stephen Hawking, Darwinian evolution, and climate warming consequences. We sat in the back of a restaurant where we ate and talked till midnight, and finally the restaurant owner asked us to leave. Back at the hotel, I arranged for a taxi to take them to the airport for an early departure on Saturday.

As usual on these trips, I had scheduled a free day to rest after the great exertions at the science meetings. I slept in on 29 January, Saturday, had a good breakfast at one pm and wandered the malls nearby. Unfortunately, I found a good chocolate shop on the river walk... and there was Almond Roca, cashews in dark chocolate, macadamia toffee and almond brittle. Oh dear, this was serious, and I needed to have some to eat on the airplane home, but even more serious was the need to hide some in my checked luggage (so I couldn't eat it) for my dear wife Sheila back in Alligator Creek.

The malls had big sofas near the doorway, full of supremely bored Texas cattlemen with their best 'going to town' Stetson hats and polished high-heeled boots good enough for church. They were waiting for their happy wives, shopping and socialising inside the shops. These men would look around nervously, nodding at passers-by, Copenhagen under their lower lip, spittoons nearby, anxiously awaiting a release to go back home where they could deal with horses and cattle. These cattlemen behaved the same as South Dakota cattlemen where I grew up, like my father and brothers. They were completely out of their element and they knew it, and it was hard to escape.

Then it was the taxi to the airport with a plastic Jesus and a rosary on the dashboard, departing San Antonio 1640, to Dallas, then to smoggy Los Angeles, and then a fourteen hour flight across the Pacific with Qantas to Sydney. Eat chocolates, films, sleep, and 30 January was lost to the date line. I arrived a little after noon on 31 January, grateful for the clean perfumed warm air, and there was tropical lady brown Sheila beaming with a red hibiscus in her hair...and off to Alligator Creek where Sir Charles loudly signaled the welcome. I got my new bouzouki tuned ADAD and invented two new 5/4 tunes for it. Sheila, Charles and I walked the creek valley orchard, picked passionfruit, pomegranates and mangoes. Ian MacGreggor visited, and we watched the St. Louis Rams barely win the Super Bowl.

Our home computer was by now hooked up to AIMS, so I could keep track of email and web news, which was probably a mistake for peace of mind. I spent the next day soothing upset TROPICS collaborators who were having problems with visas. Their shipping containers appeared to have vanished, and there were rumors of a volcanic eruption near Rabaul, where we expected to board the R/V *Lady Franklin* in two days. This was a very brief down time.

Soon Irena, John and I were off yet again, in the pouring rain in the evening of 1 February 2000, flying north in bad weather to Cairns to meet collaborator Bob Aller at a Thai restaurant with a very good Latin jazz group. Our laptop emails told us that some of Nittrouer's science crew were already in Rabaul but there was no sign of his shipping container, which contained all of his sampling gear. The next morning we were up at 0430 to reach the airport by 0500 where all flights were delayed. This meant we could have a peaceful bacon and eggs breakfast, with Aller's grad students Vanessa and Angelos (from Chile), Irena and John.

We finally got into Port Moresby's newly renovated airport, where the communications and computers were not working. Customs and quarantine service officers wanted to know what was a 'Gafenhoto'? This was apparently the reason for the delay in PNG entrance of Nittrouer's shipping containers, as the names of the equipment inside it were a big mystery to the quarantine agents, who were not big fans of oceanography. The Gafenhoto was Andrea Ogston's benthic lander rigged to settle on the sea floor and measure temperature, salinity, current direction and speed, and suspended sediment, for the purpose of measuring sediment sliding along the seafloor into deeper water. The name came from Brazilian Portuguese for 'grasshopper' during Dick Sternberg, Gail Kineke and Andrea's work on the Amazon estuary and continental shelf. We got help from AIMS and the University of Washington to fax a description and value of this gadget.

We escaped the madness of Port Moresby on a Fokker F28-1000 propeller airplane, where we were strictly forbidden to chew betel nut, and which stopped frequently at coconut plantations (Hoskins) on New Britain Island. We were trapped in our seats while refueling, and entertained with the music of Haydn, Bach and Stravinsky (*Pulcinella*!). Weird things happen in PNG.

We arrived in Rabaul at 1600. The Travelodge van was there to haul us to our hotel past drifts of black volcanic sand, where JP Walsh, Dick Sternberg and Andrea Ogsten were waiting. They had located their shipping container, covered with PNG quarantine stickers, and thankfully the padlocks were not broken. Chuck Nittrouer was still in Port Moresby, polishing our relationship with one of our sponsors, Dr. Hugh Davies of the University of Papua New Guinea Department of Geology. The hotel had cockroaches, lots of geckoes, black volcanic dust everywhere, and the restaurant wasn't very good. The air smelled of hot sulfur and the hotel desk lady told us the sulfur fumes were good for your eyes, it healed sick eyeballs.

The next day, our research ship R/V *Lady Franklin* steamed into Rabaul harbor at 0800. The ship had been in the Bismarck and Solomon Seas for two weeks with CSIRO economic geologist Ray Binns and crew, who were exploring the submarine vents at 1-2 km depth south of New Britain and up near Nauru. All of his equipment was stashed in the hold, and our shipping containers were loaded on the aft deck with fifty tonne cranes. There was tension between the CSIRO ship managers and the Pacific and Orient crew, and the smoko breaks and lunch breaks were longer than usual. It took a full day to get all the gear shifted, bolted down, and set up in the labs. This ship and CSIRO science crew then went to the airport, and a new ship and TROPICS science crew boarded the ship, leaving time for the purser to change the bunk sheets and clean the toilets. I was grateful that our ship Master was our friend Neil Cheshire and the first mate was Arthur Staron, who knew about how we were going to operate in the Gulf of Papua.

The next day, PNG Customs came aboard to inspect our passports and equipment inventory. We must have explained four times that we wanted to bring mud samples back to our labs in Australia and the USA, and that these mud samples should not be classified as 'farm soil samples'.

We wanted to have two winch wires and blocks on the aft A Frame (so we could operate both the Kasten Corer and the big box corer). This meant one of the crew needed to go up in the bosun chair to shift the blocks, while the ship slowly steamed southward across the calm Solomon Sea, full of beautiful islands covered with rainforest and white beaches. As usual, we had a ship safety drill, evacuation and medical procedures, and a lecture by the scientists to let the ship crew what we planned to do in the Gulf of Papua. We had promised Ray Binns that we would do a series of CTD casts at locations where he predicted there might be hydrothermal vents on the Solomon Sea floor, which should be detectable from temperature and salinity measurements.

The first CTD cast to 2500 m depth in the afternoon gave no vent signals, and by evening the ship computer system had crashed, there was trouble with the ship sewage system, and the ship generators were working erratically. So we gratefully had a late tea and cakes, and went to bed. Two days of frantic work by the ship crew got everything working again, and we entered the Gulf of Papua at noon on 7 February.

What follows in my journals are pages of 8-12 February time schedules for mapping the thickness of the recent (post-glacial) river mud on the shelf and slope, CTD profiling, water and mud sampling for chemistry and biology, and coring to determine the river sediment deposition history and rate over the last century. The ship ran 24 hrs/day, so we tried to work in shifts. We had a 12.5 kHz sonar on the ship, which could show us the recent sediment thickness of cross shelf transects from shallow to deep water, which would suggest where we should sample. Bob and his biologists Vanessa and Angelos were busy wet sieving, sorting and preserving benthic creatures. Chuck and JP Walsh interpreted the sonar charts of sediment thickness, operated the big box corer and a smaller kasten corer, and used an X-ray machine to look at sedimentation structure in core slabs. Dick and Andrea were deploying their turbidity meters and a benthic lander (the gafanhoto) to measure sediment sliding along the seafloor into the deep Coral Sea.

Irena was in charge of taking water samples from the big Niskin rosette, filtering and preserving the samples for chlorophyll and nutrient concentrations, and was chief data recorder. Some of the notes in my ship log are in her handwriting. She, John and I operated our big Kasten Corer and carefully sliced the 1-3 m sediment cores for mud pore water chemistry, radiochemical estimates of accumulation history and rate, and sampling the mud for geochemical and microfossil analyses back at AIMS. During all this work in the hot sun, cool darkness and pouring rain, we perfected 'the Franklin Stagger' during choppy weather, 'box corer dancing' and Franklin coffee (two spoons of instant coffee, one spoon of cocoa, ½ cup whipping cream, four spoons of cane sugar, add hot water to taste).

We were able to make sense of shelf and slope transects A-G, showing that river sediment wedge thickness and shelf width was maximum from the Purari to Fly River sector. We gratefully slept in for half a day, while the ship returned to Port Moresby on 13 February, where Nittrouer, Sternberg, Ogston, and crew departed, and the seismic team led by John Milliman (with Beth Mullenbach, Dave Cacciarone, Katie Farnsworth) boarded the ship.

Milliman and Cacciarone were operating a shallow seismic 'sparker' rig that penetrated deeper into the sediment than the ship's sonar. The low frequency power generator fed into a towed array of metal filaments that created sonic booms that bounced off sediment layers 30-100 meters below the seafloor, and came back up to be collected in a hydrophone attached to the ship. The variations in strength of the reflected signal were recorded on a computer in the ship's lab, and was used to interpret the longitudinal sedimentation history. It would detect changes in density and composition of the sediment, and provide more information about the Pleistocene history of river sediment inflow to this continental shelf and slope. After completion of this mud mapping operation, Milliman hoped to see discrete layers of river sediment of different ages, marine carbonate shell zones during hiatuses in river mud deposition, tongues of sand or gravel carving canyons now filled with fine mud, and evidence for submarine landslides. While the sparker was operating, the ship was in 'silent mode', a special effort to reduce ship noise from the engines and generators.

During 13-20 February, much of this work was done in long 12-24 hr transects across and along the shelf and slope, with stops only to trim/clean the sparker wires, and to take water and sediment samples at important locations. This meant we got to sleep for a few days, and then work furiously when the seismic record showed interesting bottom topography or unexplained sediment layers.

Bob Aller and crew were performing respiration experiments on surface mud, measuring the rate of consumption of oxygen by microbes that were decomposing organic material in the mud from the box corer. J. P. Walsh and Beth Mullenbach were operating the box corer and X-ray machine.

John Pfitzner filled several large plastic cubic meter tanks with water pumped from a supposed submarine vent site, hoping to see high radon and radium concentrations (none was found). Irena, John and I also took a few Kasten cores in interesting deeper slope locations, finding abundant shell hash, white foraminifera sand, black magnetite sand, laminations of creamy white clay and black silt, coral rubble, foul H_2S rich layers of pure organic matter, and odd layers of 'coffee grounds' (which we later found to be mostly ground up mangrove tree wood).

The results of these measurements allowed us to estimate the river sediment transport and supply budget for this continental shelf, from the five km high foggy rainforest inland mountains of the Fly, Kikori, Purari Rivers, to the base of the marine continental slope at 1000 m depth. We were also able to compute some part of the carbon budget, based upon river inputs of dissolved and particulate organic matter, algal growth in estuaries and the shelf water, respiration and decomposition of organic matter in the water column and surface sediments, and then the final residual organic matter burial rate over the past century and 10,000 years of the Holocene (the time since the last glacial period).

Partly from these measurements, Bob Aller and colleagues considered these estuaries and shelf zones to be 'incinerators' of terrestrial and marine organic matter, apparently burning off both recent and old organic matter. From our radiochemical dating of sediment cores, and analyses of organic matter in the mud, we could see how nature recycles rapidly. Most of terrestrial and marine organic matter supplied to this tropical shelf was decomposed to CO_2 and returned to the atmosphere, only ~1% of this organic matter was buried, and much of this buried organic matter was over 1000 years old. This must have been old microbially resistant organic matter eroded from inland tropical soils, as experiments on mangrove mud, estuarine and shelf organic matter showed that practically all organic matter carbon and nutrients were rapidly decomposed and released to the water and/or atmosphere.

Our Tasmanian colleague Peter Harris had previously described 'mesas' rising up 100s of meters from the sea floor in the northern Coral Sea, which Milliman's sparker showed clearly as being of curious composition. We spent a day and a dark rainy night trying to core the tops of these mesas, and finally got one good core. The core barrel was full of small spherical shiny white egg shaped sand, which caused John Milliman to go into spasms of ectasy, giving lectures on the mystery of their formation and distribution in the world ocean at 0230 hrs in the rain on the aft deck in front of puzzled and amused ship deck crew. These glisteningly white 0.5-2 mm sands were called oolites, and were thought to have formed in shallow evaporitic coastal seas. If this were true, then these oolites must have been formed during times when the sea level was 100-120 m lower than at present.

Meanwhile the ship cooks were going berserk, supplying us with dark chocolate coated cream puffs at midnight tea time. I had provided them with my grandmother's recipe for sour cream raisin pie, and they made dozens of them, which lasted one afternoon and sadly I got only two pieces of those pies.

But all good times had to end. We cleaned and packed our equipment while steaming to Daru for PNG Customs, and then to Thursday Island in the Torres Straits for Australian Customs. We had rehearsed the phrase 'These are not farm soil samples...' for when the Quarantine Officers inspected the hold full of nally bins full of packaged mud samples. For some strange reason, the Quarantine officers got suspicious of Nittrouer's X-ray apparatus and lead shielding, so they phoned their bosses in Canberra, before releasing us from bondage.

And then as we steamed for Cairns, everybody went to sleep, or began the card games, or went to the ship's library to watch films. We woke up two days later in the Cairns port swamp full of mosquitoes, and by 0900 on 22 February there were two wharf cranes and four forklifts moving equipment and crates from the ship to two semi-trailer trucks (one for TROPICS gear going to AIMS, and one for Ray Binns of CSIRO Canberra). AIMS supertechnician Bruce Parker had our semi there, loading gear destined for AIMS, Seattle, Virginia Institue of Marine Science, and Woods Hole Oceanographic Institution.

The ship was clear of us by 1500. Bruce took off in the semi for AIMS, and Irena drove a ute with our zodiac on a trailer back to AIMS. It felt odd to be walking on land again. The ground seemed to move like the deck of a ship. John and I went to the Cairns airport and were in Townsville by 1900, sweaty and tired. Bright and shining Sheila and Ian were there to rescue us, and John's dogs were excited to see their master returning home to Wulguru...and then down to Alligator Creek where Sir Charles Byrd the Cockatoo had to check me out. It has been raining here for most of February, over 800 mm of precipitation so far, a good wet season.

Two days later the trucks arrived with all our gear from the ship. We used cranes and two fork lifts to shift the gear and scientific samples from the trucks to our sheds and the institute labs. There were more urgent meetings about conservation of local wetlands, due to local industry wanting to mine clean carbonate sand from Cape Cleveland.

In March, CSIRO scientist David Post and I drove out to Pentland in the interior of western Queensland to give presentations to cattlemen about how to reduce erosion and preserve wetlands for protection of the Great Barrier Reef lagoon. The road ditches through this normally arid land were full of rainwater, there was tall grass, the eucalypt trees had pale green new leaves, but there were patches of dead trees that hadn't recovered from the previous drought. The meetings were held in the Pentland Soldiers Memorial Hall, established 1926, muddy utes with dogs in the back parked nearby, the hall roof thick with galahs, flag flying at the front, boot scrapers by the entrance, a 1950 photo of the Queen in a heavy frame on the wall, and one wall full of photos of World War I and II, Korean, Vietnam soldiers from this region that didn't come home. For some reason, this felt like my hometown, and I relaxed and made small talk with these old cattlemen. I could speak to these quiet cattlemen better than most scientists. There were overhead fans that moved slowly, and the polished/waxed wooden floors creaked with the traffic. The ladies provided ice water, sandwiches and pickles before and after the presentations.

By late March the rains intensified, roads and parts of town were flooded for almost a week. The little creek flowing through our orchard was roaring over the banks and onto the floodplain, which doesn't happen very often. Ian's girlfriend from Toronto arrived, Tiffany Hsieh, a concert pianist. Ian and I were working out arrangements for tabla and bouzouki pieces in changing time signatures (5/4, 7/8).

My performance with cattlemen in Pentland must have been satisfactory, as I got a phone call at home from the Ravenswood Future Profits Club, asking for my attendance on 1 April. Sheila and I drove out this this old gold mining town, where the setting was almost identical to Pentland, a community hall with photo of the Queen and lists of soldiers killed in action, with the ladies providing food and drink. This being a gold mining town (the mine now abandoned), I told the story about the use of mercury in the stamping mills, and how this mercury was used to amalgamate the gold in the crushed ore. Some large fraction of this mercury escaped to travel downstream in the Burdekin River to Upstart Bay and the Great Barrier Reef lagoon. Stuart Walker and my research on Bowling Bay Core 1260 showed the 1870-1900 history of 20X background mercury deposition, and the decline in Hg pollution with the advent of the Scottish method of cyanide extraction of the ore. The Ravenswood and Charters Towers region was a large gold producing region during these pioneering times, and abandoned mine pits, chimneys, and tailings litter the landscape. The central message was that inappropriate methods of agriculture and mining in inland regions results in sediment and contaminants reaching the Great Barrier Reef rather quickly, and is usually deleterious for the environment of marine biota and reef corals.

When we got home from this excursion, the TV news was warning us of Cyclone Tessie approaching our coast. We had strong wind and lashing rain for 2-5 April, the electricity failed, and the candles and kerosene lamp came out. We had a gas stove, so we could still cook, and we used my Ellesmere/Lake Hazen insulated box with ice slabs to keep food cool. Schools and my institute were closed, roads were flooded again, but the main cyclone was moving west to the inland.

In early May, good friends Bob and Josie Aller, with son Eddie and daughter Debbie arrived from Stony Brook, New York, for a month's visit to AIMS. Eddie discovered he could catch big mud crabs in nearby Cocoa Creek, and this contributed well to our feasting. We had them over to our house for feasts, parrot admiration sessions, and introduced them to the Mennonite game of crokinole. Bob and Josie had visiting scientist offices, and were trying to write papers while they were here.

Ian had applied for his Australian residency papers, so he could legally get a job, but in the meantime he was playing with a pub band. Jesse and Sarah were out in Emerald, Sarah working at a veterinarian clinic and Jesse working with Main Roads trying to build a new road on black swelling soil during this unusual wet time. He demanded the calculation of Gibbs Free Energy (heat of reaction) for slaked lime on this wet gumbo clay. Sheila's two tame magpies were still living in the house, begging for treats, sleeping on the shower spigot. Our orchard was growing fast, we had our own mangoes and lychees now, but the demand for whippersnipping the tall guinea grass was every two weeks with all this rain. The citrus trees were covered with flowers, and the scent was overpoweringly like that of my father's aftershave lotion. The frangipani tree flowers fall on the ground and glow in the dark. The moist air was sweet and warm, and life was good.

Chapter 41. Belgian Waffles and The Very English Cotswolds

Because I was one of the four international leaders of Project TROPICS, I was invited to present an overview of our research on coastal New Guinea at the 32nd International Colloquium on Ocean Dynamics in Liege, Belgium.

On May 6th Sheila hauled my groggy carcass to the Townsville Airport for an early departure to Cairns, which had an airport full of Japanese tourists. I got a laptop charger adapter plug for UK and Belgium, inspected the bookshop, found some Senior Service cigarettes and then escaped to the outdoor bench for a nap. After a laksa and beer lunch, a 1455 hr flight took me to Darwin, where the land was strangely green and not on fire, and they had good espresso and a smoking lounge. Then an 1800 hr flight took me across the Arafura Sea, Timor Sea, Java, Sumatra (which was on fire, as usual) to reach Singapore by 2100 in the rain, smog, and smoke.

They have an excellent bookshop there and wonderful Asian restaurants where a live band (white piano, drumset, F hole guitar, fiddle and bass) was trying to sound like Stephane Grappelli and Django Reinhardt. There was an exotic smoking area on the roof with cactus gardens, crowded with Dutch and German people smoking cigars and pipes. I left Singapore at 2300 hrs in the pouring rain, and I was surprised to find my Qantas ticket had been upgraded to First Class, featuring a big wide plush recliner seat with twelve buttons for TV, films in many languages, a tablecloth with platters of cheese and biscuits and wine in tall glasses. Most of this was wasted on me, as I immediately went to sleep.

I woke up over the Mediterranean for a nice breakfast of soft boiled eggs and sausage, with good coffee and bowls of fruit. We landed in London on 7 May at 0455, where the sky was gloomy grey and the city lights seem to stretch to the horizon, all except for the Thames and the sea.

Heathrow smelled of musty carpets, hot metal and diesel fumes. Grandmotherly ladies in long skirts clucked over my transfer ticket, appearing surprised that I didn't want to go to London. 'Where are you from, young man?' with raised eyebrows, as they gave me a special security pass to terminal 1. The airport was on security alert, and there were rolls of barbed wire on top of chain link fences at the transfer bus station, where guards carried pistols and shoulder rifles. They demanded to see the security pass and our passports before allowing us to board the bus. The sleepy British Air desk clerk was not impressed with us, grumbling something about '...nowt on him going to bloody Brussels, eh lad'. I got to watch some CNN news in the internet and smoker bin, and then Salman Rushdie and twelve security guards walked in, and it got very quiet all of a sudden. One of the security guards walked over to an oblivious Yankee student in worn denim shirt and pants who was playing video games on his laptop, the sound card at full blast. He turned it off quickly.

The smoker bin was full of Indians in saris and Nehru shirts, dark-suited pasty white European businessmen, worried Japanese executives on laptops with assistants bringing them tea in plastic cups. I was with a small group of Australian oceanographers *en route* to Liege, and I'm sure we looked out of place, being tanned, wearing sandals and shorts, completely unrepentant and relaxed, most of us smoking cigars. The airport shops finally woke up, and I walked past duty free shops full of fancy clothing (£600 sport jackets), booze, a lot of stuff I didn't need, but no good book shops. I got a terrible lukewarm cappuccino and an awful cold meatless sandwich. I found a money changer, and got 4000 Belgian Francs for US$100. The smog and grime, gloomy grey London skies didn't brighten much as I boarded a British Airways flight at 0940, but at least the grass was green, and the mustard fields were bright yellow.

It was a bumpy ride over the North Sea. The European sea coast was all of a muddy froth, but the land was a startling array of rectangular green crops with a few unhappy rows of trees planted amongst them. All airplane announcements were now in Walloon (sort of French) and Flemish (sort of Dutch) as we landed in Brussels. The railway station was in the basement of the airport, where I paid BF2000 for a ticket to Liege. The black-suited train conductors were straight out of *Tintin*, with mustaches, pot bellies and polished brass and leather hand-cranked ticket stamping machine. The train rolled through the farm country with Holstein dairy cattle grazing with skyscrapers in the background.

I had to switch trains in Brussels Nord (Loeuvin) to join a train going to Germany (Kohln, Aachen), and the language switched to high German. Many of the small towns consisted of grubby rows of industrial cement block buildings with gaping broken windows and partially demolished old stone barns with roof fallen in...the whole works being converted into shopping malls and apartment buildings. Away from the towns, the farms had neatly stacked wood piles, barley and oats heading out, hay in windrows, rows of trimmed grapes, rows of Lombardy Poplar trees, neat gardens behind the house with flowers in window boxes, all villages with steepled churches, usually a clock tower in the town square. There didn't appear to be a square inch of land that wasn't under cultivation. But the trains ran on 12-16 parallel tracks, fast, clean, quiet, electric, and there were groves of lilac and crabapple trees that pleased the eye. I was reading Bill Bryson's *Neither Here Nor There, Travels in Europe.*

Now if you are getting tired of all these small details about airports and train stations, I'm sorry; I will not apologise. Part of the learning experience of these trips was the people watching, language and cultural shifts from country to country, how to find decent food and snooze time, and seeking presents for Sheila on my return.

I arrived in Liege-Guillemins at 1430, staggering out to the street with my luggage, and I was savagely assaulted by the very strong smell of hot waffles and croissants and hot chocolate. I watched the natives buy a slab of waffle, take it to the chocolate shop to have hot chocolate poured over it, and then sit on a bench under flowering crabapples in the sun to enjoy it with a cup of strong coffee. Ok, I like Liege already. Before World War II, Liege was a major coal and steel city, and was the capital of Wallonia, French speaking but Protestant. The city never recovered from the war, there being only one steel mill operating now, amongst the ruins of dozens of other industrial buildings. But the city had a grand history, the birthplace of the Emperor Charlemagne, and many medieval churches and statues.

A taxi took me to a disgustingly normal Holiday Inn, across the street from a 'Quick' burger and fish and chips shop. I had a TV in my room, where the French and Italians were playing basketball against each other. I fell asleep watching CNN, and woke up hungry. This was Sunday, and nothing was open close by. I warily walked across the Meuse River to a busy café zone, and ordered from a Walloon menu without a clue...and ended up with a plate full of huge sausages and some strong cheese-stuffed pasta. I also found an ice cream shop that had *pamplemousse glacée* which was excellent. I was wary of getting lost in this city, and I wrote down the street names and statues to guide me back to my hotel; one landmark statue was a huge bronze bull with a brilliant red dangling scrotum. I went to sleep early in a good bed.

The purpose of all this travel was to attend a big oceanographic science meeting over the week at the University of Liege, high above the town. There were lots of good speakers in every language of Europe and north Asia. The Russians and Turks showed up with good stories about the Aral Sea drying up due to cotton irrigation withdrawal of river water. Several papers were given on the mysteries of the Black Sea deep water, devoid of oxygen and rich in metals. Coastal oceanography of the Mediterranean and the European coast from Gibraltar to Finland was presented. The Nordic nations were especially focused on the eutrophication of the Baltic Seas, due to pollution via the inflowing rivers. Two sessions on the validation of sediment trap results were important to me. A group of Australian oceanographers gave a session on the Southern Ocean and the Antarctic ice sheet.

My presentation on our TROPICS Project was especially interesting to the Europeans, who were frustrated by the difficulties of running international research programs. I had explained that our research consortium was entirely operated by participating scientists, and was not supervised or administered by government bosses. Several important EU officials appeared, being agents responsible for funding EU science (they were bosses of the OMEX3 project currently in operation). My Belgian, German, and French colleagues were petrified, terrified of these officials, and were horrified that they had heard my presentation, which appeared to negate the importance of top government supervision of research collaboration. However, at lunch with Michel Frankingnol, Roland Wollast, Alberto Borges, Bruno de Lille, and session chairman Jacques Nihoul, they all proposed a toast to TROPICS.

Throughout all of this heavy scientific load, we suffered mightily during lunch and supper at some of the local restaurants. My European colleagues assumed that I was ignorant of their delicacies, and they insisted on ordering for me. I would never have volunteered to eat flounder in hot strawberry soup with thick cream! Some evening feasts were held in a seventeenth century Frankish castle that had been restored for feasts and pageants. We had baked prawns in pastry, squids stuffed with ham and cheese, too much smoked salmon in cream sauce, roast duck with pears, blood sausages baked in loaves of bread, spicy salads with octopus and squid, enormous amounts of wine and good Belgian ales.

Dear wife Sheila had emailed me, demanding that I bring home some dark Belgian chocolate for her, and requesting that I not eat it on the plane coming home. My journal says that the city smelled like my Grandpa Brunskill's fish trap bait barrel (strong cheeses), flowers everywhere, smoked fish, horse manure, and chocolate everywhere. This is different from most other big cities, which smell of diesel fumes and sewage. The horse manure came from the huge Belgian draft horses (Brabanter), pulling beer wagons to all the pubs. These gigantic and gentle brown work horses must have weighed a tonne and were maybe seventeen hands high. When I was a child on the White River floodplains of western South Dakota I remember my grandfather Brunskill's team of Belgian work horses that pulled a hay wagon on skis after a winter blizzard, when the tractors couldn't work in the deep snow. But I also noticed that horse meat was featured as a delicacy in some of the best restaurants.

Alberto Borges and Bruno de Lille became good friends. They were from the University of Liege, and knew the town well. They took me to a *Tintin* shop, where all issues of the comic books by Georges Remi (Hergé) were displayed. I had read these comic books to my sons when they were little. They had seen my CV, so they knew I was not an Australian, but an escaped American; therefore, they took me to the World War II Cemetery in Henri-Chappelle, where thousands of American soldiers were buried from the US First Army drive of 1944-45.

My uncle Gerald had been here as a sergeant with the Infantry during the Battle of the Bulge, and some of those buried here were probably his comrades. Walking through the memorial cemetery affected me deeply, and I somberly returned to my hotel room to write a letter to my uncle Gerald, who was back on the White River ranch in South Dakota. As I wrote this letter to my homeland family, I remembered playing 'Taps' at the graveyards of fallen soldiers in little South Dakota prairie towns. Sometimes the Veterans of Foreign Wars would hand out poppies on Memorial Day, and young people would wear them stuck in their shirt button holes, not really understanding what they represented.

I was not mentally prepared for this shock of history: this part of Europe had been a slaughterhouse during both World War I and II. Museums in town featured maps of where all the mustard and chlorine gas, explosive and arsenic bombs, ships, and battlefield trench soldiers were buried on land and in the coastal seas. These chemical and explosive munitions were not just from the Nazi Germans, but also the British and French, and the fallen are buried where they died along 800 km of trenches from the North Sea to Calais in France.

You might wonder how the grown children of Belgian parents and grandparents can smile and have pleasant conversations with a naïve visitor like me.

I had heard the Scots/Australian folk singer Eric Bogle at the Winnipeg Folk Festival. Some of his songs were about the insanity of war and the somber visits to World War I and II graveyards in Belgium, Holland and France. Some of his lyrics came into my brain as I was trying to ignore deep emotions rising up in me.

And then it was all over: Check out of the hotel, taxi to train station, switch trains, underground shuttle train in Brussels. I must admit that train travel allows some introduction to the countryside. Under cloudy grey skies, there were church spires and domes in the smoky mist, river boats growling along with their cargo of barrels and hay bales, crowded housing districts with identical red tile roofs and multiple chimneys. The train was bloody fast, clean, with espresso and waffles being served from the mobile cart, and good newspaper and magazine racks in perhaps six languages. Many of the passengers appeared to be farmers with calloused hands and scuffed boots, but there were also wealthy French women with poodles, and a group of tourists from Nebraska wearing cowboy boots and talking too loud. I sat near two elderly ladies from Los Angeles, who were 'tickled pink' to find a South Dakota boy on this train, and they rattled on about the animated rodeo music in Wall Drug, western South Dakota. It is a small world.

By some miracle, I arrived at the Brussels airport at noon on 13 May 2000, with just enough time to buy some Liege Genievre Filliers (a spicy wine, maybe) as a gift for Sheila's uncle Freddie and aunt Ester, my next stop in England. By special arrangement, Ester's daughter Jackie and her husband Jerry hoisted a red bandana for me to locate them in the London Airport lobby. There was a quick exit to some busy four lane highway, and in an hour I was in The Honeycombe, the Chipping Camden cottage where Freddie and Ester lived. I was put on the sofa and subjected to a grilling of exactly where I'd been, while drinking a cooled ale from an antique mug. There were horses in the back yard, and Oswald the cat wanted to sleep in my lap. Ester began fussing about preparing supper, and she gave orders to Freddie to take me to the Baker's Arms for more ale, and a tour of various castles and properties.

We arrived back to find the table set elegantly with a roast leg of lamb, asparagus and roast potatoes, followed by sliced strawberries with clotted cream. We retired to the living room for the TV news and a serious discussion about Freddie's next book, more or less about the failure of the British Christian churches to inspire the young, the lack of challenging jobs for young people, and despair at England being a backwater (after having been a world leading nation).

You have to understand that uncle Freddie was a Scotsman born in India, was accustomed to British colonial service in India, Uganda and the Seychelles. Before and during World War II, he was in the 1st Infantry Division in France (Dunkirk), was awarded a MBE for his antics on the Anzio Beachhead landing, and was involved in sea landings in North Africa (Morocco, Libya) and Italy. Being in Transport, he had opportunity to travel around by jeep to visit the Vatican, and he once gave a ride to a hitchhiker who turned out to be a grandson of the King of Italy, of the Borghese family. He was subsequently invited to dinner at the royal family home, where truffle dogs were used to find the meal's first course. In 1945 he was in Palestine to assist the evacuation of the French from the Middle East. For someone who had lived such a dangerous and adventurous life, he felt that the current lives of his daughters and grandchildren were rather tame.

Sheila must have warned Aunt Ester about my needs at breakfast time, as the next morning I found the sideboard replete with soft boiled eggs, omelettes, bacon, blood sausage, small lamb chops, fried mushrooms and tomatoes, a warm loaf of whole wheat bread, three kinds of marmalade, two kinds of hot mango chutney, two pots of tea and one pot of strong black coffee. I was informed that this was a traditional English breakfast, and Freddie was enormously excited to join me in this feast (I suspect he didn't get it very often). Ester was delighted to watch us eat it all.

Meanwhile, I was given lectures on family history, much about Ester's grandfather being an administrator of Ross-Cromarty and Lewis, and then the maps of Scotland came out, with pencil markings on Invergordon Castle (now fallen to pieces), the Firth of Cromarty just north of Aberdeen where they have a big university and wonderful golf courses. The big ships would come into harbor in the firth, and the sailors would dance the hornpipes on the jetty. Aunt Ester was spouting bits of her family history that I was supposed to bring to Sheila. For example, 'Sheila's grandmother Annie Rooney was a bit strange, but animals were drawn to her'. Ester showed me a cane back chair in her sewing room, which had been in her family for at least four generations. Ester's rules of life: 1. One crossword puzzle daily, 2. Three glasses of red wine daily, and 3. Eat oily fish daily, all to prevent the onset of Alzheimer's disease.

Then I was put to work. They were very upset about their shower curtain, and I took the old one down and put up a frame and a curtain rod for the new curtain. I also helped to prepare a trellis for a recently planted clematis vine, which was curiously named President Billy Clinton. For mysterious reasons, Aunt Ester had Freddie bring cigar ashes from the pub, which she used to fertilise her yard plants. I also fixed several of their garden hose spigot fittings.

Freddie decided we needed another historical tour of the region: more stately estates in Burton, Blockley House, Chipping Camden, a history of the wool churches, World War II barracks being converted to craft centers, some Pictish or Celtic stone rings, and finally the Horse and Groom Pub, where they had cold Hook Ale, a meal in itself. The barmaid had just returned from holidays in Perth, Western Australia.

I had earlier explained that I had been in Belgium for a science meeting, and that I had visited a memorial and cemetery for the Allied Forces killed in action there during World War II. In a quiet corner table of the pub, uncle Freddie proceeded to tell me about his role in the Dunkirk Evacuation in late May/early June 1940, him being in the transport section of the British Infantry. This was not a tale of glory and heroes, but rather a royal military disaster. This operation consisted of a flotilla of fishing boats, pleasure craft, life boats, military barges, half-crippled war ships, and poorly led rescue troops trying to evacuate poorly trained English, French, Belgian and Dutch soldiers that were surrounded by the German military forces. In a speech to the House of Commons, British Prime Minister Winston Churchill called the events in France 'a colossal military disaster', saying 'the whole root and core and brain of the British Army' had been stranded at Dunkirk and seemed about to perish or be captured. The British Expeditionary Force lost 68,000 soldiers during the French campaign and had to abandon nearly all of their tanks, vehicles, and other equipment. But in eight days this ragtag flotilla of boats rescued 338,226 soldiers from the 'fires of Calais'. Freddie held strong views of the important and gallant role of UK in history, but this episode was a disgrace in his eyes. He went on to describe the failure of the British spirit, the stiff upper lip, during this part of the war. He appreciated the late US involvement in the war, and I gave him a short resume of my uncle Gerald's tour of duty in 1944-45. His comment was that the USA and the Russians turned the tide against the Nazi military machine.

Then it was back to The Honeycombe for smoked haddock chowder and a fruit salad with heavy cream. I was to catch a 10:30 flight to go home that night, so I packed up, with Ester already in tears and Freddie angry at the train being two minutes late. After some anxious moments switching from trains to bus to shuttles, I found my British Air gate at Heathrow. I had to eat again, medallions of overcooked cardboard beef with a cold beer, and then sleep until Singapore, where I had some rubbery scrambled eggs with lukewarm coffee. But they did have a nice cactus garden in the smoking area on the lobby roof, where the damp air smelt of tobacco and cloves. And then I'm on another airplane flying over Indonesian islands, the north coast of Australia, Alice Springs, and finally Brisbane at 0440 on 15 May, where it was too cold. For some reason, there were a lot of people with balloons, some of which were loudly popping, causing the airport security police to lock down the place. After shifting to the domestic terminal, I flew to Townsville where it was hot and sunny, and my brown beaming Sheila was there to meet me.

Chapter 42. Adventures in New Caledonia

June and July were now the middle of our winter. It was chilly for us, accustomed to the tropical heat, but this cool spell was essential for our citrus trees. We had wonderful white and red-meated grapefruits, mandarins, Seville and Valencia oranges, and two kinds of limes. We had a good wet season previously, so the grass and weeds were over my head in places. I spent a lot of time whippersnipping paths through the tall grass and piling the cut grass around my trees for mulch. It was dry now, so I was irrigating every weekend. There were several wildfires up on Mount Elliot, practically in our back yard. Sheila was feeding a baby fruit pigeon (called Froot Loop), an injured lorikeet (named Mulder), and the tame magpies were always in the house begging for treats. This made our house parrots Caspar and Charles upset, as they were becoming quite possessive and jealous of interlopers. Our old timey string band Meadowlark played at a Bluegrass Festival and two concerts in town.

At work in the office I was trying to prepare the data and slides for two upcoming science meetings, and drafting a journal paper on the carbon budget for the Herbert River catchment and continental shelf and slope. I managed to escape the inevitable catfights about funding between AIMS program leaders and the mahogany row managers by going to a meeting in New Caledonia with John Pfitzner. This was the annual meeting of the South Pacific Environmental Radiation Association (SPERA), and I was an invited speaker.

We left Townsville at 1130 on Saturday, 17 June 2000, while I was reading Salman Rushdie's *The Moor's Last Sigh.* We arrived at the Noumea Airport at 1700, where I could get some Senior Service cigarettes and French Francs (78F = Aus$1). SPERA vans whisked us off to the Hotel Lantana, where we had a good Indian curry supper. I walked the beach, found a shop that served mango ice cream, watched young kids playing soccer amidst roaring motorbikes and booming cabaret music coming from the nightclubs nearby. I tried to watch the CNN international news in French, but it made me go to sleep.

Small diversion: when I was in high school, we had a retired jazz musician for a school bandleader, Mr. Tommy Mathews. He kept playing for local dances with his jazz band, playing mostly saxes and clarinet. One of the tunes they played (and sang) was a real stomper blues called 'Caldonia'. But it was pronounced 'Caledonia', and I never did know who she was. It could have meant Scotland, too?? Anyway, here I was in New Caledonia, and apparently this island has nothing to do with this blues song.

We had come a day early so that we could have some collegial organisation meetings about TROPICS and other joint projects. I got to sleep until noon, had a wonderful 'Americain breakfast' at a beach shack in the warm sunny air, ripped apart by Rastafarian taxi vans with speakers playing some sort of French/Indian hip hop with sitar and tabla. There was a jet-ski slalom race competition roaring away in the nearby lagoon, and sail boarders farther out. And of course there were pasty white skinned Europeans, Chinese, Japanese and Americans sitting in the shade clutching their guide books and cameras.

About mid afternoon, Ron Szymczak, Paul Martin, Gary Hancock, Ross Jaffree, and John Twining (all Australians) arrived with booze demanding a bigger beach table. John Robbins and David Edgington (USA) and John Smith (Canada) arrived, and we all moved on to a good French restaurant where we ordered a haunch of local venison, big loaves of French bread and way too much wine. All of these people had helped John Pfitzner and me to establish the equipment, radiotracer spikes, and methods for our radiochemistry laboratory at AIMS. Robbins was an old friend who had helped me interpret ^{210}Pb profiles in Canadian Arctic lake and estuary sediment cores in the 1980s, and we were now using his computer models to interpret the chronology of sediment cores from northern Queensland and the coastal regions of Papua New Guinea.

This was one of those delicious and relaxed times when good friends and interesting scientists come together in strange places to trade stories of good and bad times in the lab and on ships, to discuss funding agencies and journal reviewers, and to acknowledge our success and failure in our work.

Well of course, we had to do some work. We had a week of science presentations, workshops and arguments about interpretations of weird results and recommendations for future activities. We met an Indian fellow from Suva in Fiji, who wanted us to help him with a coral reef coring problem. We heard too many welcome addresses from French and

New Caledonian government people in the French language, where with headphones and translations, we could only determine that they were embarrassed about the nickel mine pollution. There were papers from all of my colleagues above, as well as studies on Korean wetlands, Mexican coastal fisheries uptake of radionuclides, the latest biochemical aspects of nuclide uptake from the English lab at Lowestofft, contaminant supply history for the St Lawrence Great Lakes, John Smith's epic expedition in Canadian submarines underneath Arctic Archipelago ice, trying to track the nuclear power plant and fuel reprocessing wastes from Sellafield/Windscale in Cumbria, England. Several Russian papers described some inhumane working conditions on Arctic ice breakers, as well as trying to track the radioactive wastes from sunken nuclear powered military ships.

Strangely, there were few papers on the local coral reefs, which were under stress from nickel mine refinery wastes. Gary Hancock led an important sediment core dating workshop, which was dominated by my colleagues John Robbins and David Edgington. I tried to honour the early pioneers of this kind of work: Willard Libby on radiocarbon, and Ed Goldberg on using the uranium series nuclides in recent sediment accumulation studies.

There was a series of sessions about the chemistry and biology of the French nuclear bomb test radionuclides on Mururoa and Fangataufa atolls from 1966 to 1996. These south Pacific atolls are just wisps of coral gravel, 1-2 m above sea level, with a few coconuts and shrubs, surrounding a shallow lagoon (the submerged caldera of an undersea volcano). The first bomb test was named Aldebaran, a plutonium fission bomb, that sucked all the water from the lagoon, and it rained fish and molluscs and coral debris for half a day. Between 1966 and 1974, the French detonated forty-one atmospheric nuclear test bombs on Mururoa. Some of these bombs were exploded on floating barges, some were dropped from airplanes, some were exploded on steel towers, and some were suspended from balloons. The fission products of these bombs ($^{238,239}Pu$, ^{241}Am, ^{90}Sr, $^{134,137}Cs$) were distributed all over the south Pacific Ocean, in marine plants, fish and whales, from New Zealand to Peru. Due to protests from the Polynesian Territorial Assembly, New Zealand, and Australia, the French stopped atmospheric testing in 1974, and continued to test bombs in shafts drilled into the volcanic rock of these islands up to 1996, when the international nuclear test ban was signed. These 140 underground bomb tests generated several wide cracks in the atoll rim, some submarine landslides, and several tsunamis. Meanwhile, the Americans were busy exploding their bombs in the north Pacific Marshall Islands, to the northeast of Papua New Guinea. The British were happy to explode their bombs in the Australian desert and on the NW coast.

In the several decades since cessation of nuclear bomb testing, ecologists, oceanographers and radiochemists have monitored the distribution of the bomb nuclides in rainfall, on land, in the lagoon water and coral reef biota, in lake mud and the deep sea sediments. Physical oceanographers used several of these bomb isotopes as time tracers of ocean currents.

Our session on New Caledonia contributed dozens of presentations on all of this, as well as accounts of nuclide activity down the kilometer deep drilled test shafts on these atolls. Organisms that lived and fed in the coral lagoon sediments accumulated high activities of nuclides. The distribution of bomb fallout nuclides down cores of lake and ocean sediments, ice, and tree rings is now being used to help define a new era in the geological and ecological history of our planet, to be called the Anthropocene.

I got sleepy in the hot afternoons and would skip some session papers to walk downtown Noumea to find an electrical plug adaptor for my laptop, so I could write some science journal reviews (and our last *R/V Franklin* TROPICS cruise report) late at night. I also got a case of beer, tea, and some chocolate coated croissants, as I had a little fridge in my hotel room.

There was the inevitable tour of the nickel mine, the main business of the island. From the tropical paradise of the coastal beaches, the journey up to the mines in the chilly damp dreary mountains made me think of Tolkien's dwarves in the mines of Mordor.

The French had found the deep water harbor in the mid-1800s, and they found the settlement useful as a penal/slave colony and later on for mining of gold and nickel. The US Military had a South Pacific HQ here during World War II, which partly explains the wartime military influence on building and street names, and the sometimes embarrassing preference of the local Melanesian people for the Yankee culture, rather than the French. Melanesian shopkeepers correctly identified my accent as being from the midwest USA, and spoke kindly of US soldiers that had befriended their parents during the war. The upper class people of the government, university, police, church, and mining executives were stiffly proper snooty French, but the people in the streets were a wonderful mix of Melanesian, Polynesian and Indonesian.

I thought there was a special festival happening in one of the town squares, where bamboo marimbas, guitars, *garamudas*, dancers and singers were performing. There was hot bannock with butter and honey, pralines, and fruit in the vendor stalls, and the young people were in low slung baggy trousers with orange or green hair, boom boxes on their shoulders, and silver rings in their noses and bellybuttons. Very large *kanak* women in garishly coloured muumuu dresses were dancing with big smiles on their faces. Very stiff uniformed French police observed this riot of color and music and dance, and ramrod-backboned Catholic nuns looked on with disapproval as they marched through the food market with wicker baskets.

I asked what the holiday was, but the locals said it was like this all of the time!

There was also a Chinatown nearby, where I had a good spicy Hunan supper with cold beer. I bravely climbed on a garishly painted bus to get back to my hotel, this was like PNG...

there were crying babies, crates of poultry, dogs, old people clinging to the railings, with loud reggae music on the speakers. The driver would call out the stops in several languages, none of them English, and he had a string attached to the side door latch, where a big sign indicated 'Automatic Opening Door'.

French Colonial is not a furniture style here, and the native people do all they can to forget the history of their French masters. Downtown, the French street names are ignored, as the sections of town are actually part of the pre-French fifteen native kingdoms, which have their own dialects, music, dance, food style, and marriage customs. The young people have of course moved on, mostly bored, with greased braided shiny black hair underneath knitted wool skull caps, smoking little cigars, imitating some mixture of Jamaican Rastafarian and New York City gangsta rap swagger.

Finally the official science meeting was over, complete with a banquet under tents in one of the only green grass parks on the island. I was grateful to sleep in on Saturday, 24 June 2000, as it was cloudy and pouring rain. I had an excellent noon breakfast at a beach fruit stand under canvas, bacon and eggs, green tomato, hot peppers, plantain and mushrooms all in the same frying pan.

I was happily gazing at a sailboat race amongst the reef islands on Bae Citroelle when a fellow came up to my table and introduced himself as a geologist from the nickel mine (SLN = Societe le Nickel, a subsidiary of the large French Company Eramet). He wanted to know more about the Ok Tedi Mine and the Porgera Mine in PNG as he had heard my presentation about the radionuclide dating of the Fly River Delta sediments.

I pumped him for information on the effects of the New Caledonia mines on the local vegetation and nearby reef. He said that nickel mining has been the major sector of the island French economy since 1875, and the ground is full of it, 2-15% nickel over most of the red dusty land, and New Caledonia is the world's fifth largest supplier of nickel to European industry. The early settlers ground up the nickel-rich green garnierite and added it to whitewash or paint for their buildings, to protect against termites and other insects as a powerful pesticide. The whitewash was made from local outcrops of tremolite asbestos, which later studies demonstrated to be carcinogenic. Much of the upland soil is sterile and toxic to most plants, with high concentrations of iron, nickel, cobalt, lead and arsenic. There were two species of scrubby trees that were tolerant of these metals, but the sap of these trees contained ~20% nickel and cobalt, which was toxic to goats and cattle. There were a few patches of land with volcanic soil, where Charolais cattle and a few horses grazed on green grass. Mine and refinery wastes clogged the few river beds with metal and hydrocarbon rich contaminants, some of which was discharged into the coral reef lagoons surrounding these islands.

Previous to mining activities, this dry island probably discharged little sediment and metals into the lagoon, which was likely then a high biodiversity tropical reef wonderland.

Ok, so the island is not a tropical paradise today, despite the pasty white tourists on the beaches in front of the fancy hotels. These luxury hotels have shops with clocks imbedded in nautilus shells and other tourist junk with relatively high prices. My Hotel Lantana (like the weed) is cheap, low class, and had a shop full of toiletry and sun cream goods. The place was full of high school students and their teachers/chaperones trying to learn the French language. Our desk boy has a French Catholic chain and cross around his neck and he was reading a missionary newsletter for the entire week I was there. I was deliriously happy to notice a small group of cattlemen at my hotel, wearing high boots, wide belts, Levis, and walking with a horseman's swagger.

John Pfitzner and I (and many other scientist participants) took the 4 pm bus to the airport, and we left in the pouring rain to arrive in chilly Brisbane two hours later. We had to overnight in a little hotel near the train station, where it was a wonderful comfort to have the English language on the TV. The next morning we boarded a flight to Townsville, arriving in the warmth and perfumed tropical paradise of home.

Chapter 43. The Sierra Nevadas, Lake Tahoe and Dangerous Dan McGrew

June, July and August 2000 were unusually chilly for my tropical blood. At home, I was wearing my old ragged moose hunting black and red checked quilted shirt, track pants, and woolly ankle boots in the evenings. The bananas and pawpaws refused to ripen, and there were wildfires on the mountains near us. Ian MacGreggor and I attended part of the Australian Festival of Chamber Music in July, where we listened to some stunning modern music by Vine, Sculthorpe, and Sitsky.

At the office, Irena and John were helping me understand and graph the data from our Gulf of Papua sediment cores, in preparation for two conference presentations in the next few months.

My colleagues Nic Harvey (University of Adelaide, Mawson Centre) and Tony Belperio (Minotaur Exploration Ltd., Adelaide) arrived in July for another effort to drill core the Cocoa Creek saltpans and chenier ridges. They were interested mainly in the timing and variations of sea level rise since the last deglaciation about 12,000 years ago. This area near AIMS is part of a national park and aboriginal reserve. We already had a permit to drill from Queensland Parks and Wildlife, but our AIMS administrator Peter Isdale asked us to visit the local offices of the aboriginal people. The Bindal and Wulgurukaba people visited us on the salt flats, approved of our proposed saltpan drill sites, and explained that the slopes of nearby Mount Cleveland had many sacred caves with rock paintings and burial crevasses. The next morning we had phone calls from two other aboriginal tribes, who also claimed ownership of that region, and did not approve of our work. We had contracted a bore driller from Ayr, who apparently vanished and didn't answer his phone. So we surveyed several sites for some future drilling and gave up on this project, frustrated, with research money and time wasted.

Meantime, my AIMS colleague Kathy Burns was the chief organiser of an international meeting of organic geochemists in a downtown hotel. I put in a poster about our research on dioxins in one of my sediment cores from Bowling Green Bay. My part in this work was to provide the sediment core chronology, and my expert collaborators Caroline Gaus and Jochen Muller (University of Queensland, Brisbane) did all the difficult measurements of these highly toxic dioxins. It was (and still is) a mystery as to where these complex organic contaminants came from, as they are usually associated with the plastics industry and combustion of rubbish in big cities. I also gave a paper on the organic carbon budget for the Gulf of Papua at this meeting.

Sheila and I had been married for thirty years on 8 September 2000. With our son Ian, we borrowed a small boat from Cungulla friends Sylvia and Kevin Harney and puttered up the Haughton River/Doughboy Creek into Burrumbush Creek. We spotted a 3+ m crocodile sleeping on the creek bank, and lots of mud crabs and fish. After nearly getting lost on return, Sylvia and Kevin opened up a bottle of champagne for us, as they had been married for thirty-one years. We went into town for supper at a fancy Indian restaurant, Sheila with hibiscus flowers in her hair.

Our TROPICS research program had worked very well for four years for scientists in Australia, Papua New Guinea, Indonesia, and the USA. The USA funding agency NSF now requested a big reporting session from all of us, including what we wanted to do next. Chuck Nittrouer called the meeting to be held at a resort on Lake Tahoe. Clever fellow. And so on 9 September 2000 at 0500 Sheila drove me to the airport through smoke and fog. The 0600 scheduled flight was delayed for three hours, due to the poor visibility from the smoke and sea fog. And then off we go again, with all the delays and frustrations of travel, to Brisbane, Sydney, and then an arrival in Los Angeles at 0905 on Saturday, 9 September 2000, an hour before I left Townsville.

The city of angels was covered with smog, a rusty grey sickly sky, and the air outside the airport was mostly diesel fumes. Inside the airport were Arabs in long flowing white gowns and *keffiyeh* headdresses, teenagers with orange hair and baggy pants, Japanese sport teams, and slick black suited business men with cell phones glued to the side of their heads. After a lunch of fried chicken and Seattle Pale Ale, and exchanging Aussie dollars for US dollars, I snoozed at the American Airlines gate. Then off at 1550 to Reno, Nevada, reading Salman Rushdie's *The Ground Beneath her Feet*. The California lowland was all irrigated cropland, giving way to the bare naked deforested Sierra Nevada mountain tops and the arid desert plateau of Nevada.

Arriving in Reno at seven pm, the path to the luggage carousel was lined with slot machines and gaming tables. The blue tint brigade was busy gambling, and there were Nevada cowboys with hats and high heeled boots that I'm sure had never ridden a horse. There were dozens of stretch limos ready to pick up the highroller gamblers. As the sun went down, the garish hotel lights came on, covering up the shoddy and fake decorations. I wandered into the bus station, bought a $34 RT ticket to South Tahoe, and was grateful to see clean pine forests and clean air up in these Sierra Nevada mountains. This is also a tourist town with big billboards advertising the casinos, country and western music bands, and very posh resorts. Chuck had booked me into a fancy hotel, where I had an excellent steak and my room TV featured country music by Lyle Lovett, Michael Martin Murphy, Guy Clark, Townes van Zandt, and the Sam Bush bluegrass band.

I woke up on my birthday with a headache and a cold. But I did find a decent breakfast of bacon and eggs and white gravy on hot baking powder biscuits, down beside a Lake Tahoe boat launch, where fancy boats were tied up. This place is at 1897 m elevation, with snow on the mountain tops. The lake is 500 m deep, and the water was chilly.

The Truckee River is the lake's only outlet. The Truckee flows northeast through Reno, Nevada, and into Pyramid Lake, Nevada, which has no outlet. Only one third of the water that leaves the lake leaves via the Truckee, however, the rest evaporates from the lake's vast surface. Charles Goldman of the University of California at Davis led research on this lake for decades. He was a close friend (and naïve gambler buddy) of my Cornell Graduate School professor Jack Vallentyne. Dear Sheila sings the ballad 'Darcy Farrow' about this part of the world.

There were Bombardier motor ski races, Canada geese, and wealthy people fishing with big rods and reels. The *Tahoe Queen* was a big paddle wheeler in the New Orleans style, with a Dixieland band playing on the aft deck. Many of my colleagues were here, another gathering of the oceanography crowd, and we were all bussed to Camp Richardson for the big meeting with funding agencies.

Most scientists are not accustomed to luxury and wealth. Camp Richardson was mostly 1950s style cheap tarpaper log cabins and a small town community centre. For our purposes, it was excellent, being away from phones, computers, TV, pleasure sports and diversions. Our purpose was to invent a new research program that would be seen as a responsible and useful continuation of our TROPICS research and several other big NSF research programs, and that fit the desires of the big funding agencies in the USA, Europe, and SE Asia/Oceania. A dozen representatives from the big funding agencies were in our audience, and often grilled us on how practical different tasks and locations would be.

We also made plans for joint publication of work done to date, and to form a series of educational lectures for schools. The general plan was to create a 'Source to Sink' model, the sediment and organic matter sources being the inland mountains, and the sink being the river deltas, continental shelves and slopes. Our TROPICS group was augmented by Hugh Davies from the University of Papua New Guinea, Joe Beluka from the PNG Geological Survey, Arnold Moi from Ok Tedi Mining Ltd., Bill Dietrich from Berkeley, Gary Parker from University of Minnesota, Dave Prior from Texas, which covered most of the big river catchments, whereas our TROPICS group covered the river deltas and shelves. John Milliman, J. P. Walsh and Steve Kuehl made the strong point that the island of New Guinea and the Fly, Kikori, Purari, Sepik, Ramu, Mamberamo, and Digul carry the largest flux of sediment in the southern hemisphere, some 80-100 million tonnes/yr. We had lots of very good competition from groups working in the Gulf of Alaska, and the east coast (Waipaoa River and shelf) of New Zealand.

Of course there was a break-away rebel group of biogeochemists, led by Bob Aller, Kathleen Ruttenberg, Miquel Goni and myself. We filled blackboards with flow diagrams of all the different kinds of organic matter and chemical elements that are transported (slowly and quickly, in dissolved and particulate form) from the mountain ranges to the river deltas and coastal shelves. We made grand plans and funding proposals for the next decade, with periodic reporting sessions at the big AGU meetings, and sequential attempts to create geochemical mass balance models for big rivers draining continents with different rainfall, vegetation, tectonics, and geology. We didn't find out till a few months later, but NSF decided to fund our Gulf of Papua work and the New Zealand Waipaoa River study under the umbrella of a pre-existing NSF funded MARGINS project. All of this might be boring to you, dear reader, but this is how large scale scientific projects get organised and funded.

The serious work was over on 14 September, and we were bussed back to South Tahoe and the paddle wheeler 'Tahoe Queen'. We feasted on hot chili guacamole, pale ale, grilled fish and smoked BBQ pork belly while the Dixieland band played on into the night and the paddle wheels churned the lake water. Sir Charles Goldman appeared, all dressed in black with a black cape and an English style duffer cap. He gave a wonderful lecture about the history of research on Lake Tahoe, and his own studies 1958-2000. He knew me, so he sat with my TROPICS group for supper, and told stories of his 1970s consultancy work about the proposed Mitsubishi hydroelectric dam and aluminum refinery on the upstream Purari River. The local Purari people had no interest in this development, and Charles claimed he stopped the project before the river could be damaged. He also told stories of Bill Clinton and Al Gore visiting Lake Tahoe for some secret meeting with foreign gents. Charles was openly supporting Al Gore for President, and told several Clinton jokes.

Sir Charles always did have a good sense of humour. He got our table to participate in a pantomime on Robert Service's 'The Shooting of Dan McGrew'. We used my Alligator

Creek Roadhouse cigarette lighter and a wine bottle cork to make dark face paint. Kathleen Ruttenberg had some lipstick for our cheeks and lips, and Sir Charles took the microphone to recite the poem by memory, with a decided gleam in his eyes. He had a little tape recorder with some terrible detuned piano beer parlour music on it, and some paper bags inflated and slapped to simulate pistol shots. He had a good repertoire of sound effects from his mouth and hands. We were assigned the actor roles, and we moved with the story, as the audience hooted and hollered and whistled as the lady that's known as Lou (in a tight split skirt) took the gold poke from the dead men. All of this was going on as the paddle wheeler was chugging along in the middle of the lake after midnight. After this little performance, and after the paddle wheeler was docked, most of us went to wash our faces, and accompanied Sir Charles to the parking lot where his car was parked. He popped open the trunk (boot) to get his musical fish, and did two very brief songs (Bob Marley's 'One Love', and 'The Fishing Pole Blues'). His musical fish was a glittering fish shape with mandolin strings from the head to the tail to the anus. This whole episode was so bizarre it blew my mind, and I had a hard time calming down to get some sleep.

We all lugged our gear to the roadhouse at 0300 on 16 September, for the bus to Reno and the airport, where we all split up, the scientific family dispersed once again, promising to gather again on research ships and at meetings of the future.

And then I was off to visit my South Dakota family, flying to Denver, where all flights were cancelled due to a security scare. I had my usual cowboy feast in the upstairs smokers bar and lounge, good cornmeal smothered BBQ pork ribs and beer with lots of cigar smoke and loud country music, while watching a Purdue/Notre Dame football game. Eventually I got a flight out to Rapid City, flying over the Colorado prairie, Black Hills, the South Dakota prairie brown and clean and peaceful. And there was my brother Obie and his daughter Donna Sue (now a school teacher in Custer, So. Dak.) waiting for me, and I'm six hours behind schedule. Never mind, they had orders from Sheila to go shopping to buy Levis for Jesse and a Carhartt jacket for Ian MacGreggor. They whisked me along to the back roads beside the Bad River to reach their ranch house, where an enormous amount of food was being prepared by two more of Obie's daughters, Marcy and Julie. We ate fresh picked sweet corn, an enormous roast of beef, large home garden tomatoes for supper with cinnamon apple pies for dessert. And then we all sat on the porch talking until midnight. It is really hard on the brain to have such huge shifts in conversation and locational settings.

They kindly let me sleep in, but I woke up on Sunday to find the house and yard and barn filled with relatives, all having different kinds of breakfast. I had homemade sausages and fried eggs with home made still-warm bread with chokecherry jelly. Marcy, Donna, and Julie had taken over Obie and Annie's kitchen, and very serious food production operation was happening. All of Obie's daughters are very forceful, efficient, quick, and knowledgeable! They forcefully, kindly, efficiently put me out of the way in the tractor shed to cut up a wheelbarrow load of muskmelon (and save the seeds for next year). By noon, the place was packed with uncles and aunts and their kid's families, Toadie (my mother's distant cousin from Minnesota), Obie's three daughters' families, and my brother John's family, and it was a comfortable heart-warming madhouse! There were at least a dozen small kids there, offspring of my nieces and nephews. Now I begin to understand the troubles of my grandmothers in identifying the names and orgin of all those gremlins playing in the dirt. There were twenty-eight people here, a major reunion! Outdoors it was blowing hard and 110°F.

Again I was assigned to the tractor shed to supervise too many kids around two electric powered ice cream makers and two large grills covered with odd slabs of beef, antelope chitlins, venison sausages and burgers. Part of this was for a birthday party for Jackson (son of Julie) and myself. I got presents of six fly swatters and a Custer School t-shirt. As it cooled down, I gave the kids some Australian boomerangs, which they threw far into Obie's horse pasture, frightening them, causing snorting and bucking fits. They demanded explanations of kangaroos and possums and Australian habits. One of Annie's sons, John Paul, phoned the house party from Fort Houachouka in Arizona, where he was in 'spook school' for a military intelligence unit of the US Army. He had found his niche for his interest in computers and information technology.

The next day most people packed up and left before noon. Obie and Jackson fixed the float in the stock tank in the cattle pasture, the nearby stock dam being nearly dry. The water here comes in a pipeline from a deep artesian well north of the town of Wall. The well water was warm and high in iron content. Obie got me into his high suspension pickup truck and we drove to Philip for some business, and then to Murdo for lunch at the Skelly Truck Stop, where most of the waitresses and coffee drinkers knew Obie, and were a bit puzzled to know that his older brother now lived in Australia. 'Why?' they asked. We visited my youngest brother John at his Jones Country Court House Auditor's desk, where he actually had a big desktop computer working on a spreadsheet. I was surprised at this, as John always said he hated computers. It took us about twenty minutes to tour the town, observing many main street businesses shut down, and many homes of families I used to know being empty.

We then drove twenty miles south (past Don Hight's longhorn cattle herd) to visit my uncle Gerald and aunt Joyce on the White River ranch, where I grew up. Gerald was now seventy-nine years old, frail and thin, and had survived several serious surgical operations. Joyce provided tea and Boston date bread with butter and honey. Gerald was upset by Ted

Turner (of CNN fame) who had bought many of the big ranch properties in western North and South Dakota, intending to run buffalo on them, all the way to Texas. 'That guy hasn't the slightest ability to take care of livestock!' Joyce was mad at Governor Janklow...'He's so dumb he couldn't pour piss out of his boot if there wasn't directions on the heel.' Despite them being old and frail, they had a well weeded vegetable and melon garden, and the windmill kept the stock tank full of cool clear water for three saddle horses in the corral.

By evening we were back at the England/Brunskill ranch near Murdo, where my brother John had his grill loaded with twenty-one day aged corn-fed Angus beef steaks, about two inches thick. Normal people might consider these slabs of meat to be small roasts! Everybody was certainly well fed in this part of the world!

The brothers sat on the front porch talking about old times while being served plates of lemon cake, grasshopper pie (limes and cream), and tea. My brother John had an Australian cattle dog named Gypsy Rose, which was apparently very useful to move cattle where wanted. John was stiff with arthritis and had a hard time stepping into the saddle stirrup to get onto a horse. Gypsy came to lick my hands and laid down at my feet: this upset John, as this dog was supposed to be a one man dog.

We talked about our grandparents and parents, and how the England and Brunskill properties were not properly handed down to the family. The town lawyer C. D. Kell had split the ownership, and it was still confusing to me. I was surprised to hear that my Grandpa George not only owned the White River ranch, but two farm properties in Iowa, which were now inherited by my uncle Gerald and aunt Margie.

There was a big brass cuspidor on the porch, which would give a rich 'ping' when a stream of tobacco juice was spat toward it.

The next day John, Obie, Marcy and I drove down to visit my aunt Margie and uncle Bernard at their ranch house on Prosperity Flats on the south side of the White River. Margie was well prepared for us, with two cakes, sour cream raisin pie, tea and coffee, and a big box of old photos she had rescued from relatives in Iowa. One photo showed my grandpa George Brunskill in 1906 with his Haywarden (Iowa) high school classmates labelled 'hooligans', with cigars and beer bottles. There was the usual talk of cattle auction prices, genetically modified grain and beef, and how the prairie grass was changing in nutrient value. Marcy talked Margie into giving me some of her precious buffalo berry jam, handpicked by Bernard on horseback down by the river.

And then it was time to pack up and go: Obie and Marcy helped me to pack the buffalo berry and chokecherry jam in several sacks of bubble plastic, and some books and other stuff for Sheila, Ian and Jesse, all in two nested cardboard boxes...as if I couldn't possibly do this myself. Obie was the rope man, insisted on tying up the box with fancy knots. I hoped Customs would not ask me to open it.

I spent some nostalgic time in my grandmother (and mother's) piano room, which was kept like a shrine behind closed doors most times. Grandmother's piano sat there as always (now drifting out of tune), with little white ceramic busts of the major composers lined up neatly on the piano top. The piano room closet still had piles of music books for teaching, and some of the antique 78 rpm records that I remembered from my youth. My mother had hung high school graduation photos of her four sons on the wall beside the bookcase, and there we all were, looking so innocent and honest and clean. The bookcase held my mother's collection of classic literature and art, some early obscure volumes on South Dakota exploration and history, and some of my Augustana College books on art history and creative writing. Down in the basement, John had built major book shelves to store most of our mother's collection of family treasures, boxes of old photos, and some of my Canadian cold weather gear. Over in a corner of the basement was my father's little cupboard and writing table, where he kept his livestock records, vaccination vials and syringes, the bloat tool, sharpening stones and jackknives. As usual, this moved me deeply. I choked up, as I wasn't sure I'd ever see this again.

John drove me to Rapid City in his pickup truck, a tin can between his legs for spitting the snuss juice. Obie's daughter Donna Sue met us at the airport in a smart elegant black dress, handing me another box of Levis and Carhartt jackets for Ian and Jesse.

And then it was back to life on airplanes. I left Rapid City at four pm on 20 September 2000 in a little twin propeller airplane on a bright and sunny day, the prairie stretched out golden brown, and the Black Hills being appropriately black. And then Denver, and then Los Angeles in the grey smog, with the normal strange collection of worried frantic people. It comforted me to find my Qantas gate where Australian accents were strong, and then it was fifteen hours flying time to Melbourne, where I had a wonderful large plate of bacon and eggs with strong coffee. The Customs man said I looked like Jesus Christ and waved me on, without opening my boxes and suitcases, even when I declared the buffalo berry jam. Melbourne airport is a nice place, with good book stores, coffee shops with pastries and big soft sofas for snoozing. The place was crawling with police and airport cops, due to some protesters marching outside. And then off to Brisbane, and finally Townsville, where the tropical air was warm and perfumed, and dear Sheila was there beaming to haul me back to the peaceful sanity of Alligator Creek.

Chapter 44. Hot and Sweaty in Papua New Guinea and Indonesia, Again!

Our tropical life was enriched by a good wet season in January and February 2000, but now in November we were getting even more rainfall. The road north and south of us was flooded, and a huge semi-trailer of Tully bananas destined for the Brisbane/Sydney market tipped over in a metre of water over the nearby Killymoon Bridge on the Bruce Highway. Tonnes of bananas were bulldozed off the road into the flooded creek water. Downstream herbivorous fish must have been happy. AIMS let staff off work for a week because the road was practically closed to traffic. This allowed me to whippersnip the tall guinea grass and spread 100 kg of NPK fertiliser on my orchard trees.

Son Ian was with us, and he was now teaching and playing with several jazz groups in town. He made enough money to buy a used diesel Toyota van, which was big enough to haul around all the drum and sound gear needed for a working band. Son Jesse was out in the dry lands west of Mount Isa, a project engineer improving the highway towards Camooweal. And of course it wasn't dry. The rain flooded much of their work.

In October 2000 I was involved in the 4th Herbert River Integrations Study Workshop, led by Christian Roth of CSIRO. The grand plan was to bring together the environment, agriculture, town, mining, tourism, and coastal marine interests, and to write a book on the social, agricultural, and environmental status of this region. We had every national and state government office as well as the local people represented in our workshop.

The Herbert River catchment includes sugar cane and banana farms on the floodplain and cattle and mining in the interior uplands. The Herbert River delta and Hinchinbrook Channel are rich in mangrove and estuarine biota, and farther offshore are healthy coral reefs. I enjoyed contributing to this effort, but it was cumbersome with so many players. Graduate students Greg Doherty and Joanne Cavanagh helped me gather historical data on the number of cattle on inland properties, fuel, fertiliser, and crop tonnage of sugar cane farming, and town business growth and population variations.

For the past decade, there had been disputes between the PNG villages on the floodplain of the Fly River, the PNG Government in Port Moresby, and the Ok Tedi Mining Ltd operations by BHP. This was due to the 1984 failure of tailings dams in the Star Mountains below the mine, causing the flooding of ~1000 km of the river with mine waste sediment. According to BHP records, they dumped 60-80 million tonnes per year of mine waste into the river for several decades. This elevated the river bed, so that more of the high water overflowed the banks, and delivered mine sediment and contaminants to the floodplain village gardens, water supply and fisheries. Approximately 1600 km^2 of floodplain grassland/forest died from the sediment and metal stress.

Some of the villagers got legal help from the Australian law firm Slater and Gordon in the 1990s, as PNG land is held under a system called 'native title', and the villages claim to own this land. The national government in Port Moresby retained control over resources under the ground, and claimed taxes from the miners.

This was the complex situation to which I was called to assist both the PNG Government and the Ok Tedi Mine (OTML, BHP). They wanted a report from our TROPICS research project on the impact of the mine wastes that reached the estuary and continental shelf of the Gulf of Papua. As usual, the cautious AIMS bosses and the Australian Department of Foreign Affairs and Trade were reluctant for me to do this, as this could be seen as interference in the affairs of a foreign country. However, both the PNG Geological Survey and OTML provided a letter of invitation to me, which calmed the politicians somewhat.

So on 14 November 2000, off I went with a power point projector and three CD copies of my presentation. I arrived in the Port Moresby sweatbox airport at two pm. Ladies were selling mud crabs on the sidewalk just in front of the taxi station. My hotel room was tiny, hot, with an overhead fan and a shower but no towels. The little hotel was surrounded by a high heavy iron fence with one guarded gate. Good friends Patricia Pepina and Joe Buleka came to take me out for supper of wonderful reef fish and giant prawns, hot baked bread, and cold beer.

The conference was sponsored by the National Research Committee (Pepina) and the PNG Geological Survey (Buleka), and it was scheduled to begin at eight am the next day. I was there early to set up my power point projector, but the doors were padlocked. Nobody showed up until about 10 am, at which time I was drowning in sweat. In true PNG style, many of the participants interrupted my presentation to ask pertinent questions, which was very good. There were many anti-mine green environmentalists present, some representatives from villages from along the southern coast, and a few old fellows from the Gulf of Papua fishing industry. Also present were slick suited newspaper men, Indonesians and some PNG Mineral Resources agents. My main message was that most of the OTML and Porgera Mine contaminants were in the freshwater floodplain lakes and forest floor, and that we hadn't yet detected much contaminated mud in the seaward estuary or shelf of the Gulf of Papua. This contaminated mud was likely to be a problem for the next century, and there was little or nothing the villagers could do about this problem. It seemed likely that some of this mine waste mud would move downstream slowly over the next century, and accumulate in the mangrove forests of the Fly River estuary.

My minder Kayemene—a village fisherman from Lake Murray, on the Strickland River, eastern tributary to the Fly River—hauled me to the airport, quietly telling me that he and other Fly River people were grateful for my presentation, as they don't get much solid information from the local politicians or the mining companies.

There an OTML twin propeller Fokker (the *Pride of Tabubil*) was waiting for me, and some of their employees coming back to work. We flew over the south coast, the delta of the Purari River, Lake Kotubu, several volcanoes, and up the Fly River into the Star Mountains. Here the mining town of Tabubil sits at elevation 460 m in the dense rainforest jungle, beside the decapitated mountain, and the mine waste sediment laden streams (Ok Ma). It is one of the wettest places on earth, with an average of 8 metres of rainfall per year. Higher in the mountains was the rain fog forest, home of exotic flora and fauna.

There were perhaps 12,000 people in the town, a wild mixture of the native mountain people (the Min Nation), Indonesian and Australian engineers, some Canadians and Americans, and families of the mine employees. As in all mining towns, the mine officials run everything, and the airport was guarded by armed security forces. I was quickly hauled to my 'donga', and then to the mess hall, operated by the Pong Brothers. The next morning I was led around the OTML Environmental Laboratory, where they had a large enclosure with two pet tree kangaroo, seven parrots, a big cassowary, and a lot of possums. The lab was well equipped, having some analytical equipment better than my own lab at AIMS, and a good small library. I got introduced to the geochemists, geomorphologists, hydrologists, mangrove ecologists, and two fish biologists. I gave a two hour presentation to twenty people, and found that most of the environmental lab people already knew more than I did about the freshwater river and floodplain ecosystems. My real expertise was of course the estuary and continental shelf, about which they had little information. The Environmental Lab staff were positively impressed with our work in the estuary, but the OTML bosses and lawyers were highly critical of my predictions of the fate of the mine waste sediment along the river and floodplain. I predicted that the fine grained fraction of the mine waste would reach the estuary and inner shelf within two or three decades. Both the scientific staff and legal beavers were obviously in preparatory mode to defend the mine from lawsuits and land claims.

After a good Pong Brothers supper of sweet potatoes and Goroka beef, I walked the streets of this little town. Men, women, and kids were sitting quietly in the shade of big trees, talking, singing and playing games until midnight. OTML Security trucks patrolled the streets, and there was no sign of the 'rascals' common in Port Moresby. There was a big general store for groceries and hardware, complete with a barefoot Santa Claus ringing a bell, Christmas wreaths, Madonna and David Bowie T-shirts, lots of kids with Rastafarian braided hair and a Christian book shop that looked completely out of place. The craft shop sold knitted skull caps, machetes bent in curious shapes, woven dilly bags, some very curious chess pieces, and giant woks about a meter in diameter. The local Min men wore brightly coloured skull caps and not much else. Some were proudly wearing penis gourds. The Min ladies wore bright knitted shawls, even in the blinding mid-day heat and humidity. I bought a six-pack of South Pacific #1 (the only beer on sale), and went back to my air conditioned donga to read E. O. Wilson's *Consilience*.

The next morning, 17 November, I was hustled to the airstrip at 0630, only to find that the aircraft had already left town. The next flight to Moresby was at noon, and cost me K622. It was fun to fly down the main stem of the Fly River to the copper port at Kiunga and on to Daru just outside the Fly River Delta, a quick 1000 km lesson in big river hydrology and floodplain dynamics.

We were lucky to catch another small plane out of Daru, and reached Port Moresby at 4 pm, where the airport was in its usual state of chaos. I had two hot meat pies and a warm beer to keep me going.

The local PNG people hadn't quite got the hang of their new airport facilities. The inside was neat and clean but nearly empty of people, and two large duty free shops and restaurants were closed. Outside on the sidewalk and road margin were over two hundred people happily sitting and talking, and vendors were selling food and drinks.

Finally we got a flight across the Coral Sea to Cairns, where Australian Customs and Immigration searched my bags thoroughly, being a little disappointed that they didn't find any bad stuff. I had an overnight in a stuffy little motel, where I had to roust the kitchen staff to get a meal and a beer. The next morning I boarded a plane for Townsville in the pouring rain, arriving in more pouring rain at noon, and there was my sweetheart Sheila beaming and damp, ready to take me home.

The wet season in Alligator Creek was truly in full blast this year. Our little creek was flowing a banker, and the guinea grass in my orchard was taller than me. Roger Green and his wife Celina visited us in early December. Roger was a mathematical ecologist, and an older classmate of mine in graduate school at Cornell University. He gave lectures at JCU and AIMS, and he was here to honour a retirement party for Dr. Howard Choat. It was startling to see him with white hair and a sore leg, having retired from his professorship at the University of Western Ontario. He was a bit cranky after our galah Caspar bit him on the cheek, but he was interested in son Ian's drum collection and information about the Evergreen Club Gamelan in Canada.

I was also busy with two Ph.D. thesis exams and exit seminars from Greg Doherty (Cleveland Bay trace element chemistry of sediments) and Joanne Cavanagh (pesticides in floodplain soil, mangrove mud, and Hinchinbrook Channel sediment cores).

For exactly the same reasons as the trip to Lake Tahoe, I had to go to Jakarta to crank up the INDOTROPICS research program, so on 12 December in the pouring rain, I boarded flights for Cairns, Darwin, Singapore and Jakarta. I met several of my Australian TROPICS colleagues on the roof top smoker lounge in Singapore, where we made plans for our presentations. It was sad to hear from the Hobart Marine scientists that our Australian National Facility research ship R/V *Lady Franklin* had been sold. I had twisted my ankle in one of these airports, and my ankle swelled up alarmingly. Our pilot had a strong Irish accent and we bounced around through some storms approaching Jakarta in the dark. Luggage from the aircraft cabin overhead bins were flying through the air, brief cases flung open, spilling contents down the walkway. We were two hours late from our schedule, so our LIPI reception transport was absent, and we got to our hotel by cab.

Our hotel had a high steel fence around it and the cab was not allowed to go through the guarded gate. It was a lavish hotel full of carved wood structures with smoky clove and sandalwood atmosphere. Our Indonesian friends Gani and Meriwether rushed in to find us at the supper table enjoying Bintang and prawns. Later that night we listened to CNN TV News describing the US Supreme Court ruling on the USA election tie in Florida, giving the Presidency to George W. Bush instead of Al Gore. The elevation of Bush to the Presidency was a terrible mistake, as the following decades would reveal. Many have wondered about what might have happened on the alternate path under intelligent Al Gore, which would probably have avoided the next three decades of war and refugee problems in the Middle East.

The next morning we were up early for a 0700 Indonesian breakfast of spicy rice porridge, bonito tuna chips, curry soup, and rice. Gani hauled us to LIPI (the main science organisation of Indonesia) and a big room with humming ceiling fans. Several important dignitaries, probably the BPPT officers that would fund future research cruises, sat at a magnificently carved table and set of chairs on a high platform. About ten INDOTROPICS scientists gave their presentations for continued work on the Digul and Mamberamo River estuaries in Irian Jaya. Much of the discussion was in Baha Indonesian, and we didn't have a clue what was being said. Gani later explained that, as usual, the funding officers wanted to know when they could see published papers and books. Ron Szymczak and Denis Mackay gave short presentations on their work in the Gulf of Papua and the Sepik River estuary in the Bismarck Sea, and I tried to summarise the work of the USA oceanographers. There was a large audience, including some newspaper and TV reporters. All of this went on to four pm, with no break for tea or lunch, as it was Ramadan. Wives of the LIPI scientists were hosting a bazaar in the lobby, selling crafts, sweets, books and art works.

We were escorted to the Jakarta shipyards/wharfs, where the LIPI oceanographic ships were tied up. These ships were operated by the Navy, and were bigger and better than our Australian research ships.

We toured the *Baruna Jaya 8*, a new ship recently arrived from Norway. The ship captain spoke excellent English, and had been educated in England. This was a 53 X 12 metre ship outfitted with equipment for geophysical and physical oceanography with the latest and best seismic and gravity mapping gear, seafloor imaging and mapping gear, a moon pool, port and aft A frames and big piston coring gear. I was having difficulty walking up and down the deck stairs on this big ship, because my swollen ankle was painful and sore. A huge galley and mess room for a crew of forty served hot coffee and little cinnamon pastries to us Australian infidels. There were big computer rooms, but the computers were still in their shipping crates. The Moslem call for prayer came out over loudspeakers on the ship.

By the time I got back to my hotel room I was starving, and I ate two separate suppers: honey pigeon and rice, and then another plate of satay seafood, all with pickles, sauces, bread, and a two litre pitcher of Bintang pilsner. This hotel was obviously accustomed to providing services for non-Muslims while the Muslims were fasting for Ramadan.

The next day I was back at LIPI, only for more informal discussions with individual scientists in their offices and labs. We exchanged reprints and CD data disks, and I offered to help with getting their work published in some of the big journals. I got back to my hotel to find two of my books had been stolen. I ate another big supper of chopped sea cucumber (Holothuroidea), a spicy concoction stuffed in a fish stomach, a few giant prawns, and several Bintang. And then off to the airport for a 6 pm flight to Singapore, where a group of Indian ladies in sari garb with baseball caps on backward were singing Christmas carols. Then more living in airplanes, to Darwin, Cairns, and Townsville by 10 am on 15 December, my ankle ballooned up inside tensor bandages.

At home for the Christmas and New Year's holidays, Jesse and Sarah joined us for several wonderful feasts by Sheila in our pergola. We made grapefruit ice cream (sherbet), a new concoction, and were treated to the spectacle of Ian and Jesse playing cricket. Neither of them has the slightest idea of how to play cricket, but they did have a bat and the wicket.

Chapter 45. Adventures in Port Moresby, and The Twin Towers Come Down

The *JOIDES Resolution* tied up on a Townsville wharf in January 2001, the biggest oceanographic ship I had ever seen. The chief scientist was my colleague Ray Binns of the Economic Geology Project of CSIRO in Sydney and it carried a crew of 106, which dwarfed our research ships.

The ship had been at sea continuously from 1985 to the present, and was just returning from an effort to drill core active hydrothermal vents north of Papua New Guinea, south of New Britain, near Manus Island. The high drill tower in the middle of the ship and four powerful bow and stern thrusters allowed the ship to maintain position and drill kilometers into the seafloor. However, in this cruise they had lost the drill bit, the temperature logger, and most of the drill casing, with only 16 m of core recovery.

Over the past two years, I had been visited by university and church groups from the north coast of Papua New Guinea concerning the problem of waste disposal from a large nickel and cobalt mine and refinery being proposed for Basamuk Bay, a small native village 40 km SE of Madang. I was useful to them, as we had good oceanographic and sedimentary information about the estuary and submarine canyon of the nearby Sepik River. JCU biologist Dr. Marcus Sheaves, Flinders University physical oceanographer John Luick and myself were contracted by the NGO Mineral Policy Institute of Sydney and the German Lutheran Church of Madang to provide expert review of a large pile of documents about the mine, refinery, and waste disposal engineering proposals, and some of the mining company consultants estimates of social and environmental impacts of the project.

Highland Pacific (a Canadian mining company, later bought out by a Chinese state mining company) proposed to dump all the refinery wastes down a pipeline to 150 meter depth in a coastal canyon offshore from Basamuk Bay. They hoped that the metal rich wastes would slide down the canyon to 1200 meters depth, and not be stirred or moved by turbulence and currents on the way down. In the mining engineering trade, this is called 'Deep Sea Tailings Placement', or DSTP. The normal practice of storing mining and refinery wastes in tailings ponds was not feasible, as this was an area of frequent earthquakes and flooding. This is in the Coral Triangle, a marine region of maximum global biodiversity of corals and fishes, of importance to local fishermen and international tuna fishing vessels. I suspect that if we had known the large volume of material to be read, and the trouble we were about to get into over the next decade, we would never have joined this effort.

Having read all the documents and written our critiques, we flew to Port Moresby on 6-7 February 2001 to be greeted in the pouring rain by World Wildlife Fund and MPI coordinator Phil Shearman. Some of the town streets were washed out, and Phil had to carefully find a way to our hotel.

We had lunch with some of the church elders from Madang and Lae, who claimed that they had the legal power to block the mining project if we indicated important environmental damage to the land and sea. They knew that the PNG government was anxious to support large new developments that would generate large tax income, and that there were few social, economic, legal or environmental impediments within the PNG government. The church maintained that the constitution of PNG gave first right of control over any developments to native villagers, the native land holders.

Highland Pacific had already requested a thirty year moratorium on changes in mining and labour laws, and an exemption from taxation for a decade. Their main environmental consultant for the project was an Australian company, Natural Systems Research (NSR), which had been involved in mining operations in PNG for the past twenty years, and they were major proponents of dumping wastes into the sea, rather than trying to expensively manage these wastes on land.

We gave our presentation to about twenty of the Madang and Lae church elders and coastal village environmental activists, in which we claimed that there would be environmental troubles with the proposed mining and refining wastes. Luick talked about the likely turbulence to be expected as the slug of refinery slurry wastes were discharged from the orifice of a pipeline at 150 m depth. Surface and deep water landward tides and internal waves were likely to mix, slow and disrupt the descent of the slurry into deep water. This coast was known to have episodes of monsoonal deep water upwelling and surface and deep water currents with velocities of 0.5–1 cm/second. There was also a problem of maintaining a waste pipeline stability on a steeply sloping seafloor in a region of earthquakes and tsunamis.

I explained how the Sepik River discharge of water, dissolved minerals and suspended sediments into the Bismarck Sea was partitioned into a surface low salinity sheet of fine colloidal mud, and a much larger thick descending plume of muddy freshwater sliding down a deep canyon. The currents in the Sepik River canyon were strong enough to move our sampling equipment far from the desired location. I applied this knowledge to the proposed refinery waste slurry offshore from Basamuk Bay, and predicted a lot of upwelling of muddy water into the surface layers of the water column. Some of the trace elements (Ni, Cr, Co, Cd, Hg) and nutrients (ammonia, phosphorus) in the refinery iron-rich slurry would probably be released to seawater, and this would likely be harmful to corals and the fishery.

From our research on the Sepik River coastal canyon, my group predicted that the refinery mud would not slide smoothly down to 1200 m depth (out of sight, out of mind), but would get 'hung up' in shallow and mid depth parts of the canyon, creating a temporary dam of refinery mud.

My fisheries biologist colleague Marcus Sheaves found the mining consultant NSR biological oceanography reports to be naïve and inadequate to allow estimation of tailings impacts on this ecosystem. Nor were surveys of Basamuk Bay village fishermen done, to estimate the importance of domestic and/or commercial fisheries in this region. The fauna of the deep water mud was poorly studied, and in any case would be obliterated by huge piles of refinery mud in an area of previously low sedimentation.

We all knew how our language of science might appear to non-English speakers with little or no science education. We had been asked to speak and show slides in one paragraph bits, and then to stop while the translators tried to explain what we said, and then continue with the next paragraph. Some of these PNG people spoke English very well, but it was the language of commerce, trade (tok pisin), and the church. There were at least six official translators working this meeting, with the same questions coming up time after time, indicating that we were not understanding or answering their question directly. Part of the problem was that we were making predictions of the waste disposal impact before it had happened. The idea of how we framed our statements in probability and likelihoods led to concern that we really didn't know what was going to happen for 100% certainty. If we were not 100% sure of what would happen to the coastal fisheries, it was easy to scoff at our predictions. Our long list of science journal publication references from our TROPICS research program was not very useful here.

The church elders were correct of course, that we couldn't predict the impact with 100% accuracy. But it worked both ways. Some of the Sepik fishermen from Wewak stood up and claimed they knew all about this 'upwelling' described by John Luick. They knew that when the monsoon winds changed, sometimes the water cooled rapidly, and it was easier to catch deep water fishes. They also talked about how they avoided the turbid muddy surface plume of the Sepik River far out in the Bismarck Sea (as I had shown), as they rarely caught fish in this muddy water.

Some of the north coast fishermen were alarmed by Marcus Sheaves' estimation of the impact of the refinery waste in Astrolabe Bay to be about 150 km^2, or about thirty km seaward and NW of the Basamuk Bay coral reefs. We later heard that some of these fishermen got on the phone to warn the Madang tuna fisheries officials of potentially detrimental impacts of the seafloor waste disposal scheme.

In the afternoon, we repeated our performances for an assembly of thirty-two Lutherans and the Government agencies for business financing, mining, resources and environmental affairs. The Department of Environment and Conservation (DEC) was led by our TROPICS supporter First Secretary Gaikuvina Kula. It was obvious that the government supported the miners, and didn't really want to listen to marine scientists from a foreign country. The Lutheran Church and some of the villagers were criticised by some of the higher government minders as being obstacles to developments, progress, and jobs for PNG people.

The next day Phil Shearman gave us a tour of the mountains NW of Port Moresby, the Sogeri Plateau, the location of the beginning of the Kokoda Trail. This muddy Kokoda Trail is part of the link between the people of Australia and PNG. During World War II, Australian troops were led by PNG natives through this difficult mountainous country to attack the Japanese forces, which had camped on the eastern shores of PNG. Many wounded Aussies were also carried by PNG natives back to Moresby along this trail. And true to the myth, it was foggy, boggy, misty forest with muddy trails over rough ground. We saw pitcher plants, lots of parrots, hornbills and a large echidna, and visited a small museum there. Phil took us to the airport at noon, and we all flew across the Coral Sea to our respective homes.

In May I got a phone call from Jochen Mueller of the University of Queensland laboratory focused on organic contaminants. A publication last year about dioxins in north Queensland coastal sediments and dugongs was completely unexpected, as these complex dangerous toxic compounds were normally sourced from large city rubbish burning, the plastics industry, and some manufacturing industries. Jochen asked for a meeting at a cafe in Townsville, and we organised ourselves into a study team to explore the time history of dioxins in my sediment cores.

Caroline Gaus joined our team, and with help from German laboratories she found dioxins in my Bowling Green Bay sediment core down to a mud horizon of approximately 1900 in age. This puzzled all of us, as there were no obvious sources or precursors for this man-made contaminant in our region, and surely not in pre-1950s times. It seemed as if this stuff must come from atmospheric transport from SE Asia, China and Japan?

Another colleague, Jan Skjemsted of CSIRO labs in Adelaide, measured charcoal and soot compounds in the same core, which showed oscillations of sooty stuff over two centuries, probably related to seasons of aridity and increased wildfire in the river catchment areas.

I got a surprise 60th birthday party at AIMS on 10 September. I was invited by Irena and Frank Tirendi to 'an important noon meeting', to find an open balcony filled with tables full of very spicy curries, fresh fruits, two cakes, wine, beer and cigars.

At home Sheila and Ian had rigged another party feast, and I got the reasonable presents of a long-handled tree trimmer, a big jug of mosquito repellent, a Gerry Mulligan CD, and rude birthday cards from relatives in North America and England. Son Jesse and Sarah arrived the next day, carrying a baby roo, a possum, and a one week old emu for Sheila. The emu promptly took ownership of our house and settled down on the sofa in the living room. The emu imprinted on Sheila, and followed her around in the house.

That night, Ian and I were watching a US football game on our satellite TV ESPN. At about midnight, a groggy Sheila got up from her bed to tell us to switch to the CNN news, to find that a big airplane had crashed into one of the World Trade Towers in New York City.

As we were watching, behind the CNN reporter in New York another plane ran into the second Trade Centre Tower.

This all felt like a fictional horror film, it couldn't be true, and we waited for the hoax to be revealed. But no, it got worse, as another airplane hit the Pentagon, and yet another crashed in a field in Pennsylvania, supposedly *en route* to Washington, D. C. and the White House. It was a very unsettling time, as nobody on the news channels knew what was happening and why.

Chapter 46. A Visit to Taipei

I was up by 0400 hrs on 25 September 2001, taking a shower with the assistance of two magpies and a curious emu chick. In this house, I was only surprised that there was no pygmy rhinoceros watching. Sheila hauled my sleepy carcass to the airport for a trip to Taiwan for an international meeting about carbon and nutrient fluxes to the coastal zone, sponsored by the International Geophysical and Biological Program (IGBP), the Joint Geophysical Ocean Flux Study (JGOFS), The Chinese Geoscience Union, and the Land Ocean Interactions in the Coastal Zone (LOICZ). In this business, you have to get used to acronyms.

The airport in Brisbane was in chaos due to the terrorist threat, with security guards everywhere. All luggage was opened and examined in detail, causing all flights to be delayed. I had a Taiwanese airline (EVA) lunch of beef and noodles, and then slept as the plane flew over PNG, the Philippines, the China Sea, and finally the stormy coast of Taiwan.

I arrived in Taipei at 1730 in very strong wind and rain. Typhoon Lekima was now threatening the SW coast. Taiwan just had a major typhoon last week, so the sandbags were still in place.. There were armed police and military guards at gates and doors. I took a bus to Howard Plaza, two hrs in massive traffic, three tiers of roadways and trains, then walked slowly through rain to the Dynasty Hotel, which was a very civilised place with flush toilets and a restaurant with an alternate menu in English. I had crab cakes and seafood noodles for my supper, reading an English newspaper full of terrorist war details, with an emphasis on the Philippine troubles in Muslim Zamboanga. The city is very noisy and crowded with scooters, bikes, crowded streets, people with cell phone in one hand and umbrella in the other, roaring motorcycles, smells of food, diesel fumes, flower fragrance, sewage, and city grunge. But my room 1402 was very clean and quiet, with CNN/English on the TV and a huge bed.

My friend Kon-Kee Lui (National Centre for Ocean Research) and several colleagues had initiated this conference as a result of previous meetings and publications that described the stress of human civilisation upon estuaries and the coastal ocean, the repository for most of the wastes of agriculture, cities and industry. We wanted to stimulate the creation of a book that summarised global fluxes of nutrients and contaminants from land to the ocean. So we invited the best scientists on this topic from Italy, France, USA, Germany, South Korea, Poland, China, Greece, the Philippines, Thailand, Australia, and Taiwan, with travel funding from four different international oceanographic organisations. Our symposium was aimed at creating a group of experts to write chapters in this book, and to make sure that the data, style and content of each chapter were similar and consistent. Our meeting was held at the same time as the Joint Asian Geoscience Conference, so we had the benefit of large input from SE Asian countries.

The next day we still had dangerous strong wind and rain, and I found my hotel restaurant had a 'Mr. Stone's Australian Room', where my bacon and eggs breakfast was just fine. Back in my room, I practiced my power point presentation and read K. K. Lui's paper on continental shelf carbon mass balance again.

The city was preparing for typhoon Lekima, with police and sandbag crews at work. My waitress said they had 50 cm of water flowing down the street outside this hotel during the last typhoon.

CNN says Saudi Arabia has severed diplomatic relations with the Taliban, and we appear to have a full scale war happening in Afghanistan. My room has 'broadband in-room internet access', and a very large TV screen. Even on travel status I am swamped with information and news.

I walked through the pouring rain to the University to hear some papers from the Joint Asian Geoscience Conference, past endless rows of scooter-bikes, 7-11 stores, Starbucks, pastry shops, stereo/phone/computer game shops, one shop with Celtic fiddle tunes on loudspeakers, everything very noisy and busy. I wandered the campus sidewalks, where classes were in session, visited the Institute of Oceanography and talked with graduate students standing in front of posters of Ph.D. thesis work. I bought some tea, pastries and juice, very large Snickers bars with almonds, and retired to my dry hotel room with wet sore feet and wet clothes. I had supper at a Shanghai restaurant of shredded pig ear and cabbage, scallops and noodles. I haven't yet found pickled wasps and duck blood popsicles, which KK had recommended to me.

My Australian colleague Brad Opdyke phoned at 2130, just off the plane from Japan, desperately needing some beer and English conversation, as no USA scientists showed up at the Palaeoceanography meeting in Japan. Bradley hadn't got much news about the World Trade Centre disaster, and he had several friends who worked in the WTC, but none of them were in the building that Tuesday morning. He is now on sabbatical leave from Australian National University in Canberra, and is hiding out at Southampton University in England, trying to write papers for the journals. Our local Taiwanese news channel shows flooding in the country, city subway flooded and being pumped continuously, and the typhoon moving northward toward China.

The next day we were off by bus to a large Convention Centre, where our session starts at 0900 with our leader K. K. Liu speaking with a dead PowerPoint projection system. We also heard speakers from the Mediterranean, Baltic, North Sea, East China Sea, Gulf of Mexico, Caribbean, Okhotsk and Japan Seas, Bay of Bengal, coastal Brazil, and Australian tropical seas.

I was delighted to find and buy Yuan-Hui Li's book *Compendium of Geochemistry*. Yuan-Hui was at Lamont Doherty Geological Observatory when I was there in 1978. KK Liu's minions provided an envelope with NT$10,000 (= A$600) cash, and a check for the air fare. Then we were off to supper with KK Liu and people from Turkey, Netherlands, Japan, China, Germany, Korea, France, Italy and Taiwan. I enjoyed a lot of stuff of uncertain origin, fish heads, eel, fish eggs, shark fin, seaweeds, date soup, abalone, lotus seed cakes, melon, guava, oolong tea and beer.

Our symposium party were in a separate banquet room with loudspeakers blaring Taiwanese singers and rappers. Taiwanese oceanographer Frank Shiah had been in Texas and Montana recently, where he bought a cowboy hat, whip and saddle, which he keeps in his office here in Taipei. He loves cowboy movies. Brad told him I was a South Dakota cowboy, so I had to make up some stories about the Wild West. This was not so difficult. I made friends with Sarin from India, and we argued about how to determine the best tasting mangoes. He had been to Australia, and had tasted Australian varieties in Melbourne (you can't grow mangoes in Melbourne!!) and considered them to be poor. He strongly recommended several varieties of south Indian mangoes to correct my brazen assertion about Queensland Bowen mangoes being the best. He spoke with religious fervour about the Alphonso mango from Maharashtra, Goa, Gujarat, and Kerala, being available in the city markets for only a few weeks every year. I later found out that Alphonso mangoes were actually growing near Katherine, Northern Territories, Australia, but they were mostly shipped to SE Asian markets because of their high value.

On Friday it was still pouring rain with strong wind. We were loaded into vans for transport to our workshop at Institute of Oceanography, National Taiwan University. This was a well-organised workshop, with espresso and pastry breaks on a balcony overlooking a duck pond.

Most scientific meetings are relatively free from political differences, but we observed some erupt here. People from Taiwan, Japan, and China made a point of disagreeing with each other. There were heated exchanges between Arthur Chen from Kaiohsung and Tsuogi from Japan. The Turkish fellow said he couldn't get permission to work in Crete controlled areas of the Aegean Sea, so the Russians said they'd get the data for him.

With a large effort of diplomacy and politeness, KK Liu managed to steer the symposium in useful directions. Over a few days we heard presentations on the estuarine and continental shelf conditions from the European North Sea, the Baltic Seas, the Sea of Okhotsk and the Japanese East Sea, the Black and Caspian Seas, Mediterranean Sea, SE Asian Archipelago seas, the South China Sea, the Bay of Bengal, the tropical coasts of the Americas, the coral reefs of the world ocean, and my work about the Australian/PNG shelf seas (Coral, Solomon, Bismarck Seas). We agreed to seek out experts on the coastal regions that were not covered, such as the polar regions. There was the usual argument about the geographic boundaries, the physical and chemical units of measurement, and the time frame for annual budgets. Previous international consortia were discussed (JGOFS, LOICZ, IMBER), some doing good and some being incomplete.

Our goal was to collectively write and edit a book, so we set some ambitious deadlines for first drafts, another symposium for group editing, and hopes for a final publication date. I ended up as chairman for the tropical margins part of the book, and I wanted to have access to all SeaWiFS and other satellite imagery for the major estuaries of the tropical world.

It was a long wet walk back to my hotel, like walking in hot sweat. We walked past oxygen bars, lots of people in respiration masks, loud muzak everywhere, wet shoes, street drains backing up into water fountains. I gratefully took a nap in my quiet room and then had supper alone at Mr. Stone's, which had wall photos of Uluru, billabongs, and roos running across the desert

After a bleary-eyed 0800 breakfast with Mr. Stone, we were all hauled off to the University for more workshops. An interesting presentation was given by Bing-Sun on pCO_2 transects across the East China Sea, showing about half of it to be above atmospheric pressure, presumably due to degradation of huge inputs of organic matter. This meant it was degassing huge amounts of greenhouse gases CO_2 and CH_4 to the atmosphere. This was much to the discomfort of Chen and Tsunogai, and the argument raged over pizza, hot meat pies, bean and nut paste rolls, beer and Coke for lunch.

There was some discussion of the future of shelf research groups after JGOFS and LOICZ had ceased operations. The Netherlands, Germany, France, Japan and Taiwan were funded for continuing large research programs, whereas USA, Australia, Indonesia, India and Mediterranean countries were not well funded. We were unlikely to create a unified synthesis in the same manner as JGOFS for open ocean vertical flux work, because we have both horizontal and vertical fluxes to consider.

Tim Jennerjahn (Germany) was separately writing a book on the global fluxes of biogenic silica, and he wanted our data for the Fly and Sepik River of PNG. Magically, we got on a computer and got inside the AIMS website to see all of our data neatly arranged by Irena Zagorskis. Tim then downloaded what he wanted to a flash stick. These are modern times!

More food and drink, toasts and thank you speeches ended the workshop, with a giant slide screen behind us showing deadlines for our chapter outlines, first draft, and final text being due by the end of June 02. [As you can see from the reference section, this book was finally published in 2010!]

I returned to my hotel room for a nap and CNN news about Al Qaeda. Brad and I tried to find a roast duck supper, but city navigation was treacherous, very foul smells from sewers backing up and running down streets, flood water in some parts of the city, as it had been pouring rain all day. We had a good discussion about joint research with Brad and John Marshall on Sahul Shelf *Halimeda* mounds in the Indonesian and Timor Seas, which was of direct interest to me. Brad is the son of the famous LDGO Opdyke who travelled the world over for tectonic and midocean ridge research and always came home with neat presents for his kids. Brad was feeling guilty because he hadn't found a present for his two kids. We did find a good roast duck shack, and then it was back to my hotel room for Earl Grey tea and sweet buns.

On 30 September, Sunday, I dragged my tired carcass down to Mr Stone's for breakfast with Tim Jennerhahn from Bremen, Helmut Thomas the Dutchman, Pino from Italy, a Polish Gentleman, Brad, and Olivier of the Antarctic, and then we were off on a bus to the National Museum at 0900 in the relentless pouring rain, past overhead trains, the underground subways being flooded. The museum was on huge grounds, ten buildings stuffed with Chinese treasures brought to the brave Republic of Formosa during Mao's purges and secretly stored in caves and train stations. They display only 10% of the collection at any one time. We saw enormous rooms full of jade carvings, bronze artefacts from 5000 BC, bamboo carvings, balls with seventeen balls rotating inside made of ivory, silkscreen and tapestry, endless rows of cooking pots and bronze hand bells. We strolled through the botanical gardens with umbrellas, where swans and duck ponds contained sculptures, where little rain shelters with benches were wonderfully positioned beside stream waterfalls, carved limestone terraces, tea houses, Coca Cola drink machines, cranes and egrets and very large angleworms crawling about the grass.

We had lunch with the symposium shelf gang, hot spicy noodles, long discussions about how to create a new shelf group to replace JGOFS and CMTT, which will extinguish in 2003. Then it was over, the collegial adrenalin declines, and the symposium people start the migration to the airport. We arrived after three hours of suffering the fumes of a very congested freeway, trucks and maniac drivers, horns and sirens blaring. I was happy to leave behind the stink of the city, wet shoes, oxygen bars, respiration masks and loud muzak. There were velvety cedars along the way in the misty rain, mountains looming in the background just like the silkscreen paintings, tunnels under mountains long and lit with grubby yellow lights.

Chiang Kai-Shek Airport is neat, clean and empty, with high security military police everywhere. I had one more Taiwanese snack of noodles and BBQ pork and then left Taipei about 10 pm, to eat more and then go to sleep until Australia happens. I arrived in Townsville 1400 on 1 October where there were no (visible) security police with guns, and there was my sweetheart Sheila with a hibiscus flower in her hair, and it was sunny, warm breeze, and blissfully quiet. And then home to Alligator Creek, Moses the emu, Charles and Caspar's loud greetings, and I have sore feet.

These international travels are educational and rewarding, but they also teach me how much I like being home, where I am not expected to be intelligent, I don't have to speak to three hundred people, and I don't have to exert a large effort to communicate in a country where English is a third or fourth language. After the house pet birds check me out, tasting my skin to make sure I am not an imposter, they agree to allow my presence in the house. And now I have good jobs like mowing the orchard and taking care of my fruit trees.

I had become entrapped somehow in the post-doctoral work of Thomas Stieglitz, with Peter Ridd at JCU Marine Geophysics, whose main interests were water circulation in crab tunnels in mangrove sediments. This research led him into the Billy Moore hypothesis that subterranean fresh and saline groundwater inputs to estuaries were important but largely unknown, but could be measured with some of the short-lived radon radioisotopes of radium, and various satellite remote sensing of sea water salinity.

I had been involved in radium-radon research in ground water at the Experimental Lakes Area in Canada in 1968-72, and radiochemist John Pfitzner helped Thomas get geared up for coastal surveys of north Queensland. This work had important links to sugar cane agriculture in the Herbert and Burdekin River floodplain and delta, where they usually have either too much or not enough freshwater. In addition, the impact of sea level rise (from greenhouse gas warming of the ocean) was slowly pushing more salt water landward, bringing mangroves into the cane fields, and changing the coastal vegetation of the floodplain.

In an attempt to get funding to support this work, we latched onto the Australian Freshwater Research Centre (AFWRC) at JCU for a workshop on the topic. By this time, I had some experience in the continuum from river floodplains, freshwater wetlands, salt pans (sabhkas), intertidal mangroves, and estuarine hydrology and chemistry of coastal north Queensland and Papua New Guinea. Our workshop gathered university, engineering, agricultural, state and federal experts on groundwater and surface water cycles in the coastal zone, and we reviewed past research and made recommendations for new studies, some of this costed and scheduled. My notebooks have several pages of intricate diagrams, models, water and greenhouse gas budgets, and cost estimates for the next generation of research.

Despite the good vibrations at the workshop, absolutely zero funding came about from our effort. And so it goes, sometimes we win, sometimes we fail. But Thomas went on to measure radium/radon and other tracer elements in groundwater input to our estuaries.

Chapter 47. Old Friends in Winnipeg

In May 2002 Jesse and Sarah informed us that they were going to get married in September. Jesse was now employed as an engineer for the state road building company, and they were living in Cloncurry, interior Queensland. Of course, this resulted in major upheavals of organisation amongst the mothers of both families...fussing about a celebrant, the reception, guest lists, music, and details of the ceremony. After serious discussions, Sheila and I decided the best wedding present would be a White Mountain hand crank ice cream freezer, which we ordered from the company in Vermont, USA. Homemade ice cream was one of our family traditions, and we were on our second freezer at this time, much rusted and repaired many times. The wooden bucket would dry out, and had to be soaked for a week before it would hold the salty ice water slurry.

Son Ian was now playing with several jazz groups in town, including a JCU band that played Pat Metheny charts in unusual time signatures (7/8, 5/4). His girlfriend Tiffany Hsieh from Canada was living with us, a very good classical musician and piano player, and she could play much of the traditional wedding music if we could put a decent piano in Queen's Park under the palm trees.

Our orchard was now seriously producing fruit. I was picking sacks full of oranges, grapefruits, kumquats, and limes, the excess being converted to juice and frozen for use later in the year. We had a poor wet season, so it was essential to irrigate frequently, but still some trees had yellow leaves and were unhappy.

Because of my 1969 work on the good ship *Bradbury* on Lake Winnipeg, I was invited to give a keynote lecture at a meeting of the International Association of Great Lakes Research in Winnipeg. My expenses were paid by Fisheries and Oceans Canada, so Sheila decided to come along to visit her family. So it is once more into the travel game, flying to Brisbane, Los Angeles, cold rainy gloomy Minneapolis, and then Winnipeg at midnight, where it was a chilly 6°C, and Sheila's sister Peabody and husband Al were there to meet us.

It seems that every one of our Canadian friends and family knew we were coming, and had arranged feasts and music parties for the whole week. Never mind that I was supposed to be a serious scientist at a big science meeting at the University of Manitoba.

Al's father was also visiting, and in discussions it came out that he had been a navigator on a de Havilland DH.98 Mosquito combat/bomber aircraft in World War II. His wife was a Morse code operator for a coastal weather station. He said everything after the war was anti-climax, and he often worried that he had used up all of his adrenalin. So I had to sing Tony Kaduck's ballad about the 'Princes of the Clouds', which sobered up everybody for a little while.

Dozens of our friends asked us out for supper, showing us their new musical instruments and new tunes. Peabody's son Gordon was playing with a dance band on a cruise ship off the coast of Puerto Rico, which caused Sheila and me to perform 'The Road to Bayamon', a very crazy song about a circus being rained out in Puerto Rico. Peabody's daughter Janet was lead trumpet in the school band, and she demanded that I play 'Stardust' one night at the supper table. My chops were definitely out of practice, but I did manage to struggle with her trumpet through this 1927 Hoagy Carmichael jazz standard.

Apparently I was also famous for making breakfast, and most of Peabody's kids would gather round when I was at the stove, making my standard bacon, sausage and eggs, begging for more.

But during the daytime, I was supposed to be a scientist. At the meetings (3-4 June 2002), there were all my scientific friends from the old days of ELA, Lake Winnipeg, and our work in the Arctic. My colleague David Schindler gave the plenary presentation, an overview of the history of nutrient enrichment of this huge lake. Since my work on Lake Winnipeg in 1969, Manitoba Hydro had engineered control structures on the outlet of the lake to regulate Nelson River flow appropriate for electrical power production. This allowed the regulation of lake level as a reservoir for flow to the turbines. There were objections to all this, some cottage owners on the South Basin shoreline, claiming that lake level regulation caused increased shoreline erosion. This was shown to be unlikely. Others claimed that moderation of lake level variations diminished the productivity of shoreline wetlands and the Red River Delta, mainly by eliminating the extreme high and low lake levels. Others were claiming that Lake Winnipeg was a net sink for CO_2 and an important agent of carbon sequestration in the climate change game.

I rescued some old and new measurements of algal growth rates and respiration/decomposition rates that had not been published, to show that the lake was a giant organic matter incineration plant, a good sewage digestion pond, and was emitting huge amounts of CO_2 to the atmosphere, perhaps 912,000 tonnes of carbon/yr. If there were a carbon tax on emissions, then this would be equivalent to about US$8 million/yr, at US$10/tonne carbon. Most of the organic matter from the soils and farms and towns of the huge catchment area (953,250 km^2) was delivered to the lake by rivers, most of it was decomposed in the lake waters rich in oxygen, and very little of this organic matter survived to be buried in the lake sediments. But on the whole, lake regulation and Manitoba Hydro electricity production was a good thing, as there were few coal or diesel power stations in Manitoba, and greenhouse gas production was mainly from agriculture.

Good friend Jim Teller (from the University of Manitoba) led a session on the geological and post-glacial history of Lake Winnipeg. As the glacial ice was melting and receding northward 10-12 kya, Glacial Lake Agassiz formed and was larger than the area of the state of Texas. This proto-Lake Winnipeg had at least three outlets, one to the north via the proto-Mackenzie River, one to the east via the proto-St Lawrence Great Lakes, and one to the south via Big Stone Lake in North Dakota, and the proto-Mississippi River. Seismic soundings indicated that there was at least 50-100 m of glacial debris underlying the more recent 10-20 m of lake mud. During some parts of the early post-glacial time (the Holocene), the south basin of the lake (and its inflowing rivers) may have been dry, as there were several soil horizons in long sediment cores. Other studies showed how palaeomagnetism, radioisotopes, pollen, and stable isotopes of carbon and nitrogen allowed interpretations of the environment of Lake Winnipeg over the past several thousand years. Shorter sediment cores were used to show peaks in contaminant mercury and polycyclic aromatic hydrocarbons (PAHs) during 1960-80. A variety of satellite imagery was used to show that much of the turbidity in the lake was caused by shore erosion and bottom sediment resuspension, and was less related to muddy river water inputs.

Old colleagues Mike Stainton and Ray Hesslein described their work to measure the partial pressure of CO_2 in the Lake Winnipeg water (compared to atmospheric CO_2) over several seasons. These data generally supported my calculations that the lake was emitting large amounts of CO_2 to the atmosphere (perhaps 388,000 tonnes carbon/yr during summer only, whereas my calculations from old data suggested ~900,000 tonnes/yr).

Greg McCullough, Mike and Hedy Kling showed that the rivers were discharging more water and nutrients to the lake (compared to my 1969-74 data), the lake biomass of algae were increasing, more algal blooms were occurring, even under the ice. Algal and zooplankton biomass was 10 times that in 1969, but the diversity of species was declining.

Good musical friend Walter Lysack and colleagues reported on the past century of the Lake Winnipeg fishery, and the spectacular increase in walleye catch in the last decades. Other fish studies showed that mercury in commercial fish tissues was declining, but all big fish had higher Hg, some exceeding guidelines for commercial sale and consumption. Some of these presentations resulted in published papers, which are given in the reference section.

It was rewarding to me to be in the midst of all this...seeing old friends after twenty years, most of them knowing of my work on the lake in 1969. We drove up to Selkirk to view the good ship *Bradbury* in dry dock in the museum, now converted to a tourist site, polished and painted. I visited my Chief Scientist cabin, and the parts of the ship we had converted to laboratory work. But I was even more rewarded to have new scientists come up to me, handing me reprints of their work on Lake Winnipeg, and saying how important it was to have our 1969 research information. There was a movement stirring, a scientific, public, and political desire to pay more attention to Lake Winnipeg, one of the Canadian Great Lakes. The Lake Winnipeg Research Consortium had been formed in 1998, and was now reporting the first results of their work. Several colleagues were trying to form a new organisation to shine more political light on the lake. Neither the federal or provincial government had been paying attention to Lake Winnipeg for decades. This concern led to the formation of the Lake Winnipeg Foundation in 2005.

GJB on CGS Bradbury *in dry dock, Selkirk Manitoba*

And then it was all over. Peabody and her son John drove us to the Winnipeg airport through the pouring rain on 6 June. We arrived in Minneapolis at noon, where my nieces Marcy and Julie picked us up and whisked us off to a restaurant for another family lunch. My mother's cousin Toadie was there, demanding attention, photographs, and account of what's going on in our lives. We were solemnly given a jar of chokecherry jelly, knowing that such stuff does not exist in Australia. I was happy to see a Butterfield Florist shop in the Minneapolis airport: most of the Butterfields in Minnesota are part of my grandmother Gertrude Butterfield England's family.

Sheila and I got on another airplane to fly to Los Angeles, Sydney, Brisbane, Townsville, where son Ian rescued our tired carcasses and took us home to Alligator Creek, where the house parrots Sir Charles and Caspar registered their appreciation that we finally came back home. Moses the Emu was now very large, over two metres tall and sometimes feisty. Moses liked to dance to Ian's darbuka drumming, puffing up his feathers and jumping into the air. As described below, we took Moses inland to a big cattle station and released him into the wild.

THE SAGA OF MOSES THE EMU

Sarah brought a one week old emu to Sheila on 10 September '01. He got separated from his emu family out near Emerald somewhere, and was brought to the veterinarian in Emerald, where Sarah works. Sheila gathered up into her arms this little white and brown striped chick, about 1 kg in weight, the size of a laying hen. He liked walking into the yard ponds, eating aquatic vegetation, and noodling around the yard chewing on grass and Sheila's yard flowers. He slept in the bathroom on a blue fuzzy rug, making little 'meap meap' sounds, right next to the door closest to Sheila's bedroom. He imprinted on Sheila quickly, and followed her everywhere.

As the emu got bigger, a small plastic kid's wading pool was supplied for his bath, and he enjoyed getting soaked under the spray of a water hose. All of Sheila's yard plants were nibbled, and precious plants migrated higher and higher on chain suspended planter pots. The grass gradually disappeared from the yard. As he grew to about a metre tall, our yard table of plant pots had to have high wire mesh around the seedling fruit trees, parsley, spices, etc. He was a menace around a toolbox, running away with screwdrivers and sockets, and he ate several piles of screws, some bottle caps, used Kleenex, a small padlock, and parts of newspapers. He grandly strolled through the yard, chasing away the brush turkeys and wallabies and terrorizing the neighbor's dog who dared come onto our property. He helped Sheila paint our house by eating paint-saturated rags and stealing paintbrushes. His normal food was horse pellets, but he liked to eat snow peas, sliced apples, lettuce, and onion tops. He would also eat chopped liver thrown off the verandah for the kookaburras. It was nearly impossible to eat in our pergola during this time, as he would devour everything on our plates, especially my breakfast bacon and toast. At Christmas time he was only mildly amused by the Christmas tree, but we had to make several extra pieces of French toast for him.

When he was large, almost two metres in height, I started calling him by the name of Moses, and he would come down to the creek valley with me to help irrigate my fruit trees (only if Sheila would hide and be quiet). When I was bent over pulling weeds, he would stick his neck between my legs to nibble at the weeds. He liked to eat the leaves from Chinee apples, a prickly weed tree I'm trying to eliminate from our grounds. He would happily run around on the creek bottom at full speed, chasing the wallabies and running through the tall grass. When there was water in our creek bed, he would walk into the billabong and sit down, scooping up water with his very large beak. The sound of our neighbor's horses running along the opposite side of our creek would scare him, and he'd run home. At any sound from our house up on the hill, indicating that Sheila was home, he would zoom straight up the hill to be with Sheila.

Sheila was away twice during Moses's upbringing. Ian MacGreggor took care of Moses while she was gone. This involved scheduling paid emu sitters to stay with Moses while Ian went to town for various duties. These emu sitters were a varied lot, including a Filipino drum maker and a couple from Rio de Janerio, and their experience in taking care of an emu was certainly limited. Ian came home one night to find the emu sitter watching TV and Moses AWOL. This was his first late night, but he finally appeared in the house at 10:30 p.m. and went straight to bed on his blue fuzzy in the bathroom. In order to adjust his musician's life style to Moses' customary dawn waking hour, Ian slept in a cot in the pergola, so that the local birds would wake him at dawn.

We built him a ramp so he could walk in and out of the house by himself. He would come into the house at dusk, and sit down in front of Sheila's feet to watch the evening news and *Jeopardy*. He still slept in the bathroom on his blue fuzzy rug, and we could hear him in the night thumping the floor with his long bony legs, trying to get comfortable. The bathroom floor was covered with newspapers to contain the rather large droppings, and these were dumped in a special compost container. He would be ready to move outdoors at dawn, and his feet would sound 'tick tick tick' from his huge toenails striking the floor in the house. Sometimes when I would come back from a research cruise, having been away several weeks, he would do his 'war dance', fluffing up his neck feathers, standing near two metres vertical, jumping and kicking sideways, flopping down, and then jumping up again. He would go through three or four-wire fences somehow, and visit the neighbours, chasing their dogs, nibbling their yard plants. Most of our visitors were treated kindly by Moses, especially children. He would often flop down in the pergola beside the table, and be the object of wonder and admiration. When Ian would play his big low pitched drum, Moses would go into a war dance, standing on his tip-toes, jumping and kicking, fluffing up his neck feathers and vocalizing 'Haaaaaa'.

GJB serenading Emu, Alligator Creek, 2003

By mid-August Moses began exploring further and further away from our house, and not returning at night. Although most neighbours were kind to Moses, he was in danger of finding the road traffic below our house. Sheila decided it was time to move Moses out to a safe place in the bush. On the evening of 23 August '02, we prepared a 60 kg concrete slab, and pushed Moses through the gauntlet, making feet marks in the wet concrete. This concrete slab is now embedded in our front yard as a memento of Moses. Sarah provided 1.5 ml of Xylazil, a sedative drug. We waited till 0200 on 24 August, and Ian MacGreggor went into the dark bathroom to restrain Moses, Sheila swabbed his thigh with alcohol, and I injected the sedative into his thigh muscle. Moses was a bit upset, and got into the bathtub while Ian was trying to restrain him. He kicked several large holes in the bathtub, and bent the tap upwards with his very strong legs and feet. Sheila herded him to our front verandah, where Ian's van was backed up and open. Sheila had prepared the back end of the van with foamies, blankets and the blue fuzzy rug, with the back window blacked out. When the back door was closed, it was dark in the van, and Ian and Sheila got Moses settled down.

We then drove like fury west to Charters Towers, and north to Greenvale and Jervoise Station, about 400 km into the Valley of the Lagoons, the upper Burdekin River. It was a cold winter night, with some fog and smoke over the road. We refueled in Charters Towers at 0330, narrowly avoiding a police car inspection, and bounced and swerved along progressively narrower and poorer roads, cattle and dead roos in the road. The sedative made Moses vomit a few times, and he would try to stand up now and then. Sheila spent most of the trip sitting beside Moses, trying to comfort him. She tried to put a sock on his head, but this didn't work, making him more upset. We arrived at Jervoise Station at 0700, with Greg and Kerry Jonsson there to meet us in the yard. There were several wild emus in the pastures near the cattle station, open grass country with ironbarks and white gums. I opened the back door of the van, Sheila got out, and Moses stood up, blinking at the new situation. After some hesitation, and seeing the station dogs and pet pig, he jumped down to the ground and chased the dog and pig. He then strolled around the yard, inspecting the house, the fenced garden, the big mechanical workshop shed, the chook pen, and the river behind the house. He enjoyed running beside some of the station children on their bicycles, the kids screaming with delight, calling him 'Big Bird'. If he was stressed by this wild night ride, he seemed to have got over it quickly.

Sheila and I had a good bacon and eggs breakfast, and met some of the station son and daughter families, the station governess, and other transients. Jervoise Station produces organic beef, and has been a depot for disturbed children in the past. I found a bed and slept for four hours, but Sheila watched the emu walking around the yard. About noon he vanished, probably found a quiet spot to take a nap. We packed up the van and left, as Sheila was getting weepy. This was a good place for Moses, as the families understood and enjoyed animals, and the property was already occupied by wild emus.

We returned to our home in Alligator Creek to view the damaged bathtub and call a local plumber to get estimates on the cost of fixing the bathroom. I phoned our insurance agency to report the damage, and of course they rejected the claim. They laughed loudly at my description of our pet emu in our bathtub, and told that was a grand story! But no worries, the rejection was not because it was emu damage, but because the bathtub was plastic, not ceramic.The pergola ramp was taken down, and the house seems a bit quiet at night without Moses thumping away in the bathroom. Sheila was very sad to have another of her children leave our home. In addition to our sons Ian and Jesse, this included a Canadian pet fox, two horned owls, various turtles, a Canada goose, and a lot of repaired birds.

Several weeks later, Kerry Jonsson of Jervoise Station phoned to say that Moses would come to the house some evenings to have some horse pellets and lay down on his blue fuzzy blanket. The children riding bikes would bring Moses out of the bush to play with them. But mostly he was with the wild emus out in the cattle paddock.

In late March 2003, we got a very late Christmas letter from Greg and Kerry Jonsson of Jervoise Station, with a hand-written note at the bottom, indicating that an emu had come up to the house a few weeks ago, with two babies walking behind it. This was probably Moses, showing off his babies and looking for his blue fuzzy blanket. Kerry said that they had lived there for thirty-three years, and wild emus had never come close to their house.

So Emu had a happy ending.

Chapter 48. The Timor and Arafura Sea

One of my pet arguments with colleagues at AIMS was that *Halimeda* mounds were at least as important as coral reefs in our coastal ecosystems.

Halimeda is the genus of a macroscopic green alga with heavy calcium carbonate encrustations, which tends to occur in shallow water in meadows, sometimes in the interior protected lagoons of atoll coral reefs. As *Halimeda*'s life cycles go around, the distinctive calcareous encrustations pile up over thousands of years, forming a dome shaped mound. I had seen these amongst the northern ribbon reefs of the GBR, and I had read about giant *Halimeda* mounds at the shelf edge of the Timor and Arafura Seas.

One of my TROPICS collaborators, John Milliman, also pestered me to explore these poorly known geological and ecological systems, which were sometimes associated with submarine seeps and gas vents. In addition to my curiosity, my colleague Kathy Burns wanted to explore the Timor Sea to locate the distribution and sources of unique hydrocarbons in the water and sediment, to help explain what she had found on our R/V *Franklin* cruises north of NW Cape and the sediment traps in Exmouth Plateau in November 1996. We were also directed to find ship time funding from the petroleum industry, as much of the north coast shelf is thick with drilling platforms. We managed to infect some of our colleagues with this *Halimeda* mound project, Andrew Heyward (corals and oil drilling platform impacts), Mark Meekan (sharks), Janice Lough (AIMS) and Michael Gagan at ANU (coral core history of the Indonesian Throughflow), Bradley Opdyke at ANU (foram isotope climate history) in informal workshops, and we did manage to organise a few cruises with other collaborators to gather preliminary evidence for the importance of *Halimeda* mounds (also called the Karmt Shoals, Sahul Shoals.

And so, barely out of jet lag from our trip to Canada and USA, on 9 June 2002 I flew from Townsville to Cairns to Darwin. Kathy, Irena, and John were already at the Rydges Plaza Hotel in Darwin, with several AIMS ship crew personnel, eating oysters Thai, mud crabs, and Gulf of Carpentaria prawns. The R/V *Harry Messel* was already in port, due to extremely rough weather (five metre swells) during the previous research cruise. They had returned to the port of Darwin exhausted and without accomplishing their research plan.

The next day the wind was blowing even stronger, and Master Allan Reif suggested that we relax in port for a few days. Even the Navy border patrol ships were tied up in port. We picked up deck rigger Cary McLean and Bradley Opdyke at the airport and moved from the hotel into accommodations on the good ship R/V *Harry Messel.* We all had supper at a tiny Thai street restaurant. John Pfitzner selected some Barossa Valley wine, and we ate good food till about midnight, watching a World Cup Soccer match between USA and Korea.

It was obvious we were not going anywhere in the next two days. We prepared our deck and lab gear for extremely rough conditions, and Cary tested the new CTD attachments (methane, photosynthetic light, fluorescence, oxygen, and % light transmission) and calibrated the depth/pressure, temperature, and salinity sensors. Kathy visited the Northern Territories University to get some clean solvents for her hydrocarbon work, and Bradley bought several sets of the electrode leads for his alkalinity titration apparatus. Cary and John tried to test the new 3-D sonar software (Max Sea) that we wanted to use to map the *Halimeda* mounds. John was repeatedly calibrating his sampling sorbent column and lab apparatus for the measurement of the isotopes of radium and radon, which we hoped would be a good indicator of submarine seeps feeding nutrients to the *Halimeda* mounds. The ship crew were busy sandblasting the hull, painting, and welding. There were about thirty prawn trawlers tied up here, all of them busy welding, refitting, drying nets, making good use of the wild weather safe in port. I spent most of the time writing an introduction to a whole journal issue of TROPICS papers.

An important fax arrived from the Australian Department of Foreign Affairs and Trade, giving us the Australian view of the position of the boundary between Australian and East Timor's sea boundary. At this time, this boundary was in dispute, as new oil and gas deposits had been recently mapped by Geoscience Australia, and the Australian government was anxious to claim more than its share of these resources. As it happened our ship track plan went right up to this boundary, and AIMS was a bit nervous about this. You need to understand that at the stroke of midnight on 20 May 2002, the Democratic Republic of Timor-Leste became the first new nation of the twenty-first century. From that moment, those who fought for independence have faced a challenge even bigger than shaking off Indonesian occupation: running a country of their own. See the book by Peake in the reference section.

Finally, on 12 June, the Master moved the R/V *Messel* out of the safe harbour Duck Pond, past the big Navy ships, livestock ships, fancy sailboats, oil rig tenders, container ships, pearling ships and wharf cranes, so we could anchor in Darwin harbour, ready to take off if the wind died down. By 1400, the wind was only twenty kts, so we up anchor, and off we go. It would take us two days to reach our study site, and we regularly tightened the ropes and chains on our aft deck and lab gear, as our little ship was moving wildly.

On 13 June we had engine trouble, a ruptured seal, and we pulled in behind a coastal island for repairs. When the ship stopped moving, we had to pull in the aft deck mackerel lines. Cook Pascal performed miracles with his tiny galley, and we ate very well, appetizers of spicy cheese and spinach stuffed flaky pastries, a lot of fresh mackerel stuffed with squids, pork loin, cuttlefish and hot peppers.

All along this coast are pillars of wildfire smoke, an almost grey wall along the coast, in spite of the clear blue sky. Bradley entertained us with wild stories about his father, a Columbia University/Lamont Doherty Oceanographic Observatory geologist, who mapped palaeomagnetic structures in Afghanistan and Iraq some forty years ago, and to whom is credited the origin of our new field adventure motto...'Ho Ho, she cried! And waved her wooden leg aloft'.

Suddenly the wind went down, so we steamed fearlessly due north to the shelf edge, where there would be no protection if the bad weather came back. We arrived at our first target at 0400 on 14 June, in the midst of some small isolated submarine pinnacles near Pea Shoals. After a hearty breakfast of pancakes, bacon and eggs and strong coffee, we began our exploration of these massive mounds, with CTD/methane/light/chlorophyll profiles, grab samples of the seafloor from the tops (10-20 m) and margins (200-1000 m) of the mounds along the shelf edge.

You need to appreciate that some of these mounds originated in 600-700 m water depths and rose almost vertically to within 20 m of the surface, about the same height as the World Trade Centre skyscrapers. We used the MaxSea sonar facility to make 3-D representations of some of these mounds. Sure enough, the core of the surface of these mounds were almost pure *Halimeda* carbonate fragments (almost like white bran cereal, corn flakes, or oatmeal), with only small patches of living *Halimeda* and some apparently dead coral bommies, some bryozoan and ophiuroid skeletal parts, a few mollusk and red algal fragments. Cary did a few SCUBA dives to calibrate our sonar map identifications of live *Halimeda* on top of the mounds, and to obtain a big sack full of live *Halimeda* for identification by Ed Drew back in Townsville. John found no strong radium signal from seep water near the base of the mounds, and Kathy's methane sensor detected nothing above background. Seaward and northward from the shoals was the Timor Trench at 1-2 km depth, where we found fine clay mixed with pteropod spines, crinoid stems, scaphopod and worm tubes, and small amounts of *Halimeda* fragments. The deep water was cold (~10°C) and highly charged with excess CO_2, which probably dissolved most of the carbonate material supplied from the shelf edge. The fine clay must have come from the erosion of Indonesian islands and transport via the Indonesian Throughflow.

We were lucky with the weather, five to ten kt winds, but boiling hot and humid. The deck was very hot underfoot even with safety boots. For the next week we puttered along the shelf edge, working the tops of mounds of the Sahul, Big Bank, and Karmt Shoals and into the margin of the deep Timor Trench. Most of the mounds had near vertical sides, and our grab sampler could not sample this material. My coring equipment would not work (and was damaged by the hard gravel) in this kind of sediment, so I made sure to bring a big tub of this *Halimeda* sediment to the AIMS workshop engineers, in hope that they could build a drill corer that might penetrate this material.

We woke up on 19 June to twenty-five kt winds and three metre swells breaking over the margins of the shoals. The Master quietly recommended that we return to Darwin for our safety, as most of the other oil rig tender ships in the region were putting out distress signals. One yacht, *Lost World*, had been abandoned with a broken mast near Penguin Shoals, and two people were in a life raft asking for assistance.

So we again chained down our aft deck equipment, put away all lab gear in their shipping containers, and steamed directly into the bad weather for the port of Darwin. Seabirds were wheeling over the ship backwash, there were rainbows from waves making spray off the bow, and the aft deck was sometimes knee deep in seawater froth. I went back to bed to read Phillip Hensher's *The Mulberry Empire*, a good lesson in the futility of military expeditions against the Pashtun mountain men of Afghanistan. Despite the rough conditions, Pascal performed miracles in his tiny galley, providing seafood pizza lunch and a roast rack of lamb for supper. The next day we reached the shelter of Bathhurst and Melville Islands, and it was possible to walk the deck without hanging on for dear life. For some reason, the ship air conditioning was on full blast, and we all dressed warmly for meals.

We were safely in Darwin harbour by 0700 on 21 June 2002, with a good breakfast of bacon and eggs, and the ship was not rolling and pitching. Then all hell broke loose, with the phones ringing and the fax machine whirring. AIMS was establishing a new substation in Darwin, and we had purchased a big rail container to store our equipment for the following planned cruises. For some reason, AIMS administration phoned to say that we should sell the container for $2,000 and discontinue our plans for future research in the Timor Sea.

Then my New York colleague Bob Aller phoned to say that NSF failed to fund our planned MARGINS cruises in the Gulf of Papua. This was all bad news.

After cleaning the ship, we all went to a beach restaurant for a seafood feast, returning in the wee hours where we found the aft deck door hinge pins had been removed. We were ready to call the police to report robbers, but we found Cary McLean asleep in his bunk. He had come back early, didn't have a key, so he improvised with his jackknife to get inside the ship.

Most of our science and ship crew flew back to Townsville, but I needed to stay in Darwin for a few days to help settle the deal about new AIMS facilities and our warehouse, and establish collaboration with Niels Munksgaard at NTU and Andreas Bollhoefer at Jabiru. Irena went off to visit her brother who lived in Darwin.

John Pfitzner demanded that I visit his old stomping ground at the Jabiru field station on the Alligator River. John had worked at this federal government establishment (Office of the Supervising Scientist, Department of Environment) for over a decade before I hired him. This lab did all of the alpha and gamma spectrometry for samples from water and mud runoff from the big uranium mine into the billabongs, swamps and Alligator Rivers. I had read about a decade of research at the Ranger Uranium Mine, affectionately referred to as the 'RUM Jungle', on the uranium series elements in file snakes, crocodiles, birds and the cloaca of freshwater turtles. The mine is now called Energy Resources Australia, and is being cleaned up and rehabilitated for closure.

After some phone calls, we bought a ute load of groceries for John's friends out at Jabiru and the Kakadu Park, half a day's drive SE of Darwin. And here we entered a different world of small rivers with muddy banks of salinity stunted mangroves, mimosa thickets, fine black silt soils 'burnt to buggery' with black stumps smoldering and ash blowing in the wind. I was not surprised to find that the roadside café waitress called John by his first name. We had to visit an albino buffalo (named Bertram) in the paddock at the rear, a frill-necked lizard in an aquarium, and a cage full of ratty looking parrots. Farther on, the gravel road featured a lot of road killed feral pigs and wallabies, which the dingoes and black kites were happily devouring.

I could tell John was enjoying this drive. As we drove through the Kakadu Park buffalo pasture to Dave Lindner's camp, the magpie geese and ducks flew up from the lagoons by the hundreds.

Lindner was about all that remained of the park 'buffalo rangers' whose job was to cull the population of Indonesian water buffalo from these swamps. These introduced buffalo were thought to upset the floodplain, estuarine and intertidal ecosystems by making deep furrowed trails from the tidal channels back into the freshwater wetlands, thereby allowing salt water into the more productive freshwater ecosystems. To me, it seemed this problem was partly due to greenhouse warming and sea level rise, but John cautioned me not to bring this into the discussion. Over several decades, the rangers shot or captured the buffalo for export, and the present residual population was now limited to one area of the park, mainly to feed aboriginal communities nearby. We drove past the old corral and loading chutes, an old time example of bush craft, made from railway sleepers, steel rails, and rusty steel cable. The corral fence was decorated with hundreds of bleached white buffalo skulls with their huge horns.

Lindner's camp was an untidy cluster of old railway containers focused around a large 'tea table' perched on big melaleuca stumps that might have served ten or fifteen hunters not so long ago. He proudly led me through the old rusty containers, to show off his pet 'cats' (chuditchs, actually a native marsupial quoll), two emus in a pen, several goannas, a lot of big ravens and black kites sampling the buffalo gut and hide pile. One shed was a rifle and shotgun cartridge reloading station. Another was full of skinning equipment, skinning knives, equipment and materials to make skinning knives, and tanned hides, another shed was full of ammunition boxes full of groceries and cooking gear, one shed was full of loins and hind quarters from this morning's buffalo kill, hanging on hooks, with a two hundred litre barrel of plucked and gutted magpie geese in a salt water brine. I walked behind the camp, beside the electric fence that kept the buffalo out of his camp, alongside the lagoon, muddy from buffalo traffic, a buffalo dust wallow up the slope, a melaleuca/billy goat plum/*Eleocharis*/*Hymenachne* wetland swamp. I should have written a ballad about this place.

When the sun went down Dave Lindner's aboriginal wife Patsy arrived, after having operated a nearby tourist ride through the swamps. Dave had two big fat magpie geese roasted in a homemade pressure cooker for our supper, with boiled potatoes, pickled onions and beer... couldn't be better. We were told not to clean up our plates, as the camp quolls would take all the goose scraps. Patsy fed the goannas and ravens chopped buffalo liver.

Then the stories were told, and I wished I had a tape recorder. It turns out their favourite homemade skinning knife had been designed by John Pfitzner. Dave had been unemployed for four years, had lost most of his teeth, a few more to go, and he will get dentures. Dave told stories about the early days of Kakadu, before the park was established, crocodile biology adventures of Sydney physics professor Harry Messel (after which our ship was named, a Sydney University nuclear physicist), Tom Cole's crocodile and buffalo hunting tales, and of course all the bad situations that happen from buffalo hunting. Dave tried to sing 'The Buffalo Shooter's Song', but without most of his front teeth, I couldn't understand most of the words. Dave told of the cranky bull buffalo named 'Bin Laden', who had two .475 calibre Weatherby magnums in his neck, and was still out there somewhere, angry.

The next morning we had damper, cold goose meat and coffee for breakfast, and we were off to 'Goose Camp' to shoot a ute load of geese for the aboriginal community. It was chilly and misty in the tall grass, and the sky was full of geese, so no problem getting the geese with two shotguns. Dave gave three of the geese to 'Rough Nuts', a local crocodile who demanded his tithe.

After another lunch of goose meat, we drove back to Jabiru, a miniature Canberra, complete with an artificial lake in the middle of town. Uranium mining began here in 1979, and the tailings pile was about thirty-five metres high, beside a tailings pond where feral pigs were happily rooting around in the mud. Nearly all of the yellowcake is exported to France, Germany and Japan. We drove farther east to Cahill's Crossing on the East Alligator River, where there were several caves/overhangs with very old rock art. The view of the coastal wetlands was stunning. We drove back to Darwin to overnight with one of John's buddies, a conservation biologist, who served us rusa deer meat for supper. It was good to have a shower.

We barely had time the next day to eat a hearty breakfast of venison sausage and eggs, when our cell phones began to riot. Sheila phoned to say Bob Aller from New York had phoned to indicate that NSF had funded the MARGINS cruises in the Gulf of Papua, the AIMS ship manager phoned to indicate that we should keep our storage container and rent a larger warehouse, and Frank Tirendi phoned to indicate that the Timor Sea project was fully funded by AIMS, and that we should leave our gear in Darwin for future cruises.

Well, this was a complete about-face!

Frank was the AIMS point man for ATSEF (Arafura Timor Sea Expert Forum, including scientists from the Indonesian Institute of Sciences, Research Centre for Oceanography, Geoscience Australia, James Cook University, Northern Territories University, and AIMS biologists and chemists). In addition, AIMS was planning to establish a field station in Darwin to manage research along the north coast. My notebooks were filled with urgent plans, equipment lists and operation costings for the next six years.

Then we were on another airplane flying over Cape York dotted with bushfires, to Cairns and Townsville, where my beaming Sheila collected our tired carcasses.

As usual in this business, we seemed to oscillate between despair and exciting research opportunities.

The path for extended research in the Timor Sea was now open, with opportunities for collaboration with Indonesia, Irian Jaya, and several agencies/universities in Australia. We held another big workshop in Darwin on 25 July, with the intent of attracting some of the oil and gas companies (Woodside, Shell) that worked in the Timor Sea. We had proposals from the geoscience groups interested in ancient and presently active gas and oil seeps along the shelf edge, sea floor mappers wanted to run seismic transects across this complicated region, the microbiologists who wanted to study oil decomposition in the sea, the physical and chemical oceanographers and the fisheries and benthic biologists. The importance of submarine seeps in fueling chemosynthetic communities with methane and hydrocarbons was hot in our minds.

Part of the geological theories proposed that the Karmt and Sahul Shoals and the *Halimeda* mounds were initiated and maintained by methane seeps on the flank of the Timor Trough, beginning in Pliocene times. The giant petroleum companies were interested in Kathy Burns, Nicolle Webster, and Jo Johnson's work at AIMS on hydrocarbons in the sea and sediments, and Geoff O'Brien's ideas about how the submarine methane seeps might have built the 400-600 m high towers of Karmt and Sahul Shoals. Fisheries biologists were worried about depauperate shark, tuna, and marlin populations, perhaps due to Indonesian over harvesting. Coral and *Halimeda* biologists wanted to know how tolerant these ecosystems were to methane and low molecular weight hydrocarbons. Proposals were made and costed for Indonesian, Australian, and oil company ships, and we were all naïvely optimistic about the next six years of Timor Sea research.

Without giving me a chance to catch my breath, the next day most of my USA collaborators in the MARGINS research project appeared at AIMS for a Gulf of Papua workshop. This was fairly high level planning with Chuck Nittrouer, Bob Aller, John Milliman, John Crockett, Mike Ridd and Peter Harris discussing the big US ships, visa problems, storage of US gear at AIMS, and planning for 2003-04 cruises. This was all the spinoff from our TROPICS Project which was now defunct and had not been funded by the US National Science Foundation. AIMS backed off from funding more PNG research, and would only support this work if NSF paid for ship time.

Chapter 49. GBR *Halimeda* Atolls, and Jesse and Sarah's Wedding

A group of us were infected with a craze for research on *Halimeda* sp., a macroscopic green alga common in tropical reef seas. My colleague John Milliman from the Virginia Institute of Marine Science, who wrote the book on carbonate sediments around the world, had been pestering me to organise a cruise to the northern GBR and Torres Strait, where atolls were thought to be built mostly from *Halimeda* carbonate crusts. John wanted to use various seafloor mapping and seismic equipment to determine the thickness and extent of the *Halimeda* beds. He needed Irena and myself for sediment grab sampling and coring work as ground truth for his computer images.

John had a pile of NSF money to support the use of the AIMS R/V *Lady Basten* and so on 2 August 2002 we loaded the ship with all sorts of seismic gear, computers, extra diesel generators and power capacitors. John brought along his super technicians Dave Mucciarone and Jesse McNinch. I loaded my own office computer and two big monitor screens to assist with complicated data capture and storage. Master John Staib was suspicious that the Yanks were actually spying on the GBR, with four desktop computers in the lab. In testing the seismic gear, the power unit and capacitors kept overheating, so we stopped in Cairns to buy two big fans to cool them. Leon Cook, who was the cook, exceeded all standards with his supper of rosemary spiced roast lamb chops and a passionfruit junket for dessert, as we steamed northward on calm seas. We saw three big whales spouting and breeching somewhere north of Cairns. The mackerel lines were full out, and we brought in a few medium sized mackerel and bonito before midnight.

Well, the kind weather didn't last long, as the wind came up and the swells were 2-3 m the next day, but it was a following sea which helped. We reached the shelter of Lizard Island where we put the ship stabilisers out and deployed four gadgets off the stern. These were shallow seismic 'sparkers', Chirp (a kind of multiple frequency sonar), side-scan sonar, and a towed high power sonar, in addition to our ship having two different frequencies of sonar transducers in the hull of the ship. After Dave and Jesse got all the computers and electronics working and the sonars calibrated, the plan was to sneak outside the barrier reef into the Coral Sea, where Ashmore Reef and Great Detached Reef were perched as atolls in 500-1500 m depth on the continental slope. *Halimeda* experts and my advisors Ed Drew and Richard Orme suggested that the interior of these atolls would be almost entirely *Halimeda*. The interior of these atolls was only 20-30 m deep, and we could anchor safely with the protection of the coral rim of the atoll. Some of us had a quick snorkel dive to view the colourful reef ecosystem around Lizard Island, perhaps the least impacted by human developments. We settled down at anchor to wait for better weather and had a wonderful supper of fresh caught squid, small octopi, barramundi and mackerel.

For reasons that South Dakota boys cannot comprehend, the ship Master Staib decided to pull anchor and 'thread the needle' through the many reefs at 0130 on 4 August, as the tide was ebbing (better to spotlight the reefs), the wind was down to twenty kts, and the whitecaps were less vicious. Two deck hands in safety harnesses were on the roof of the wheelhouse with spotlights and radio communication with the Master, guiding the ship through the narrow channels between reefs. We reached the Great Detached Reef at 1945, and everyone was exhausted from the tension.

This atoll was about twenty km long north to south, and maybe ten km wide east west. The rim of the atoll is composed of coral and hard red algal colonies that are level with the water surface at low tide, with four ten metre deep openings into the irregular oval shaped atoll. The atoll sits on top of blocks of continental Cape York that fell away from northern Australia, and these many offshore atolls (Osprey, Shark, Ashmore, Boot, Eastern Fields, Portloch, Lagoon) are pinnacles that arise from deep (200-2000 m) water, and are considered outside of the Great Barrier Reef proper. We gratefully anchored inside the atoll, with good protection from the roaring waves crashing on the coral margins to the seaward.

The next day all the sounding gadgets were soothed, caressed and deployed behind the ship chugging along at slowest speed. The ship runs at 1800 rpm at full speed, and the engines did not like running slow. There was a frantic scramble between computers, and the aft deck was completely covered with sensor cables to the sounders. The side-scan fish needed cleaning, the sparker power source was blowing fuses, and the sparker hydrophone needed repairs. And then the cook brought out sugared ginger muffins with tea and espresso for smoko, and everybody relaxed.

All day we ran transects across the inside of the atoll, and observed the thickness of the recent (Holocene, maybe ten thousand years) sediment being 20-40 m. Finally in late afternoon they pulled in all the seismic gear, and I set up the Smith-MacIntyre Grab sampler to gather some ground truth for the sounder signals. And sure enough, most of the sediment was nearly pure *Halimeda* flakes, live *Halimeda*, some huge calcareous sponges, bryozoa and some coral rubble. Irena catalogued the locations of grab samples in accordance to the seismic transect locations. We also noted the locations of several ship wrecks, anchors, and odd piles of steel cables on the seafloor.

The next day it was blowing hard and raining, and I tried to obtain some cores with my big Kasten Corer with a tonne of lead attached to drive it. The *Halimeda* gravel was resistant to penetration, and my corer just tipped over in four attempts, damaging my core catcher device. My notebooks show ideas and designs for a drill corer that would work in this hard bottom. I gave up doing this before noon, and we steamed at full throttle back inside the barrier reef, as the wind was thirty kts and rising.

Inside the safety of the barrier reef, we relaxed for lunch, but we couldn't find Leon the Cook. We found him in his bunk, pale and having serious chest pains. Master Staib immediately began a medical evacuation procedure by radio and satellite phone to hospitals in Cairns and Thursday Island. It was sixteen hours steaming to Thursday Island, and twelve hours to the settlement at Lockhart River, where there was a nursing station, but the weather was getting worse. We were fortunate to have Irena Zagorsis (high level first aid training) and John Crockett (USA Emergency Medical Officer) in our crew, and they firmly took charge of the situation on the ship. The diagnosis was that it was a mild heart attack. I didn't realize the full nature of the ship's several first aid kits. There was even a defibrillator, an oxygen tank and mask, and several kinds of angina pills, all of which Irena and John knew how to use.

The Royal Australian Navy ship *Hawkesbury* heard our Mayday radio calls, and indicated that they were anchoring out the storm behind Clerke Island to our NW. They had a medical officer and helicopter on board, and were coming to help us. This ship was three times the size of the *Lady Basten*, and they came close to our anchorage, amidst white caps, rain and thick fog. There were also two prawn trawlers hiding near our anchorage. The Navy sent two large zodiacs to our ship, with paramedics and stretcher boards, to shift the cook to their ship. We later heard he was helicoptered to the Cairns hospital. He survived the heart attack, but gave up his career as a ship's cook.

Skipper John Staib had made arrangements for another cook to be flown to Thursday Island in Torres Strait. There was a thirty kt wind blowing with torrential rain, so we stayed at anchor for the night in the peaceful protection of several reefs, within sight of the big Navy ship.

We watched a *Mad Max* film, and the crew had fun catching fish and squid from under the aft deck lights. All night we could hear the big storm breakers roaring against the reef to the seaward of us.

At dawn Master Staib and crew raised anchor and steamed carefully inside the barrier reef into Thursday Island for fuel, food, freshwater, and a new cook. I gratefully slept till 10 am, and as usual Irena was mad at me for being able to sleep blissfully when the ship was in rough weather. Irena and Jesse McNinch made lunch of squid rings, stuffed squid, left-over stir-fry, and a gigantic spicy fish chowder that should last several days. We anchored in Thursday Island harbor at 1630 with a twenty kt wind. The huge windmill turbines on the hill behind the little town were spinning very fast. We drew straws to determine who had to remain on the ship, while everybody else took off in zodiacs to walk the streets, visit shops, and have supper at Jardine's Hotel balcony.

The ship crew had vanished, and it was only days later we found out they were dancing with strippers at some hidden night club. By midnight, the wind was up to thirty kts and bending the coconuts and sea almond trees to the NW. The next day was worse, thirty-five kt wind and rain. Our new cook arrived at the Horn Island airport, but the airport/town ferry decided it was too rough to travel, so Carmel Clark had to sit on Horn Island for most of the afternoon waiting for the wind to die down. She eventually arrived on Thursday Island, took inventory of the galley, and went to town to select more fresh fruit and vegetables. Of course this was a good excuse for another night on the town for the ship and science crew.

Master Staib got fed up with being safe and quiet at anchor, roused all of us up at 0400 to help put out the stabiliser fins, and steamed NE through a nightmare of reefs, atolls, islands in amongst white caps and thirty kt wind. At least the sun eventually came out and visibility was good. Most of the little cays around here had prawn trawlers anchored in the lee of the wind, the crew asleep, as they usually work at night. We were aiming for the Murray Islands (Mer, the home island of Eddie Mabo), which appeared to be a large bowl shaped lagoon, maybe full of *Halimeda* carbonate. We reached Mer the next day with only twenty-five kt wind and clear skies, to find that the island lagoon was not full of *Halimeda*, but coral and red algal fragments, with no living *Halimeda* at all. But the islands provided shelter from the big swells, the fishing was good, Carmen made pizza, and we watched a Zorro film and played card games. It was dangerous to open the fridge, as it had to be strapped to the galley bulkhead.

The next morning we lashed everything down on the aft deck, labs and galley, and threaded the needle of the Fly Entrance in thirty kt winds to dash southeastward to Ashmore Reef. This was about three hours of open water (500-600 m depth) steaming, without the protection of the barrier reef, but we made it to the NW entrance of Ashmore Reef. The entrance to this atoll was a deep blue fifty metre depth notch in an irregular oval coral reef rim. We were safe inside the atoll, but outside the big swells crashed into the coral rim and spume and spray was ten metres high in the air. All the seismic gear was quickly deployed, and we began a sounding grid across the 40-60 metre deep atoll lagoon. Then there was a computer crash, the ship's generator had to be tuned up, and all sensor calibration had to be done over again.

I did a grid of grab samples, showing that none of the bottom sediment was *Halimeda* flakes, but rather a curious mixture of coralline algae, foraminifera, hat corals, oolites (little egg shaped carbonate grains, normally associated with evaporate deposits), and some terrestrial sand and silt. Milliman was now puzzled, as this did not fit our theory. But I pointed out that the huge PNG Fly River was only a hundred km to the north of us, and its estuary and delta must have been close to Ashmore Reef during low sea level stand. It took the rest of the afternoon and evening to get the computers running again. Carmen made us some spicy prawn balls for supper. Whoever selected the entertainment DVD films for this cruise was a genius. Now we had *The Year of Living Dangerously* on the big screen, an excellent account of the Suharto revolution in Indonesia in the early 1960s. Somebody also had a CD of The Chieftains, so we had loud Celtic music on the aft deck, and I wished I had brought my fiddle.

The next day we completed the Ashmore Reef atoll seismic transects, confirming that the sediments of this atoll were hard, cemented, marine and terrestrial material with little or no *Halimeda* flakes. We left this atoll midafternoon, after lashing down everything again, in preparation for a long three day run to Cairns for a crew change. Just to the south of Lizard Island, where my associate Ed Drew had worked on *Halimeda* mounds decades ago, Milliman and team obtained seismic data that was perfectly in agreement with previous ideas: the classic *Halimeda* mound was nearly pure *Halimeda* flakes to 30-40 m sediment depth, where 'Reflector A' indicated the low sea level stand surface some 12,000 years ago. To the delight of John Milliman, there was a series of Clint Eastwood films, some of which he wanted to see more than once! I caught up on my sleep and faxed Bruce Parker at AIMS about the next action schedule on this cruise. I wanted to take a few Kasten cores outside of Hinchinbrook Island, a few days' work on the way back to Townsville and AIMS.

We arrived at the port of Cairns (Trinity Harbor) at 0915 on 15 August. John Milliman and his crew had been furiously packing up all their seismic gear, computers, tools, and a lot of external hard drives. They rented a small truck to haul their gear to the airport for shipment back to their labs in USA. The ship took on fresh food, water, fuel, and some parts for the ship electrical generator. The Milliman crew got hotel rooms, and most of us went to party with them, including our new science crew from JCU, John Crockett (University of Washington,

Seattle) and an AIMS post-doc. There were gale force winds on the islands just seaward from Cairns, so we decided to stay in port for the day.

We spent the next three days trying to obtain sediment cores seaward of Hinchinbrook Island, but our equipment would not work in this hard skeletal carbonate gravel and sand. We sort of expected this, but we had to try, to confirm our maps of terrestrial river sediment being largely trapped in mangrove estuaries and the inner shelf bays. We were happily back in Townsville marina on 18 August, the ship Master Staib and John Pfitzner's birthday.

Two AIMS trucks awaited us, with a crane to unload our equipment for hauling it all back to AIMS for repairs and storage. Beaming Sheila gathered me up and hauled me back to Alligator Creek, where I slept for about three days.

And suddenly it was September 2002. The preparations for Jesse and Sarah's wedding were in place, so what could possibly go wrong? Well, Jesse's bachelor (buck) party went wild. It was a pub crawl for lubrication, then cheeseburgers at Hungry Jack's, some beach volleyball, a go-kart race, get into tux for a nightclub prance, Jesse being tackled by a very large bouncer and being badly bruised on his neck. Sarah was not happy about this...but then we never heard anything about her hens party.

Ian and Tiffany moved a piano to Queens Garden so Tiffany could play the wedding march underneath a rain tree. The wedding processional was the 'Theme from Man from Snowy River', and the recessional was a piece written by Sheila's mother. Sarah was half an hour late, so Tiffany had to scramble to keep music going. Flocks of parrots were screeching in the palm trees around the wedding site, and a big white ibis pulled discarded sandwiches from a rubbish bin behind the ceremony celebrant. Sheila and I sang two songs for the ceremony. It was about the right amount of chaos, and Sheila and I were proud to watch this good-looking couple take their vows.

At the post-wedding reception and dinner, brother Ian provided a PowerPoint slide show of the history of Jesse's childhood, as usual with a few embarrassing scenes. The fathers of the bride and groom gave little speeches. At the reception dance, I got to play some bouzouki tunes with Ian and a bass player. Sheila and I did some dance music as well. Our Alligator Creek neighbor Colleen's grandmother made the cake, covered with Alligator Creek orchids, with a horse on top. The wedding papers were signed with a pen made from a Moses the Emu feather. Good friend Rhonda Lyons bought Jess and Sarah a fancy room at Jupiter's Casino Hotel, but they spent the night partying in the casino till five am...their close friends from Emerald, Aaron and Carly with them...didn't get to sleep as they had to go to the airport for a 0600 flight to Vanuatu for their honeymoon.

It was rather restful at home after they left for their honeymoon. We had a good 2002 Christmas, and Ian's Canadian pianist girlfriend Tiffany was with us, delighted to have her own stocking full of toys. We spent a lot of time in our neighbour's swimming pool, as it was blistering hot and humid. Ian was building a bamboo meditation temple down in our creek valley orchard. At the demand of Sheila's older brother Ian, we recorded some music for him, Ian MacGreggor being the recording engineer. Somehow he obtained a hard disk recorder and a half a dozen microphones, which worked very well on our kitchen table. At this time Ian was giving lectures in the music department at JCU.

Chapter 50. Meadowlark CDs, MARGINS Mud and Black Sand

AIMS got a new ship, the R/V *Cape Ferguson*, and I had to revise my deck coring gear to fit on it in January 2003. We took the ship and the new gear out in Bowling Green Bay to test the new gravity corer, extrusion device and core handling deck table.

Bob Aller (State University of NY) and three graduate students (Madrid, Panzeca, Dantzer) arrived at AIMS. They were feasted at our home table, and we made ice cream in the hand crank freezer bucket. This was part of the new MARGINS research program funded by US NSF. We loaded our gear into the ship's hold on 16 January 2003, as the ship would service two other research cruises along the north Queensland coast before it reached Thursday Island in Torres Strait, where we would join it in February.

Early the morning of 8 February, we drove through rain and fog down the coast to get me to the Townsville airport, where Irena, John and Paul Dixon were waiting for me. We flew to Thursday Island in Torres Strait, where Justin McCallum and the R/V *Ferguson* awaited us. We pulled our equipment out of the hold, put everything else back into the hold for storage, and prepared the deck and lab for our work. Bob Aller and students arrived the next morning, and they spent the day calibrating their lab chemical equipment. John Pfitzner spent the day standardising and testing his radium and radon isotope methods, Irena set up her filtration gear and data recording computers, and I helped the ship engineer weld attachments to the new aft deck coring table, so it could be securely bolted to the deck. Australian Customs and Immigration visited the ship, obtained an inventory of our supplies and equipment, and cleared us for international travel.

And so, merrily off we go, steaming into the Sunday night of 9 February, a gentle misty rain cooling the deck. John pumped 1.6 m^3 of Torres Strait seawater through a small plastic tube filled with MnOOH coated acrylic fibres (from 1900 to 0100 the next day) to obtain a sample for radium determination. Radium atoms are sorbed on the MnOOH, and the radioactive emissions of radium daughters radon 223 and 224 were measured to estimate radium activity. Part of this research voyage purpose was to seek evidence for submarine fresh groundwater (usually rich in radium and other trace metals) flowing from the New Guinea mountains into the Gulf of Papua continental shelf.

We were at anchor in Daru by 1100 hrs for PNG Customs and Immigration, where officers searched the ship for illicit drugs, pornography and weapons. For reasons we knew not, they also emptied and inventoried our food freezers, asking the cook to open packages wrapped in opaque paper. We later found that there had been several disputes between PNG and Australia about apparently Australian fishing boats carrying large amounts of cash, drugs and guns.

Our job as part of the USA funded MARGINS program was to get into the small river mouths and shallow waters of the south coast of the Gulf of Papua, where the big US research ships could not go. In my earlier work here, I had avoided most of these areas, as they were not well mapped or sounded, and were mainly complexes of sandbars and tidal channels that probably move with the season. But the AIMS research vessels *Harry Messel* and R/V *Ferguson* were well suited for inshore work, as they had a 'skeg', or a steel cage protected screw (propeller) and rudder, and a shallow draft of about 4.5 m.

Nevertheless we grounded out in front of the Kikkori River delta in four metres of milky turbid water. First Mate Justin McCallum tried to pull the ship off the sandbank for several hours, but we eventually relaxed and waited for the tide to come in and lift us off. We had our evening meal tipped at an angle of about 20° as we sat on the sand bank. Brett the cook had a savage sense of humour, and he provided Viennese Coffee and a big cake with a ship decoration tipped on its side.

This area was full of black sand and huge mangrove trees with gigantic roots uplifted floating by the ship. This was a bad place to be at night, but we found safe anchorage for the nights behind mangrove islands. We steamed up and down our cross-shelf map coordinate lines taking mostly black sand grab samples and mapping salinity and chlorophyll. Most shallow water sites could not be cored for sedimentation history, as it was all hard packed black sand. On the outer shelf and slope we did find soft mud, and my new gravity corer worked well to capture several centuries of mud deposition history. Our radiochemical sampling for radium isotopes worked well, but we found no evidence for large amounts of groundwater with exotic chemical input, as we expected. Twice the wind and rain came up, and we had to run up the Kikkori River delta for shelter.

There were several villages in amongst this maze of deltaic islands, and some of the village people came out in their dugout canoes to trade for salt, sugar and cigarettes, while we got big bunches of bananas, mud crabs and smoked fish. During this down time, my notebooks show that I was designing a better aft deck core slicing table, to make the work easier and safer during rough weather.

We made friends with some of the people at Gauri Village, a logging station on one of the maze of deltaic islands. Nick Kaubai, a sawmill winch operator who spoke good English, came on board to give us a copy of their homemade delta channel map. He figured we were dumb white people completely lost, and hadn't a clue where we were. But he had friends on the Kumar oil pumping platform (ChevronTexaco, ExxonMobile, Santos) on the outer shelf south of us, and he got them on the radio to arrange a visit for our ship's company when the weather improved.

The Kumar oil platform had a crane with a winch wire caboose that could lift visitors off our ship deck and up onto the oil pumping platform, which was an amazing small city in a different world. We got treated well, served coffee and pastries, and given a tour of the computer rooms that control the PNG inland pipelines and the deep sea tanker vessel loading operations. The oil comes from the inland Lake Kutubu Oil Field, which went into production in 1992, and oil flows through pipelines in the Kikkori River channels and forty km offshore to Kumar pumping terminal where it is transferred to deep sea transport ships.

After a difficult voyage with some weather and equipment trouble, we returned to Thursday Island on 3 March 2003, and everybody went to the Grand Hotel for supper and beer. That night there was trouble on the wharf, as both Australian, Indonesian, and PNG police were again searching vessels and demanding passports. Several Indonesian *prahas* were set alight on the beach, probably for illegal fishing in Australian territory. The *Harry Messel* was cleaned up and all our gear was packed and lashed down for the voyage back to AIMS. The next day the science crew flew back to Townsville. Irena and Paul Dixon carefully brought two Eskies full of our frozen filters and water samples.

In March, Jesse's Canadian high school friend Kevin Wieler came for a visit, him being now a successful hog farmer. Jesse and Sarah treated Kevin to visits to farms and cattle properties of inland Queensland, the coastal regions up to Tully, and some wide-eyed appreciation of the last of our home-grown mangoes.

Mainly due to demands from relatives on both sides of the family, we recorded our first CD in our kitchen over several sessions from January to July. Our good musical friend Jennifer Hultgren joined Sheila and me for the Australian version of Meadowlark. All I remember about this was that it was almost too easy. Sheila and I had been performing most of these tunes for over twenty years. We rarely had to do a second take, and most of the editing was easy and quick. Ian set up the microphones and hard disk recorder (Roland VS1680) in our kitchen, so that snacks, drinks, beer and cider were within easy reach between tunes. Meadowlark, the band that plays on its stomach. And of course the CD title was *Kitchen Table Music*.

Our life changed at 11:30 am on 22 July 2003, when Sheila and I became grandparents. Georgia Rose Brunskill was born, loud and strong, as expected. Parents Sarah and Jesse were calmly radiant, strangely transformed by this little baby. So were we.

Our garden came into serious food production mode, and we were swamped with eggplants, fennel, garlic chives, leeks, tomatoes, sweet corn and a few big watermelons. Son Ian was busy performing with a West African drumming group (Wassa) and a jazz trio in town and at Magnetic Island tourist resorts.

We had been planning various research ship cruises for the US National Science Foundation funded MARGINS Project for two years in workshops in the USA and Australia. This involved some of the big USA ships that were involved in international world ocean operations. The R/V *Thompson* was in Vietnamese coastal waters in February '03, and arrived in Townsville in early March to load the Milliman/Nittrouer USA equipment that had been stored at AIMS.

This NSF funded USA ship was capable of seismic mapping of recent sediment thickness, swath mapping, and large piston coring to obtain a ^{14}C dated record of glacial and post glacial sedimentation. All of this work was far beyond the capability of AIMS ships and labs. But the big ship could not work safely in shallow water or in the rivers of the PNG coast. Nittrouer and graduate students hired the Ok Tedi Mines twin-hulled catamaran supply ship to explore the upper reaches of the Fly River Delta. Another UNOLS ship R/V *Melville* was scheduled to work in the Gulf of Papua in September. Much of my time was spent smoothing the path for various ship crews and international scientists to obtain the necessary clearances and visas for Australia and Papua New Guinea, and I became a familiar voice on the telephone to the embassies of Australia, the USA and Papua New Guinea. Some of this international diplomatic work was made difficult by armed military PNG seizures of Australian fishing vessels in Daru and Fly River Delta communities in October and November 2003. We never got an explanation of the outcome of these disputes.

Despite these complications, I led another Gulf of Papua shallow water expedition (3-24 November 2003) for the USA partners in MARGINS in November, using the new R/V *Ferguson*. Our target was to get estuarine and shelf salinity gradient samples of dissolved and particulate elements, zooplankton and benthic ecosystems, sedimentation rates and geochemistry, mainly in the poorly explored estuary of the Purari River. We found a lot of muddy water and a whole lot of platy black quartz, hornblende, biotite sand. The consortium wanted me to take sediment cores along the salinity gradient, but the freshwater and inshore areas had very hard packed black sand, ground-up mangrove wood (coffee grounds), and a lot of buried mangrove trees. Over this cruise, I snapped four of my steel core barrels and lost two core catchers, using a half a tonne of lead weight to drive the core barrel into the sediment. This was all hand-made AIMS workshop equipment that I admired very much. The ship engineer Ian Henderson tried to repair some of this gear on the ship, with pretty crude results.

The best part of this research cruise was the end of it in Thursday Island, after escaping a very serious search by PNG Customs and Police in Daru. We had damaged a lot of our equipment, and some of our plans were altered by rough weather and personnel problems. But we all got together at a Thursday Island hotel for supper, before everybody flew off in different directions. Bob Aller had brought a new hand-crank wooden ice cream bucket, to replace our badly rusted one. He had it packed in with his science gear, so we had it on the ship, and made ice cream with whole vanilla beans smashed into the mix. Everybody was telling stories about adventures at sea, or in strange sea ports. Several of the ship crew vanished with Bob Aller's two grad students (attractive young ladies), Irena, and our cook Brett. We later got the story that Brett and the two USA ladies entered a karaoke contest in a pub (an AC/DC song), but lost out to a Thursday Island girl. Afterwards, our ship Master, Ray Thorns, told me this following true story in the dark with cold beer on the aft deck of the R/V *Ferguson*. I had worked the Exmouth Gulf (NW Shelf) in 1996-98, so I knew most of the place names he mentioned.

Master Ray Thorns, Exmouth Gulf Story about Cyclone Beverly.

Ray told me this story on the upper deck of the R/V *Cape Ferguson* in the Thursday Island harbour on 21 November 2003, at the conclusion of a rough research expedition to the tidal rivers of southern Papua New Guinea, mainly the Pia-a, Turama, and Bamu Inlets. We had just finished a blow-out Torres Strait celebratory feast at the Jardine Motel, all fifteen of us eating roast duck, prawns and scallops, with large amounts of beer and wine, a good time was had by all. This below is Ray Thorns' story, mostly in his words: he is talking about the fisheries on Exmouth Gulf, on the far NW corner of Australia.

It was towards the end of November in '74 or '75 when I was a nineteen year old deckie working for M. G. Kalis Fish Company on the old mongrel rusty prawn trawler *Kachula*, Gary Flynn, Master. Gary, he died of an overdose of heroin a few years ago. The other deckie was Glenn, even younger than me. Kalis had ten other ships working in the Gulf, one of them called the *Murat*, skippered by Rick Rowe, big bugger, good bloke. The Fleet Master was a fellow called King, a real prick whose main job was to cut costs and increase profits.

We'd been warned there was a big blow coming, so we went to the Kalis Factory in Exmouth to get some better chain for our anchors. They pointed to a row of rusty barrels full of old World War II rusty chain, the best they had. We hauled this to the ship, along with some six inch hauser (tiger rope, black and yellow), and rigged the ship as best we could. We drove the ship over to Eva and Why Islands, the best place to anchor in a storm, and waited there drinking tea. We watched the sky turn pink as the wind picked up the red dirt from the land. Pretty soon the rain came, and the wheelhouse windows and decks were covered with dripping blood red chocolate mud coming out of the sky. The wind blew off the airscoops on top of the wheelhouse, the red rain poured onto the radar, and we had bolts of lightning arcing across the wheelhouse, and the smell of burning rubber and plastic. The wind was so loud we couldn't hear each other, even when shouting in your ear. There was nothing you could do but grit your teeth and hang on. The front windows in the wheelhouse were thick glass, but the side door windows were thick plastic and they were bulging inward from the excess pressure, three of us were pushing on the window to keep it from breaking. Your ears hurt from the pressure and the howling shrieking noise. My deckie friend Glenn wanted to go out to pee, but the master forbade him to open the wheelhouse door, told him to pee in a bucket.

We had both our anchors out on this old chain, one of them reinforced with the six inch hauser. The single chain snapped in the first few hours, leaving only one anchor on the chain and rope. The second chain snapped and threw itself back at the ship, across the top of the wheelhouse and aft deck. The big rope stretched, going from 6' to about 1' in diameter before it broke, with a loud crack, whipping back at the ship and tangling about the deck gear. The ship was now swinging wildly, we had no control at all, the lid on the fish hold brine tank flew up and became a sail, pushing the boat downwind stern first, water and foam rising over the stern, moving us into a pile of rocky islands. The brine tank cover was lashed down with cray-rope, was heavy hardwood and fibreglassed, and it jammed against the mast ladder and could not be brought down again. The ship engine was a Rolls Royce 240 hp and was useless against the wind at full throttle. The ship was perched for a while on top of some rocks, and we thought we were finished. But a big wave pushed us into a pool of protected deep water, where the Master told us to put down the net boards to help keep us there, as we had no more anchors. He also told us to abandon ship, but nobody wanted to go in the dingy, as there was about four feet of bloody froth on top of the water, and you couldn't see anything. Hard to tell which way was up. You would've suffocated in the foam I think. The seafoam was like the froth on top of a cappuccino, except it was bloody red. We must have lost track of time, because you couldn't tell if it was night or day.

The wind got even worse and blew us up on the shore, in amongst some sorry mangrove sticks and into the saltpan, the wind blowing the water SE above the high tide mark, and left us lying on our side. We woke up in the morning, the air still, quiet and hot, our faces pressed against the wheelhouse windows, all of us covered in red mud, but we were miles from seawater. The sand flies and mosquitoes were bad, so we rubbed 30 grade oil and salty mud on our skin. We had no food and only a little drinking water, so we burned some tires as a signal for a rescue signal for three days. An airplane flew over, circled and came down low, we thought we were rescued and there would be a food drop. A fellow in a white shirt leaned out the aircraft door with a TV camera, and then left us waving our hands on the sandflat.

Finally on the evening dusk, we could hear cursing across the sandflats. The skipper of the *Murat*, Rick Rowe, was dragging a zodiac (with an outboard attached) across the sandflats. He was cursing because he had gone up the wrong mangrove creek hunting for us. He gathered us up and we went out in the zodiac on the flood tide in the zodiac, and got back to the Kalis Factory. There were ten trawlers in the fleet, most of them had better anchors and chain, and they survived the cyclone. At the Kalis Factory everybody stared at us, because we were covered with grease and red mud. We got a shower and ate a lot of food. We had no ID, clothing or money, and no job due to the loss of the ship. We got little sympathy from the Kalis Fish Company, and we were some pissed off that the Fleet Master King did not come looking for us. We heard later that they dragged the *Kachula* out of the sandflats on a spring tide, refloated her for many more seasons of trawling.

So I signed on with 'Chuckles' Wilkins (his father was an American, his mother an Australian, he was always chuckling) for three months work in the Gulf of Carpentaria. He was a pisshead on land, but a good skipper when he dried out at sea. Chuckles often fished in the 'off-limits' areas during bad weather at night, because he knew the fisheries patrol vessels wouldn't come out during bad weather. The Gulf of Carpentaria fleet was owned by Henry Ott, and the ship I worked on was called *The Betty Ott*, after Henry's wife. The Otts had two sons named Frank and Bill, they weren't much good, so we called them Silly Ott and Idi Ott. The deckie was called Bum Nut, because he had a broken nose. Life was a little better in the Gulf of Carpentaria.

Chapter 51. Science in San Francisco and a Visit to South Dakota

After sleeping for a few days, to recover from a tough time on ships in the Gulf of Papua, Sheila, Ian and I drove inland to Cloncurry to surprise Jess on his birthday. We also toured the nearby abandoned mine sites, the Fountain Springs. Little Georgia Rose smiled a lot at the sound of my banjo.

Later, in December, the AIMS Christmas Party in the pouring rain featured skits celebrating the departure of our last CEO.

Some very disturbed person scheduled the American Geophysical Union Marine Science Meeting in San Francisco 8-12 December 2003, when most decent people would be thinking about Christmas holidays. We were in the midst of a good wet season, and I left Townsville 0600 in the early morning rain. I gratefully found a decent bacon and eggs breakfast in the International Brisbane Airport, where I also encountered my friend Peter Dymond Ramplin, a fellow folkie and banjo player of some repute.

He was bleary eyed and confused, sitting with his banjo and guitar case and no other luggage at a table with a big cup of coffee. He explained that he doesn't sleep well on airplanes, and he'd been travelling for a day and a half. He was returning from a badly organised concert tour of Germany, where he went bust in Munich but scored well in Berlin. He was *en route* home to Townsville now, fearing that his wife had bought a lot of stuff assuming he made a lot of money on this tour. In fact, he barely broke even. He wearily explained that the last time he was on tour in the UK, his wife bought a new car and a dog, and his daughter became pregnant. He said he was about to write a humorous ballad about all this.

And then it was Brisbane to Auckland, more food, and then a long sleep before landing in Los Angeles, the City of Angels surrounded by smoky amber-grey smog, on 6 December, the same Saturday I left home. The whole airport was blaring Bing Crosby, Andy Williams, and Jose Feliciano Christmas carols. Customs and Quarantine Service and Security agents were on high alert, and little old ladies were required to take their shoes off and dump their purses on the table. I usually carry a stockman's jackknife, and I had put it in my suitcase, but the security x-rays found it. They opened it, asked me where I was from and where was I going, shook their heads wearily and gave it back to me.

And then up to another departure gate lounge occupied by scientists heading to San Francisco. I sat down beside some Australians, most of them bitching about the recent loss of geophysical and geochemical science departments at La Trobe, Melbourne, and Monash Universities. We flew down the coast of California, over the ash grey denuded coastal mountains, to the chilly grey smog covered city of San Francisco. My luggage miraculously appeared with the latches broken, but a luggage agent had put a rope and masking tape around it. A hotel van took me to the top of a hill where I was bunked in the Cathedral Hill Hotel. The TV in my room was already on for the Army-Navy football game. I walked down the steep hill to a Thai noodle bar where I got a very good duck laksa. On the return I found a pastry shop just closing, and got some baklava and elephant ears for hotel room tea later. As I watched bad gangsta movies, Sheila phoned to announce the birth of Arthur Barry Gordon Dennett in Winnipeg, son of her brother Alistair and wife Sherri.

The next day I slept till two pm. It was Sunday, and the AGU registration and cocktail afternoon were in full swing down at the Moscone Convention Center in Union Square. This was culture shock for an Alligator Creek boy: too many people, too much noise. Here were street vendors peddling pot pipes, jewelry, hair decorations, pretzels, Cuban cigars, hot doughnuts with cocoa, all to the loud background of Peggy Lee and Bing Crosby Christmas songs and a particularly obnoxious country and western version of 'Jingle Bells'. A brilliantly tattooed and costumed Santa Claus roared down the sidewalks on a Harley-Davidson motorbike, throwing handfuls of Tootsie Rolls at gaggles of kids on the street. I innocently wandered through a Saks Fifth Avenue store and a Macy's store, looking for exotic Christmas tree ornaments for Sheila. I escaped to a bench in the square where I rolled a cigarette. Two young kids came to sit beside me asking if I was smoking weed, and could they have some. Yup, this was La La Land. On the way back to my hotel I found an Indian restaurant that fed me some very hot spiced butter chicken, and then I needed to sleep again.

The next day the real work began, helping to orchestrate, enhance and manage a two day session on our work in TROPICS and MARGINS in Papua New Guinea. My good colleague Bob Aller and I were co-chairmen, and we had to juggle the schedule due to some speakers being sick or absent. This was a popular session, as not too many big research projects were in remote, little-disturbed, tropical locations. There were presentations with three slide screens going at the same time, videos and interviews with local people. Australian, US and Indonesian speakers gave information on the Pleistocene evolution of the Fly and Sepik and Mamberamo River channels over the past eight ice ages, engineering models of deltaic evolution with rising and falling sea levels, past and present submarine channels offshore from the deltas, different modes of sediment transport landward (up into the estuary) and seaward (usually via submarine canyons). Bob gave his interpretation of an organic matter budget for the south coast riverine inputs to the gulf: virtually all modern organic matter was being burned off (metabolised, decomposed by microbes), this tropical and natural incinerator being fueled by highly nutritious marine organic matter. Even more surprising was that much of the river supplied organic matter was ^{14}C dated to be several thousand years old. Bob's wife Josie, who is a microbiologist, expanded upon this theme using SeaWifs satellite imagery for chlorophyll, macrobenthos and microbial abundance, and the short lifetime of organic matter along the estuarine salinity gradient.

I gave a paper on the river and shelf budget for barium and radium. There were lots of poster presentations by graduate students about various remote sensing tools being used in the Gulf of Papua and Coral Sea. This was all an enormous amount of information, and I hoped to gather it into a whole issue of a science journal.

Of course the collegial spirit was in evidence during lunch and suppers. We were always hunting for a more exotic kind of food. We found several 1950s style diners with booths and punch button juke boxes featuring The Platters, The Four Freshmen and Elvis music, just like high school at Mack's Café in Murdo!. You could also eat on spinning stools at the bar, where busty girls provided free coffee if you had a piece of their pie.

Our TROPICS/MARGINS team got to relax and attend sessions on submarine vents by Russian and US experts, the distribution of methane hydrates in the mud of continental shelves of the Arctic to the Tropics, and the danger this represented for rapid increases in greenhouse gases in the atmosphere (the clathrate gun hypothesis). Jordan Clarke described the Coal Oil Point methane seep on the California coast, a massive 100,000 m^3/day injection of powerful greenhouse gases from over 3 km^2 of the coastal sea. Several papers showed swath and seismic mapping of massive methane vent chimneys in the northern Gulf of Mexico. There were German oceanographers reporting on methane gas hydrate volcanoes offshore from the Congo, some of these mud volcanoes being decorated with colorful colonies of tube worms, molluscs, acanthocephalans, pogonophorans and crustacea. All of this fired up my desire to explore the Timor Sea for methane vents.

We attended all sorts of other palaeoclimate research from ice, sediment, and coral bommie cores, French work on Lake Chad, cave stalagmite isotopic history of Mayan collapse and the Medieval Climate Anomaly in Central America, and the history of dust inputs to the

Pacific. Good friend Taro Takahashi gave a magnificent summary of twenty-five years of study of the partial pressure of CO_2 in the Pacific, showing that the current rise in anthropogenic atmospheric greenhouse gases resulted in an increase of 1.8 μatm/yr in the surface ocean, and how this was a large sink for the combustion products of coal, oil and wood over the last century. Colleagues Steve Emerson and Roberta Hamme were testing the solubility and distribution of neon, krypton, argon, and nitrogen in surface and deep seas, trying to understand how rapidly the ocean can come to equilibrium with changing atmospheric gases. Several large studies reported on how the continental shelf sediments were a major sink for the troublesome greenhouse gases let loose on the planet by humans. There were big sessions on POPs (persistent organic pollutants, pesticides and industrial poisons) in the deep ocean and polar zones, often in association with plastic particles in the water and sediments. There was frequent mention of the idea of the 'Anthropocene' as a new geological planetary era, caused by misguided actions by human societies.

MARGINS leaders Chuck Nittrouer and Bill Dietrich called a meeting out on the University of California campus in Berkeley. There was great enthusiasm to continue the 'Source to Sink' study on the PNG and New Zealand rivers and estuaries. We spent the whole day trying to identify the next big thrusts in our consortium research program, the missing links, and how much money we would need to do this. Many of the reports at this meeting displayed how important it was to understand the past cycles of glacial and interglacial sea level, physical and chemical erosion rates on land, distribution and burial rate of organic matter, and how this affected climate. It was rewarding and heartwarming to see how our small Australian contingent was well regarded by our colleagues in the USA and Europe.

And then it was all over. The adrenalin declined, and exhaustion seeped into our brains. I had to wake up at 0300 on 12 December to catch a shuttle bus at 0400 to reach the airport for a 0600 flight. The shuttle bus driver must have been a close relative of Steve McQueen, as he enjoyed steep hills, fast corners, evading trolley cars, and shuddering fast stops. My departure gate at the San Francisco airport seemed to be full of ornery looking cowboys with big hats, high heeled boots and large belt buckles en route to Denver. It was snowing in Denver, and the smoking lounge was full of cowboys with cell phones glued to their ears and tall glasses of whiskey beside their big plates of bacon and eggs. I flew over the dry brown prairie with patches of snow wherever there was enough grass and shrubs to stop it blowing away. There was not very much grass and shrubs...most of it was brown dirt blowing in the wind.

On the outskirts of Rapid City, there was a bit of tall grass prairie around the airport, which felt good. My brother Obie was there to meet me in Levi jacket, cowboy hat and boots, with his big diesel pickup truck. We were off to Wall Drug for cheeseburgers and beer, donuts and coffee, and I could absorb the Dakotan language again...the lack of rain, cattle prices, etc. Out at the ranch north of Philip, parts of the family were gathering, and the phone was ringing. Toady in Minnesota wanted to know about our new granddaughter Georgia Rose. Toady also spoke of Cherry Stubbs dying in the hospital there of bone cancer, having just returned from genealogical research in Scotland. She was one of the major sources for my England and Butterfield family genealogies. Niece Marcy phoned from Rochester, getting ready for a caesarean birth of her son Peter, but bubbling with Christmas cheer. Obie's wife Annie and niece Donna arrived, and the feasting began, as well as recitations of all the progress of my relatives (three pages in my journal, and the fridge was covered with photos of grandchildren). It was a little embarrassing when some of the neighbors drove into the yard: they wanted to see this strange brother of Obie, the one that lives in Australia. 'How come you are living down there?'

The next morning, Obie took a pickup load of kids to the stock dam to get two gunny sacks full of ice. He also pestered Donna to make the ice cream mix. I teased my brother for having an electric ice cream freezer, as the kids didn't have to crank the freezer. We had a big noon dinner feast, stuffed turkey with cranberry sauce, roast potatoes, acorn squash from the garden, sweet corn swimming in tubs of salty butter. We had to put two extensions in the dining room table. Ross brought Australian Foster's beer to keep me happy. Afterward there was a football game on the living room TV, and a rodeo on the basement TV. There was a big wood stove in the basement, stuffed with cedar, elm and cottonwood fence posts, to keep the house warm despite the screaming wind outdoors. I helped do chores with Obie and Ross, and we could see maybe fifty pronghorn antelope on the ridge to the south.

Donna, Obie and I drove to Murdo to visit my brother John and his wife Pat, also bringing John a few tubs of dehorning paste, which he couldn't get in Murdo. The old England house was nicely decorated, the big spruce trees lit up, and John's son Dan was building three bedrooms in the basement. After a home-grown steak supper we watched NRA rodeo, where three of the top bronc riders were South Dakotans. The next day, John drove me down to the White River ranch, where my uncle Gerald and Aunt Joyce were living. Joyce was spry as usual, but Gerald was very thin and white haired. Gerald was a civil engineer, and he wanted to talk about my Australian son Jesse who was a civil engineer, my attempt at the Brunskill and Gregg genealogy, and how GPS and ArcView works on a computer. I had seen some of his hand-drawn engineering designs for some of his projects, and I'm glad I didn't have a computer with me. The old ranch house felt good to me, familiar furniture in the living room beside the fuel oil stove. We drove across the river and up onto Prosperity Flats to visit my uncle Bernard and aunt Margie, where they served us BBQ beef with cheese coated potatoes.

Some of the downside of this visit was to learn about John's arthritis. He is on a no-carbohydrate Atkins diet and wears high lace boots to keep his ankles from twisting, but he still uses Copenhagen snuff. He took one look at me, and said I'd be shot/culled if I was a steer. I'm too skinny he said; his steers gain four lbs/day. Brother Obie was struggling with heart and ear balance problems, and was taking a lot of pills. Several of my high school classmates were not in good health, and the little town of Murdo had a main street that was mostly boarded up and closed. Philip, however, seemed to have more town business and good useful shops.

I was up at 0700 on 15 December 2003, and the South Dakota wind was roaring and howling around the house. I packed up Ian MacGreggor's box of musical items (shipped from Pennsylvania) and my gear in plastic bags for the back of John's pickup truck. We loaded a ton of livestock salt blocks in the pickup tray for ballast, in case the road was icy. John had two big parkas, tire chains, towing chain, a handyman jack, a shovel, two cartons of Copenhagen, and two big sacks of salted peanuts for his winter emergency gear. We fueled up at the Sinclair station in Murdo, where Howard Peters was still in control. He was eighty-four yrs old, had a big wad of Copenhagen snuss under his lip, a big brass spittoon beside the cash register, and he had some of his old buddies in the back drinking coffee. They were all puzzled to have a look at 'Gregg and Virginia's oldest boy, the one that lives down in Australia.' 'What did you want to live down there for, anyway, how come?' We left Murdo at 0900 in blizzard conditions with an icy road, but it all cleared up at Wall Drug, where John slyly drove up to the restaurant. He made me promise that I would not tell anybody that he ate a hot sugared raised doughnut with coffee. It was good to drive through the Badlands, the Cheyenne River breaks, where the deer and the antelope play. And then we are at the Rapid City airport and John had to leave quickly, as the blizzard was getting worse. The airport parking lot seemed to have mostly pickup trucks with bales of hay, saddles, salt blocks, and big rolls of barbed wire in the back. [If these South Dakota visit paragraphs above sound familiar, you can visit earlier chapters to see almost the same sequence of events for a family visit.]

Now I have to shift my brain from South Dakota mode to that of an international traveler. The Rapid City airport is very civilised. You can smoke in the restaurant, where I had a hot Reuben sandwich and cold beer, with a large piece of warm rhubarb-strawberry pie smothered with clotted cream and washed down with endless cups of coffee. I was reading some of Larry McMurtry's books, which fit right in. The blizzard was howling outside the airport, ice covered flagpole rope clanging, and there was a good view of the encircling truly black Black Hills from the upstairs pub windows. At the back of the pub were some big round tables where about a dozen good old boys sat playing cards, smoking and drinking Wild Turkey whiskey from big water glasses. There were huge brass spittoons beside this table. Standing at a high table at the other end of this pub were two gentlemen in dark suits, white shirts and neckties, shiny briefcases, plugging their laptops into the wall circuit. There was an abyss between the small town prairie cattlemen and the world of CNN, the megacorporations and global affairs. The super patriotism, imperial/military/gun attitudes, and the idea of US exceptionalism began to grate on my nerves. The airport TV was showing the capture of Saddam Hussein and the booming financial market news. This was all somehow unsettling.

The Dakota wind was strong enough to rock the small airplane on the tarmac in Rapid City. The stewardess asked us all to move to the back of the plane, so the pilot could get a quicker takeoff. Denver was ok. I found some honey soaked baklava served on paper plates by prancing cowgirl waitresses in the smoking lounge. Los Angeles airport was nearly in lockdown at Terminal 4 Tom Bradley International, and I survived three separate security gates, x-rays and pat-down body searches. They were a bit puzzled with my two passports (Australian and USA), and wanted to know if I was a medical doctor: they had a hard time with the idea that I studied river mud in the coastal ocean. I was grateful to escape to the Qantas lounge, where a nearby Mexican restaurant served hot chili, spicy chicken wings and beer, and there was football on the wide screen TV.

I left L.A. at midnight, went to sleep, and woke up as we landed in Sydney. Customs and Quarantine wanted to clean my shoes, as I admitted I had been on a cattle ranch, but they had no interest in several jars of horseradish in my suitcase. And on to Brisbane, where it was warm and sunny, and on to Townsville, where the air was warm and perfumed, and there was my bright and beaming and shining Sheila, hibiscus in her hair. There were brilliant orange red poinciana flowers outside the airport door...what a change from blizzards in South Dakota! And then we drove down to Alligator Creek, where the house parrots Sir Charles and Deputy Caspar had to check me out, suspicious of me being away. Ian came down for a supper of green curry prawns, so he could get his box of sheet music and drum bits.

I picked our own mangoes over the Christmas holidays, and made twenty litres of puree out of the excess. We had a good wet season, our little creek was flowing, the orchard grass was over my head, and the big fruit bats were very happy with all the fruit. We had a good Christmas and New Years with the family home with us. As usual, the house parrots Charles and Caspar had their perches in the midst of the Christmas tree presents, and they had their own stockings filled with toys and seed sticks. We had to take Sheila to the AIMS beach on Tickle Belly Bay, so she could fly her Christmas kite while up to her neck in the Pacific Ocean.

GJB and Sheila, AIMS Beach

Chapter 52. A Good Wet Season, the Big US Ships, and Another Timor Sea Adventure, 2004

In early January Sheila and I drove through pouring rain northward to the little sugar cane town of Tully, where Jesse's wife Sarah's parents lived. Doug and Dinah Hazzard had their garage set up for a family feast, and Doug had his grill covered with snags, chops, eggs and onions. Sheila provided a mango cake, and I ran the ice cream freezer, with some help from the kids. Doug had asked for music, so Sheila and I brought our instruments and played for about an hour, as long as the brandy and tea was served. The next day, we performed a two hour show for the Blue Nun Old Folks Home in Tully.

We drove to Cairns on 5 January, where the big USA ship R/V *Melville* (120 metres long) had arrived from Vietnam/Thailand seas, and was being loaded with railway containers of equipment from MARGINS leaders Chuck Nittrouer, Dick Sternberg, and Bob Aller. Chuck had his wife Debby along, a ballet teacher, and Dick's wife Lois was a piano player, all of these people had been to our home for feasts during TROPICS. The ship was a small city, and would be in operations in the Gulf of Papua through May of this year. I wasn't personally involved in this, but I had helped organise ship clearance, visas, and links to the University of PNG.

A few days later, my Bretagne post-doctoral fellow Philippe Monbet arrived with his wife Caroline. We put on a big house party for Philippe and Caroline, who were infatuated with our house parrots. Caspar seemed to like eating off Philippe's supper plate. And thus began a year of lessons in a version of the French language spoken in Brittany, and many difficulties in finding equivalences for Australian slang. For example, we had a Russian tea cosy, in the form of a Leningrad babushka with heavy quilted skirts, which we called 'Hot Crotch'. The French version of this was not very polite. We also had Kathy Burns and her Belgian-French student

Cecile Kohn, and my Ph. D. students Thomas Steiglitz and Severine Thomas around the table. Sheila made tandoori chicken and a wonderful coffee/chocolate ice cream.

We had a lot of rain and the insects, mosquitoes, deer flies and rhinoceros beetles were seriously out of control. The rain flooded large areas of inland Queensland, causing a lot of damage to Jesse's roads west of Mount Isa. For Sheila's Valentine's Day present, I had brought in three trailer loads of good dirt to help make a lawn in front of our house (where stony poor soil exists). The torrential rain washed most of it down the hill. Sheila had rescued a baby scaly breasted lorikeet from under the house, and was feeding it crushed mango and porridge. Sarah is a licensed carer, so we drove west to Hughenden to meet them and give up the bird. By this time, our granddaughter Georgia Rose was completely familiar with parrots, snakes, wallabies, Brahman cattle and big lizards, and gleefully pulled on her Grandpa's whiskers.

In February , I flew with Irena, John, Dave McKinnon, Paul Dixon, Kathy Burns, Cecile Kohnn to Cairns and Darwin. I was reading Joseph Conrad's *Typhoon* and it made my research partners nervous, hoping this was not a signal for the weather up ahead. This expedition was part of our effort to locate and quantify the seafloor source of methane and oil sheens in some parts of the Timor Sea. Kathy was determined to find out if the Timor Sea hydrocarbons were similar or different from those found on the Exmouth Plateau of the NW Shelf, from our 1996 R/V *Franklin* expedition. Cecile was a Belgian-French expert in hydrocarbon gas analysis by gas chromatography, and she had installed a special sensor in Kathy's GC for methane. Dave was measuring zooplankton abundance, algal growth and respiration rates, wondering how methane seeps would influence sea productivity. Paul was trying to measure microbial metabolism in the seafloor mud, and John was our radiochemical expert on radium and radon fluxes from the seafloor.

We were delayed in Darwin, as an important part of our specially designed water sampling equipment had not been delivered. This was a special rosette frame for six 30 litre Niskin bottles, which could sit on the seafloor to sample water from near the sediment surface. I used this week of down time to write my introduction to the TROPICS issue of the science journal *Continental Shelf Research*. This delay was also good for us, as there were two cyclones wandering around over the area we wanted to visit with our ship.

Kathy and I had also helped organise a 3 March workshop in Darwin on the objectives and priorities for oceanographic research on the Timor Sea. This was part of the beginning of ATSRF (Arafura Timor Sea Research Facility) which was being formed by the Australian National University (ANU), AIMS, and the Northern Territory University in Darwin. There were also speakers from LIPI in Indonesia, Timor-Leste Irian Jaya, Torres Strait people, and Papua New Guinea, all being concerned about their coastal seas. The CEO and most of the project leaders of AIMS, ANU, and NTU gave little speeches. Kathy and I presented information from our past research, and the petroleum companies were vocal in the audience.

Much of this discussion was the passion of Bob and Marilyn Wasson of ANU, and the main message was that there was a great need to push oceanographic and fisheries studies in this region, fully loaded with oil rigs and plans for natural gas extraction. Marilyn was a social scientist, and was concerned more about the native people and their food sources, so she called the consortium ATSEF, for the Arafura Timor Sea Environmental Facility. There were already several international treaties and agreements between these nations, but there was little practical protocol for funding or support for the poor villages bordering these seas. This region had been ruled and contested since World War II by the Portuguese, Dutch, Germany, the UK, Japanese, and now Indonesia and Australia. Timor-Leste had recently gained an uncertain independence from Indonesia, and Irian Jaya was a troubled colonial outpost of Indonesian military police. For example, the boundaries that exist now on the Arafura and Timor Sea are contested by Indonesia, Timor-Leste, and Australia, a situation which inhibits free movement of research ships, personnel, equipment, moored sensors and petroleum/gas exploration and extraction. Big plans for future symposia, conferences, UN support, and funding from the oil industry were earnestly discussed, but we never got much benefit from all of this well-intentioned hot air.

Agreements between Timor-Leste and Australia were never respected, and were abandoned in about a decade.

Meanwhile, the *Cape Ferguson* and crew had been tied up at the Stokes Hill Wharf at the Port of Darwin. Cyclones Monty and Evan were causing some damage to the oil production platforms and their support ships in the Timor Sea. The ship Master decided to go to sea on 5 March, even though the sea would probably be still rough. We tested our new water sampling carousel niskins, and a sneaky way to sample the gas in the water, just outside of Darwin, where we did not expect to find seepage gas. Our new CTD was also balky, but we found that the new laptop that received the CTD signal had the wrong software, so we had to adapt an older laptop to the job.

Two days later we were 400 km NW of Darwin, at the edge of the continental shelf near Karmt Shoals, in water depths of 500 m. We had been here before, thinking that these shoals were giant *Halimeda* mounds, initiated and fed by submarine seep nutrients over the past several million years. The wind had calmed, we had only one metre swells to worry about, and it was bloody hot on the aft deck. We began our routine sampling, a CTD profile, water sampling down the salinity/temperature/oxygen gradient, pumping several cubic meters of water from the surface and near the bottom for radon and radium isotopes, samples of zooplankton, water for algal productivity and respiration, and several different kinds of sampling for methane in the water column.

For example, surface water was at 28°C., salinity was 34.4 psu, oxygen concentration was 160 μM, and chlorophyll maximum was at 70-80 m depth. Water at 500+ m was at 7°C., salinity was 34.6 psu, O_2 was <90 μM, and radium/radon isotope activities were at background levels. Kathy's small sediment traps were deployed for one day at 100 and 400 m depth, and successfully recovered the next day. By measuring oxygen concentration changes in sealed bottles incubated in a screened water bath, Dave was able to determine that these waters were strongly heterotrophic, and that respiration and organic decomposition was much faster than algal growth rates. This would be expected if there were large amounts of easily oxidised organic matter (hydrocarbon gases) being supplied to the water column.

Kathy has especially modified one of the Niskin bottles so that she could sample by syringe the gas released inside the bottle as it rose to the top, due to the release of 50 atmospheres of pressure. I didn't think my modified Kasten coring rig would work well in these deep water sites, but we were lucky to get a 1.7 m core on the third try. The green/grey sediment had exotic carbonate shells of pteropods, scaphopods, polychaete tubes, crinoid and coral fragments, sponge spicules, and foraminifera. We followed this routine for the next two weeks. We used our special Niskin carousel with 8-30 L Niskins to obtain 240 L water samples for methane and radium/radon, but we did not find evidence of submarine seeps around the base of the giant *Halimeda* mounds. David's bongo net was designed to allow sampling of deep plankton, and he found more plankton than would normally be expected for these algal poor tropical seas. Several of Kathy and Cecile's samples indicated both oily residues and low molecular weight hydrocarbon gases at concentrations higher than would be expected for clean tropical ocean water.

We also had MaxSea mapping sonar ability on this ship, and we amused ourselves by putting names on mounds and submarine trenches: there was Justin's Mount, McKinnon's Ridge, Hendo's Mount, Carmel's Valley, Pfitzner's Fissure, and Irena's Canyon. Most of this work was in deep (400 m+) water, but we could always steam to one of the big shoals for anchorage for some sleep and rest. The big *Halimeda* mound shoals were within 20-30 m of the water surface, and the shoals reduced the big swells. The tops of the *Halimeda* mounds were largely vast areas of white *Halimeda* flakes, with a few stony and red corals.

At peace at anchorage on Karmt Shoal top in the evening, our cook Carmel made filo pastry coated seafood, freshly caught by the crew the previous night. Cecile made chocolate ice cream puffs, and the ship crew applauded and whistled loudly. After the hot sun went down,

Justin McCallum, Ian Henderson and David McKinnon jumped ship, swimming around with the salps and chambered nautilus.

At anchorage on top of the shoals some nights the sunset made the quiet water look like red molten metal. The pitch of the hydraulic pump and the winch noise were a perfect fifth interval, but when the winch stops the pump motor pitch rises up a whole tone. We could sometimes see the drilling platforms at night, flaring off the natural gas in wild orange-scarlet flames, lit up like Christmas trees over the sea. The wind would whistle through the ship air vents like wind chimes when I tried to get some sleep, and I could hear at least four tone scales. We had dolphins around us most days, and sea birds circling and diving into the propeller foam. North of the shoals we often saw Indonesian (Bugi) *prahas*, colourful shade cloth over the deck, latine sails sewn and patched many times. Steaming toward Darwin we could see columns of smoke rising from the distant land; the shoreline was on fire, and we had woodsmoke and ash all over the ship.

We were safely back in Darwin harbor on 15 March, and we loaded our gear and samples on a semi-trailer for shipment back to AIMS. Then we all went off to a good seafood restaurant, where my Tooheys New beer had a fly in it, but we had good laksa and mud crab. I discovered that our engineer Ian Henderson was a Miles Davis jazz maniac, and I made a note to bring some CDs on the next *Cape Ferguson* expedition. Irena and Cecile were in charge of getting dry ice for some of the samples being airlifted back to AIMS. The next day we flew to Cairns and Townsville, back to a different kind of civilisation. The *Cape Ferguson* and her crew would steam eastward around the Gulf of Carpentaria, Torres Straits, Cape York, and south to return to her home port at AIMS.

At home I spent several weekends whippersnipping the tall grass in our orchard. By June and July we began to harvest a decent crop of bananas, citrus fruit, Seville oranges, mandarins, tangelo and two kinds of limes. We recorded our second Meadowlark CD (*Kitchen Sink Music*) in June-December, with Jennifer Hultgren and Ian MacGreggor the recording engineer.

We went out to Mount Isa for our granddaughter's first birthday in July, and she had teeth and no fear, liked to eat mud, and rode a Shetland pony named Mischief. In September, my Father's Day treat was to help son Ian MacGreggor to scrape, stretch, and frame a Brahma steer skin, for the purpose of making drum heads. By this time, Ian was teaching at the JCU music department and in the Townsville school system. Jesse was a civil engineer and project manager for state road and bridge building to the west of Mount Isa, toward the town of Camooweal on the Queensland/Northern Territory border. He moved his family out to a cattle station named Yelvertoft, where they lived quite happily in a huge shed with all the conveniences, and so he could be closer to the construction crews. He learned a lot about the local aboriginal people, some of them descendants of the Kalkadoons, a rather unusually organised group who demanded some voice in selection of the route of the road. Several sacred sites required moving the road move to the north or south.

Jess, Ian, and I organised a helicopter ride for Sheila for her birthday out of the Ayr Aerodome. She got to ride a little two person bubble chopper over the sea coast, salt pans, our home, and to fly with the kites and eagles. We had two Christmas parties, one with Jesse at Yelvertoft, and one in Alligator Creek with Ian and the parrots.

I had been working and editing for the past two years to assemble all the TROPICS science reports into one whole issue of a science journal. This finally got published in the 2004 December issue of *Continental Shelf Research*. It felt good to see that our work spurred on the continuing USA supported research in the PNG Gulf of Papua for the next several years.

I had done plenty of work in the Gulf of Papua, but AIMS did not want to continue to support myself and international scientists on our ships outside Australian territory seas. In addition, I was being sucked away from this PNG program by the tantalising vision of submarine seeps in the Timor Sea, their importance in supplying greenhouse gases to the atmosphere, and how this might be involved in the geological history of climate change. AIMS wasn't very hot on this direction of research either, but I wrangled enough logistic money support from CSIRO and universities to keep going. Kathy Burns and I wrote an ambitious proposal for a research cruise in the big CSIRO ship R/V *Southern Surveyor* for next year, but the chances of success seemed dim. The National Facility of CSIRO Ships required input of ~$30,000 from the applicant's institution, as a partial support for the much larger cost of running the ship.

The world expert guru on submarine seeps was Allan Judd, who was part of a CSIRO workshop that I attended in Canberra in early October, organised by Graham Logan of Geoscience Australia. I attended this workshop to learn more about how these seafloor volcanoes, pockmarks, black smoker chimney vents operate. Judd was from UK and had worked with Martin Hovland on seeps in the North, Irish and Norwegian Seas, where seeps were sometimes associated with deep sea coral reefs and/or exotic chemosynthetic ecosystems. Judd actually 'wrote the book' on submarine seeps, and he told lots of stories about bad experiences with ships and equipment in those nasty cold rough northern seas. One of his UK ships caught fire and destroyed ten years of equipment development. He recommended using 100-200 kHz sonar and low frequency swath mappers to detect and monitor gas seeps, and to avoid trying to drill core the seafloor vent location, for fear of letting loose a volcano.

There were also speakers concerned about the need for protection of the exotic biodiversity associated with these seeps. Part of the support for this workshop came of course from the petroleum industry, under a banner title 'Wealth from Oceans', and they wanted to find large (commercial) deposits of methane hydrate deep in the sediments, using seismic and sounding methods. I had a good Greek restaurant supper with Alan Judd and Wayne Baily (Perth CSIRO), spicy scallops, espresso and baklava. I was grateful to have input from these experienced fellows on our *Southern Surveyor* cruise plan for 2005.

Chapter 53. More Hot Air on Climate Change, and Finally We Locate a Bubbling Methane Seep in the Timor Sea, 2005

We had good rain in early 2005, and the Burdekin River was in flood by late January. We had been working on the chemical and sedimentary oceanography of Upstart, Bowling Green, and Cleveland Bays for the past decade. These bays are fed by the Burdekin River at high flow, so we wanted to sample the dissolved and particulate chemistry of the river at high flow and track the flow path of this muddy water along the inner shelf northward to Hinchinbrook Channel, our other study site.

To gain access to the Burdekin Delta channels required a week of phone calls and meetings with the Burdekin Water Board, the Burdekin Delta cane farmers, and Sun Water (the agency regulating city and irrigation water supply). John Pfitzner and I drove an AIMS flatbed truck onto the Rita Island bridge over one of the main channels, and we filled three one cubic metre tanks for our work. The flood water was nearly touching the underside of the bridge, and we heard that the next day the bridge had been washed away by the water over the top of the bridge railing.

The activity of bomb fallout ^{137}Cs and natural atmospheric fallout ^{210}Pb and phosphorus concentrations on the suspended sediment told us that most of the sediment was topsoil, rich in fertiliser, and not deep gully erosion. The dissolved elements suggested that the flood waters were removing a lot of the fertiliser used on the cane fields.

THE CASTROL BARREL STORY

GJB, June '05

My gun licence expired a few years ago, and I was instructed to go to the police station to renew my existing licence for my little .22 rifle. This expedition was combined with a trip to the dump (tip) with our domestic rubbish. As it turned out, the trash barrels were badly rusted, so I took the old barrels and the rubbish in the back of Ruby, our old battered Subaru station wagon, to the dump. At the dump, I bought two Castrol barrels in good shape, for our home rubbish burning.

Then I went on into Townsville to find the police station. I waited in the lobby for a long time, as there were other people with their own problems trying to get the attention of the staff. All of a sudden, three very large police officers in flak jackets came into the lobby, asking who was the owner of the Subaru parked outside. I stood up and answered, and was whisked into a back room for a rather scary interrogation. They wanted to know where I was born, where I lived, what I did for a living, and what was my business here. I told them I was here to get my gun licence renewed, and they wanted details of my gun collection. They wanted to know what was in the Castrol barrels in my car. I innocently told them that they were empty, and that I planned to use them for burning rubbish. They asked me to go to the car and demonstrate that they were empty. I walked to the car and the policemen stood far back as I opened the back door, pulled the barrel out, took the bung out, and tipped it upside down on the asphalt. The barrels had some residual oil in them, which dripped out. But they could tell that the barrels were empty, as I could lift them easily. They began to relax, and eventually smiled and apologised to me.

All this occurred shortly after the Oklahoma Federal Building bombing, where a van had been parked in front of the building loaded with barrels of ammonium nitrate soaked in diesel fuel. Apparently the police had spotted the old beat up car with two barrels in the back, and assumed the worst.

I eventually did get my shooter's licence renewal, with special attention from several of the head honchos at the police station. One of them was from Alligator Creek, a good old boy who liked to talk about hunting with big rifles, and had hunted buffalo in the Northern Territories when he was young.

In April, we got word from CSIRO in Hobart that the *Southern Surveyor*/Timor Sea submarine seep proposal by Kathy Burns and myself was approved, subject to payment of the fee and a detailed cruise plan. AIMS was not happy with this. The research director thought it was an April Fool's prank, as there was not enough money in our project funds to pay the fee. For a while, the CEO threatened to cancel the project, but he eventually paid the bill after I got some of our collaborators at universities to pitch in some money.

To many scientific institutions, having an approved National Facility ship operation is a mark of success and prestige, and it would have been embarrassing to AIMS' reputation to cancel the project. But it also meant that we had to get our act together, operate on a low level of funding, bring in collaborators, and organise shipment of our gear to Darwin one more time.

AIMS had been reluctant to join the climate change bandwagon, probably because it was a federally funded institution, and we were being governed by politically conservative people who didn't really believe in this 'nonsense.' I talked up AIMS attendance at a Hobart (Tasmania) climate change workshop in April with CSIRO Marine Science, Geoscience Australia, and several universities. We finally got four people to attend, with some of the GBR coral core climate signal group. And so off we go, down to Brisbane (the Qantas flight had too many chattering teenage girls with toe rings, tattoos and pierced bellybuttons), Melbourne, and into Hobart by nine pm where it was chilly and raining.

It was an excellent workshop, and the AIMS speakers were well appreciated. Madeleine van Oppen from AIMS gave a synthesis of evidence for global warming as a cause of coral bleaching around the world. Scott Wooldridge and Craig Steinberg gave excellent presentations on managing a World Heritage Area, and the distribution of hot water along the GBR coast. CSIRO Marine and Geoscience Australia were much larger institutions than AIMS, but they were anxious to establish collaborative projects with biologists, chemists, and physical oceanographers in tropical regions. Much of the planned work by CSIRO Marine and Geoscience Australia was in the Southern Ocean, the polar seas, where greenhouse gas heat was being absorbed in surface water and transferred to deep Pacific currents. Several colleagues wanted funding to study the titration of seawater alkalinity by carbonic acid from the increasing concentrations of CO_2 in the atmosphere, and preliminary data suggested that the Southern Ocean had already been titrated down the omega scale. Pteropods (sort of a small marine molluscan butterfly) were abundant in the Southern Ocean, their shells were made of aragonite (a crystallographic form of $CaCO_3$), and pteropod shells were dissolving due to the decline in pH. There were several very good proposals for an integrated series of automated moored sensors all around Australia, but the cost of this was in the billions of dollars. Examples of this moored sensing operation were the GOOS, HOTS, and BATS programs in the USA. And again, there were advocates of 'Wealth of Oceans' themes, emphasising options of how to make money from the sea. There were several social scientists that presented important advice about the need to convince the public that climate change was a real threat, and they provided examples of the complete failure of scientists to speak to the public in this way. We spent half a day drafting a press release and a workshop summary, in hopes that this might lubricate the money wagons.

Of even more value was the opportunity to visit the good ship R/V *Southern Surveyor*, and to discuss my cruise plan with the National Facility Ship Captain Fred Stein, Director Ron Plaschke and ship coordinator Steve Thomas. This ship was 66 m in length had fifteen bunks for the science crew, and we needed to support the CSIRO voyage manager, a computer technician, a lab chemist, and an expert on swath mapping from Geoscience Australia. We had to arrange for a superclean lab and steady power supply for Kathy Burns and her gas chromatographs, and a constant temperature room for Dave McKinnon's lab algal incubations. I anticipated damage to some of our gear, so we needed welding and metal bending skills. They gave me maps of the three decks of this ship, so that I could plan the loading of the ship in Darwin and the placement of our crew operations. I had a good supper (roast duck, Boag beer) with Captain Fred, who liked to tell stories about his experiences in the cold and icy Southern Ocean. He said he always dreamed of running a ship in the warm tropical regions.

By 19 May 2005, we had loaded two 12 m semi-trailer trucks with our equipment, some of it in railway containers. Cary McLean and John Pfitzner were master packers, and were also responsible for the sequence and position of our equipment on the decks of the *Southern Surveyor*. There was a media storm 26-27 May, probably due to our press release in Hobart. We were interviewed by the ABC, Queensland University News, and the *Catalyst* science show on TV.

Our freight ready to be loaded onto the Southern Surveyor

We later heard that AIMS was embarrassed by all this publicity, as they had not arranged it. AIMS has an evaluation procedure, PPE (personal performance evaluation), done yearly, an irritation to me. The AIMS management were not impressed with my performance, wanted to reduce my research budget, and wanted me to pay more attention to the Great Barrier Reef. They asked me to focus on fewer tasks, avoid the ATSEF connection in Darwin, and to stop my collaborations with Indonesia and PNG. They really didn't want to hear any more about *Halimeda* mounds, submarine seeps and climate change.

As usual, I completely ignored these suggestions, but I should have realised I was pushing the managers too far.

On 29 May, Kathy and her microbiologist student Ken Wasmund, Dave McKinnon and his zooplankton student Philippe Gusmao, Max Rees (underwater camera expert), Cary McLean and Barry Bennett (our ship riggers for the big sediment traps), Irena, John and myself flew to Darwin, where the R/V *Southern Surveyor* was already tied up at the wharf. The next morning, the big trucks arrived and we were busy all day unloading the railway containers for forklifting our gear onto the ship. Andrea Cortese (Geoscience Australia) and Bradley Opdyke (ANU) arrived, and amongst other things, they were responsible for the seismic and swath mapping of the submarine seep sites.

We had our gear neatly installed and secure on the ship by five pm, and we all went to Nirvana, a Thai/Indian restaurant by the sea, where I ate another butter chicken and red duck curry. We walked back to the ship, had tea in the galley and inspected our lab gear again in case of rough weather.

As the ship steamed NW towards the Sahul Shoals and Cartier Trough, both the science and the ship crew were scrambling to get organised for some difficult operations. I had meetings with the ship Master Arthur Staron and Voyage Manager Steve Thomas to review the cruise plan and to agree on daily duty sheets to be posted around the ship. We were carrying more than the usual quantity of hazardous chemicals, such as radioisotopes, liquid nitrogen, strong acids and bases, and all these had to find designated safe storage locations on the ship. I met with our rigger Cary McLean, bosun Tony Hearne, and the winch/A Frame operators to review how our grab and coring equipment worked, how the big sediment traps were to be deployed and recovered, how we planned to collect bubbles of methane rising through seawater using divers, and how we planned to drill cores of coral bommies on the tops of the Sahul Shoals.

Irena and John worked with the ship chemist Dave Terhell to work out the procedure for the normal CTD profiles and how our big caged Niskin carousel was to be used. John and I set up some big tanks on an upper deck for collecting cubic metre sized water samples from the ship cooling water pumps, for radium, radon, caesium, and plutonium. Dave calibrated his oxygen titration gear for algal growth and respiration, zooplankton sampling, and chlorophyll calibration of the sensor on the CTD carousel. Andrea Cortese and Brad Opdyke tested the swath mapping gear and the system to record and display the massive amounts of digital information. Kathy found that the centrifuge needed rebalancing and the ship electrical power supply to her gas chromatographs needed additional control. Ken explored the ability of the fume hoods to handle his microbial oil degradation experiments. Irena and Bernadette Heaney set up a routine for data management and backup for all records, measurements and samples. This is where you want everybody to be expert at what they are doing, and to anticipate what can go wrong before it happens.

This below is copied directly from the Voyage Narrative, a report to the CSIRO ship committee:

> 'After departing Darwin at 1045 on 31 May 05, we began work at our first sediment trap site (11° 09' S. and 125° 00' E. at 500 m water depth) at 0800 on 2 June 2005, on the northern margin of the Cartier Trough. The previous night had been spent swath mapping and sub-bottom profiling the central part of Cartier Trough. Our intent was to obtain useful information about the water column and sediments that would be needed to interpret our results from the sediment trap deployment. This included a CTD profile, including % transmission of light, photosynthetically available radiation (PAR), oxygen, fluorescence, and the normal pressure, salinity, and temperature sensors. Samples were taken from the rosette Niskin bottles for dissolved and particulate nutrient elements (and other major/minor elements), methane and hydrocarbons, and samples for measurements of oxygen consumption and production rates (= net community primary production in the euphotic zone). Three 1000 litre tanks and six 200 litre barrels were filled from the ship seawater supply hose for measurements of plutonium and radium isotopes in surface water. A Smith-MacIntyre Grab sample of surface sediment was taken for microbiological culture work and hydrocarbon chemistry. Triplicate zooplankton net tows with a bongo net were obtained for estimation of copepod growth rates. Our benthic camera sled was deployed on the seafloor for 3 hours, to determine the biological diversity of macrobenthos and fish at this site. We had previously obtained sediment cores and water column data from this site from AIMS research cruises in 2002 and 2004. After this work was done, the aft deck was cleared for deployment of two sediment trap arrays on one mooring wire, which was accomplished smoothly in two hours. Upon completion of work at this station, we steamed for our next work station to the west of Sahul Shoals, running the swath mapper and sub-bottom profiler to test our assumptions about sediment cover in this region.
>
> During 3-4 June, we repeated the procedure above at 11° 26.8'S. and 124° 00 E. at 500 m water depth, to the west of Sahul Shoals. After the deployment of two sediment trap arrays on one mooring wire, we obtained two Kasten cores from within a mile of the sediment trap site. One core was for hydrocarbon chemistry, and the second core was for

inorganic elemental analyses and radiochemical estimation of sediment accumulation rate. During this sediment trap deployment time, we had good weather, 5-15 kt winds and clear skies, gentle swells, and all of the ship and scientific equipment worked well. When this work was completed, we departed for the Cornea Seep site to the south.

Deploying large sediment traps in the Timor Sea, from the Southern Surveyor, *June 2005*

During 5-10 June, we explored the Cornea Seep site region centred around 13° 38' 34.69' S. and 124° 43' 5.75', where we found large bubble plumes coming from the seafloor, through the water column, and into the atmosphere. These bubble plumes were easy to view on the ship 120 and 200 kHz sounders, and were obvious from swath mapping and sub-bottom profile records of 'hardgrounds'. Full use was made of the acoustic Doppler current profiler, to determine the direction and strength of the tide, because the bubble plumes were most active at low neap tides. After sufficient mapping work to determine the shape and extent of the hardground, we deployed a set of small sediment traps near the largest bubble plume, and put a lot of effort into water column sampling as described above. A special AIMS CTD and Large Niskin cage designed for sitting on the sediment surface was deployed to obtain large volumes (200-400 litres) of sediment surface water for chemical and radiochemical measurements, looking for signals of submarine seep fluids. Helium balloons were attached to the 30 litre Niskin bottle top vents to obtain seep gases not dissolved in water. Our benthic camera sled was deployed several times in towed mode and stationary mode, to inspect the seafloor near the gas vents. Net tows for zooplankton were done, with some copepods being sampled for stable isotopic measurements to relate vent gas carbon to the food chain system. The Smith-MacIntyre Grab sampler gave us samples of gas vent tubes seen on the sled cameras.

The AIMS bubble catcher was deployed from the ship's workboat, and was successful at obtaining 10-100 ml samples of bubble gas near the sea surface, and this gas was determined to be mostly methane by shipboard gas chromatography. Samples of the sea surface slick and apparent oil globules were taken from the workboat. We were not successful at capturing any of the red crabs swimming at the sea surface. We were not successful in our attempts to use the AIMS ROV camera system. Attempts to core the hardground seep site with our Kasten Corer failed, but soft sediment in palaeochannels to the north were successful, and these sediments were found to have much higher hydrocarbon concentrations than at the main bubble seep site. During the night and early mornings of this period, the swath mapping and sub-bottom profiler (operated by Andrea Cortese and Bradley Opdyke) were used to create a detailed map of the region, in coordination with the bubble plume maps detected by the 120 and 200 kHz sounders. Bernadette Heaney kindly spent time putting the coordinates of the seafloor sources of the bubble plumes (from saved records of the 120 kHz sounder) into a spreadsheet, so that we could generate a map of the locations of weak, medium, and strong bubble plumes.

During this work time at the Cornea Site, we had excellent calm weather, and optimized work time during low tides, to maximize the bubble plume signal. Two mishaps occurred on 6 June, when the workboat was nearly swamped by the bow and stern thrusters, and the ship apparently drifted over the grapple buoy for the sediment traps, cutting the buoy rope line. The hydraulic system for the A Frame got too hot on 7 June, so we diverted our work to activities that did not need aft deck services. The AIMS lightweight ROV utilising a new ballasted cage deployment system was trialled at the Cornea site. All ROV operations were eventually suspended despite efforts to respond to fibre optic data transmission problems and camera tilt mechanism failure. We greatly appreciated the use of the CSIRO Sonar dyne pinger which was attached (by Steve Thomas) to the AIMS benthic camera sled, the big Niskin cage, and the towed bongo nets, to allow GPS and depth control for these devices. After repeated measurements of the methane bubble plume fluxes, we steamed north to recover the sediment traps, and departed for our next work station on the southern margin of Cartier Trough around Pee Shoals.'

Back in Mount Isa, Sarah and Jesse welcomed their son Daniel, born 7 June 2005.

During 11-18 June we were working on the southern rim of Cartier Trough and Sahul Shoals region, alternating between coral coring operations and exploring the new deep (200-300 m) mounds found by the swath mapping and seismic sounding equipment on this ship and the R/V *Melville* in July 2003. Coral coring at 18-20 m water depth was successful at Pee Shoals, 11° 45.49' S., 124° 49.65' E., where triplicate cores 30-50 cm were obtained, and at 'Kathy's Shoals' nearby at 11° 42.5' S. and 125° 01' E.

During this time, coral coring was delayed by troubles with the drilling bit, a slow drill drive motor, and slow and overheating scuba tank compressor. During gaps in the coral coring program, the AIMS sled camera was deployed in various modes (towed, moored for 1-2 days) to obtain shallow and deep water views of macrobenthic biodiversity. Our sled camera was twice damaged during tows over the deep hard mounds, but this was repaired by Cary McLean and his welding gear.

Bubble Catcher. Timor Sea

Sediment Trap Deployment, RV Southern Surveyor*, 2005*

Samples were routinely taken almost daily for oxygen productivity and zooplankton growth rate measurements. Kasten cores were obtained along a previously established transect from 250 to 500 m depth zones to the north, in the late afternoon, after the diving work was done for the day. We used the ship's rock dredges to recover living and dead benthos and hard rock samples from the moats and tops of the deep mounds (Max's Mound at 11° 38.102' S. and 125° 06.08' E., top at 222 m, and Melville Mound at 11° 37.282'S. and 125° 07.4' E., top rim of mound at 225 m). We were excited to discover vent tubes similar to those found at the Cornea Site on top of these deep mounds. Coral coring efforts on top of the larger Sahul Shoals were made difficult by strong currents, but several good cores were obtained. During this time there were several failures of the ship's electrical system when the Kasten Corer rope winch spool controls were dead, and when the ship radar was out of action. The lab water supply pump failed on 14 June. This did not seriously impede our work, as John Morton's engineers fixed things as quickly as possible. After recovering our last sled camera mooring on 350 m deep hardgrounds near the Sahul Shoals, we steamed west to recover our sediment traps that were deployed at the beginning of this voyage. The Kasten coring railway, core slicing table, and other aft deck gear were taken apart and consolidated to increase deck space for transit storage of sediment traps.

Removing Core Tube from Weights, Timor Sea

During 19-20 June, the two large AIMS sediment trap moorings were recovered successfully without incident. This is due to the good work of AIMS mooring team Cary McLean and Barry Bennett, Bosun Tony Hearne, winch operators Graham McDougall and

Paddy Chamberlain, and Master Arthur Staron's team on the bridge. After obtaining CTD and Niskin water sample profiles, oxygen productivity and zooplankton samples, and near bottom samples with the AIMS big Niskin Cage, the ship was put to final swath mapping and seismic sounding to complete the Cartier Trough map. Meanwhile, I had informed the cooks that it was Cary McLean's birthday, and a major feast was prepared including several desserts with candles. Irena gave me about six hours notice that she wanted to have a special birthday song for Cary's supper party, which I wrote on the back deck. I sang it, along with my harmonica, and the crews chimed in nicely for the chorus.

The above photo is of the science crew on the Southern Surveyor, *a surprise gift for deck rigger Cary McLean. Left to right: Barry Bennett, Sir Alexander David McKinnon, Ken Wasmund in front of Philippe Gusmao, Max Rees, Kathy Burns, Gregg Brunskill, Bradley Opdyke behind Irena Zagorskis, Andrea Cortese, David Tirel, and John Pfitzner. Missing is the Birthday Boy, Cary McLean.*

We departed for Cootamundra Shoals en route to Darwin on 21 June, with a mid-day station at 11° 22.5' S. and 126° 52.1' E. that was supposed to be a 'control station' for the Cornea Site. We found as much UVF hydrocarbon signal in grab sample sediments, and as much methane in Niskin water samples, as we had found on the inner shelf north of the Cornea Site. We also obtained a CTD profile, large volume radium samples, and complete chemical sampling from the rosette Niskins at this location. At Cootamundra Shoals (10° 50' S., 129° 10' E.), the AIMS dive team found that visibility was poor and most large bommies were too deep for coral coring. Bongo net tows were done in a nearby deep channel, amidst swarms of small fish, and another large volume sample for radium isotopes was done. Most of the science crew were busy packing equipment and samples for the past few days.

We arrived at Stokes Hill Wharf at 0700 on 23 June, and were able to begin unloading our equipment at 0900. Our equipment was off the ship by noon, and most scientific personnel left the ship before 1500 hrs. Our equipment occupied one semi-trailer and one 22 ft container for shipment to AIMS near Townsville. Frozen samples were carried by scientific staff as excess baggage on the aircraft.

This was a huge effort by everybody, and we all realised that this was probably our last chance to operate research programs on a large scale. The management and funding climate at AIMS were in retreat, and it was unlikely that we'd be able to do this again.

Chapter 54. Family Visits in USA and Canada

There was a need for Sheila and me to visit family in Canada and USA. We left Townsville at 0600 on 15 July 2005, flew to Brisbane, and then nine hours to Taipei, where the spotless airport provided a restful orchid garden, a National Museum display of Formosan art works and ceramics, two good bookshops, and a good laksa supper with cold beer. The flight from Taipei to San Francisco was eleven hours, and I could sleep, eat, sleep, and watch films. There is a note in my journals in Sheila's handwriting: 'There is a hairy Amoeba Lady sitting beside me, her feet in my space, her pillows and elbows in my ribs, I can't sleep.' We arrived in San Francisco several hours before we left and had an overnight in a hotel on Brisbane Bay (confusing), part of San Francisco Bay, about which Jesse Fuller sang his song with a fotdella (foot piano).

The next morning we took an early airport bus through a hideous tangle of cloverleafs and under/overpasses to reach a crowded airport where security was on high alert. They took my cigarette lighter, but I hid my jackknife inside my suitcase luggage. They did have good chocolate croissants and espresso. Then on to Denver (Prairie Martini = beer with olives, advertised in the upstairs smoking lounge), where we had a fast turnover to fly to Rapid City. Leaving Denver was like entering a different planet: the land did not have big cities or lots of roads, just dry alkali flats and dry brown prairie, empty country. I guess city people might be frightened by this, but it was strongly nostalgic to me, coming back once again to the home of my youth.

If you have read previous chapters in this book, the following description of visiting our families in South Dakota and Winnipeg will be almost identical. This is what fuels nostalgia, the continued tradition of feasting, storytelling, music, and review of who got born, who's still alive and who died.

As we stepped into the lobby of the Rapid City airport, standing there was my brother Obie in his usual Stetson hat and high heeled boots, grinning at us. He whisked us off to town where he was determined to find suitable drinks for us: cider and cranberry juice for Sheila, and Sam Adams Ale and Moose Drool beer for me. We drove east on the big interstate highway, with an angry hot July wind, lightning, thunder, and prairie fires to the south of the Badlands, heading north of Philip for Annie and Obie's ranch. Sheila spotted some white-tailed deer, pheasants, and lots of meadowlarks sitting on fence posts. We drove then through wheat fields, windrowed oats, combines and grain trucks, rows of hay bales, and a lot of happily grazing Hereford cattle. Obie's wife Annie had a large roast beef supper waiting for us, with a dining table full of their daughters' families. The angry wind blew all night with the strong smell of prairie fire somewhere to the south. The next morning Annie found all of her staked garden tomato plants flattened by the wind. Such is life on the lone prairie, where the coyotes howl and wind blows free, where the old night owl hoots so mournfully, and the deer and the antelope play.

The next day, all of us drove to my brother John's house near Murdo, where another noon day feast had been prepared for even more relatives. John had his BBQ grill operating full blast, and the big dining room table had all the extra panels in place. There were many small nieces and nephews running around, and I couldn't quite remember any of their names. And then all three brothers sat on the porch near the juniper trees, telling stories mostly about people that are now dead and gone, reviewing all of the Gregg and England relatives, Annie's Philip Library Project, and all the medication for Obie and John. They snorted a little about me, having no medications to brag about. Brother Gerald didn't come up from Arizona this time. My uncle Gerald couldn't be there, because he'd been crunched by a bull in a horse trailer, and had been in the Pierre hospital. He had just returned to the ranch on the White River, bruised and sore at eighty-four years of age.

One of my favourite things to do when visiting the home ground was to drive down Highway 84 to Westover, and then the dirt road along the White River. Sheila spotted bald eagles, wild turkeys and lots of meadowlarks. The prairie went on forever, with a sea of green and brown grass, yucca and sweet clover with yellow flowers, willow and cottonwood trees in the creek beds. They had good rains and the White River had good flow, somebody should have been setting a fish trap.

There was the old ranch house where I grew up, getting smaller every time I visit, with my uncle Gerald and Aunt Joyce expecting us, ready with lunch. My uncle Gerald now has long white whiskers: Aunt Joyce explained that he wouldn't trim his whiskers because he wanted to show me how a beard should look! Gerald had a pile of genealogical material to give to me, including some hand-written wartime letters from my grandmother Edna Mae Brunskill. Gerald had the floodplain bottom land planted to alfalfa, which was now in second crop purple flower. We drove across the river to visit my Aunt Margie, who was out in the garden watering her tomatoes. We got sour cream raisin pie and coffee! Uncle Bernard and sons Chuck and Mark were combining wheat with three big John Deere combines, but Bernard said it was 'scabby' wheat, worth maybe $2.48/bushel, about the cost of growing it, about the same as wheat prices in 1922, whereas new combines cost $300,000.

Obie took us back west the next day, down through the edge of the badlands, where we could see lots of deer, antelope, mountain goats, and spectacular white, pink, cream, purple, blue and gray cliff/spire scenery in this Sioux Indian country. We had good muffins, cinnamon buns and coffee at the A and M Café in Interior, where the waitress called my brother Obie by his name. At the cowboy store in Wall, we bought presents for our sons Ian and Jesse and some South Dakota gold earrings for Jesse's wife Sarah. Back at Obie's ranch house we sat on the porch in the shade with ice water. Most of Obie's kids and their children were there. We made ice cream again and Annie mushed up garden strawberries for the topping.

Niece Marcy was absent, due to pregnancy troubles in Rochester, Minnesota, but she phoned to let me know she was still alive. She demanded that I phone Toady, which I did, and this old lady was fully loaded for conversation with her wandering Australian relative. She was mainly interested in Australian green tree frogs, our SE Asian tsunami, correcting some bits of my genealogy, and a general notice that she was too old and would die soon. She wanted to be cremated and her ashes put into an old coffee pot to be buried with her husband. She was really pissed off to be growing old, was losing her vision, memory, and couldn't write anymore.

Obie took us to his neighbor's Prairie Moon Ranch, where we could watch a herd of buffalo slowly grazing the prairie, and there were a few long shaggy coated yaks too.

Next morning we were up at 0430 to drive with Obie in his pickup truck to the Rapid City airport. There were flocks of pheasants and small groups of white tailed deer along the way. We had a good South Dakota breakfast with Obie in the airport café, and then we were back in the air again on a United Air flight. This time it was a propeller plane to Denver, where we had 'not the best landing ever', according to the pilot. And then Denver to Minneapolis, where the stewardess examined me carefully and decided that I was 'a nice sort of guy, and I looked just like Jesus'. 'What you going to Canada for anyway?'

Finally we flew north to Winnipeg, over green prairie and forests of southern Manitoba, lots of standing water in the fields, and the Assiniboine and Red River diversion around the city of Winnipeg were in full flow operation. There in the airport was Sheila's sister Peabody and family, beaming and shrieking with the sight of us coming down the stairs. We were whipped off to their home in north Winnipeg, where their grandaughter Audrey was peeking at us shyly from the bedroom window. Al and I cooked steak and onions on the back yard BBQ grill, and the most of the old Meadlowlark band showed up with fiddles, guitars, banjo, autoharp and mandolins. Peabody's daughter Janet gave us a demonstration of her juggling skills, and it was a good family party. Over 22 July to 2 August, there was a blur of feasts and music parties with Peabody, Susan, David and Barbara Ediger, Walter and Bean Lysack, and Buster and Cathy Welch. We planted six cedar trees in the front and back yard of Peabody's house, and I severely trimmed their hedge, chokecherry and elm trees, and expanded and fertilised their raspberry patch.

We left the Winnipeg airport two hours late on 3 August, and arrived in Denver after our connecting flight to LA had left. We hustled to rebook with United/New Zealand Air, as we were going to miss our Los Angeles China Air flight to Brisbane, but found that China Air had no intention of cashing in our ticket, and New Zealand Air would not honour the China Air ticket. We spent eight hours trying to solve this problem, and eventually we spent $1450 to buy a new ticket. Once again we ate Asian Beef with cashews and beer and went to sleep over the Pacific while August 4th didn't happen at all. Twelve hours later we found Auckland Airport had good lime pie and cappuccino, cool clean air, hoop pines and frangipani flowers. Sheila was starving and sleepy and disgusted with this trip. We got to Brisbane where it was winter, frosty cold, and our luggage was missing, but promised delivery to Townsville in a few days. And forty-eight hours from leaving Peabody's house in Winnipeg, we arrived 8 pm at the Townsville Airport, where the air was warm and perfumed and our son Ian MacGreggor was there to collect his dead tired parents. He was busy performing at the Palmer Street Jazz Festival, and he urged us to come to at least a few of the shows there. And yes indeed, our luggage did eventually appear on our doorstep.

In October, Sheila and I drove 1008 km out to Yelvertoft/Johnson Yards, a cattle station shed on the road between Mount Isa and Camooweal in western Queensland. Jesse was a Queensland State Government Main Roads project engineer for upgrading the old road and building some new sections with better bridges for the rare times when this dry country was underwater from flooding. Jesse was dealing with aboriginal tribes who had ancestral history with parts of this land, as well as white cattlemen who wanted to use the bores drilled by the road builders.

Sarah and Jesse turned this huge open shed into a nice place to live, and a wonderful space for their children Georgia Rose and Daniel to play. I spent a fair amount of time on the floor, building towers and bridges from play blocks, so that Georgia could knock them down laughing and shouting 'eego'! Nearby were windmills, stock water tanks, long feeding troughs, and small pens for calves, the pet pig and a Shetland pony. Evening suppers were attended by

Penelope the black duck, with the pig expecting the food scraps. Jesse would be finishing this job soon, and he was considering several job options for the future. The local cattlemen wanted to hire him as a cattle property manager, but jobs in southern Queensland were also available.. We drove further west to the Queensland/Northern Territory border at Camooweal and had lunch at a typical outback hotel, where the pub had all the thirty-six mailboxes labeled with family names, not numbers. This area had the same feel as western South Dakota.

Chapter 55. Adventures in Borneo

A New Zealander dendrochronologist had visited AIMS a few years ago, hunting for mangrove ecologist Norm Duke. Duke had left AIMS and was working at the time at the University of Queensland. This Kiwi gave a seminar about estimating climate, precipitation and evaporation from tree rings, and he wanted to find out if mangrove trees had annual rings. This was Jonathan Palmer, and I quickly snagged him for discussions about a paper we were trying to write with several tree ringers and radiochemists, especially Gary Hancock at CSIRO. We had measured radium and radiolead in two species of mangrove, and found that the rings in the heartwood were roughly annual in one species but not others, based upon the decay of the natural radioisotopes. This was interesting, as most tropical trees do not appear to have annual tree rings. I had been in correspondence with Ed Cook and Gordon Jacoby at Lamont-Doherty Earth Oceanography tree ring laboratories of Columbia University in New York, and Ed also visited me at AIMS. He looked at dozens of big Fly River Delta mangrove tree slab sections that we had cut with chainsaws over the past years. He pronounced the observed rings to be seasonal but not regular enough to be annual and took some of my slabs with him, back to LDEO.

For some serendipitous reasons, Jonathan Palmer asked me to accompany him on a consultancy job in Malaysian Borneo, for a German palm oil plantation company, the Golden Hope Plantation, and the Feindt Foundation. Our job was to make expert observations on the impact of deforestation, palm oil plantations and developments on the coastal ecosystems of Sarawak. They wanted general information concerning the palm oil plantation impact of river erosion, mud and nutrient transport by rivers into the coastal mangroves, coral reefs and shelf fisheries, compared to sections of the Sarawak coast that had 'pristine rainforest.' I volunteered the expertise of Sheila in vertebrate wildlife, especially birds. AIMS was not happy for me to take this job, partly because there was no large sum of money being paid into institution coffers, except for my travel, food and accommodations.

So here we go again, an expedition to yet another wild part of the earth. We left Townsville on 30 October 2005 to reach Sydney in the early afternoon. Our Sydney colleague Ron Szymczak collected us at the airport for a tour of his favourite beaches, pubs and Thai restaurants, where red curry duck was encountered. We left Sydney airport at 2300 hrs on Malaysian Airlines, and they gave us a nice blanket, hot wet towel, and I slept for the 10.5 hr flight to Kuala Lumpur.

Kuala Lumpur airport was dark, neat and clean and empty in the very early morning, with a huge smoking lounge provided by British American Tobacco. We were enticed by lots of Belgian dark chocolate, durian and mangosteen candy, and a barista singing away right beside Burger King and Starbucks.

It was still dark as we left on a Malaysian Airlines flight for Kuching in Sarawak, where we could see the steam rising from the mangrove intertidal zone swamps near the airport. Jonathan was there to meet us, and he whisked us off to the inevitable Holiday Inn, complete with a properly dressed doorman and luggage porter.

After a shower, we met Jonathan and Kerry-Jayne Wilson (NZ Lincoln University biologist) for lunch. Kerry-Jayne was an expert in birds and reptiles, especially wood pigeons, but she seemed most concerned as to whether Jonathan had brought along a case of gin. She said this was her anti-malarial treatment, gin and tonic. Jonathan assured her that he had lots of gin in his room. We were introduced to Steve Oakley, a marine fisheries and coral reef lecturer at the local university and the Sarawak Biodiversity Centre. His favourite research sites were the oil drilling rig platforms off the coast, around which there was a 200 m exclusion zone for fish and marine biota. Outside these exclusion zones, there were many large and small fishing boats, and this part of the Southern China Sea was heavily used for seafood harvest. We had brief reviews of the various Malaysian organisations that have data that we need, the National Parks and Forestry Departments, several wildlife NGOs, a Chinese Forestry Corporation (timber barons), and the various agencies that measure river discharge and marine resource harvests.

Kuching is the administrative capital of Sarawak, a state of the nation of Malaysia. There were over 300,000 people in town, a ferment of Malay, Chinese, Indian, Iban, Dayaks and stiff upper lip UK cultures stirred up with Muslim, Buddhist, Taoist, Hindu, Christian faiths, and blessed with some wonderful pagan animism. We were solemnly told that some of the white people in the University were 'secularists', which we eventually discovered meant that they didn't believe in any of those religions...oh dear me. There were lots of variations of languages here, but most shop, restaurant, and science people spoke excellent 'very British' English to us. It was also humiliating to hear shop keepers flipping between four or five languages, depending upon the customer. This melting pot culture was historically enriched and ruled by UK white Rajah James Brooke and descendants since the mid-1800s. Japan cruelly occupied Kuching during World War II, and the last Rajah of Sarawak, Sir Charles Vyner Brooke decided to cede Sarawak as part of British Crown Colony in 1946. See the references section for more of this Borneo history.

Sheila and I escaped the science meetings to wander along the Sarawak Riverside shops. The market stalls were selling fish and fruit, but we couldn't find any mangosteens. There were McDonald's, 7-11, and endless tourist shops selling fabric, carvings, fashion and Michael Jackson T-shirts. We rented flippers for our coral reef work. We had to wait till well after sundown for supper, as this was Ramadan. We staggered through the misty heat to have supper with Jonathan and Kerry-Jayne at Benson's Seafood, which shoveled huge prawns, chunks of venison, fish, fiddleheads, okra, and Tiger Beer at us. All the while there were firecrackers and rockets shooting over the river toward a big castle. We learned to say *Satu Lagi* for same again, for more food, and *Trima Kasi* for thank you.

The next day, 1 November 2005, the Forestry Corporation drove us out to Semenggoh Nature Reserve, Wildlife Centre, and Orangutan Rehabilitation Centre. We drove through a long winding road surrounded by suburbs, high rise apartments, banks, trading companies, flower stalls, new construction sites, logged plots, logging regrowth, and not a quiver of rainforest. But finally we came to a wall of big trees with a narrow muddy road entering it, and there inside the gloomy hot wet forest, we saw orangutans feeding at a table suspended high in the trees. There were all sorts of birds, hornbills, squirrels, fern and bamboo gardens, and a special garden for medicinal plants. These were public and tourist viewing locations, and we were part of a crowd of twenty or thirty people.

Jonathan rented us a van with a driver to haul us and our luggage northward along the coast to Sematan, where we found rooms at the Palm Beach Resort Hotel on a long grey sand beach. Tough life! I was happy to hear that the agricultural crop for this area was watermelon. He served us anti-malarial gin and tonics, and provided cigars after supper.

We had an early 'American' breakfast at the hotel, and then we were off to Semantan village to shop for food and supplies for four people for five days, at a Research Station camp further west down the coast called Semansan. The shops were crowded with halal tinned goods, vegetables and fruit spread out on canvas sheets on the floor, 200 kg sacks of rice, chillies, and unknown rooty spices. Just after lunch we loaded our gear and food into two small boats at the dock and slowly putted down the river past sandbars and tree snags out into the South China Sea, where there were 1-2 m long and gentle swells. Frigate birds were flying above us and floating demurely on cut sections of mangrove logs.

We arrived at the Samunsam Wildlife Sanctuary Research Station midafternoon, and the Station Manager was not amused to see us a day early. He pointed to four wheelbarrows for us to move our gear from the dock across the camp grounds to our cabin. He apologised for the cabin being closed down, all windows and vents shut tight, and no mattresses or bedding on the bunks. There might have been a dozen unoccupied cabins scattered around the central green area, and our cabin was called 'Ramin Villa'.

Jonathan must have pulled some strings at high levels of Malay government to get permission to visit this location, as this Research Station was not open to the public, but rather it was carefully protected against development and settlement, even by native people. The tap water was brown, there was no fridge, and most surfaces were covered with mould, wasp cocoons or ants. Our cabin had a one burner gas stove, a wok, kettle, big bowls, glasses and tea cups. But we had a nice lunch of corned beef, French bread, fried green banana, warm Tiger Beer and tea. The rain poured down, and we sat on the verandah.

The camp was on the Samunsam River estuary, and there was good access to a sequence of salt-to-freshwater mangroves (even a boardwalk!), huge undisturbed rainforest trees, and insane levels of exotic and rare biodiversity. Some of these rainforest trees were fig and *Dipterocarp* species, buttress rooted and over fifty metres high. After a good supper of highly (& unknown) spiced corned beef, rice, fiddleheads, and warm Tiger Beer, we walked the mangroves with torches, and Kerry-Jayne tried out her high frequency hand-held bat monitor. There was no good reason to go into the mangroves to see or hear the bats, as we had maybe a dozen of them hanging on the eaves of our cabin verandah.

The next day, after a strangely spiced scrambled egg and fresh-grown coffee breakfast, the bird crew (Kerry-Jayne and Sheila) did a bird count survey of the beaches to the NE of Samunsan.

Jonathan and I ran several transects through the mangroves, recording variation of species with salinity, girth, and a few hand boring to inspect the tree rings. There were several mangrove species I hadn't seen before, and two saltwater palms with spines all over the base. The fetid odor and difficult walking conditions through the mangrove roots caused a lot of sweat and bruised feet and legs, and we spent some time washing up from a rainwater tub beside our cabin. After sardine sandwiches, Tiger Beer, and Durian Pops lunch, we tried to liberate a heavy boat sitting on the mud beside the dock.

The Station Manager had not been informed we needed a boat and tried to tell us we could not use it, it was unsafe, it would sink, the crocodiles would get us, and poisonous snakes would drop from the trees into the boat and kill us instantly. Most of this might have been true! There was no outboard motor in the shed, so Jonathan used a chainsaw to make some big heavy oars and some small light paddles. It did leak a bit, so we caulked it with old canvas, and we had a bailing can.

After an hour's work trying to prepare the boat for our use, the tide had come in, and floated the heavy boat off the mud bank. We rode the tide landward, doing bird counts, mangrove species, noting sand and mud bars, the howler gibbons swinging across the river on overhanging tree branches. Here there was no evidence of deforestation, with the minor exception of a few fisher cabin clearings along the river. The river water was milky coffee coloured with little suspended mud, and we spotted no mud banks. We tied up at the camp docks at dusk with the monkeys screaming at us. We later found out that the river was essentially sea water 14 km inland at high tide, whereas it was freshwater by our camp during low tide. We cooked up a good supper of giant slabs of mackerel and garlic noodles, with some huge paw paws for desert. The rain came down, and we smoked cigars and drank more warm beer on the verandah. We had to check the ropes on our boat at the dock at midnight and three am, as the boat was filling up with water from the torrential rain.

Jonathan Palmer, GJB, & Sheila Borneo 2005

We tried to repeat these adventures the next day, and the bird crew ladies took the boat across the river on an ebbing tide, this strong tide pushing them nearly out to sea. Jonathan wanted more tree cores for his tree ring growth band studies, so we climbed through the mangrove roots and made standing or sitting platforms for the coring work. We staggered back to the camp for lunch in early afternoon, and noted that Sheila and Kerri-Jayne were standing forlornly on the far shore of the strong high tide river. But they had their packed lunch, so we abandoned them to hike up into the high ground rainforest where the camp water supply waterfall and pool was located. We came across patches of grey-white (sometimes lemony-yellow) clay-silt soil, which we had noted in some of the nearshore embayments. We later found that this clay was important for the kiln manufacture of ceramic tiles, which were exported all around the world. We also found calamansi and mirianda trees in fruit, which were very tasty for snacks while walking the trail.

Sampling River Mud, Borneo

Sheila had jumped out to walk up the river to the jetty at low tide, pulling the boat behind them. The Camp Manager was furious, claiming the crocodiles were watching them. But 'No Worries, Mate', we had a supper feast of wok hash of corned beef, onions, raisins, rice and some dreadful Kiwi Biscuits (would have made better skipping stones). And the rain came down in buckets, so we sat on the verandah drinking gin and tonic, Baileys Irish Cream, smoking cigars, and telling stories of past adventures.

Kerry-Jayne had been threatening us with the idea of wokked oatmeal porridge for breakfast, and the next morning, 5 November, it happened. We survived this, using lots of raisins, cashews, evaporated milk and honey, but it was a near thing. We were told to get up extra early, because a boat was coming to take us out to the Talang Talang Islands, so we could see some coral reefs and island biota of the South China Sea. As we arrived on the jetty, the camp manager pointed out a bow wave proceeding upstream: 'That's old grampa—an eighteen foot croc. He lives here'. Magically, an outboard motor appeared, and the very boat that we repaired was made ready at the camp jetty. We loaded all our gear, food, beer and extra freshwater, as we were told there were no services on the island.

We arrived at Pulau Talang Besar at noon, to see a logging ship anchored near the camp jetty, with a big crane scoop dredging the approach to the island. The ship was fully loaded with large trimmed logs, and there were large *Porites*-like brain corals piled on the rocky shoreline. This did not bode well for the 'Talang-Satang National Park', also called the Turtle Islands. Once again we hauled our gear and supplies up the beach and across the shoreline to a neat and tidy two story building, where we were surprised to hear Malay pop tunes from a radio. This building was also temporarily occupied by five young park biologists who were waiting for the turtle hatch to happen. This being Sheila's birthday, I helped prepare some exciting spicy scrambled eggs on toast for our lunch. We were deliriously happy to see two gas refrigerators in the pantry, where we stashed our beer and drinking water. The grounds around the building were planted with short rows of mango, citrus, coconut, cheese fruit, jackfruit, date palms, banana, plantain and pawpaw, with an irrigation system from rainwater tanks higher up on the hills somewhere.

The bird survey crew took off, walking on the beaches and rocky shores. Jonathan and I walked out onto the ~4,000 year old remnants of a coral reef, which had been decapitated by sea level fall and exposure to too much hot sun. We used snorkel, face mask, and fins to explore the deeper (5-15 m) corals, which seemed to be abundant, diverse, and happy, despite very high water temperatures and salinity. The water was a bit murky from the dredging however, but there were no patches of terrestrial mud around the reef. Now that we had an outboard motor, we toured completely around the island to the delight of Kerry-Jayne, who found a group of Pied Imperial Pigeons, her main target. I expected these islands to be coral cays, but they were volcanic in origin, lots of black gabbro and pink granodiorite (same as at ELA!). Back at the camp house, the rangers sold us some big crays (similar to lobster) which were boiled in seawater, and we had these with mountains of rice, wokked greens, cold beer, egg jam and honey on toast, and endless cups of tea on the verandah listening to the bats chittering. We had some party poppers and singing on the beach for Sheila's birthday party. The turtle rangers started a generator so they could watch TV pop singers and game shows, which seemed totally out of place on this island.

The next day was a Sunday, a day of rest. Yah, sure, no chance of that. Eza Edwin, the boat driver, hauled us about three km away to Palau Talang Kecil at dawn, another of the Turtle Islands where tourists and visitors were forbidden, and even research biologists visits were carefully restricted and recorded in a log book. The Park Ranger Kamin showed us a group of eagle nests, his fruit tree orchard, and some of the high hill rainforest trees. They had a problem with eliminating introduced cats, which preyed on some of the rare small mammals and baby turtles. The water here was very clear, the corals looked healthy and diverse, and there was no terrestrial mud anywhere. We were back at Talang Besar for a fresh fish, cray and ginger, sardine lunch. The local boys took the boat to go back to Sematan for generator fuel, but they returned without any fuel...we never knew why. We snorkeled the reefs again, under thunder, lightning, strong wind, and eventually pouring rain.

The ranger boys were worried that the emerging turtles would get swept away by the rain or eaten by lizards and eagles, so they were watching the beach sand carefully, and then it happened, just as the sun was going down. The beach in front of the camp was suddenly crawling with baby turtles, and the rangers were gathering them into twenty litre buckets. Sheila was in heaven, as she is a turtle maniac. She stood on the beach waving her arms, shooing the little turtles as they frantically paddled through the sand to the water. The ranger boys would lay down on the sand and push their arm deep into the loose beach sand, to feel and grab the eggs and baby turtles. Some of the nests were about fifty cm deep in the sand, and there were over a hundred eggs in most of the nests. The buckets of eggs were taken back to the camp for incubation in a wire mesh protected sunny location. These were Green Turtles (*Chelonia mydas*), and three other species (Hawksbills, Olive Ridleys, and Loggerheads) used the mainland beaches for their eggs. According to park officials, these turtles were often over a hundred years old, and did not breed until they reached thirty to fifty years of age. The whole reason for this camp and the rangers was to collect and protect the baby turtles and eggs from all the predators (bearded pigs, sea eagles, monitor lizards, feral cats and dogs, and at sea the sharks and big fish).

The next morning the rangers inflated three little dingy boats and paddled well outside the island shallows to release buckets of turtles, supposedly away from the attention of the predators. Even with all this help, perhaps only one out of a thousand turtles survived to reach the age of reproduction.

Eza was determined to go back to Semantan to get fuel for the generator, so we loaded up the big boat and went along to sample bottom sediment (with a small ponar grab sampler) for the purpose of determining the amount of terrestrial mud coming into the sea from deforestation. We had also booked rooms at a seaside hotel there, in anticipation of a trip up the Semantan River. We took bottom sediment samples every ten minutes across four km of seafloor, and found little or no terrestrial mud, only skeletal marine carbonate sand and gravel. Inside Semantan harbor, the grab sampler brought up some sticky white clay, which we recognised from our work in the rainforests here. This was important and good news, as most of the coastal plain and mountains inland had been deforested for agriculture or palm oil plantations. The catchment of the Samunsam River and the Wildlife Refuge had never been logged, and that was our 'control' river catchment.

From the estuary jetty, Eza Edwin vanished with an odd collection of tins and plastic jugs to find fuel for his generator. We waited for him in a busy Semantan street restaurant, where we ate *mee goreng*, mud crabs and roti (fried bread) and drank beer and strong coffee (*kopi koso*). The Chinese own the shops, but the native Malay Dyaks do all the heavy work. The street was full of motor bikes loaded with humans and cargo, Mercedes sports cars, old military trucks, motorised trolley carts, human or pedal bike taxis (rickshaw, *becak*). And then Eza came back with his fuel containers in a wheelbarrow, and we all helped him load the boat for the return to Talang Bezar.

Eza and Jonathan hired a driver and small truck to haul us all to a Chinese hotel on the soft brown sand beach. It was very hot and humid, and we all went swimming in the South China Sea. After a good supper, we again sat on the verandah drinking gin and tonic as the rain came down in buckets. The Chinese hotel manager took a liking to us, and regaled us with stories of pangolins roaming the local cemetery...and of course I began writing a ballad about this.... Jonathan and I began talking about creating an ecological consulting company consisting of retired tropical scientists, 'The Greybeards Society'. We even drafted a prospectus!

Jonathan was well lubricated and told a story about his family. His father was an ex-Malay avocado and oil palm plantation manager in 1960-80s, and now at age seventy-eight was a retired book seller in Christchurch. Jonathan's grandparents lived in the Reading, Basingstoke, Berkshire region of England, where they owned and managed five pubs pre-World War I. The beer, ales and lagers were delivered by Clydesdale teams pulling big wagons decorated with banners and advertisements from the breweries. The horses were all named and were like family pets. Kids could ride them and walk between their legs. When World War I came, the government called for all the Clydesdales to pull cannons on the battlefield in Europe. Jonathan's grandmother cried when the horses were loaded on the ship for transport to the war. None returned. The pubs went broke, never re-established, and motorised trucks replaced the horse drawn wagons for beer delivery.

The next morning before we woke, Jonathan hired a motorised banana boat and driver (Rupiah 150) for an expedition up the Semantan River, to view remnants of the coastal lowland rainforest, and patches of deforestation for agriculture and timber harvest. I tried to identify the mangrove species along the salinity gradient (high salt *Rhizophora spp.*, *Xylocarpus* sp. {The Cannonball Mangrove}, *Brownlowea, Ceriops sp., Avicennia, Soneratia, Heritiera brugosa, Nypa fruiticans* at 10 psu, and finally *Casuarina* spp. and a prickle palm in freshwater). There were lots of clearings for fish farms, fishermen's shacks, jellyfish drying racks, sawmills and lumber yards. Here we found more jelly mud shorelines and river mud bars, lots more mud crab holes, and evidence of terrestrial erosion input to riverine ecosystems. There was the usual plethora of birds and monkeys, and some of the fish in the fishermen's drying racks were strange, one of them like a giant catfish. Jonathan and I cored several big mangroves to count the tree rings, and the mosquitoes were fierce and large. We were back in Semantan by midafternoon, had another meal of mud crab, rice, roti and cold beer. We walked the streets to view the fruit and fish markets, and noted the Kentucky Fried Chicken shops, Coke and Pepsi Five, Colgate toothpaste, Captain Destruction video games, dried fish and jellyfish, cell phones, motorcycles, spinning tops, kites, gongs, music CDs, fabric shops, tubs of black peppercorns, bookshops, and crates of Snickers. We bought a sack of longans and small mangoes to bring back to our hotel. The mangoes were sour, not anything like Queensland sun ripened mangoes.

The Palm Beach Hotel vehicle arrived at the jetty to haul us back along the shoreline to the imagined safety of the hotel grounds. The lady driver told us that the hotel was owned by one of three Chinese brothers, and the other two brothers owned the local sawmill, lumber yard, two grocery stores, and several coastal freight ships. She told of the local Dyaks catching pangolins in the Chinese cemetery, for which they could sell them for a lot of money to the Chinese. Pangolin scales were used for magical medicines in China, but the flesh is way too strong to eat and smells like turpentine. After supper, the hotel Chinese lady arranged for a Dyak to drive us to the Chinese cemetery to look for pangolins. Here was bizarre! We were all standing in the back of a small utility truck, hanging on to the roll bars, with large flashlights flickering around the gravestones and flower beds. We didn't see any pangolins at all, but the night smelled of camphor, sandalwood, cooking fires, copra, sea spray, sewage, and diesel fumes. Later we had gin and tonic and Snickers on the verandah long into the night, discussing the organisation of The Greybeard Society.

The next day I slept in and Jonathan went off to buy some cross section slabs of mangrove trees for his tree ring research. The ladies went off birding along the Semantan beach. Jonathan returned with the news that he had seen a pangolin swimming in a road ditch creek, which completely ruined my naïve sense of pure tropical ecosystem health. We suffered through another noon lunch of mud crabs, noodles with meat, roti canai, cold beer, egg jam on toast and *kopi koso*, and then we boarded a taxi van for the ride back to the big city of Kuching.

We got rooms in our Holiday Inn again, where they had a large fake gingerbread house perched in the hotel lobby, covered with Christmas decorations and flickering lights, with elves and gnomes leering at us. This is a dominantly Muslim country!

In our search for evening food, we discovered a wonderful restaurant on top of a giant car parking garage, maybe eight floors high, well above the traffic noise and lights. There were booths featuring all sorts of fish and seafood, freshly grilled camel chops, huge prawns, pickled jellyfish, fiddleheads, snow peas and large mushroom caps filled with strong cheeses.

We were obviously getting close to civilisation now, as our Holiday Inn provided us with a proper English breakfast of soft boiled eggs, beef sausages, croissants, butter and marmalade... but the coffee was terrible. There was a big TV screen on the wall, giving us reports of sixty-five dead from a hotel bombing in Jordan, immigration policy riots in Paris, a US Senate review of a scandal in the Texas oil industry, and finally parts of an Austin, Texas Country Music Festival concert. Sheila went off to a cultural village where folk music was presented. Kerry-Jayne went to some nearby island to view a bird sanctuary.

Jonathan and I spent the day touring government offices to gain environmental and forestry data, and to discover more of what the Feindt Foundation wanted in our report. My journal notebook has six pages of notes here, all boring details of government department administration details, addresses and phone numbers. Many of these environmental officials were desperate for collaboration with international research institutions. I hadn't planned on it, but I did have some brochures about AIMS and our TROPICS research group to hand out to these hungry people. They wanted workshops and consortia on topics near to riverine and coastal marine ecology, all of which AIMS was ideally situated to provide. AIMS was well known and respected by some of these folks as a centre for mangrove ecological research. Several of these institution directors wanted to be invited to visit AIMS and JCU, but they wanted us to pay for their travel and accommodation. The forestry, agricultural, fisheries, and oil industries were all interested in funding research work. [On return to AIMS, my CEO wanted nothing to do with any of these proposals. Very sad.] We were also searching for guidance in what these people wanted in our report, which was partly intended to stimulate internal and international research studies on the rainforest and coastal ecosystems, and at least partly to enable intelligent progress of the forestry, palm oil, tourism and fisheries industries.

Jonathan invited several of these people to join us on the parking garage roof for our evening meal. We could see the *sampans* crossing the river below us, some of them with 'Lipton' painted in large letters on the umbrella shade cloth over the ferry boat. I shared a huge roast of camel meat with Steve and Sally Oakley (and their son Jack), and I learned a lot about life in Sarawak. Steve had been a Professor at the University of Malaysia, had surveyed coral reefs, fish and benthic biology all along the coast of Sarawak, Brunei, and Sabah, and claimed that there had been no bleaching since the 1998 event. He found maximum biodiversity around and on the oil producing platforms, where the commercial fishermen were forbidden to harvest fish. For this work he was given the Duke of Edinburgh Award a week ago. He now had been funded to build a huge aquarium in Kinabalei, Brunei. His wife Sally was a school teacher in Kuching, and she spoke of the difficulties of teaching in English to kids that already speak three to five Asian languages, are Muslims, and know very strange, warped and bizarre aspects of European history or culture.

We had a free day on 11 November, so Sheila and I joined our guide Khalik for an overnight expedition to Bako National Park. As soon as we got in the banana boat ferry for this trip, Khalik whipped out his cell phone to order our rooms and meals. As we were underway along the mangrove coast of the Bako peninsula, we could see large bearded pigs (*Babi-Hutan*) and lots of fishing pole platforms and gill net floats. At the Telok Asam HQ for tourists, we were guided to a very nice cabin, and were told to keep the doors and windows closed, because macaque monkeys were accustomed to raiding these cabins and running away with personal items. We walked the beach and did some clambering up a rocky path to the high ground, where we could see some of the famed *Nepenthes* (large pitcher plant), some large green pit vipers in the trees, and Khalik lectured us on many varieties of medicinal plants. Khalik was Iban, a native Malay, and I overheard him speaking both German and French to other tourists. He had an extra thumb on his right hand.

Back on the boardwalk through the high rocky coast, a few beaches, and some muddy mangroves, we watched the endemic proboscis monkeys stuffing themselves with mangrove (*Avicennia*) leaves. They were so bizarre in their twenty kg body form that it was hard to stop watching them. Their long nose is the reason the locals call them 'Dutchman's Monkey'. Their long and brilliant red penis flopped around as they moved through the trees. After a supper of fish, beef with rice, eggplant and cuttlefish chips, we had another night walk through the rainforest. We were told to notice the smells, sandalwood, durian, and several foul-smelling medicinal plants. When we got back to our cabin, it was a mess, as we had left one window open. The long tailed macaque monkeys had got into our fridge and food cupboard, and the pillows had been pulled away from our bed. We live and learn; we had been warned.

I slept in the next morning, while Sheila went off on a rainforest walk and experienced the wonderful smells of a durian grove (durian is a tropical fruit that smells like rotting flesh). I had my English breakfast at a nearby tourist restaurant, at an outdoor table where a macaque monkey jumped from the verandah railing to my table and stole my toast. Khalik gathered our group together for the boat trip back to the mainland, all the while lecturing us on Dyak/Iban marriage customs, the job market, *Nypa* sugar, his brother's motorcycle repair shop, and housing development scandals.

Back in Kuching and our Holiday Inn rooms, we packed up for our trip home. I bought a Muslim prayer rug and a *parang* in a leather sheath. The *parang* was the main weapon of the Dyaks during the World War II fighting against the Japanese. It is like a short machete, with a heavy steel blade, and very sharp. We had a late laksa supper in the basement of an insanely posh high rise shopping centre, loud with western muzak and Christmas carols. And then it was all over, taxi to the airport, departed Kuching 1755, bowls of fish curry in Kuala Lumpur at midnight, blissfully asleep until our arrival in Sydney in the dark at 0440 on 13 November, then Brisbane for blueberry tarts and cappuccino, and finally back to Townsville by 1700 hrs. The air was clean and good.

Jonathan and I spent about a year trying to send repeatedly revised reports of our overview to the Malay and Sarawak governments and the German Feindt (Oil Palm Plantation) Foundation. We were hired to provide plans for a 'forest to the sea' study to remedy impacts of human population, industry and resource harvests on biodiversity and ecosystem health. Our brief visit allowed limited conclusions based upon tours of new and old logged land and coast (Semantan) and relatively undisturbed rainforest, river and coastal shelf (Samunsam Wildlife Sanctuary and the Turtle Islands). Logging of the magnificent rainforest certainly reduced plant and animal diversity, but the recovery rate appeared to be rapid (decadal?) and biologically different from the original ecosystem. Deforestation almost certainly caused increased soil erosion and river transport of mud to the coastal mangroves, but we couldn't find any evidence of terrestrial soil transport into the coral reef zones of the continental shelf. Perhaps the huge monsoonal flooding episodes carry lots of mud into and across the coastal zone to the continental slope, where we were unable to sample.

From our survey of human and scientific resources for a 'forest to the sea' study, we had to argue that the State of Sarawak was not equipped to manage and execute research and regulatory programs to protect and enhance ecosystem services for the next century. Almost all past and present ecological research was done by scientific staff from western Europe, the USA, and Australia, but this was not managed or regulated by the State of Sarawak. While there were small commercial fishing vessels, no oceanographic or marine biological research vessels were available. We proposed that a panel of experts be assembled for series of workshops to further explore pathways to long-term ecosystem protection. Our 'Greybeards Society' concept was our personal contribution. We had little feedback from the fisheries, forestry, natural resources, environment and biodiversity group at the University and State Government, and to my knowledge the 'forest to the sea' study was never attempted.

To some extent, I was a little embarrassed by the way the academic, parks, and research people in Sarawak treated us as experts. The local environmentalists and academics knew our group was funded by a German palm oil plantation company, and they must have wondered how biased our report would be. To a large extent, I would expect the native people living in undisturbed rainforest and coastal settlements would be more expert than I was. Certainly the coastal fishermen and the marine island parks staff knew more about the local marine ecology than I did. I did have more expertise about the ability of those lazy tropical rivers to carry erosion products from deforested land into the sea, and where it might go along the coast.

Back at my desk at AIMS, I was disappointed that my institution and government were not interested in providing expertise and organisational skills to assist Sarawak, especially if there were not huge amounts of money attached.

I also spent a lot of time drafting proposals for Coral Sea research on climate change, sea level change, and ocean acidification. There was a movement in Europe and North America for an 'integrated ocean observing system' (IOOS), and some of us attended workshops on Australia's contribution to this international program. The IOOS program proposed to install automatically profiling measurements of temperature, algal pigments, pH and parts of the CO_2 carbonate system at moorings across the coastal shelves and continental seas. The data would be transmitted to research institutions from surface buoys with radio transmitters to internet connections onshore. As you might guess, all of this required considerable investment in ships, mooring ability, monitoring, and computer skills.

According to my journals, senior management at AIMS made most of this effort vanish with no trace. I spoke out to the press about anthropogenic greenhouse gas emission history, and how this had influenced the carbonate chemistry of sea water, slowing the building of carbonate mineral skeletons of corals, molluscs, echinoderms, bleaching of corals, changing the transmission of sound in the sea, heating the land and ocean, and changing the strength and direction of major ocean currents of the world. This was not appreciated by senior management, and this contributed to a movement to restrict contact with the media to senior management and the graphics department. Very odd.

I was witnessing the decay of a research institution into a money machine, linked to commerce and trade, overtaken by occupational health and safety inspections, and too many managers. It was becoming important to commercialise science products to satisfy consultant airheads with mission statements, key performance goals and milestones. In the past few years, I had seen the loss of dozens of experienced older leading scientists to 'redundancy', when their skills were needed more than ever. The symptoms of the Management Disease were excessive concern over performance goals, targeted research, safety, accountability, intellectual property, research targets being justified by contribution to economic wealth, and media smarts. On the other hand, scientists were probably guilty of poor communications with the managers and taxpaying public. How can we explain that we cannot predict the value of the results of esoteric research to the taxpayer? Is science and technology really the driver of civilisation, or a social parasite, like the financial institutions? As Einstein said, 'If we knew what it was we were doing, it would not be called research, would it?'

Chapter 56. Honolulu Beaches and South Dakota Prairie Wind in 2006

The Ocean Sciences section of the American Geophysical Union was held in late February 2006 in Honolulu. I was one of the session convenors for the TROPICS/MARGINS research in coastal Papua New Guinea and New Zealand. We were delayed en route in Sydney due to airplane problems, and there must have been twenty Australian scientists in the international airport beer lounge and smoking parlour. We eventually got stuck into some very good duck laksa, with dark ales, espresso and cinnamon donuts. Thomas Stieglitz of JCU was my seat mate, and we ate another beef curry supper over the dark Pacific Ocean.

Of course we arrived in Honolulu before we left Australia. And it was an early grey and wet morning bus ride to our hotel, through six lane highways, grubby docks and industrial sectors where Coast Guard and military ships loomed above the jetties, and then past the famed Waikiki beaches already full of tourists slathered with coconut oil. I was bunked in a third-rate hotel amidst a row of tattoo, acupuncture, motorcycle repair and fast food shops, but they were not ready for me yet and I left my luggage behind the concierge's desk to walk the streets until 3 pm. I found a bench to watch the surfers, old men playing chess, bikini babes, and the big canoes out beyond the surf waves. Finally got into my hotel room, where I collapsed in bed to sleep till supper time.

The next day I woke with a sore throat and raging headache, but a good bacon and eggs breakfast made it all go away. The Honolulu Convention Centre was full of about seven thousand scientists from all over the world, with equipment and book suppliers showing their wares, and a large active media desk hunting for interviews.

I soon was immersed with my gang, the scientific consortium related to TROPICS and now MARGINS. The MARGINS project had groups working on rivers supplying water and mud to continental shelves of PNG, New Zealand, Alaska, China, Gulf of Mexico, and the west coast of Europe. I gave a presentation on the annual budget of the Gulf of Papua river inputs of ^{226}Ra, ^{210}Pb, and ^{137}Cs to the continental shelf. Over the next five days, sessions were held on all of these projects, and the impact of Hurricane Katrina and Rita on the coastal waters of Louisiana to Florida.

This was the usual overload of new methods and data, new friends and old friends. My brain was on adrenalin most of the time. There were organisational and funding agency meetings for these groups of scientists, and a special party to honour my friend Billy Moore, who had almost singlehandedly created the concept of submarine groundwater inputs to the coastal zone, utilising his methods for measuring and modeling with the four isotopes of radium. I frequently had supper with good friends Bob and Josie Aller at Chinese and Thai restaurants nearby; Korean BBQ scallops and prawns, duck, Portuguese sardines, roast suckling pigs and cappuccino ice cream, carefully avoiding the Hawaiian delicacy of Spam Salad. I sat down at a table with Steve Emerson and Paul Quay, who had been Wally Broecker's graduate students working at ELA in 1970. Steve was telling ELA stories about Schindler and myself, some of which was highly classified: Steve had my 1970 radon measurement notebook which contained a note about the night I proposed marriage to summer student Sheila Dennett. He also told stories about Sheila and her pet grackle at ELA.

And then the South Dakota mafia struck again: Lois Norris, a friend of my mother, now widowed and living in an old folks home in Honolulu, phoned me in my hotel room. She simply ordered me to come to visit her, and she would feed me South Dakota food. Her husband John had died in 2002, and she kept all his books, music, piano and cello in her tiny little apartment. I was fed medium rare T-bone steak with baked potato, thankfully without the jello salad, while Lois gave a non-stop history of her life. She repeated several stories about Murdo high school activities, and friendships with my mother and father as teenagers. Lois and John were both South Dakotan school teachers, but John went on to university, studying linguistics, Greek, Latin, Hebrew, theology, and field medical services. They worked in Ghana, Nigeria, Morocco, Washington, D.C., several European cites, and finally Hawaii, where John was drafted and joined the Navy after Pearl Harbour in 1941. They had four daughters and a son, and daughter Kathleen was now touring Texas giving lectures on prairie social customs. Lois was a lady not to be messed around with, and she commanded the discussion with strength and power.

After supper I was invited to join a discussion group which happened to contain the seventy year old widow of Keith Chavez (who lectured to my group at the Bermuda Biological Station in 1964), and an ornithologist who had studied the big palm cockatoos and curlews in north Queensland in the 1970s.

And then it was all over. The Convention Hall booths were taken apart, and everybody made bus and taxi arrangements to get to the airport. I was up at 0530 to join the hotel shuttle to the airport. I was leaving another land of frangipani, where there was an excess of everything: multiple TV screens on different channels, flower baskets on the street signs and lights, surfers on the big waves, *wahini* clothed only in body paint, pigeons in the restaurants, ukulele singers on the streets, joggers with heart-rate monitors taped to their chests, Hummers and SUVs and Vespa bikes, street vendors selling fake watches, stretch limos, pasty white tourists, strip joints and hookers and tattoo parlours, orchids in the bathrooms, US Marines in combat gear, US flags everywhere, but also anti-George Bush posters, gay rodeos, and autistic basketball games behind the Taco Bell. At the airport on 25 February, I'm with Australian friends Bette Willis, Kathy Burns, Thomas Stieglitz, and Malcolm McCullough, and we ate and slept nine hours and twenty-three minutes over the Pacific, coffee and croissants, beef and chicken curry, arriving at Sydney for an overnight stay, and then flying north on 27 February to reach home at 1140,with bright and shining Sheila there to meet me.

At home in Alligator Creek, we had a good wet season, over a metre of rainfall, and our little creek had flowing water into June. In our tropical fruit cycle, we were picking oranges, mandarins, tangelo, limes, grapefruits, kumquats in the relative cool of our winter (May-September). I had planted banana shoots in all the wet spots of our little floodplain, and many of them grew fast and provided big bunches of lady finger and sugar bananas which hung from a rope to a ceiling beam in our kitchen and pergola. We also had some severe wind storms that tipped over some of our yard and creek valley trees. Most of these trees were propped back up, fertilised, and they grew back just fine.

Partly due to strange management choices, and partly due to government funding support, AIMS was in financial and political trouble. Our CEO and Research Director were frantically hunting for the magic language that would please the conservative government agencies in Canberra, without too much concern for the quality of science. There were too many sour and bitter workshops trying to identify our 'key providers' and the difference between the cost and value of our work, the ships, support services, and administrative reporting practice.

My colleague Rick Willis quit in disgust at having a scientific simpleton being put in charge of the analytical laboratory. In an eerie way, I had seen this institutional desperation and self-immolation before, with the Canadian Freshwater Institute being gutted by a Conservative government.

I had been urged by my family and high school classmates to attend the Centennial Celebration of the little prairie town of Murdo, including a reunion of the Murdo High School Class of 1959, of which I was the class President.

Murdo had been established as a railway station in 1906. The town name was due to the Scottish cattle baron Murdo MacKenzie, who needed railway transport of cattle from his 1885-1916 Prairie Cattle Company. Later he was the general manager of the Matador Land and Cattle Company. When I was very young, the remnants of an elevated water tank and metal discharge arm still stood beside the huge wooden railway station building. This building, well constructed of huge beams of solid timber, was no longer needed by the 1960s, and my uncle Gerald had moved the roof down to our ranch on the White River. He put it under the trees beside the river, for winter protection of cattle, and himself during calving time. He had built a snug little insulated room in the corner for himself, with a coal stove, a bunk, a gas cooker and coffee pot, and a shelf of tinned food to keep him going during blizzards in calving season. When I last visited this timber shed, there were also a lot of empty Copenhagen snus containers in the wastebasket.

Anyway, on 12 June Sheila drove me to the airport for an early flight to Brisbane, where a terrible breakfast was endured. But I got two more breakfasts on the Qantas flight to Los Angeles, arriving 0700 on the day I left... The USA Customs and Immigration man gave me a hassle because I couldn't remember the street address of my brother Obie's ranch, and they looked puzzled at my Australian passport. Then an American Airlines stewardess gave me a wonderful soft red blanket and I slept across the salt pans, Sierra Nevadas, Rocky Mountains and Colorado circular irrigation patches in otherwise brown grassland to Denver. There I found the smokers lounge up high, with Sam Adams Pale Ale, bratwurst sausages and whole wheat buns. I sat beside a large man with a tiny poodle in his satchel, who ate an alarming number of bratwurst with Miller's Lite Beer. He fed the butt ends of his sausages to his poodle. My flight to Rapid City was delayed twice, but we landed in the dark cool prairie at 2330 to find my brother Obie standing beside the luggage rack in cowboy hat, boots and rodeo belt buckle with Copenhagen in his shirt pocket. Once again, I was entering a different planet.

Obie drove me through the dark rolling hills of the western South Dakota prairie, arriving at his ranch house at 0200, way past his bedtime. I had a cup of tea with toast smeared with a concoction made by Obie's school teacher daughter Donna, made from crushed almonds, cashews, pecans, maple syrup, and dark chocolate. This was pretty damned good, so I had to have another one. I woke up the next day at 1030, listening to the sound of meadowlarks and killdeers. Obie's wife Annie was working at the Pierre library, so Obie took over the kitchen, producing a giant breakfast of fried bacon, pork chops, sausage, eggs, toast and cinnamon buns. Apparently he is not normally allowed this kind of food, due to some 'good heart' diet prescribed by his doctor, but he took advantage of my visit and certainly enjoyed it all. He had let his driver's license lapse, took the written test and failed it, and was now studying the booklet for a second try at the test. He took off to take the test again, leaving me to sit on the verandah to watch the birds and smell the hot prairie wind. I walked through a cattle pasture to the nearby stock dam and saw red and yellow wing blackbirds, crows, Baltimore orioles, meadowlarks and killdeers. Obie came home about six pm and immediately put a big rack of ribs in the oven and produced his 'home fries' grinder for potatoes. This, and some Sam Adams ale was just fine for supper. We took the big diesel pickup to check the water tanks, settled down to watch the basketball playoffs between Miami and Dallas, and both fell asleep in our recliner chairs.

We messed around on 14 June getting ready to go to Murdo for all the Centennial festivities. Obie brought a borrowed trailer house to brother John's ranch to accommodate all the relatives that would appear shortly. We arrived at John's place in time to witness the artificial insemination of some of his black Angus cows, his son Bill being the expert at this process. They had a big liquid nitrogen Dewar for the semen, contained in long plastic tubes. As usual, there was the relentless strong wind blowing dirt into my teeth. John's wife Pat made a giant kettle of hamburger and hash brown potatoes and hot buns for our supper.

The next day I again woke up to meadowlarks singing and a wonderful South Dakota breakfast. Again I walked down to the stockdam by the house, where maybe a dozen big turtles were sunning amidst the ducks and geese. The Jaycees had a big pit BBQ going in town, and John brought us a bucketful for lunch. I borrowed Pat's town car to drive down to the White River ranch to visit with my uncle Gerald and aunt Joyce. The river floodplain had alfalfa in bloom and lots of hay stacks from last year. The mule deer were in the foothills and the whitetail deer were in the river bottom timber, where pheasants and wild turkeys were abundant. Gerald was down in the timber, fixing a well pump for stock water tanks beside the corrals of the old Murdo train station. We came back to the ranchhouse for coffee and warmed up banana bread, and Gerald launched into his latest research on the Brunskill genealogy. His father George Harrison's mother was Marion Elizabeth Diamond (who was born and grew up in England). Marion had a sister, Eva, who had imported a fancy piano from England, and Gerald had somehow obtained a small print photo of this intricately wood carved piano. It looked like a small portable organ with roller wheels. On the way back to Murdo I picked some yucca flowers, and admired the longhorn cattle and buffalo against the fences.

Then absolute mayhem ensued, as Obie's daughters Marcy (with husband Steve and three kids) and Julie (and three children) arrived, as well as John's two sons and daughter. We all sat on the back porch and visited quietly, reviewing carefully who we might encounter at the Centennial Celebrations, who was still alive and who had died since I was last home.

As usual for these small town celebrations, there was a parade, and there were antique tractors and cars, the school marching band, a lot of decorated horses with riders and flags, motorcycle groups, a car dealership float throwing lollipops to the kids, and some convertible cars with really really old people waving at us. Some of them were my classmates!

My high school class of 1959 met at the local truck stop coffee shop, where we were served hamburgers and hot dogs and some of the worst jello salads anyone could imagine: green, yellow, orange slabs quivering in a threatening way. I was President wielding a gavel. Anna Louise the class secretary read the minutes of our 2001 meeting, and Pat the treasurer reported that we had no money at all. Of eighteen classmates in the 1959 school album, eleven showed up for the reunion. Two different classmates stood up to thank my grandmother Gertrude Butterfield England for educating so many of them in piano music. The story goes that local piano competitions in this region seem to have a high proportion of winning pianists. The judges wondered why this might be. Linda Haugland Brost explained that nearly all of the piano teachers in this region had been taught by Gertrude England. We made tentative plans for a 50th year reunion in 2009.

Later we all gathered again in the high school gymnasium, where a lot of my parents' friends appeared, and I had to explain several times where I was living, what was my job, and when I might be coming home. The next two days were full of fireworks, visits at home and in town, a town dance with a local cowboy band, hot biscuits and pit BBQ roast beef, aunts and uncles and various relatives.

By 18 June everybody was packing up to go home. I had to carry a lot of stuff I rescued from the family treasures in the basement, and some of my mother's precious books. Sheila had ordered a crokinole board to be sent to John's house; we repacked it with stiff cardboard for airplane transport to Australia. There were a lot of teary hugs and scuffing of cowboy boots in the dirt. John hauled me to the Rapid City airport for a rough flight to Denver, then the diesel fume smog of Los Angeles, then a lime chicken supper, and I slept for thirteen hours yet again across the Pacific Ocean. I was in the arms of my darling Sheila in Townsville by 1120 on 20 June, one day having vanished as usual.

Back home in Alligator Creek, our citrus trees were loaded with fruit in June and July, and we had lots of feasts with scientific friends and students. Bob and Josie Aller from SUNY on Long Island, NY,with their children Eddie and Deb, were living at AIMS for a month. Eddie was catching mud crabs in Cocoa Creek, so Josie made Maryland style crab cakes for several of our feasts. Later, Chuck and Debbie Nittrouer (and Deb's brother and wife, from Seattle, Washington) demanded one of Sheila's famous feasts, in payment for a series of seminars at AIMS. We walked the orchard and had heavy scientific discussions under the bamboo and frangipani trees beside the chicken pen.

Kathy Burns and I drove an AIMS car loaded with AIMS posters and video equipment to the Cairns Convention Centre, for a part of a public show of Queensland culture. There were aboriginal dancers, industrial booths, museum artefacts, tourism and trade and commerce displays, several university booths, and our show about Great Barrier Reef science and technology. Outside the Centre were some street musicians, body painters, hair braiding, and lots of exotic finger food, wine and beer. We were surprised by demands for information on north Queensland wetlands and mangroves. Fortunately I had some experience in the salt pans of Cocoa Creek.

Chapter 57. Professional Crash Landing, and then Santa Fe

On 3 December 2006, I was called to the office of Ian Poiner, the CEO of the Australian Institute of Marine Science.

He informed me that my expertise was 'excess to requirements' (an outrageous lie), and that I and six others had been selected for involuntary redundancy.

I hadn't seen this coming. I was angry, especially since the CEO (with interests in sea grass biology) and the Research Director Peter Doherty (a fish scientist) had little or no ability to judge my research skills and contributions to marine science. My radiochemical technician John Pfitzner was also declared redundant, even though there was no one else capable of operating the radiochemical lab.

My son Ian now had a van, so he and Sheila helped me clean out my office, move some of my books, two (out of eleven) file cabinets, my computer, two monitors, and a backup hard drive. I donated over half of my personal books and journals to the AIMS Library. After a few weeks, I began to lose the anger, and felt a sense of relief at escaping from the harness of government employment. I got a decent severance pay, and we consolidated this with savings from Canada.

Within a few months, AIMS had practically given away our half million dollar gamma spectrometry radiochemical laboratory, and this capability was lost to AIMS. I had commitments to scientific meetings in the USA and PNG, which AIMS agreed to fund. I was sixty-five years old, and my professional scientific life had come to an end. This was a painful paragraph to write.

But we had an excellent mango crop, and I was bringing twenty kg sacks of sun ripened mangoes up the hill daily for the rest of December and January. We used a blender to make mango puree, which was stored in 1.5 litre jugs in our veranda freezer. Son Ian and I toured the Townsville shops to find Christmas presents for Sheila, and we found a large timber pergola table, made of PNG merbau, a highly valued SE Asian rainforest wood. Our dear sweet Sheila also got a wheelbarrow and a magnifying desk lamp for Christmas from her sensitive gentlemen. Ian, Jesse, Sarah, Georgia Rose and Dan were with us for Christmas and New Year's. We gave Jesse and Sarah our inherited big kitchen table (from down on the South Dakota ranch), which was taken apart and loaded in their ute for transport to their Amamoor home in SE Queensland. The centre support for the table had White River silt inside, for which I provided a written account of what happened when the ice jammed on the White River in spring. I was deeply moved to see that old kitchen table go down the driveway to support another generation of Brunskill kids.

Kitchen Table

Our home kitchen table came from 'downt the ranch', the homestead of my grandparents George Harrison Brunskill and his bride Edna May Gregg Brunskill. The story is that Simon Brunskill, a fairly successful businessman in Haywarden, Iowa, helped my newly married grandparents to buy a ranch out in western South Dakota. The homestead was built out of cottonwood logs by a fellow who couldn't pay the bank on a loan, so the purchase price was low ($600), sometime before World War I. Grandpa came out west in 1915 with a rail car full of cattle, horses, and some pigs, and grandmother followed with all the furniture and household goods. The train only went as far as Chamberlain, so Grandpa moved the cattle to the homestead on horseback, and Grandma drove a Model T car with a trailer for their household goods. In this trailer was a big round wooden kitchen table with five extension leaves. My father Gregg George and uncle Gerald Paul and aunt Marjorie grew up around this table eating meals and doing homework. After Grandpa and Grandma died, my mother gave Sheila and me this big round table for our new home at Tourond, Manitoba. For nearly twenty years we hosted feasts and music parties around this table. When we took the table apart to move it, we could see White River silt in the central pedestal cavity, from the river bottom floods. Our sons Ian and Jesse grew up around this table. We brought it with us when we moved to northern Queensland in 1991. Our grandkids ate at this table in our house. When Jesse got married to Sarah and had kids, they started a hobby farm home near Amamoor, Queensland. We gave the big round table to Jesse and Sarah. They moved to Tasmania, where the table was refinished and varnished one more time, and our three grandchildren do their school homework on this table.

In late January 2007, Sheila and I flew down to Brisbane, and rented a car to visit son Jesse and Sarah and family at Amamoor. I got to play with grandchildren Georgia Rose and Daniel, shovelled a lot of Russian thistles in their paddock, helped build a horse corral, and some fencing for poddy calves. We did the home-made ice cream ceremony with kids squealing, played UNO and Seven-Up, drank a lot of beer and ate chili-sour cream chips. Neighbours had peach orchards, and it was nice to have fresh peaches and cream again.

Back home in early February, we had torrential rain for a week, and there was standing water in parts of the orchard. All sorts of bugs and mosquitoes were thick in the air, there was slippery mould on every surface, and our house reeked of DEET, citronella and sandalwood mosquito repellents. The Alligator Creek bridge was under water for a few days. There were several cyclones in the Gulf of Carpentaria and north coast during this month.

Nevertheless, brave Sheila and I drove to town and the airport at 0430 in early February for my trip to an American Society of Limnology and Oceanography meeting in Sante Fe, New Mexico. I was a convener of two sessions related to the MARGINS Project, and I had two presentation papers to give. Once again in Brisbane, there were three international airport security checks on the way to my departure gate, and they took away my nasal spray and eye drops in case they were explosive. I had the usual nice twelve hr sleep over the Pacific, woke up in Los Angeles, and transferred to a flight to Phoenix over bare desert, mesas and mountains. I took the bus to a rental car depot, and got lost twice trying to escape from the suburbs of Phoenix to drive to Sante Fe. I had to stop for guidance in Sante Fe, as Alligator Creeker travel agent John Dale had booked me into a strange accommodation.

It was an abandoned hospital and nunnery, all done up in fake adobe (concrete) style, and each room was obviously a recovery unit for patients at the hospital. There was an outdoor heated pool (steam rising into the chilly air), around which the maids and nuns were smoking black cheroots. Next door was a College of Acupuncture, a small shopping mall, and some basketball and tennis courts covered with snow. The main hospital building was now a restaurant and museum. I had a very nice little room with a kitchenette, crucifixes in the bathroom, a Jesus picture over the writing desk, washer and dryer, and a small library of shoot-em-up Western Cowboy novels. There was interesting Mexican music on the radio, and the TV was going nuts about the drug death of Anna Nicole Smith, the NFL Superbowl (Indiana Colts beat Chicago), the spreading of bird flu, and two women fighting over an astronaut. I bought some bacon, sausage, eggs, cheese, tea, butter, and a lot of pastries to stock my little kitchen nicely. I especially appreciated large slabs of corn bread and honey with tea.

As usual, I got lost driving to find the location of the science meeting, and I ended up on the road to Taos, which I probably should have continued on, as this is where all the hippies and artists and philosophers live. But no, I had to turn around and struggle through the city streets to find the El Diablo Hotel, where seventy-six year old Sir Charles Goldman and Gene Likens shouted to me from the registration table. Woods Hole and Scripps Oceanography people were lurking at the bar beside a crackling fireplace, and the place was full of the usual grubby graduate students and post-docs with their leather elbow patched jackets.

This was also a tourist hangout, as the nearby shops were full of very expensive electronic doodads, $2000 saddle blankets, $900 leather jackets, and high end restaurants with mesquite grills out front. Some of the side rooms were galleries for display of photographs and art work related to the SW desert cultures, Indian and cowboy, bleached bones on the desert sand, strangely Marilyn Monroe, chili peppers, Spanish architecture, and some jazz groups I never heard of. Many of the street names and building names came straight out of Zorro (Don Diego) and *The Cisco Kid*, with Pancho close by.

The week went by in a blur of morning, afternoon, and evening sessions about the physics, chemistry and biology of rivers, lakes and oceans around the world. I met old friends and made some new friends, many of which were telling wild (and partly true) stories about my past career in Canada and Australia. Some of my colleagues had quit science to join business ventures in solar, geothermal, and ocean wave energy production. There were several series of sessions from various consortia of research labs, universities, and government institutions, all of them hustling money from the National Science Foundation, the Office of Naval Research, and several private foundations. There was an increasing effort and ability to synthesise and extrapolate from new and old studies to make estimates of global processes and fluxes of carbon, nitrogen, phosphorus, metal and organic contaminants, and water circulation over past centuries and millennia.

Many sessions emphasised the use of satellite sensors of ocean heat, turbidity, chlorophyll, which generated huge data sets for large sections of the world ocean. Some continental shelves were clearly affected by anthropogenic contaminants, climate change and greenhouse warming of the oceans, and some regions did not display any impact. The timing and magnitude of world river discharge into the world ocean was changing, partly due to dams and irrigation projects, and partly due to changes in the distribution of precipitation. Most of the world's estuaries were acting like giant incinerators, oxidising (burning) most of the riverine organic matter that flows into the ocean. There were some disturbing sessions on century scale chemical ocean history from the composition of the annual growth bands in corals, with the observation that many of the locations in coastal

Puerto Rico, Gulf of Mexico, and the Caribbean displayed sick, bleached, and crumbling coral reefs. Several research groups had sessions on how anthropogenic CO_2 was dissolving into the ocean as carbonic acid, and causing problems with calcification of many groups of organisms.

One of the new research consortia being formed was GEOTRACES, mainly concerned about the supply and removal of heavy (trace) metals and radionuclides from the world ocean. I was interested to hear from the physiologists that copper, zinc, cadmium, nickel were all essential elements for algal growth in natural seawater low concentrations, but rapidly became deleterious or toxic at 10-100 X increase in concentration. Mercury and arsenic had no biological function, and were usually a negative factor in ecosystem functions.

Other large consortia competing for funding were the alphabet soup acronyms JGOFS, IMBER, LOICZ, SOLAS, CLIVAR, IOOS, all of which had their special sessions. I was especially intrigued to attend two sessions on submarine vents, mud volcanoes, and black smoker chimneys in various parts of the world ocean - we had not yet published our work on the Cornea vent site in the Timor Sea.

Some of my close friends were shocked to hear of my involuntary retirement at AIMS, and they offered funereal condolences and expressions of joy for my future career in tropical fruit production, music, and plenty of time to write up all the unfinished projects in the office file cabinets. Peter Santschi showed up at one of my supper table groups, dressed immaculately in Texas cowboy gear with a fossil ammonite tie clasp. Peter was a European multi-lingual Swiss post-doctoral student at Wally Broecker's lab at Lamont when I spent a year there. Taro Takahashi demanded a chair at my table. He was my office mate at Lamont in 1978. John Crusius took me out to a hot spicy Indian restaurant to explain his Woods Hole USGS North Atlantic cross shelf work on radium and radon to detect submarine discharge of freshwater, and iron as an essential nutrient for algal growth. John was one of the Broecker students at ELA in 1970-72, when he wrote and sang folk songs. Atsushi Suzuki from Japan had spent a year at AIMS, and I had helped him in some of his work there. Joanie Kleypas now had a good job in Colorado, but she did her Ph.D. in coral reef oceanography at James Cook University in Townsville. She remembered attending a Meadowlark concert, my banjo music, and Caspar the Galah. Sir Charles Goldman and wife Nancy took me to a sushi bar where we drank too much sake, talked about his consultancy in Papua New Guinea, the legacy of Jack Vallentyne, Richard Vollenweider and Bob Wetzel, some of the giants in freshwater science. Sir Charles was a close friend of Jack Vallentyne, and he told stories of trying to swim in Lake Baikal in the USSR. Charles was seventy-six years old and was still lecturing at the University of California at Davis.

I was saddened to realise that I would no longer be attending these international science meetings, an intense social and cultural treat for the soul and brain.

Saturday 10 February 2007 was a free day, and I slept till noon. After a green chili quiche breakfast, I walked through a wonderful North American Indian museum, which had a surprisingly good section on the Dakota Sioux people and a good bookshop. I drove northward to visit the mountains, the Bandelier National Park, where a canyon cliff dwelling Indian village was open to tourists. There were large haciendas back in the mountains where rich people lived.

The next day I was up early to drive back to Albuquerque, through snow patched mountains, little adobe towns with casinos and huge race tracks, pickup trucks with hay bales, and empty desert all around. I was half an hour early for my American West flight to Phoenix, but Homeland Security was paranoid - they were concerned about the big slab of cornbread in my suitcase, but I let them have a bite to verify that it was not explosive. On the airplane I could see rodeo cowboys sitting next to Arabs with keffiyeh headdress, hippies with nose rings, Hari Krishna people in robes, nuns, military fellows in full uniform, and the usual suits with laptops and cell phones.

And there beside the Phoenix, Arizona airport luggage carousel was my brother Gerald grinning his fool head off. I hadn't seen him in a long time. He whisked me off to several houses in Mesa where his grown children lived: Sarah Melissa, Becky, Christopher, and Carrie, all of them with their own children. We ended up in San Manuel (up above the San Pedro River valley) for a supper of fried cheesy chicken. We collapsed in the living room in recliner chairs in front of a huge screen TV to watch two separate football games and the news of the murder of nine people in a Salt Lake City mall. Much to Gerald's delight, I had brought a big slab of cornbread from the bakery in Sante Fe. Late at night, we warmed up pieces of corn bread and slathered on the butter and honey to have with tea. Gerald and Doreen have four Yorkshire terriers, all of which pile onto Gerald's lap when he is in his recliner chair. They tried to teach me the dog names: Molly, Ginger, Lizzie, and MacGreggor, but I couldn't tell the difference between them. Gerald got onto a family history binge, telling me about our grandfather George Brunskill, who told the engineers building the bridge over the White River near Westover in the early 1950s that the bridge was too low, and the spring ice would tear it apart. They wouldn't listen to him, so he drove to Pierre and told this to the state governor. They all ignored him. And twenty years later, the spring ice tore the bridge from its concrete pillars, and it was never replaced.

Gerald had graduated as a metallurgical engineer from the Black Hills School of Mines in Rapid City, South Dakota, in 1971. He worked for Caterpillar in Peoria, Illinois for a year, got bored with that and came to Yellowknife to live with us briefly in the early 1970s. He worked for Giant Mines in Yellowknife, and also for my colleague Mike Healey's fisheries experiments in lakes north of YK. He was also the captain of a car ferry at Fort Simpson, where he befriended some of the Dene people and apparently lived on bear meat. Enough of that, and he hitchhiked to Florida to work in a lumberyard, then a printing factory, while he lived in a tent on the beach north of Miami. He hitchhiked to California and worked for a roofing company in San Diego, and a hotel in Long Beach for released prison inmates. He returned to South Dakota to help his brother John and cousin Chuck take care of their livestock. He also briefly had an interest in a clothing store in Rapid City. He married Doreen in 1976, and worked for a uranium mine near Edgemont, So. Dak. driving a GeoTeck seismic sounding truck. He certainly had some adventures before he settled down!

Gerald had been a mining engineer at Magma Copper Mine since 1979, but the mine had recently been closed by BHP, and Gerald was kept on to manage the rehabilitation of the mine property, the massive tailings and facilities. Gerald had three of the 1920s England and England Construction Company photographs in his BHP office. Gerald's wife Doreen runs a small care home for children, many of which call her 'Gramma'.

Over the next few days, Gerald drove me around the region to visit the Tonto National Forest, the Superstition Mountains, some botanical gardens, saddle shops, several Mexican restaurants, a Tucson music store to buy banjo strings (for Gerald's banjo, hidden underneath his bed), some bookshops and old time cowboy museums. Doreen's son Josh (by a previous marriage) and Gerald packed me up on 16 February to drive 2.5 hours to the Phoenix airport, through whirligig cloverleaf roads and a car park eleven stories high. Josh latched onto Gerald's sleeve, as he was afraid of getting lost. There must have been some crazy people around, as Homeland Security was uptight and slow to allow me into the airport.

Once again I entered a different world of international airport travel. Tom Bradley airport in Los Angles was dirty, loud, crowded and reeking of sweat and diesel fumes, Punjabis in turban headdress, cowboys, teenyboppers in low-slung baggy pants, hookers, pan-handlers and beggars, Muslim ladies in hijabs, groups of little old ladies in wheel chairs, Japanese tourists with cameras.... a wonderful place for people watching. At my security check, a spitting mad angry airline pilot was led off by two burly policemen. I was grateful to find the Qantas gate peaceful and quiet, and I boarded the plane at midnight, ate a nice supper with Aussie Victoria Bitter, and slept across the Pacific into 18 February. Brisbane was warm and sunny, neat and clean compared to Los Angeles. We landed in Townsville at noon where it was pouring rain, and a John Deere slasher was bogged on the margin of the runway. Sheila and Ian were there to meet me, with a pair of shorts and flip-flops so I wouldn't suffer on the way home to Alligator Creek.

So here I was at home. I didn't have to go to work, and I could sleep till noon. I was feeling a little lost/vacant/underutilised/useless, but then Sheila came to my rescue, as usual. She hornswoggled our neighbour John Pozzebon into helping me build a chicken and duck pen down in the creek valley . We spent a little over $1500 to build a poultry fortress against the attacks of dingoes, feral dogs and eagles. The design came from John, who had worked in the construction industry, and he had mates that loaned us posthole diggers, cement mixers, generators, and a small front end bucket loader.

We could buy laying hen mash feed from local farmers from the Burdekin Delta south of us, for $18-20 for a 20 kg sack. Several friends began bringing us chickens and ducks and in no time at all, we had fresh eggs and some of the broody hens hatched both ducklings and chicks. Of course, Sheila named all the new chicks and ducklings. We wanted to raise some of these for meat, but then we had to learn not to let the poultry out if we were not there. Feral dogs killed several batches of our poultry and each time I reinforced the pen, and reduced the time that the chooks were out grazing. Over the next few years we butchered some of the ducks and chickens, but eventually gave up the meat production and kept only egg layers.

Sheila also encouraged me to write up my life story, as inspiration to young people to follow a career in science. Took me ten years to put it all together, going back through all my journals. I guess I did have a few adventures... .

Our friends, biodiversity scientist John Endler and wife Lorna left north Queensland for a job down south. They wanted to leave behind their big Toyota Troop Carrier, which they had used for expeditions into the jungle inland from Cairns. They bought a steel pole frame with a canvas tarp overthrow, to keep their vehicle out of the sun, and this was constructed and cyclone anchored in an unused part of our orchard. When they returned to claim their vehicle, they left the frame pole shelter for me; this got turned into a 3 X 6 metre wire mesh enclosure for my vegetable garden. I brought in six cubic meters of aged chicken manure which I enclosed in timber planks above the hard red soil of our creek valley. This was richly augmented with wheelbarrow loads of horse manure—brumbies (wild horses) invaded our orchard because of our stocktank water, and left a bonanza of garden fertiliser. I really liked several of the brumby colts, and they didn't do much damage to my fruit trees...but they did like to eat the young sprouting banana plants. In no time at all, we had more tomatoes, eggplants, zucchini and potatoes than we could eat, but there wasn't room enough for large plantations of melons and sweet corn. So the following year I bought two more of these pole cages, and then we had melons and lotsa sweet corn. We had bore water to irrigate the orchard and garden. Some of our ambitious tomato plants grew to over two metres in height, peeking out the mesh roof of our garden enclosure. Over the next years we grew squash/pumpkins, leek, garlic chives, shallots, strawberries, snow peas, cucumber, climbing spinach, sweet potato, various kinds of beans, sweet and very hot habanero peppers. The chickens produced lots of manure for the garden and nearby citrus trees, and the huge bamboo forest produced a thick carpet of fallen leaves which worked well for chicken nest boxes and garden mulch during the inactive wet season. Well, I guess life was sweet, and I didn't have to go to work.

Our son Ian had been teaching percussion part time at schools in Townsville, but in June he was finally hired as a full time employee by the Education Department of the State of Queensland. This gave him a little more security than is usual in the professional music game. He also was a guest timpanist with a symphony orchestra that performed Carl Orff's *Carmina Burana*, which was very exciting for me. We watched Ian conduct a percussion ensemble performance in a theatre, which blew our minds...Ian the long-haired rebel and black leather jacket heavy metal kid was now 'The Teach'. Sheila was now performing with her Tea Bag Ladies folk band and giving hammered dulcimer lessons. Even more miraculous, Sarah gave birth to Scarlett Grace on 2 July 2007 in Gympie, Queensland, our third grandchild. She started out wrinkled like a prune, but improved quickly into a beautiful and spunky little girl.

I was invited to be a consultant to a workshop on how to reduce the carbon footprint of AIMS by using alternate energy. Before I left AIMS, I had a website page giving information about how to shift from hydrocarbon fuels to equivalent energy production from solar, wind, geothermal, tides, ocean currents and waves. So in late November 2007, we sat around a big polished boardroom table to review the situation. In 1996, AIMS was spending over $600,000 for power and air conditioning. The AIMS engineers calculated that AIMS buildings, ships, and commuter cars were generating over 9000 tonnes of CO_2/year, and there was about fifty-eight tonnes of landfill waste per year. Estimation of cost of wind and solar panels was made and then ignored, as far as I know. This was a waste of time.

Chapter 58. Tasmania to the North Coast of Papua New Guinea in 2008, and On and On...

We had a wonderful wet season in January 2008 and half of our orchard and garden in the creek valley was under 50 cm of floodwater for a few days. It felt like it was going to rain hard for a month, and the air was so humid you could almost drink it.

Sheila the fierce grandmother went down to visit Jesse, Sarah and her grandchildren in southern Queensland with two suitcases...one of them full of toys, books and clothing for the kids. We had also purchased a good piano for Jesse's kids, and this was installed and tuned. It rained so hard here that Sheila couldn't get home, because airports and roads were flooded due to over a metre of rain in a week. The vegetation responded in a tropical way, filling the air with fragrances, stinks, flowers, mushrooms and mould on the walls and my South Dakota cowboy boots and batwing chaps. We had a splendid mango harvest, and Sheila went to town to buy a good Braun blender to make puree for the freezer. We also harvested five gallon buckets of passion fruit, which were packaged in one litre containers for the freezer. There was also abundant watermelon and sweet corn. Despite being in the tropics, where fruit and vegetables grew year around, my South Dakota prairie boy upbringing urged me to put away food for the 'winter'.

Our good musical friend John Weier arrived in February, a musical instrument maker and old time folky banjo/guitar player with the famous Winnipeg Duck Donald and Cathy Fink band. He told stories about his father, who built wagons in the little Mennonite town of Niverville, close to our home near Tourond. He was mainly interested in seeing tropical birds and spent a lot of time peering through his field glasses and writing poetry. When he left to return home in Winnipeg, he took my old (1842) black fiddle to rebuild the back end of it, which has sort of melted over the years.

Over the next few months our guest room was very busy, and Sheila made many homegrown chicken and duck feasts in her new kitchen. Her younger sister Lindsay (Noddy) and her husband Ted appeared, and were treated to the local tourist sites of interest. Then her niece Rachel Dennett appeared, followed by old Canadian friend and scientist Andy Hamilton and wife Betty Fello. Limnologist Gene and Phyllis Likens appeared at our table, full of stories about adventures at Cornell University after I left the campus in 1967.

Ian went down to Brisbane to take a conducting course, as required for his instrumental music teaching diploma. Meadowlark was still rehearsing and performing in town and at the local Palm Creek Folk Music Festival. Ian performed with an ensemble of musicians and dancers at the July Australian Festival of Chamber Music. This was the premiere performance of *Remember Me*, an affectionate reminiscence of old time dance halls in Charters Towers in the 1920s. The composer Iain Grandage was part of the ensemble, performing on piano and cello. Both Sheila and I were deeply moved by this composition, which reminded city people of rural and small town social cohesion.

Kathy Burns asked me to take her place on the National Facility Ship Committee (NFS) that operates and schedules the big deep water research ships in Australia. Unfortunately I accepted, and I nearly froze to death down there in Hobart, way too close to Antarctica. In late June 2008, Hobart was under grey skies at 10°C and there was snow on the mountains. I was dressed in strange clothing, full long pants with socks and shoes, and a woolly long sleeved shirt with a long DrizaBone coat. I walked the docks to view the *Aurora Australis*, the big icebreaker ship used for Southern Ocean research, the lobster and crab boats, and many full masted sailboats. I was lucky to stumble into a good restaurant that served good rare roast beef as well as a wonderful cream rhubarb pie.

Like most government committees, NFS is composed of some honest oceanographers and some administrative appointments from research institutions around the country. There were nine of us around the big table, each of us with a large binder of information for background and decisions to be made. The National Facility and CSIRO had sold the R/V *Franklin*, and was now managing the R/V *Southern Surveyor*, a 1971 built 66 m refitted North Sea stern trawler. This committee was pushing the budget limits to keep the ship working for 180 days/yr, but they were running out of money to keep the ship up to modern oceanographic standards and equipment and crew. So they were trying to propose that we needed a new and bigger purpose built vessel for deep water research at an approximate cost of $120 million. It had cost CSIRO about $4 million for last year's *Southern Surveyor* operations. Part of the problem was to design a ship that we could afford to operate for the next three decades. I proposed that we consider nuclear, solar, and sail power, rather than diesel. This was not considered realistic by the committee. The committee was jealous of my institution, AIMS, which had recently got a new (much smaller) coastal research vessel (R/V *Solander*). There was a lot of discussion about how the USA and European oceanographic institutions built and managed their ships (UNOLS), and some puzzlement about how my consortium TROPICS and MARGINS had brought UNOLS ships (R/V *Melville*) into Australian and PNG waters. [Note well...The new ship, called the *Investigator*, was finally launched in late 2014).

The meat of the work was to evaluate two dozen proposals for use of the R/V *Southern Surveyor*. Some of the proposals were easy to eliminate, as the ship couldn't do the proposed work. Many proposals were poorly prepared, and some institutions could not afford to pay the ship fees. But still, there were good proposals for over four hundred days of ship time for next year, when only one hundred and eighty days of shiptime were available.

I got yelled at because I refused to vote on some proposals that were out of my area of expertise (seismic geology, 3-D mapping of the sea floor). After this was all done, Captain (retired) Fred Stein grabbed me and took me to his favourite restaurant for seafood. I was enormously proud that he indicated all of my cruise reports for the R/V *Franklin* and R/V *Southern Surveyor* were greatly appreciated and were being used as examples of how to write these reports. We had a very hot seafood curry, some scallops in filo pastry, abalone, some sort of pickled fish, and way too much Tasmanian cider and beer.

I escaped on 25 June. The shuttle to the airport passed a small herd of Hereford cattle, a hell of a place to raise cattle! The airport was cold and grey and there were galahs on the lawn. Australian Federal Police were swarming the place and two paddy wagons were sitting with engines running outside the main doors. But they didn't get me, and I flew to Sydney, Brisbane, and Townsville, where the air was warm and fragrant, and I was home again.

Over the past few years, I had helped with the creation of a Canadian Lake Winnipeg Foundation, an NGO for the purpose of providing information and lobbying for improvement of the health of this big lake. In October, I was startled to get an email from the LWF Director indicating that I was the first to be awarded the Bajkov Prize, for my 1969-74 work on this big muddy lake. Alexander Bajkov was a pioneer ecologist who visited the lake in a rowboat in the 1930s. I could not attend the ceremony in Winnipeg but my colleague David Schindler spoke for me, and I heard later that he told a lot of wild stories that included just a little exaggeration about the storms and waves. Many of my Canadian friends were at this ceremony, and I heard that there were many hoops and hollers. Sheila's sister Heather (Peabody) accepted the award for me.

I was also polishing two book chapters and advising a graduate student (now Dr. Jasmine Jaffreys) who studied the heat budget and physical oceanography of the Coral Sea, with respect to greenhouse warming of the planet.

In late October I began taking lariam, the pills to slow down malaria infections because I had been invited by the Minerals Policy Institute (MPI) to participate in a conference on deep sea tailings disposal in Madang, PNG in early November.

The background to this was that Highland Pacific (a Canadian-Australian joint venture mining company) had sold the Ramu NiCo mining project to a Chinese consortium, and they wanted to mine for nickel and cobalt ore inland near the Ramu River, pipeline the ore slurry to Basamuk Bay on the north coast, where they wanted to build a refinery and a port, for shipping the refined nickel and cobalt to the Chinese markets. The mining company had hired the Scottish Association of Marine Science (SAMS) to help collect information useful for their marine tailings disposal proposals to the PNG government mining and environment departments. The waste from the refinery was to be disposed through a submarine canyon pipeline to water depths of 150 metres. The mining company and their consultants hoped this waste slurry would slide down the submarine canyon to over 1000 m depth and stay in one place, and not appear in surface water where the local people fished and stunningly beautiful coral reefs were thriving. This region, from the Solomon Islands to northern Indonesia, was called 'The Coral Triangle', where over six hundred species of corals were found, the world's greatest coral biodiversity. These coastal seas were also relatively free of industrial and population development.

I was invited because our TROPICS publications described the outflow of the naturally muddy Sepik River into the crystal blue Bismarck Sea, where the muddy water split into a tan colored surface plume, and a larger heavy mud laden turbidity current flowing down the seafloor of a large submarine canyon leading to over 1000 m water depths. We tracked the fine mud from the Sepik River hundreds of km to the west into Irian Jaya, being carried by the strong deep currents.

Despite being retired, I spent a month building a powerpoint presentation of our work (with Irena's help), and submitted this to present at the conference. I smelt a rat early on, as my presentation was rejected because the program was already full of the mining consultants and SAMS presentations. It became clear that this conference was intended to promote local jobs, the mining operations, and deep sea tailings disposal. However, MPI said they would pay expenses if I would attend as part of their team of environmental experts, and to help with a critique of the proposed mine refinery waste disposal into the Bismarck Sea.

By 1 November, the Alligator Creek poinciana and frangipani were in full flower, and the perfume was strong in the warm air. I had scuttled around furtively hiding some birthday cards and presents for Sheila because I would not be home on 5 November. Sheila was actually still in Halloween mode, with a jack-o'-lantern on the verandah. She drove me to the airport at 0400 and off we go to Brisbane, where colleagues Amanda Reichelt-Brushett (marine toxicologist, Southern Cross University) and John Luick (physical oceanographer, Flinders University, SARDI) joined me on the Air Nuiguini PX flight to Port Moresby, where we were greeted by Phil Shearman (Geographer, University of Papua New Guinea).

Port Moresby airport is outside of town, we could see the wild fires in the hills inland and smoke was in the air. We flew across the green mountains to the north coast of PNG, landing in Madang, a town that smells good, not like a city. MPI organiser Techa Beaumont (from Kolkata, London, and now rural NSW) and Wrenn met us, and hauled us to the town hotel for a good buffet supper of rare roast beef and plantain. There we met other members of our team, including Jim Tait (fisheries biologist), Bob Moran (retired US Geological Survey geochemist), Catherine Coumins (Canadian anthropologist with NGO Miningwatch), Brian Brunton (lawyer and retired PNG Judge), and Gert Schriever (BIOLAB, Marine Consultant, Germany). All of us were formally lined up to shake hands with several PNG politicians and Tracy Schimmield, the conference organiser from SAMS.

We were whisked away for a bumpy ride to Jais Aben (Peaceful Place), a wonderful resort beside a lagoon just outside of Madang, where the offices of the environmental activist Bismarck Ramu Group and MPI were found. We did not suffer at Jais Abend, as our rooms were big and clean with satellite TV, good computer/internet connections, merbau furniture, a full restaurant and bar, swimming pool, and a very good PNG artefact shop. The lagoon had a raft a little offshore, where you could swim out to sit in the hot tropical sun. We later were told there was a World War II Japanese aircraft on the bottom of this lagoon, in 25 m water depth.

The next day, after a hearty bacon and eggs and strong PNG coffee breakfast, we were hauled off to a nearby community hall where the Lutheran Church elders and the Bismarck Solomon Sea Indigenous People's Council had gathered. They owned the land around Basamuk Bay, where the miners wanted to build their refinery and harbour. Many of them were neatly dressed in suits with neckties, and our motley crew felt a little embarrassed to be in shorts and short-sleeved shirts, not a necktie amongst us. These intelligent PNG native people were familiar with some of us, as they had met us in Port Moresby some years ago. One fellow pointed at me and said that I was a cowboy, and that I knew the name of Roy Rogers' horse when he had last met me. These people interspersed their questions in English with several other languages describing their customs of calling sharks, hunting crocodiles, and deep water fishing for the bright pinky-red snappers (*Lutjanidae*). They lectured us on the Rai Wind (SE Trades that blow the water west) and the Taleo Wind (a Northwest wind that blows the water east). We had three translators up on our platform, and we often got different versions of their questions. They wanted us to explain the purpose of the conference, to explain deep sea tailings disposal and its likely consequences, how the nickel/cobalt mine would affect the Ramu River inland, would there be lots of jobs, and what would happen when the pipeline was broken.

Good questions, but not easy to answer quickly, especially since the mine, pipeline, and waste disposal system had not yet been built. We tried to warn them of the difference between the opinions of consultants working for the mining company, the town business council that wanted new jobs, the environmental advisors from MPI and BRG, and the PNG government departments of mine development and environment. Each of us gave an account of our expertise, and how it related to deep sea tailings disposal.

Over the next two days we were hauled in utes to other villages that wanted to know about the conference, all as above. The previous evening, we made a dot point handout summary of our group expertise, our purpose at the conference, and an explanation of what would happen if they did go ahead with the mine, refinery, and waste disposal into the coastal ocean. We had not been warned of these village pre-conference sessions. We had lunch with these friendly people on the beach with driftwood fires, cooking whole fish, plantain, chunks of meat of unknown origins, and buckets of mangoes. Techa was chewing betel nut and her teeth were all red, like half the people in the village. We talked with these people all day, and it was bloody hot and humid.

On the return trip, we drove through Madang and the port, where huge Chinese ships were unloading bulldozers, long steel I-beams, railway containers and crates, huge plastic and metal tanks, and a variety of Chinese made vehicles. We were told that the local people did not like the Chinese, and that the PNG people sabotaged their work when possible. Two Chinese had been killed at the Basamuk Bay refinery site, about twenty km SE of Madang, bashed to a pulp because they had insulted the local Basamuk Bay people. As a consequence, the Chinese now have their own police and security guards that occasionally treat the local people badly. It became apparent that there were some hot tempers in town, some in favour of the jobs from the mine, some against the pollution from the mine. Now it became clear why MPI had hustled us quickly out to Jais Aben, to keep us away from the town disputes. We were slightly alarmed to discover that there were quiet armed guards around the Jais Aben grounds. We were very grateful to get back to Jais Aben to swim in the lagoon, have a shower and a good supper with cold beer, and to discuss what we had learned from these sessions with the local people.

The conference began in the big Madang Hotel on 5 November. The morning was spent listening to various local governors and mayors and speakers from the Moresby federal government mining and environment agencies. Tracy gave an overview of the proposed mining operations, with details of the business and employment opportunities from the mining project. It quickly became apparent that this conference was mainly for convincing the local people to approve of the mining operations. The speakers all assumed that the mining operation would go ahead, but Catherine, John and I asked some embarrassing questions, and the fight was on.

In the afternoon, a good presentation was given about the gold mining operation on Lihir Island in the Solomon Sea, where the mine waste was intentionally dumped nearshore on top of coral reefs. During one of these presentations by the mining consultants, a PNG Minister interrupted the talk to announce that John McCain had conceded defeat, and that Barack Obama had been elected as President of the USA. Whistles and cheers came from the audience, and everybody went out to the lobby to get a beer and talk about this. I didn't realize that these people paid attention to US politics, but Obama was very popular here (and still is).

That night, we were feasted in a tropical garden beside the hotel. We had the French Ambassador and an EU Commissioner speaking on and on about the importance of industrial development for PNG and the Pacific islands. The *kundu* (drums) began throbbing, and a lot of PNG dancers pranced through the grounds, some of the ladies bare breasted, and some of the men with penis gourds, some had boar tusks through their noses, all with feather headdresses, all greased up with coconut oil. They were singing too, and one of the songs was *Obama Obama*. All of our environmental group was seated at a table farthest away from the guest and speaker table, but the dancers seemed to spend a lot of time around our table.

The next day the environmental NGOs were given an hour to speak. None of the mining consultants, SAMS, or government ministers were there, but we spoke to maybe two hundred local native people. There was open hostility in the local people for the arrogant miners and pro-mining business interests. The rest of the day was for the more polished presentations by the mining consultants, and more examples of deep sea tailings disposal methods in Indonesia, the Nordic countries and Hungary. We got to know Samantha Smith, who was spruiking for the Canadian (Vancouver) Nautilus Minerals Inc. Company, who were proposing to mine submarine chimney vents at 1500 m water depths SW of New Britain Island in the Solomon Sea. Several consultants reported on coastal and deep water recolonisation of mine waste polluted sediment by bottom organisms, implying that it would all go back to natural conditions once the waste disposal was over. James Wang represented the Chinese MCC company that was moving heavy equipment into Basamuk Bay village. There were several industrial lawyers given speaking time to indicate that the mining proposal was legal. Later we had two suckling pigs with a mountain of sweet potatoes at our evening meal, with speakers from town business leaders, and more local dancers with *kundus*.

PNG Dancers with their kundus at Deep Sea Tailings Conference, Madang

The next morning there were some good presentations from the SAMS biologists who had surveyed the Basamuk Bay and Canyon. They made a rough bathymetric map of this coast, and sampled the bottom organisms that would be in the path of many tonnes/day of refinery waste stream. They also did a quick survey of the nearby Lihir and Misima Island coastal zone, where gold mine wastes had been dumped nearshore for almost a decade. We were approaching the end of the scheduled conference in the afternoon, and the SAMS organisers wanted us to help draft some draft guidelines to allow the Chinese miners to dump their refinery waste into the clear blue Bismarck Sea.

Almost in unison, most of the black people and NGO environmentalist rose up and walked out of the conference room, wanting nothing to do with guidelines drafted by the miners themselves. Tracy called us back, loudly saying we were not being responsible, and that we were endangering the success of the conference and the mining operations. We were all sitting on the ground outside the hotel in a big circle and Tracy was standing there crying, angrily shouting at us. Our MPI group answered questions in the blistering hot sun until the evening. That night at the banquet, several PNG Ministers gave addresses stressing the importance of mining development for town development. They told the story of the nearby Bougainville Island trouble, where the local militant landowners revolted and sabotaged the Rio Tinto mining equipment. Much of the rusting equipment still sits around gaping holes in the mountains, and there has been no rehabilitation of the damage from the miners. They were also pointing out that the NGO environmentalist team were not from PNG, but from Australia, Canada and the USA. Retired PNG Judge Brian Brunten stood up to indicate that he was a citizen of PNG.

Then the conference was over, and I got to sleep late on Saturday, 8 November. I had breakfast with some of the Scottish marine biologists, who claimed they were not aware of how their rapid biological surveys were now being used as 'before mining' data, to be looked at after the mine waste began flowing down the coastal canyon. They considered themselves to be in a hornet's nest of intrigue, not suitable for good scientists. They too were upset at the arrogant Chinese and mining consultants.

Our MPI team were hauled to the Divine Word University just outside of town, where 350 students were waiting for us in their gymnasium. The University was opposed to the Chinese and coastal pollution, and they had been banned from the conference. These students were from many different tribal communities along the north coast of PNG, and we had at least six translators busy sorting out the questions and answers in *tok pisin* and English. Some of these people came from both distant and nearby tribes that did not trust the tribes of the Madang people, and they felt betrayed that Madang wanted the mining operations to proceed. That evening the Jais Aben cooks led us to their favourite country club down the coast.

There was a dance floor, live music, kids and grandparents laughing, and a very unusual Chinese kitchen that specialised in local squid, octopus, sea urchin gonads, and jellyfish pickles. They were handing out PNG cigarettes, strong tobacco wrapped tightly in newspaper. I had to demonstrate my ability in 1950s jitterbugging when the band played Bill Haley and the Comets' 'Rock around the Clock'.

We had pouring rain, strong wind, and a lightning storm the next morning, as we were waiting to fly from Madang to Moresby. Some of the Townsville military units were at the airport with their Blackhawk helicopters. The airport reeked of smoked fish smells, seaweed, and big ship diesel smoke. The airport was packed with native PNG big mommas carrying babies waiting for their sons to return home, Chinese miners frightened by these big mommas, and a few sunburned Japanese tourists with cameras around their necks.

We finally got out of Madang four hours late, arrived in Moresby in the pouring rain, and were met by an Australian Broadcasting Company reporter (Stephen Marshall) with microphone and cameraman. He wanted an account of our experience at the Madang Conference, which we gave in our best cautious and neutral way.

We just barely caught the last flight to Cairns at eight pm, no time for supper. Most of my Australian colleagues and I missed our ongoing flights, so we overnighted in Cairns at a strange computer operated hotel with no front desk humans or restaurant. Severe starvation. Up the next morning for breakfast of instant coffee and PNG tobacco cigarettes, taxi to airport where I finally found a cheeseburger and a beer at 0800. My flight to Townsville was on a small propeller airplane that looked like a museum piece. And then Sheila gathered me up and took me home to sanity, peace and near ripe mangoes in our creek valley.

Within a week, I had an irate phone call from the media officer at AIMS, complaining that I had not informed him about my ABC interview in Port Moresby. They had been ambushed by somebody from the Canberra Department of Foreign Affairs and Trade, saying that they didn't want government scientists speaking (without permission from DFAT) to the press about sensitive matters in foreign countries. I never heard the radio or TV interview, but I guess somebody in Canberra was paying attention. Since I was no longer a government agent and not in the employ of AIMS, I paid no attention to their concern.

Chapter 59. Family Adventures at Home and Abroad

In May, Sheila flew to England for her Auntie Ester's ninetieth birthday.

While she was gone I was batching. John van Bockern, girlfriend Janice and lumberjack Ehren visited me from Colorado. They had been touring the east coast of Australia in a hippie van, eating fast food and enjoying the beaches. John is the son of my cousin Gail England van Bockern. They could really eat! I roasted a duck...they ate it all and looked hungry. So the next night I roasted two ducks, and they ate it all. I finally conquered them by roasting two of our homegrown chickens and a duck, and they left a little on the platter. Janice had to explain that I needed to make large pots of mashed potatoes to fill them up before the meat. Ian brought a large tub of ice cream and some coffee liqueur, which was quickly consumed.

John and Ehren had operated a dead tree timber harvesting company on the east slope of the Colorado Rockies, the pines having been killed by bark beetles which could over-winter these days due to climate warming. They rapidly sharpened my chainsaw, and helped me cut down some big gum trees that were too close to the house. They also helped me rescue my bogged orchard mower tractor. John could play the piano well, and was university educated. He was interested in how to convert redneck Midwestern people to green energy and sustainable life styles. He realised that the science and engineering was all ready to go, but the social and political conversion was going to be difficult. Damn right! I was gratified that they recognised and applauded my skills in making liqueurs from passionfruit, kumquat, and jackfruit. Ian invited them to a Wassa concert on the Strand, which they enjoyed greatly.

I doubled the area of our vegetable gardens, so I could grow more melons and sweet corn in 50% aged chicken manure augmented soil. We had a massive tomato crop for almost eight months...they grew almost two metres high. I was pleased to be called 'The King of Eggplants' by our neighbours, who had to eat some of them.

In addition to this gardening, I managed to write a science journal paper on submarine methane seeps in the Timor Sea. After sending the paper off to the editors, a Thai-Australian oil drilling platform sprung a leak in our Timor Sea study area. The Montara blowout leaked more than 2000 barrels/day (320 m^3/day) over seventy-four days. It was on fire for several weeks, and Timorese and Indonesian fishermen were finding 'cookies' of slippery oil in the gill nets far to the northeast of the rig location.

You need to understand that this remote region had crystal clear blue water, and you could see *Halimeda* patches on top of sea mounds 25 m below the sea surface. What a mess, but for some reason, Australia was very quiet about this disaster and we were not consulted, despite having spent two weeks of ship time in this region in 2006, with Kathy Burns' hydrocarbon gas chromatograph working full time.

In August we visited families of Sheila's Canadian brothers and sisters in Winnipeg and Ontario. I had earlier sent my second fiddle (the Luke Brooks fiddle) back to Winnipeg with luthier John Weier. He repaired the collapsed tail peg and provided a new case. It was good to see this old fiddle again. A party at Dave and Barb Ediger's house gathered most of the old Meadowlark and Chicken Flats String band, and we had four fiddles going strong for much of the night. Susan Israel and Sam Baardman hauled us up to visit my Lake Winnipeg research vessel (CGS *Bradbury*) which was now part of a museum, up on dry dock. Sam had been involved in the photographic history of Lake Winnipeg and the old ships, and there was a nice website display of poetry, artworks, carvings, photos, and interviews with old ship masters. After a Vietnamese supper, we went to a nightclub to hear the soul/funk band The Solutions. This band was led by Susan's son Isaac on guitar and his school buddies, as well as Crowbar (trombone) and Janet on trumpet (offspring of Al and Peabody).

As explained in an earlier chapter, I was honoured to be awarded the first Bajkov Prize for my 1969 research cruises on Lake Winnipeg, but I couldn't come to accept the award then. Now all my Lake Winnipeg colleagues gathered for a party, and to give me the award plaque. We also roasted an eighty lb pig in Al and Peabody's back yard to feed twenty-five people. Here were most of the people I had worked with in 1969-74, most of them retired. The next day, Alex Salki, Chairman of the Lake Winnipeg Foundation, drove Sheila and me to the little lakeside town of Gimli, where the new R/V *Namao* was tied up at dock, with crew waiting for me. All lines were cast off, and the ship proceeded to the middle of the south basin of Lake Winnipeg for a few water samples for chemistry, incubations for nitrogen fixation, testing of trawl nets, and filtering of the water to obtain the suspended sediment chemistry. Lake Winnipeg was still very muddy, like coffee with lots of cream. The First Mate on the old CGS *Bradbury* in 1969 was a young Métis named Merv Sinclair...and now he was a retired ship captain, in the wheelhouse telling stories about our early work. He snorted in amusement when I show some deck photos of him in 1969 on the *Bradbury*.

Peabody's family spent several days eating the leftovers of the roasted pig. They also were kind enough to have a large bowl of Eccles cakes in the kitchen, which was one of my favourite pastries when we lived here. We had several bad bad games of 'Extreme Croquet', starting out in Peabody's front yard, but ending up on the street and in the gutters in the dark. Of course, Sheila's nephew, delightfully nicknamed 'Crowbar', won all the games. On 6 August, we flew down to Toronto where Sheila's youngest brother Bill helped us escape the city for his little village north of Toronto.

Here we ate planked salmon, sweet corn, local sausages and lobster, and watched Bill fly model airplanes. Apparently nobody in this family eats breakfast, so I proceeded to cook a dozen eggs and three lbs of bacon, and it all vanished quickly. You need to understand that Bill and Carla's house porch had a row of 100 litre barrels labelled as being dangerous and highly flammable. One of Bill's hobbies was rocket technology. Bill took us to the Edenvale Aerodrome, where old airplanes and gliders were restored and flown. One of these airplane maniacs (Dan the Honeyman) had a Tiger Moth canvas biplane fired up, and Sheila (in a black leather jacket) got to ride in the back passenger seat for a joyride over the nearby flax and onion fields, and on to wing rolls over Georgian Bay. We then drove to the home of Sheila's sister Lindsay (Noddy), where again I made gigantic breakfasts for everybody, Sheila and I played old familiar family music, and many relatives showed up for supper.

Mike Stainton, Alex Salki, GJB, Hedy Kling–presentation of Bajkov award to GJB, Winnipeg 2009.
Photo: Lyle Lockhart

We left Toronto on 13 August in the chilly early morning, flying to Minneapolis for a brief visit with my nieces Marcy and Donna. Marcy took me to an old folks' home where my mother's cousin Toadie was now living. It nearly broke my heart seeing this vital and bouncy lady mute in a wheelchair, after all the wild adventures she had with the England and England Road Building Company pre-World War II. Here she was, trapped in the clutches of a medical facility, with none of her family left to care for her. I played a few fiddle tunes for Toadie in this hospice in Minneapolis, but I'm not sure she heard anything. But she perked up when Sheila gave her a ceramic green tree frog from north Queensland.

We flew on down to Sioux Falls, South Dakota, where two of my classmates now lived. Terry was a retired bank inspector, and Tom was a retired gyn/obstetrics doctor. They had organised an amazing fiftieth reunion of my Murdo South Dakota High School Class of 1959 in Tom's back yard, under a circus tent, where chokecherry wine, bbq pig [right tolerable, fit to eat], sour cream raisin pie, and lots of cold beer was consumed. Exploits of our school days were fondly remembered by over a dozen people. Tom had planted a row of buffalo berry trees, thinking of his (and my) early days on the White River foothills. There were late night sessions with eighteen year old whiskey and cigars. I drove out to the Augustana College campus to meet Augie classmates Annalee Rantapaa and Larry Tieszen, whom I hadn't seen since graduation in 1963.

We drove west to Murdo, happily crossing the wide Missouri at Chamberlain, into the rolling west river prairie, where all four Brunskill brothers and families were together again. Quite a few stories were told over brother John's excellent home-grown beef steaks, pulled pork, melons, fresh peaches and cream. We had to make a cheat sheet for the names and parentage of the squillions of nieces and nephews. We sat on the back porch telling stories. Brother Obie said ...'The government ought to give everybody a cutting torch and pay them to cut up their tractors and ploughs and discs, so we could get back to the prairie cattle and horse business.' We drove down to the White River ranch, where the house and barn were sadly empty. We rescued some of the old harness buckles and bell collars for our sons. As we entered Australian customs, the officers fingered our strange collection of horseshoes, sheep bells, chokecherry and buffalo berry jams, and a rusty narrow tined manure pitch fork. But they let us back in the country, and we were home and safe again.

Back home in Queensland, son Ian continued mighty efforts toward his Graduate Diploma from Griffith University in Brisbane. He was also busy teaching at a dozen schools in town, and performed at the July CD launch of his new group Torakina. He was in several jazz groups that performed at beachfront restaurants, played with West African drum ensemble Wassa, and toured with school bands, orchestras and choirs to Charters Towers, Brisbane and New Zealand.

Sheila kept increasing the number of orchids hanging on our verandas, feeding voracious semi-tame wildlife in our yard, and tried to keep me aware of the day of the week. She was writing for the website of our local wildlife sanctuary, and is a founding member of The Old Time Music Teabag Band. After years of delay by the male members of the family, she took charge of her sixtieth birthday this year, and ordered a swimming pool to be built in our yard (financed by my severance pay from AIMS). This development required some major trimming of large gum trees, generating a small mountain of woodchips for mulch. One of my Ph.D. students, Jasmine Jaffres, from Switzerland, handed in a good thesis about warming of the Coral Sea, and how this might affect cyclone strength and coastal coral reef bleaching.

And now the decorative Christmas beetles were flying into our house at night, the geckoes and green tree frogs were fat and happy with more insects to eat, the cicadas were singing a seventh interval drone in the big gum trees, the leafless poinciana trees were aflame with bright orange-red flowers, and our creek valley frangipani and citrus trees were in flower. The fragrance makes the humid hot air almost too heavy to breathe. We have been walking and swimming at the beach, where turtle tracks show trails to their supratidal nesting sites. The water is very warm now. All this was interrupted by the news from Canada that Sheila younger sister Noddy was diagnosed with lung cancer, and was now undergoing radiotherapy in Toronto.

Sheila has a potted cycad to decorate for our Xmas tree. I've made a fresh batch of my famous fermented banana jam, and butchered a large Xmas jackfruit. We now have ripe mangoes and sacks of lychees in the kitchen. Right Tolerable, yessiree Bob!

Chapter 60. More Adventures in Madang, North Coast of PNG

We had a wet 2010 and I had trouble keeping the big guinea grass down in my orchard, even into the dry season. We had over 500 mm of rain in two days, and our little creek came up out of its banks to flood my chicken pen and garden again. Sheila was getting her swimming pool to work, and spent a lot of time in the new pool.

Ian was still in his university courses, and he brought oboes, bassoons, trombones, flutes and tenor sax down to our house for practice. He was supposed to learn how to play a few scales on each of these instruments, as well as my old Rafael Mendez trumpet, and I was happy to help with this.

I was hornswoggled into being a guest editor for a tribute volume for Dr. Jack Vallentyne (my Ph.D. thesis advisor) and Dr. Richard Vollenweider (an important Swiss limnologist who created a useful model for eutrophication), both of whom had died in 2007. This involved long hours at the computer late at night, on email, trying to contact old colleagues and scientific associates to elicit commemorative articles about these two gentlemen, who had greatly influenced the careers of many aquatic scientists over three quarters of a century.

I was mowing my orchard with my John Deere in early Feb '10, after some good rain. I came to a corner and the rotor blade stopped suddenly. I shut down the engine and got off to inspect the drive belt that turns the mower blades. The greasy black drive belt appeared to be broken in several places. I rolled a cigarette and wished my neighbour Pozz would stroll by to help fix this mess. Then I noticed that there were too many pieces of the broken drive belt, and that some of them were still moving. On closer inspection, I found that two yellow-bellied black snakes were entangled in the drive wheels, cut into many pieces. I untangled the snakes and hung them on a nearby peach tree, and then discovered that the drive belt was still intact, and only needed to be fitted back onto the drive wheels. These drive wheels were pretty slippery with snake guts. A day later, the magpies and ravens had consumed most of the snake parts hanging on the peach tree.

Sheila went down to south Queensland to stir up her grandchildren at Jesse and Sarah's home in Amamoor in April and September. This was partly arranged to coincide with my upcoming expeditions to Madang in Papua New Guinea.

Over the past year, the legal kettle had been furiously boiling about the pollution of the Bismarck Sea by the Chinese Ramu NiCo MCC refinery, and one of the local PNG NGOs got serious about trying to stop this outrage. [See previous chapters.] I had been providing scientific references and summary documents about this problem for the past year. The Bismarck Ramu Group (BRG) got funding from various local and international sources (mostly the Christianson Foundation), and they hired Australian-PNG lawyer Tiffany Nonggorr, who was also the wife of a PNG Constitutional legal expert. Tiffany phoned me and most of our team from previous efforts in Madang, offering funds to pay for our travel, food and accommodation in Madang to help plan the court case.

I flew down to Brisbane to meet the new head of the Minerals Policy Institute, Charles Roche, and we had good coffee and croissants while waiting for the Air Niugini flight to get organised. In Port Moresby, good friend Phil Shearman met us to haul us to a cheap hotel that smelled like smoked fish and wood smoke. It was very hot and there was no air conditioning, but the seafood was good as we ate by candlelight, as the electrical power failed for the night.

The next morning, 14 April, the headlines in the local newspaper were '12 bodies hacked to death found in the Purari River delta mangroves', which was related to unpopular LNG plants and pipelines in that region. This gives a taste of the volatility of the PNG people when they feel excluded from their own land. We flew across the rainforested mountains and valleys of the Markham and Ramu River into the massive confusion of the Madang airport. Tiffany and her children and two law students met us in a big van, and hauled us to Jais Aben. We later found out that the 'two law students' were actually bodyguards from the Mount Hagen highlands.

Over the next four days we presented our information to Tiffany and her legal team. Tiffany explained that the PNG Constitution gave local tribes land ownership (customary property rights), and they were entitled to stop development if their land and coastal seas were damaged. The funding agencies put some ground rules on expenses and personal safety. It was explained that there were a dozen security guards around us at Jais Aben, but we were not protected in town.

More important were several interviews that were filmed by Toby and his Sacred Grounds TV crew. Sama was a member of the Basamuk Bay village community, where the Ramu NiCo MCC refinery was located, and he told the story of the Chinese bulldozers flattening the community cemetery without consultation or warning. He explained the complicated politics of his village, where his cousin had been hired by the Chinese to be a village policeman, and two magistrates and a village councillor were members of his extended family. Sama came from a family that had lived in Basamuk Village for more than ten generations, and he told stories of his grandfather teaching him how to tie the magic knot on a rock at the 400 m end of a fishing line. As the rope tension relaxed upon hitting the bottom, the knot would slip and release the heavy rock, and the rope was pulled up slowly to catch the big red snappers in the deep cool water. Sama told of open sewer lines from the refinery to the sea, and coral and fish that were no longer in Basamuk Bay. Sama had invited Toby's crew to film the devastation of their village cemetery by the Chinese, but they had been arrested by Chinese police, confined in the railway container, and guarded by Chinese police armed with assault rifles. Some of this got surreptitiously filmed by GoPro cameras.

Another afternoon was devoted to conversation and filming of interviews with Melchior, who was a Ramu River floodplain big man, a well-educated leader in his community. He had a diploma in Agricultural and Fisheries Science, had travelled to the USA and UK, and had been a teacher of fisheries and forestry conservation at the University of Papua New Guinea. He explained the life style of his people, the cycle of floods and dry seasons, the importance of hunting, fisheries and gardens. He especially liked the eel fishing and groaned about the wonderful taste of smoked eel. The mine had already produced salty water, which the freshwater fish would not like, and the waste water killed their gardens. He had little respect for any benefits of the mine, and he implied that if the court case could not shut them down, he would help to dynamite the pipeline.

During this time, Tiffany was successful in obtaining a court order and injunction to stop the Chinese mine and refinery from operating. Tiffany explained that there would be an appeal, and we would have to go back to court sometime in September, with much more evidence to support her case.

Tiffany led a discussion group at the downtown Madang Hotel, with all of us behind her. In the audience were several governors and ministers, and Sir Michael Somare, the PM. She reviewed the PNG Constitutional customary property rights law, and showed photos of blatant pollution on the land and sea in Basamuk Village. There was no environmental monitoring or control. A Madang medical officer gave a short presentation on the proper design of a human and food toxicology procedure, based on samples of Basamuk Village people's blood and tissue, garden vegetables and fish. There was no evidence that this was in the government plans for Ramu development. He worried that the local Madang tuna packing plant might already be contaminated with toxic metals and organic reagents from the mine and refinery.

This was a direct confrontation of the environmental side vs the local business people and miners, and tensions were obvious. We were all quickly gathered up and hauled back to the safety of Jais Aben, outside of town. I had supper with a tobacco and cigarette businessman, who introduced me to *girigiri,* a kind of coconut curry seafood cooked in a big iron pot, bush camp style.

This part of the battle was now over, and we prepared to go home on 18 April, hot and tired, but Tiffany demanded that we all support her at a press conference in Port Morseby the next day. Our Environmental Team arrived in Port Moresby the next day at 1240, and Tiffany whisked us off to the Gateway Hotel, where the bodyguards and media were abundant. The press were about equally divided between the pro-mine development and pro-environment camps, and we were peppered with questions into the evening.

The next day, our airplane developed engine trouble halfway to Australia and returned to Moresby for repairs. This meant that our Australian contingent was late for our connecting flights from Brisbane, and we had to find a nearby hotel for the evening. We all had a late smoked duck supper with US basketball on the big screen TV. Early the next day we caught our flights home, and there was Sheila waiting for me, all brown and with a hibiscus flower in her hair.

There was a brief period of sanity and peace from May to August. I was busy editing and corresponding with authors about our tribute publication to Vallentyne and Vollenweider. I was also kept busy writing briefs and opinions for Tiffany on the developing court case about the Ramu NiCo MCC refinery waste disposal into the Bismarck Sea. Our little creek had flowing water even during our winter season. Consequently, there was an abundant crop of insects, especially mosquitoes. My garden was producing large amounts of tomato, eggplant, snow peas, shallots and leeks, and we had lots of bananas, grapefruits and limes.

In July, Sheila flew to Toronto to have a last visit with her sister Lindsay (Noddy), who was dying of lung cancer.

Both the PNG and Australian governments were unsettled: Sir Michael Somare's cabinet resigned, and Australian Prime Minister Kevin Rudd was demoted by his own Labour Party and replaced with Julia Gillard.

I wrote some letters to the editors of newspapers of coastal towns of North Queensland, trying to explain that expansion of coal ports was a bad idea; fossil fuel combustion causes greenhouse warming of the planet and the Coral Sea, where the Great Barrier Reef grows unhappily in hotter and hotter water. In addition, dredging of coal ports from Gladstone to Cape York would add terrestrial mud to the clean water habitat of the GBR. The North Queensland Conservation Council called in the national TV and radio networks for interviews, and I spent two days in our yard talking with several other experts about this problem. I enjoyed having our galah Caspar on my shoulder during most of these discussions.

On 20 September I flew to Cairns, Port Moresby, and Madang for another session with the BRG lawyers and the Ramu court case to obtain an injunction to stop the pollution. Our team was met with a chaotic situation of apparent legal trickery. Dr. Tracy Shimmield (the scientific leader of the Scottish marine scientists who had done an oceanographic survey of the Basamuk Canyon disposal site) was apparently alone and angry in a Port Moresby hotel room, because the Chinese miners and the PNG Mineral Resources Agency decided not to call her to the witness stand. We had relied on her science team reports about the certain damage that would result from construction and waste disposal.

Four of Tiffany's witnesses (residents of the village of Basamuk, including Sama and Melchoir mentioned in April) had vanished, and we later heard the gossip that one of the Chinese helicopters took them to Port Moresby and hid them somewhere.

As usual, we were hustled out of town to Jais Aben Resort, where we were treated very well, guarded by PNG police. My air conditioned room was beside the beach, and a bouquet of hibiscus and frangipani flowers was on my pillow with a note that said 'For Dr. Gregg'. I later discovered that the owners of Jais Aben Resort were Jerry and Sheila Scott, who had previously worked at the Kenricia Hotel in Kenora, Ontario, one of the places I had courted my Sheila in 1970. They took a liking to me, as they discovered that I was one of the crazy Canadian scientists at ELA, not too far from Kenora. One morning Tiffany's mother 'Boo Boo' Beryl, a Yorkshire actress, joined me at breakfast, and she talked my ear off.

The next few days I had breakfast and lunch with Tiffany and John Nonggorr, and Ray Williams, all lawyers, and all despondent because of their missing plaintiffs. They now had to rebuild their court case without the testimony of four PNG citizens of the village of Basamuk, and they secured twelve new plaintiffs from the coastal villages.

Other members of our science team (Amanda Reichelt-Brushett, John Luick, Phil Shearman, Gavin Mudd) joined us for long discussions about how we should behave as environmental experts while in the witness chair in court. Several of us were kept busy writing opinion pieces and science summaries for the lawyers, and affidavits for our new plaintiffs. We had our suppers beside the beach and talked till midnight, as the sea waves crunched up on the shore. Little bats were eating figs in the trees above our supper table. Local fishermen were in the lagoon beside us, with kerosene lanterns at the bow, and their bow and arrows ready for fish that came to the light.

Local PNG lawyer Ray Williams informed me that I would never be attacked by the town people, whereas other members of our white skinned environmental team should be very careful to stay close to our police guards. Ray said I looked like Jesus, and also like one of the respected priests in town, and I didn't behave like an outsider, because I was friendly, polite, tanned and not afraid. He explained that many native people were clever at reading body language, and they especially despised the arrogant and disrespectful Chinese.

The next day we were hauled to the courtroom in the back of a big utility truck, with police cars in front and behind us. We observed Tiffany and Ray arguing for the admission of new plaintiffs, and submitted twelve new affidavits. The judge was not amused at this whole affair and declared that more court proceedings had to be scheduled for all these new witnesses.

And so on Sunday 26 September, we didn't have to be serious anymore and the goofy behaviour began. John Luick took it upon himself to present a religious service beachside on this holy day. We had two drunk Enga men with us, roaring out hymns. John read some Sufi poetry about men with two penises, and the ecological consequences of the phases of the moon in this part of the world. We were joined by Debbie Chapman (who worshiped Che Guevara), Dr. Mackerel (a medical doctor interested in toxic metal poisoning from the mine works), and Canadian Nancy Sullivan (an anthropologist living in Madang, originally from Vancouver). John, Amanda and Gavin left early the next morning, but Tiffany asked me to stay two more days to help prepare the next case. We had phone calls from SAMS Chief Scientist Tracy Shimmield, who was sitting in a hotel in Brisbane, about to return to Scotland, confused and angry.

Barry, Rosa, and Gerry Scott finally got a flight booking for me, and I got to Port Moresby at noon on 28 September, where the airport was full of PNG people with cell phones anxiously texting and speaking loudly, all the while chewing betel nut with their teeth and mouth a blood red colour. A young girl from Perth had been in Madang, was stung by a bad jellyfish in the lagoon, and was here in a wheelchair trying to get home for hospital treatment.

I had to overnight in Cairns, where the airport was in chaos the next day, front end loaders and graders were on the runway, and hang gliders and parachute skiers were all over the beach. There was a very nice bevy of nude full body tanned women prancing along the miserable muddy beach, being filmed by several cameramen. To complete the brain busting contrasts, I was reading Garrison Keillor's *Life Among the Lutherans*, which was very close to the kind of people I knew in South Dakota. We finally got out of Cairns on a wonky propeller Fokker aircraft in late afternoon, which got me back to Townsville in the rain at 1825 dead tired. Son Ian was there to haul me back to Alligator Creek, where two upset parrots were demanding attention and treats.

GJB, Casper and Charles, Alligator Creek

And now December was upon us and Sheila was spending time in the granny flat with the sewing machine humming, something related to Christmas I'll bet. Ian was in rehearsals for a January production of *West Side Story*. The cicadas were thrumming especially loudly, and the curlews and nightjars hoot softly in the evenings. The poinciana trees are full of orange/red blossoms, and the frangipani flowers fall on the ground and are luminescent in the evening. Meanwhile, during one of our Christmas feasts with friends, I listened to my darling Sheila being proud of her musician son Ian, having fed him sandwiches, pushed him into university and grad school, and how we watched him playing with a rock band on stage, wearing a caftan, playing his mother's pots and pans, the university graduate. (He would survey the sound of his mother's metal pots and lids, ice cream canister lid, some of which had a good long ringing pleasant tone). But now he was teaching music at twelve schools in Townsville, playing in jazz bands and symphony orchestras. And then there was our son Jesse the engineer, building roads and bridges, using Gantt charts and math...he was the boy who could put a protractor in his mouth in high school, and taped the door shut on his science teacher. We continue to be amazed and grateful that our ornery little boys grew up to be intelligent gentlemen.

Chapter 61. Floods and Cyclones in 2011

We continue to live in interesting times. January gave us almost two metres of rain in one week, and SE Queensland experienced some vicious flash flooding that washed away small towns, floodplain barns, livestock, and badly flooded the big city of Brisbane. Our son Jesse and family lived down near Brisbane, and they had to evacuate the family and livestock to higher ground for several days. We live on a high ridge in the far north, so no damage here. We drove through minor flooded highways to attend a local production of *West Side Story*, where our son Ian was in the pit orchestra. They only had one show, the opening night, because the next day Category 5 Cyclone Yasi (2-3 Feb) came roaring out of the Coral Sea and smashed into the coast 200 km north of us, pulverising several small sugar towns. We were on southern margin of the 700 km diameter cyclone, 200 km/hr winds, 4 m storm surge, 7 m waves in Cleveland Bay, where it knocked down navigation buoys and lights.

Sheila had prepared well for this, battened down the hatches, and stockpiled batteries and food. We lost some yard and orchard trees, and chainsaws whined up and down our community for weeks afterward. With the help of our neighbour John Pozzebon, we cleared our house yard and driveways. The strong wind blew away my grapefruit, lime and avocado crop for the year. Power was off for two weeks, but we had a little 800W generator to keep our fridge and freezer cold. It was very hot and humid, lots of mosquitoes and snakes, and we furiously tried to eat up the frozen food that could not be saved. Avocado trees were tipped over but I propped them up again, and our laying hens were ok, high and dry. Some confused rainforest birds were blown off-course and spent time in our yard.

I abandoned Sheila at home to go to Madang 7-17 Feb for a big court trial about Chinese mine waste dumping into the Bismarck Sea. Airports en route were confused and still battened down from cyclone threats, and there were few scheduled flights. Our Australian group found hotel rooms in Cairns and Port Moresby where the Super Bowl was on the large screen TV and the beer was cold.

We arrived in Madang in the pouring rain, where Barry Lalley hauled us to the BRG command centre and computer room at Jais Aben. Barry was originally from New York City, but left the USA when Ronald Reagan got elected President (he said it made him sick to his stomach to watch that clown talk). He studied at Florida State University for an advanced degree in Asian Studies, worked with the Peace Corps in the Solomon Islands, and was now a major facilitator for BRG and Tiffany's science team. Unfortunately, he was also a vegetarian who was irritated by my insistence on mostly meat for my meals.

Since our last effort to close down this Chinese NiCo mine and refinery, Tiffany and her lawyers had obtained over a thousand affidavits from local people who lived along the Basamuk Bay and Ramu River coast. We were instructed to read all of these affidavits, to be aware of the local conditions near the mine and refinery. Tiffany coached us on behaviour in the witness chair, and we reviewed the mining consultants' expertise and their printed documents. Our scientific advisory team was guarded by police in a fenced accommodation compound, where the pub provided cold beer and big chunks of meat. We were hauled to court under police escort, to testify with six other scientists supporting the plea for an injunction from a thousand native village plaintiffs.

A half a dozen mine consultants were also called for testimony, some of them from the USA and Europe. The MCC Chinese mining lawyers were from Melbourne and they grilled us with no mercy, trying to find evidence of lack of expertise in this specific region. Twelve boxes of affidavits and written testimony were presented to the court from the Chinese MCC company. This was very tiring, as we had Mandarin and *Tok Pisin* translators who were not familiar with most of our scientific words. The court room was not air conditioned but had ceiling fans, and the little room was packed solid with concerned people.

You haven't lived until you've been interviewed by the local PNG and Australian media in the courtyard full of fist fights, shouting combatants, and a betel nut juice spitting contest. Tiffany and BRG treated us to the best restaurants in town for our evening suppers, including a visit to the exclusive Madang Boat Club (Members Only), where we had a massive Hunan Chinese feast, underneath a very loud thunderstorm with cracking lightning and pouring rain.

During our few off days, we snorkelled the lagoon near Jais Aben. I had an interesting discussion with a pasty white fundamentalist missionary on the lagoon raft at Jais Aben. He and his church troops were going inland to convert little villages to their religion, build churches, and demand that they wear clothes and not pay any attention to their pagan taboos and totems. We were also treated to meals by the owners of Jais Aben Resort, who supported our cause.

Five months later the judge revealed that we had won on three of the four counts, but the fourth count was about financial and contractual stuff, which won the judgement in favour of the Chinese miners in a PNG Supreme Court decision. An appeal was now in place, so there had been no action so far.

I came home to find Sheila exhausted from heavy work cleaning up the cyclone debris at home and at Billabong Sanctuary. It rained on through March, when we had a Richter 5 earthquake that trembled our house. During the big earthquake off the coast of northern Japan, Fukushima Daiichi, the 8 m tsunami, and six nuclear reactors malfunctioned. I had been on a Canadian response team for Chernobyl (twenty-five years ago), and the team was briefly called back into advisory roles, all via email.

Takeshi Kanno is a medical doctor in the Shizugawa public hospital at Minami Sanriku town. After the quake, the tsunami warning siren, and before the tsunami wave hit, he moved many of his patients to the top floor. After the wave went through, all lower floor patients were gone. Kanno stayed with his surviving patients for two days before the helicopters arrived to evacuate them. After much finger pointing and blame game, measurements indicate that the release of radioactive element impact from the Fukushima Daiichi nuclear power plant was about 50 millisieverts, or about 10% of that from Chernobyl, according to a World Health Organization report. So far there have been no reports of death or sickness from nuclear power plant radiation. Japanese debris piled up around Hawaii and the BC/Oregon coast, from the tsunami erosion. Small amounts of Fukushima nuclides appeared in tuna caught on the west coast of the USA.

My vegetable gardens were under 50 cm of water for weeks, and most plants perished, except for the loofahs. For several months afterward, I was the 'King of Loofahs' in our region. The cyclone flattened a lot of our banana patches, but a few survived to ripen suddenly. This gave me opportunity to do further research on my prized fermented banana jam product. It took several months for the gardens to dry out, but by July we had abundant eggplants, tomatoes, garlic chives, leeks, and too many habanero peppers. The latter were extremely dangerous, but Sheila made a wonderful pepper jelly with them.

Throughout all of this excitement, Sheila was rehearsing weekly with the Teabag Ladies Orchestra and performing for veterans, retirement homes, and old folks home dementia wards. She also supervised a major renovation to her bathroom and continued to help with the Billabong website, writing, taking photos and managing the Facebook page. Gramma made a couple of trips to Amamoor to stir up the grandkids and watch platypus in the creek. In August, Sheila's younger sister Lindsay (Noddy) died from cancer, after bravely fighting for over a year. It had been very dry, so the bush fires began and smoke was in the air for a few months.

Sheila and I celebrated our forty-first wedding anniversary on 8 September, and I had my seventieth birthday on 10 September. Jesse and his son Daniel surprised me by showing up to assist with the celebrations and feasts, and Ian was photographed in the swimming pool with Caspar the galah on his shoulder. He had just returned from a school ensemble tour in Melbourne, where he focused his phone camera on graffiti and good coffee bars. Sheila became famous around Alligator Creek for winning first prize at the Billabong Halloween Party as a green witch, the prize being the honour of feeding a crocodile named Nipper. Her birthday came on Guy Fawkes Day as usual, and her menfolk gathered around to provide a spectacular feast with a Tia Maria-ice cream-donut birthday cake. Jesse arrived with daughter Georgia Rose to help belatedly celebrate Ian's fourtieth, and much of the birthday weekend was spent petting emus, tickling wombats, walking the beach and carousing in the swimming pool.

We finally got our tribute journal volume of papers published, honouring Jack Vallentyne and Richard Vollenweider.

Chapter 62. The Hippy Goes Metal in 2012

It is a madhouse here, as usual. *Dementia praecox*. Yes, we had demented peacocks wandering in our house, demanding treats. And then the turtles demanded crickets, we had low-flying bananas in the kitchen, and the cicadas sang their guts out, like a thousand demented highland bagpipers on a temperature dependent seventh interval diminished drone. Our orchard now stank of frangipani flower fragrance, the poinciana trees were covered in brilliant orange flowers, and our laying hens hatched four new chicks. Sir Charles the Cockatoo and Chairman Casper the Galah were in full control, but Sheila was feeding way too many wallabies in our yard.

In March, I got a new right hip joint, a titanium steel shaft and ball, and ceramic socket. I was happy to have lots of opioid drugs for three months, oxycontin and endone. I was home four days after surgery, on crutches for three months, but walking normally by July. A steel hip makes for interesting attention in airport security stations...the bells go off loudly.

In June, Ian and Sheila trucked our old Buster-built Canadian cedar strip-fibreglass canoe down to Jesse's family at Amamoor. Jess rebuilt some of the rotten and termite damaged bits, applied a new epoxy finish, and it looks like new. This 'moose master' canoe was built by our friend Buster Welch in the early 1980s. Ian returned home with his new marimba, made in Japan. Ian was featured on percussion and the marimba in several events at the Australian Festival of Chamber Music here. He also completed the university work for his music teaching diploma, and is a bit more firmly embedded in his day job teaching at schools in town. Meanwhile, civil engineer son Jesse put 5 kW of solar panels on his farm shed roof, and is generating about two thirds of his whole farm electrical usage on sunny clear days. Everybody should be doing this.

Our gardens provided tomatoes, sweet corn, leeks, garlic chives, sweet and white potato, sweet and hot peppers, eggplants, fennel, and even some strawberries from June to November. We also had an excellent year for grapefruits, oranges, limes and mandarins. The garden production went into late November, and then it gets too hot...about the time the mangoes are ripe, I ignore the garden until May next year. In September, Sheila's Canadian niece Ana Baginski visited for a month, giving her a good excuse to go jungle surfing, snorkel the reef and prowl the local beaches.

In early December I flew down to Sydney for a big PNG mining meeting of fat cats, me being an unpaid consultant to lawyers concerned about pollution of the Fly River and Coral Sea. This trip reminded me that I don't like big cities or the smell of vehicle exhaust. And so I was whisked up to the forty-third floor of the Sydney Hilton for a quick briefing from Tiffany and some other lawyers, and Charles Roche from the Mineral Policy Institute, and then into a large conference room where we had a large banquet meal with speeches by the PM of Papua New Guinea Peter O'Neill, who begged the audience to invest in PNG.

The task for us was to prepare an environmental view of the future plans by the industrial proposals that were presented. Over the next few days, we listened to industrial representatives from PNG, Australia, the USA, China, Canada, Germany and France as they presented flashy PowerPoint and video shows about their ambitions in PNG for the next decade. Most of the multinational corporations were well aware of the lack of environmental and financial regulations in PNG, and sought to develop their business in ways they could not operate in their home countries. With few exceptions, these people were concerned with growth of their business, wealth creation, resource exploitation, and unsustainable development for the local people on the land. Most of the promotion was irresponsible hype to generate maximum profits for business investors back in the USA and Europe. There were maybe a half dozen investment firms, banks, and policy advisors lurking, all anxious to push development ahead.

There was no mention of greenhouse emissions, reclamation of disturbed land and sea ecosystems, health and safety training, education, problems with HIV, malaria, tuberculosis, and the options for alternate energy sources for the local people. Most of the planned operations involved oil and gas production and export, metal mining, timber exports, seafloor mining for precious metals, with precious little attention to the many poor PNG villages in the mountainous regions where highly technological exploitation was proposed.

Several of the hotel floors were devoted to booths hustling the audience with trinkets, bumper stickers, toys, and pamphlets about their businesses. I walked among Texas oilmen, black suited Chinese, turbaned Punjabis, a few Arabs in white robes, Japanese and Filipinos, and some very proper English and Scots people, including Samantha Smith from Nautilus Minerals of Toronto.

The mining companies must have hired PNG people to attend: there were tall Mount Hagen men, small barrel shaped coastal men, shiny black people from Bougainville and the Soloman Islands people, large women with flowers in their hair, all carrying dillybags of handout treats for all.

Most lunches and evening meals were lavish and expensive, sometimes called 'degustation', where raw beef and fish were served in alcoholic sauces with white and black carrots, Tasmanian and Barossa Valley wines, cheese plates and big chunks of dark chocolate. By 6 December I was sick of all this hustle, and Charles Roche and I escaped to a taxi driving us down tunnels of four lane traffic, breathing exhaust fumes on the way to the airport. There was no peace in the airport, where teeny boppers with electric blue toenail polish were shouting on their cell phones and taking selfies, and big haired Rastafarians were singing in a corner. All this was complicated by the airport metal detectors which discovered my steel hip. I was so glad to get back home in north Queensland, where I could spend the next few weeks gathering my magnificent mangoes from our orchard.

I eventually had to write a report of my views of this conference, which did not include much hope for a clean and green PNG future.

Chapter 63. 2013...And On We Go

Sheila volunteers for Billabong Sanctuary, keeping their website and Facebook page up to date about staff, wombats, emus, crocodile pens and parrot training exercises. She helped with several other wildlife education organisations in our region, including duties such as holding wombats and pythons for kids' shows in shopping malls. She is one of the pillars of the Teabags folk/country music band, who get together weekly for fun and tea and cakes. I do my daily chores, usually with a parrot on my shoulder, taking care of our laying hens and veggie gardens, and try to stay out of trouble. In January, I proudly put twenty-six litres of mango puree in our freezer.

In January, Sheila helped son Ian empty his house for renovations, mainly to create more room for instruments, including his new marimba. Son Jesse got flooded again in late January, horse sheds and corrals underwater down near Brisbane. Sheila fed a baby currawong for several weeks, which she named Zorro. Zorro was released in February, but returned frequently to our verhandah for chopped beef heart treats. Later it appears one of our yard owls ate Zorro, and thus the food chain spirals upward, and Sheila was devastated. Jesse and six year old daughter Scarlett visited us in February, so we had lots of family feasts and loud kids in the swimming pool. Ian traded off his old van for a new Hyundai van, both of which are usually full of drums. Our usual wet season failed and we had poor rainfall for most of the year.

An old friend from the Great Barrier Reef Marine Park Authority drove into our yard one evening, demanded a beer, and proceeded to push me to write critiques on the environmental impact statements for six different coal port expansion projects up and down our North Queensland coast.

Over February-May, I wrote critiques of plans for two coal port expansions and dredge spoil disposal in the Great Barrier Reef lagoon. There are large coal reserves in Queensland, and the Indian and Chinese miners want it badly. I made the very unpopular claim that the coal industry should vanish, due to the resulting horrible local pollution, black lung disease, and greenhouse gas warming of the planet. In addition, they should not be dumping contaminated harbour/port mud into the clear clean Great Barrier Reef lagoon. But we have a Conservative state and national government, so big business will probably win the debate.

[Addendum: 10 Dec 2013, the National government approved one of the coal port expansion/dredging/dredge spoil disposal projects in the GBR. Maybe I'll give up being an environmental scientist.]

In September, giant crocodiles were moved around from North Queensland community beaches to Billabong and other croc farms. This was the result of a government policy on croc removal from public areas, supposedly to keep the crocs from eating tourists and the family dogs. Sheila was called upon to photograph crocs being loaded and unloaded and shifted from one pen to another.

Ian went on tour to Perth with students, where he got to spend time at the Western Australian Academy of Performing Arts at Edith Cowan University.

In October we began to get little dribbles of rain, and the hills reluctantly turned dusty green, after a very long dry spell.

ABC TV came to our home to film my critical comments on dredge spoil dumping in the GBR lagoon, as the state and national government are agonising over a decision to permit pollution of a World Heritage Area.

December has been very hot and sunny, and I picked mangos every day. Most of my strawberry plants frizzled up and died from the heat, but I still have a good supply of Habanero peppers, a few tomatoes, eggplants and some fennel (what on earth am I supposed to do with fennel?). The poinciana trees are orange and red, their sepals falling on the ground, our back path down to the creek is coloured red. Frangipani and citrus flowers in my orchard make it smell like a perfume factory.

Has our research work helped improve the human condition?

Yes and No. Lead from gasoline was banned and lead in the ocean is declining, but they keep finding too much lead in municipal water supplies. DDT and some other organochlorine pesticides were banned in the wealthy western countries (thanks to Rachel Carson), but they are still floating around the world in polar ice and seafloor mud. The ozone hole might be healing, due to banning of the chlorofluorocarbons from aerosol cans and refrigerators. In some ways it is worse, as we keep manufacturing new chemical products without knowing their effects on the ecosystem. Now we have oestrogenic hormones in the rivers, new pesticides and antibiotics every decade, UV tanning filters and cosmetic ingredients in fish tissues, fire retardants, wood preservatives, and volatile components of plastic bottles in bird eggs, dioxins in seafood, coal tar residues in our livers, and enormous islands of plastic debris floating in the central Pacific Ocean. These organic contaminants are now found in biota at the bottom of the oceans and in Arctic snow. Human mother's milk in Europe and Asia still carries a lot of man-made chemical substances to the sons and daughters of the next generation. The Minamata Convention was signed and adopted by 140 countries on 13 October 2013 in Japan, home of several bad incidents of mercury poisoning of local human populations, nevermind the wildlife.

Chapter 64. The Last Chapter

As I write this, we are surrounded by leafless poinciana trees aflame with orange/red flowers. And in the evenings with full moon, the frangipani flowers fallen on the ground are iridescent. The seasons ebb and flow—long, bone dry winters turn the hills brittle and brown, till summer rains spark an explosion of fresh green growth and the air becomes heavy with the perfume of jasmine and eucalyptus.

In 2014 I helped to defeat state and national governments in their bid to dump coal port dredge spoil in the Great Barrier Reef World Heritage Area.

With encouragement from the various conservation groups along our GBR coast, I wrote several documents for state and national parliamentary inquiries, some letters to editors in coastal towns, and the tourism industry. Although the original coal port dredging and expansion proposal was intended to cover a dozen ports from Gladstone to Cape York, the debate focussed upon Abbott Point and Cleveland Bay, where I did have some scientific expertise in how the coastal mud moves. I contributed to several North Queensland Conservation Council meetings, and got to know dozens of strong citizen conservation groups that wanted to oppose the pollution of the Great Barrier Reef Lagoon.

In July, I was invited to participate in a Senate Inquiry about the health of the GBR, with specific reference to the coal port terminal expansion and dredging. This was held in the 'Sugar Shaker', the highest hotel in Townsville, on the fifteenth floor. There were a dozen senators present, with a sound system and TV crews watching us. There were presentations from the port officials, the coal companies, the tourism industry, local businesses, Great Barrier Reef Marine Park Authority, AIMS and JCU scientists, several NGO environmental groups, representatives of the non-fossil fuel alternate energy options, and some legal firms that were likely to support the environmental movements. Somehow, this felt good, a kind of democracy in action, with very different points of view being described, without anger and spite. Of course there was a conflict between the business, economic and jobs view, and the environmental, tourism, greenhouse gas/climate change, and World Heritage Area criteria. After a lot of political debate, the state and national governments decided to prohibit dredge spoil dumping into the GBR, but rather to dump the spoil into the RAMSAR Protected Caley Wetlands, adjacent to the Abbott Point port coal storage area.

This was a small victory.

Sheila continues to volunteer at Billabong Sanctuary, taking photos and writing stories for their Facebook page and website, lending a hand to rangers, and occasionally babysitting a cockatoo. She created a slogan for their 30th birthday celebration and wrote a ballad for owners Bob and Del. Sheila enjoyed being part of the Tea Bags musical group that gets together weekly for music and tea and cakes, and performs for residents in aged care homes.

Son Jesse, wife Sarah, and our grandchildren Scarlett Grace, Dan, and Georgia Rose packed up their belongings and animals and moved to Tasmania in February 2014. Their southern Queensland farm was on a floodplain, and had been flooded too many times. Jesse works as construction manager for a civil engineering firm. He and Sarah live on a farm, close to the southern coast where they can get fresh tuna and oysters. They raise horses, sheep, dogs, cats and chickens, supervised by two alpacas, Vala and Joop.

Ian travelled to South Korea in 2014 to visit Ed Choi, one of his university musical colleagues who is now principal percussionist in the Seoul Philharmonic Orchestra. He returned in 2018 to perform with the orchestra and percussion ensemble. He also attended an international percussion conference in Indianapolis, Indiana, where he got to refresh contacts with his mentors, heroes, and his uncle Alistair, who inspired him to play the drums in the first place.

Ian continues to work hard teaching percussion in Townsville schools. Performance highlights of the year included the Australian Festival of Chamber Music and a concert with African drumming group Wassa, featuring the amazing Epizo Bangoura from Guinea.

We made it through 2016, a year of drought and heat. We also managed to avoid zika, dengue, leprosy, psittacosis...our area was part of an important microbial and genetic experiment to make several species of mosquitoes locally extinct, especially carriers of zika and dengue. Lack of rain helped to keep the mosquito population down during these dangerous times. There was no wet season for the past two years, and we are still dry here. The nearby town (which supplies our house water) is on water restrictions. Our garden and orchard are suffering, but we managed to get a few tomatoes and habanero peppers, and I'm now daily bringing a sack of ripe mangoes up the hill from our stressed trees. The mangoes are few and small, but delicious, and I'm grateful we get to have any at all. In this hot weather, I'm also grateful for Fat Yak Ale.

In July 2016, our niece Marcy and family from Minnesota visited, and we fed them fresh picked mandarins and tangelos from our orchard. Sheila and Ian wore them out up here, with adventures at Billabong, Magnetic Island, and local beaches. They went on down to Tasmania with Sheila and Ian, to combine their four kids with the three kids from Jesse and Sarah: 'What a riot! We did one boat trip up the east coast of the Tasman Peninsula to look for whales—plunging up and down three metre swells against bitterly cold winds. Spectacular views of coastal cliffs—among the highest in the world, one whale tail in the distance, and a busload of Asians on board puking over the rails... .—SMB

Some family history: forty-six years ago, Canadian university summer student Sheila got her first paycheck at a remote research camp in boreal forest Ontario, where I was chief scientist at the time. She asked me to drive her into Winnipeg so she could buy a steel string guitar. She had been playing a cheap nylon string guitar. Well now, she got a 00018 Martin guitar, top of the line, but didn't have enough money for a case, so I bought it for her. We had supper at Hy's Steak Loft, and then I delivered her to the YWCA for overnight accommodations, but the door was locked. So she came to my little single room apartment, and things got out of hand from there on... And now, on our 46th wedding anniversary, 8 September, we drove out to Charters Towers to Taffy Evan's house and workshop, where Sheila picked up her new guitar, named 'Cockatoo'.

Where we live, the coastal coral reefs of the Coral, Timor, Indonesian, South China Seas, and Indian Ocean are bleaching, due to anthropogenic greenhouse heating of the ocean. This was the fourth major global coral bleaching event over the past decade, and it is probable that the coral community will be permanently stunted. Mangroves are dying back into the black sulphurous mud. And in early 2017, we have our city, state, and national government begging the Indian Adani Energy Company to open one of the world's largest coal mines in interior Queensland, as well as provide a railway and coastal port facility so they can ship coal to India. Despite this, Australia has taken up solar, wind, and tidal energy production, with coal and gas becoming less desirable even to the bankers and investors.

In March 2017 I began having more and more difficulty walking up our steep path from our garden and orchard. I was not getting enough oxygen, and puffing badly during easy physical tasks. Sheila finally got me to a doctor, who demanded X-rays, lung tests, PET Scans, where a small cell carcinoma was found on the right lung. This was a rapid growth cancer, and may have developed in the last month. Chemotherapy began almost immediately. Sons Ian and Jesse came to most of the medical sessions with Sheila and me, and they tried to slow me down when doing chores. The chemotherapy helped, as long as I didn't try to do heavy physical work. I started using our Kubota ride-on mower to go down the hill for my chores, but then in May, Ian and Jesse bought me a Polaris Ranger field buggy with a tray in the back. I began to enjoy even more the serenity and peace of our home and orchard and the chooks and ducks. And now I pick buckets of limes from our little West Indian lime tree. Jesse bought me an Aussie Akubra hat, as my hair is falling out now, and cognitive dysfunction is setting in.

Dr Gregg John Brunskill

Gregg Brunskill, cowboy, musician and scientist, lived out the rest of his days with Sheila at Alligator Creek. He died after a brief illness in September 2022, days before his 81st birthday, with his family by his side.

Acknowledgements: My wife Sheila Margaret Dennett Brunskill and our two sons Ian and Jesse have made comfortable our homes in Canada and Australia, while I was travelling around the world on airplanes and ships. My parents Gregg George Brunskill and Virginia Carol England Brunskill raised me up better than expected, I could have been a lot worse. And all of the other partners in family life, science, music, travel, who helped me deviate from ordinary pathways, and to pursue relentless and serendipitous adventures.

www.ingramcontent.com/pod-product-compliance
Ingram Content Group UK Ltd.
Pitfield, Milton Keynes, MK11 3LW, UK
UKHW062010041026
15591UKWH00017B/1255

9 781922 603227